T0397292

ROADS AND AIRPORTS PAVEMENT SURFACE CHARACTERISTICS

Roads and Airports Pavement Surface Characteristics contains the papers presented at the 9th International Symposium on Pavement Surface Characteristics (SURF 2022, Milan, Italy, 12–14 September 2022). The symposium was jointly organized by the Italian company that manages Italy's National Roads (ANAS–Ferrovie dello Stato Italiane Group), the World Road Association (PIARC) and Politecnico di Milano. The contributions aim to improve the quality of pavement surface characteristics while accomplishing efficiency, safety, sustainability, and new generation mobility needs. The book covers topics from emerging research to engineering practice is divided in the following sections:

- Advanced and performing construction methods and equipment
- Next generation mobility
- Data monitoring and performance assessment
- Surface features and performances
- Maintenance and preservation treatments
- Pavement management
- Economic and political strategies
- Safety and risk issues
- Minimizing road impacts
- Sustainability and performances issues about materials and design
- Pavements surfaces and urban heat islands
- Weather conditions impact
- Airport pavements

Roads and Airports Pavement Surface Characteristics is of interest to academics, engineers and professionals in the fields of pavement engineering, transport infrastructure, and related disciplines.

PROCEEDINGS OF THE 9TH INTERNATIONAL SYMPOSIUM ON PAVEMENT SURFACE CHARACTERISTICS (SURF 2022), 12–14 SEPTEMBER 2022, MILAN, ITALY

Roads and Airports Pavement Surface Characteristics

Edited by

Maurizio Crispino and Emanuele Toraldo

Department of Civil and Environmental Engineering, Politecnico di Milano, Milan, Italy

CRC Press
Taylor & Francis Group
Boca Raton London New York Leiden

CRC Press is an imprint of the
Taylor & Francis Group, an **informa** business

A BALKEMA BOOK

First published 2023
by CRC Press/Balkema
4 Park Square, Milton Park, Abingdon, Oxon, OX14 4RN
e-mail: enquiries@taylorandfrancis.com
www.routledge.com – www.taylorandfrancis.com

CRC Press/Balkema is an imprint of the Taylor & Francis Group, an informa business

Library of Congress Cataloging-in-Publication Data
A catalog record has been requested for this book

ISBN: 978-1-032-55149-4 (hbk)
ISBN: 978-1-032-55150-0 (pbk)
ISBN: 978-1-003-42925-8 (ebk)

DOI: 10.1201/9781003429258

Typeset in Times New Roman
by MPS Limited, Chennai, India

Table of Contents

Advanced and performing construction methods and equipments

Next generation mobility (light, electrical, autonomous)

Data monitoring and performance assessment

Surface features and performances

Maintenance and preservation treatments

Pavement management

Economic and political strategies

Safety and risk issues

Minimizing road impacts (noise, vibration, pollution, etc.)

Sustainability and performances issues about materials and design

Preface

SURF 2022 was the 9th event in the series of highly successful International Symposiums on Pavement Surface Characteristics, previously held in Australia (2018), USA (2012), Slovenia (2008), Canada (2004), France (2000), New Zealand (1996), Germany (1992) and USA (1988).

The main scope of SURF 2022 was to improve the quality of pavement surface characteristics while accomplishing efficiency, safety, sustainability, new generation mobility needs, and users, managers, and social expectations. The Symposium covered the current related topics not only in Italy, but also all over the world.

SURF 2022 welcomed transport infrastructure managers, practitioners, researchers, academics, industry professionals, road agencies, and all other interested parties to present papers (research, case studies, state of art, discussions, etc.).

In this context, SURF2022 presented the occasion to develop a profitable debate, comparison and exchange of scientific, technical, and managerial knowledge and experiences in worldwide scale. The papers presented in SURF2022 covered topics from emerging research to engineering practice. For the proceedings the contributions are grouped under the following themes:

- ADVANCED AND PERFORMING CONSTRUCTION METHODS AND EQUIPMENT
- NEXT GENERATION MOBILITY (LIGHT, ELECTRICAL, AUTONOMOUS)
- DATA MONITORING AND PERFORMANCE ASSESSMENT
- SURFACE FEATURES AND PERFORMANCES
- MAINTENANCE AND PRESERVATION TREATMENTS
- PAVEMENT MANAGEMENT
- ECONOMIC AND POLITICAL STRATEGIES
- SAFETY AND RISK ISSUES
- MINIMIZING ROAD IMPACTS (NOISE, VIBRATION, POLLUTION, ETC.)
- SUSTAINABILITY AND PERFORMANCES ISSUES ABOUT MATERIALS AND DESIGN
- PAVEMENTS SURFACES AND URBAN HEAT ISLANDS (COOL PAVEMENTS, ETC.)
- WEATHER CONDITIONS IMPACT (SNOW, ICE, ETC.)
- AIRPORT PAVEMENTS

Chairman's Note

As Chair of the Symposium, and on behalf of Politecnico di Milano, I was delighted to welcome all the participants to the 9th Symposium on Pavement Surface Characteristics, SURF 2022.

Politecnico di Milano, the leading Italian Technical University in Engineering, Architecture and Design, was honored to host this Symposium.

For the first time, the Italian company that manages the National Roads (ANAS – Ferrovie dello Stato Italiane Group) and the World Road Association (PIARC), together with Politecnico di Milano were joining forces for a global event that aimed at involving people from all over the world.

The Symposium was directed not only towards the academic environment, and researchers in general, but was also pertinent to managers, practitioners, companies, authorities, associations, engineers, and technicians dealing with everyday tasks and responsibilities related to transport infrastructures.

The topics of the Symposium dealt with data monitoring and performance assessment, innovation in asset management to meet next generation mobility needs, economic and political strategies, life-cycle cost analysis and assessment, safety and risk issues, sustainability issues (noise, vibration, pollution, fuel consumption, etc.), surface layer design, maintenance, and preservation treatments, as well as materials and design.

SURF 2022 program included plenary sessions, concurrent technical sessions, poster sessions, workshops, and lectures by worldwide experts.

The main scope of SURF 2022 was to improve the quality of pavement surface characteristics while accomplishing efficiency, safety, sustainability, new generation mobility needs, as well as users, managers, and social expectations. For this reason, all participants were actively involved in discussions and workshops to improve their knowledge and to take the know-how back to their own businesses.

Last but not least, I would like to thank the Symposium patrons, the sponsors, and all the members of the Symposium Committees for their valuable contributions to the organization of this international event.

Prof. Maurizio Crispino
Chair of SURF 2022

Chair & Committees

Symposium Chair
Maurizio Crispino
Politecnico di Milano | Italy

Symposium Co-Chair
Aldo Isi
ANAS | Italy

Sponsoring Committee
PIARC International TC 4.1 Pavements
PIARC Italian TC 4.1 Pavements

Supporting Committee
PIARC TC 3.3 Asset Management

Scientific Committee
Members from PIARC TC 4.1 and TC 3.3
coordinated by M. Crispino, E. Toraldo, M. Briessinck, R. Wix

Local Organizing & Professional/Student Challenge Committee
M. Crispino, E. Toraldo, M. Ketabdari, A. Antoniazzi, E. Mariani, C. Nodari
Politecnico di Milano | Italy

Roads and Airports Pavement Surface Characteristics – Crispino & Toraldo (Eds)
© 2023 The Editor(s), ISBN 978-1-032-55149-4

Hosting Institution

POLITECNICO
MILANO 1863

Politecnico di Milano is a scientific-technological University which trains engineers, archi-tects, and industrial designers. The University has always focused on the quality and inno-vation of its teaching and research, developing a fruitful relationship with business and productive world by means of experimental research and technological transfer.

Research has always been linked to didactics and is a priority commitment which has allowed Politecnico di Milano to achieve high quality results at an international level as well as connecting University to the business world.

Moreover, research activities constitute a parallel path to that formed by cooperation and alliances with the industrial system. Knowing the worldwide working environment is a vital requirement for training students approaching their future careers. By referring to the needs of the productive, industrial world and public administration, research is facilitated in fol-lowing new paths and dealing with the need for constant and rapid innovation. The alliance with the industrial world, in many cases favored by Fondazione Politecnico and by con-sortiums to which Politecnico belongs, allows the University to follow the vocation of the territories in which it operates and to be a stimulus for their development.

The challenge which is being met today projects this tradition which is strongly rooted in the territory beyond the borders of the country, in a relationship which is developing first at the European level with the objective of contributing to the creation of a "single professional training market".

Politecnico takes part in several research and training projects collaborating with the most qualified European universities. Politecnico's contribution is increasingly being extended to other countries: from North America to Southeast Asia to Eastern Europe. Today the road to internationalization sees Politecnico di Milano taking part in the European and world network of leading technical universities and offering several exchange and double-degree programs beside many programs which are entirely taught in English.

Sponsors

Gold Sponsor

Silver Sponsor

Bronze Sponsor

Exhibitors

Media Patner

Patronages

Organizing Secretariat

Scientific Secretariat

POLITECNICO
MILANO 1863

Advanced and performing construction methods and equipments

Roads and Airports Pavement Surface Characteristics – Crispino & Toraldo (Eds)
© 2023 The Author(s), ISBN 978-1-032-55149-4

Improved accelerated ageing of asphalt samples in the laboratory

Greg White* & Ahmed Abouelsaad

School of Science, Technology and Engineering, University of the Sunshine Coast, Queensland, Australia

ABSTRACT: There are currently two established approaches to the accelerated ageing of asphalt mixture and bituminous binder samples in the laboratory. The first is to expose a film of the bituminous binder to heat and air, under controlled conditions. Different protocols are intended to represent the ageing associated with asphalt mixture production, as well as the long-term field ageing of an asphalt surface. The second approach is to expose the asphalt mixture to elevated temperature in an oven, intended to represent long-term field ageing. The current protocols for ageing bituminous binder samples do not take into account the interactions between the binder and the aggregate. The binder ageing protocols do not include UV irradiation and do not use representative binder film thicknesses. Furthermore, the current protocols for ageing asphalt mixture samples do not include UV irradiation, and the heat is applied equally to all sides of the sample, which is not representative of field conditions. The limitations associated with the current protocols for the accelerated ageing of asphalt binder and mixture samples in the laboratory is a significant issue. Any research into the effect of incorporating recycled waste materials in asphalt mixtures relies on a robust accelerated ageing protocol. Similarly, research on asphalt rejuvenation and preservation treatments also requires reliable accelerated ageing protocols, to avoid the need for 10 or more years of field ageing to measure the benefits. Therefore, more reliable accelerated asphalt ageing protocols are critical to better pavement surfacing technology. This paper explains the limitations of the current asphalt binder and mixture laboratory ageing protocols and summarises efforts to develop a more reliable approach. The proposed approach incorporates UV and heat exposure, which is more representative of field conditions. Examples demonstrate that the proposed approach results in more realistically aged asphalt surfaces and the preferred indicators of sample age are described.

Keywords: Asphalt, Accelerated, Ageing, Laboratory

1 INTRODUCTION

Asphalt mixtures, which are a conglomerate of bituminous binder, fine aggregate and coarse aggregate, are critical for the construction of road, port and airport pavements around the world. Asphalt mixtures come in many forms, with different aggregate gradations and different maximum aggregate sizes, which represents the variety of applications for which asphalt mixtures are used in pavement construction (Shell 2015). The bituminous binder also varies from lower to higher viscosity grades of bitumen, or bitumen modified with acid, polymer or other additives (Porto *et al.* 2019). However, one issue that affects all asphalt mixtures is the change in bitumen properties with time, a phenomenon known as 'ageing'.

---

*Corresponding Author

DOI: 10.1201/9781003429258-1

Aggregate does not age significantly over the life of a typical asphalt mixture, except where aggregate that is prone to breakdown or chemical attack is used (White & Fergusson 2021). However, with exposure to radiation and oxygen, bitumen molecules change, and the resulting bituminous binder becomes harder and more brittle (Abouelsaad & White 2021). As the binder ages, it becomes more prone to cracking and cohesive failure, as well as adhesive failure with the aggregate particles. As this occurs, the bituminous mastic, which is the combination of the bituminous binder and the fine aggregate particles smaller than 75 μm in size (Pérez-Jiménez et al. 2008) can erode, particularly at the surface of the pavement (White 2018). This is a distress known as fretting, which can lead to ravelling (FAA 2014).

Ravelling is a significant distress for pavement surfaces. Where load-related cracking and rutting are avoided, age-related fretting and ravelling often lead to pavement resurfacing, either by asphalt overlay or asphalt removal and replacement (Abouelsaad & White 2021). This is particularly the case for airport pavements, which are generally designed to be thick and strong to avoid rutting and fatigue cracking and where the tolerance for loose aggregate from asphalt ravelling is low (White 2018).

To research the properties and performance of aged asphalt mixtures, researchers rely on accelerate laboratory ageing of asphalt mixtures and specimens (Abouelsaad & White 2021). As a result, reliable accelerated ageing protocols, that adequately represent the field ageing of asphalt mixtures, is important to asphalt mixture research (Rad et al. 2017). Furthermore, as the desire for more sustainable pavements results in an increased use of recycled materials in asphalt mixture, their effect on mixture performance will become important (Jamshidi & White 2020). Although there are established and reliable tests for the relative effects on asphalt mixture stiffness, deformation resistance, crack resistance and moisture damage resistance, there remains questions regarding the effect of these recycled materials on the expected service life or durability of the resulting asphalt mixtures (Van Den Heuvel & White 2021). Similarly, the use of surface preservation treatments to extend the life of asphalt surfaces is increasingly attractive (White & Thompson 2016). If an economical preservation treatment can add just one year to the service life of an asphalt surface, the whole of life benefit is substantial. However, practical research into the effects of asphalt preservation treatments relies on accelerated ageing of asphalt mixtures before and after treatment application. Consequently, reliable and representative accelerated ageing processes are important for future research on the effects of recycled materials in asphalt material production, as well as surface preservation treatments for asphalt mixtures. This paper reviews the issues associated with the accelerated ageing of asphalt mixtures and bituminous binders in the laboratory. The issues considered are the testing of bituminous binder or asphalt mixtures, the replication of realistic field ageing, developing realistic ageing processes and the preferred indicators of ageing.

1.1 Background

1.1.1 Asphalt and binder ageing

The ageing if bituminous binders and asphalt mixtures is universally acknowledged (Abouelsaad & White 2021). It is actually only the bituminous binder that ages, with the properties changing with exposure to oxygen, radiation and weather. However, the interaction between the bituminous binder, the fine aggregate and the coarse aggregate skeleton means that the change in the bituminous binder has significant influence on the properties and performance of the associated asphalt mixture (Wu & Airey 2009).

Bitumen ageing has two aspects; thermal-oxidative ageing, which is mainly caused by heat and oxygen (Lu & Isacsson 2000; Xiao et al. 2009), and photo-oxidative ageing, which is caused by ultraviolet (UV) irradiation and oxygen (Wu et al. 2010). Both heat and UV degrade the bituminous binder properties, but their effects on ageing are distinctly different (Feng et al. 2012). Furthermore, more complicated and more severe ageing is observed when heat and UV irradiation exposure occur simultaneously (Wang et al. 2012).

As bituminous binder ages, the ductility reduces, the penetration reduces, while the softening point, viscosity and flash point all increase (Siddiqui & Ali 1999). The effect of these binder changes on asphalt mixtures is similar, with Marshall stability increasing and Marshall flow decreasing with age (Hainin *et al.* 2019). Furthermore, asphalt mixture stiffness increases (Briliak & Remišová 2021), fracture and crack propagation resistance decreases (Mohammadafzali *et al.* 2017), while deformation resistance increases (Yu *et al.* 2021).

Although not considered by most researchers, it is clear that asphalt surfaces aged in the field also exhibit an increase in the surface macro-texture, as a result of bituminous mastic erosion and fretting, which can eventually lead to ravelling (FAA 2014). To confirm this, Abouelsaad & White (2020) measured the surface texture of aged airport runways and taxiways. It was found that there was no statistically significant difference between the wheel paths and the untrafficked pavements edges, indicating that surface texture increases with age, leading to fretting and ravelling, as a result of age-related weathering. An example is shown in Figure 1, which contrasts a typical new airport asphalt surface (macro-texture ~0.5 mm) and an aged surface, near the end of the service life (macro-texture ~1.5 mm). Both surfaces are 14 mm nominal maximum size dense graded asphalt mixtures produced to the same volumetric specification requirements (White 2018).

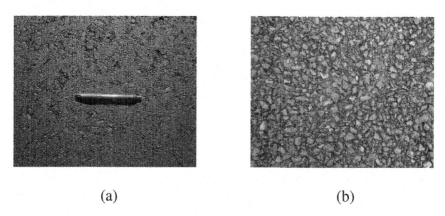

(a)   (b)

Figure 1.   Typical (a) new and (b) aged dense graded airport asphalt surfaces.

These changes in bituminous binder and asphalt mixture properties are important for understanding the effect of recycled materials, such as plastic and crumbed tyre rubber, which are intended to partially or fully dissolve into the binder phase during production (White 2020). However, the importance of asphalt mixture ageing also depends on the application of the asphalt mixture and its location within the pavement.

1.1.2   *Asphalt mixture applications*
As stated above, asphalt mixtures are used for many different applications in pavement construction. As well having different performance and functional requirements for road, port and airport pavement (Huang 1993) different asphalt mixtures are required for different functions within a pavement structure (Figure 2).

For example, high modulus asphalt mixtures with large maximum aggregate size and lower bituminous binder contents, provide good base layers in full-depth or deep asphalt pavement structures (Shell 2015). For these mixtures, ageing is less critical because the main purpose of the layer is to provide structural stiffness to the pavement. Surface erosion is not

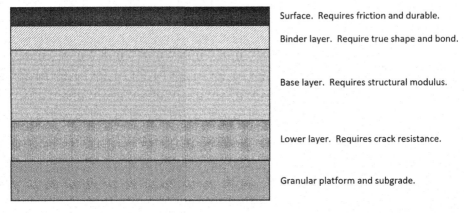

Figure 2.   Typical asphalt mixture applications in a full depth asphalt pavement.

an issue (Widyatmoko *et al.* 2009) and fatigue cracking is less critical than for layers near the bottom of the pavement where the tensile strains are highest (Shell 2015). Similarly, the bottom layer in a full depth asphalt pavement will need to resist fatigue cracking, meaning a higher bituminous binder content and a modified binder to provide elasticity, is more attractive. Ageing is less of an issue at this depth in the pavement, because the overlying layers protect the lower layers from oxidation and radiation exposure. In contrast, the surface is exposed to the weather, including high temperatures and UV irradiation (Rad *et al.* 2017). Consequently, ageing occurs faster at the surface and ageing is slower and potentially insignificant deeper in the pavement structure. The surface is also where any loose aggregate can impact vehicle safety.

Asphalt surfaces are generally expected to last more than ten years. In airport applications, where the tolerance for loose stones and cracking is low, ten years is a typical surface life expectancy (White 2018). In contrast, in local roads where structural failure is less likely and loose stones are less critical, a twenty year surface life is more realistic (Kidd *et al.* 2021). For the deeper asphalt layers, the service life is generally longer than the surface, because the upper layers protect the lower layers from the weather and UV radiation, as well as from high temperatures. Regardless of the expected asphalt life, asphalt researchers can not generally wait for the full life cycle to assess the effect of incorporating recycled materials or surface preservation treatment. Consequently, most research is performed on asphalt samples that are subject to accelerated ageing in the laboratory.

1.1.3   *Accelerated ageing protocols*
To allow asphalt ageing to be investigated in a viable timeframe, various accelerated laboratory ageing protocols have been developed over time. For bituminous binders, the rolling thin film oven (RTFO) is intended to simulate the short-term ageing associated with asphalt production, haulage and paving processes (Hossain & Wasiuddin 2019). The pressure ageing vessel (PAV) is then used to represent long-term ageing, with 7–10 years of ageing in the field intended to be represented by one cycle of PAV ageing (Yan *et al.* 2019). Different jurisdictions have different standards, but ASTM D2872-19 (RTFO) and ASTM D6521-19 (PAV) are typical examples. For compacted asphalt mixture samples, the only known standard ageing protocol is to place compacted asphalt specimens in an oven at 85°C for a period of 5 days, as detailed in AASHTO R30, which is intended to replicate 5–10 years of ageing in the field (AASHTO 2002). However, it is not clear whether that is surface ageing or lower layer ageing.

2   DISCUSSION

Challenges for the accelerated laboratory replication of realistic field ageing of asphalt mixtures are discussed in the following sections. The requirements for improved accelerated laboratory asphalt mixture ageing are identified. The resolution of these issues is critical to achieving improved accelerated ageing of asphalt mixtures in the laboratory.

2.1   *Ageing of binders or mixtures*

Different researchers have investigated the ageing of bituminous binders, while others have investigated the ageing of asphalt mixtures (Abouelsaad & White 2021). In general, aged binders have been compared to unaged binder using rheological tests such as temperature-frequency sweeps from the dynamic shear rheometer (DSR), and via chemical analysis of the saturates-aromatics-resins-asphaltenes (SARA) composition of the binders. Some have taken a simpler approach by comparing the penetration or softening point. In contrast, aged asphalt mixture investigations have used asphalt mixture properties, such as modulus. Few researchers have aged asphalt mixtures and then extracted the bituminous binder, to allow a direct comparison of binder properties from aged asphalt mixtures (Abouelsaad & White 2022a).

Bituminous binders are convenient for researchers. Binder testing avoids the need for asphalt production and binder extraction methods, which makes research faster and less expensive. However, the interaction between the bituminous binder and the fine aggregate is important (Wu & Airey 2009) and can not be considered by bituminous binder ageing. Although asphalt mixture testing is more complex and expensive, it is more representative of field ageing. Despite this benefit, ensuring the binder extraction process does not adversely affect the bituminous binder properties is also important. Consequently, extracted binder from aged asphalt mixtures must be compared to extracted binder from unaged asphalt samples, so that both the aged and unaged binder has been exposed to the same asphalt production and binder extraction processes, which will reflect the short-term ageing. Furthermore, the binder extraction and solvent evaporation processes, which must be designed to minimise the impact on the binder properties, but must fully extract the binder from the asphalt mixtures and evaporate off any cutters used in the extraction process. Rotary evaporators and low flash point solvents are recommended for this purpose (Mikhailenko *et al.* 2019). This issue is likely to be a challenge for all asphalt mixtures, regardless of their type and depth within the pavement. However, binders modified with crumbed rubber and plastic are likely to be particularly challenging because of the potential for only partial digestion of the rubber/plastic into the bituminous phase (White & Reid 2022).

2.2   *Replicating field ageing conditions*

Asphalt mixtures located deeper in a pavement are expected to age more slowly than surface layers, primarily as a result of less intense thermal loading. In most practical circumstances, the overlaying asphalt layers will protect the lower layers from high temperatures, UV radiation and oxygen. Consequently, the lower asphalt layers of any pavement will age uniformly. In contrast, the surface layer of a pavement is more exposed to high temperatures and oxygen, while the pavement surface is exposed to UV radiation, high oxygen circulation and rainfall (White & Abouelsaad 2021). This creates an ageing profile, with the top of the surface layer far more aged than the bottom, as shown schematically in Figure 3, compared to the profile generated by simply oven ageing, as required by AASHTO R30. This ageing profile has been demonstrated by slicing aged asphalt cores into 10 mm thick discs, extracting the binder from each disc and testing the chemical and rheological properties of the extracted binder (Abouelsaad & White 2022a). The uppermost disc exhibited

7

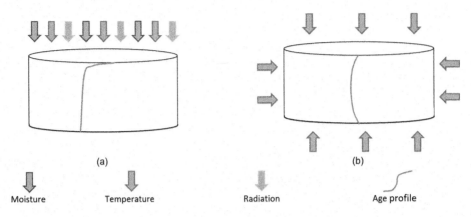

Figure 3.   Schematic comparison of (a) field and (b) oven ageing profiles.

significantly greater ageing than the lower discs, with a DSR measured complex modulus (G*) four to seven times higher for the uppermost discs, as well as a significant decrease in aromatics and a decrease in resins based on the SARA analysis (Mirwald *et al.* 2020). Furthermore, some researchers have indicated that most of the surface ageing only penetrates 5–10 mm from the pavement surface (Zeng *et al.* 2018), meaning the true slope of the ageing profile is even greater than that shown in Figure 3. The profile of ageing is most important to asphalt ageing research focussed on the resistance of surface asphalt mixtures to age-related weathering, and by association, to technologies intended to slow or reverse ageing, such as surface preservation treatments. Because a significant portion of all asphalt production and construction is related to pavement resurfacing, this is an important element of asphalt ageing research. In contrast, replicating the ageing profile is less critical for cracking and fatigue research associated with asphalt layers located deeper in the pavement, where ageing is slower and more uniform. However, regardless of the ageing profile, the amount of accelerated ageing that is required to replicate field conditions is another challenge.

2.3   *Representative ageing protocols*

The aim of laboratory ageing protocols is to replicate field ageing in an accelerated timeframe, in a controlled laboratory environment. Although the existing oven-based asphalt mixture ageing protocol (ie. AASHTO R30) provides thermal loading, which increases the asphalt mixture modulus with ageing time, the 85°C for five days protocol was based on only limited field data from cores in the USA (AASHTO 2002). Consequently, the representativeness of this approach has been questioned because different climates are likely to age asphalt mixtures differently in the field (Sirin *et al.* 2017) and because it omits UV irradiation, which is a catalyst for thermal ageing (Hossain *et al.* 2018). Regarding the differences in field ageing around the world, it is likely that field ageing of asphalt mixtures in northern Europe is quite different to ageing in central Africa. Furthermore, ageing in winter is also certainly slower than ageing in summer, particularly in locations such as New Zealand, where a large ozone layer hole develops each summer, allowing a significant increase in UV irradiation (Kataria *et al.* 2014). Regarding the importance of UV as an ageing catalyst, this is more important for surface layers than for asphalt mixtures deeper in the pavement structure. This adds another complexity to the development of a universal accelerated ageing protocol.

To avoid these complexities, accelerated ageing research is often performed in a relative manner. That is, a control asphalt mixture is aged and alternate asphalt mixtures, which are identical apart from the factor of interest, are aged and the two are compared after various ageing intervals. For example, Abouelsaad & White (2022b) compared the resilient (indirect tensile) modulus of the same asphalt mixture compacted in the field and compacted in the laboratory. Both sample types were aged in a conventional oven and in a UV-equipped weathering chamber. The resilient modulus of triplicate samples was measured every 14 days up to 98 days, and the results are summarised in Figure 4. For consistence with the UV-equipped weathering chamber, the oven was set to 70°C rather than the AASHTO R30 recommended 85°C.

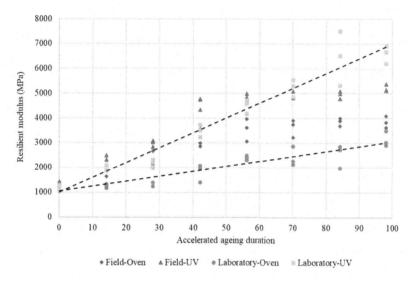

Figure 4.   Schematic comparison of (a) field and (b) oven ageing profiles.

From Figure 4, the relative effect of ageing can be readily determined. For example, after 98 days of accelerated ageing, the modulus was approximately three to seven times the initial modulus values. Similarly, it can be calculated that, on average, the laboratory compacted sample modulus was 26% higher than the modulus of the field compacted samples, and that samples aged in the UV-equipped chamber had a modulus 61% higher than samples aged in the oven. However, relating this to field ageing remains a challenge.

To calibrate the accelerated laboratory ageing protocols to field ageing, a large number of field cores must be extracted from surfaces of different ages. A locally representative average field ageing profile can then be developed against which the accelerated ageing temperatures and durations can be calibrated. However, ageing is associated with an increase or percentage change in the initial modulus, so the initial modulus of the mixtures would need to be known. Furthermore, different asphalt mixtures are expected to age differently, meaning that the mixture type and volumetric composition, as well as the bituminous binder type, would also need to be known. Finally, as discussed above, asphalt mixtures will age more slowly if located deeper in the pavement structure, so different accelerated ageing protocols would need to be developed for different mixtures, with different binders, and for different depths in the pavement. Consequently, this is a substantial undertaking, and it is likely that generic or relative accelerated protocols will remain dominant. The above discussion used resilient modulus as the indicator of ageing because that is commonly used by asphalt mixture ageing researchers and because it is routinely tested in many jurisdictions. However, there are many other indicators of ageing that may be more appropriate.

2.4 *Indicators of ageing*

There are many potential indicators of ageing. For bituminous binders, the DSR frequency sweeps and characteristic properties such as the complex modulus are commonly tested (Tarsi *et al.* 2018). However, some researchers have used SARA fractions, particularly the asphaltenes content, as the bitumen ageing is generally associated with an increase in asphaltene content (Mirwald *et al.* 2020). Other researchers have used either the Gaestal index (GI) or the colloidal instability index (CI), which are calculated as various ratios between various combinations of the SARA fractions, effectively the proportion of solid components divided by the proportion of liquid components (Gawel & Baginska 2004). Other researchers have used Fourier-transform infrared spectroscopy (FT-IR) to chemically fingerprint bituminous binders (Cong *et al.* 2019). Peaks at different wavelengths are taken to indicator the relative degree of ageing.

Because these indicators of ageing use different units and have vastly different initial values, indices of ageing are preferred. Any ageing index can be calculated as the ratio of the aged value to the unaged value of any particular property. Most properties increase with age, such as modulus, resulting in an ageing index greater than 1.0. However, indicators such as CI decrease with age, so an inverse index (1/CI) is preferred, so that the index increased with age, making comparison to other indices simpler. All these indicators of ageing change differently with accelerated ageing time, the inclusion of UV irradiation and whether the bituminous binder or the asphalt mixture is aged. For example, Table 1 indicates that the ageing index associated with various indicators of ageing for samples aged in both an oven and a UV-equipped chamber (White & Abouelsaad 2022a, 2022b). From Table 1 it is clear that the different ageing indices vary greatly, with 1/CI increasing by 60–70% for extracted binder, compared to the G* from the top 10 mm of compacted asphalt samples, which increased by more than nine times. It is also clear that the UV affected some properties substantially, such as the G* in the top 10 mm, more than other properties, such as the surface macro-texture. Finally, the effect on binder properties was different, depending on whether the asphalt mixture was aged for 98 days, or whether the binder was aged as a thin film for 8 days, which was intended to be approximately equivalent. Based on G* and FT-IR, the 98 days of asphalt mixture ageing in the UV equipped chamber was more severe than the 8 days of thin film bituminous binder ageing. However, for the CI, the same bituminous binder ageing was more severe than the asphalt mixture ageing.

Table 1. Asphalt and binder ageing indices.

Sample type	Property	Index for oven ageing	Index for UV-chamber ageing
Aged asphalt mixture (98 days)			
	Resilient modulus	2.76	5.54
	Surface texture	1.14	1.22
Bituminous binder extracted from aged asphalt mixtures			
	G* (top 10 mm)	9.01	56.0
	1/CI (top 10 mm)	1.56	1.69
	FT-IR (top 10 mm)	2.55	2.51
Aged 1 mm thin film of binder (8 days)			
	G*	1.98	2.24
	FT-IR	1.10	1.20
	1/CI	1.26	2.89

The preferred indicator of ageing for any particular research effort is expected to be based on convenience. Where asphalt mixtures are aged and a resilient modulus test apparatus is locally available, then mixture resilient modulus is likely to be preferred. In contrast, when binder is aged and a DSR is available, the G* value from a temperature-frequency sweep is likely to be adopted. However, the most appropriate indicator of ageing must also consider the intent of the research and the type/application of the asphalt mixture being considered. For example, when the research is intended to determine the effect of recycled plastic on expected service life of asphalt surfaces for airport pavements, where fretting and ravelling generally trigger resurfacing, surface macro-texture is likely to be preferred, perhaps supported by extracted binder G*. A universal approach is unlikely to be identified and the use of different indicators of ageing will lead to different conclusions.

3 CONCLUSION

There are significant challenges that need to be addressed to improve accelerated ageing of asphalt mixtures and bituminous binders in the laboratory. These include the balancing the convenience of binder ageing with the increased representativeness of mixture ageing. Depending on whether the research is focussed on structural layers that age slowly and evenly, or asphalt surfaces, which age quickly and with an extreme profile or gradient of ageing, the need for UV irradiation is different. Furthermore, different applications and environments warrant different ageing temperature and duration combinations. Despite these complexities, the current asphalt ageing protocols are simple and are applied universally. Consequently, local and application-specific ageing protocols must be developed. Similarly, the tendency to select the most convenient indicator of ageing must be replaced by the selection of an ageing indicator that reflects the focus of the research being undertaken. In conclusion, there are many improvements required for the reliable accelerated ageing of asphalt mixtures and bituminous binders, and this will become a more important topic as researchers and practitioners become increasingly interested in the effect of recycled materials and preservation treatments on the expected life of asphalt mixtures, particularly in pavement surfaces where ageing is most severe.

REFERENCES

AASHTO 2002 'Standard Practice for Mixture Conditioning of Hot Mix Asphalt (HMA) R30', Provisional Standards, *American Association of State Highway and Transportation Officials*, Washington, District of Columbia, USA.

Abouelsaad A. & White G. 2020, 'Fretting and Ravelling of Asphalt Surfaces for Airport Pavements: A Load or environmental distress?', *Nineteenth Annual International Conference on Highways and Airport Pavement Engineering, Asphalt Technology, and Infrastructure*, Liverpool, England, United Kingdom, 11–12 March.

Abouelsaad A. & White G. 2021, 'Review of Asphalt Mixture Ravelling Mechanisms, Causes and Testing', *International Journal of Pavement Research and Technology*, doi. 10.1007/s42947-021-00100-7.

Abouelsaad A. & White G. 2022a, 'Comparing the Effect of Thermal-oxidation and Photo-oxidation of Asphalt Mixtures on the Rheological and Chemical Properties of Extracted Bituminous Binder', *Road Materials and Pavement Design*, Article-under-review.

Abouelsaad A. & White G. 2022b, 'The Combined Effect of Ultraviolet Irradiation and Temperature on Hot Mix Asphalt Mixture Aging', *Sustainability*, vol. 14, no. 5942, pp.1–19.

Briliak D. & Remišová E. 2021 'Research Into Effect of Asphalt Mixture Aging on Stiffness', *Transportation Research Procedia*, vol. 55, pp. 1251–1257.

Cong P., Hao H. & Luo W. 2019 'Investigation of Carbonyl of Asphalt Binders Containing Antiaging Agents and Waste Cooking Oil using FTIR Spectroscopy' *Journal of Testing and Evaluation*, vol. 47, no. 2, pp. 1147–1162.

FAA 2014, *Pavement Surface Evaluation and Rating: Asphalt Airfield Pavements*, Advisory Circular AC 150/5320-17A, Federal Aviation Administration, Washington, District of Columbia, USA, 10 September.

Feng Z.G., Yu J.Y. & Liang Y.S. 2012 'The Relationship Between Colloidal Chemistry and Ageing Properties of Bitumen', *Petroleum Science and Technology*, vol. 30, no. 14, pp. 1453–1460.

Gawel I. & Baginska K. 2004 'Effect of Chemical Nature on the Susceptibility of Asphalt to Aging', *Petroleum Science and Technology*, vol. 22, no. 9, pp. 1261–1271.

Jamshidi A. & White G. 2020, 'Evaluation of Performance of Challenges of use of Waste Materials in Pavement Construction: A Critical Review', *Applied Sciences*, vol. 10, no. 226, pp. 1–13.

Hainin M.R., Ramadhansyah P.J., Awang H., Khairil Azman M., Intan Suhana M.R., Nordiana M., Norhidayah A.H., Haryati Y. & Che Ros I. 2019, 'Marshall Stability Properties of Asphaltic Concrete with Kaolin Clay Under Aging', *IOP Conference Series: Earth and Environmental Science*, vol. 220, no. 1.

Huang Y.H. 1993, *Pavement Analysis and Design*, Prentice-Hall, New Jersey, USA.

Hossain K., Das P. & Karakas A. 2018 'Effect of Ultraviolet Aging on Rheological Properties of Asphalt Cement', *Canadian Technical Asphalt Association*, Toronto, Ontario, Canada.

Hossain R. & Wasiuddin N.M. 2019 'Evaluation of Degradation of SBS Modified Asphalt Binder Because of RTFO, PAV, and UV aging using A Novel Extensional Deformation Test' *Transportation Research Record*, vol. 2673. no. 6, pp. 447–457.

Kidd A., Stephenson G. & White G. 2021 'Towards the use of Crumb Rubber Modified Asphalt for Local Government Roads', *AfPA International Flexible Pavements Symposium*, A Virtual Event, 3–5 August.

Lu X. & Isacsson U. 2000 'Artificial Aging of Polymer Modified Bitumens', *Journal of Applied Polymer Science*, vol. 76, no. 12, pp. 1811–1824.

Mikhailenko P., Ataeian P. & Baaj H. 2019, 'Extraction and Recovery of Asphalt Binder: A Literature Review', *International Journal of Pavement Research and Technology*, vol. 13, pp. 20–31.

Mirwald J.,Werkovits S., Camargo I., Maschauer D., Hofko B.,Grothe H. 2020 'Investigating Bitumen Long-term-ageing in the Laboratory Byspectroscopic Analysis of the SARA Fractions', *Construction and Building Materials*, vol. 258.

Mohammadafzali M., Ali H., Musselman J.A., Sholar G.A. & Massahi A. 2017, 'The Effect of Aging on the Cracking Resistance of Recycled Asphalt', Advances in Civil Engineering, vol. 2017.

Pérez-Jiménez F.P., Miró Recasens R. & Martínez A. 2008, 'Effect of the Nature and Filler Content on the Behavior of the Bituminous Mastics', *Road Materials and Pavement Design*, vol. 9, pp. 417–431.

Porto M., Caputo P., Loise V., Eskandarsefat S., Teltayev B. & Oliviero C. 2019, 'Bitumen and Bitumen Modification: A Review on Latest Advances', *Applied Sciences*, vol. 9, no. 4.

Rad F., Elwardany M., Castorena C. & Kim Y. 2017, 'Investigation of Proper Long-term Laboratory Aging Temperature for Performance Testing of Asphalt Concrete', *Construction and Building Materials*, vol. 147, pp. 616–629.

Shell 2015 *The Shell Bitumen Handbook*, 6th edn, ICE Publishing, Italy.

Siddiqui M.N. & Ali M.F. 1999, 'Investigation of Chemical Transformations by NMR and GPC During the Laboratory Aging of Arabian Asphalt', *Fuel*, vol. 78, no. 12, pp. 1407–1416.

Sirin O., Paul D.K. & Kassem E. 2017, 'State of the Art Study on Aging of Asphalt Mixtures and Use of Antioxidant Additives', *Advances in Civil Engineering*, vol. 2018.

Tarsi G., Aikaterini V., Lanttieri C., Scarpas A. & Sangiorgi C. 2018, 'Effects of Different Aging Methods on Chemical and Rheological Properties of Bitumen', *Journal of Materials in Civil Engineering*, vol. 30, no. 3.

Van Den Heuvel D. & White G. 2021, 'Objective Comparison of Sustainable Asphalt Concrete Solutions for Airport Pavement Surfacing', *International Conference on Sustainable Infrastructure*, a virtual event, 6–10 December.

Wang H., Feng Z., Zhou B., & Yu J. 2012 'A Study on Photo-thermal Coupled Aging Kinetics of Bitumen', *Journal of Testing and Evaluation*, vol. 40 no. 5, pp. 724–727.

White G. 2018, 'State of the Art: Asphalt for Airport Pavement Surfacing', *International Journal of Pavement Research and Technology*, vol. 11, no. 1, pp. 77–98.

White G. 2020, 'A Synthesis of the Effects of Two Commercial Recycled Plastics on the Properties of Bitumen and asphalt', *Sustainability*, vol. 12, no. 8594, pp. 1–20.

White, G. & Abouelsaad A. 2021 'Towards More Realistic Accelerated Laboratory Aging of Asphalt Samples', *International Symposium on Frontiers of Road and Airport Engineering*, Delft, Netherlands, 12–14 July.

White G. & Fergusson K. 2021, 'Exploring the Durability Specification of Coarse Aggregate used in Airport Asphalt Mixtures', *Sustainable Issues in Infrastructure Engineering*, Shehata, H. & El-Badawy, S, (Eds.), vol. 2, pp. 207–224.

White G. & Reid G. 2022, 'Challenges for the Development and Implementation of Asphalt Mixtures Containing Recycled Waste Plastic for Pavement Surfacing', *9th Symposium on Pavement Surface Characteristics*, Milan, Italy, 12–14 September.

White G. & Thompson M. 2016, 'Australian Airport Asphalt Surface Treatments', *6th Eurasphalt & Eurobitume Congress*, Prague, Czech Republic, 1–3 June.

Widyatmoko I., Hakim B., Fergusson C., Richardson J. & Cant S. 2009, 'Sustainable Airport Pavement using French Asphaltic Materials (BBA and EME2)', *2nd European Airport Pavement Workshop*, Amsterdam, Netherlands, 13–14 May.

Wu J. & Airey G.D. 2009, 'The Influence of Aggregate Interaction and aging Procedure on Bitumen Aging', *Journal of Testing and Evaluation*, vol. 37, no. 5, pp. 402–409.

Wu S., Pang L., Liu G. & Zhu J. 2010 'Laboratory Study on Ultraviolet Radiation Aging of Bitumen', *Journal of Materials in Civil Engineering*, vol. 22, no. 8.

Xiao, F., Amirkhanian, A. & Shen, J. 2009 'Effects of various long-term aging procedures on the rheological properties of laboratory prepared rubberized asphalt binders', *Journal of Testing and Evaluation*, vol. 37, no. 4, pp. 329–336.

Yan, C., Huang, W., Lin, P., Zhang, Y., & Lv, Q. 2019 'Chemical and rheological evaluation of aging properties of high content SBS polymer modified asphalt', *Fuel*, vol. 252, pp. 417–426.

Yu, C., Hu, K., Yang, Q., Chen, Y. 2021, 'Multi–scale observation of oxidative aging on the enhancement of high–temperature property of SBS–modified asphalt', *Construction and Building Materials*, vol. 313.

Zeng, W., Wu, S., Pang, L., Chen, H., Hu, J., Sun, Y., Chen, Z. 2018, 'Research on ultra violet (UV) aging depth of asphalts', *Construction and Building Materials*, vol. 160. pp. 620–627.

Optimization of reinstating drill core holes – laboratory investigation and accelerated pavement testing

Tim Schroedter, Pahirangan Sivapatham*, Stefan Koppers & Hartmut Beckedahl
Faculty of Architect und Civil Engineer, Pavement Construction and Maintenance, University of Wuppertal, Wuppertal, Germany

ABSTRACT: In order to assess construction work on roads, drill cores are taken which tend to cause a weakening of the road construction. Inferior closings of drill core extraction points often result in open holes in the pavement or damage to the surrounding original construction. For the purpose of achieving a professional and high-quality execution of the closing, research was carried out on different methods and materials, and a selection of 14 closing variants were evaluated. The closings consist of a surface layer replacement and a substructure. They were tested using modified and adapted test methods for leakage and durability under cyclical load. Hot mix closings (HM) produced in the laboratory are waterproof and enable a good bond with the original construction, whereas cold mix closings (CM) produced in the laboratory were permeable. Despite similar compaction, the surface images differ between the closings produced in the laboratory and those produced in situ. The durability tests (cyclical load) deform the core fillings in different ways during the consolidation phase and correlate with the rigidity of the substructure or the compactibility of the materials used. Two closure variants (HM, CM) were installed on the demonstration, investigation and reference site of the German Federal Highway Research Institute (duraBASt) at a test site and loaded with the Mobile Load Simulator MLS30. Traverse evenness and Falling Weight Deflectometer (FWD) measurements were performed continuously. The results show that given a proper closing of the drill core sampling point, no impairment of load capacity is to be expected. Finally, the insights gained were summarized in a draft manufacturing procedure specification.

Keywords: Drill core reinstating, laboratory testing, Accelerated Pavement Testing APT, Mobile Load Simulator MLS30

1 INTRODUCTION

Drilling cores are already being taken today to check road construction work. The technical requirements for taking cores are contained in the Technical Test Specifications for Asphalt (TP Asphalt-StB) (FGSV 2012a) and the Technical Test Specifications for Base Courses with Hydraulic Binders and Concrete Pavements (TP Beton-StB) (FGSV 2010). In contrast to the extraction of drill cores, the reinstating and sealing of drill core holes has not yet been regulated.

With the completion and introduction of the guidelines for structural assessment in Germany (RSO Asphalt and RSO Concrete), numerous drill cores will be necessary at the respective object during the period of use. For this purpose, homogeneous sections will be

---

*Corresponding Author: psivapatham@uni-wuppertal.de

DOI: 10.1201/9781003429258-2

identified and subsequently sampled with 16 drill cores in the first kilometre and five further drill cores for each additional kilometre within the homogeneous section.

In practice, it has been shown that the areas of the drill core holes often tend to premature damage (cracks, etc.) and thus cause a weakening of the road construction. The individual reasons for this premature damage are not always comprehensible. Taking drill cores is described in the relevant guidelines, but the procedure for long lasting reinstatement of the holes is not specified. A sustainable control of the quality of the closure of a drill core hole is also not taking place at present because no corresponding specifications exist.

In order to maintain the required quality level for the reinstatement of drill core holes, a professional and high-quality execution is necessary. This means that all damages to or in the immediate surroundings of the core sampling points should be reduced and avoided. For this purpose, both conventional and new methods are to be used and systematized. From this, a draft for a process description for the professional reinstatement and the quality control of drill core holes is to be compiled, which can be supplemented to the road construction-technical guidelines.

The research was divided into three parts. The first part includes the compilation and quantified comparison of procedures for the professional reinstatement of drill core holes. This includes conventional methods such as reinstating with cold mix, mastic asphalt and fast cement mortar but also with existing old drill cores that are glued in. In addition to the drill core closure, the direct surrounding of the road construction is also a decisive point of investigation.

In the second part, two drill core closure variants were installed and examined on a 1:1 scale on a test section of the German Federal Highway Research Institute (BASt). The drill cores were positioned in a coordinated scheme in order to consider two loading situations (full or partial loading) by rolling truck wheels. The load was applied within two months with the Mobile Load Simulator MLS30.

Within the third part, the findings from the previous parts were combined for practical implementation, especially on roads with a high volume of traffic. In the process, a draft for a process description for durable reinstatement was prepared, which can be incorporated into the road construction rules and regulations.

2 DIFFERENT DRILL CORE REINSTATEMENT VARIANTS

Basically, the different reinstatement variants can be distinguished by the asphalt or concrete surface course replacement (fillup material for top layer) and the fillup material for the substructure. Within the scope of the research, five different reinstatement materials and surface course replacement materials were used to replace the surface course on asphalt roads. These include hot mix (HM), emulsion-bound cold mix (CM-E) and reactive cold mix (CM-R), as well as two solid body reinstatement variants (bitumen plug and repair drill core). A road repair concrete C30/37 was used to replace the concrete slabs. For the substructure, unbound aggregates (AG) and hydraulically bound aggregates (hbAG) as well as cold mix and hot mix were used for the asphalt construction. This results in 14 closure variants, which are described in Table 1.

2.1 *Laboratory evaluation*

A laboratory evaluation of closure variants should provide an objective opportunity for testing and comparing different closure methods and materials. In particular, the main properties – durability and tightness – are to be tested under conditions that are as realistic as possible. Existing test methods were modified and adapted for this purpose and a test system was developed. This procedure was necessary because no prescribed test methods are available.

15

Table 1. Overview of the reinstatement variants.

Variant	Fillup material for top layer	Fillup material for substructure	Construction
A	Hot mix (HM)		Asphalt
B	Concrete (C 30/37)		Concrete
C	Hot mix (HM) MA 5 S	Hydraulically bound aggregates (hbAG)	Asphalt
D	Reactive cold mix CM-R 0/4	Hydraulically bound aggregates (hbAG)	Asphalt
E	Cold mix CM-E 0/8	Hydraulically bound aggregates (hbAG)	Asphalt
F	Cold mix CM-E 0/8	Cold mix CM-E 0/11	Asphalt
G	Cold mix CM-E 0/8	Unbound aggregates (AG) 8/11	Asphalt
H	Cold mix CM-E 0/8	Unbound aggregates (AG) 5/11	Asphalt
I	Reactive cold mix CM-R 0/4	Unbound aggregates (AG) 8/11	Asphalt
J	Reactive cold mix CM-R 0/4	Unbound aggregates (AG) 5/11	Asphalt
K	Hot mix (HM) MA 5 S	Unbound aggregates (AG) 8/11	Asphalt
L	Hot mix (HM) MA 5 S	Unbound aggregates (AG) 5/11	Asphalt
M	Bitumen plug	Unbound aggregates (AG) 8/11	Asphalt
N	Repair drill core	Unbound aggregates (AG) 5/11	Asphalt

2.1.1 *Concept*

Durability testing was performed on drill core closure variants manufactured in the laboratory and subjected to dynamic loading in a testing machine. The durability test is based on the test principle of the uniaxial pressure-swelling test with restricted lateral elongation. The aim of the test is to show how the various closure variants behave under cyclic loading. Particular attention must be paid to possible settlement / post-compaction and possible damage occurring in the joint area between the surface course replacement material and the original construction.

For this purpose, appropriate test specimens with the various closure variants had to be produced. Since most of the closure variants consist of two layers, the upper layer with the surface course replacement material and the lower layer with the substructure, a correspondingly deep drill hole had to be drilled or produced in a laboratory prepared asphalt specimen.

Therefore, asphalt specimens were assembled from individual asphalt slabs. The asphalt specimen thus consists of an approx. 30 mm thick base slab without drill hole (AC 5 D N) and three asphalt slabs (AC 16 B S), each 80 mm thick, as well as a 40 mm high surface course slab (SMA 11 S), in each of which a drill core was removed centrically. The individual asphalt slabs were then bonded with a 50/70 road bitumen.

These laboratory-drilled holes, with a diameter of 150 mm and a depth of approx. 280 mm, were then reinstated with the 14 different closure variants. The reinstatement materials were compacted using an electric percussion hammer with tamper plate (Ø 140 mm).

A production of test specimens in the laboratory for variant M (bitumen plug) could not take place. The heating of the drill core hole reduced the stiffness of the asphalt so considerably that the test specimens cracked radially to the drill core hole when the bitumen plug was driven in. It could be observed that the bitumen plugs displace the asphalt considerably due to their dimensions and the surrounding warm asphalt. The resulting clamping and fusion of the bitumen plug with the asphalt is probably very effective, but could not yet be realistically simulated in the laboratory.

A method based on DIN EN 12697-19 "Asphalt – Test method for hot asphalt (Part 19: Permeability of test specimens)" (DIN 2012) was developed for testing the impermeability of the surface course substitute material. This method was used to test whether water could penetrate the substructure of the drill core hole closure through the surface course replacement material. The substructure was not tested.

For the test, so-called carrier slab were first produced, which were used to accommodate the surface course substitute material to be tested. For this, rectangular asphalt or concrete slabs with a height of 50 mm were produced. Subsequently, a centric drill core removal (Ø 150 mm) was carried out from these test slabs to produce the drill hole.

The drill core hole in the carrier slab was then closed with various surface course replacement materials (CM-R, CM-E, HM, Repair drill core, bitumen plug and Concrete). For the final test for tightness of the surface course replacement materials, compaction was carried out using an electric percussion hammer with tamper plate (Ø 140 mm). This should result in a more even and, if necessary, more intensive compaction. The self-compacting HM used did not require additional compaction.

2.1.2 Durability tests

The test was carried out according to the principle of the uniaxial pressure-swelling test based on the TP Asphalt-StB Part 25 B1 (FGSV 2012b). The loading of the drill core hole closure was carried out dynamically via a load plate (Ø = 96 mm) by means of impulse loading. The applied cyclic pulse load of 0.35 N/mm^2 corresponds to the high tension from the TP Asphalt StB Part 25 B1 (4). The duration of the test was fixed at 30,000 load cycles and the test temperature at 20°C. The long duration of the test was intended to ensure that any settlement behaviour or post-compaction that might occur was sufficiently well documented.

In preliminary tests, higher stresses were experimented with, but damage was found in the area of the surface course slab. In order to avoid such damage, a higher overstress was dispensed with.

The quantitative results of the durability test are shown in Table 2, where the deformation in mm after 10,000 loading cycles and after 30,000 loading cycles as well as the gradient between the 10,000th and 30,000th loading cycle are shown parts per million [ppm]. The

Table 2. Results of the durability test.

| Variant | Deformation | | Slope | Rank |
	After 10,000 cycles [mm]	After 30,000 cycles [mm]	Between cycles 10,000 and 30,000 [ppm]	According to the slope
A	−0,29	−0,35	−3,19	6
B	−0,12	−0,14	−1,20	1
C	−0,41	−0,49	−4,03	8
D	−0,41	−0,47	−3,17	5
E	−0,23	−0,29	−2,93	4
F.1	−0,77	−1,06	−14,67	18
F.2	−0,66	−0,97	−15,53	19
G.1	−0,63	−0,78	−7,43	15
G.2	−0,33	−0,45	−5,81	13
H.1	−0,42	−0,54	−6,04	14
H.2	−0,46	−0,63	−8,30	16
I	−0,26	−0,32	−2,57	2
J	−0,69	−0,86	−8,48	17
K.1	−0,93	−1,03	−5,07	10
K.2	−0,60	−0,70	−4,98	9
L.1	−0,56	−0,66	−5,12	11
L.2	−0,47	−0,58	−5,46	12
N.1	−0,12	−0,18	−2,79	3
N.2	−0,44	−0,51	−3,88	7

results as creep curves of the tested closure variants are shown in Figure 1. The section for determining the slope dimension of the creep curve was selected so that there is a sufficient phase for consolidation and the further course is nevertheless representative of long-term behaviour with an approximately linear slope. In addition, the rank to the other variants is determined according to the slope dimension.

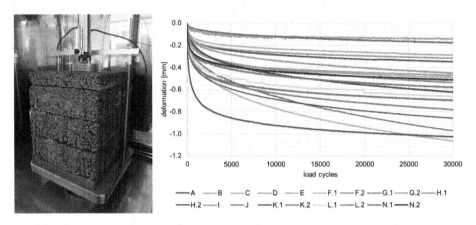

Figure 1.   Durability test setup (left) and results (right).

2.1.3   *Leakage tests*

After closing the drill core removal point, a plastic pipe (Ø 200 mm) with a silicone sealing was attached to the carrier slab in a watertight manner. A defined amount of water was filled into the plastic tube to check the tightness and to check whether water was escaping from the underside of the reinstated drill core hole.

The termination criteria of the test were either the complete flow of water or an unchanged water level after 60 minutes compared to the water level after 30 minutes (saturation period of the test specimen) or a dry underside after 60 minutes. These termination criteria resulted in "leaky" or "tight" test results. If after 60 minutes, the complete amount of water has not flowed through but the water level has changed significantly or the underside of the drill core hole is wet, the test result is "moderately leak-proof". In the case of leaky or moderately leaky test results, it shall also be clarified whether a justification for the test result is discernible (e.g. defective test specimen production).

2.1.4   *Findings*

In conclusion, it can be stated that the results of the durability test largely correspond to expectations and that there was no failure of a variant. With regard to the compactibility or compression willingness of the substructure, it was observed that the use of slightly graded aggregate (5/11) had a clearly positive effect on the spans of the test results. The spans of the variants H and L after 30,000 cycles are 0.09 mm and 0.08 mm. A similar span width (0.09 mm) could be determined for variant F. The variants with a single grain mixture (8/11) as substructure (variants G, K and L) each provided a span of 0.33 mm between the test results.

The spans between the test results of the durability test show a possible reproducibility or uniformity in the compaction of the substructure. Due to the manufacturing process, the compaction with a motor hammer and the small number of tests, however, no statistically reliable statement can be made.

Overall, the method developed here for testing durability provides a very small span for the measured deformations for all investigated variants. The maximum measured deformation on the surface of the backfill for the variant F is 1.06 mm. Due to the low upper stress under cyclic loading, the effective depth and thus the influence of the substructure on the test results is unknown. It should also be noted that influences resulting from the ageing of surface course replacement materials, particularly CM, could not be taken into account. In addition, the production of test specimens is extremely time-consuming and reproducibility cannot be proven.

The results of the tightness tests show that with the road repair concrete, the hot mix and the repair drill core a tight surface course replacement or concrete ceiling replacement can be produced. These results are plausible and correspond to in-situ observations. The situation is different for both types of cold mix (CM-E and CM-R), for which the leakage of the investigated materials was determined. However, despite mechanical compaction using a hammer drill and plate, the surface appearance was noticeably different and more open-porous than with in-situ compaction. Considering the surface appearance when evaluating the results, it must be mentioned that no realistic compaction was possible in the laboratory.

For the load tests with the mobile load simulator MLS30 on the duraBASt, the variants H and L were selected based on the findings. The decision for the substructure 5/11 was based on the small range of the test results and the resulting reproducible compaction. The variants with HM, CM-E and CM-R were taken into consideration for the selection of the surface course replacement material due to their frequent use and low technical expenditure. For this purpose, the span of the test results of all variants with the substructures made of AG 5/11 and AG 8/11 was determined and evaluated. The hydraulically bound aggregates were not considered due to the technical complexity. In this analysis, the smallest span widths were determined for the variants with HM and CM-E surface course replacement materials.

2.1.5 *Full-scale test*

Within the scope of the project, laboratory tests were carried out as well as full-scale operations. For this purpose, drill cores were taken from the public road space, reinstated with various reinstatement variants and observed over a period of one year. In addition, closure variants were installed on the demonstration, investigation and reference site of the BASt (duraBASt) for an Accelerated Pavement Test. These two investigations are described and evaluated below.

2.1.6 *Long-time evaluation*

In addition to the laboratory evaluation of closure variants, field tests were carried out on roads in the urban area of Wuppertal (Germany). Different closure variants were installed at different locations and continuously observed and documented over a period of approx. one year during the project. As quality control, a photo documentation of the drill core closures was accomplished.

Within the scope of control tests in Wuppertal, eight different closure variants were installed in four test fields in 2017. For comparison purposes, these were produced in two versions. In test fields 1 and 2, variants C and K (HM) and variants D and I (CM-R) were installed. In test fields 3 and 4, variants E, F, G and H (CM-E) were used (Figure 2).

An image documentation of the test fields for the detection of possible damages and the recording of the progress was carried out. No damage was found at the drill core holes of both test fields.

2.1.7 *Accelerated pavement testing*

To test the closure variants H and L, ten drill core reinstatement points were closed with two different closure materials in a test area on the demonstration, investigation and reference site of the BASt (duraBASt). This was followed by an Accelerated Pavement Test with the Mobile Load Simulator MLS30, a mobile large-scale test facility on a scale of 1:1, with a total of 1.6 million overruns with a wheel load of 50 kN. The test field and the closures of the drill core reinstatement points were continuously examined using a Falling Weight

19

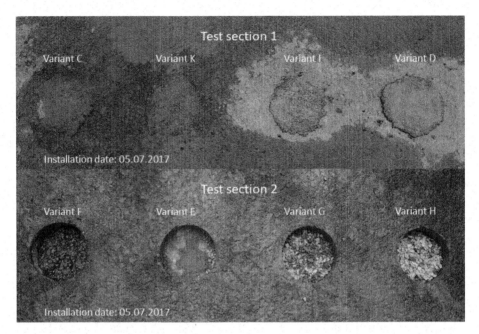

Figure 2. Test section 1 and 4 in public road.

Deflectometer (FWD) and transverse flatness meter. In addition, the test was accompanied by photographic documentation.

The loading area is 3.50 m long and 1.00 m wide. The rolling track of the MLS30 has a width of 0.46 m. The two closure variants were each used for five sampling points (Ø 150 mm). Of these, two are located in the rolling track, which was rolled over completely, and three each at the edge, which were rolled over approximately halfway (Figure 3).

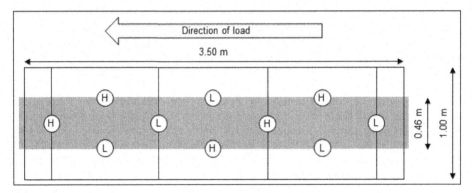

Figure 3. Scheme of the loading area with closure variants H and L as well as the rolling track of the MLS30.

In summary, a positive conclusion can be drawn from the APT program with the MLS30 on duraBASt. The closures of both variants have withstood the high stresses of 1.6 million loading cycles at partly high air temperatures. There were no eruptions or total failure of the surface course replacement materials. The settlements took place approximately at the same height as the surrounding original construction (Figure 4). Only the tread marks of the Super-Single tire on the surface of the variant L are negatively significant, but can be explained by the loading without lateral wanderer and the extreme boundary conditions. At the edge of the rolling track, the L variant delivers better results than the H variant. In the roll-over area, both closure variants also have a settlement of nearly the same height as the surrounding original construction. Outside of the loading area, strong compressions can be observed especially in the H variant. The results of the FWD measurements do not indicate any direct influence of the closed drill core holes on the load-bearing capacity of the surrounding road construction.

Figure 4.   Closure variant H in the rolling track after 1,600,000 loading cycles.

3   SUMMARY

The laboratory analysis paired with a lot of practical experience in this project detected 14 different closure variants (variants A to N) and divided them into different application areas. With the help of various tests (leakage and durability tests), objective results could be achieved in order to classify the different variants. These investigations also enabled important findings to be obtained for the laboratory analysis of new closure variants. With the help of these investigations, practical results before implementation into the road network will also be possible in the future. Further important findings could be gathered by the parallel use of different variants at one point in the road network. In addition, two selected variants could be monitored in an Accelerated Pavement Test Program on duraBASt. Very good results were achieved especially with the direct rollover and the L variant also performed very well with the half-sided load. All findings were compiled in a first process description. In this process, the most realistic and applicable variants possible were used in order to optimize the use of the machine as much as possible. A rock mixture with a grain size of 5/11 is therefore to be paved in layers and compacted with a motor hammer. For reinstating, the uppermost 4 cm must be reinstated with the cold mix asphalt (variant H) or with hot mix asphalt (variant L). Based on the research results and in special on the long-time-evaluation, the road construction department of the city of Wuppertal decided to reinstate drill core holes with the tested emulsion-bound cold mix asphalt. So far, no damage characteristics have been encountered.

ACKNOWLEDGEMENTS

This report is based on parts of the research work carried out on behalf of the Federal Ministry of Transport and Infrastructure, represented by the Federal Highway Research Institute, under FE-No. 07.0279/2014 (Beckedahl *et al.* 2020). The responsibility for the content lies solely with the authors.

REFERENCES

Beckedahl, H. J., Koppers, S. and Schrödter, T. (2020): *Untersuchungen zur Verbesserung der Methode zum fachgerechten Schließen von Bohrkernentnahmestellen* (engl.: *Investigations to Improve the Method Used for Professional Closing of Drill Core Extraction Points*). Schünemann Verlag, Bremen.

Deutsches Institut für Normung e. V. (2012): DIN EN 12697-19:2012, *Asphalt – Prüfverfahren für Heißasphalt: Teil 19: Durchlässigkeit der Probekörper (engl.: Asphalt – Test methods for Hot Mix Asphalt: Part 19: Permeability of Test Specimens*). Beuth Verlag, Berlin.

Forschungsgesellschaft für Straßen- und Verkehrswesen e. V. (2010): *Technische Prüfvorschriften für Tragschichten mit hydraulischen Bindemitteln und Fahrbahndecken aus Beton (TP Beton-StB)*, Ausgabe 2010 (engl.: *Technical Test Specifications for Base Courses with Hydraulic Binders and Concrete Pavements*). FGSV Verlag, Köln.

Forschungsgesellschaft für Straßen- und Verkehrswesen e. V. (2012a): *Technische Prüfvorschriften für Asphalt (TP Asphalt-StB) Teil 27: Probenahme*, Ausgabe 2012 (engl.: *Technical Test Specifications for Asphalt Part 27: Sampling*). FGSV Verlag, Köln.

Forschungsgesellschaft für Straßen- und Verkehrswesen e. V. (2012b): *Technische Prüfvorschriften für Asphalt (TP Asphalt-StB) Teil 25 B1: Einaxialer Duck-Schwellversuch – Bestimmung des Verformungsverhaltens von Walzasphalt bei Wärme*, Ausgabe 2012 (engl.: *Technical Test Specifications for Asphalt Part 25 B1: Uniaxial pressure-swelling test – Determination of the deformation behavior of asphalt under heat*). FGSV Verlag, Köln.

Roads and Airports Pavement Surface Characteristics – Crispino & Toraldo (Eds)
© 2023 The Author(s), ISBN 978-1-032-55149-4

Comparative investigation on the use of fibers into porous bituminous mixtures

Edoardo Mariani* & Emanuele Toraldo
Department of Civil and Environmental Engineering, Politecnico di Milano, Milan, Italy

ABSTRACT: Porous hot mix asphalts (PHMAs) are widely used for safety purposes as wearing courses of bituminous pavements. PHMAs are characterized by high air void content, allowing high permeability and rainwater drainage, thus increasing traffic safety trough hydroplaning, and splash-spray phenomena reduction.

The high content of voids is obtained by an aggregates sieve size distribution without a significant portion of intermediate-fine particles, especially sand. The final voids content is almost five times higher than the one of a standard wearing course. This implies the reduction of the stone-to-stone contacts into mixtures, and the consequent decrease of the course structural performance. Polymer-modified bitumen is used as binder for increasing the structural response and cellulose fibers are added for reducing loss of binder during transportation and laying operations, caused by the lack of the intermediate-fine part of the aggregate gradation. However, it is possible to replace the polymer-modified bitumen with neat bitumen if the fibers are composed by cellulose and other components able to replace the role of the polymer into the binder. This is the focus of the laboratory investigation described in this paper, in which the role of three cellulose-based fibers into neat bitumen PHMAs was assessed in terms of compaction properties, volumetric characteristics, mechanical behavior, and functional performance, also in comparison with the same mixture in which polymer-modified bitumen was used.

Keywords: Pavement Materials, Pavement Surface Properties, Road and Airfield Safety

1 INTRODUCTION

Porous Hot Mix Asphalts (PHMAs) are a particular kind of hot mix asphalts currently used in Europe as wearing course of road pavements. Such PHMAs are characterized by high air voids content, and then high permeability, which enhances the drainage of rainwater, thus reducing hydroplaning and splash and spray phenomena produced by the vehicles' movement. Consequently, PHMAs increase traffic safety because their ability in reducing the water veil between pavement and tires, thus guarantying satisfactory levels of friction also in case of rainfalls.

PHMAs are mainly composed by coarse crushed aggregates without a significant portion of fine particles, especially sand, producing an interconnected void system in which the water can drain during a rain event. From a mechanical point of view, this imply the reduction of stone-to-stone contacts into mixtures, and the consequent decrease of the structural performance of the course. For this reason, modified bitumen is currently used as binder for PHMA. The need of high working temperatures for this bitumen (because to reduce its viscosity) and the lack of intermediate-fine part of the aggregates cause loss of binder during

*Corresponding Author: edoardo.mariani@polimi.it

DOI: 10.1201/9781003429258-3

transportation and laying operations. For reducing this drain-off effect, cellulose-based fibers are currently used, because their capability of absorbing bitumen. Acting as a sponge, these fibers increase the thickness of the mastic film enveloping the aggregates. In other words, the binding system is composed by modified bitumen, filler, together forming the so-called bituminous mastic, and cellulose fibers, the role of which is strictly related to the working phase. However, it is possible to replace the modified bitumen with a standard bitumen if the fibers are composed by cellulose and other components able to replace the role of the polymer into the modified bitumen. This is the focus of the paper. In fact, the research herein described was aimed at evaluating the role of three cellulose-based fibers into PHMAs bonded using standard bitumen, comparing their compaction properties, volumetric characteristics, mechanical behaviour and functional performance to the same mixture in which modified bitumen is used.

2 LITERATURE REVIEW AND SCOPE

In the past, several researchers addressed their efforts on the study of PHMAs (Alvarez *et al.* 2011; Scholz & Grabowiecki 2007). Moreover, the PHMAs' permeability is the core of several recent studies aimed at defining the most appropriate features of such materials in order to guarantee an efficient drainage system and a filter from pollution, together with a good clogging resistance (Cedergren *et al.* 1973; Klenzendorf *et al.* 2011; Pagotto *et al.* 2000; Sansalone *et al.* 2012).

Many researches are focused on the influence of binder type on the performance of PHMAs (Chen J. S. *et al.* 2012; Liu & Cao 2009; Molenaar J. & Molenaar A. 2000; Suresha *et al.* 2009a), other investigate PHMAs volumetric features (Alvarez *et al.* 2010; Praticò & Moro 2007), top down cracking performance (Chen Y. *et al.* 2012) or compaction technique and raveling resistance (Mo *et al.* 2009; Suresha *et al.* 2009b). As far as fibers, some researches demonstrate the positive effect of such additives as reinforcement in both HMAs (Abtahi *et al.* 2010; Toraldo *et al.* 2015) and PHMAs (Mallick *et al.* 2000). For the latter, available literature is not agree about the effects of fibers on the drain-down effects of PHMAs (Hassan *et al.* 2005; Punith *et al.* 2011), as also demonstrated by a number of recent papers, in which organic, i.e. date-palm, textile (Hassan & Al-Jabri 2005), polymeric (Punith & Veeraragavan 2010) and cellulose fibers (Cooley *et al.* 2000; Wu *et al.* 2006) are investigated.

Given the literature review briefly mentioned above, the research herein described focused on the evaluation of the use of cellulose-based fibers as reinforcement of PHMAs bonded using standard bitumen. For this purpose, a laboratory comparison was performed, considering three types of cellulose-based fibers currently available on the European market. The fibers were composed by cellulose, cellulose and glass, cellulose and recycled polymer. The investigation included a number of tests to highlight the effects of fibers in terms of compaction properties (Workability and Self-compaction from the Gyratory Shear Compactor), volumetric characteristics (Void Contents, Voids in the Mineral Aggregates and Voids Filled with Bitumen of the compacted specimens), mechanical behavior (Stiffness S and Indirect Tensile Strength ITS, both at three different temperatures) and functional performance (Permeability and Surface Friction). The effects of bitumen content were also considered in the study. Finally, for comparative purposes, four PHMAs bounded with modified bitumen (two dosages) were analyzed: (I) using the sole modified bitumen and (II) adding glass cellulose fibers to the same modified binder. The latter mixture was included in the investigation because it is currently used in Europe to prevent the loss of bitumen during transportation and laying operations.

3 EXPERIMENTATION

3.1 *Key materials*

The key materials used during the research described in this paper were:

- 50–70 standard bitumen (EN 12591), herein named SB;
- 25-55/60 polymer modified bitumen (EN 14023), containing 4% of SBS polymers and named MB;
- calcareous aggregates;
- calcareous filler completely passing through a 75μm sieve;
- three cellulose-based fibers: a pure cellulose (named Fiber A), a cellulose and glass (Fiber B) and a cellulose and recycled polymer (Fiber C).

More details of both bitumen and fibers are given in Tables 1 and 2, respectively; Figure 1 reports the aggregates gradation; Los Angeles Coefficient, Shape Index and Flakiness Index of the aggregates are presented in Table 3.

Table 1. Main characteristics of the bitumen.

Test	Standard Specification	Standard Bitumen -SB-	Modified Bitumen -MB-
Needle penetration [dmm]	EN 1426	51.3	35.2
Ring and Ball [°C]	EN 1427	46.5	61.7
Rotational Viscosity [cP]	EN 13302	116	424

Table 2. Main characteristics of the fibers.

Fiber	Source	Length [μm]	Diameter [μm]	Temperature resistance [°C]
A	Cellulose	~200	~7	>200
B	Cellulose and glass	~200	~7	>200
C	Cellulose and recycled polymer	~200	~10	>200

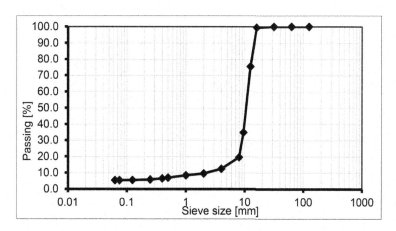

Figure 1. Aggregates sieve size distribution.

Table 3. Main characteristics of the aggregates.

Test	Standard Specification	Result
Los Angeles Coefficient [%]	EN 1097-2	13
Shape Index [%]	EN 933-4	11
Flakiness Index [%]	EN 933-3	12

3.2 *Experimental plan*

As mentioned above, the research described in this paper was aimed at evaluating the role of fibers in reinforcing PHMAs bonded by standard bitumen in substitution of modified one. Twelve mixtures were prepared in the laboratory, considering two bitumen contents (4.0% and 4.5% with respect to the weight of aggregates). For each bitumen content, two reference mixtures were prepared using either modified or standard bitumen, three mixtures were obtained adding 9% of the above-mentioned fibers (with respect to the weight of standard bitumen), and one mixture was obtained combining modified bitumen and cellulose-glass fibers. The fiber content (9% by the weight of bitumen) was selected according to the range (1.5% and 12% by the weight of bitumen) currently used for road applications (Bonica *et al.* 2016; Crispino *et al.* 2013; Toraldo *et al.* 2015).

In this paper, the mixtures are identified by an alphanumeric code formed by content (4.0 or 4.5) and type of bitumen (SB for standard bitumen or MB for the modified one) followed by the fiber identification (A, B, or C); e.g., the code of a bituminous mixture containing standard bitumen at 4.0% and fiber A will be 4.0-SB-A.

A heated laboratory mixer was used to prepare the mixtures. Mixing and compaction temperatures were set according to the results of a previous investigation carried out by the Authors (Crispino *et al.* 2013) focused on the effects of fibers on corresponding bituminous mastics at working temperatures. This was crucial in order to maintain almost constant the volumetric characteristics of the mixtures, thus to compare their mechanical and functional performances at the same volume of voids. Both the above-mentioned temperatures are given in Table 4.

Table 4. PHMAs mixing and compaction temperatures.

Mixture Code	Reference Temperature* [°C]	
	Mixing	Compaction
4.0-SB	155	145
4.5-SB		
4.0-SB-A	187	177
4.5-SB-A		
4.0-SB-B	159	149
4.5-SB-B		
4.0-SB-C	190	180
4.5-SB-C		
4.0-MB	175	165
4.5-MB		
4.0-MB-B	178	168
4.5-MB-B		

Notes
*Reference Temperature ± 5°C

26

Each mixing batch was 25 kg. After mixing, a 3.0 kg sample of each mixture was used to control both aggregates gradation and bitumen content and to measure the Theoretical Maximum Density (EN 12697-5). The other part of the batch was used to compact cylindrical specimens by using the Gyratory Shear Compactor – GSC (EN 12697-31), obtaining nine nominally alike specimens (150 mm in diameter and 70 mm in height; number of gyrations of 130) for each mixture.

The use of the GSC allowed to record the degree of compaction of a specimen at each rotation, and thus to calculate: Self-compaction (C_I), that is able to describe the mixtures' proneness to reduce its internal voids under its own weight, and Workability (k), that gives information about the mixtures' ability to be compacted under the rollers on site (Cominsky *et al.* 1994; Santagata & Bassani 2002).

The GSC compacted specimens were subjected to a complete volumetric characterization (EN 12697-8), that included the evaluation of voids content v (that provides the measure of the compaction level achieved by a given mixture), voids in the mineral aggregates *VMA* (that assesses the degree of packing of the aggregate particles) and voids filled with bitumen *VFB* (that gives the degree of void filling caused by the effective bituminous binder).

In order to investigate the contribution of fibers to the PHMAs' performance, the mixtures' mechanical behavior was evaluated by means Stiffness tests (EN 12697-26 – Annex C), and Indirect Tensile Strength (EN 12697-23) at three different temperatures (5°C, 20°C and 40°C). In detail, Stiffness test was selected because it reveals the binding effects due to both bitumen and fibers among the stone particles into the mixtures in a range of temperatures suitable to cover the typical climatic variation in the Mediterranean regions. Similar findings can be obtained by Indirect Tensile Stress Tests, even if in this case, at the failure limits of the mixtures. Moreover, both tests allow highlighting the performance of the PHMA with respect the expectation in the field: rutting potential at high temperatures, bearing capacity at intermediate temperatures, and cracking at low temperatures.

Functional performance were evaluated measuring both surface friction resistance and permeability on GSC specimens. Surface friction was evaluated by measuring the Pendulum Test Value *PTV* (EN 13036-4) on both faces of the specimens. On each face, two perpendicular measures were taken. A falling head permeameter was used to measure the permeability of the PHMA specimens. The test was performed in duplicate, considering both faces of each specimen. Both tests allowed to assess the fibers' filling effects on the surface of a PHMA.

4 RESULTS AND DISCUSSION

Figures 2, 3, 4, 5 show the results obtained by the laboratory investigation. Histograms in the figures give the average results; the error bars display the corresponding standard deviation.

4.1 *Compaction properties*

In Figure 2 the compaction properties (Self-compaction C_I and Workability k) measured during the GSC compaction are given.

Regarding the Self-compaction (C_I), all the mixtures exhibit similar values of the investigated parameter, regardless of type and content of both bitumen and fibers, revealing the proneness of the mixtures to reduce their internal voids under their own weight is not affected by the investigated variables.

The ability of the mixtures to be compacted under the rollers in the field (graph of Workability) is slightly influenced by the above-mentioned variables (type and content of both bitumen and fibers). Although, it is not easy to define a trend of the results, particularly considering the data dispersion, as shown by the error bars in the graph.

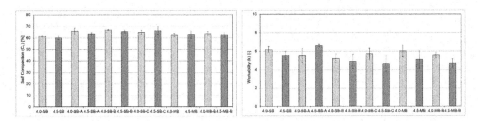

Figure 2. Compaction properties.

4.2 *Volumetric characteristics*

Volumetric characteristics of the investigated PHMAs are shown in Figure 3. Analyzing the graphs in the figure it can be noticed that no significant effects come from the use of fibers. Obviously, it derives from the compaction method used; in fact, the compaction temperature was changed (Table 4) in order to maintain the mastic's viscosity almost constant, as proven by the Authors in previous studies (Bonica *et al.* 2016; Toraldo *et al.* 2015; Crispino *et al.* 2013). Because of this compaction approach and according to the compaction properties above presented and discussed, the content of voids is similar for all the investigated mixtures (between 23% and 26%). Same findings derive from *VMA* (between 30% and 35%) and *VFB* (between 21% and 23%) parameters.

To conclude, both compaction and volumetric parameters reveal that the samples have similar volumetric characteristics, allowing to compare the mechanical and functional performance of mixtures in the correct manner, avoiding the effects of compaction properties variations.

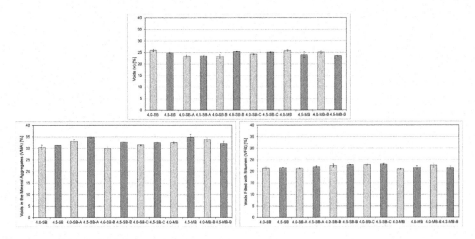

Figure 3. Volumetric characteristics.

4.3 Mechanical performances

In order to highlight the fibers effects on the typical distress expected in the field, it is useful to discuss both the investigated mechanical performance (Stiffness S and Indirect Tensile Strength ITS) of the mixtures (Figure 4) according to the temperature.

As far as low temperatures, longitudinal cracks, parallel to the center line, and transverse cracks are the typical distresses that occur on the surface layers of a road pavement in the field during winter, normally due to a quick reduction of temperatures between daytime and night. To prevent these distresses, low stiffness and high resistance to tensile tension would characterize a surface layer. It is more important in the case of porous mixtures, because the stone-to-stone contact into the bituminous matrix are reduced to increase the content of voids (for drainage purposes), if compared to a standard wearing course.

On this basis and considering the objective difficulty to produce PHMA specimens for Direct Tensile tests according to EN 12697-46 because their shape and dimension, it is possible to have an idea of the proneness of PHMA to form low temperature cracking in the filed by the mechanical tests selected in this research at 5°C, as shown in the graphs in the upper part of Figure 4.

Analysing both graphs, it is possible to note that the mixtures in which Fibers B (cellulose and glass) and C (cellulose and recycled polymer) are used in combination with standard bitumen show the same behaviour of the reference ones, in which modified bitumen or modified bitumen and cellulose and glass fibers are used.

As far as mixtures reinforced by Fiber A (cellulose), a reduction of ITS can be observed, resulting in a diminishing of the resistance of the mixture to low temperature cracking distresses.

Obviously, more prone to this phenomenon appear to be the mixtures in which the sole standard bitumen is used; in fact, the ITS of these mixtures is the lowest.

A final consideration about low temperature is related to the fact that all the investigated mixtures exhibit high values of S because of the thermo-dependency of the bitumen, which become stiffer according to the temperature decrease. That is the reason why the mixtures containing the higher amount of bitumen (4.5%) exhibit better performance (that means lower S) compared to the ones with the lower content of binder (4.0%). The contrary is true if ITS is considered.

Regarding intermediate temperatures, pavement engineers commonly use mechanical performance of the bituminous mixtures at these temperatures as input for road pavement design. In general, there is not a desirable value or range of S or ITS because it depends on the number, thickness and bearing capacity of the layers composing the road pavement. Due to that, the comparison among the investigated mixtures can be done by means the results obtained by the reference ones (modified bitumen or modified bitumen and cellulose and glass fibers) and the ones with standard bitumen reinforced with fibers.

Thus, as shown by the graphs in the middle of Figure 4, the mixtures with Fiber C (cellulose and recycled polymer) guarantee the same performance of the reference mixtures. The worst performance is exhibited by the mixtures with Fiber B (cellulose and glass) and, as expected, by the mixtures with the sole standard bitumen. Mixtures with Fiber A (cellulose) show and intermediate behaviour.

Rutting is a road surface depression in the wheel path. It is a typical pavement distress that can be caused by plastic movement of the bituminous matrix in hot weather conditions and static (or quasi-static) loads. Because PHMAs have high voids (for drainage purposes), the rutting potential of this mixtures is primary due to high temperature conditions, which reduce the stiffness of the binder into the mixture and allow it to move under wheel loads. Even if there are specific tests to measure the rutting behaviour of an asphalt mixture (EN

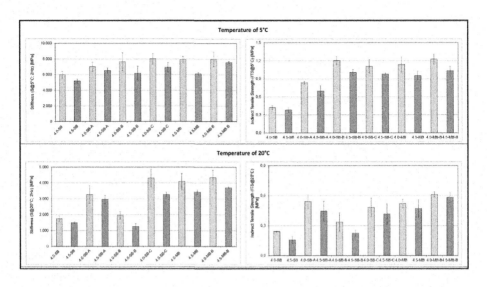

Figure 4. Mechanical performances.

12697-22), which will be included in the upcoming part of this research, both *S* and *ITS* can be used for this purpose. In general terms, both the investigated parameters (*S* and *ITS*) should be as higher as possible. The results in the right part of Figure 4 demonstrate that mixtures with Fiber C (cellulose and recycled polymer) have the same behaviour of the reference ones (modified bitumen or modified bitumen and cellulose and glass fibers), confirming that Fiber C can be used with standard bitumen in replacing the modified one. Mixtures with Fiber B (cellulose and glass) and, as expected, the ones with the sole standard bitumen exhibit the worst performance. Mixtures with Fiber A (cellulose and glass) show and good behaviour in terms of *ITS*, but the *S* is acceptable only if the lower amount of bitumen is used.

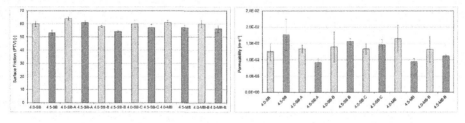

Figure 5. Functional performances.

4.4 *Functional performances*

In Figure 5 the functional performances (Surface Friction and Permeability) are shown.

Regarding the Surface Friction (evaluated by PTV), all the mixtures exhibit similar values of the investigated parameter, regardless of the type of fibers, revealing that fibers have no effects on the surface characteristics of the PHMAs. Obviously, at a certain content of fibers, the investigated parameter decreases according to the increase of bitumen.

As far as Permeability, it is not easy to define the role of fibers. However, the results suggest that these additives not negatively influence the draining ability of the pavement.

To conclude, both the investigated parameters confirm that fibers not negatively affect the surface characteristics of a PHMA, revealing that there are no significant filling effects due to the use of these additives.

5 CONCLUSIONS

This paper reports the results of a laboratory investigation aimed at evaluating the role of three cellulose-based fibers into PHMAs bonded using standard bitumen, comparing their compaction properties, volumetric characteristics, mechanical behavior, and functional performance to the same mixture in which modified bitumen is used, according to the European practice.

The obtained results reveal that:

- mixtures' compaction properties and volumetric characteristics are not affected by the use of fibers if both mixing and compaction temperatures are carefully defined; thus, field laying procedures must be defined and supported by the results of a field test;
- standard bitumen porous mixtures containing cellulose and recycled polymer fibers proved to guarantee the same mechanical performance of the ones in which modified bitumen (or modified bitumen and cellulose glass fibers) is used, even if it is necessary to continue the investigation in order to assess their rutting potential and the fatigue resistance;
- all the investigated fibers were not detrimental in terms of Surface Friction and Permeability in comparison with the reference ones (modified bitumen or modified bitumen and cellulose glass fibers), so there are not significant surface filling effects deriving from the use of fibers; however, a real scale test is necessary to definitively assess the effects of fibers on the functional performance of PHMAs.

Finally, this research demonstrated that it is possible to replace the modified bitumen currently used in Europe into Porous Hot Mix Asphalts with standard bitumen and fibers, identifying the cellulose and recycled polymer fibers as the best among the three types of fibers considered in the study.

ACKNOWLEDGEMENTS

The Authors acknowledge the Experimental Laboratory of Transportation Infrastructures of Politecnico di Milano (InfraLab@PoliMI) for the technical support and the contribution during the laboratory activities.

REFERENCES

Abtahi S.M., Sheikhzadeh M., and Hejazi S.M., 2010. "Fiber-reinforced Asphalt-concrete–a Review.", *Construction and Building Materials*, 24.6: 871–877.

Alvarez A.E., Martin A.E., and Estakhri C, 2010. Internal structure of compacted permeable friction course mixtures. *Construction and Building Materials*, 24(6), 1027–1035.

Alvarez A., Martin A.E., and Estakhri C., 2011. A Review of Mix Design and Evaluation Research for Permeable Friction Course Mixtures. *Construction and Building Materials*, 25(3), 1159–1166.

Bonica C., Toraldo E., Andena Marano C., and Mariani E., 2016. The Effects of Fibers on the Performance of Bituminous Mastics for Road Pavements. *Composites Part B: Engineering*, 95, 76–81.

Cedergren H.R., Arman J.A., and O'Brien K.H., 1973. Development of Guideline for the Design of Subsurface Drainage Systems for Highway Pavement Structural Sections. *Federal Highway Administration*, Report No. FHWA-RD-73-14 Final Rpt..

Chen Y., Tebaldi G., Roque R., Lopp G., and Su Y., 2012. Effects of Interface Condition Characteristics on Open-graded Friction Course Top-down Cracking Performance. *Road Materials and Pavement Design*, 13 (sup1), 56–75.

Chen J.S., Sun Y.C., Liao M.C., and Huang C.C., 2012. Effect of Binder Types on Engineering Properties and Performance of Porous Asphalt Concrete. *Transportation Research Record: Journal of the Transportation Research Board*, 2293(1), 55–62.

Cominsky R., Leahy R.B., and Harrigan E.T., 1994. Level One Mix Design: Materials Selection, Compaction, and Conditioning. *Strategic Highway Research Program*, Report No. SHRP-A-408.

Cooley L.A., Brown E.R., and Watson D.E., 2000. Evaluation of Open-graded Friction Course Mixtures Containing Cellulose Fibers. *Transportation Research Record: Journal of the Transportation Research Board*, 1723(1), 19–25.

Crispino M., Mariani E., and Toraldo E., 2013. Assessment of Fiber-reinforced Bituminous Mixtures' Compaction Temperatures Through Mastics Viscosity Tests. *Construction and Building Materials*, 38, 1031–1039.

Hassan H.F. and Al-Jabri K.S., 2005. Effect of Organic Fibers on Open-graded Friction Course Mixture Properties. *International Journal of Pavement Engineering*, 6(1), 67–75.

Hassan H.F., Al-Oraimi S., and Taha R., 2005. Evaluation of Open-graded Friction Course Mixtures Containing Cellulose Fibers and Styrene Butadiene Rubber Polymer. *Journal of Materials in Civil Engineering*, 17(4), 416–422.

Klenzendorf J.B., Charbeneau R., Eck B., and Barrett M., 2011. Measurement and Modeling of Hydraulic Characteristics of Permeable Friction Course (PFC). *American Society of Civil Engineers, Proceedings of the First Transportation and Development Institute Congress*, March 13–16, Chicago, Illinois, 803–813.

Liu Q. and Cao D., 2009. Research on Material Composition and Performance of Porous Asphalt Pavement. *Journal of Materials in Civil Engineering*, 21(4), 135–140.

Mallick R.B., Kandhal P., Cooley L.A., and Watson D., 2000. *Design, Construction, and Performance of New-generation Open-graded Friction Courses*. Asphalt Paving Technology, 69, 391–423.

Mo L., Huurman M., Wu S., and Molenaar A.A.A., 2009. Ravelling Investigation of Porous Asphalt Concrete based on Fatigue Characteristics of Bitumen–stone Adhesion and Mortar. *Materials & Design*, 30 (1), 170–179.

Molenaar J.M.M., and Molenaar A.A.A., 2000. An Investigation into the Contribution of the Bituminous Binder to the Resistance to Ravelling of Porous Asphalt. European Asphalt Pavement Association and European Bitumen Association, *Proceedings of the papers submitted for review at 2nd Eurasphalt & Eurobitume Congress*, September 20–22, Barcelona, Spain, Book 1 – Session 1, 500–508.

Pagotto C., Legret M., and Le Cloirec P., 2000. Comparison of the Hydraulic Behaviour and the Quality of Highway Runoff Water According to the Type of Pavement. *Water Research*, 34(18), 4446–4454.

Praticò F.G. and Moro A., 2007. Permeability and Volumetrics of Porous Asphalt Concrete: A Theoretical and Experimental Investigation. *Road Materials and Pavement Design*, 8(4), 799–817.

Punith V.S. and Veeraragavan A., 2010. Characterization of OGFC Mixtures Containing Reclaimed Polyethylene Fibers. *Journal of Materials in Civil Engineering*, 23(3), 335–341.

Punith V.S., Suresha S.N., Raju S., Bose S., and Veeraragavan A., 2011. Laboratory Investigation of Open-Graded Friction-Course Mixtures Containing Polymers and Cellulose Fibers. *Journal of Transportation Engineering*, 138(1), 67–74.

Sansalone J., Kuang X., Ying G., and Ranieri V., 2012. Filtration and Clogging of Permeable Pavement Loaded by Urban Drainage. *Water Research*, 46(20), 6763–6774.

Santagata E. and Bassani M., 2002. Full-scale Investigation on the Volumetric Relationship Between Laboratory and in-situ Compaction. *Proceedings of the 3rd International Conference on Bituminous Mixtures and Pavements*, November, Thessaloniki, Greece, Vol. 1, 237–259.

Scholz M. and Grabowiecki P., 2007. Review Of Permeable Pavement Systems. *Building and Environment*, 42 (11), 3830–3836.

Suresha S.N., Varghese G., and Shankar A.R., 2009a. "A Comparative Study on Properties of Porous Friction Course Mixes with Neat Bitumen and Modified Binders.", *Construction and Building Materials*, 23(3), 1211–1217.

Suresha S.N., Varghese G., and Shankar A.R., 2009b. Characterization of Porous Friction Course Mixes for Different Marshall Compaction Efforts. *Construction and Building Materials*, 23(8), 2887–2893.

Toraldo E., Mariani E., and Malvicini S., 2015. Laboratory Investigation into the Effects of Fibers on Bituminous Mixtures. *Journal of Civil and Environmental Management*, 21(1), 45–53.

Wu S.P., Gang L, Mo L.T., Zheng C, and Ye Q.S., 2006. Effect of Fiber Types on Relevant Properties of Porous Asphalt. *Transactions of Nonferrous Metals Society of China*, 16, s791–s795.

Innovative cement-based material as environmental and efficient technical alternative for pavements

Stimilli Arianna* & Caraffa Tullio
Direzione Operation e Coordinamento Territoriale, Anas S.p.A., Rome, Italy

Alessandro Marradi
Civil and Industrial Engineering Department, University of Pisa, Pisa, Italy

ABSTRACT: Over time, pavements must guarantee smooth and safe driving surface. Often, considering the remarkable age of the Italian road network and the increasing damage rate due to the increase in traffic over years, reaching adequate safety and comfort standards requires maintenance actions that involve also deep pavement layers. Consequently, the intervention becomes too burdensome and costly. In this context, ANAS, the main Italian road agency, is constantly promoting innovative design solutions aimed at maximizing technical efficiency and environmental sustainability (longer service life and reduced maintenance needs).

For these purposes, particularly performing and eco-sustainable materials for binder and base layers can substitute traditional HMA guaranteeing thinner layers with, at least, same final performance.

Specifically, a cement-based material, developed through laboratory studies and real-scale experiments, has been tested. It is a mixture of aggregates, potentially even entirely composed of RAP, mixed at room temperature with a specific bituminous emulsion and reactive filler. It combines rigid and flexible pavements benefits, with environmental and technical advantages:

- reduction of execution time (fluency, no need of compaction efforts);
- less air pollution (cold mix);
- high stiffness modulus (reduction in layer thickness);
- high elasticity (high fatigue resistance) and rutting resistance;
- resistance to chemical aggression and atmospheric agents;
- combination with high percentages of RAP (reduction of virgin materials);
- no temperature sensitivity of the layer.

All these peculiarities mean longer service life compared to traditional pavements and make this material particularly suitable for accomplish environmental goals established by the new regulations in the context of the European Green Deal.

First laboratory studies were started in 2014. Afterwards, full-scale trial sections were realized also involving an Anas road segment still under monitoring (i.e. SS3 "Flaminia", pk. 136 + 100 to 136 + 600). Optimum performance were detected in terms of bearing capacity, fatigue, rutting and cracking.

Other trial sections are scheduled in the coming year to confirm the potential of this material, particularly useful where deep and time consuming maintenance actions cannot be carried out.

---

*Corresponding Author: a.stimilli@stradeanas.it

DOI: 10.1201/9781003429258-4

This paper describes the material potential, with particular reference to the results obtained with Accelerated Pavement Testing with FastFWD on the trial sections involved in the project.

Keywords: Cement-Based Material, Semi-Flexible Pavement, Reactive Filler, Long-Term Performance

1 BACKGROUND

Hard surfaced pavements, which constitute the majority of Italian national roads, are typically categorized into flexible, rigid and semi-rigid pavements (surfaced with bituminous materials, Portland cement concrete and a combination of both, respectively). Each of these pavement types distribute loads over the subgrade trusting on different mechanisms. Rigid pavements, thanks to its high stiffness (high elastic modulus), tends to distribute the load through the cement concrete layer over a wide area with a consequent low stress transferred to the subgrade. On the contrary, flexible pavements distributes loads over a relatively smaller area relying on a combination of several bituminous ductile layers for transmitting the load to the subgrade. Semi-rigid pavements represent a solution in between which combines asphalt and cement bound aggregates.

An additional pavement type can be referred to the so-called semi-flexible pavements (SFP), which simultaneously exhibit flexible properties of asphalt pavements and high bearing capacity (stiffness) of concrete pavements (Hassan *et al.* 2002; Hassani *et al.* 2020). They combine the advantages of both pavement types with significant improvements in terms of durability, especially in those areas subjected to heavy and slow-moving loads. SFPs consist of a surface layer made of open-graded hot mix asphalt (HMA) with high air void content (i.e. 20–25%) filled by injecting special highly flowable cementitious grouting materials (i.e. Grouted Macadam) (Zoorob *et al.* 2002). However, the construction complexity of this technology, which implies a two-stage operation for laying down with many manpower efforts, leads to a reduction in the construction speed (normally two consecutive days) and requires high initial production costs (at least four times than a traditional wearing layer), so limiting the application and the advantages of this solution to few areas of modest extension (e.g. airports, ports, container terminals) (De Oliveira *et al.* 2006; Zoorob *et al.* 2002). Moreover, the benefits of this material are deeply related to the void connectivity (that guarantees the possibility for the grout to flow through them) and to a suitable grout workability for completely filling all the voids available. Both cases could mean unfilled voids that may cause premature pavement failure (Hassani *et al.* 2020).

To solve the abovementioned issues without renouncing to the advantages guaranteed by SFPs, a new technology has been recently implemented as a valid performing and eco-sustainable alternative of Grouted Macadam.

This material, called Ready to Mix (RTM), can substitute traditional HMAs guaranteeing less thick layers with, at least, same final performance at lower costs and reduced construction times compared to grouting materials (Pratelli *et al.* 2018).

It is worth noting that, differently to the Grouted Macadam, RTM showed the potential to be used also as a structural layer and not just as an improved wearing course. Moreover, this mixture has been developed taking into account also sustainability in terms of recycling and energy saving, so enlarging its potential applications as well as ensuring additional environmental benefits. In fact, RTM is a cement-based material made of aggregates, potentially even entirely composed of reclaimed asphalt pavement (RAP), mixed at room temperature (i.e. cold mix) with a specifically designed slow-setting cationic bituminous emulsion (ECL 60-type C60B4, dosage between 4,6% and 6,8% by mixture weight), a special cementitious reactive filler (around 30,0% by mixture weight) and water (around 5,0% by mixture weight depending on the desired flow ability).

The results of preliminary laboratory studies and real-scale experiments seem demonstrate the potential of this material to combine rigid and flexible pavements benefits, with environmental and technical advantages:

- reduction of execution time (fluency, no need of compaction efforts);
- room temperature production (less air pollution);
- high stiffness modulus (reduction in layer thickness);
- high elasticity (high fatigue resistance);
- resistance to chemical aggression and atmospheric agents;
- combination with high percentages of RAP (reduction of virgin materials)
- no temperature influence on stiffness.

The next paragraphs summarize the main implementation steps and in-situ trial sections experienced with this mixture in order to provide an overview of its potential applications.

2 PRELIMINARY LABORATORY EVALUATION

RTM was optimized trying to develop a mixture that allows the laying down process in a single layer even with high layer thickness (around 20 cm), guaranteeing high bearing capability, good fatigue behaviour and resistance to chemical aggression and intense heat. The determination of the optimum mix composition required an iterative optimization process, which in each phase evaluated the effect of different components (e.g. aggregates, reactive filler, bituminous emulsion, water) and dosages on mixture performance.

A wide range of mechanical tests was carried out in laboratory according to well-defined criteria and the main technical standards for pavement materials characterization (Figure 1).

Compressive Strength [N/mm^2-MPa] UNI EN 13286-41	Indirect Tensile Strenght [MPa] UNI EN 13286-42	Stiffness Modulus IT-CY UNI EN 12697-26 Annex C	Fatigue Testing IT-CY UNI EN 12697-24 Annex E
22.12	3.32	27,773	With $\varepsilon = 38\mu\varepsilon$ NO FAILURE

Figure 1. RTM preliminary laboratory characterization (Pratelli *et al.* 2018)

Compared to traditional mixtures and open-graded mixtures prepared with grouting materials, a significant increase in stiffness modulus (i.e. more than double values, 27,000 MPa instead of 12,000 MPa of Grouted Macadam) and ductility was observed, as well as a remarkable improvement in terms of fatigue resistance (no failure was observed even by applying the highest load achievable through the standard asphalt testing machines). The material showed also high Indirect Tensile Strength (ITS) and resilient modulus, high resistance to chemical agents and fuels and low sensitivity to temperature (Pratelli *et al.* 2018).

3 IN-SITU EXPERIENCES

Based on the promising laboratory findings, the research activity has deepened the performance analysis to verify the potential of large scale in plant productions of this cement-based material as well as its long-term performance. After the laboratory optimization of the mixture composition and preliminary in-situ attempts realized on specific and controlled production sites, additional trial segments on Italian national roads were constructed and monitored also by means of in-situ tests (i.e. Accelerated Pavement Testing, APT – Metcalf 1996).

The technical specifications provided by the Italian National Road Authorities (Anas 2016; Autostrade 2004) for asphalt layers constituted the reference for defining the aggregate distribution of the mixture (Figure 2) and verifying other characteristics and performance requested for coarse and fine aggregates.

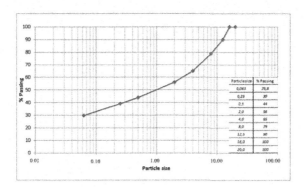

Figure 2. RTM aggregate (Pratelli *et al.* 2018).

3.1 *In plant production and laying down operation*

Considering the high impact of material components' properties on the development of RTM performance, high attention must be paid on the production process (e.g. materials dosages, mixing time and adding sequences, accurate reactive filler entry operation, moisture content).

Due to the presence of cementitious components and bituminous emulsion, to properly activate cementitious and bituminous bonds the mixture must be appropriately hydrated. At the same time, the amount of water must provide a mixture fluency suitable for laying down the material with traditional paver machines. To this aim, the plant production is equipped with a continuous "loss in weight" dosing system, capable of controlling in real time the flow rate of water, emulsion and reactive filler and, hence, the quality of the mixture produced. Once all the aggregate components have been inserted into the mixer, the right amount of water is added for the complete hydration of the filler allowing the activation of all its components.

At this point, the RTM mixture is loaded directly onto the truck and transported to the construction site where is laid down with standard road pavers with a procedure similar to a traditional HMA (Figure 3). It is worth noting that the dosage of mixture components can be adapted for obtaining self-compacting mixtures or less fluid mixtures to be compacted with suitable mechanical rollers.

3.2 *Trial section 1 and 2*

The first in-situ full-scale trial sections were realized on April and June of 2016, respectively. They were located on a production site characterized by the daily and controlled passage of a number of heavy vehicles. Figure 4 describes the pavement structure analysed in the two sections. The first one consists of a 12-cm thick wearing layer of RTM, realized on a pre-existing 7-cm thick bituminous layer. The deepeest layer is a 50-cm thick compacted base course laid on a cohesive subgrade. The second section was specifically designed to emphasise the behaviour of RTM when subjected to repeated dynamic loads under critical

Figure 3.   Laying down operations of RTM by means of a traditional road paver.

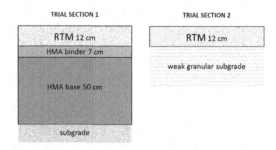

Figure 4.   Pavement structure scheme for trial section 1 and trial section 2.

stress conditions (to lead the material to failure in a short period). To this aim, the pavement structure involved only a wearing layer of 12 cm laid on a pre-existing granular subgrade.

Along with the optimization of the in-plant production process and the laying down activities, the primary goal of trial sections 1 and 2 was to study the fatigue behaviour of the material.

In this sense, the field modulus variation trend under dynamic load repetitions was used as indicator of the fatigue performance. It was investigated by means of an APT test performed by means of the Fast Falling Weight Deflectometer (FastFWD – Briggs *et al.* 2016; Sundgaard *et al.* 2015) configured with a 300-mm diameter loading plate and with nine deflector geophones used to record vertical displacements caused by the impulsive load (positioned under the loading plate and up to 1800 mm from the plate). During the test, the RTM layer temperature was hourly monitored to verify any temperature influence on test results. Air and surface temperatures were recorded as well.

The recorded deflection data were backcalculated on the base of the pavement structure. The modulus evolution of each layer was consequently determined and plotted as function of the number of load cycles (Isola *et al.* 2013; Pratelli *et al.* 2018).

Trial section 1 was subjected to 120 kN load level (highest load level configuration) for three days. The test was interrupted after 30,000 load cycles since the recorded deflections did not show any significant variation and the pavement structure was far from failure conditions.

Although this limited number of load repetitions were not significant enough for judging long-term performance of the material, test results gave a preliminary indication of the potential behaviour under repeated load cycles. Moreover, on the base of the results recorded, it was observed a good ability of the material to recover part of its stiffness when not stressed. It was also reasonable to assume that RTM stiffness is not influenced by the temperature within the investigated range (around 16°C and 32°C). This low temperature sensitivity is mainly attributable to the cementitious bonds that lead material behaviour (Pratelli *et al.* 2018).

The second trial section was specifically designed for overcoming the limitations observed in the first attempt and to reach a failure condition in a short period. Several test points were investigated with a total of 60,000 load cycles applied for each one.

Primary, despite the material high stiffness, RTM confirmed promising fatigue properties as demonstrated by the high resistance of the material to dynamic repeated loads. A specific performance relationship to correlate number of loads to failure and initial deformation induced in the material by the test load was identified. Figure 5 describes RTM performance criteria based on in-situ data results. It is worth noting that the strain level applied during the test to verify the material response under repeated load cycles were much higher than what usually recommended for HMAs. This shrewdness was adopted to emphasise RTM fatigue behaviour since for standard strain levels no evidence of cracks or permanent deformation was observed on the tested surface at the end of the investigation. Based on this observation, reasonably, this relationship can be considered conservative suggesting the presence of a fatigue endurance limit (the material never breaks) under traffic-induced strain levels.

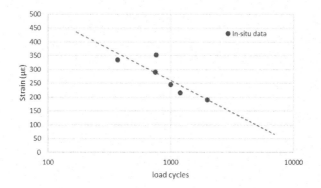

Figure 5.   RTM fatigue performance criteria based on in-situ data results.

Moreover, results showed again a certain self-healing capability (Figure 6) and low temperature sensitivity of the material.

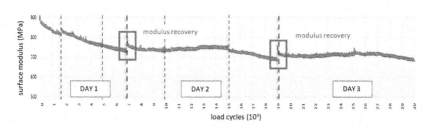

Figure 6.   RTM self-healing aptitude.

3.3 Trial section 3

On July 2017, a third trial section was realized on the Italian national road SS3 "Flaminia" (pk. 136 + 131 and 136 + 207, direction SUD). Unlike trial sections 1 and 2, in this case RTM was used as structural layer (base layer) for substituting a traditional HMA at equal thickness as shown in Figure 7.

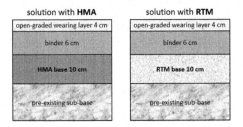

Figure 7.   Pavement structure scheme for trial section 3: solution with HMA and with RTM.

The main objective of this trial section was to analyse in-situ pavement performance in terms of bearing capacity and resistance to repeated dynamic loads in comparison to a traditional flexible pavement under real traffic conditions.

Two testing campaigns, on September 2017 and June 2018, were carried out by means of the FFWD (Marradi *et al.* 2018). For each one, the number of load cycles exceeded 50,000 repetitions with a load level of 120 kN. Based on the results obtained in both experimental phases, some observations can be drawn as follows:

- the overall bearing capacity of the pavement made with the RTM base layer was much higher than that one of the traditional pavement structure with asphalt base layer.
- As a result of the total load application, the surface modulus of the measurements points with RTM base layer decreased by 15%, whereas the measurement point with the asphalt base layer suffered a much higher decrease in the surface modulus (Figure 8).
- Even after the application of over 15,000 load cycles with a constant load level of 120 kN, the surface modulus of the pavement structure with RTM base layer remained approximately 50% higher than the initial value (measured before the load application; Marradi *et al.* 2018). This finding is a consequence of the low deflection measured under the load plate, along with a significantly lower increasing rate of the deflection compared to HMA

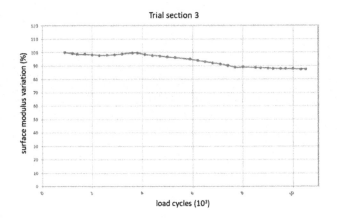

Figure 8.   RTM surface modulus variation under repeated load cycles.

(Figure 9). That implies not only better initial performance of RTM, but also the ability to preserve optimum properties over time with a higher and higher performance gap compared to HMA in a long term perspective.

- Low temperature sensitivity was confirmed.

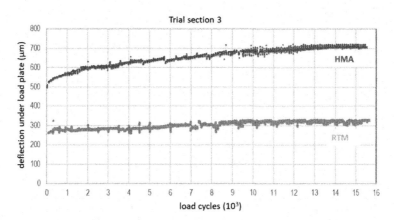

Figure 9.   Evolution of the deflection under the load plate after 15.000 load cycles: RTM *vs* HMA.

Moreover, elaborating the evolution over time of applied loads and deflections (measured under the loading plate), it was possible to calculate the energy absorbed by the pavement during each load cycle (i.e. energy loss – Rowe 1996). The graph in Figure 10 shows how, in the case of traditional HMA pavements, the energy increases as the test proceeds, whereas for the pavement with RTM the dissipated energy does not increase over time, indicating a substantial stability of RTM under dynamic loads.

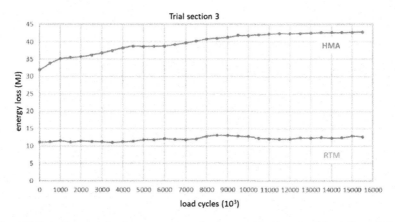

Figure 10.   Dissipated energy by the pavement under repeated load cycles (Marradi *et al.* 2018).

Summarizing, the RTM base layer was charcaterized by a stiffness modulus significantly higher than the traditional HMA base layer. The APT results have also highlighted a much higher resistance to repeated load cycles and low temperature sensitivity for the pavement structure with RTM base layer confirming laboratory and first in-situ results.

It is worth noting that, to further monitor RTM performance evolution, an additional investigation was performed in January 2021 with same loading conditions but different test temperature (around 10°C in 2021 *vs* 20–30°C in 2017 and 2018). Test results, currently under elaboration and that will be discussed in details in future papers, suggest the consistency of what observed with the previous analyses and confirmed the optimum behaviour of RTM in a long term perspective (i.e. bearing capacity and fatigue resistance). The comparison between test data acquired in 2017–2018 and 2021 also allowed the identification of specific relationships that describe the deflection as function of temperature.

Moreover, correlating FFWD test data with the traffic volume actually experienced on the test road segment in the time gap between 2018 and 2021, the damage equivalence factor that allow the correlation between different load axes was recalibrated in case of pavements with RTM base layers. That will allow a more feasible forecast of the evolution of RTM bearing capacity over the entire road service life (i.e. 20 years), obtaining a precise long-term evaluation of the material modulus, essential aspect for reliably design pavement structure in case of RTM layers.

3.4   *Trial section 4 – work in progress*

A fourth trial section is currently under development. It will be realized as part of maintenance activities on a segment of the Italian national road SS3bis "Tiberina" (pk. 122 + 900 and 133 + 755, direction NORTH) subjected to high heavy traffic volume and, analogously to trial section 3, it will provide the use of the RTM as structural layer. A scheme of the project solution identified as alternative to traditional flexible pavements is depicted in Figure 11. The proposed solution assumes, for the driving lane, a 14-cm thick RTM base layer that will substitute 18-cm HMA layers and will be laid down through a two-stage operation. Straddling the joint between the driving lane and the fast lane, in the latter (where the pre-existing asphalt base layer will not be removed) a reinforcement geogrid will be interpose between the pre-existing asphalt base layer and the second RTM layer in order to minimize the possibility of fracture formation due to the high difference in stiffness between materials.

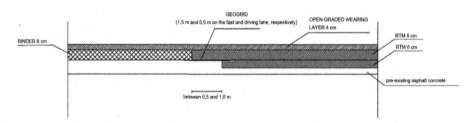

Figure 11.   Pavement structure scheme for trial section 4 with RTM.

As the previous experiences, the trial section will be continuously monitored during construction, to check the in plant production feasibility of the RTM mixture, the material composition consistency and the adequacy of laying down operations, as well as during the service life to verify the resistance of the material under traffic in a long-term perspective. Results will be discussed in future papers.

4 CONCLUSIONS AND DISCUSSION

Laboratory and in-situ experiences undertaken for optimizing and evaluating RTM properties have shown its potential for various applications. RTM, initially developed in laboratory as a cheaper and easier system to realize wearing layer as substitute of Grouted Macadam, demonstrates high performance also for deeper maintenance solutions. The promising outcomes obtained with the preliminary laboratory evaluation largely encouraged the investigation of the material when produced in large-scale amounts. To this purpose, various trial sections were realized to fine-tune all the operational phases and to accurately monitor material performance when subjected to heavy traffic volume.

In-situ trial sections validated laboratory results and certified similar or even superior characteristics compared to Grouted Macadam or traditional HMAs for base layers, as well as excellent properties in terms of stiffness and ductility, which mean optimum bearing capacity and fatigue resistance. Specifically, the latter has been investigated directly in-situ by means of FastFWD. Moduli evolution and surface deflections trends have been analysed and evaluated for all the tested trial sections. Results confirmed a great resistance of the material to repeated loading cycles (no evidence of cracks or permanent deformation was produced on the tested surface at the end of the surveys even by applying the maximum load for a number of loading cycles up to 60,000). This finding, combined with the high RTM stiffness due to the cementitious bonds created by the specifically designed active filler, means superior performance when compared to traditional HMAs. The presence of RTM affects the pavement response when subjected to heavy traffic showing a clear reduction in deflections with no regard to the test temperature (i.e. temperature insensitivity), so allowing a more homogeneous structure in the operating temperature range. A certain self-healing aptitude was also observed.

According to the technical evaluation presented in the previous paragraphs and considering other practical aspects related to the environmental sustainability as well as the ease of construction (e.g. potential of incorporating extreme percentages of RAP, room temperature for mixing and laying down, possibility of using traditional mixers and common road pavers, reduced construction time and materials' amounts), a number of relevant benefits make this material an optimum and desirable alternative for pavement maintenance activities.

Finally, in-situ validation of preliminary laboratory results allowed also the identification of feasible correlation laws and parameters to properly arrange the design of pavement structures that incorporate RTM as wearing or deeper layer. On this matter, studies are still ongoing and results will be discussed in future works.

All the trial sections realized are still under monitoring and additional in-situ sections are planned to further investigate RTM behaviour and validate the material response under different boundary conditions (e.g. loading, temperature, high RAP amount, pavement structure).

REFERENCES

Briggs R.C., Navarro Comes A. and Ullidtz P., 2016. Accelerated Pavement Testing Using the Fast Falling Weight Deflectometer. *Proceedings of the Transportation Research Board 95th Annual Meeting,* Washington, DC, USA, 10–14 January 2016; The Transportation Research Board: Washington, DC, USA.

Hassan K.E., Setyawan A., Zoorob S.E., 2002. Effect of Cementitious Grouts on the Properties of Semi-Flexible Bituminous Pavements. *In Proceedings of the Performance of Bituminous and Hydraulic Materials in Pavements, Nottingham,* UK, 11–12 April 2002; CRC Press: Boca Raton, FL, USA.

Hassani A., Taghipoor M. and Karimi M.M., 2020. A State of the Art of Semi-flexible Pavements: Introduction, Design, and Performance. *Construction and Building Materials* 253.

Isola M., Betti G., Marradi A. and Tebaldi G., 2013. Evaluation of Cement Treated Mixtures with High Percentage of Reclaimed Asphalt Pavement. *Construction and Building Material* 48, 238–247.

Marradi A., Flores L. and Rossi M., 2018. *Indagini ad alto rendimento per il rilievo delle condizioni di portanza dei campi prove con strato di base in RTM® e in Conglomerato Bituminoso Tradizionale S.S.3 – Flaminia.* Dynatest, Report Version 01 Revision 00, 02/11/2018.

Metcalf J.B., 1996. Application of Full Scale Accelerated Pavement Testing. *NCHRP Synth. Highw. Pract.* 235 (3).

Pratelli C., Betti G., Giuffrè T., Marradi A., 2018. Preliminary In-Situ Evaluation of an Innovative, Semi-Flexible Pavement Wearing Course Mixture Using Fast Falling Weight Deflectometer. *Materials* 11 (611).

Rowe G.M., 1996. *Application of Dissipated Energy Concept to Fatigue Cracking in Asphalt Pavements.* Ph.D. Thesis, University of Nottingham, Nottingham, UK.

Sundgaard J., Navarro Comes A., Lund B., Resca L., Ullidtz P. and Carvalho R., 2015. *Application of FastFWD for Accelerated Pavement Testing.* Dynatest Denmark A/S: Søborg, Denmark.

Zoorob S.E., Hassan K.E. and Setyawan A., 2002. Cold mix, Cold Laid Semi-flexible Grouted Macadams Mix Design and Properties. *In Proceedings of the Performance of Bituminous and Hydraulic Materials in Pavements*, Nottingham, UK, 11–12 April 2002; CRC Press: Boca Raton, FL, USA.

Roads and Airports Pavement Surface Characteristics – Crispino & Toraldo (Eds)
© 2023 The Author(s), ISBN 978-1-032-55149-4

Field evaluation of different rollers effect on the compaction of hot mix asphalt overlays using an electromagnetic densitometer

Seyed Rasool Fazeli* & Hamidreza Sahebzamani
School of Civil Engineering, University of Tehran, Tehran, Iran

Omid Ferdosian
School of Civil Engineering, Iran University of Science and Technology, Tehran, Iran

ABSTRACT: Paving asphalt overlays, from raw materials extraction, production of asphalt mixtures, transporting the asphalt mixture to the job site, and laying hot mix asphalt with paving machines should constantly be tested. One of the main controls after asphalt paving is the compaction of the new layer and its accurate density. Inaccurate compaction will reduce the pavement life.

Core sampling is the most known method for determining newly paved roads. Using this method has its defects. Destruction of the new paved layer for sampling will be the weakness of the new layer. As a result, there will be a limitation for measuring the asphalt pavement density. On the other hand, contractors cannot achieve the accurate number of rollers passes and density of the implemented HMA layer. Samples should be tested at the laboratory, and insufficient density will face contractors to penalties or even reconstructions.

These concerns led to non-destructive methods to measure in-place density. There are two non-destructive methods for measuring the in-place density of asphalt overlays. The electromagnetic and the nuclear densitometer. The main concern of using these non-destructive methods is the accuracy of measurements.

In this research, authors tried to evaluate the effectiveness of using an electromagnetic densitometer for measuring the density of new asphalt overlays in two sites. Device results after site calibration were compared to core samples. Determination of sufficient roller pass using an electromagnetic densitometer was the other purpose of this study.

The most common rollers that use for hot mix asphalt compaction are pneumatic and steel wheel rollers. The effectiveness of using pneumatic and steel wheel rollers on HMA layer compaction was the other issue evaluated in this study.

Results showed that electromagnetic densitometer is a not reliable non-destructive method for quality assurance of asphalt layer in-place density. Although, a sufficient number of roller passes could be determined for any project by this method. The other important conclusion was that the steel wheel roller has a greater role in gaining density than the pneumatic roller. Moreover, in the case of using a pneumatic roller without a steel wheel roller, the need for passes will increase.

Keywords: Electromagnetic Densitometer, Asphalt Compaction, Pneumatic and Steel Wheel Roller

1 INTRODUCTION

In the first of the 90s, there were two major methods for determining Bulk specific gravity of newly paved asphalt pavement, core sampling, which was a destructive method, and the

---

*Corresponding Author: rasoolfazeli@alumni.ut.ac.ir

DOI: 10.1201/9781003429258-5

nuclear densitometer, which was the non-destructive method. In the nineties, a new method was proposed. This new method was non-destructive, and it was much safer, faster, lighter, and cheaper than the nuclear method. The primary model of this apparatus was fabricated in 1991 as a pavement quality indicator (PQI), and Troxler made the next model in 1998.

This apparatus measures the compaction of HMA by measuring the dielectric constant of the components of the asphalt layer. It passes the electromagnetic waves through the asphalt layer. These waves move from transmitters to receivers in an isolated field. Impedance or electrical resistance will be used for dielectric constant by passing through the asphalt layer. Accumulation of dielectric constants is a proportion of density (Hausman & Buttlar 2002).

Hurley et al. (2004) evaluated non-destructive methods including electromagnetic densitometer (pavement quality indicator and Pave Tracker apparatus) and nuclear densitometer in different asphalt pavements. PQI electromagnetic densitometer showed a medium to high consonant with core sample results in 16 out of 20 projects. The Pave Tracker results were acceptable in 7 out of 10 projects. The results showed that the nuclear densitometer still correlates better to core samples than the electromagnetic densitometer.

Kvasnak et al. (2007) investigated Pave Tracker and PQI electromagnetic densitometer on field and laboratory in 2007. The results showed that core sample results differed from electromagnetic densitometers data. Although, the final results obtained from electromagnetic densitometers for the pavement's quality assurance (QA) were close to the core sampling outcome.

Koudous Kabassi et al. (2011) Performed similar research on PQI 301 and nuclear densitometer on 13 sites. Regression statistical assessments were used for the evaluation of results. The coefficient of determination (R^2) was 0.42 for the nuclear densitometer and 0.18 for the non-nuclear densitometer versus core samples. This study showed a better performance for nuclear densitometer than electromagnetic densitometer.

2 OBJECTIVE

In this study, an electromagnetic densitometer was primarily evaluated for its accuracy. For this purpose, two sites were investigated, and core samples were extracted at the specified points. The other goal was to use this apparatus to determine sufficient roller passes for each project. Moreover, both pneumatic and steel wheel rollers have their own effect on the compaction of newly paved asphalt pavement. Investigation of the effectiveness of each of these rollers for gaining enough compaction by use of an electromagnetic densitometer was the other purpose of this study.

3 INSTRUMENTS

Pave tracker model 2701B made by Troxler company was used in this study as an electromagnetic densitometer. Hilti DD200 was used for core sampling, and the density of core samples extracted by this apparatus was determined by the surface saturated density (SSD) method in the laboratory. SAKAI TS290 pneumatic roller and HAMM D75 steel wheel roller were the rollers used in this study for compaction of asphalt pavement.

4 MATERIALS

As mentioned before, two sites were chosen in this study, Meygoon road and Babaei Highway in Tehran province. In both projects, asphalt mixtures were prepared using PG 64-22 asphalt binder and 19mm Nominal Maximum Aggregate Size (NMAS) of limestone crushed aggregate.

5 TEST METHODS

5.1 *Saturated Surface Dry (SSD)*

The surface Saturated-Dry (SSD) method (AASHTO T166) was used to determine the Bulk Specific Gravity (Gmb) of the core samples. In this method, the mass of the specimen is calculated at first. Then, the specimen is submerged in the water bath at 25°C for four minutes. The submerged weight of the specimen should determine. Then the specimen surface would dry with a towel immediately. The mass of the specimen after drying with a towel would determine. The bulk specific gravity of samples should be determined by Equation (1). It should be noted that this method can be used for samples with less than 2 percent water absorption.

$$Gmb = \frac{A}{B - C} \tag{1}$$

In which:

Gmb: Bulk Specific Gravity
A: dry specimen mass
B: surface dried submerged specimen mass
C: specimen mass in water

5.2 *Calibration of electromagnetic densitometer*

The calibration process for this study was performed according to ASTM D7113. Based on this standard, it is recommended to have three to ten points for the calibration process. In this study, five points were chosen in an area of three meters in length and 1.5meters wide (Figure 1). The apparatus needs theoretical maximum specific gravity (Gmm) and bulk specific gravity of the mixture, which is imported as raw data.

Pave tracker model	Hilti DD200	SAKAI TS290	HAMM D75
2701B		(Weight: 29 tons)	(Weight: 10 tons)

Figure 1.   Equipment's.

Five readings perform by the electromagnetic densitometer in each circle, one at the center and four in each circle at 2, 4, 8, and 10 o'clock. Core samples will be extracted from the center of each circle. Bulk Specific Gravity of each core sample will be determined. The average of core samples and electromagnetic densitometer will be calculated. The numerical difference between the average of core samples and the electromagnetic densitometer will be imported to the apparatus as an offset.

5.3 *Main operation procedure*

A test section was needed to measure the accuracy of the Troxler electromagnetic densitometer and each roller effect on gaining compaction. In this test section, 30 points are

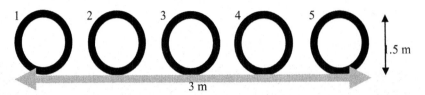

Figure 2.   Schematic of calibration area.

determined for reading by an electromagnetic densitometer, and core samples are extracted from six points (black circles) out of 30 points. This pattern covers the whole paved lane, including marginal points. The number of roller passes and the type of roller used for compaction was similar for all points.

For the first project (Babaei Highway), 5 points were chosen for calibration on each test site (Figure 4), and after calibration, the pattern is marked on the newly paved asphalt layer for main evaluation (Figure 5).

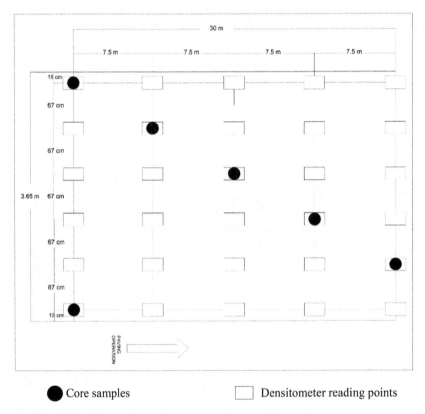

Figure 3.   Schematic of test section.

As mentioned before, the other purpose of this study was to evaluate the effect of rollers on compaction. Three points were chosen other than the pattern. After each pass of rollers, the apparatus measurements were recorded. A steel wheel roller (HAMM D75) was primarily used until the readings of the apparatus converged to the highest number, and then a pneumatic roller (SAKAI TS290) was used (Figure 6).

Figure 4. Calibration of electromagnetic densitometer.

Figure 5. Test pattern on Babaei Highway.

Figure 6. P1 point for evaluation of roller effect (Babaei Highway).

The same process was used for densitometer accuracy evaluation for the second project, the Maygoon road. Four points were chosen in this project to assess the roller's effect on compaction. Based on previous studies and Babaei Highway results and a more accurate evaluation of the steel wheel roller effect, the pneumatic roller was prevented from passing in the aforementioned four points. It should be noted that the asphalt mixture and execution team was constant in both projects (Figure 7).

Figure 7. Test pattern on Meygoon road.

6 RESULT AND DISCUSSION

6.1 *Babaei Highway results*

Table 1 shows the results obtained from Babaei Highway. The regression analysis was used to compare the accuracy of the electromagnetic densitometer versus core samples (Figure 8).

Table 1.   Compaction result of densitometer and core samples at Babaei Highway.

Point	Compaction percentage obtained by core samples	Compaction percentage obtained by electromagnetic densitometer
A1	96	97
B2	95	97
C3	93	93
D4	94	91
E5	95	92
F1	95	91

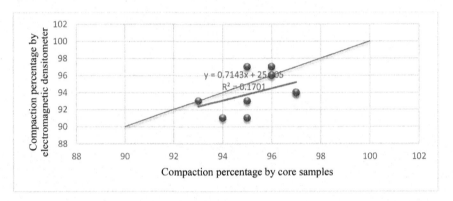

Figure 8.   Comparing compaction of core samples and electromagnetic densitometer on Babaei Highway.

Rollers' effect on compaction for the three points is shown in Figure 9. The blue line is related to steel wheel roller passes, and the red line is related to pneumatic roller passes. According to the data achieved from Babaei Highway, the results are as follows:

- In three points, densitometer measurements had up to four percent difference which is noticeable. Only one point showed the same results for densitometer and core samples measurements. R2 value was 0.17, which expresses the inaccuracy of this device. Points positions (middle or edge of the paving lane) has no effect on densitometers evaluations.
- Steel wheel and pneumatic rollers are used for P1, P2, and P3. It was supposed to use an electromagnetic densitometer after each rollers pass to achieve maximum compaction. Core samples of these three points showed 96, 97, and 97 percent of compaction. It could be concluded that the device is reliable for determining the number of roller passes.
- Figure 9 shows that the density of Hot Mix Asphalt (HMA) decreases by passing pneumatic roller after steel wheel roller. This phenomenon could be related to the shearing force imposed by pneumatic roller passes. This effect would disappear after more passes of the pneumatic roller. Nevertheless, the total density achieved after pneumatic roller passes was less than steel wheel roller passes in two out of three. The authors decided to eliminate using the pneumatic rollers on the next site for a better conclusion.

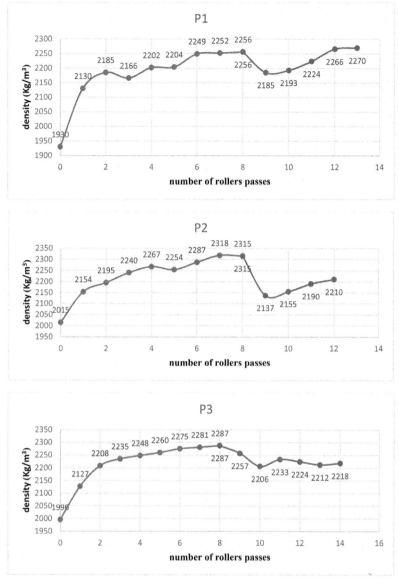

Figure 9.   Effect of rollers on compaction in Babaei Highway.

6.2 *Meygoon road results*

Results from Meygoon road are shown in Table 2. The regression method was used to evaluate the accuracy of the electromagnetic densitometer compared to core samples (Figure 10).

According to the data achieved from Meygoon Road, the results are as follows:

• Electromagnetic densitometer device measurements were close to core samples in this project. Although there was a two percent difference in some points gained, and just two points were exactly the same, the R-squared was much better than the previous site. This means that the device performance was much better on Meygoon road.

- Points positions (middle or edge of the paving lane) does not affect densitometers evaluations. This result was the same for Babaei Highway.
- Just steel wheel roller was used for P1, P2, P3, and P4. Core samples of these four points showed 96, 98, 96, and 97 percent of compaction. The accuracy of using the electromagnetic devices to the determination of the number of roller passes was considerable.
- Results obtained by four points prove that using pneumatic for the compaction of the top layer of HMA overlays is not necessary. It could decrease the final compaction in some cases, and it is better to use a steel wheel roller solely.

Table 2. Compaction result of densitometer and core samples at Meygoon road.

Point	Compaction percentage obtained by core samples	Compaction percentage obtained by electromagnetic densitometer
A1	97	99
B2	95	95
C3	98	100
D4	97	98
E5	97	97
F1	96	97

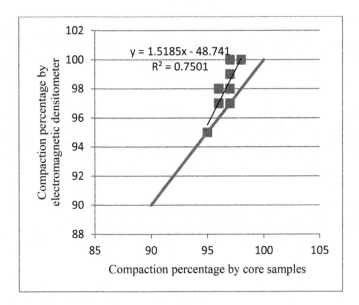

Figure 10. Comparing compaction of core samples and electromagnetic densitometer on Meygoon road.

7 CONCLUSION

Based on the study conducted, the results could be summarized as below:

1. The R^2 value was 0.17 for the first site and 0.75 for the second site. Using an electromagnetic densitometer could not be a reliable method for quality assurance of HMA layer compaction, and core sampling is still required. Although, using this method can decrease the number of core samples which could cause weakness in the newly paved

asphalt layer. Contractors can also use this method for quality control of their job and decrease their job errors that could cause penalties and future reconstructions.

2. Results from using an electromagnetic densitometer at seven specific points to determine the number of roller passes was successful. By using this device, contractors can specify the number of sufficient rollers to gain enough compaction at any HMA project.

3. The study shows that a steel wheel roller can be used solely to achieve maximum density in HMA layers, and utilizing a pneumatic roller terminates the compaction process in a lower compaction percentage. An insufficient number of pneumatic roller passes will decrease the compaction attained by the steel wheel roller.

4. The electromagnetic device is not affected by the position of measuring points (middle or edge of pavement).

According to the importance of asphalt compaction on pavements life, using an electromagnetic densitometer as a device for quality control of pavements would be helpful for asphalt contractors. This device is a non-destructive method that measures the density of asphalt pavement much easier and faster than the conventional core sampling method. Determining the number of passes for rollers at any time and in any project's section is another advantage of utilizing this device.

The effect of different rollers on the compaction of asphalt pavements was also noticeable in this study. Pneumatic rollers on their insufficient use after steel wheel rollers may lower the efficient compaction.

REFERENCES

Hausman J.J., and Buttlar W.G., 2002. Analysis of TransTech Model 300 Pavement Quality Indicator: Laboratory and Field Studies for Determining Asphalt Pavement Density. *Transportation Research Record* 1813(1): 191–200.

Hurley G.C., Prowell B.D., and Cooley Jr L.A., 2004. "Evaluating Nonnuclear Measurement Devices to Determine in-place Pavement Density." *Transportation Research Record* 1900(1): 56–64.

Kabassi K., Im H., Bode T., Zhuang Z., and Cho Y., 2011. Non-Nuclear Method for HMA Density Measurements. *Associated Schools of Construction (ASC) 47th Annual International Conference in Omaha, NE.*

Kvasnak A.N., Williams R.C., Ceylan H., and Gopalakrishnan K., 2007. *Investigation of Electromagnetic Gauges for Determining in-place HMA Density*, Iowa State University. Center for Transportation Research and Education.

The use of aggregates with different resistance to polishing in wearing courses

Pavla Nekulová* & Jaroslava Dašková
Brno University of Technology, Institute of Road Structures, Brno, Czech Republic

Leoš Nekula
Měrení PVV – Leoš Nekula, Vyškov, Czech Republic

Jiří Kašpar, Petr Bureš & Michal Sýkora
EUROVIA CS a.s., Prague, Czech Republic

ABSTRACT: The paper presents conclusions of a research project, which dealt with the use of aggregates with high and low resistance to polishing in asphalt mixtures for wearing courses. As part of the project, asphalt mixtures (stone mastic asphalt and asphalt concrete for wearing courses) were designed with various combinations of aggregates - basalt and greywacke. Basalt is a local, cheap aggregate with a low resistance to polishing, while greywacke is highly resistant to polishing but must be imported from a distant quarry. Several test sections were made from these asphalt mixtures. Friction after polishing was measured on cores taken from these sections. Based on the results, the asphalt mixture with the highest amount of basalt but still sufficient friction coefficient value was selected for paving a test section whose longitudinal friction coefficient is measured regularly. The paper presents the results of these measurements and evaluates the service life of the road skid resistance in the test section.

Keywords: Skid Resistance, Friction After Polishing, Polished Stone Value

1   INTRODUCTION

The research project TH02030194, supported by the Technology Agency of the Czech Republic, dealt with the possibility of using aggregates with a lower Polished Stone Value (PSV) in asphalt mixtures for wearing courses than usually required. The topic of the project was chosen with the view that a shortage of aggregates in road construction can be expected soon in the Czech Republic because no new quarry has been opened in the last thirty years, and the existing deposits are almost depleted (Czech Mining Union 2020). Because the skid resistance properties of the road surface have a significant impact on road safety (Mayora & Piña 2009), there are high requirements for used aggregates with PSV, which can ensure long-term satisfactory skid resistance properties of the road surface. Such aggregates are already unavailable in some parts of the Czech Republic, and with the increase in new constructions and road reconstruction, the shortage of aggregates will only increase. Using aggregates or recycled materials with unsatisfactory PSV to wearing courses with a small amount of aggregate with satisfactory PSV would lead to the economical use of materials

---

*Corresponding Author: pavla.nekulova@vutbr.cz

DOI: 10.1201/9781003429258-6

while maintaining the required lifespan of the skid resistance properties of the road surface (Kane & Edmondson 2020; Pomoni *et al.* 2021; Xiong *et al.* 2021).

2 DESIGNED ASPHALT MIXTURES

The most frequently used types of asphalt mixtures, namely (i) AC 11 (asphalt concrete for wearing courses with the aggregate up to 11 mm), and (ii) SMA 11 (Stone Mastic Asphalt with the aggregate up to 11 mm), were selected for the project. A basalt (PSV 51) from a quarry in northern Bohemia was chosen for asphalt mixture design. The availability of aggregates with a sufficient PSV is a general issue for road construction Czech Republic. The second aggregate was greywacke (PSV 62) imported from the distant quarry. A modified bitumen was used for both asphalt mixtures.

2.1 *SMA 11*

In the asphalt mixture type SMA 11, the aggregate was used only for the 8/11 fraction because this fraction is represented in the mixture by more than 50% and therefore has a significant effect on the skid resistance properties of the road surface. Basalt was always used in all the other fractions. First, three asphalt mixtures with 100% basalt (for comparison without the aggregate with higher PSV), 50%, and 100% greywacke in the 8/11 fraction were designed. Because the laboratory measurement delivered promising results, asphalt mixtures with 34% and 20% of the greywacke were added. All details about test specimens of the SMA 11 type are listed in Table 1.

Table 1. SMA 11.

Asphalt mixture	Test specimen label	Aggregate weight ratio, fraction 8/11	Test section
SMA 11	127655 A–E	100% basalt	Asphalt plant
	127656 A–E	100% greywacke	None
	127657 A–E	50% basalt	
		50% greywacke	Asphalt plant
	129752 A–E	66% basalt	
		34% greywacke	Asphalt plant, road
	130073 A–E	80% basalt	
		20% greywacke	None

2.2 *AC 11*

AC 11 type asphalt mixtures were initially designed analogously to the three SMA 11 mixtures, i.e., 100% basalt, 100%, and 50% greywacke in the 8/11 fraction. However, because the 8/11 fraction is significantly less represented in this type of asphalt mixture, the estimated lifespan of the skid resistance properties of the road surface was similar for all designed AC 11 mixtures. In addition, there was an issue with removing the modified asphalt binder from the surface of the test specimens made in the laboratory. Therefore, the work on the AC 11 type mixture was temporarily interrupted and resumed only at the end of the research project. In the newly designed AC 11 asphalt mixtures, the aggregate with low PSV was used for the 4/8 fraction to the same extent as in the 8/11 fraction. The exact aggregate ratios and test specimen labels are given in Table 2.

Figure 1. SMA 11 - core from a slab made in the laboratory, cleaned with benzene solvent (left) and core from a test section on the road (right).

Table 2.   AC 11.

Asphalt mixture type	Test specimen label	Aggregate weight ratio, fraction 4/8 and 8/11	Test section
AC 11	9, 10	100% basalt	Asphalt plant
	13, 14	50% basalt	
		50% greywacke	Asphalt plant

Figure 2.   AC 11 - core from the test section at the asphalt plant (left) and detail of its surface (right).

3 FRICTION AFTER POLISHING AND LONGITUDINAL FRICTION COEFFICIENT MEASUREMENTS

Slabs were made from designed asphalt mixtures in the laboratory from which cores (test specimens) of 225 mm diameter were taken. The specimens then underwent a test of accelerated polishing in accordance with the EN 12697-49, see Figure 3. During this test, the friction coefficient is initially measured on the test specimen surface. After that the specimen surface is polished by rubber cones rolling over; a mixture of water and quartz powder is added during polishing. This process simulates the traffic load. The friction coefficient measurement is repeated after a given number of passes of rubber cones. These two operations alternate until the required number of passes or a stable value of the friction coefficient is attained. The result is a graph of the development of the friction after polishing (FAP – an average value for the asphalt mixture from the measurements on individual cores from the same mixture) depending on the number of cones passes. Given this outcome, it is possible to approximately estimate the lifespan of skid resistance properties of the specific asphalt mixture under laboratory conditions (Daskova *et al.* 2016).

Figure 3. Equipment for determining the coefficient of friction after polishing according to EN 12697-49 (left) and dynamic measuring device TRT (right).

The skid resistance properties measurement on test sections, made from selected SMA 11 asphalt mixtures, was performed by a dynamic measuring device TRT (CEN/TS 15901-4) according to national standard CSN 73 6177. The measurement output is the longitudinal friction coefficient F_p value.

3.1 *SMA 11*

The surface of test specimens made in the laboratory was covered with a thick layer of modified bitumen binder and could not be removed by polishing cones. In this case, the EN 12697-49 standard recommends removing the binder by blasting corundum on the specimen surface, but this procedure significantly damaged the coarse aggregate. Cleaning of the specimen surface by benzene solvent was tested. Cleaning by benzene solvent also does not correspond to normal wear by traffic load. Still, it has proven to be sufficient for an indicative laboratory determination of the lifespan of the skid resistance properties.

Figure 4 shows the course of the relationship between FAP values and the number of cones passes for all asphalt mixtures from Table 1. The asphalt mixture containing only basalt aggregate has a FAP value of approximately 0.14 lower after 390,000 cone passes than asphalt mixtures containing varying amounts of greywacke. In contrast, there was no significant difference between asphalt mixtures containing greywacke, regardless of the

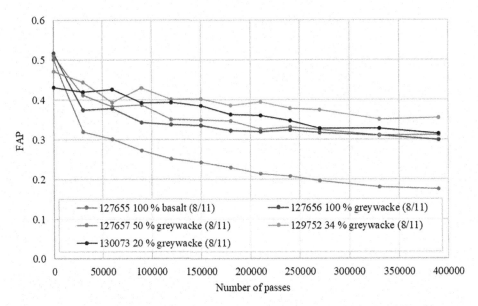

Figure 4.   FAP values as functions of the number of cones passes for individual asphalt mixtures type SMA 11 – test specimens made in the laboratory, cleaned by benzene solvent.

quantity contained (20-100% in the 8/11 fraction). The asphalt mixture with 34% greywacke in the 8/11 fraction had the highest FAP value after 390,000 passes.

Due to the ambiguity of the results of asphalt mixtures with greywacke, it was finally decided to pave the test sections at the asphalt plant, see Figure 5. The motivation was to reveal whether the laboratory results were distorted by the different surface textures of specimens made in laboratory conditions. For these purposes, the asphalt plant was selected, in which all used types of aggregates were stored for project solution purposes. Three asphalt mixtures were used: 100% basalt as a reference, 50%, and 34% greywacke (127655, 127657, 129752). The surface of the sections was not covered by chippings, i.e., the technological procedure for increasing the initial skid resistance of SMA was skipped.

Figure 5.   Paving of test sections in the area of the asphalt plant.

Cores for measuring the FAP values were taken from the test sections at the asphalt plant. It was not necessary to clean the surface of the cores with a benzene solvent. When paving with conventional techniques (road paver, compaction roller), there was no excess binder on the road surface, and it was possible to remove it by polishing cones. The surface texture also corresponded to the usual wearing course, which could not be achieved in the laboratory.

The graphs in Figure 6 show the FAP values of cores taken at the asphalt plant compared to the FAP results of test specimens made in the laboratory. It can be seen from the graph that the test specimens made in the laboratory had a higher initial FAP value, which is due to the cleaning of the binder with the benzene solvent and not gradually by the traffic load. Therefore, the initial values were shifted by 180,000 cone passes when the binder was removed, even on the cores from the test sections. Nevertheless, the FAP values of the test specimens from the laboratory and the cores from the test sections are different. It is due to the differences in the binder removal methods and the differences in the textures and com-paction methods. However, the difference in the FAP values between asphalt mixtures with and without greywacke is still significant: approximately 0.09.

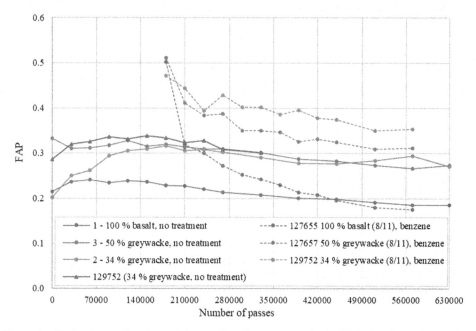

Figure 6.   FAP values as functions of the number of cones passes for individual asphalt mixture of the SMA 11 type – test specimens made in the laboratory (dashed line), cores from test sections in the area of the asphalt plant (solid line), and cores from test section on the local road (red line).

The asphalt mixture with 34% of greywacke was selected for the paving of the test section on the operated road. The test section was made on a local road between two villages. Unfortunately, it was impossible to obtain a permission to pave a test section on the road with a higher traffic load. The section was paved in the autumn of 2019, and the coefficient of longitudinal friction has been regularly monitored by the dynamic measuring device TRT. As a reference, a follow-up section was made of SMA 11, commonly used as a wearing course on motorways. A core was also taken from the test section to verify the FAP value of the asphalt mixture. The results agree with the FAP values measured on cores taken from the test section on the asphalt plant (red line in Figure 6).

The graphs in Figure 8 show the time histories of the longitudinal friction coefficient F_p for both test sections on the local road (SMA 11 with 34% of greywacke, SMA 11 used for motorways). Measurements immediately after paving exhibited the lowest values of F_p because the road surface was not treated by chippings, and the coarse aggregate was covered

Figure 7.   Paving of test sections on the local road.

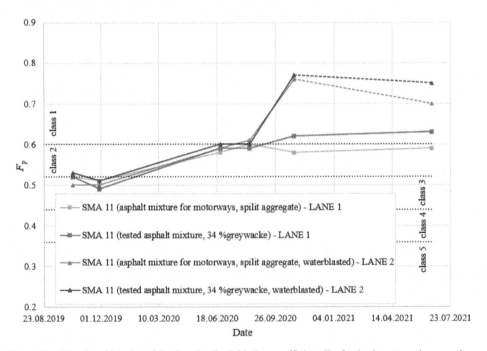

Figure 8.   The time histories of the longitudinal friction coefficient Fp for both test sections on the local road (SMA 11 with 34% of greywacke, SMA 11 used for motorways) and skid resistance classification levels 1 to 5 according to the national standard CSN 73 6177.

with a layer of binder, which is gradually worn by the effect of traffic. The F_p values in both sections are still rising because the binder is still being removed, and the aggregate has not yet been polished by traffic. To verify the lifespan of the skid resistance properties of the test sections quickly, the binder was removed in part of both sections by water blasting, which caused an increase in the F_p values. From this moment on, the aggregate wears in the asphalt mixtures. From the measurements of the F_p values performed so far, it follows that the asphalt mixture SMA 11 with 34% of greywacke is at least comparable with the asphalt mixture SMA 11 with spilite aggregate used on motorways. In the last measurement, it even attained 0.05 higher F_p value.

3.2 *AC 11*

None of the test specimens listed in Table 2, which was made using the AC 11 type asphalt mixtures, was produced in laboratory. The reason to skip the laboratory specimens was that it was not possible to achieve a surface texture that would mimic the texture of the road paved by a conventional technology. Therefore, test sections were paved at the asphalt plant, from which cores were taken to measure the FAP values. The measurement results are shown in Figure 9. From the course of the FAP values, it is evident that the greywacke again has a favorable effect on the long-term development of the FAP values. After 330,000 passes, the difference between the FAP values of asphalt mixtures is 0.08.

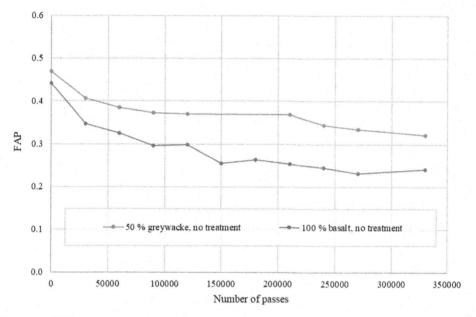

Figure 9.   FAP values as functions of the number of cones passes for individual asphalt mixtures type AC 11 – cores from test sections at the asphalt plant.

4   CONCLUSION

The results obtained within the research project and presented in this paper confirmed the assumption that if a part of the aggregate in the asphalt mixture with insufficient PSV is replaced by aggregate with high PSV, the lifespan of skid resistance properties of the road surface can be significantly extended.

For the asphalt mixture type SMA 11, a total of 5 mixtures with different proportions of aggregate 8/11 fraction with higher PSV - greywacke (0%, 20%, 34%, 50%, and 100%) were designed. To implement the test section on the local road, an asphalt mixture with 34% greywacke was finally selected based on laboratory measurement of the FAP values. According to the F_p results measured in the test section, the level of skid resistance properties of the road surface is so far evaluated with classification level 1 according to national standard CSN 73 6177. This asphalt mixture is also comparable with SMA 11 asphalt mixture, which was used on the motorways and to make a follow-up test section on the local road.

The measurements of the FAP values on cores from test sections made of asphalt mixture type AC 11 also showed a positive effect of aggregate with higher PSV on the lifespan of skid resistance properties of the road surface. Compared to the asphalt mixture type SMA 11, where the 8/11 fraction makes up approximately 50% of the aggregate volume, this percentage is significantly lower for AC 11. Therefore, it is necessary to use aggregates with a higher PSV even in the 4/8 fraction.

A key finding was also that producing slabs from asphalt mixtures with a surface whose texture would sufficiently match the surface of a paved road is problematic. To achieve the most reliable results in the project, the production of slabs was replaced by the paving of test sections at the asphalt plant, which is costly, and some asphalt plants do not have enough space. A potential solution for future research may be the use of a different type of compactor (segmental instead of lamellar) and a lower compaction rate in the production of the slab.

The calculation of the costs for the production of SMA 11 mixture showed that, apart from the variant where aggregates replaced all aggregates of fraction 8/11 with higher PSV (greywacke), a small amount of such aggregates is financially advantageous. The rest of the aggregate in the mixture was a low PSV aggregate (basalt), but it was from a local quarry. The costs were compared with the SMA 11 asphalt mixture used for paving motorways for which the skid resistance properties of the road surface were similar. Due to the current shortage of aggregates, the use of combinations of aggregates with different PSV may be a solution to reduce the cost of construction and also to manage the better-quality aggregates resources economically.

Unfortunately, it was not possible to prove in any general way which combination of aggregates to use (e.g., a weighted average of the PSV of aggregates). Each aggregate has a different shape index and creates a different surface texture in combination with another aggregate (Xiong *et al.* 2021). So far, the only solution is to make test specimens from the designed asphalt mixture and measure the FAP.

ACKNOWLEDGMENTS

The findings presented in this paper were gathered with the financial support of the Technology Agency of the Czech Republic, project TH02030194.

REFERENCES

CEN/TS 15901-4 Road and Airfield Surface Characteristics – Part 4: *Procedure for Determining the Skid Resistance of Pavements Using a Device with Longitudinal Controlled Slip* (LFCT): Tatra Runway Tester (TRT). November 2010.

CSN 73 6177 Measuring and Evaluation of the Skid Resistance of Road Pavement Surfaces (in Czech). December 2015.

Daskova J., Nekulova P. and Nekula L., *Implementation of EN 12697-49 standard into practice*, E&E Congress, 2016.

EN 12697-49 *Bituminous Mixtures – Test Methods for Hot Mix Asphalt – Part 49: Determination of Friction After Polishing*. February 2022.

Kane M. and Edmondson V., Long-term Skid Resistance of Asphalt Surfacings and Aggregates' Mineralogical Composition: Generalisation to Pavements Made of Different Aggregate Types, Wear, Volumes 454–455, 2020, 203339, ISSN 0043-1648.

Mayora J.M.P., Piña R.J., An Assessment of the Skid Resistance Effect on Traffic Safety Under Wet-pavement Conditions, *Accident Analysis & Prevention*, Volume 41, Issue 4, 2009, Pages 881–886, ISSN 0001-4575.

Pomoni M., Plati C., Kane M. and Loizos A., Polishing Behaviour of Asphalt Surface Course Containing Recycled Materials, *International Journal of Transportation Science and Technology*, 2021, ISSN 2046–0430.

Press release from the Czech Mining Union. September 2020. Available at (in Czech): https://tezebni-unie.cz/2020/09/30/stavebnictvi-hrozi-nedostatek-tezeneho-stavebniho-materialu/

Xiong R., Zong Y., Lv H., Sheng Y., Guan B., Niu D. and Wang H., Investigation on Anti-skid Performance of Asphalt Mixture Composed of Calcined Bauxite and Limestone Aggregate, Construction and Building Materials, Volume 306, 2021, 124932, ISSN 0950-0618.

Roads and Airports Pavement Surface Characteristics – Crispino & Toraldo (Eds)
© 2023 The Author(s), ISBN 978-1-032-55149-4

Experimental analysis of asphalt concrete modified with polymeric compounds including steel slags as an alternative to basaltic aggregates

G. Bosurgi & G. Bertino
Department of Engineering, University of Messina, Messina, Italy

A. Ciarlitti
Iterchimica S.p.A., Suisio (Bergamo), Italy

C. Celauro
Department of Engineering, University of Palermo, Palermo, Italy

M. Fumagalli
Iterchimica S.p.A., Suisio (Bergamo), Italy

O. Pellegrino, A. Ruggeri & G. Sollazzo*
Department of Engineering, University of Messina, Messina, Italy

ABSTRACT: Studying sustainable solutions in the infrastructure sector represents a strategical opportunity for reducing CO_2 emissions and other relevant issues on the environment. Beyond innovations in vehicles, the construction sector is fully involved in this process for defining and promoting novel and more sustainable materials and technologies, but with adequate mechanical performance. In this regard, since the use of "non-traditional" materials may determine huge benefits, several studies confirmed the possibility (and opportunity) of use waste materials after appropriate recycling actions (aimed to reach performance and environmental requirements) even in novel constructions. In this study, an experimental analysis in laboratory was carried out to evaluate the possibility of use steel slag aggregates to be used in innovative surface asphalt mixtures. The experimental phases relied on the comparison of the performance offered by an innovative mixture with steel slags, with those offered by a traditional one, with natural limestone and basaltic aggregates only. For both mixtures, a polymeric compound was used together with an adhesion activator for balancing effects of the high percentages of silica that hinder satisfactory adhesion between binder and aggregates. The experimental results evidenced an overall equivalence between the two mixtures in terms of mechanical performance, with clear environmental benefits for mixture including steel slags, due to the reduction of virgin materials and bitumen percentages. Further, the experiments showed higher crushing resistance of steel slags, a relevant advantage in laying and compaction phases since this property favours avoiding or minimizing effects of grain size distribution variation due to aggregate crushing.

Keywords: Steel slag, sustainable pavements, mechanical performance, polymeric compound

---

*Corresponding Author: gsollazzo@unime.it

DOI: 10.1201/9781003429258-7

1 INTRODUCTION

Environmental sustainability and the related issues, into the planning, construction and management of various works, belonging to multiple fields of civil engineering, play a fundamental role in order to assure a more stability of ecosystem and biodiversity in a future.

For this reason, in last decades, in the civil sector is preferable the use of reused materials to replace virgin ones with a high environmental impact and high economic charge. In the infrastructure field, the construction and maintenance of pavements has an environmental impact in terms of sustainability. Different studies are present in literature related to eco-compatible technologies for pavements (Abukhettala et al. 2016; Ahmedzade et al. 2009; Euroslag 2006); among these, the reuse of materials deriving from industrial processes, including steel slag, has proved to be a particularly effective practice over the years.

Studies carried out by the European Commission in 2009 showed that the percentage of slag per ton of steel produced in the EU is in the range of 12–15%; moreover, the research conducted in Germany in 2010 and 2015 confirmed these values. Into the EU, the position of Italy is of considerable importance, being second only to Germany, with 22 million tons/year of steel produced in year 2015. The adoption of slag in the production of road pavements contributes not only to the disposal of this material, but also to making the steel production process more eco-sustainable, allowing to be disposed and a new destination.

The first studies on the reuse of steel slags in asphalt mixtures note in the literature date back to the early 1980s (Emery 1984). Most of the research in this field has focused on chemical, physical and geometric properties (Asi et al. 2007; Pasetto et al. 2010, 2017; Xue et al. 2006). Motz & Geiseler (2001) investigated the possibility of reusing steel slag in road pavements, carrying out leaching and expansion tests to evaluate environmental behaviour and volumetric stability, respectively. Expanding the results of a previous research (Del Fabbro et al. 2001), Sorlini et al. (2003) carried out a study to evaluate the recovery of steel slags in road subgrade, in which, to investigate the release of pollutants and the volumetric expansion, a chemical characterization of the slags was carried out. Furthermore, several laboratory and field experiments conducted from previous studies show that the presence of steel slags in the asphalt mix can improve its durability, the stability of the slip resistance and the affinity with bitumen (Goli et al. 2017; Wen et al. 2016).

The choice of using recycled material is also useful for asphalt producers, who could take advantage of materials with a lower cost than those normally used, such as basalt, widely adopted in the Southern Italian context. Basalt, indeed, has a glassy porphyritic micro-crystalline structure or fine-grained that guarantees high mechanical resistance, making it very useful in road applications for the construction the most stressed layers of the pavement. At the same time, however, this material ensures low affinity with bitumen. Its composition mainly based on silicon, indeed, causes adhesion problems between aggregates and binder (Zhou 2021). In order to compensate the relative problematics to the percentage of residual voids normally encountered with the use of basalt (Iskender 2013), the basalt material is combined with limestone. In addition, the extraction and crushing of these stone aggregates entail very high costs and environmental impacts, very critical for sustainability evaluation of asphalt mixtures (Sollazzo et al. 2020).

This work reports an experimental laboratory study aimed at a comparative analysis between two mixtures, the first including a mix of ordinary aggregates (and, thus, basalt), and the second including steel slags as an alternative to basalt aggregates. In order to resolve the problems relative to adhesion with the binder and ensure high mechanical and durability performance to the final mixtures, some specific additives were equally added to the mixtures. The mixtures were investigated thorough an experimental laboratory campaign including physical and mechanical tests. The sought and achieved objectives with this study are to be traced back purely to the expected performance of the mixture containing the slag, which respected and satisfied structural, economic, environmental and functional aspects.

2 MATERIALS AND METHODS

In this paper, two different asphalt mixtures, suitable for surface layers, were considered for the experimental analysis. The first mixture (called "BM") includes ordinary aggregates (basalt and limestone) widely adopted in the Southern Italian context. The other mixture (called "SM"), instead, includes different coarse aggregates as an alternative to basalt, i.e. steel slags, obtained as a by-product of different industrial processes in the steel production.

The first part of the experimentation regarded tests on the aggregates; then, a mix-design, with Marshall and volumetric methods, of the asphalt mixtures for surface layers was performed to achieve a comparative analysis through different test procedures.

2.1 Aggregates

As a first characterization of the aggregates, a gradation analysis was performed (UNI EN 933-2). For each of the two compared aggregate types, two different classes were identified, respectively Slags 4/6-Slags 8/14 e Basalt 4/8- Basalt 8/12. The gradation curves of the various aggregates are reported in Figure 1.

In Table 1, the specific gravities of the 4 reference classes of aggregates are listed.

Moreover, both the shape (UNI EN 933-4) and the flakiness (UNI EN 933-3) indices were calculated. The abrasion and degradation resistance was evaluated trough the "Los Angeles" test, according to the UNI EN 1097-2. Finally, the affinity of the aggregates with bitumen was evaluated according to EN 12697-11 standard, as there are significant percentages of silicon in the composition of both basalt and slags (equal to 15% for the latter).

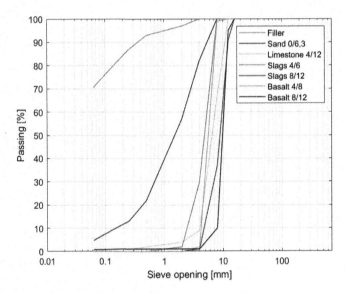

Figure 1.   Gradation curves of the various aggregates.

Table 1.   Specific gravity of the various aggregates.

	Filler	Sand 0/6,3	Limestone 4/12	Slags 4/6	Slags 8/12	Basalt 4/8	Basalt 8/12
Specific gravity [g/cm3]	2.701	2.671	2.645	3.558	3.561	2.941	2.900

2.2 *Bitumen*

In Table 2, the characteristics of the adopted bitumen (class 50/70) are reported, with the classification thresholds indicated by ANAS (Azienda Nazionale Autonoma delle Strade).

Table 2. Characteristics of the adopted bitumen.

		ANAS thresholds		
		Bit. 50/70	Bit. 70/100	Adopted bitumen
Penetration test 25°C (UNI EN 1426 – CNR BU 24/71)	dmm	50/70	70/100	68.9
Softening point (UNI EN 1427 – CNR BU 35/73)	°C	45–60	40–60	49.9

2.3 *Polymeric compound and additive*

The two mixtures investigated in this research were modified using a polymeric compound (PC), a mix of low-density polyethylene (LDPE), ethylene-vinyl acetate (EVA), and others polymers with low molecular weight and medium melting point. It was proved to ensure improved performance to asphalt mixtures in dry method modification, especially in terms of stiffness moduli and rutting resistance (Celauro *et al.* 2019). According to literature and product indications, the PC was added to the mixture as 0.3% of the aggregate weight.

Another anti-stripping agent (AS) was added to both mixtures, as 0.3% of the bitumen weight, to guarantee a perfect adhesion of bitumen to the aggregates. It has been proven to modify the chemical structure of the bitumen extending the service life of the pavement and allowing to obtain better workability, compaction, and high mechanical resistance. In this regard, specific tests were performed on the aggregates for evaluating improvements (EN 16297-11).

2.4 *Mix-design and performance tests*

The tests for mix-design described in the following were performed on mixtures including different types of aggregates (limestone plus basalt or slags) and various percentages of bitumen. For each mix, 3 different cylindrical specimens were tested.

The optimal percentage of bitumen was derived from the Marshall test (UNI EN 12697-34), by varying this percentage between 4.5% and 6.3% on the aggregate weight. All the mixtures include 0.3% on the aggregate weight of PC and 0.3% on bitumen weight of AS.

The volumetric characterization of the mixtures, even to evaluate the void ratio (UNI EN 12697-8) to compare with ANAS' specifications, was performed considering the theoretical maximum density (TMD) of the material (UNI EN 12697-5) and the bulk density calculated through the hydrostatic weight method (UNI EN 12697-6).

Then, a further analysis of the volumetric properties of the mixtures was performed using a Superpave Giratory Compactor (UNI EN 12697-31). During the compaction process, the results obtained after 10 cycles (representative of the first compaction, N1), 140 cycles (in situ compaction during construction phases, N2) and, 230 cycles (end-of-life compaction, N3) were analysed.

Finally, the stiffness modulus of the mixtures was evaluated according to the UNI EN 12697-26C, by applying a cyclic stress into the specimen. Results of these test were reported in form of ITSM (Indirect Tensile Strength Modulus).

3   EXPERIMENTAL RESULTS

In the following, the results of the various tests on both mixtures are provided. The test results for the flakiness index and the shape index are provided in Table 3. The results of the Los Angeles test are reported in Table 4. It is easy to evidence that both materials are in compliance with ANAS' specifications.

Table 3.   Flakiness and shape index for both mixtures.

	Slags		Basalt		ANAS' specifications
Flakiness index		10%		9%	< 20%
Shape index	4/6	8/14	4/8	8/12	—
	1%	5%	5%	7%	

Table 4.   Los Angeles test results.

Aggregate	LA (Los Angeles) [%]	ANAS' specifications [%]
Slags	13	< 20
Basalt	17	< 20

In Table 5, results show that both the aggregates exhibited a weak affinity with the virgin bitumen (< 15% even for the limestone aggregate, characterized by higher porosity). Then, the addition of AS (0.3% of the bitumen weight) was similarly evaluated. The related results showed a clear improvement, with values up to 95% for basalt and 90% for slags, widely in compliance with production requirements.

Table 5.   Result of the affinity between aggregates and bitumen test.

Material	AS [%]	Result [%]
Basalt + bit	0.0	10
Basalt + bit & AS	0.3	95
Slags + bit	0.0	5
Slags + bit & AS	0.3	90
Limestone + bit	0.0	15
Limestone + bit & AS	0.3	80

The gradation of the compared mixtures was evaluated in terms of both weight and volume. In both cases, as clear in Figure 2, the design curves stay into the specification fuse defined by ANAS. The composition of the final aggregate mixtures is reported in Table 6.

In Table 7, then, the volumetric characteristics derived for the various mixtures and different bitumen percentages are reported. In particular, the table reports the bulk specific gravity of the compacted mixture (Gmb), the theoretical maximum density (TMD), the void ratio (v), the voids in mineral aggregates (VMA) and the voids filled with bitumen (VFB).

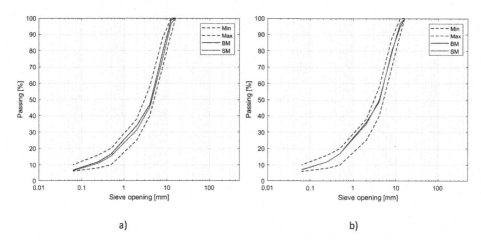

a)

b)

Figure 2.   Design gradation curves of the two mixtures: a) passing in weight; b) passing in volume.

Table 6.   Composition of the aggregate mixtures.

	Filler	Sand 0/6,3	Limestone 4/12	Slags 4/6	Slags 8/12	Basalt 4/8	Basalt 8/12
% for BM	6	49	0	0	0	22	23
% for SM	6	44	5	8	37	0	0

Table 7.   Volumetric features of the mixtures.

	b [%]	Gmb [g/cm3]	TMD [g/cm3]	v [%]	VMA [%]	VFB [%]
BM	4.7	2.418	2.567	5.8	17.1	66.2
	5.2	2.422	2.548	4.9	17.4	71.5
	5.7	2.425	2.53	4.2	17.7	76.5
	6.2	2.419	2.512	3.7	18.3	79.6
SM	4.5	2.655	2.761	3.9	15.8	75.6
	5.0	2.657	2.739	3	16.2	81.5
	5.5	2.654	2.717	2.3	16.7	86.1
	6.0	2.652	2.696	1.6	17.1	90.5

According to ANAS' specifications, to find the optimum percentage of bitumen, the Marshall test methods must be considered. Then, Marshall stability, flow, and ratio values for each mixture were calculated (Figure 3).

In Table 8, then, the results of the compactions performed using the Gyratory compactor are provided.

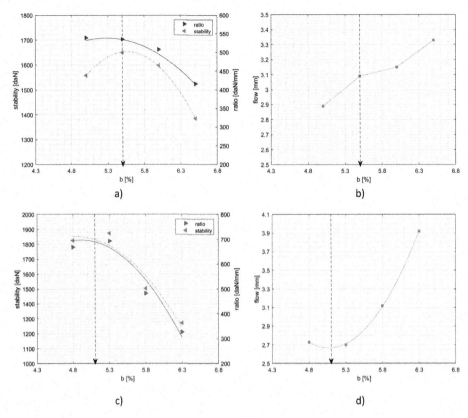

Figure 3.   Graphical results of the Marshal test method for both mixtures: a) stability and ratio for BM; b) flow for BM; c) stability and ratio for SM; d) flow for SM.

Table 8.   Volumetric features for specimens obtained using Giratory compactor.

	Gmb@N1 [g/cm3]	Gmb@N2 [g/cm3]	Gmb@N3 [g/cm3]	v@N1 [%]	v@N2 [%]	v@N3 [%]	TMD [g/cm3]
BM	2.079	2.381	2.422	18.4	6.6	5.0	2.549
SM	2.341	2.632	2.674	14.8	4.2	2.7	2.748
				11%–15%	3%–6%	>2%	

	VMA@N1 [%]	VMA@N2 [%]	VMA@N3 [%]	VFB@N1 [%]	VFB@N2 [%]	VFB@N3 [%]
BM	29.1	18.8	17.4	36.7	65.0	71.5
SM	26.0	16.8	15.5	43.0	74.9	82.7

Finally, the stiffness modulus in terms of ITSM was calculated at 5°, 20°, and 40°C. The maximum deformation level for each pulse was equal to 4 µm at 5°C, 7 µm at 20°C, and then 9 µm at 40°C. A peak time equal to 124 ms and 5 pulses, after 10 preconditioning pulses, were fixed. The results of these tests are provided in Figure 4.

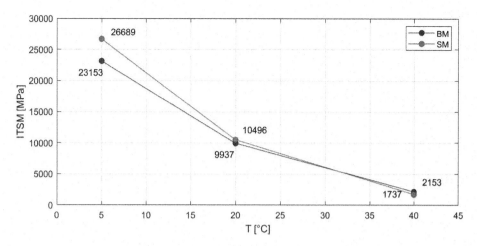

Figure 4. ITSM at various temperatures for the two mixtures.

4 DISCUSSIONS AND CONCLUSIONS

The laboratory tests presented in the previous section evidenced almost similar and comparable performance for the two mixtures investigated in this study. In detail, considering the acceptable results in terms of both mechanical and volumetric features of the mixture, steel slags appear as a reliable alternative to basalt aggregates to be used also in surface layers. This opportunity may determine relevant benefit both in terms of economic savings and sustainability improvements of asphalt pavements.

The analysis of the various results evidenced some interesting aspects:

- In terms of shake and flakiness of the aggregates, for the considered classes, there are not differences among the two materials (Table 3). These results may prove that the slags can ensure the same granular interlocking than basalt aggregates, i.e. a good attitude to compaction.
- The toughness ensured by the slags is stronger than basaltic aggregates, despite basalt is still characterized by good values (Table 4). Both values are in compliance with standard specifications for practical applications.
- Considering affinity with bitumen, results of the water boiling test evidenced that both aggregates have low affinity values (Table 5). This is due to the presence (in both materials) of remarkable percentages of silicon. This issue was solved by adding to both mixtures an anti-stripping agent, that may guarantee more durability to the mixtures.
- Gradation curves of the two mixtures were fully comparable in both weight and volume methods (Figure 2), in compliance with existing technical regulations. Both BM and SM are thus suitable for the production of adequate asphalt layers.
- By analysing the volumetric features of the mixtures (Table 7), BM exhibits reduced density values and higher percentage of voids than SM. Moreover, since the max value of Gmb is reached with 5% of bitumen in SM, while BM requires 5.7% of bitumen, the mixture with slags ensures higher density values together with beneficial reduction in bitumen needs.
- The results of the mix-design (Figure 3) evidence that both mixtures guarantee good performance and fully comply with the standard specifications. However, SM reached better values for stability, ratio, and flow. Furthermore, the analysis of the optimum bitumen percentage for the two mixtures showed that SM requires lower percentage of bitumen

(5.1%) than BM (5.5%), for maximizing the performance. For these bitumen percentages, the mixtures guarantee values of void ratio in the admissible range [3%–6%] (Table 7).

- The results of the volumetric characteristics of the gyratory compactor tests (Table 8) evidenced that SM may ensure a compaction level in compliance with specification limits for the void ratio. SM, in fact, exhibited good values at all investigated cycles (N1, N2, N3). On the contrary, the results for BM were not acceptable for the compaction levels corresponding to the initial in-situ compaction of the pavement (N1) and the exercise compaction (N2). Moreover, SM again showed higher density value than BM.
- Also the results in terms of stiffness modulus (ITSM, Figure 4) confirmed the good performance of both mixtures at different temperatures. In all cases, the observed values cannot evidence any relevant sensitivity to low or high exercise temperatures, critical in terms of low temperature fragility or permanent deformation accumulation. In particular, SM, at 5°C, exhibited a higher modulus than BM. Similarly, at 20°C, SM overpassed BM, but the difference is very slight. At 40°C, instead, the trend is inverted (although the difference is still very small). As said, the obtained value (1737 MPa) is still compatible for every practical application.

Then, the steel slags may be considered reliable alternative aggregates for asphalt mixtures to be adopted in the surface layer of road pavements. The results offered by the two mixtures are fully comparable and their alternative application may favour the satisfaction of construction quality performance, even respect to typically adopted ordinary mixtures. Similar optimized mixtures including steel slags, SA, and PC may represent a strategic alternative for pavement construction in order to reduce economic and environmental impacts due to basalt aggregates, that are currently widely considered in current practice in the South of Italy.

In fact, it should be considered that, as evidenced also in previous studies, the adoption of steel slags in asphalt concrete may ensure advantages in economic terms and from a sustainable perspective. Considering costs, in fact, since slags are a by-product of industrial processes (they were initially considered as waste), they have for sure lower costs than basalt aggregates. In terms of impacts, similarly, their use is convenient for at least two reasons: from one side, they should be treated and/or disposed by the steel industries, with consequent huge environmental issues; from the other, the production of basalt aggregates determines huge impacts (for example, in terms of abiotic depletion or energy consumption). All these issues, then, may be positively reduced by extending the adoption of asphalt mixtures including slags, favouring circular economy processes that may ensure preservation of resources and the ecosystem quality.

In further studies, the authors will try to deepen the performance investigation and the convenience analysis: first, other mechanical tests will be performed on both the aggregates and the mixtures to confirm what emerged in this research and to investigate sensitivity of slag mixtures at higher temperatures; further, a comparative economic analysis of the production and supply costs will provide other useful information for supporting decision makers and technicians in planning and design phases; finally, other interesting topics may be derived from a comparative assessment of the impacts (trough Life Cycle Assessment approaches) on the environment determined by the two materials.

REFERENCES

Abukhettala M. and Fall M. 2016. Literature review on Using Recycled Materials in Road Construction, *2nd Intern. Conf. on Civil, Structural and Transportation Engineering, Ottawa, Canada.*, May 2016

Ahmedzade P. and Sengoz B. 2009. Evaluation of Steel Slag Coarse Aggregate in Hot Mix Asphalt Concrete, *Journal of Hazardous Materials*, 165, 300–305.

Asi M., Qasrawi H.Y. and Shalabi F.I. 2007. *Use of Steel Slag Aggregate in Asphalt Concrete Mixes, Canadian Journal of Civil Engineering*, 34, 902–911.

Celauro C., Bosurgi G., Sollazzo G. and Ranieri M. 2019. Laboratory and In-situ Tests for Estimating Improvements in Asphalt Concrete with the Addition of an LDPE and EVA Polymeric Compound. *Construction and Building Materials* 196, 714–726

Emery J. 1984. Steel Slag Utilization in Asphalt Mixes, In: National Slag Association, *MF*. 186-1.

Euroslag. 2006. Legal Status of Slags Position Paper, *European Slag Association*. http://www.euroslag.org/.

Goli H., Hesami S. and Ameri M. 2017. Laboratory Evaluation of Damage Behavior of Warm Mix Asphalt Containing Steel Slag Aggregates, *Journal of Material in Civil Engineering*, 29 (6), 04017009-1-9.

İskender E. 2013. Rutting Evaluation of Stone Mastic Asphalt for Basalt and Basalt–limestone Aggregate Combinations. *Composites Part B: Engineering,* Nov 1;54: 255–64.

Motz H. and Geiseler J. 2001. Products of Steel Slags. *An Opportunity to Save Natural Resources, Waste Management* 21, 285–293.

Pasetto M. and Baldo N. 2010. Experimental Evaluation of High-performance Base Course and Road Base Asphalt Concrete with Electric Arc Furnace Steel Slags, *Journal of Hazardous Materials*, 181, 938–948.

Pasetto M., Baliello A., Giacomello G. and Pasquini E. 2017. Sustainable Solutions for Road Pavements: A MULTI-scale Characterization of Warm Mix Asphalts Containing Steel Slags, *Journal of Cleaner Production*, 166, 835–843.

Sollazzo G., Longo S., Cellura M. and Celauro C. 2020. *Impact Analysis Using Life Cycle Assessment of Asphalt Production from Primary Data, Sustainability*, 12 (24), 10171.

Sorlini S., Collivignarelli C. and Vezzola M. 2003. *Recupero di Scorie Nei Rilevati Stradali, Recycling*, 1, 101–106. (in Italian)

Wen H., Wu S. and Bhusal S. 2016. Performance Evaluation of Asphalt Mixes Containing Steel Slag Aggregate as a Measure to Resist Studded Tire Wear, *Journal of Material in Civil Engineering*, 28 (5), 04015191-1-7.

Xue Y., Wu S., Hou H. and Zha J. 2006. Experimental Investigation of Basic Oxygen Furnace Slag used as Aggregate in Asphalt Mixture, *Journal of Hazardous Materials*, B138, 261–268.

Zhou L., Huang W., Zhang Y., Lv Q. and Sun L. 2021. Mechanical Evaluation and Mechanism Analysis of the Stripping Resistance and Healing Performance of Modified Asphalt-basalt Aggregate Combinations. *Construction and Building Materials*, 273.

Roads and Airports Pavement Surface Characteristics – Crispino & Toraldo (Eds)
© 2023 The Author(s), ISBN 978-1-032-55149-4

Grip mastic asphalt – high friction asphalt for wearing course with the addition of steel slag

Cristina Tozzo*
Tecne S.p.A., Rome, Italy

Davide Chiola
Movyon S.p.A., Rome, Italy

Nicoletta Gasbarro & Paolo Spinelli
Autostrade per l'Italia S.p.A., Rome, Italy

ABSTRACT: Adherence between tire and pavement surface is essential to ensure traffic safety and driving comfortability. The phenomenon is very complex and there are many impact factors affecting the friction performance, including the tire type, pavement structure characteristics, pavement material characteristics, aggregate gradation, wet and dry condition, and vehicle speed. For this reason, significative efforts were made to understand the macroscopic friction based on continuum mechanics (Liu et al. 2021; Zorowski 1973) and the correlation between the micro texture of asphalt pavement that determines the actual contact area between tire and pavement and influences the intermolecular force and adhesion force (Chu et al. 2017). In this field the most evolutive model is represented by the study in micro-friction area (Yue Hou et al. 2018) analysed using Molecular Dynamics (MD) simulation. From an experimental perspective, several combinations of asphalt mixes and aggregates were investigated in terms of skid resistance to tyre. High value of skid resistance was easily related to the use of aggregate with high polished stone value (PSV) and mixes designed with high texture. An innovative test, the German Wehner Schulze test, was proposed within Europe as a replacement of the PSV and was developed with the aim to simulate the trafficking and assess the change of skid resistance with time of samples or cores extracted from pavement surface (Huschek 2004). Important comparison of PSV and WS methods are available in the literature (Allen et al. 2008). Moreover, Highway Administrations are every day committed in the promotion of innovations and technologies that enhance road safety. Among those, the most important one is related to the identification of materials or treatment able to significantly enhance skid resistance. These solutions reach a high level of importance when applied in locations with high friction demand with the aim to reduce crashes. With the aim to define innovative solutions for asphalt pavement to be used along the highway network, the research center of Fiano Romano (RM) investigated the friction properties of asphalt mixtures made with the addition of steel slag. The project belongs to an innovative approach aimed at the development of a sustainable engineering where the supply of natural sources is limited, and the use of recycled materials is encouraged.

Keywords: Pavement Temperature, Temperature Distribution, Prediction Models.

---

*Corresponding Author: cristina.tozzo@tecneautostrade.it

DOI: 10.1201/9781003429258-8

1 INTRODUCTION

Adherence between tire and pavement surface is essential to ensure traffic safety and driving comfortability. The phenomenon is very complex and there are many impact factors affecting the friction performance, including the tire type, pavement structure characteristics, pavement material characteristics, aggregate gradation, wet and dry condition and vehicle speed.

For this reason, significative efforts were made to understand the macroscopic friction based on continuum mechanics (Liu et al. 2021; Zorowski 1973) and also the correlation between the micro texture of asphalt pavement that determines the actual contact area between tire and pavement and influences the intermolecular force and adhesion force (Chu et al. 2017). In this field the most evolutive model is represented by the study in micro-friction area (Yue Hou et al. 2018) analysed using Molecular Dynamics (MD) simulation.

From an experimental perspective, several combinations of asphalt mixes and aggregates were investigated in terms of skid resistance to tyre. High value of skid resistance were easily related to the use of aggregate with high polished stone value (PSV) and mixes designed with high texture. An innovative test, the German Wehner Schulze test, was proposed within Europe as a replacement of the PSV and was developed with the aim to simulate the trafficking and assess the change of skid resistance with time of samples or cores extracted from pavement surface (Huschek 2004). Important comparison of PSV and WS methods are available in the literature (Allen et al. 2008).

Moreover, Highway Administrations are every day committed in the promotion of innovations and technologies that enhance road safety. Among those, the most important one is related to the identification of materials or treatment able to significantly enhance skid resistance. These solutions reach a high level of importance when applied in locations with high friction demand with the aim to reduce crashes.

With the aim to define innovative solutions for asphalt pavement to be used along the highway network, the research center of Fiano Romano (RM) investigated the friction properties of asphalt mixtures made with the addition of steel slag. The project belongs to an innovative approach aimed at the development of a sustainable engineering where the supply of natural sources is limited, and the use of recycled materials is encouraged.

2 OVERVIEW OF THE AVAILABLE SOLUTIONS TO IMPROVE ADHERENCE

According to the ASPI Technical Specification (TS), the treatments to improve the adherence of a pavement surface include all those construction technologies or applications intended to restore and maintain pavement friction to reduce crashes. Generally, those treatments refer to the application of very high-quality polish-resistant aggregate to the pavement using a polymer binder. They can be a temporary or even permanent solution.

This is the case of the Micro-surfacing, also known as micro-texturing, macro-seal, or macro-pavement, a latex-modified asphalt emulsion slurry paving system. This system was originally developed in West Germany in the 1970's and has been used in the U.S. since 1980. The total system generally consists of a latex-modified asphalt emulsion, chemical additives, high quality crushed aggregate and mineral filler (usually Type 1 Portland cement), and water (Guyer 2015).

As alternative, the TS include the use of pavement surface treatments, a thin layer of asphalt concrete formed by the application of emulsified asphalt or emulsified asphalt plus aggregate to protect or restore an existing roadway surface. The treatment may be placed over dense-graded asphalt pavement and rigid pavement.

In the recent years, an innovative and high adherence surface treatment called GRIPROAD (as the German company that has subcontracted this specialized work) was presented. According to the FDOT High Friction Surface Treatment Guidelines, this commercial product is composed of hard, polish and abrasion resistance aggregate, the calcined bauxite, bonded to the pavement using a polymer resin binder. Calcined bauxite is an imported product and the China region is the most important supplier.

Since 2005, ASPI decided to use this surface treatment on the highway network. This type of aggregates recorded a very high roughness and abrasion resistance with an initial CAT value more than 80 and a macro-texture above 0.30 mm typical of a traditional wearing course. The aggregates were used in the size of 1-4 mm, by crushing the calcined bauxite aggregates, and glued over the top pf the wearing course to be treated using a two-component epoxy adhesive with a mixing value from 2,600 to 3,000 g / sqm for a gross cost of the over 40 € / sqm. Particular attention was made in the execution of this surface treatment; all the surface applications were carried out in good environmental conditions (medium-high temperatures and low humidity, typically summer conditions necessary, if provided in night shift, for a rapid reopening to traffic the next morning) and on pavements in good structural condition having a surface layer consisting of a mixture of bituminous asphalt, closed graded. Paving operations were carried out using a machine specially designed and set up by GRIPROAD (Figure 1).

Figure 1.   Paving operation with GRIPROAD.

Preliminary results after few years highlighted that the GRIPROAD, while maintaining very high values of adherence (CAT still above 70), is affected by the different thermal susceptibility between the resin and the asphalt that leads to a progressive deterioration of the pavement with the appearance of surface cracks more or less widespread. After 4-5 years the deterioration showed an important evolution, with irregularities on the surface that can evolve more or less rapidly in structural problems depending on traffic conditions, environmental conditions and the state of the deeper layers.

3   ALTERNATIVE SOLUTIONS FOR HIGH ADHESION WEARING COURSE

According to the above, with the aim to define innovative solutions for asphalt pavement to be used along the ASPI Network, the research centre of Fiano Romano (RM) investigated the friction properties of asphalt surface treatments with high-performance in terms of adhesion, to be use in place of Griproad. The research was focused on a solution with a high level of affinity with the support layer (asphalt binder) and a high workability that allow the use also in less favourable conditions.

Moreover, the problem related to the high cost of supply of the calcined bauxite and its not easy availability (it comes only from China through a single European importer located in Belgium) led to seek more "viable" solutions. Following an innovative approach of the ASPI group aimed at the development of a sustainable engineering where the supply of natural sources is limited and the use of recycled materials is encouraged, it was decided to experiment the use of other "artificial" aggregates deriving from industrial processes with suitable characteristics for their reuse in the production of bituminous conglomerates.

To date, the market offers several opportunities for the use of these aggregates coming from iron and steel production processes in civil engineering, in response to the increasing attention to a sustainable development, based on the reduction of the consumption of natural resources and the minimization of waste production. In particular, the solution has been identified in the use of the granular material obtained from the crushing and granulometric selection of steel mill slag produced in plants located in Italy. This material is generated from the recovery of black steel mill slag and, thanks to the high temperature of the melting process that originates them, has composition and chemical characteristics similar to the rocks of volcanic origin commonly used in asphalt materials.

Moreover, the use of steel slag in small percentages (usually 10%) was already considered in the wearing course asphalt mix (both open graded and dense graded) from asphalt plants of the ASPI network, in all the cases an integration of the natural aggregates was needed because the availability of the latter did not fully cover the production needs. Steel slag commonly reported a good affinity with the bituminous binder.

In these proportions, however, the beneficial effect on the performance offered by the mix as a whole cannot be appreciated.

4   THE GRIP MASTIC ASPHALT (GMA)

Referring to the above, it was decided to experiment an asphalt surface treatment where the aggregate component is essentially made of Steel mill slag.

Given the fact that the application of such treatment is limited to the surface with a special attention to the adherence performance, a proper mix design was studied considering a mixture for a layer of thickness 2.5-3 cm with discontinuous grain size, typical of Splittmastix Asphalt or Stone mastic Asphalt (SMA).

4.1   *The field section*

The trial section, laid on July 14th and 15th, 2020, was located in the A1 North carr. in proximity of Arezzo. Paving operations were performed at night, after the restoration of 7 cm of the old binder course with a previous GRIPROAD on the surface with an advanced state of deterioration. The area of intervention is located on a "right" curve with a radius of about 410 m (longitudinal slope between 2 and 3%).

The percentage in weight of the steel slag on the total amount of aggregates used for the mix is between 78% and 97 and represents the lithoid skeleton of the layer. As can be seen from the cores extracted from the field section, the surface has a good roughness, and the slag aggregate shows a good porosity and some metal intrusions as showed in the Figure 2.

4.2   *Field measurements of the adherence with GMA*

4.2.1   *First analysis*
In order to confirm the suitability of the proposed GMA surface treatments, the CAT values were acquired with the equipment SuMMS (Survey Machine for Macrotexture and Skid). The acquisitions were performed 3 days after paving and reported an average CAT of 76.

Figure 2. Pictures of the cores extracted.

After 5 months the results achieved a value slightly above 60. The Figure 3a shows that after 650 days the CAT values are between 58 and 61.

Regarding the macrotexture, excellent values of the TEX were recorded, with a variation between 0.8 and 0.9 (Figure 3b). If this performance is maintained over time, as expected thanks to the excellent properties of durability of the steel slag, would ensure an alternative solution to GRIPROAD.

Moreover, for a better comparison of the GMA performance, a traditional wearing course was laid down in a near field section after a week for approximately 300 m starting at km 357 + 000. As can be seen from the Figure below (Figure 3c), after an initial fluctuation (due to the stripping of the binder covering the aggregates or due to the particular initial measurement occurred in the summer period) is showing the better performance of GMA compared to the traditional mix. After 650 days the CAT value of the GMA shows better performance than the traditional wearing course, with transversal adherence values of 61 and 49,7 respectively.

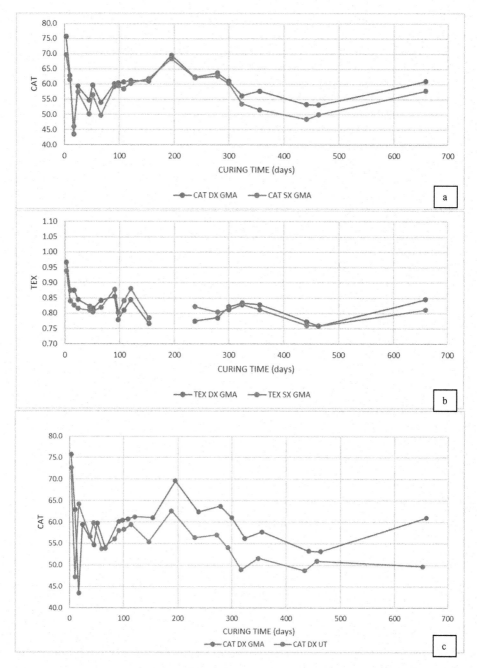

Figure 3. Field results of GMA in terms of TEX (a), CAT (b) and comparison of the GMA results in terms of CAT with a traditional wearing course (UT).

4.3 *Cost analysis of the GMA*

After a preliminary evaluation of the GMA costs, it is possible to assume that they are comparable with an open graded wearing course of the same thickness.

Regarding the GRIPROAD, the GMA costs are significantly lower (the latter has a cost 3.5 times higher) because the resin and the calcinate bauxite is more expensive if compared with recycling aggregates such as the steel slag.

5 CONCLUSIONS

According to the results of the field section described in the previous chapter, the following conclusions could be achieved:

- GMA costs are equivalent to an open graded wearing course of the same thickness;
- field measurements confirm that the performance of GMA in terms of adherence are higher than a traditional wearing course;
- GMA mix design is in line with the most recent principles of sustainable engineer and environmental criteria with the use of recycling products;
- it is possible to lay the GMA in more extreme environmental conditions than GRIPROAD (e.g. in case of night time construction sites in spring or autumn);
- the performance of GMA in terms of adherence are comparable to that offered by GRIPROAD, since the lower micro-roughness (lower CAT) is partly compensated by an excellent macro-roughness (important in wet pavement conditions) obtained from the particular particle size of the GMA mixture;
- the GMA is compatible with the underlying layers of traditional bituminous asphalt pavement, differently to GRIPROAD where the resin-induced cracking lead the deterioration of the support;
- the GMA could be easily integrated in the production of a traditional asphalt plant commonly used in the maintenance works.

Additional tests will be performed in order to confirm the preservation of such high performance and to monitor additional pavement indicators during the design life of the structure when the GMA is used.

ACKNOWLEDGMENTS

The activities presented in this paper were sponsored by Autostrade per l'Italia S.p.A. (Italy), which gave both financial and technical support within the framework of the Highway Pavement Evolutive Research (HiPER) project. The results and opinions presented are those of the authors.

REFERENCES

Allen B., Phillips P., Woodward D., & Woodside A. (2008, May). Prediction of UK Surfacing Skid Resistance Using Wehner Schulze and PSV. *In* International Conference Managing Road and Runway Surfaces to Improve Safety (pp. 11–14). SaferRoads. org Cheltenham, England.

Chu X.M., Li Y., Yan X.P., Wan J., Development of Skid Resistance Evaluation Based on Asphalt Pavement Micro-texture, Comput. Commun., 25 (1) (2007), pp. 61–65

FDOT *High Friction Surface Treatment Guidelines*

Hou Y., Zhang H., Wu J., Wang L., Xiong H., Study on the Microscopic Friction Between Tire and Asphalt Pavement Based on Molecular Dynamics Simulation, International Journal of Pavement Research and Technology, Volume 11, Issue 2, 2018, Pages 205–212, ISSN 1996-6814, https://doi.org/10.1016/j.ijprt.2017.09.001.

Huschek E.S. (2004). Experience with Skid Resistance Prediction Based on Traffic Simulation. *In Symposium on Pavement Surface Characteristics of Roads and Airports*, 5th, 2004, Toronto, Ontario, Canada.

Liu Z.J., Liu H.Q., Zhang Z.Y. Characterization of Pavement Texture and Research Progress on Testing Technology of Anti-sliding Performance, Road Eng., 4 (2012), pp. 62–65

Paul Guyer J. 2015, *An Introduction to Hot Mix Asphalt Spray and Surface Applications*, Continuing Education and Development, Inc.

Zorowski C.F. (1973). Mathematical Prediction of Dynamic Tire Behavior. Tire Science and Technology, 1 (1), 99–117.

Development of new method for detecting partial deformation and subsidence of porous asphalt surface layer and examination of new management criteria

Kiyohito Yamaguchi*
Deputy Division Chief Pavement Division Road Research Department
Nippon Expressway Research Institute Company Limited, Machida-shi Tokyo, Japan

Ryo Kato
Division Chief Pavement Division Road Research Department
Nippon Expressway Research Institute Company Limited, Machida-shi Tokyo, Japan

ABSTRACT: Japanese highways use the porous asphalt mixture on the surface layer of pavements. This mixture allows water to penetrate inside the pavement and prevents puddles from developing on the road surface. As a result, driving safety in rainy weather is improved and the number of traffic accidents has been significantly reduced. In fact, there was a case where the number of traffic accidents plummeted by 80% compared with before the mixture was used at a section of a Japanese highway. Furthermore, since the mixture also absorbs noise generated by vehicles, it makes it possible for drivers and passengers to travel comfortably and to reduce noise for people living along the road. As stated above, while the porous asphalt pavement has many advantages, it also has its disadvantages. Damages caused by the asphalt peeling off from the aggregate due to water infiltrating through the surface, called "stripping," may develop in the base course layer beneath the porous-asphalt surface layer. If this type of damage occurs, partial deformation or subsidence will appear on the surface. Road inspections on highways in Japan check the amount of rutting, crack rate, and IRI. However, it is difficult to quantitatively capture partial deformation and subsidence of the road surface caused by stripping of the base course layer with these check items. Therefore, we have developed a method to detect partial deformation or subsidence by utilizing the transverse profile data of the road surface, obtained from the road surface condition measurement vehicle, and examined the management criteria for this partial deformation or subsidence. * IRI: International Roughness Index: An international index for travelling comfort of pavement.

Keywords: Porous Asphalt Pavement, Partial Deformation, Transverse Profile, Management Criteria

1 CHARACTERISTICS OF POROUS ASPHALT MIXTURE

The porous asphalt mixture has been used on Japanese expressways since 1998 to provide a safer and more comfortable road pavement for users. As the porous asphalt mixture has a high porosity (general areas: 20%, snowy and cold areas: 17%) (Design Guidelines, 2020), it functions to drain rainwater from the surface layer quickly (Figure 1) and reduce noise generated when tires of travelling vehicles excite the road surface.

---

*Corresponding Author: k.yamaguchi.ac@ri-nexco.co.jp

DOI: 10.1201/9781003429258-9

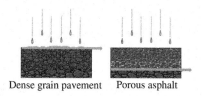

Dense grain pavement   Porous asphalt

Figure 1.   Drainage function of porous asphalt mixture.

2   EFFECTS AND PECULIAR DEFORMATION OF POROUS ASPHALT PAVEMENT

The effects of drainage and noise reducing functions of porous asphalt pavements are introduced in detail in this section.

2.1   *Traffic safety effect*

Investigations on the number of accidents at 213 locations on expressways where many accidents have occurred during heavy rainfalls one year before and after the porous asphalt pavement was laid showed that the number during rain has lowered to below 20% compared with before the pavement was laid. (Expressways and Automobiles 2002) (Figure 2(a).)

2.2   *Noise reducing effect*

According to the results of investigations at expressway sections paved with the porous asphalt mixture, the noise level was some 3 dB lower compared with sections using dense grain pavement. It was also found that the difference was maintained for some 3 years after laying the porous asphalt (Figure 2(b).). This noise-lowering effect of some 3 dB is equal to that of halving the traffic volume at the sections. (Expressways and Automobiles 2002)

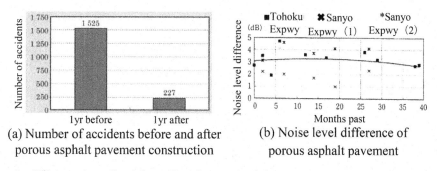

(a) Number of accidents before and after porous asphalt pavement construction

(b) Noise level difference of porous asphalt pavement

Figure 2.   Effects and peculiar deformation of porous asphalt pavement.

In this way the porous asphalt pavement benefits the road user as a safe and comfortable pavement to travel on. On the other hand, if the stripping resistance of the base course layer, which is the drainage basal plane under the surface layer, is low, stripping may cause the top face of the base course layer to turn fragile, leading to partial deformation of the surface porous asphalt pavement. This type of deformation is distinctive to porous asphalt pavements.

3 DAMAGE CAUSED BY STRIPPING OF BASE COURSE LAYER

An example of partial deformation of porous asphalt pavement caused by stripping of the base course layer is introduced in this section. Figure 3 shows subsidence, longitudinal cracks at the outer wheel pass (hereinafter OWP) and swelling at the outer lane line. The piece cut out from the pavement, on the left side of the Figure, is a sample from the swelled part, and the sample piece on the right is the subsided part of the OWP. In the samples, the asphalt stabilization roadbed layer, the upper layer of the two samples, and the porous asphalt mixture surface layer, are about same thickness, but the thicknesses of the base course layers are completely different. In the sample on the right, you can see a gap at the border of the surface layer and base course layer (top face of base course layer). It was assumed that stripping of asphalt, which is the base course layer mixture, and aggregate reduced the adhesive strength of the aggregate, causing the base course layer to drift to the side. And the longitudinal crack developed because the surface porous asphalt mixture could not adhere to the deformation of the base course layer mixture. This partial deformation is distinctive to porous asphalt pavement that was not seen when dense grain pavement was the mainstream of surface layer pavements (Motomatsu *et al.* 2004).

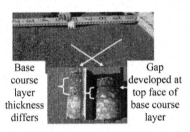

Figure 3.   Base course layer drifts to side due to stripping.

4 PAVEMENT MANAGEMENT CRITERIA

The management criteria for expressway pavements in Japan are set as shown in Table 1 for the purpose of maintaining the management standard at a level that allows operators to provide safe and comfortable roads to users. The management criteria indicate that it is preferable to carry out repairs before each deformation reaches the set criterion. Of the management criteria, other than that for flatness (IRI), indices set in the "road maintenance and repair guidelines (Outline of road maintenance and repair,1978)" in 1973, taking into consideration the results of surveys and research on road surface properties and vehicle performance at the time, were adopted. From around 1975, measurements of typical deformations, like rutting, flatness (σ before revising to IRI), and crack rates, obtained by the road surface condition measurement vehicle, have been compared with the management

Table 1.   Management criteria of pavement (Survey Guidelines 2020).

		Level difference (mm)				
Item	Rutting (mm)	Bridge mount	Cross structure mount	Skid friction coefficient	Flatness (IRI)	Crack rate (%)
Management criteria	25	20	30	0.25	3.5	20

criteria to rationally grasp the present state of road surface properties, and the method has not changed to date.

5 DEFORMATIONS REQUIRING REPAIRS AND MANAGEMENT METHOD OF POROUS ASPHALT PAVEMENT

Deformations requiring repairs of porous asphalt pavement in recent years are shown in Figure 4. Figure 4 is a summary of damages that were repaired on a total 28.7 km of lanes in one year at various locations in Japan (Eguchi *et al.*).

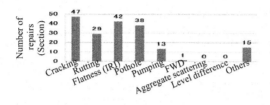

Figure 4.   Deformations on porous asphalt pavement requiring repair.

Repairs for rutting, flatness, and cracking, which are items of the management criteria in Table 1, take up some 65% of the total number of repair works, followed by those for potholes and pumping which take up 30%. Of the deformations needing repair, rutting, flatness and crack rate, are inspected every 2 to 3 years using the previously mentioned road surface condition measurement vehicle and thus it was possible to sufficiently grasp their current states. Therefore, pavement management for them has been carried out without any big problem in planning and conducting repair works. As for potholes and pumping, the current status is grasped by regular inspections (Maintenance and Inspection Guidelines 2021), which are basically visual inspections from a travelling vehicle carried out at least two times a week to rank the deformations per scale. And repair works are carried out following the priority ranking of the damage.

6 ISSUES FACING ROAD SURFACE MANAGEMENT OF POROUS ASPHALT PAVEMENT

When stripping occurs in the base course layer and penetrating rainwater cause ejection (pumping) of fine grains of the lower base course material, partial deformation and cracking occur that may lead to potholes. This is one way potholes develop in porous asphalt pavements. Even when only one pothole develops through this process, it is known that in many cases potholes will develop successively in the area in a short period of time. In such cases, there is usually an increase in the maintenance work rate for porous asphalt pavements with repair works not cannot catching up with the number of potholes developing and long hours needed to grasp the condition of the pavement through regular inspections, and it often becomes difficult to carry out maintenance as scheduled.

This issue may be solved if regular inspections to grasp the current road surface condition, which are mostly carried out after a deformation develops, may be performed beforehand and if the deformation may be understood rationally and quantitatively. But predicting potholes and grasping the state of their precursors, such as pumping, partial deformation and cracking, that develop within a small area, extending about 5 m, are completely different tasks compared with those for rutting, flatness and crack rate. These deformations, which develop in a wide area at a certain uniformity, are evaluated at 100 m intervals, using the road surface condition measurement results. This is why, currently, it is difficult to predict potholes or even grasp their precursors.

7 EXAMINING EVALUATION METHOD OF PARTIAL DEFORMATION OF POROUS ASPHALT PAVEMENT

If deformations like pumping, partial deformation, and cracking, which are precursors to potholes developing in an area extending some 5 m, may be grasped in advance, many issues may be solved and effective pavement management would become possible. This will lead to operators providing a safer, securer and more comfortable expressways to users. Thus an evaluation method for these deformations is examined.

In studying evaluation methods, it is preferable to examine a rational and quantitative method for precursor deformations that develop earlier than others, as the purpose of the study is to grasp the deformation in advance. One precursor to potholes, pumping, is the ejection of minute grains of the lower layer roadbed material accompanying the infiltration of rainwater. But it is unrealistic to quantitatively and rationally grasp traces of ejection. Another precursor, partial deformation, is a damage of the road surface and is measured every 2 to 3 years using the road surface condition measuring vehicle. So it may be possible to evaluate it rationally and quantitively using the measurements taken by the vehicle. Therefore, we decided to study an evaluation method by looking at the transvers profile used to calculate rutting.

7.1 Creating three-dimensional point cloud data with transvers profile

The transverse profile used in the study of the evaluation method was measured by the road surface condition measuring vehicle at 10 mm intervals in the transverse direction and at 50 mm intervals in the longitudinal direction, applying three-dimensional shape measurement using the optical cutting method.

To evaluate partial deformations of the porous asphalt pavement, the change in height compared with the reference height needs to be calculated. So it is necessary to convert the transverse profile into three-dimensional point cloud data, shown in Figure 5. But three-dimensional point cloud data is influenced by the gradient, therefore, the longitudinal and horizontal gradients of the transverse profile need to be corrected to 0%.

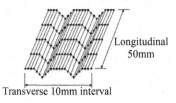

Figure 5.   Image of transverse profile point cloud data (Owaki & Takahashi 2021).

Figure 6(a) and 6(b) are images of the corrected transverse profile. Assuming that the incline of the transverse profile's approximate line is the horizontal gradient, the incline is corrected to 0 (Figure 6(a).). Next, to put the longitudinal gradient to 0%, the height (intercept) of the approximate line is set to 0 mm (Figure 6(b).). By performing this process on all transverse profiles, transverse profiles of reference height 0 mm will run in the longitudinal direction, resulting in three-dimensional point cloud data with the road surface gradient corrected to 0%.

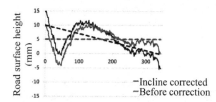

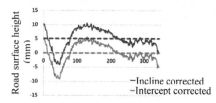

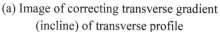

(a) Image of correcting transverse gradient (incline) of transverse profile

(b) Image of correcting longitudinal gradient (intercept) of transverse profile

Figure 6.   Images of the corrected transverse profile (Owaki & Takahashi 2021).

7.2   Method for aggregating the amount of change in three-dimensional point cloud data

The aggregation method was examined (Eguchi *et al.*) to quantitatively evaluate the amount of change in height compared with the reference height using the three-dimensional point cloud data with the road surface gradient corrected to 0%.

A partial deformation of porous asphalt pavements is the change in the height of the road surface which propagates to a certain range. Therefore, indices that can easily grasp the change were checked and the following characteristics were found.

1. Used the amount of change as is

   The difference in the surface level can be checked directly by the eye. But because the measurement intervals of the transverse profile are narrow, the level difference turns into minute noise and it is difficult to distinguish the detected difference.
2. The mean (μ)

   The mean (μ) is expressed as height distribution which was converted from the mean amount of a change in the noise.
3. Standard deviation (σ)

Road surface roughness is expressed by the differences in the height distribution.

From the above, it was decided that 2. The mean (μ) and 3. Standard deviation (σ) would be compared in detail.

To aggregate the amount of change in the three-dimensional point cloud data with the road surface gradient corrected to 0%, the mean (μ) and standard deviation (σ) of the transverse profile in the range (50 cm $\times$ 50 cm) marked by a square frame in Figure 7. were calculated. The position of the square frame is moved for each transverse profile data. The

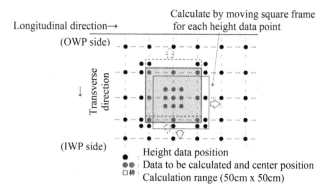

Figure 7.   Image of calculating the mean (μ) and standard deviation (σ) (Owaki & Takahashi 2021).

aggregation range is set to 50 cm × 50 cm because when the range is too large the amount of change is evened out, and when the range is too small, the number of data decreases and the reliability of the amount of change lowers.

In addition, to make it easier to visually grasp the partial deformation, it was decided to add color according to the size of the calculated mean value (μ) and standard deviation (σ). The positive value (high value) of the amount of change in the three-dimensional point cloud data was colored red, and the negative value (low) was colored blue. And the colors were made darker as the absolute values became larger.

The mean (μ) and standard deviation (σ) colored for better visualization are shown in Figure 8. Places circled in the road surface image indicate traces of ejection of the fine grain of the lower roadbed material caused by pumping. It can be confirmed that the mean (μ) and standard deviation (σ) are marked at the same positions. However, with the mean (μ), you see portions marked in red alongside places that are circled, while with standard deviation (σ), only the circled portion is colored. This means that partial deformations can be more clearly distinguished using standard deviation compared with the mean (μ).

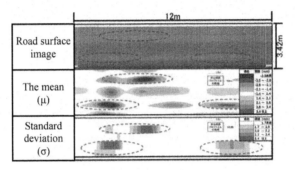

Figure 8.   Comparison of visualized mean (μ) and standard deviation (σ) (Eguchi *et al.*).

Based on the above results, it was decided that the aggregation method for capturing partial deformation from the amount of change in height with respect to the reference height of the three-dimensional point cloud data would be evaluated using standard deviation (σ).

7.3   *Examination of visualization method*

For the above-described partial deformation of porous asphalt pavement to be more easily visualized, the coloring for standard deviation (σ) of road surface height was examined in detail.

Figure 9. shows the threshold values of standard deviation (σ), considering the state, etc. of partial deformations, and colors used to indicate them. The classification of the threshold values is as follows.

1. □: Standard deviation (σ) of road surface height is less than 3 mm and there is no pumping and cracking
2. ▧: Standard deviation (σ) of road surface height is 3 mm to less than 4 mm and there is pumping
3. ▨: Standard deviation of road surface height (σ) is 4 mm to less than5mm, and there are pumping and cracking (medium degree)
4. ■: Standard deviation (σ) of road surface height is 5 mm or more and there are pumping and cracking (large degree)

Standard deviation of road surface height		Image	Pumping	Cracking
Color	Threshold			
□	Less than 3mm		none	none
▦	3mm to less than 4mm		yes	none
▦	4mm to less than 5mm		yes	medium
▦	5mm or more		yes	large

Figure 9.   Coloring of threshold values (standard deviation (σ)) (Owaki & Takahashi 2021).

As a result of confirming the standard deviation (σ), calculated from the three-dimensional point cloud data, and the state of partial and other deformations in the road surface image, a close relationship was confirmed between the size of the standard deviation (σ) and the state of the partial and other deformations.

7.4   Confirmation of coloring pattern according to standard deviation (σ) of road surface height

Partial deformations of porous asphalt pavements can be visualized and captured by the afore-mentioned method. But because colors to indicate rutting and partial deformation are placed alongside each other, it was decided to analyze deformations by focusing on the differences between the deformed state of the two.

As shown in Figure 10, it was found that in places where rutting has occurred, only one wheel pass is colored continuously in the longitudinal direction, while the section where partial deformation has occurred, two wheel passes in a range of about 5 m are colored. Next, when comparing the shapes of the transverse profiles, it was found that the section where partial deformation has occurred has a gentle shape at the V-shape bottom compared with that where rutting has occurred, and that standard deviation (σ) is lower and shown without coloring, as indicated by the ▲ in the figure.

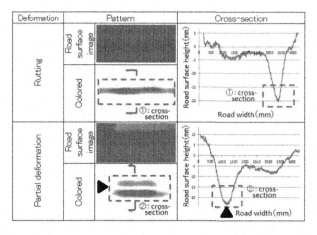

Figure 10.   Color pattern for rutting and partial deformation (Owaki & Takahashi 2021).

8   VERIFICATION OF EVALUATION METHOD FOR PARTIAL DEFORMATION
OF POROUS ASPHALT PAVEMENT

It was decided to verify the *in situ* applicability of the evaluation method of partial deformations of the porous asphalt pavement examined so far. The verification was carried out using data obtained by the road surface condition measuring vehicle in its monthly inspections carried out for two years from December 2014 at a 9.8 km section of the Joshinetsu Expressway (outbound, cruising lane). The evaluation procedure explained in the preceding sections is summarized as follows, and it was decided to verify the applicability following this procedure.

1. Calculate an approximate line for each transverse profile measured by the road surface condition measuring vehicle at 10 mm intervals in the transverse direction and 50 mm intervals in the longitudinal direction, and create three-dimensional point cloud data by setting the intercept and tilt at 0 and eliminating the longitudinal and transverse gradients of the road surface.
2. For the three-dimensional point cloud data, calculate the standard deviation (σ) of road surface height for the transverse profile data measured at intervals of 50 cm in the longitudinal and transverse directions.
3. Create a color diagram based on the threshold values of the standard deviation (σ) considering the state of partial deformation, etc.

As a result of confirming partial deformations using the color diagram created following the above process and the road surface image, the road surface image showed pumping had occurred at 74 places and the color diagram showed that there were 69 places. This means that the results agree at a high percentage (93%). It was also verified that improvement in the road surface may be confirmed by applying this evaluation method on road surfaces after maintenance and repair works of partial deformation were carried out.

9   STUDY OF MANAGEMENT CRITERIA (DRAFT) BY STANDARD
DEVIATION (Σ) OF ROAD SURFACE HEIGHT

It was decided to study new management criteria (draft) using the evaluation method for partial deformations of the porous asphalt pavement. In the study, to avoid establishing management criteria (draft) that deviate from the actual conditions of road pavement management, it was decided to confirm the repair records of regular inspections on the extent of standard deviation (σ) of the road surface height at places where potholes and their precursors pumping, partial deformation, and cracking developed within a range of about 5 m to see the degree of deformation that required maintenance and repair. Standard deviations (σ) from twenty-three 5-meter sections from five routes (Tohoku Expressway, Takamatsu Expressway, Kochi Expressway, Sanyo Expressway, Chugoku Expressway), where records and data of regular inspections, maintenance and repair dates of the above-described deformation, and transverse profiles of road surface conditions for about two years before maintenance works were carried out were available, were selected for the study (Figure 11).

From Figure 11, it could be confirmed that standard deviations (σ) were largely a distribution of 5.0 or more in 21 sections of the 23 sections. Therefore, it was understood that standard deviation (σ) 5.0 would likely be the management criterion (draft) based on the actual conditions of road pavement management. Next, after confirming deformations with standard deviation (σ) 5.0 or more using the results of regular inspections described in Figure 12 and the color diagram, it became clear that these deformations are pumping, partial deformation, and cracking that can be visually confirmed. And it was judged suitable to set standard deviation (σ) 5.0 as the management criterion (draft), the target value which repairs should be carried out before the deformation reaches it.

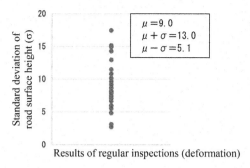

Results of regular inspections (deformation)

$\mu = 9.0$
$\mu + \sigma = 13.0$
$\mu - \sigma = 5.1$

Figure 11.   Relationship between regular inspections and standard deviation (σ).

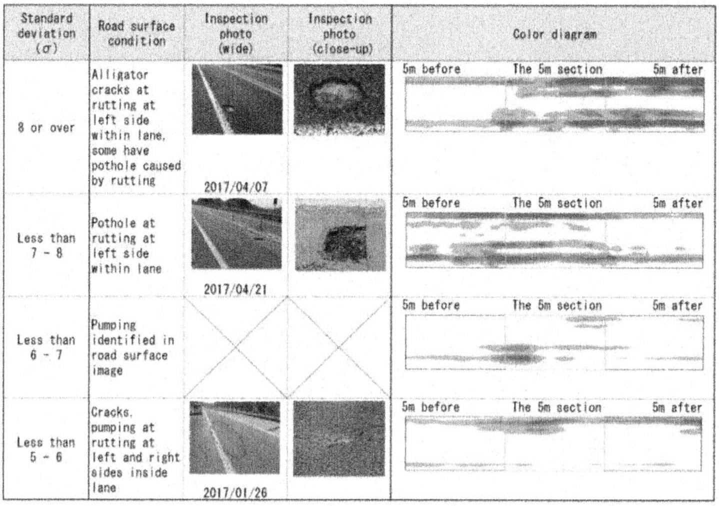

Standard deviation (σ)	Road surface condition	Inspection photo (wide)	Inspection photo (close-up)	Color diagram		
8 or over	Alligator cracks at rutting at left side within lane, some have caused pothole caused by rutting	2017/04/07		5m before	The 5m section	5m after
Less than 7 - 8	Pothole at rutting at left side within lane	2017/04/21		5m before	The 5m section	5m after
Less than 6 - 7	Pumping identified in road surface image			5m before	The 5m section	5m after
Less than 5 - 6	Cracks, pumping at rutting at left and right sides inside lane	2017/01/26		5m before	The 5m section	5m after

Figure 12.   Relationship between regular inspection results, standard deviation (σ) and color diagram.

If standard deviation (σ) 5.0 is set as the management criterion (draft), it would be necessary to grasp and study in advance deformed places with standard deviation (σ) of less than 5.0 that would be candidates of repair works and reflect what was found in the repair plan. Therefore, it was decided to analyze in detail the changes (Figure 13(a).) in the

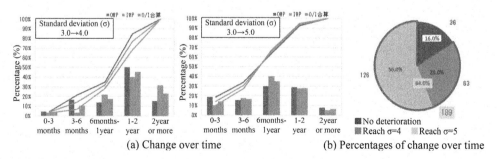

(a) Change over time   (b) Percentages of change over time

Figure 13.   Standard deviation (σ).

standard deviation (σ) over time and its ratio (Figure 13(b).) using data described in "Verification of evaluation method for partial deformation of porous asphalt pavement."

Figure 13 (a) shows that the time it takes for a standard deviation (σ) of a road section to exceed 3.0 and reach 4.0 is approximately within 2 years, and it takes less than 2 years or more to reach 5.0. The progression ratio of standard deviation (σ) is as follows and shown in Figure 13(b).

- The percentage of standard deviation (σ) deteriorating from 3.0 to 4.0: 84%
- The percentage of standard deviation (σ) deteriorating from 3.0 to 5.0: 56%
- The percentage of standard deviation (σ) not deteriorating to below 3.0: 16%

The above results show that a comparatively high percentage (56%) of standard deviation (σ) deteriorate from 3.0 to 5.0. In addition, it takes about within 2 years for the deterioration to progress. That period is about the same as the inspection interval of the road surface condition measurement vehicle. Therefore, it was decided that it would be preferable to plan repairs when the standard deviation (σ) reaches 3.0.

10   EXAMINATION OF REPAIR PLAN (DRAFT) APPLYING STANDARD DEVIATION (Σ) OF ROAD SURFACE HEIGHT

The above-described standard deviation (σ) 3.0 is suitable for planning repair works for potholes that occur in a range of about 5 m extension and their precursors pumping, partial deformation, and cracking, as it can be easily applied to small-scale replacements which are typical of maintenance works. However, deformations that develop continuously require large-scale repair employing cutting overlay works, and it is reasonable to plan repairs in 100 m sections like when planning repairs applying the existing management criteria (rutting, flatness, crack rate). Therefore, it was decided to examine the repair plan (draft) using standard deviation (σ) 3.0. The examination, using the repair data of cutting overlay works, was carried out to find the minimum number of 5 m sections within a 100 m length that have reached standard deviation (σ) 3.0, the measure to start planning repairs of the 100 m section.

Figure 14 shows the number of 5 m sections within a 100 m section in which the standard deviation (σ) exceeded 3.0, from the repair data used for the previously-described "Study of management criteria (draft) by standard deviation (σ) of road surface height." Figure 14 shows that if the number of 5 m sections, whose standard deviation (σ) exceeded 3.0, was minimum 8 sections (40 m), repair works were carried out by cutting overlay. The average was 14 sections (70 m) and the maximum 20 sections (100 m), and the range varies. But from the viewpoint of providing a more safe, secure, and comfortable expressway by managing the road pavement based on a certain management level, it was decided that repairs would be planned if the number of 5 m sections reached minimum 8 sections (40 m).

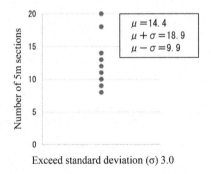

Figure 14.   Number of 5 m sections exceeding standard deviation (σ) 3.0.

11 SUMMARY

1. Partial deformations that are precursors to potholes occurring within the range of about 5 m in extension, peculiar to porous asphalt pavement, may be evaluated by creating a three-dimensional point cloud data by correcting the transverse profile obtained by the road surface condition measuring vehicle, then calculating the standard deviation (σ) of the road surface height, and determining the shape using the color diagram.
2. It is desirable to detect deformations through regular inspections, and repair pumping, partial deformation, and cracks that are precursors to potholes occurring in the range of about 5 m in extension before the standard deviation (σ) reaches 5.0.
3. Analysis of change over time of the standard deviation (σ) of road surface height show that the standard deviation (σ) deteriorates from 3.0 to 5.0 within about two years.
4. Judging from the data of repairs carried out by cutting overlay works, repairs were carried out if there were minimum 8 sections (40 m) of 5 m sections where the standard deviation (σ) of road surface height had reached 3.0.

From the above, partial deformation, a precursor to potholes that occur in the range of about 5 m in extension, which is peculiar to porous asphalt pavements, can be evaluated by calculating the standard deviation (σ) of road surface height from the three-dimensional point cloud data created by correcting the transverse profile obtained by the road surface condition measurement vehicle and determining the shape using the colored diagram. Further, it is desirable to carry out repairs by the time the standard deviation (σ) of the road surface height reaches 5.0. In planning repair works by cutting overlay for repairs of 100 m extensions, it is desirable for places with 8 (40 m) or more of 5 m sections that have reached standard deviation (σ) 3.0 to be extracted for repairs.

12 CONCLUSION

We plan to apply the evaluation method and management criteria (draft) developed this time for partial deformations which are precursors to potholes occurring in the range of about 5 m in extension, peculiar to porous asphalt pavements, after verifying the results of trial implementations in various parts of the country.

REFERENCES

Design Guidelines Vol.1 *Pavement Maintenance Edition*, July 2020, East Nippon Expressway Co., Ltd.
Eguchi M., Kawamura A., Tomiyama K., Takahashi S. and Endo K., Basic Study on the Evaluation of Road Surface Deformations by Three-dimensional Point Cloud Data Conversion of Transverse Profile, *Proceedings of Japan Society of Civil Engineering E1 (Pavement Engineering)*, Vol.73, No.3 (Pavement Engineering Papers Vol.22)
Maintenance and Effects of Porous Asphalt Pavements, Expressways and Automobiles Vol.45 No.3, March 2002
Maintenance and Inspection Guidelines Structures Edition, April 2021, East Nippon Expressway Co., Ltd.
Motomatsu S., Kamiya K., Matsumoto D. and Yamada M., Research on the Evaluation Method of Stripping Resistance of Existing Base Layer Mixtures, *Proceedings of Pavement Engineering, Japan Society of Civil Engineers*, Volume No.9, Dec. 2004
Outline of Road Maintenance and Repair, Japan Road Association (1978)
Owaki S. and Takahashi S., *Development of Evaluation Method Suitable for Damage to Porous Asphalt Pavement Magazine Pavement* 56–11 (2021)
Survey Guidelines, July 2020, Expressway Co., Ltd.

Next generation mobility (light, electrical, autonomous)

© 2023 The Author(s), ISBN 978-1-032-55149-4

Rutting effects on electrified asphalt roads embedding electric vehicles charging units

Claudia Nodari*, Maurizio Crispino & Emanuele Toraldo
Department of Civil and Environmental Engineering, Politecnico di Milano, Milan, Italy

ABSTRACT: Battery Electric Vehicles (BEV) are a potential solution to enhance the environmental sustainability of the transportation sector over the incoming years. Among the arising technologies, the current trends seem to be addressed by the dynamic inductive charging system as a promising solution. However, since this system requires a new charging infrastructure network composed by prefabricated Charging Units (CUs) embedded into the road pavement, some issues related to the structural life of the so-called electrified road (e-road) pavements must be addressed. Moreover, the available literature reports few studies on the structural response of e-road, but the long-term performances of pavements are not fully investigated, including fatigue and rutting effects. Regarding the latter, this paper shows the results of an investigation in which the e-roads and the traditional ones (t-roads) are compared in terms of rutting proneness. In particular, the study is devoted to identifying the number of critical load cycles leading to rut an e-road pavement from a theoretical point of view. As a second step, the theoretical results are verified applying them to a real case study: Viale Forlanini in Milan (Italy), which is an important two carriageway road connecting the city center to the Linate airport. As a result of these investigations, interesting achievements were obtained regarding the rutting effects of embedding CU into road pavement. Therefore, CU seems to be compatible with the on-site high temperature effectiveness of the pavement, as also confirmed by the case study results.

Keywords: Electrified Road, Dynamic Charging, Charging Unit (CU), Finite Element Modelling (FEM), Rutting Resistance

1 INTRODUCTION

The vehicles electrification is a feasible answer to increase environmental sustainability of road transport system (Azad *et al.* 2019; Gruetzmacher *et al.* 2021; Mahesh *et al.* 2021; Schiavon *et al.* 2022), which is responsible for the 18.34% of the total Green House Gases emissions (European Commission 2021).

Dynamic inductive charging is one of the emerging technologies for Electric Vehicle (EV) battery recharge, but it requires a proper infrastructure network. Therefore, the future roads have to satisfy the EV needs by transporting electricity (e-roads) as well as the traditional road (t-road) requirements. For this on-the-road charging system, a possible network solution consists of prefabricated concrete Charging Units (CU) embedded into the road pavement (Chabot & Deep 2019; Chen *et al.* 2019; Nodari *et al.* 2021a). This generates further questions related to both structural response and long-term performances of e-road pavements, in terms of fatigue life and rutting resistance. Even if in the available scientific literature, few studies examine the e-road structural behavior, the rutting resistance of

---

*Corresponding Author: claudia.nodari@polimi.it

DOI: 10.1201/9781003429258-10

pavements is not fully investigated. This is why, the current research analyzes rutting proneness of e-road compared to t-road, using a Finite Element Model (FEM) to calculate stresses and strains in different cross-sectional geometries and for various load positions.

2 LITERATURE REVIEW

Rutting is the phenomenon that leads to rut depth development. The rut depth is a depression on the road surface in the wheel path resulting from the permanent deformations in each pavement layer due to traffic repetitions (AASHTO 2015). It is representative of the vertical difference between the transverse profile of the pavement surface and a wire-line across the lane width (AASHTO 2015). The rutting resistance assessment consists in the evaluation of the rut depth at the end of the pavement service life. The maximum rut depth is variable: in (Lister & Addis 1977) rut does not exceed 12.5 mm; in (ASTM 2020) rut depth lies between 6 and 13 mm (low severity level) and in (Direzione Generale Infrastrutture e Mobilità - Regione Lombardia 2005) rut should be lower than 15 mm (low severity level). Scientific literature provides numerous formulas to calculate permanent vertical deformations in each pavement layer, as reported in the following.

In (AASHTO 2015), for example, a specific relationship allows to calculate the permanent deformation into asphalt concrete layer (HMA), as shown in Equations from (1) to (4).

$$\Delta_{p(HMA)} = \varepsilon_{p(HMA)} \cdot h_{HMA} = \beta_{1r} \cdot k_z \cdot \varepsilon_{r(HMA)} \cdot 10^{k_{1r}} \cdot n^{k_{2r} \cdot \beta_{2r}} \cdot T^{k_{3r} \cdot \beta_{3r}} \tag{1}$$

$$k_z = (C_1 + C_2 \cdot D) \cdot 0.328196^D \tag{2}$$

$$C_1 = -0.1039 \cdot H_{HMA}^2 + 2.4868 \cdot H_{HMA} - 17.342 \tag{3}$$

$$C_2 = 0.0172 \cdot H_{HMA}^2 - 1.7331 \cdot H_{HMA} + 27.428 \tag{4}$$

Where:

$\Delta_{p(HMA)}$ is the accumulated permanent vertical deformation in the HMA layer/sublayer [inch];

$\varepsilon_{p(HMA)}$ is the accumulated permanent axial strain in the HMA layer/sublayer [-];

$\varepsilon_{r(HMA)}$ is the elastic strain calculated by the structural response model at the mid-depth of each HMA sublayer [-];

h_{HMA} is the thickness of the HMA layer/sublayer [inch];

n is the number of traffic repetitions;

T is the pavement temperature [°F];

k_{1r}, k_{2r}, k_{3r} are the global field calibration parameters, equal to -3.35412; 0.4791 and 1.5606;

$\beta_{1r}, \beta_{2r}, \beta_{3r}$ are the local or mixture field calibration constants equal to 1;

k_z is the depth confinement factor, defined in Equations from (2) to (4);

D is the depth below the surface [inch];

H_{HMA} is the total HMA thickness [inch].

In (AASHTO 2015), a definite equation is suggested to calculate permanent vertical deformation in all unbound pavement subbase layers and foundations. Moreover, among the various equations available from the scientific literature, Equation (5) allows to calculate the permanent deformation in granular material (GM), as shown in (Veverka 1979).

$$\varepsilon_{p,\text{GM}} = 2 \cdot \varepsilon_{yy} \cdot n^b \tag{5}$$

Where:

ε_{yy} is the vertical compressive strain calculated by the structural model;

n is the actual number of traffic repetitions;

b is a coefficient equal to 0.2 (thickness higher than 0.12 m) or 0.3 (thickness lower than or equal to 0.12 m).

The previous formula is adapted properly to HMA, considering that permanent strains are proportional to elastic strains (Monismith 1992).

$$\varepsilon_{p,HMA} = 4.49 \cdot \varepsilon_{yy} \cdot n^{0.25} \tag{6}$$

Where:

ε_{yy} is the vertical compressive strain calculated by the structural model;

n is the actual number of traffic repetitions.

Regarding subgrade (S), Heukelom and Klomp formula – Equation (7) – allows to calculate permanent deformation based on the maximum acceptable vertical stress.

$$\varepsilon_{p,S} = \varepsilon_{yy} \cdot (1 + 0.7 \cdot \log n) \tag{7}$$

Where:

ε_{yy} is the vertical compressive strain calculated by the structural model;

n is the actual number of traffic repetitions.

3 OBJECTIVES

The current research is divided in two phases.

The first one is devoted to identifying the number of critical load cycles leading to rut an e-road pavement from a theoretical/general point of view. Therefore, the aim is the assessment of CU effect on the pavement structural performances.

In the second phase, the theoretical outcomes are verified applying them to a real case study: Viale Forlanini in Milan (Italy), which is an important two carriageway road connecting the city center to the Linate airport. In this step, the actual traffic (load cycles) is known and therefore the rut depth for each cross-sectional geometry can be calculated.

The paper is planned as follow. Section "Rutting resistance: general approach" describes the rutting resistance of pavement from a general perspective, comparing e-road results with the ones obtained for t-road. The case study is detailed in Section "Rutting resistance: a case study", considering real traffic loads and bituminous mixtures (previous studied by the Authors). In the end, conclusions and future investigations are drawn.

4 RUTTING RESISTANCE: GENERAL APPROACH

4.1 *Methodology*

Three cross-sectional geometries are investigated, as shown in Figure 1: t-road, e-road_solid CU and e-road_void CU. As regard the latter, CU cavity dimensions (0.24 m in width and 0.03 m in height) are obtained by using an optimization process (Nodari *et al.* 2021a).

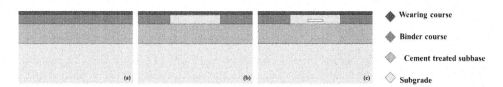

Figure 1.   Cross-sectional geometries (a) t-road; (b) e-road_solid CU; (c) e-road_void CU.

Concerning materials, the key characteristics of asphalt layers are reported in Table 1, while the details of cement treated subbase, subgrade and concrete are given in (Nodari *et al.* 2021a). In order to evaluate the asphalt layer behavior during all seasons - in line with the fatigue analysis of (Nodari *et al.* 2022) - Young's modulus at 20°C (spring/autumn) is increased (winter) and decreased (summer) by 50%, as shown in Table 1.

Table 1. Key characteristics of asphalt layers.

	Thickness [m]	Bulk density [N/m^3]	Young's Modulus [N/mm^2]			Poisson's Ration [-]
			Winter (E + 50%)	Spring/Autumn (E)	Summer (E-50%)	
Wearing Course	0.05	24,000	8,250	5,500	2,750	0,35
Binder course	0.14	23,500	5,250	3,500	1,750	0,35

Additionally, to relate cross wander distance to CU location, three load positions are examined: *centered on CU* (load is centered on CU, in the middle of the lane); *CU edge* (the left wheel load of EV is located along CU left edge) and *CU center* (the left load is located along CU center). More details about loads and 2D FE model are given in (Nodari *et al.* 2021a).

Based on these conditions (cross-sectional geometries, load positions and temperatures), 27 FEM simulations are performed using COMSOL Multiphysics 6.0 software, in order to calculate stresses and strains in the entire pavement domain.

Among the obtained model outputs, the maximum vertical compressive strains along proper analysis sections (Table 2) are required to calculate the permanent strains in each layer -equations from (5) to (7)-. Regarding the analysis sections, maximum strains are evaluated at the interface between layers and into subgrade, as reported in Table 2. This approach is in line with the one used in (Nodari *et al.* 2022) to assess fatigue life of e-roads. It is important to note that along *binder course-cement treated subbase* interface, vertical strains related to CU concrete are not computed since they do not describe the asphalt concrete behavior. To summarize, 108 analyses (3 cross-sectional geometries, 3 load positions, 3 temperatures, 4 analysis sections) are performed in order to obtain the maximum vertical compressive strains in each layer.

Table 2. Analysis sections.

Layer		Analysis sections	Y-coordinate [m]
Layer	Wearing course	Wearing course-binder course interface_top	-0.049
	Binder course	Binder course-cement treated subbase interface_top	-0.189
	Cement treated subbase	Cement treated subbase-subgrade interface_top	-0.489
	Subgrade	Into the layer	-0.590

Concerning permanent strain models, equations from (5) to (7) are used to calculate strains in cement treated subbase, asphalt concrete and subgrade, respectively. Since in these FEM simulations cement treated subbase is considered as a cracked material, a model dedicated to calculating permanent strain in granular material is adopted -equation (5)-.

Moreover, *b*-coefficient of equation (5) is equal to 0.2, considering that the thickness of cement stabilized subbase layer is equal to 0.30 m.

As already said, the goal of this first part is the identification of the critical load cycles number leading to rut an e-road pavement for a depth of 1.5 cm (ASTM 2020; Direzione Generale Infrastrutture e Mobilità - Regione Lombardia 2005; Lister & Addis 1977). Thus, based on the aforementioned conditions, the critical load cycles are calculated adopting the Microsoft Excel Solver Tool, in which:

• The *objective cell* is the rut depth equal to 1.5 cm;
• The *changing variable cell* is the critical load cycle number.

The experimental plan of the first part is summarized in Figure 2 for the sake of clarity.

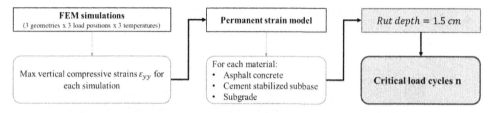

Figure 2.   Flow chart of the general approach.

4.2   *Results and discussion*

The number of critical load cycles *n*, that leads to a rut depth of 1.5 cm, is reported in Figure 3, according to cross-sectional geometries and load positions. The higher the load cycles number, the higher the rutting resistance.

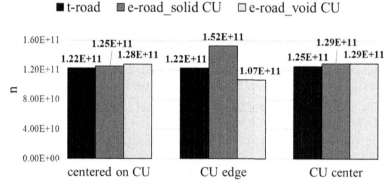

Figure 3.   Critical load cycles [ESAL] leading to rut depth of 1.5 cm.

When load is *centered on CU* or along *CU center*, all the graphs are characterized by the same trend/similar value of load cycles. In particular, the highest difference between load numbers is equal to 3% (considering *solid CU-CU center* compared to *solid CU-centered on CU*). Comparing different cross-sectional geometries, with the same load position, CU effect is weak. In fact, respect to t-road, the maximum increase in critical load cycles value of e-road is equal to 4% and to 3%, for load *centered on CU* and load along *CU center* respectively.

On the contrary, when load is along *CU edge*, the effect of CU is evident. The critical cycles number increases of 24% (solid CU) and decreases of 13% (void CU) respect to t-road. Considering the same cross-sectional geometries, if load is along *CU edge*, an increase of

101

20% (solid CU) and a decrease of 17% (void CU) are registered respect to load *centered on CU*.

Based on these results, the lowest critical load cycles number is obtained for e-road_void CU and load along *CU edge* $(1.07 \cdot 10^{11}$ ESAL).

Figure 4 compares the critical cycles number for rutting (that lead to a rut depth of 1.5 cm) to the critical cycles number obtained for fatigue phenomenon (that lead to a Damage Index DI equal to one, as discussed in previous Authors' study (Nodari *et al.* 2022)). For the sake of simplicity, the y-axis (critical cycles numbers) is represented on a logarithmic scale.

As reported in Figure 4, the fatigue traffic repetitions are significantly lower (three orders of magnitude) than the ones obtained considering rutting. Thus, for the analyzed cross-sectional geometry, it can be stated that the most critical phenomenon is fatigue.

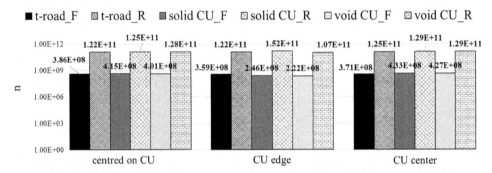

Figure 4.   Fatigue critical load cycles (F) *vs* Rutting critical load cycles (R) for different cross-sectional geometries and load positions [ESAL].

Based on the achieved results, once again, the most important role is played by the load position. In fact, assuming e-road_solid CU:

- When load is on *CU edge*, the increase of rutting critical cycles is 619 times higher than the critical number of fatigue traffic repetitions (the highest increment);
- When load is along *CU center*, the increase of rutting critical cycles is 297 times higher than the critical number of fatigue traffic repetition (the lowest increment).

Considering both the phenomena, the lowest number of critical load cycles is obtained for e-road_void CU and load along *CU edge*.

5   RUTTING RESISTANCE: A CASE STUDY

5.1 *Methodology*

In this section, the theoretical results of the previous part are verified applying them to a real case study. Viale Forlanini in Milan (Italy), which is an important two carriageway road connecting the city center to the Linate airport, is selected as the case study, in line with (Nodari *et al.* 2022).

The conditions, considered in the case study for rut depth calculation, are listed in the following:

- Three cross-sectional geometries (t-road, e-road_solid CU and e-road_void CU), as reported in the general part (Figure 1);
- Three load positions (*centered on CU*, *CU edge* and *CU center*), as described in the general part;

- Three pavement temperatures (8°C, 21°C and 34°C corresponding to winter, spring/autumn and summer, respectively), defined according to the Linate Airport weather station, as discussed in (Nodari *et al.* 2022);
- Two bituminous mixtures - one for wearing course and one for binder layer - studied by the Authors in previous researches (Nodari *et al.* 2021b, 2020) are considered, in line with the analysis conducted in (Nodari *et al.* 2022). The key main properties of these mixtures are reported in Table 3. Regarding the other layers, the characteristics of concrete CU, cement stabilized subbase and subgrade are in line with the ones presented in the general part;

Table 3. Key characteristics of bituminous mixtures for wearing course and binder course, derived from (Nodari *et al.* 2021b, 2020).

| | Thickness [m] | Bulk density [N/m³] | Young's Modulus [N/mm²] | | | Poisson's Ration [-] |
			8°C	21°C	34°C	
Wearing Course	0.05	24,240	17,502	8,092	2,822	0.35
Binder course	0.14	23,970	17,242	8,610	3,363	0.35

- Two traffic conditions (considering a pavement service life of 20 years): vehicles travel on one lane (5.75 · 10⁷ ESAL) and vehicles travel on two lanes (2.88 · 10⁷ ESAL). More details about the procedure used to determine the traffic can be found in (Nodari *et al.* 2022);
- Four analysis sections, as presented in the general part (Table 2), are evaluated for the calculation of vertical compressive strains;
- Three permanent strain models are adopted, as detailed in the general part – equations from (5) to (7) -.

The experimental plan of the second part is reported in Figure 5.

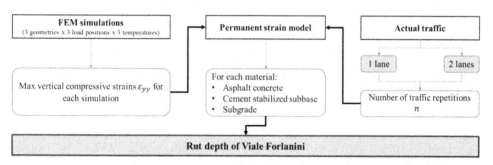

Figure 5. Flow chart of the case study.

5.2 *Results and discussion*

Table 4 shows the rut depths for the two traffic conditions, according to cross-sectional geometries and load positions.

Comparing the traffic conditions, all the graphs are characterized by the same trend of rut depth. As expected, the results considering traffic on two lanes are lower than the ones obtained for vehicles travelling on one lane (decrease of 15%). Even if it is not possible to

Table 4. Rut depth considering the two traffic conditions.

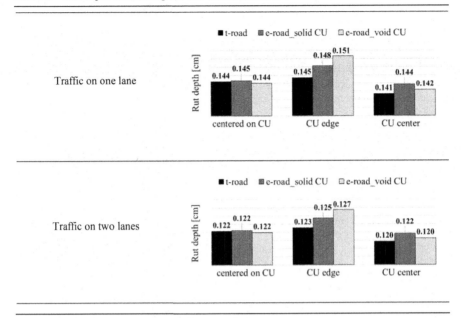

identify a correlation between rut depths and traffic repetitions, it can be stated that an increase of traffic involves an increase of rut depth.

Regarding the effects of load positions on rutting results, the main findings are listed above:

- Load *centered on CU*: no significant differences are appreciable between cross-sectional geometries. Thus, CU effect is very weak;
- Load along *CU edge*: void CU is characterized by the highest values, with increase of 4.14% and 3.25% compared to t-road for traffic on one lane and on two lanes, respectively. This load position guarantees the lowest pavements rutting resistance;
- Load along *CU center*: solid CU shows the highest values of rut depth (respect to the other cross-sectional geometries at equal load position), with increase of 2.13% and 1.64% compared to t-road for traffic on one lane and on two lanes, respectively.

E-road_void CU is the geometry characterized by the highest difference between load positions. In particular, when load is on *CU edge*, void CU rut depth increases of 6.34% (traffic on one lane) and 5.83% (traffic on two lanes) respect to load in *CU center*.

For both traffic conditions, rut depth values are significantly lower than the limit of 1.5 cm, previously defined (ASTM 2020; Direzione Generale Infrastrutture e Mobilità - Regione Lombardia 2005; Lister & Addis 1977). These results are in line with the ones obtained in the general section, in which the traffic cycles that lead to rut a pavement are in the order of magnitude of 10^{11} ESAL.

6 CONCLUSIONS

Dynamic inductive charging is one of the emerging technologies for Electric Vehicle (EV) battery recharge. Among the alternatives, the use of prefabricated concrete Charging Units (CUs) embedded into the road pavement seems to be a promising infrastructure network.

However, the CU effects on pavement long-term performances are not fully investigated. This is the reason why, the current study analyzes rutting resistance of electrified road (e-road) compared to traditional-road (t-road), using a Finite Element Model (FEM). The research is divided in two steps. The first one is devoted to identifying the number of critical load cycles leading to rut an e-road pavement from a theoretical point of view. In the second step, the theoretical outcomes are verified applying them to a real case study.

Based on the achieved results, the main conclusions arising from the research described in this paper can be drawn as follow:

- In the first part, the rutting resistance of e-road is affected by load position. When load is *centered on CU* or on *CU center*, CU effect is weak. In fact, comparing different cross-sectional geometries with the same load position, the number of traffic repetitions for e-roads is similar to the one characterized t-road. Otherwise, when load is along *CU edge*, the negative effect of CU is evident;
- The most critical configuration in terms of rutting resistance is defined by e-road_void CU and load along *CU edge*;
- The case study outcomes validate the results of the theoretical part, also in the worst traffic condition (traffic on one lane);
- Comparing long-term performances of pavements, fatigue phenomenon is more critical than the rutting one. The configuration described by e-road_void CU and load along *CU edge* leads to the lower results.

Finally, as further research steps, real scale tests are required in order to both validate the achieved results and optimize the construction phase.

REFERENCES

AASHTO, 2015. *Mechanistic-Empirical Pavement Design Guide - A Manual of Practice, Concrete Pavement Design, Construction, and Performance.*

ASTM, 2020. D6433 – 20 - *Standard Practice for Roads and Parking Lots Pavement Condition Index Surveys.*

Azad A.N., Echols A., Kulyukin V.A., Zane R., Pantic Z., 2019. Analysis, Optimization, and Demonstration of a Vehicular Detection System Intended for Dynamic Wireless Charging Applications. *IEEE Trans. Transp. Electrif.* 5, 147–161.

Chabot A., Deep P., 2019. 2D Multilayer Solution for an Electrified Road with a Built-in Charging Box. *Road Mater. Pavement Des.* 20, S590–S603.

Chen F., Balieu R., Córdoba E., Kringos N., 2019. Towards an Understanding of the Structural Performance of Future Electrified Roads: a Finite Element Simulation Study. *Int. J. Pavement Eng.* 20, 204–215.

Direzione Generale Infrastrutture e Mobilità - Regione Lombardia, 2005. Catalogo dei Dissesti Delle Pavimentazioni Stradali.

European Commission, 2021. *EU Transport in Figures - Statistical Pocketbook 2021, Notes.*

Gruetzmacher S.B., Vaz C.B., Ferreira A.P., 2021. Assessing the Deployment of Electric Mobility: A Review. In: *Lecture Notes in Computer Science (Including Subseries Lecture Notes in Artificial Intelligence and Lecture Notes in Bioinformatics)* 12953 LNCS. pp. 350–365.

Mahesh A., Chokkalingam B., Mihet-Popa L., 2021. Inductive Wireless Power Transfer Charging for Electric Vehicles-A Review. *IEEE Access* 9, 137667–137713.

Monismith C.L., 1992. Analytically Based Asphalt Pavement Design and Rehabilitation: *Theory to Practice*, 1962-1992. *Transp. Res. Rec.* 1354 5–26.

Lister N.W. and Addis R.R., 1977. Field Observations of Rutting and Their Practical Implications. *Transp. Res. Rec.*

Nodari C., Crispino M., Pernetti M., Toraldo E., 2021a. Structural Analysis of Bituminous Road Pavements Embedding Charging Units for Electric Vehicles. *Comput. Sci. Its Appl. – ICCSA 2021. Lect. Notes Comput. Sci.* vol. 12952 1, 149–162.

Nodari C., Crispino M., Toraldo E., 2020. Bituminous Mixtures with High Environmental Compatibility: Laboratory Investigation on the Use of Reclaimed Asphalt and Steel Slag Aggregates. *Lect. Notes Civ. Eng.* 76, 433–442.

Nodari C., Crispino M., Toraldo E., 2021b. Laboratory Investigation on the Use of Recycled Materials in Bituminous Mixtures for Dense-graded Wearing Course. *Case Stud. Constr. Mater.* 15, e00556.

Nodari C., Crispino M., Toraldo E., 2022. Fatigue Effects of Embedding Electric Vehicles Charging Units into Electrified Road. *Case Stud. Constr. Mater.* 16, e00848.

Schiavon M., Adami L., Ragazzi M., 2022. Private Electric Mobility and Expected Impacts on Climate and Air Quality. *Int. J. Transp. Dev. Integr.* 6, 25–36.

Veverka V., 1979. Raming Van de Spoordiepte bij Wegen met een Bitumineuze Verhanding. *De Wegentechniek* Vol. XXIV, 25–45.

Roads and Airports Pavement Surface Characteristics – Crispino & Toraldo (Eds)
© 2023 The Author(s), ISBN 978-1-032-55149-4

Transparent communication and improvement of cycle path maintenance for an increase in bicycle use

Rainer Hess & Carsten Mahnel*
Department of Civil Engineering, Mainz University of Applied Sciences, Mainz, Germany

Liliane Stein
Department of Communication Design, Mainz University of Applied Sciences, Mainz, Germany

ABSTRACT: A sustainable modal mix is being promoted by encouraging the use of bicycles. In this context, it is important to consider the choice of transport mode, which depends on objective factors such as the presence of infrastructure and personal attitudes as the individual valuation of it. The empirical research design of the «AllRad» project includes an application in three German cities with preceding demand- and following impact-analysis. Therefore, the subjective perception of the population in relation to the local cycling infrastructure was collected with an online survey. Results illustrated the importance of cycle lane maintenance in relation to bicycle use, as respondents perceived the condition after the quality of the cycling infrastructure as the highest potential barrier to bicycle use. Besides the survey revealed a greater need of information. Based on the findings technical measures to improve maintenance and communication measures to generate comprehension were implemented.

Keywords: Political Strategies, Communication, Maintenance, Monitoring and Inspection

1 INTRODUCTION

Germany has set itself the target of a transport transition, consisting of a mobility transition and a propulsion transition, in order to shape cities that are more liveable and simultaneously protect the environment. A sustainable modal mix is the core objective, for which cycling and public transport have been promoted and supported for years, so that the share of cycling in the modal mix has risen to 11 % (Nobis 2019).

The approximately 1.46 billion € in funding for the years 2020 to 2023 from the Federal Ministry of Digital and Transport serves the development of cycling and is preferably used for the expansion of the cycling infrastructure (BMDV 2021). The availability of infrastructure is only one factor among many that can have a lasting influence on the individual's choice of transport mode. Quality, condition and general comfort are further objective evaluation factors that have to be taken into account in everyday mobility.

As a complement to the expansion of the cycling infrastructure, it is important to find out, through long-term quality assurance, how local authorities can improve cycle path maintenance in order to increase the attractiveness of the mode of transport and thus everyday bicycle use.

---

*Corresponding Author: Carsten.Mahnel@hs-mainz.de

DOI: 10.1201/9781003429258-11

2 METHODS

2.1 *Setting*

In order to develop potential improvement measures for cycle path maintenance, quantitative data was collected by means of an online survey in order to obtain the subjective requirements of users for the cycle infrastructure, and guided interviews were conducted with those responsible for cycle path maintenance in three German cities in order to identify potential for improvement and scope for action.

The findings from the survey were used to design the improvement measures together with the selected pilot cities and to test them in ongoing operations. The direct evaluation of the measures revealed optimisation potential in the implementation and further need for action.

Within the framework of the project, the individual measures of the pilot application are subject to a final evaluation, which is structured symmetrically to the first survey. The effects of the individual measures on the subjective perception of the users will be surveyed, which will allow conclusions to be drawn about bicycle use.

Since the empirically designed project includes a pilot phase with technical and organisational measures, a practical application in the cycling infrastructure is advisable. Therefore, three cities were selected that have a significant share of cycling in the modal split, but at the same time show differences in structure, type of space and expansion of cycling

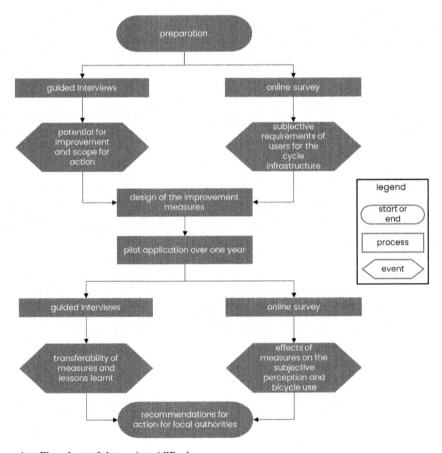

Figure 1. Flowchart of the project AllRad.

infrastructure. This ensures that unchangeable influences are included and a variation of measures is applicable. Pilot cities are:

- Mainz (capital of Rhineland-Palatinate) Part of the Rhine-Main Metropolitan Region
- Muenster "Cycling Capital" with a particularly high share of cycling and well-developed cycling infrastructure
- Munich (capital of Bavaria) Large city, centre of region

2.2 *Guided interviews with those in charge of cycle path maintenance*

In Germany, the responsibility for maintaining of roads and paths, which includes cycling infrastructure, is divided among four levels depending on their importance and function.

- Federal Government
- Federal state
- County
- Municipality / City

The project focuses on municipalities and cities, as this is, where most of the cycle infrastructure is located. The tasks of cycle path maintenance are divided into different departments, which in most cases, as in the selected pilot cities, are subordinate to the civil engineering department. As an intermediary between the population and traffic planning, there is the role of the bicycle commissioner or a bicycle office in Germany, which thus overlaps with the tasks of the civil engineering department.

In the period 10/2020-12/2020, the responsible persons of the two mentioned offices from the three pilot cities were interviewed in a one-hour conversation via telephone or video conference.

The guided interviews with the civil engineering offices focused on surface repairs, green maintenance, cleaning, winter services and prioritisation of the cycle network in maintenance. The processes, planning, necessary expenditure and possible weak points were openly discussed.

The discussions with the cycling representatives and the bicycle offices focused on the interfaces between the services of the municipalities/city and the population. The establishment and operation of communication platforms for reporting defects in the infrastructure and public relations work were key issues. Here, too, the potential was discussed openly.

2.3 *Online user survey*

In the period from 10/2022 to 12/2022, the online survey was conducted using the software "LimeSurvey". The survey was primarily advertised in the selected cities, but was not limited to them. The answers could be assigned regionally by specifying the postcode.

In terms of content, the questions and answer options were based on the recurring survey of the Federal Ministry of Digital and Transport "Mobility in Germany". This enables comparability with the participant characteristics as well as their assessments of the superordinate survey.

After asking about personal characteristics and mobility behaviour, which were structured similarly to the "Mobility in Germany", the subjective perception of the cycling infrastructure was then asked. For this purpose, the satisfaction and safety perception of the regional infrastructure was queried depending on the infrastructure development as well as on the cycle path maintenance. A differentiation was made according to individual subject areas, which contain the areas of maintenance as well as infrastructure development. In addition, specific questions were asked about the need and potential for improvement in the area of cycle path maintenance.

In addition to the technical measures, the use of and satisfaction with the communication offers in relation to the cycling infrastructure were surveyed. For this purpose, the channels used, desired information and need for communication on the topic were surveyed.

3   RESULTS

3.1   *Results of the interviews*

Based on the interviews the service depots responsible for cycle path maintenance are appropriately equipped with material and personnel for their tasks. For the prioritisation and priority maintenance of cycle routes, smaller sweeping and clearing vehicles or cleaning by hand are used in contrast to the road. In the case of priority management of the cycle network, the three pilot cities have different levels of expansion.

Structural measures in the case of surface damage, for example, only fall within the scope of cycle path maintenance if the monetary and spatial scope is small. In the case of larger and multiple damages on a route, the renewal is managed as an investment, so that this must be brought into the municipal household individually. Since this requires several higher-level approvals, also at the political level, it can take several years from the occurrence of the damage to the construction measure.

The weak points in the maintenance of cycle paths mentioned by the responsible persons are the above-mentioned inertia in the repair of existing cycle paths. Furthermore, trees that raise the pavement often cause severe damage due to their roots. The repair is very costly due to tree protection, where the roots generally must not be damaged, which is why it is rarely carried out. There is a need for a cost-effective and pragmatic solution here.

In addition, a discrepancy has been identified between the public's expectations of cycle path maintenance and the implementation of an economical maintenance service. In this context, the maintenance services of the municipal operations are usually underrepresented in public relations and the discrepancy is further intensified.

3.2   *Results of the online user survey*

Through a broad public relations campaign, 1.018 participants in the online survey were reached. Through targeted communication in the pilot cities, primarily the local population was recruited. As the survey focused on cycling infrastructure issues, people who are more likely to cycle participated, as was to be expected. Compared to the modal split in Germany, the cycling share is overrepresented. However, it can be assumed that the participants have a certain amount of experience with cycling infrastructure, which is necessary for an evaluation of the same.

The evaluation of the survey showed that 43% of the respondents would like to use the bicycle more often but are prevented from doing so for various reasons. Infrastructural problems such as quality (39,4 %), number of cycling facilities (24.1 %) or route connections (20,2 %) were mentioned most frequently. Among other things, this is reflected in the perception of safety. 81 % of respondents regularly experience danger from turning vehicles and 68 % from overtaking manoeuvres by other road users. In addition to infrastructural problems, the poor condition of the cycling infrastructure (26,4 %) is seen as an obstacle to cycling.

The responses from the different cities show mostly identical trends. There are major differences between the pilot municipalities in the area of transport modality, the average distance cycled and the assessment of the quality of the cycling infrastructure. For the city of Muenster, the problem with scrap bikes has also become more apparent, while in Munich the users are more dissatisfied with the winter service and communication than in the other

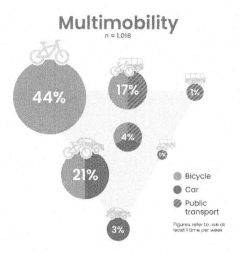

Figure 2.   Mobility of participants.

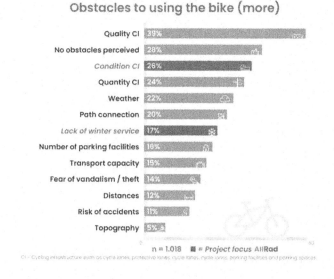

Figure 3.   Obstacles to using the bicycle (more).

cities. The latter also applies to the city of Mainz, which also received more negative ratings for the quality of its cycling infrastructure.

The survey was able to identify four essential fields of action, addressing of which will result in an improvement of the maintenance and thus an increase in the attractiveness of the cycling infrastructure.

The evaluation of several questions in the survey shows a significant need for action for the surface condition of the cycling infrastructure. According to this, 74.8 % rate the surface condition in parts of the cycling infrastructure as insufficient or in need of improvement. Uneven surfaces also represent a regular danger (40.3 %) or at least an inconvenience (49.3 %) for the majority. With regard to the maintenance of the cycling infrastructure, the removal of pavement damage (71.9 %) is also most frequently desired. This is followed by the removal of surface irregularities (65.0 %).

There is general satisfaction among cyclists about the cleaning of cycling infrastructure, but broken glass is often perceived as a danger (33.1 %) or an inconvenience (55.4 %) for users. About half of the respondents would like to see an improvement in the cleaning of broken glass (54.6 %).

A large proportion of participants are regularly confronted with obstacles such as bollards, road signs, parked objects (cars, bikes, e-scooters, bulky waste, bins) on the cycling infrastructure (91.2%), necessitating unpleasant evasive manoeuvres with potential consequential hazards.

Despite increased efforts to ensure a good winter service, many respondents rate the winter service as inadequate or at least in need of improvement (59.7 %). This value is significantly higher (74.4 %) in the pilot city of Munich, which usually experiences a more severe winter than the others as experienced in 2019 with 63 frost days and 11 ice days (City Munich 2020). The situation is similar with the cleaning of fallen leaves (48.2 %). Here, the result of the pilot city Muenster is higher (57.8 %).

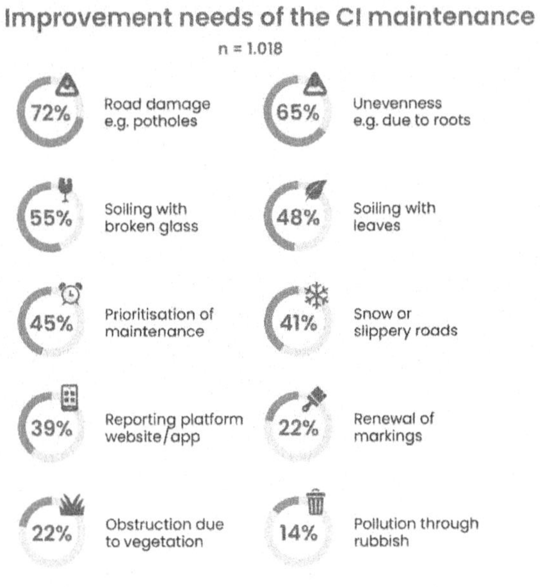

Improvement needs of the CI maintenance

n = 1.018

72%	Road damage e.g. potholes	65%	Unevenness e.g. due to roots
55%	Soiling with broken glass	48%	Soiling with leaves
45%	Prioritisation of maintenance	41%	Snow or slippery roads
39%	Reporting platform website/app	22%	Renewal of markings
22%	Obstruction due to vegetation	14%	Pollution through rubbish

Figure 4. Need for improvement of cycle infrastructure maintenance.

In the area of communication, the majority of respondents, 55.8 % in total, stated a need for more information on cycling infrastructure. While 68.6 % of them stated that the current communication offer of those responsible is at least expandable or even insufficient.

3.3 *Lessons learnt from both surveys*

The maintenance operators complain that the public expects too much of their services, while they claim to be well equipped technically and in terms of personnel for their tasks in order to fulfil the legal requirements for maintenance. The negative feedback from the population on maintenance suggests that the legal requirements for maintenance are clearly below the quality expectations of the population. This shows a discrepancy that can be closed by increasing the maintenance service or promoting the transparency of the tasks and performance of the cities. However, the current poor communication by the public authorities further increases this discrepancy. This ultimately affects the experienced quality of cycling and thus the modal split.

The long time required for comprehensive structural maintenance are reflected in the obstacles for the population to cycle (more). After all, road damage (e.g. potholes) and unevenness (e.g. roots) are demanded as the most urgent improvements. This indicates that the removal of such damages is not ensured quickly enough and thus has an impact on the quality of infrastructure and thus on the use of bicycles.

3.4 *Results of the measures applied until now*

In the ongoing project, findings for two applied measures could already be obtained before the second survey.

In order to address the lack of public communication, selected topics of maintenance were prepared for social media posts and the website. The objective was to provide a transparent view of the services provided by the public sector and thus shape expectations and understanding in favour of the cities. However, the speed of processing and decision-making required for the fast-moving communication channels on the internet could not be achieved due to the long-lasting processes in the administration. In addition, after the start of the work an internal decision was made against transparent communication, so that the measure was not implemented. It is to be expected that similar reasons lead to a lack of communication in the public sector in many places. However, this passive approach permanently worsens public opinion.

On the other hand, a high potential for improvement could be worked on with the measure around bicycle parking, which is to manage the preservation of parking capacities and the prevention of wild parking of bicycles with creative solutions.

It has become clear that the legal situation in Germany for bicycle parking is not yet adequately developed (Koerdt 2010). In the case of bicycle parking, the result is that a large number of bicycles left behind are not removed or are removed only after a long period of time and in many places the parking of bicycles is carried out without any means of control, that paths are narrowed by parked bicycles or that disorderly parking is carried out in green areas / open spaces (e.g. station forecourts), which leads to a reduction in the quality of all traffic, especially bicycle and pedestrian traffic.

4 DISCUSSION

The data shows that, in addition to the development of the cycling infrastructure, its condition has an essential influence on the people's choice of transport mode. Thus, cycle path maintenance, which is responsible for the condition, is a decisive factor in making cycling more attractive in means of transport and increasing its share in the modal split in the long term.

In concrete terms, the potential for improvement through structural repair work to deal with surface damage or unevenness has become clear. However, the study also showed that tackling such damage in Germany is a lengthy process and that municipalities have no quick-action method and usually too few resources to take action against it area-wide. If this potential is recognised at the upper federal and state level and funds are made available accordingly, municipalities could strengthen the cycling infrastructure and advance the mobility transition more quickly.

In addition to the technical findings, the need for better and more frequent communication on the part of those responsible for infrastructure and cycle path maintenance became clear. This can increase the population's knowledge about the processes of cycle path maintenance and about the infrastructure in general. The resulting understanding among the population has a positive influence on the subjective evaluation of the infrastructure, which can increase the acceptance of the means of transport and the use of bicycles.

However, the studies cannot show how an improvement in cycle path maintenance can be implemented in concrete terms or what effort is involved.

It is recommended to take a more detailed approach to the areas of action opened up by developing, testing and evaluating concrete improvement measures. For this purpose, it makes sense to differentiate between long-term measures, as is the case with structural maintenance measures, and short- and medium-term measures, which are more within the scope of action of the municipalities. The reason for this is that different approaches, which can be divided into strategic and tactical / operational, will most likely develop.

ACKNOWLEDGEMENTS

This study was supported by the Federal Ministry for Digital and Transport within the framework of the National Cycling Plan. We would like to thank our project coordinator Robert Eltner from the Federal Office for Goods Transport for the great support.

We also like to thank the staff of the pilot cities for their willingness to cooperate and their motivation for the implementation.

REFERENCES

Koerdt A., 2010. *German Bicycle Club (ADFC), Positionspapier - Fahrradparken im öffentlichen Raum*, 2010, https://www.adfc-nrw.de/uploads/media/adfc_position_fahrradparken_01.pdf

Nobis C., 2019. *Mobilität in Deutschland (MID), Analyse zum Radverkehr und Fußverkehr*, Figure 2, S21

Federal Ministry of Digital and Transport (BMDV), 2021. *Förderung und Finanzierung des Radverkehrs*, https://www.bmvi.de/SharedDocs/DE/Artikel/StV/Radverkehr/finanzielle-foerderung-des-radverkehrs. html last Access on 29.04.2022

Statistisches Amt der Landeshauptstadt München (City Munich), 2020, *1. Quartalsheft Jahrgang* 2020, S30

Roads and Airports Pavement Surface Characteristics – Crispino & Toraldo (Eds)
© 2023 The Author(s), ISBN 978-1-032-55149-4

Freeway interchanges maintenance operations. Preliminary safety analysis of connected automated vehicles impact on road traffic

Tullio Giuffrè*, Salvatore Curto & Andrea Petralia
Università di Enna Kore, Enna, Italy

ABSTRACT: Traffic safety benefits provided by Connected Automated Vehicles (CAVs) are related to their maneuvers precision; therefore such capability would lead to a significant reduction of traffic potential conflicts and congestion phenomena. However, some CAVs categories are not still allowed to operate due to legislative reasons, but mainly because there is a limited awareness on potential effects of their cooperation with conventional (non-automated ones) vehicles (CVs), and in specific roadway geometric layout configurations. The latter issues according to geometric layout parameters, traffic data and vehicle distribution can be analyzed through microsimulators providing a network performance estimation and defining safety level through surrogate safety assessment measures (SSAM). This methodology would guarantee a useful prevention analysis technique, especially with introduction of CAVs considering that several tests highlighted critical issues related to detection phases in freeway interchanges. In addition, it is necessary to observe how due to CAVs high lane-keeping system efficiency, road pavement zones in contact with tires would be strongly stressed. These critical topics would be increased in case of work-zones, due to deviations and lanes closure that would lead to a traffic flow increase in temporary single lanes. Considering that in Italy many freeways were realized in the XX century, it is interesting to evaluate CAVs impact on the former framework. So far, this paper will discuss: an analysis of traffic network performance of Italian interchange samples carried out through VISSIM microsimulation; in addition, with safety estimation through SSAM parameters based on traffic potential conflicts values and categories. The research results will allow to identify critical areas of a freeway interchanges with various geometric layouts subjected to maintenance interventions and related to CAVs impact on traffic circulation.

Keywords: Pavement Maintenance, Road Safety, Traffic Management

1 INTRODUCTION

In the last decades maintenance interventions on roadway surface pavements increased; it can be linked to the fact that traffic volumes were subjected to significant rise (Ragnoli et al. 2018) with consequent increase in induced stresses. The latter had a major impact on outdated roadway infrastructures that were designed for reduced traffic flows. Then required work sites implementation of the aforementioned interventions lead to discomforts such as interruptions and changes in the ordinary routes, affecting traffic safety level (Zou & Qu 2018). The latter is strongly related to traffic volumes, which affect interactions among vehicles and the average speed of their manoeuvres.

A typical case where stresses induced on pavement are amplified occurs when braking manoeuvres are performed because of high speeds, as for example in freeway junctions and

---

*Corresponding Author: tullio.giuffre@unikore.it

DOI: 10.1201/9781003429258-12

interchanges. Freeway junctions are characterized by significant traffic flow speed variations, and when they are subjected to maintenance interventions the above aspects play a key role. Therefore, traffic characteristics are constantly evolving, just consider the introduction of connected automated vehicles (CAVs) whose precision of manoeuvres would guarantee a high level of traffic safety (Boualam et al. 2022; Zhao & Sun 2013), but their efficiency would require high quality standards of roadways infrastructure and specific upgrading (Liu et al. 2019).

CAVs represent one of the most relevant innovations that affected transportation sector in last decades: circulation of the former would allow significant benefits in terms of traffic flow efficiency and safety (Guériau & Dusparic 2020; Zeidler et al. 2018). The latter would be achieved thanks to CAVs high efficiency detection sensors leading to driving mistakes reduction of human drivers and consequent collision risks decrease (Zhu et al. 2022).

By the way, considerations regard road surface and pavement performances would be affected by the CAVs lane high keeping position systems efficiency that would stress mainly zones in contact with wheels. Therefore, CAVs introduction in ordinary traffic will occur leading to a traffic context characterized by presence of conventional vehicles (CVs) leading to an interaction between these two vehicle types. The main proposal of this paper consists in providing an estimation of effects on traffic safety due to maintenance intervention in addition with CAVs presence. The former is expressed in terms of surrogate safety assessment measures (SSAM) while parameters related to traffic circulation were observed through microsimulation tools.

2 RELATED RESEARCH

This study was carried out both based on real traffic data provided by ANAS and with several sources related to traffic safety. The latter was analyzed according to sources concerning traffic microsimulation and SSAM, and selection of a specific Safety Performance Function (SPF) that better fit with the case study. Microsimulations and SSAM study phases occurred following criteria provided by (Giuffrè et al. 2017) where it needs to carry out several simulations with alternative design layouts that in our case correspond to various scenarios related to work zone interventions and characterized by implementation of traffic flows routes provided by origin-destination matrices. SSAM software was used to combine micro-simulation with automated conflict analysis and calculate surrogate measures of safety for the specific junction under examination. Analysis was firstly carried out without a fleet of CVs; subsequently traffic composition was integrated with a percentage of CAVs (Giuffrè et al. 2021; Williams E. 2020). Potential conflicts estimation characterized by CVs and CAVs cooperation occurred trough SSAM; where in each examined traffic composition, CVs included 5% of heavy vehicles and a filter was applied to TTC parameter to consider CAVs skills. TTC represents a time interval difference between the end of encroachment of a turning vehicle and the time when a through vehicle (with priority) arrives at the conflict point, but it must have traveled its trajectory with the speed at the time it started deceleration to avoid a collision.

Traffic safety analysis occurring in this study belongs to a techniques field that lead to a quantitative safety estimation, identified as Accident Prediction Models (APMs). The latter allows to provide the connection between expected collisions and specific roadway segment characteristics (Yannis et al. 2016a,b); in addition APM technique results useful as a forecasting tool during decision making phase for new design solutions (Srinivasan & Bauer 2013).

The most used methods applied in road safety assessment is based on Highway Safety Manual (HSM 2010), which indicates that a specific roadway site crash frequency can be obtained through a regression model developed from past crash data just if they are collected on a wide sample. Such models are called Safety Performance Functions (SPFs) and are related to specific geometric layouts and traffic parameters; then many SPFs were developed but it is possible to identify base functions which depend just on segment length and annual average daily traffic (AADT) (AASHTO 2014). Therefore, it has to be considered that HSM models were developed for American roadway sites, and then it has to be highlighted how the choice of SPFs application for an Italian interchange was reliable (La Torre et al. 2014). The use of SPF resulted to be necessary to provide an additional estimation related to traffic safety considering that the former leads to obtain crashes quantity related to roadway contexts considered (Srinivasan R. & Carter D. 2011).

3 DATA COLLECTION AND METHODOLOGY

3.1 *Vissim analysis*

Based on what discussed in the introduction section, it was selected as a case study: the Enna (Italy) freeway interchange along Palermo-Catania route. The latter was chosen for two main reasons: Enna freeway interchange geometric layout belongs to Italian geometric layout standards which make this study functional for other existing interchanges with similar characteristics as geometrical or operational ones; secondary, in late 2021 maintenance interventions began leading to the possibility of analyzing real time data in combination with structural sensor and traffic count data recording.

First steps of this study were characterized by implementation of Enna freeway interchange model in VISSIM, where vehicles routes were set using the "dynamic assignment", this option was preferred due to the possibility to obtain dynamic itineraries between junction nodes instead of static ones (PTV GROUP 2018). For the estimation of vehicles performance such as traffic flows speed variations data collection measurements and queue counters were placed in all critical areas of the road interchange such as ramps entrance and exits, freeway run lanes and overtaking lanes as shown in Figure 1. Considering that maintenance interventions will occur in several interchange zones, five scenarios were considered based on work intervention plan design phases provided by ANAS. The latter provided traffic volumes data necessary to define the interchange traffic demand; for the research goals three main O/D matrices were set referring to three selected traffic peak hours: 8 A.M., 2 P.M. and 7 P.M.

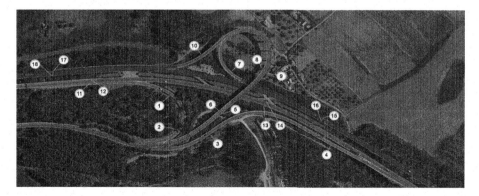

Figure 1.   Data collection positions at the interchange.

Each O/D matrix changed according to a specific scenario, which was characterized by variation of traffic flows viability due to alternating ramps closure phases. Figure 2 shows a

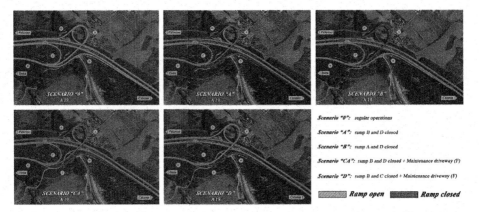

Figure 2. Simulated scenario configurations.

scheme of simulated scenarios, while in Figure 3 O/D referring to 8 A.M are reported as an example, at the same way of findings showed in section 4. In conclusion seven simulations run were carried out for each scenario considering that a higher quantity of runs did not lead to relevant changes in traffic network performance.

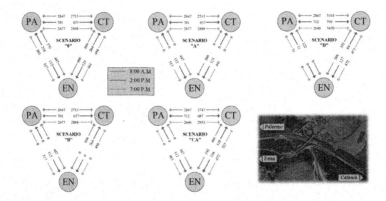

Figure 3. O/D matrices schemes.

3.2 Connected Automated Vehicles (CAVs) introduction

CAVs implementation within the road interchanges model has been modelled in VISSIM and it was carried out setting speed distribution functions and driving behavior Wiedmann 99 parameters based on (Sukennik 2020). In Table 1 below driving behavior parameters of CAVs and CVs that were applied are reported respectively, according to main technical literature and author's evaluation.

In order to relate CAVs and CVs operation with surface pavement characteristics it was assumed that in case of compromised pavement roughness, CAVs would perceive roughness variation just after significant vibrations induced in vehicles not varying their speed, while CVs would slow down a priori as a result of the driver's visual skills (Xinyi et al. 2021).

Ultimately, unsatisfied roughness was considered in VISSIM model as a reduced speed areas activated just for CVs. Speed rates were set considering values provided by Xinyi et al. 2021 considering that a higher percentage of CAVs would speed up road pavement IRI

Table 1. Driving behavior parameters.

W99	CAVs	CVs
CC0	0.75	1.5
CC1	0.45	0.9
CC2	0	4
CC3	-8	-8
CC4	-0.1	-0.35
CC5	0.1	0.35
CC6	0	11.44
CC7	0.1	0.25
CC8	3.5	3.5
CC9	1.5	1.5

increase due to reasons mentioned in section 1. Speed values were associated to international roughness index (IRI) on the basis of CAVs percentage as follow:

- 0% CAVs; IRI = 90/142;   50 mph
- 10% CAVs; IRI = 178/224; 40 mph
- 20% CAVs; IRI = 225/318; 40 mph
- 25% CAVs; IRI = 364;   35 mph

Based on previous observations following assumptions were considered:

- Compromised pavement roughness zones were modeled as reduced speed areas positioned in potential more stressed pavement stretch such as ramps characterized by significant slopes and potential braking maneuvers areas.
- Additional reduced speed areas were activated just for CVs
- In the same way of CAVs percentage increase, reduced speed areas were expanded, considering higher pavement stress due to the precise lane-keeping position capabilities.

Finally, traffic microsimulations with CAVs presence were carried out based on the scheme mentioned in section 3.1, where in each traffic volume scenario it was considered a raising percentage of: 0%, 10%, 20% and 25% of CAVs.

3.3 *Traffic conflict analysis*

Vissim simulation process provides a trajectory file (*trj* file) as output; the latter can be uploaded in SSAM application provided by Federal Highway Administration (FHWA). SSAM application allows to analyze *trj files* leading to an identification of the entire potential conflicts of the analyzed network (Li et al. 2016). For the case study examined, each trj file was analyzed through SSAM for each scenario respectively, leading to the preliminary identification of critical areas of the junction where potential conflict could be identified as three main categories: rear-end, lane change and crossings. The last are related to the angle extent of vehicles trajectories as described below:

- Rear-end conflict if trajectory angle < 30°
- Crossing conflict if trajectory angle > 80°
- Lane-change conflict if 30° < trajectory angle < 80°

Traffic conflicts data provided by SSAM were several, then in order to select just high severity traffic potential conflict it was necessary to set two main parameters thresholds as described below:

- $0.5 < TTC < 1.5$
- $0.5 < PET < 2.5$

These parameters are the time to collision (TTC) and the post encroachment time (PET) respectively, or rather surrogate safety assessment measures (Pu et al. 2008). The latter, although they characterize a dangerous interaction between vehicles of a probabilistic type, represent a functional parameter for detecting areas with a high probability of severe crashes.

The potential traffic conflict analysis outlines have been used for road safety evaluation. In other words, the potential traffic conflict was the basis of SPFs application and modeling. For the specific case study, it has been assumed the following SPF provided by Sacchi and Sayed, 2016:

$$E(Y) = c_0 \cdot AHC^{C_1} \tag{1}$$

The reported above equation was chosen due to its relationship with conflicts quantity, where:

- $E(Y)$ = expected collision frequency
- AHC = average hourly conflicts
- $ln\ c_0 = -0,024$
- $c_1 = 1,144$

4 FINDINGS AND DISCUSSION

Analysis of traffic simulations and the modeling of junction configuration scenarios with presence of CVs and CAVs can be used to correlate kinematic parameters of traffic flows, pavement state of use and road safety level. It must be premised that results evaluation related to geometric configuration of the chosen junction must be related to functioning of the freeway carriageways (PA-CT; CT-PA) based on ramps operational characteristics.

Figure 4 shows a comparison of longitudinal acceleration (left vertical axis) and speeds (right vertical axis) trend, the latter in the speed range 40-100 km / h for each of the five simulated scenarios and each main element of the interchange respectively (freeway lanes, ramp A, ramp B, ramp C, ramp D, and the Maintenance Driveway Lane). Findings obtained for the two carriageways, as well as confirming simulations realism of roadway sections geometric characteristics considered, highlight model sensitivity to CAVs incremental progression the traffic flow. As for example, outputs determined by the increase from 10% to 25% of the CAVs in terms of longitudinal acceleration for the CA and D scenarios are affected in a gradual but always consistent way with respect to the geometry of the A19 PA-CT road section.

Similarly, conditions due to increase from 10% to 25% of the CAVs show a symmetrical and consistent trend for scenario 0 in terms of travel speed, while an asymmetrical trend is determined although it is also consistent for geometric layout in scenarios A and CA.

Dynamic conditions imposed by junction ramps geometry show a significant and differentiated correspondence in relation to CAVs percentage. In ramp C (scenario B and CA) and B (scenario B) speed trends result to be significant in terms of consistency with road geometry and traffic specificities with presence of CAVs (considering reactions to performance conditions of road surfaces). Evaluation of acceleration trend is actually more problematic, both for Wiedemann 99 model calibration, and for strong correlation between vehicle position and ramps geometric configuration, with the same O / D matrix. However, such simulation behavior can be considered as a quality measure of developed model even though the case of Maintenance Driveway Lane, present only in CA and D scenarios and characterized by a modest performance of horizontal/vertical geometry as well as by a poor quality of pavement surface, determines evident fluctuations of longitudinal acceleration parameter but gradual with CAVs percentage variation.

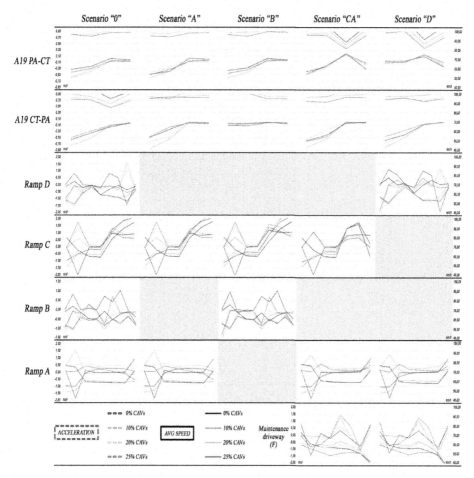

Figure 4   Acceleration and Average speed plots.

In Figure 5 simulation results in terms of potential traffic conflicts can be observed, where for Rear-end and Lateral change conflict types, just the ones that simultaneously show a value of TTC < 1.5 and PET < 2.5 are reported. Which means that are reported just conflicts probabilistically more likely to generate a high severity event between two vehicles. In addition to identify most critical zones of considered geometry, SSAM outputs showed a trend linked to CAVs percentage increase. It is possible to observe specifically how in traffic scenarios related to realization of interchange maintenance intervention, scenarios B and CA are the most problematic and most "promising" in terms of generating high probability traffic conflicts.

The ultimate aspect needs to be further investigated in future research works, in order to study extraordinary maintenance techniques of road pavements capable of guaranteeing traffic safety standards even when CAVs vehicles will be present in traffic flow.

Findings showed with the Table 2 are provided in terms of conflicts quantity obtained through thresholds mentioned in section 3.3 and expected collisions where E(Y) is referred to a single scenario while E'(y) is referred to a year.

121

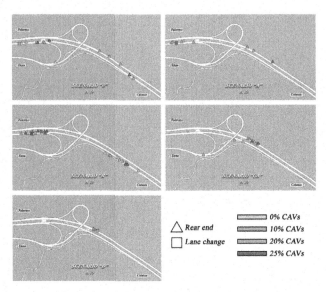

Figure 5.   Potential traffic conflicts positions and categories.

Table 2.   Driving behavior parameters.

0% CAVS – IRI = 90/142				10% CAVS – IRI = 178/224			
SCENARIO	AHC	E(Y)	E'(Y)	SCENARIO	AHC	E(Y)	E'(Y)
0	0	0	5	0	3	3	20
A	1	1		A	3	3	
B	2	2		B	3	3	
CA	1	1		CA	4	5	
D	0	0		D	1	1	
20% CAVS – IRI = 225/318				25% CAVS – IRI = 364			
SCENARIO	AHC	E(Y)	E'(Y)	SCENARIO	AHC	E(Y)	E'(Y)
0	4	5	20	0	3	3	27
A	4	5		A	2	2	
B	4	5		B	11	15	
CA	2	2		CA	2	2	
D	0	0		D	0	0	

5   CONCLUSIONS

Methodology applied in this paper would guarantee a valid prevention analysis technique, especially with introduction of CAVs considering that several tests highlighted critical issues related to detection phases in freeway interchanges. Moreover it was discussed pavement maintenance strategy related to CAVs high lane-keeping system efficiency, because road pavement zones in contact with tires would be strongly stressed. These critical topics would be increased in case of work-zones, due to deviations and lanes closure that would lead to a traffic flow increase in temporary single lanes. Results were allowed to identify most critical zones of considered geometric layout; then effects linked to CAVs percentage increase were reported. In particular, it is possible to observe how traffic scenarios related to realization of interchange maintenance intervention (scenarios B and CA) result to be most problematic and "promising" in terms of generating high probability potential traffic conflicts.

ACKNOWLEDGMENTS

This research was funded by the Italian Ministry of University and Research, grant PRIN (Research Projects of National Relevance) number PRIN 2017 USR342. Authors thank the local office of ANAS spa and all technicians that collaborated with the research team.

REFERENCES

AASHTO, 2014. *Highway Safety Manual, First Edition*. 2014 Supplement American Association of State and Highway Transportation Officials.

Boualam O., Borsos A., Koren C. and Nagy V., 2022. *Impact of Autonomous Vehicles on Roundabout Capacity*. MDPI Sustainability 2022, 14, 2203.

Eenink R., Reurings M., Elvik R., Cardoso J., Wichert S. and Christian S., 2005. *Accident Prediction Models and Road Safety Impact Assessment: Recommendations for Using These Tools*. Ripcord 506184 (January), 1–20.

Giuffrè T., Trubia S., Canale A. and Persaud B., 2017. Using Microsimulation to Evaluate Safety and Operational Implications of Newer Roundabout Layouts for European Road Networks. *MDPI Sustainability* 2017, 9(11), 2084.

Guériau M. and Dusparic I., 2020. *Quantifying the Impact of Connected and Autonomous Vehicles on Traffic Efficiency and Safety in Mixed Traffic*. IEEE 23rd International Conference on Intelligent Transportation Systems (ITSC), DOI: 10.1109/ITSC45102.2020.9294174.

Guerrero-Ibáñez J., Zeadally S. and Contreras-Castillo J., 2018. Sensor Technologies for Intelligent Transportation Systems. *MDPI Sensors* 2018, 18(4), 1212.

La Torre F., Domenichini L., Corsi F. and Fanfani F., 2014. Transferability of the Highway Safety Manual Freeway Model to the Italian Roadway Network. *Transp. Res. Rec.* (2435), 61–243571. Transportation Research Board of National Academies, Washington DC, 2014.

Li S., Xiang Q., Ma Y., Gu X. and Li H., 2016. Crash Risk Prediction Modeling Based on the Traffic Conflict Technique and a Microscopic Simulation for Freeway Interchange Merging Areas. *Int. J. Environ. Res. Public Health* 2016, 13, 1157.

Liu Y., Tight M., Sun Q. and Kang R., 2019. A Systematic Review: Road Infrastructure Requirement for Connected and Autonomous Vehicles (CAVs). *Journal of Physics: Conference Series*, Volume 1187, Issue 4.

PTV GROUP, 2018. *PTV VISSIM 10 User manual*.

Pu L., Joshi R. and Energy S., 2008. *Surrogate Safety Assessment Model (SSAM)—SOFTWARE USER MANUAL*. U.S. Department of Transportation Federal Highway Administration, Publication No. FHWA-HRT-08-050.

Ragnoli A., De Blasiis M.R. and Di Benedetto A., 2018. Pavement Distress Detection Methods: A Review. *MDPI Infrastructures* 2018, 3(4), 58.

Sacchi E. and Sayed T., 2016. Conflict-Based Safety Performance Functions for Predicting Traffic Collisions by Type. *Transportation Research Record: Journal of the Transportation Research Board*, No. 2583, Transportation Research Board, Washington, D.C., 2016, pp. 50–55. DOI: 10.3141/2583-07.

Srinivasan R. and Carter D., 2011. *Development of Safety Performance Functions for North Carolina*. FHWA/NC/2010-09.

Srinivasan R., Carter D. and Bauer K., 2013. *Safety Performance Function Decision Guide: SPF Calibration Vs SPF Development*. U.S. Department of Transportation. FHWA-SA-14-004. Federal Highway Administration.

Sukennik P., 2020. *Coexist Automation-Ready Modelling with PTV VISSIM*. PTV GROUP.

Xinyi Y., Yihao R., Liuqing H., Ying H. and Pan L., 2021. *Evaluating the Impact of Road Quality in Driving Behavior of Autonomous Vehicles*. Proc. SPIE 11591, Sensors and Smart Structures Technologies for Civil, Mechanical, and Aerospace Systems 2021, 1159106 (22 March 2021); doi: 10.1117/12.2583641.

Yannis G., Dragomanovits A., Laiou A., La Torre F., Domenichini L., Richter T., Ruhl S., Graham D. and Karathodorou N., 2016b. *Development of an Online Repository of Accident Prediction Models and Crash*

Modification Factors. 1st European Road Infrastructure Congress, 18-20 October 2016 Leeds, United Kingdom.

Yannis G., Dragomanovits A., Laiou A., Richter T., Ruhl S., La Torre F., Domenichini L., Graham D., Karathodorou N. and Li H., 2016a. Use of Accident Prediction Model in Road Safety Management–an International Inquiry. Transportation Research Procbahahaedia 14, 4257– 266.

Zeidler V., Buck S., Kautzsch L., Vortisch P. and Weyland C., 2018. *Proc. 98th Annu. Transp. Res.* Board Meeting. 2019.

Zhao L. and Sun J., 2013. Simulation Framework for Vehicle Platooning and Car-following Behaviors Under Connected-vehicle Environment. *Procedia-Social and Behavioral Sciences*, 96, 914–924.

Zhu J., Ma Y. and Lou Y., 2022. Multi-vehicle Interaction Safety of Connected Automated Vehicles in Merging Area: A Real-time Risk Assessment Approach. *Accident Analysis & Prevention*, 166, 106546.

Zou Y. and Qu X., 2018. On the Impact of Connected Automated Vehicles in Freeway Work Zones: A Cooperative Cellular Automata Model Based Approach. *Journal of Intelligent and Connected Vehicles* ISSN: 2399-9802.

Safety aspects of e-scooters in urban areas: Preliminary results on citizens' perception, users' behavior and role of pavement

Arianna Antoniazzi*, Elena Davoli, Claudia Nodari & Maurizio Crispino
Department of Civil and Environmental Engineering, Politecnico di Milano, Milan, Italy

ABSTRACT: This study is focused on safety aspects related to the use of e-scooters in urban areas. Since e-scooters recently appeared in urban areas, rapidly spreading, and increasing in popularity, several Countries and Municipalities have been not able to properly update both legislation and infrastructure. In fact, many of the enacted rules have been temporary and repeatedly modified, causing confusion among users. Moreover, road pavements are mostly inappropriate to e-scooters, equipped with small wheels and with serious stability issues, especially during braking. As a result, recently, a significant number of e-scooter-related accidents occurred, raising attention to the road-safety topic. These are the reasons why the preliminary investigation reported in this study is devoted to understanding typical user behaviors and the overall safety perception connected to these vehicles. In detail, the study is focused on three main points: (i) the Italian citizens' opinion on e-scooters' effects on urban mobility safety, by submitting an on-line questionnaire to collect opinions from both users and non-users of e-scooters; (ii) the e-scooter users behavior, by performing on-site surveys aimed at achieving a direct observation of the phenomenon in Milan (Italy); (iii) the e-scooters-pavement interaction by means field tests carried out installing an accelerometer on an e-scooter and measuring the vibration effects deriving from different pavement types and distresses. The preliminary results reported in the paper demonstrate the different safety perception of e-scooters users and non-users, some critical aspects related to e-scooters drivers' behavior, non-negligible safety effects arising from pavement-e-scooter interaction as consequence of pavement type and condition and e-scooter's speed.

Keywords: E-scooters, Safety, Survey, User Behavior, Pavement Quality, Accelerations

1   INTRODUCTION

Recently, e-scooters experienced a rapid and wide diffusion in many countries, mainly thanks to their versatility and capillarity (in many cases because of sharing services offered in several cities), which seemed to offer a low-cost solution to the so-called "last-mile problem". The USA witnessed this phenomenon since September 2017, with the introduction in Santa Monica (California) of the first dockless electric kick scooters by Bird and Lime (Verge. 2018). Some months later, Europe followed the trend: the first e-scooters appeared in Paris (France) during the summer of 2018 (Reuters 2018), while the first 20 trial e-scooters were introduced in Milan (Italy) during the Autumn of the same year (DMove 2018). Regarding the rapid increase of e-scooters in urban areas, the Italian National Observatory of Sharing Mobility (*Osservatorio Nazionale Sharing Mobility* 2020) reported that the number of e-scooters circulating within the Italian territory grew up from 4'650 units in December 2019 to 27'850 in September 2020, with a significant increase in the cities of Rome and Milan.

---

*Corresponding Author: arianna.antoniazzi@polimi.it

DOI: 10.1201/9781003429258-13

Because of the rapid introduction of e-scooters in urban mobility, a great number of e-scooter related accidents can be obviously expected. On this matter, the Italian National Institute of Statistics (Istat 2019), recorded a total number of 564 accidents in the period May 2020-July 2021. Continuing to look at the Italian situation, ASAPS (Associazione Sostenitori ed Amici della Polizia Stradale, 2021) reports that 12 fatal accidents occurred in 2021, 2 of which caused the death of a pedestrian and one of an underage driver. As far as 2022 is concerned, there have already been 4 victims of fatal accidents, 2 of which are related to the same crash involving a motorcyclist and an e-scooter. Looking at the accident consequence of e-scooter accidents, the most frequent mechanism of injury is fall from the vehicle (80.2%) followed by collision with objects (11%) and being hit by moving vehicles (8.8%) (Trivedi et al. 2019). Moreover, despite the great number of accidents implying head injury (40%), only 4.4% of riders wore the helmet, demonstrating it is a crucial safety issue. In addition, Trivedi et al. stated that a common injury pattern is fracture to body extremities (distal upper extremity - 12.5%; proximal upper extremity - 6.8%; distal lower extremity - 4.4%) and to the face (5.6%), stressing the importance of protection equipment, especially for the head (Trivedi et al. 2019). Additionally, a specific study was conducted with the aim of simulating different typologies of e-scooter-vehicle collision (frontal impact, side bonnet impact, and impact against B-pillar), in order to define the severity of the impact in terms of accelerations measured on a dummy (Fernandes et al. 2022). Results showed that an extremely important aspect for safety is the use of involuntary reflexes to protect the head with the arms, but since time is not always sufficient to develop these reflexes, other measures for head protection are needed.

Another crucial point is related to pavement-e-scooter interaction, since these vehicles are characterized by small wheels and high center of mass, making them dangerous, especially when moving on uneven pavement surfaces. Going through to the scarce scientific literature on this point, Posirisuk et al. addressed their efforts in evaluating potholes' impact on e-scooter safety, through the simulation of fall from the vehicle, demonstrating that the head-ground impact force/speed change as a consequence to a variation in pothole's depth/width, travel speed or typology of rider, in terms of weight and height. Since the vehicle's speed is proven to be a key factor (a speed increase from 10 to 30 km/h corresponds to a 67% increase in the head-ground impact speed), the importance of wearing the helmet to prevent serious head injuries has been confirmed (Posirisuk et al. 2022).

Given the aforementioned background deriving from the available scientific literature, the preliminary investigation described in this paper is devoted to highlight specific safety aspects related to the introduction of e-scooters in urban areas, focusing the attention on: (i) the Italian citizen opinion on e-scooters' effects on urban mobility safety; (ii) on-site observation of e-scooter users behavior in Milan (Italy); (iii) the e-scooters-pavement interaction by field tests, measuring the vibration effects deriving from different pavement types and distresses.

2 GOALS AND LITERATURE REVIEW

According to the main findings of the available scientific literature discussed in the previous section, the three main goals of the study reported in this paper are detailed as follows. Moreover, the scientific literature related to each specific point of the research is furtherly discussed.

1. Italian citizen opinion on the effects of e-scooters on urban mobility safety. This point is developed by submitting a questionnaire to a sample of Italian respondents, to collect opinions both from e-scooters' users and non-users, in terms of safety perception or propensity to choose these vehicles rather than other modal alternatives. On this topic, many studies have already been developed, for instance, Almannaa et al. declare that in

Saudi Arabia the major obstacles to e-scooters are the lack of adequate infrastructure (70%), followed by weather conditions (63%) and safety (49%). Because of the lack of proper infrastructure, drivers of e-scooters usually tend to behave in a hybrid manner, that combines typical pedestrians' and mopeds' habits (Almannaa *et al.* 2021). Moreover, Tuncer *et al.* investigated the tendency of e-scooters to shorten the waiting time at signalized intersection, meaning that when the traffic light is green for pedestrians, the e-scooter is more likely to cross with them, while if the signal allows the flow of vehicles then the e-scooter will probably cross as a moped (Tuncer *et al.* 2020). This extremely dangerous attitude should be definitely discouraged through the imposition of clear rules on how to behave at intersections and the identification of dedicated spaces for these vehicles, especially on critical paths.

2. On site observation of e-scooter users' behavior in Milan. 15 sites were chosen in the city of Milan in order to record, collect and analyze the most diffused behaviors and circulation modes (e.g.: number of users travelling on sidewalks, percentage of users wearing a helmet or carrying a passenger). According to literature review, most of the studies aimed at conducting behavioral or road safety analysis are based both on surveys and direct observation of the phenomenon (Sucha 2014). Moreover, there are two possible alternatives (typologies of observer) to collect data about users' typical attitudes, camera or human observer (van Haperen *et al.* 2019). The present study follows the option of the human observer, by selecting similar sites in different positions of the city, as discussed in the following Sections.

3. E-scooters pavement interaction by field tests. 3 pavement types (asphalt, cobblestones, slabs of stones) with different evenness were selected with the aim of exploring the e-scooter-pavement interaction in terms of safety and driving stability. From an operational point of view, an accelerometer is installed on the e-scooter with the goal of quantifying the impact of pavement surface characteristics on the vehicle and, consequently, on the driver. The available scientific literature on this topic highlights both the difference of stress induced on the vehicle while traveling on the sidewalk or on the vehicle lane (Ma *et al.* 2021) and the enormous difference in terms of driving dynamic comparing bicycles and e-scooters; mainly due to the fact that e-scooters require a bigger steering torque moment at handlebar and show a higher weight shift during braking phase (Paudel & Fah Yap 2021).

3 INVESTIGATION METHODS

3.1 *Italian citizens' opinion on e-scooters' effects on urban mobility safety*

The present part of the research is based on an on-line questionnaire administered to Italian citizen in the period June-July 2021. It is composed of a total of 30 questions and it is structured into two main sections. The first section is the same for all respondents and consists of general questions aimed at defining the main socio-demographic characteristics of respondents (e.g., age, gender, city). In fact, since the questionnaire is structured for being delivered to people belonging to different age groups and living in different contexts, the collected answers are essential to check if the sample could be judged sufficiently representative. Then, answers given to the key question (Q1) "Have you ever driven an e-scooter?" allow to define 3 respondents categories: (i) e-scooters frequent users, (ii) e-scooters non-frequent users and (iii) non-users. Consequently, the second section of the questionnaire consists of specific questions developed for each class of respondents, with the aim of analyzing safety perception (both for users and non-users) and the most common habits and practices diffused among users. Regarding e-scooter's users the second section includes a set of 15 questions (Q3 – Questions for users), among which the most relevant are:

- "Do you consider e-scooters dangerous for other road users?";
- "Associate a safety perception level (low, medium, high) to each circulation zone";
- "Have you ever risked falling off the e-scooter due to pavement distresses?".

Alternatively, 11 questions are submitted to non-users in section 2 (Q2 – Questions for non-users) and, within this study the most representative question is "do you consider e-scooters dangerous for other road users?" since it allows to compare safety perception of users and non-users.

The scheme depicted below in Figure 1 synthetizes the previously discussed questionnaire's structure.

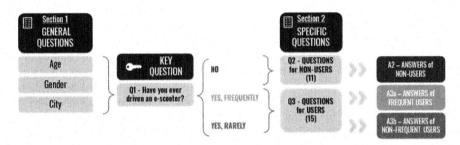

Figure 1.   Questionnaire general outline.

3.2 *On-site observation of e-scooter users' behavior in Milan*

The aim of the on-site survey is to observe and, thus, collect data on the real behavior of e-scooters' users and to analyze the most diffused practices.

Field surveys have been carried out in Milan in July 2021, for time spans of 30 or 60 minutes, mainly during periods of high traffic congestion, meaning at morning and evening rush hours. The main scope of the surveys is to gather information about: (i) number and typology of users (e.g., gender, age, sharing/private vehicle), (ii) e-scooter driving modes, (iii) use of helmet, (iv) carriage of passenger/objects and (v) unusual or improper behaviors. Observation places (15 sites located in the city of Milan, as depicted in Figure 2) are chosen

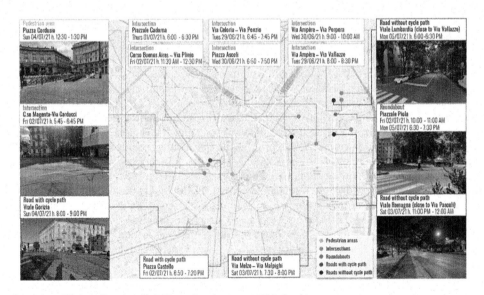

Figure 2.   Observation sites in Milan of e-scooter users' behavior.

in order to represent many different infrastructure locations and characteristics: pedestrian areas, intersections, roundabouts, roads with cycle path and roads without cycle path.

3.3 *E-scooters-pavement interaction by field tests.*

The field investigation on e-scooter-pavement interaction is carried out by equipping with an accelerometer an e-scooter, characterized by a mass of approximately 20 kg and a maximum speed of 25 km/h. The main scope consists of collecting data related to characteristics of vibrations transmitted to the driver passing on different road pavement types and distresses at various speed regimes. The instrument is installed on the e-scooter's platform, as showed in Figure 3, where the chosen coordinate system is also reported.

Figure 3.   Coordinate system for the configuration with the accelerometer installed on e-scooter's platform.

Different pavement types (asphalt pavements, cobblestones and slabs of stone) and level of distresses (new and distressed asphalt pavement) with e-scooter's speeds (ranging between 6 km/h and 25 km/h) considered in the field tests are summarized in Table 1. In particular, the e-scooter's speed is changed to compute the potential effect of a speed reduction on both safety and stability of the two-wheeled vehicle. With respect to cobblestones and stone pavements, the maximum operating speed reached during trials is 15 km/h, due to the physical impossibility to drive safely at higher speeds on extremely uneven pavements.

Table 1.   Field test configurations in terms of pavement typology and speed.

		Speed [km/h]		
		6	15	25
Pavement typology	New asphalt pavement			x
	Distressed asphalt pavement			x
	Cobblestones		x	
	Slabs of stone	x	x	

Field tests, performed by a driver of an average mass of 60 kg, allow to record accelerations with respect to different road pavements and speeds along reference axes (x, y, z). Raw data is processed according to the UNI ISO 2631 (Mechanical Vibration and Shock - Evaluation of Human Exposure to Whole-Body Vibration 1997), in order to evaluate the vibration transmitted to the whole human body. Thus, as a first step, the value of the weighted Root-Mean-Square (RMS) acceleration is computed, as defined below, where T_i is the time measurement [s] fixed at 1 second in the study (not overlapped time window).

$$a_{RMSi} = \sqrt{\frac{1}{T_i} \int_0^{T_i} a^2(t)dt} \tag{1}$$

The applicability of the basic evaluation method is verified by checking the crest factor being below or equal to 9. This parameter represents the ratio between the maximum value of acceleration registered in the time interval T_i and the weighted RMS acceleration, as reported in the following equation.

$$CF = \frac{a_{\max}(t)}{a_{RMSi}} \tag{2}$$

Once the validation is passed, the acceleration values need to be weighted according to the frequency of the signal in the time span. Therefore, the nominal frequency for each interval T_i is computed based on the number of registered peaks, as presented below.

$$f = \frac{N_{peaks}(T_i)}{T_i} \tag{3}$$

The frequency value is essential to compute the weights w_i (w_k, w_d), thanks to a specific table provided by UNI ISO 2631, to be applied to RMS acceleration values, as reported below.

$$a_w = \left[\sum_i (w_i a_{RMSi})^2 \right]^{\frac{1}{2}} \tag{4}$$

With regards to comfort evaluation, the representative parameter is the total acceleration value along 3 axes, which can be calculated as defined in the following equation, where the values k_i is defined according to the position of the driver (upright or seated posture).

$$a_v = \left(k_x^2 a_{wx}^2 + k_y^2 a_{wy}^2 + k_z^2 a_{wz}^2 \right)^{\frac{1}{2}} \tag{5}$$

The obtained value of a_v must be compared with comfort thresholds imposed by the regulation, which are defined as follows in Table 2.

Table 2. Comfort thresholds defined by ISO 2631.

Acceleration range	Level of comfort
Less than 0,315 m/s^2	Not uncomfortable
0,315 m/s^2 to 0,63 m/s^2	A little uncomfortable
0,5 m/s^2 to 1 m/s^2	Fairly uncomfortable
0,8 m/s^2 to 1,6 m/s^2	Uncomfortable
1,25 m/s^2 to 2,5 m/s^2	Very uncomfortable
Greater than 2 m/s^2	Extremely uncomfortable

4   RESULTS AND DISCUSSION

4.1   *Italian citizens' opinion on e-scooters' effects on urban mobility safety*

Answers gathered in the questionnaire belong to 273 people of different age groups and living in various Italian contexts. As a first step of the questionnaire analysis, the key question (Q1) "have you ever driven an e-scooter?", revealed that 33% of respondents have driven at least once an e-scooter, and among them 11% are frequent users, while the remaining 67% have never driven an e-scooter Figure 4a. With reference to the second section, results show an overall consciousness of the danger related to e-scooters. In fact, the majority of non-users Figure 4b and a consistent number of e-scooter's users Figure 4c is convinced these vehicles are dangerous. The risk awareness is an essential aspect for evaluating a possible reaction to the imposition of safer behaviors, also allowing the Countries' lawmakers to address new stringent rules acceptable to e-scooter users. Moreover, interesting results are obtained regarding the safety perception of e-scooters users on various urban areas. As illustrated in Figure 4d, collected answers show a clear preference for cycle paths, followed by zones 30, sidewalks, and finally roads with speed limit higher than 30 km/h. This outcome underlines the importance of promoting the diffusion of dedicated and protected paths for light mobility in the urban environment. Finally, regarding the risk of falling from the e-scooter due to distressed pavements, the collected answers from e-scooter users highlighted that both frequent (63%) and non-frequent users (57%) risked at least once to lose stability when passing over an uneven road pavement Figure 4e.

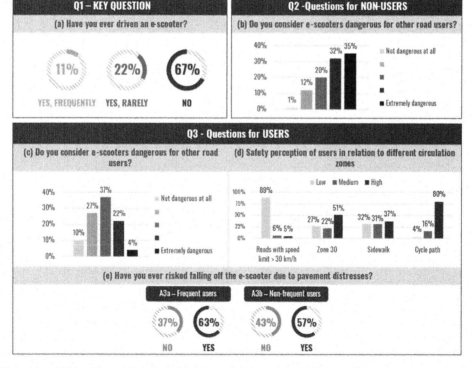

Figure 4.   Italian citizen opinion on the effects of e-scooters on urban mobility safety (Q&A).

4.2   On-site observation of e-scooter users' behavior in Milan

The most relevant results on e-scooters' behavior obtained through the on-site surveys can be summarized as follows.

With regards to the choice of urban areas Figure 5a, the on-site observation confirms the clear preference (79%) for cycle paths, where available. On the contrary, in absence of cycle paths, most e-scooters are travelling on vehicle lanes (83%). Moreover, despite being forbidden, a fair number of e-scooters drives on sidewalks, especially in absence of cycle path (17%).

Though not mandatory, 17% of private vehicle users wear a helmet, compared to only 3% of shared services users, as shown in Figure 5b. Thus, it seems that one of the greatest obstacles to the enforcement of mandatory helmet is associated to shared e-scooters, whose riders are usually incapable of providing independently the helmet. However, a possible solution could be that of integrating these devices on-board, as already done by some sharing companies.

Most private users (63%) and about a third of sharing users (36%) carry at least an object or a passenger, careless of the prohibition, as shown in Figure 5c. While the transport of light objects, such as small backpacks, can be considered non-dangerous for the vehicle's stability, the carriage of heavy or oversized objects, such as luggage placed on the platform, is definitely to be excluded.

Crossing modalities at intersections are significantly affected by user's safety perception and driving ability. However, a main distinction can be done according to the surrounding context, between high-traffic intersections and low-traffic intersections. The on-site observation allows to affirm that, a discrete number of drivers prefers to cross with pedestrians on the crosswalk (29%) in high traffic conditions, while, on the contrary, low traffic volumes induce e-scooter's drivers to cross with vehicles, as confirmed in Figure 5d.

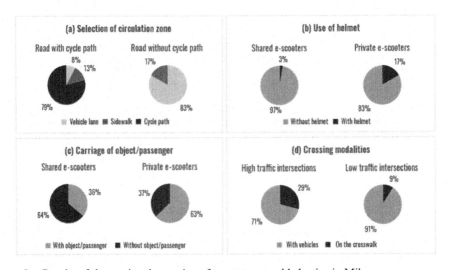

Figure 5.   Results of the on-site observation of e-scooter users' behavior in Milan.

4.3   E-scooters-pavement interaction by field tests

Field tests allow to record raw data of accelerations with respect to different road pavements, which are processed according to the UNI ISO 2631. Since the regulation ISO 2631 is referred to motor vehicle, where driver and passengers are usually seated, the evaluation cannot be considered representative in quantitative terms, but it is interesting from qualitative point of view. This is the reason why ISO 2631 procedure is applied to e-scooters for evaluating weighted Root-Mean-Square acceleration (a_{RMSi}) associated to specific typologies of pavement crossed at various speeds, and, thus, for qualitatively comparing vibrations transferred to human body in different test conditions (pavement typology and speed). Accelerations recorded during the field tests and, subsequently, processed according to the ISO 2631 are reported in Figure 6 and allow to state that:

• distresses effects on asphalt pavements while travelling at the maximum speed (25 km/h) have a great impact in terms of acceleration, as demonstrated by the achieved acceleration peaks in correspondence of potholes (time 6 s and 9 s in Figure 6);
• a speed reduction from 15 to 6 km/h on stone pavements can significantly reduce the accelerations suffered by the driver, including peaks; in particular, such a reduction of speed on stone pavements, could approximately halve acceleration values, with a great beneficial effect in terms of stability and comfort.

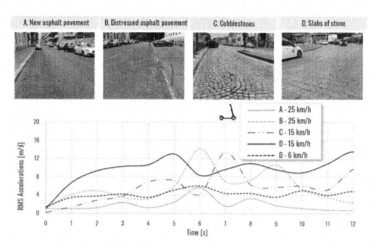

Figure 6.   E-scooters-pavement interaction by field tests - comparison of accelerations registered along vertical axis for different pavements/speeds.

5   CONCLUSIONS

E-scooters are still undergoing a period of development and urban environments are experiencing a process of continuous adjustment to rules and users' needs. However, it is evident that most of international cities are not ready yet to accommodate these new types of vehicles, with different requirements in terms of infrastructure and usually characterized by unexpected/inappropriate behavior. Starting from such a situation, this study reports the preliminary results concerning the safety effects due to the e-scooters introduction in urban areas, as listed below.

• Citizens' perception: Italian citizens' opinion on e-scooters' effects on urban mobility safety shows significant concern in relation to e-scooters. In fact, both e-scooters users and non-users show a general awareness of the danger caused by the introduction of these vehicles in urban areas. On one side, e-scooters users are generally conscious of the risk of driving these vehicles, especially on uneven pavements and tend to prefer driving on cycle paths, sidewalks or zones 30. On the other side, non-users perceive the danger due to the presence of e-scooters, when walking on sidewalks or driving other vehicles. Therefore, it is crucial to implement safe infrastructures, such as cycle paths, which allow to channel e-scooters' traffic into a proper path, with adequate pavement surface characteristics and separate from pedestrians/high-speed vehicles.

- Users' behavior: on-site observation of e-scooter users' behavior in Milan demonstrated that e-scooters' behaviors are frequently inappropriate and risky, meaning users fail to comply with the law. For instance, despite being forbidden, a quite common observed behavior is related to e-scooters travelling on sidewalks or carrying objects/passengers. From an opposite perspective, some drivers, being aware of the high risk related to these vehicles, choose to wear the helmet, even if it is not compulsory. Moreover, e-scooter's driver behavior is proven to be strictly dependent on his experience and on traffic conditions, as in the case of crossing modalities at intersections. Thus, it is evident that a proper legislative framework and the compliance with rules are essential conditions to ensure urban mobility safety.
- E-scooters-pavement interaction: the field tests confirm the great influence of speed and pavement evenness on e-scooter's safety. In fact, accelerations, registered during trials show that pavement typology and level of distress have a crucial role in e-scooter's stability: due to the vehicle design (characterized by small wheels and high center of gravity), e-scooter's driving stability is strictly affected by pavement surface conditions, that are incapable of ensuring minimum safety conditions in most circumstances. In these terms, a speed reduction on uneven pavements could be a temporary solution, but it does not work in the long-term period, thus a redesign of urban roads is strongly needed. Moreover, the speed reduction cannot be left to the user's common sense, but should be enforced through speed regulators. New technologies are always being developed and, maybe, in a near future speed regulation could be combined with geo-fencing techniques.

Based on the preliminary results, the main crucial aspects for e-scooters' safety are speed, pavement characteristics, level of traffic segregation and the use of protective devices, such as the helmet. However, since e-scooters' related risks are various and the surrounding context is continuously evolving, additional studies are required to achieve a complete insight on dangerous behaviors, vehicle infrastructure-interaction and possible crash consequences.

REFERENCES

Almannaa M.H., Alsahhaf F.A., Ashqar H., Elhenawy M., Masoud M. & Rakotonirainy A. (2021). Perception Analysis of E-scooter Riders and Non-riders in Riyadh, Saudi Arabia: Survey outputs. *Sustainability* (Switzerland).

Associazione Sostenitori ed Amici della Polizia Stradale. (2021). *ASAPS.it - il Portale Della Sicurezza Stradale.* https://www.asaps.it/tags/55-osservatorio_monopattini.html

DMove. (2018). *A Milano si Testa lo Sharing dei Monopattini Elettrici di Helbiz.* https://www.dmove.it/news/a-milano-parte-lo-sharing-dei-monopattini-elettrici-fuorilegge-se-prendete-una-multa-la-paghiamo-noi

Fernandes A.O., Id M.D., Welter C. & Id M.P. (2022). *Analysis of Electric Scooter User Kinematics After a Crash Against SUV.* https://doi.org/10.1371/journal.pone.0262682

Istat. (2019). Incidenti Stradali. *Istituto Nazionale Di Statistica,* 2019, 22.

Ma Q., Yang H., Mayhue A., Sun Y., Huang Z. & Ma Y. (2021). E-Scooter Safety: The Riding Risk Analysis Based on Mobile Sensing Data. *Accident Analysis and Prevention,* 151(December 2020).

Osservatorio Nazionale Sharing Mobility. (2020). http://osservatoriosharingmobility.it/

Paudel M. & Fah Yap F. (2021). Front Steering Design Guidelines Formulation for E-scooters Considering the Influence of Sitting and Standing Riders on Self-stability and Safety Performance. *Proceedings of the Institution of Mechanical Engineers, Part D: Journal of Automobile Engineering.*

Posirisuk P., Baker C. & Ghajari M. (2022). Computational Prediction of Head-ground Impact Kinematics in E-scooter falls. *Accident Analysis and Prevention,* December 2021. https://doi.org/10.1016/j.aap.2022.106567

Reuters. (2018). *Lime Launches Electric Scooters in Paris, Targets Europe.* https://www.reuters.com/article/us-lime-paris-bike-share-idUSKBN1JH0PK

Mechanical Vibration and Shock - Evaluation of Human Exposure To whole-body Vibration, (1997).

Sucha M. (2014). Road Users' Strategies and Communication: Driver-pedestrian Interaction. *Transport Research Arena (TRA)*, 12.

Trivedi T.K., Liu C., Antonio A.L.M., Wheaton N., Kreger V., Yap A., Schriger D. & Elmore J. G. (2019). Injuries Associated With Standing Electric Scooter Use. *JAMA Network Open*. https://doi.org/10.1001/jamanetworkopen.2018.7381

Tuncer S., Laurier E., Brown B. & Licoppe C. (2020). Notes on the Practices and Appearances of E-scooter users in Public Space. *Journal of Transport Geography*, March. https://doi.org/10.1016/j.jtrangeo.2020.102702

van Haperen W., Riaz M.S., Daniels S., Saunier N., Brijs T. & Wets G. (2019). Observing the Observation of (Vulnerable) Road User Behaviour and Traffic Safety: A Scoping Review. *Accident Analysis and Prevention*, December 2018. https://doi.org/10.1016/j.aap.2018.11.021

Verge. T. (2018). *The electric Scooter Craze is Officially One Year Old - What's Next?* https://www.theverge.com/2018/9/20/17878676/electric-scooter-bird-lime-uber-lyft

Roads and Airports Pavement Surface Characteristics – Crispino & Toraldo (Eds)
© 2023 The Author(s), ISBN 978-1-032-55149-4

Indirect IRI estimation through smart tyre technology

Gabriele Montorio* & Massimiliano Sallusti
Pirelli Tyre S.p.A., Milan, Italy

Stefano Melzi
Department of Mechanical Engineering, Politecnico di Milano, Milan, Italy

Davide Chiola
Movyon S.p.A, Roma, Italy

Federico Cheli
Department of Mechanical Engineering, Politecnico di Milano, Milan, Italy

Benedetto Carambia
Movyon S.p.A, Roma, Italy

ABSTRACT: In the last decades smart tyre technology has generated growing interest towards different fields of application. The main reason lies in the opportunity to turn passive elements like tyre into active sensors. Moreover, being the tyre a crucial element for vehicle dynamics and the only vehicle part directly in contact with the road, it provides the best monitoring location of road and vehicle interaction. This paper demonstrates how smart tyres can provide information about road surface. In particular, the work investigates how a vehicle fitted with smart tyres can enhance the monitoring of road conditions through the estimation of road unevenness, which is one of the most critical aspects for vehicle ride comfort. The presented solution represents a flexible and widely usable system capable of providing an indication of the IRI index (International Roughness Index), the most worldwide used road unevenness index. The solution is based on a generic vehicle fitted with four smart tyres, each equipped with an accelerometer sensor.

Keywords: Smart Tyre, International Roughness Index, Tyre-Road Interaction, Vehicle Comfort, Accelerometer Sensor.

1 INTRODUCTION

1.1 *IRI*

The World Bank established the International Roughness Index (IRI) in 1986 (Sayers 1986). The development was the result of an early study performed in Brazil in 1982. Since then, IRI spread worldwide, becoming one of the most widely used and well recognized index for road roughness measurement. Stability over time and transferability throughout the world are the major features that favoured the diffusion of this index.

The IRI calculation from longitudinal road profile is the preferred choice among the several possible approaches for IRI measurement, well described in Sayers 1986. As detailed in Sayers 1995, IRI is calculated from longitudinal road profile by accumulating the output

---

*Corresponding Author; gabriele.montorio@pirelli.com

DOI: 10.1201/9781003429258-14

from a quarter-car model and dividing by the profile length to obtain a summary roughness index with units of slope.

The general idea behind this approach for the IRI estimation is to feed to a simple suspension model, that represents a generic vehicle suspension system, the road longitudinal profile to obtain the effective response in terms of vehicle body vibration and filter out those road wavelengths that do not have an impact on ride comfort. To achieve the goal, a standard set of parameters for the suspension model has been provided (Gillespie *et al.* 1980), moreover, being the suspension response strictly related to the simulated vehicle speed, the standard defines a speed of 80 km/h as reference.

1.2 *Smart tyre*

The automotive industry witnessed in the last decades the growth of smart tyre technologies. The basic principles behind this solution is the transformation of the tyre from a passive element for vehicle dynamics to an active sensor able to provide information about vehicle and road status. Moreover, the integration with ADAS system could strongly enhance vehicle ride safety and comfort. A key factor for the adoption of this technology is its favourable location. The in-tyre sensor position provides a more reliable information about the interaction between road and tyre, the latter being the only contact point between vehicle and surrounding environment.

The idea of smart tyre adoption for monitoring road roughness is based upon this potentiality. In particular, the current work focuses its attention on the study of tyres deformation due to road unevenness. The height variation of the road longitudinal profile induces a sudden variation of the tyre deflection, which is then transmitted to the vehicle chassis through the suspension and perceived by the passenger as a vertical acceleration. From this point of view, the smart tyre technology offers an interesting opportunity to convert every vehicle in a potential sensor for road monitoring, delivering a continuous stream of information.

Pirelli Cyber^{TM} Tyre product line implements the smart tyre approach (Cheli *et al.* 2013; D'Alessandro *et al.* 2012). The key element is a wireless sensor based on an accelerometer mounted in the tyre.

2 METHODOLOGY

Before detailing the algorithm for the calculation of the indexes related to IRI, it is necessary to describe an important premise. As previously mentioned, IRI could be computed feeding to a simple suspension car model the road longitudinal profile. The simulated suspension car model at a reference vehicle speed of 80 km/h results to be a pass-band filter for those road wavelengths in the range from 1,25 up to 30 m, as reported in Sayers, Karamihas 1998. For this reason, a suitable instrument for IRI computation should cover this wavelength range, where the largest wave number corresponds to the distance between two consecutive samples, while the shortest is defined by the length of the acquisition itself. The Cyber^{TM} Tyre system delivers information every tyre revolution, with an approximate radius of 0,32 m, therefore every 2 m. The system has no limitations on the acquisition length, so it is possible to state that, from a theoretical point of view, it covers most of the wavelengths involved in IRI computation and thus it is suitable for the IRI estimation.

The current project is based on the study of tyre deformation. The Cyber^{TM} Tyre sensor can sample the accelerometric signal of the tyre region where it is attached to. The most interesting part of the signal is the angular sector around the contact area between tyre and road. It holds most of the information related to the vertical forces exchanged between tyre and ground, which generate tyre deformation. Figure (2) shows a generic accelerometric signal in the radial direction. The signal can be divided into three main regions, based on its

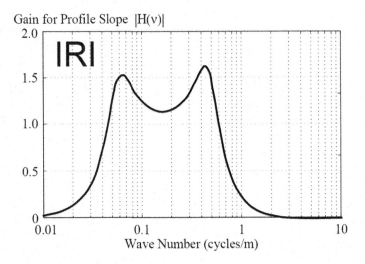

Figure 1.   Golden Car transfer function at 80 km/h.

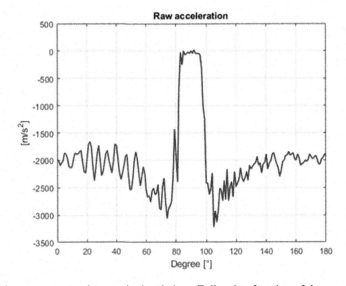

Figure 2.   Cyber tyre: raw accelerometric signal along Z direction function of the angular position.

characteristics. A first region, namely before the contact area (0°→45°), where the acceleration is approximately equal to the centrifugal acceleration and no deformation occurs. A second region, namely the contact area (45°→135°), where the acceleration reaches null value approximately where the tyre external circumference becomes flat. A third region, namely after the contact area (135°→180°), where the deformation runs out and the acceleration goes back to the centrifugal values.

From the measurement of the acceleration within the tyre it is possible to study and define multiple indicators of tyre deformation. The length of the second region previously defined is one of those indicators, strictly related to the length of the external flat circumference region of the tyre. The integration of the accelerometric signal could be another method to estimate tyre deflection. To guarantee a proper sampling of the above-described wavelengths, the implemented solution should track the tyre deformation for an adequate distance. Considering multiple tyres signals increases the reliability of the methods under the hypothesis that two tyres in the same vehicle side should follow the same path and thus should see the same road portion and tyres of the same axle would be linked by the rigid body motion of the chassis.

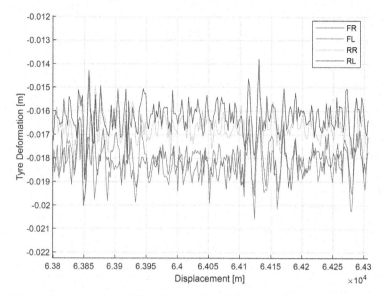

Figure 3.   Tyres deformation signals (FR: Front Right, FL: Front Left, RR: Rear Right, RL: Rear Left).

The power spectral density is a probabilistic method, which describes the power distribution of a signal into frequency components. This method is widely used to analyze random vibration signals to determine how energy is distributed over different frequencies. Tyre deformation, which depends on road unevenness, could be defined as a stochastic process with a mean value that changes with tyre operative conditions, thus this method is suitable for the study of its energetic content. Aside from the theory about PSD, we report a possible estimation formula that was adopted for our purposes:

$$PSD = 2 * \left(\frac{1}{F_s}\right)^2 * \frac{1}{L_p} * FFT(x) * conj(FFT(x)) \qquad (1)$$

where:

- F_s is sampling frequency.
- L_p is length of period.
- $FFT(x)$ is Fast Fourier Transformation of the signal.

One of the main advantages of PSD against FFT is the normalization of the amplitude by the length of the period, which allows the comparison between signals of different lengths. For the purpose of our project, PSD calculation is performed for every deformation signal from the four wheels of the vehicle. PSD is generally associated with the analysis in the time domain, but the very same approach could be used considering the distance as reference variable. This approach is more suitable for our purposes since the adopted system samples the deformation with a fixed spatial sampling. With this kind of approach, a high value of

PSD in a specific range represents a periodic oscillation in the deformation signal with a specific wavelength, which again represents road unevenness.

The next step of the methodology is to correlate the information from different tyres to enhance the robustness and filter out some random noise. The estimation of the correlation is performed through the simple equation reported below. With this equation, namely coherence (Coh) is calculated as the cross-correlation between the two input signals normalized by the square root of the PSD of each single signal.

$$Coh = \frac{real\left(2 * \left(\frac{1}{F_s}\right)^2 * \frac{1}{L_p} * FFT(x_1) * conj(FFT(x_2))\right)}{(PSD_{x_1})^{\frac{1}{2}} * (PSD_{x_2})^{\frac{1}{2}}} \tag{2}$$

Where:

- x_1 is deformation signal of first wheel
- x_2 is deformation signal of second wheel
- PSD is the power spectral density, as previously defined.

This coherence function ranges from -1 to 1. The maximum value corresponds to a synchronous contribution of the two wheels at that specific wavelength. The minimum value corresponds to half period delay between the two wheels at that specific wavelength. Moreover, a null value of coherence corresponds to null value of at least one of the two signals in the corresponding wavelength range, thus the input will be filtered out. The abovementioned coherence is computed for each couple of deformation signals for the two axes and the two sides of the vehicle.

The last step of the method is to synthetize the information into a single value. Firstly, a mean value between coherences and PSDs is computed to obtain an equivalent PSD at vehicle level.

$$EqPSD = \frac{PSD(x_{FL}) + PSD(x_{FR}) + PSD(x_{RL}) + PSD(x_{RR})}{4}$$
$$* \frac{Coh_{S_L} + Coh_{S_R} + Coh_{a_F} + Coh_{a_R}}{4} \tag{3}$$

Where:

- Coh_{S_L} is coherence at left side, thus between front left and rear left tires
- Coh_{S_R} is coherence at right side, thus between front right and rear right tires
- Coh_{a_F} is coherence at front axle, thus between front left and front right tires
- Coh_{a_R} is coherence at rear axle, thus between rear left and rear right tires

The equivalent PSD represents the energy content for each wavelength which is associated to multiple tyres. It is, therefore, an estimation of the energy transmitted from road longitudinal profile to vehicle body.

In a similar way to the methodology of IRI calculation from longitudinal profile, the final index, namely DI, is calculated as the integral value of the above-mentioned equivalent PSD in the wavelength region that, according to the golden car suspension model, mostly affects the vehicle response.

$$DI = \sum_{n=1/30}^{1/1,25} EqPSD_n \tag{4}$$

To complete the current section, it is worth to briefly describe another definition of the current methodology. IRI values are commonly estimated separately for each side of the vehicle. Considering a longitudinal profile that approximates the road input to the corresponding vehicle side. Similarly, DI index could be estimated based on deformation signal from two tyre of the same vehicle side. According to this methodology, the equivalent PSD is calculated as mean value between the PSD of the two tyres times the coherence between them.

$$EqPSD_L = \frac{PSD(x_{FL}) + PSD(x_{RL})}{2} * Coh_{s_L} \qquad (5)$$

Two more indexes are, therefore defined, DI_{Left} and DI_{Right} that refers to an IRI estimation respectively on the left and right side of the vehicle.

3 TESTING CAMPAIGN

The current research project is the result of the cooperation between Pirelli Tyre S.p.A. and MovyOn S.p.A. – Autostrade per l'Italia Group (ASPI); thanks to the sharing of the respective knowledge on smart tyres and road infrastructure it was possible to develop a proof-of-concept testing campaign carried out on a highway infrastructure segment managed by ASPI.

The proof of concept involved a VolksWagen Multivan equipped with four sensorized tyres 265/65 R16 Cinturato All Season SF2. Inside each tyre, one Cyber^{TM} Tyre sensor was installed. The sensor samples the acceleration along the three orthogonal directions for the monitoring of tyre deformation. A main component of the sensor is a MEMS accelerometer, while data and command transmission is achieved through wireless communication. Besides tyre sensors, the Cyber^{TM} Tyre system features two vehicle radio receivers, one for each axle, allowing the communication between sensors and a vehicle elaboration unit, which controls and operates the entire system. This device also implements the communication towards external devices, and, for this proof of concept, it is connected via USB to a hard disk for data storing and via Ethernet to a PC for system monitoring.

The system is coupled with a PCAN-GPS unit, which provides the information about vehicle position during the testing. Data are stored and synchronized by the vehicle

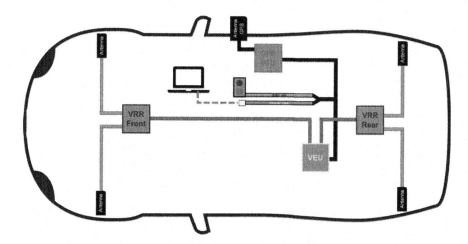

Figure 4.   On-board Cyber^{TM} Tyre acquisition system scheme.

elaboration unit. PCAN-GPS provides information with a sampling rate of 1hz, while precision of the position depends on multiple factors but mostly on available GPS satellites.

The testing methodology requires several passages on the first lane of the considered highway segment, vehicle speed ranges approximately from 80 to 90 km/h. Lane changes or speed variations outside these boundaries will not be considered for the proof-of-concept scope, in order to be as much as possible aligned with the testing methodology of the reference vehicle computing the IRI.

4 RESULTS

Before presenting the current project results, it is worth underlining that the DI unit is not relevant, since the deformation signal, which is based on, is just an estimation of the actual tyre deformation.

The graph in Figure 5 represents a first estimation of the DI directly compared to the value of IRI acquired on the same highway segment, over 15 km. The DI index is computed every 20 m, based on a deformation signal acquired for 20 m. This method guarantees a robust alignment with the implemented IRI calculation methodology.

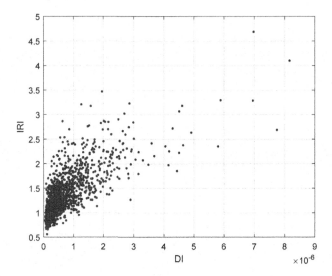

Figure 5.   DI and IRI comparison.

DI and IRI show very coherent trends with some few high values of IRI corresponding to high values of DI.

The reference vehicle provides three different IRI values. Alongside with standard IRI, IRI left, and right are also computed. These two are based on a longitudinal profile extracted respectively on the left and right side of the vehicle. Similarly, it is possible to define a variation of the current methodology based only on the deformation signals from one side of the vehicle and computing the equivalent PSD as the mean value of the two PSD tyre signals multiplied by the coherence between them.

The graph in Figure 6 shows the DI and IRI values calculated for the left and right side of the vehicle. Moreover, the graphs in the first row correspond to the right roadway, while the second row corresponds to the left roadway, named according to ASPI convention.

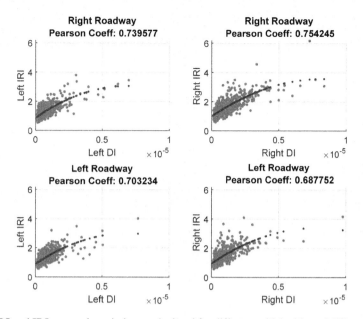

Figure 6. DI and IRI comparison, indexes calculated for different vehicle side and different roadways.

The Pearson coefficient was calculated for each data subset, representing the linear correlation between the two distinct variables DI and IRI. Values close to 1 correspond to a very strong linear correlation. The obtained coefficient underline the good correlation between IRI and DI for the whole considered dataset.

Blue dots in the graphs represent a quadratic fit of the dataset. For sake of brevity, we report only an example of the quadratic fitting performance for the right roadway on the left side of the vehicle. The formula is:

$$IRI = 0.8432 + 6.4362e5 * DI - 4.6731e10 * DI^2 \tag{6}$$

Regression performance could be estimated in terms of $R^2 = 0.574$ and residuals as reported in below in Figure 7.

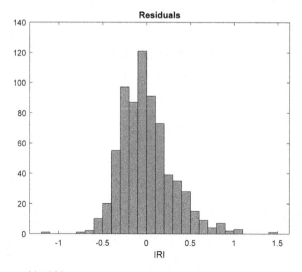

Figure 7. DI fitting residual histogram.

143

Applying the above-mentioned formulation for the calculation of IRI values from DI, it is possible to obtain a calibrated method for the estimation of the IRI, as reported in Figure 8 where the IRI value is directly compared to the estimation based on the previously defined formula. Data are reported only for left roadway on both vehicle sides.

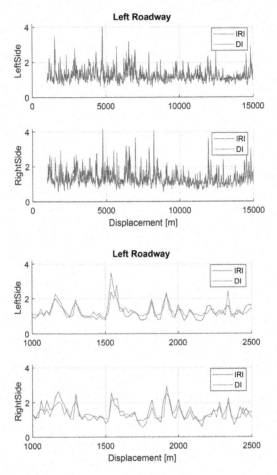

Figure 8.   Calibrated DI with respect to IRI (left) and specific portion (right).

As can be noted, the value estimated through DI well reproduces the trend of the IRI function.

5   CONCLUSION

The current project targeted the development of a new algorithm for the IRI estimation from smart tyre signals. A possible methodology was described in detail and the results were shown in the previous section. The obtained results confirm the method reliability and consistency of the final output. Moreover, several other tests on different highways and in different environmental conditions were carried out but are not reported in this paper. Under the same assumptions described above, the DI indicator always shows a good correlation level with respect to IRI. One possible point of attention for future work investigations could be the vehicle speed. This investigation could possibly allow the infrastructure manager to get an indication of IRI beyond the slow lane. This key factor, combined with the flexibility of smart tyre solutions that can be widely adopted in commercial vehicles, could become the enabler for the usage of this technology as a tool for road monitoring.

ACKNOWLEDGEMENTS

The presented methodologies was developed within Cyber^{TM} Tyre project from Pirelli Tyre S.p.A. for the development of smart tyre in collaboration with Autostrade per l'Italia S.p.A. (Italy), The activities presented in this paper were sponsored by Pirelli Tyre S.p.A. and Autostrade per l'Italia S.p.A. (Italy), which gave both financial and technical support within the framework of the Highway Pavement Evolutive Research (HiPER) project. The development of the presented work was achieved undere the thecnical supervision of Mechanical Department of Politecnico di Milano. The results and opinions presented are those of the authors.

REFERENCES

Gillespie T.D., Sayers M.W. and L. Segel 1980. *NCHRP Report 228: Calibration of Response-Type Road Roughness Measuring Systems.* TRB, National Research Council, Washington, D.C.

Sayers M.W. et al., 1986. Guidelines for Conducting and Calibrating Road Roughness Measurements. *Transportation World Bank Technical Paper* Number 46.

Sayers M.W. 1995. On the Calculation of International Roughness Index from Longitudinal Road Profile. Transportation Research Record 1501.

Sayers M.W. and Karamihas S.M. 1998. *The Little Book of Profiling.*

D'Alessandro V., Melzi S. and Sbrosi M., Brusarosco M., 2012. Phenomenological Analysis of Hydroplaning Through Intelligent Tyres, *Vehicle System Dynamics.*

Cheli F., Melzi S. and Sabbioni, E. 2013. Development of an ESP Control Logic Based on Force Measurements Provided by Smart Tires. *SAE Int. J. Passeng. Cars - Mech. Syst.*

Inductive systems for electric vehicles: Optimization of the charging units embedded into road pavement

Claudia Nodari*, Misagh Ketabdari, Maurizio Crispino & Emanuele Toraldo
Department of Civil and Environmental Engineering, Politecnico di Milano, Milan, Italy

ABSTRACT: Dynamic inductive charging system for Electric Vehicles (EV) is a promising technology to increase the environmental sustainability of road transport sector in the future years. This on-the-road charging system involves a new charging infrastructure network made of prefabricated Charging Units (CUs) embedded into the bituminous road pavement. In this way, traditional roads (t-roads) are converted in electrified roads (e-roads). Based on previous Authors' studies, CU is considered as a box with a void in which the electrical technologies are held. The present research focuses on the optimization of void CU shape and dimension in order to suggest different cross-sectional geometries. Since CU has not to negatively affect the structural performance of pavement, a specific study using a Finite Element Modelling (FEM) approach is conducted. In these FEM simulations, various load positions are considered to describe cross wander distance of Electric Vehicles compered to CU location into bituminous pavements. Therefore, as a result of this study a set of void CU geometries and dimensions is proposed, obtaining two main goals: satisfy the electrical technology needs, while preserving the pavement structural performance.

Keywords: E-Road, Concrete Charging Unit (CU) Optimization, FEM Simulation

1 INTRODUCTION AND LITERATURE REVIEW

During 2019 in EU-27, transport sector produced 25.8% of the total Green House Gases (GHG) emissions, as stated in the European Commission report 2021 (European Commission 2021). Road transport was responsible for 71.1% of the sector's emissions. As regard vehicles, cars accounted for the 60.6% of the road transport sector's emissions (11.2% of the total GHG emissions). Due to these high GHG emissions, Scientific Research investigates new solutions to increase environmental sustainability of road system. The electrification of vehicles is a possible solution (Ajanovic & Haas 2016; Azad *et al.* 2019; Bi *et al.* 2019; Mahesh *et al.* 2021; Marghani *et al.* 2019); however, the environmental advantages are effective only if renewable energy sources are used to generate electricity (Ajanovic & Haas 2016; Mahesh *et al.* 2021). Among the arising technology for electric mobility, dynamic vehicle charging seems to overcome the current limitations of Battery Electric Vehicles -BEVs- (Azad *et al.* 2019; Bi *et al.* 2019; Cirimele *et al.* 2020; Marghani *et al.* 2019; Trinko *et al.* 2022). The main BEV points of weakness are high initial cost, long recharging time, limited diffusion of static charging stations, and range anxiety (Cirimele *et al.* 2020; Mahesh *et al.* 2021; Marghani *et al.* 2019; Soares & Wang 2022). In particular, almost 70% of people do not buy electric vehicles, due to their limited range (Azad *et al.* 2019). Therefore, EV future depends on developments in infrastructure, car manufacturing industry, energy prices and electricity sector (Soares & Wang 2022). Regarding infrastructure, a key element to

---

*Corresponding Author: claudia.nodari@polimi.it

DOI: 10.1201/9781003429258-15

promote BEVs' diffusion is the so-called electrified road (e-road), that is a road able to continuously provide electricity for electric vehicles (Soares & Wang 2022). The dynamic systems use both conductive and contactless (wireless) charging techniques. In order to obtain effective wireless power transfer, the three basic principles are high power, high efficiency and a proper air gap (Soares & Wang 2022). Between the different wireless charging technologies, Inductive Coupled Power Transfer (ICPT) is the most promising solution (Azad et al. 2019; Soares & Wang 2022). The main ICPT key factors are the high power (up to 250 kW), the high charging efficiency (71%-96%) and a feasible air gap tolerance (7.5-50 cm) (Soares & Wang 2022). An addition aspect to consider for energy transfer efficiency is the alignment of the electric vehicle with the ground charging element (Cirimele et al. 2020).

Based on the inputs deriving from the aforementioned available scientific literature, the present research focuses on contactless dynamic vehicle charging, using Charging Units (CUs) embedded into the road pavement. These CUs are prefabricated cement concrete box, in which the electrical technologies are accommodated. From a pavement point of view, the introduction of the CU has not to be detrimental for the structural behavior of the entire pavement, both in short and long period. Thus, the present research focuses on structural response of e-roads by using the Finite Element Model (FEM) approach. The available scientific papers assume the CU as a solid box, as shown in Figure 1, (Chabot & Deep 2019; Chen et al. 2019; Marghani et al. 2019; Nodari et al. 2021; Soares & Wang 2022). It is obviously a limitation of the methods; that's the reason why in the current investigation CU is considered as a box with a void for the electrical devices positioning. In detail, the optimization of CU cavity dimension and shape is the goal of the present study. Regarding the constitutive model, the scientific literature proposes different sophisticated laws to describe the mechanical behavior of asphalt materials, such as the thermodynamics-based finite strain viscoelastic-viscoplastic model with damage coupled (Chen et al. 2019). However, for the sake of simplicity, in this investigation, an elastic model is assumed to be acceptable for all materials (Chabot & Deep 2019; Nodari et al. 2021).

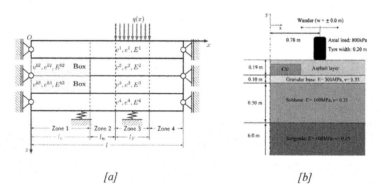

[a]   [b]

Figure 1. E-road cross-sectional geometries [a] (Chabot & Deep 2019) [b] (Chen et al. 2019).

2 OBJECTIVES

As previously stated, the goal of the present research is the study of void CU shape and dimension in order to optimize the e-road cross-sectional geometries. In other word, the study is aimed at improving the cavity characteristics (to accomplish the electrical technology needs), preserving the pavement structural performance.

The paper is planned as follows. The optimization process of void CU shape and dimension is described in Section "Optimization of void CU shape and dimension", considering different load positions and numerous analysis sections. Section "Structural response of e-road_void CU" details the structural behavior of e-road_void CU in terms of both horizontal and vertical stresses, using a Finite Element Modelling (FEM) approach. Finally, conclusions and future investigations are outlined in the last section.

3 OPTIMIZATION OF VOID CU SHAPE AND DIMENSION

CU external dimensions (0.80 m in width and 0.14 m in height) are the same for all the analyzed geometries. Four different void geometries are investigated (as shown in Figure 2):

- Void CU_R1: this geometry is characterized by a rectangular cavity (R1), as discussed and studied in previous Author's research (Nodari *et al.* 2021);
- Void CU_R2: this geometry consists of two rectangular cavities (R2);
- Void CU_C4: this geometry has four circular cavities (C4) equally distributed along CU horizontal axis;
- Void CU_C6: this geometry has six circular cavities (C6) equally distributed along CU horizontal axis.

As regard materials, bituminous pavements are analyzed as indicated in Figure 2. The characteristics of each layer/material - thickness, bulk density, Young's modulus and Poisson's ratio- are given in (Nodari *et al.* 2021).

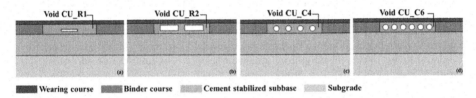

Figure 2. Cross-sectional geometry (a) e-road_void CU_R1, (b) e-road_void CU_R2, (c) e-road_void CU_C4, (d) e-road_void CU_C6.

Moreover, three load positions are studied to describe cross wander distance compared to CU location, as follow:

- *Centered on CU*: load is centered on CU. This is the perfect case (expected improvement in battery charging efficiency), in which the electric vehicle is in the middle of the lane;
- *CU edge*: the left load (corresponding to the left wheel of the electric vehicle) is located along CU left edge;
- *CU center*: the left load (corresponding to the left wheel of the electric vehicle) is located along CU center.

More details regarding the implemented 2D FE model - equations, boundary conditions, triangular mesh, load characteristics - are presented in (Nodari *et al.* 2021). COMSOL Multiphysics 6.0 software is adopted to implement the FE model.

During the study, several FEM simulations are performed progressively changing lateral/ up/down thickness of CU in order to obtain the optimum void CU geometry.

The optimization process compares FEM stress results with the failure tensile/compression stresses of each material (Table 1). Even if FEM software calculates the stresses within the domain of the entire pavements, the structural comparisons are carried out along defined analysis sections. Each geometry has typical both vertical and horizontal analysis sections, defined according to the ones in (Nodari *et al.* 2021), as follows:

Table 1. Comparison between maximum FEM stresses (both horizontal and vertical) and failure stresses for different Void CU geometries, considering load on CU center.

| | | $\sigma_{xx,tens}$ | | | $\sigma_{xx,comp}$ | | | $\sigma_{yy,comp}$ | | |
| | | Max FEM stress | | | Max FEM stress | | | Max FEM stress | | |
		Value [N/mm²]	Analysis section	Failure stress [N/mm²]	Value [N/mm²]	Analysis section	Failure stress [N/mm²]	Value [N/mm²]	Analysis section	Failure stress [N/mm²]
	R2	0.82	Cavity2 left		-1.42	Surface of wearing course		-1.16	Cavity2 left	
Wearing Course	C4	0.18	Surface of wearing course	1.25	-1.31	Surface of wearing course	-3.75	-0.83	CU center	-6.25
	C6	0.18	Surface of wearing course		-1.37	Surface of wearing course		-0.88	Surface of wearing course	
	R2	0.09	Wearing course–Binder course down		-0.40	Wearing course–Binder course down		-0.74	Wearing course–Binder course down	
Binder Course	C4	0.09	Wearing course–Binder course down	1.55	-0.42	Wearing course–Binder course down	-4.65	-0.76	Wearing course–Binder course down	-7.75
	C6	0.09	Binder course–Cem. stab. subb.top		-0.42	Wearing course–Binder course down		-0.76	Wearing course–Binder course down	
	R2	0.16	Cem. stab. subb.–Subgrade top		-0.10	Binder course–Cem. stab. subb. down		-0.69	Binder course–Cem. stab. subb. down	
Cement Stabilized Subbase	C4	0.12	Binder course–Cem. stab. subb. top	0.50	-0.11	Binder course–Cem. stab. subb. top	-1.50	-0.39	Binder course–Cem. stab. subb. top	-4.00
	C6	0.12	Cem. stab. subb.–Subgrade top		-0.12	Binder course–Cem. stab. subb. down		-0.41	Binder course–Cem. stab. subb. down	
	R2	2.16	Binder course–Cem. stab. subb. top		-6.89	Cavity down		-4.88	Cavity1 right	
Concrete	C4	1.82	Circular cavities	2.21	-2.26	Circular cavities	-20.00	-2.13	Circular cavities	-20.00
	C6	1.90	Binder course–Cem. stab. subb. top		-2.00	Circular cavities		-2.74	Circular cavities	

- E-road_void CU_R2: 11 vertical analysis sections and 9 horizontal analysis sections, as shown in Figure 3;
- E-road_void CU_C4: 7 vertical analysis sections and 8 horizontal analysis sections (including circular cavities), like the ones reported in Figure 4;
- E-road_void CU_C6: 7 vertical analysis sections and 8 horizontal analysis sections (including circular cavities), as illustrated in Figure 4.

149

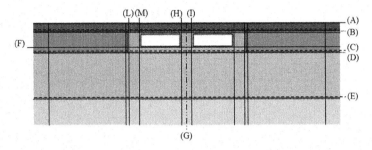

Figure 3. Analysis sections for e-road_void CU_R2. (A) Surface of wearing course; (B) Wearing course-Binder course down; (C) Binder course-Cement stabilized subbase Top; (D) Binder course-Cement stabilized subbase down; (E) Cement stabilized subbase-Subgrade top; (F) Cavity down; (G) CU center; (H) Cavity1 right; (I) Cavity2 left; (L) CU left edge in; (M) Cavity1 left.

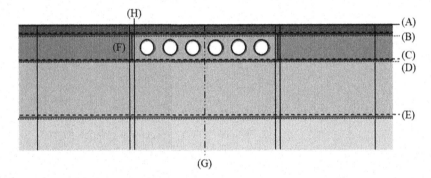

Figure 4. Analysis sections for e-road_void CU_C6. (A) Surface of wearing course; (B) Wearing course-Binder course down; (C) Binder course-Cement stabilized subbase top; (D) Binder course-Cement stabilized subbase down; (E) Cement stabilized subbase-Subgrade top; (F) Circular cavities; (G) CU center; (H) CU left edge in.

Therefore, a total of at least 150 structural comparisons are performed (considering all the sections and the three load positions for each cross-sectional geometry).

Table 1 shows the comparison between the failure tensile/compression stresses (for each layer/material) and maximum FEM stresses, calculated for each cross-sectional geometry, considering load on CU center. The failure stresses are deduced from the available scientific literature, as explained in (Nodari et al. 2021). As highlighted by the values in the table, the maximum stresses are far short of the failure limits.

The optimization process leads to three optimized CU geometries. Table 2 shows the optimized geometries and the related CU characteristics: thicknesses (up, down and lateral), cavity dimensions and cavity area.

Increments of 488.89% (void CU_R2), 179.25% (void CU_C4) and 318.87% (void CU_C6) are obtained respect to e-road_void CU_R1 cavity area. E-road_void CU_R2 is the geometry characterized by the highest value of cavity area (424 cm^2), bringing to advantages for electrical technologies accommodation.

Table 2. CU geometries and corresponding characteristics.

CU geometry	CU characteristics				
	Thickness [cm]			Cavity dimension [cm]	Cavity area [cm²]
	Up	Down	Lateral		
	7.00	4.00	28.00	24.00 x 3.00	**72.00**
	2.00	4.00	9.00	26.50 x 8.00	**424.00**
	3.00	3.00	9.60	8.00 x 8.00	**201.06**
	3.00	3.00	4.57	8.00 x 8.00	**301.59**

4 STRUCTURAL RESPONSE OF E-ROAD_VOID CU

The FEM software allows to calculate several parameters (e.g., stresses, strains, displacements, etc.) according to the characteristics of each simulation, as explained in the previous section. Among these parameters, the current paragraph focuses on stresses in both vertical (along y-axis) and horizontal (along x-axis) directions in order to evaluate the structural behavior of each e-road cross-sectional geometry. According to the FEM outputs, compression stresses are characterized by negative values; tensile stresses are indicated as positive numbers.

Table 3 shows vertical and horizontal stresses within the domain of the e-road pavements in relation to the three load positions.

Focusing on vertical stresses σ_{yy}, the following consideration can be done:

- Load centered on CU: stresses are close to the applied load, which enters and dissolves gradually from the surface to the lower layers, according to the results discussed in (Nodari et al. 2021). CU cavity shape does not affect stress distribution within the pavement. However, a slight difference in stress diffusion can be noticed in Void CU_C6. As regard stress values, the order of magnitude is 0.1 N/mm², in line with the ones observed in (Nodari et al. 2021) for traditional road (road without CU).
- Load in CU edge: higher stresses are registered compared to the ones obtained when load in centered on CU. This stress increase is clear for Void CU_R2 and Void CU_C6. In particular, in Void CU_R2, the highest value in compression increases by 525% compared to the one registered considering load centered on CU. Similarly, the tensile stress increment is equal to 400%.

As regard stress distribution within the whole pavement, CU cavity shape affects the results. This difference is evident both between cross-sectional geometries and between different load positions. Void CU_R1 is characterized by similar stress diffusion comparing both the load locations. Void CU_R2, Void CU_C4 and Void CU_C6 show evident

Table 3. Distribution of vertical/horizontal stresses σ_{yy}/σ_{xx} [N/m^2] at different cross-sectional geometries of e-road and different load positions.

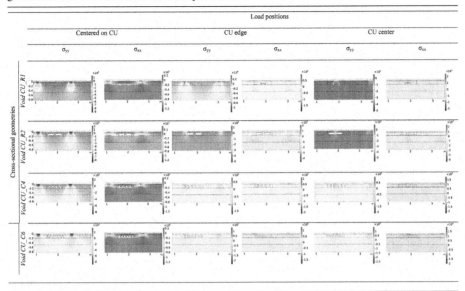

differences. Stresses are close to the applied load, but both CU presence and CU shape affect stress distribution. In particular, a compression stress concentration rises in the left bottom corner of the left rectangular cavity of Void CU_R2. The highest compression stress in CU concrete along "Cavity1 left" analysis section (Figure 3) is equal to -2.63 N/mm^2. Considering the same analysis section in Void CU_R1, the compression stress value is -0.47 N/mm^2. Therefore, the concrete compression stress of Void CU_R2 is increased up to 460% respect to Void CU_R1.

Considering "CU left edge in" analysis section (Figure 3 and Figure 4), the highest compression stresses in CU concrete are equal to -1.51 N/mm^2 (Void CU_R2), -1.15 N/mm^2 (Void CU_C4) and -1.35 N/mm^2 (Void CU_C6). Comparing these results with the one obtained in Void CU_R1 (-1.19 N/mm^2), it is possible to note stress increase of 27%, 3% and 13%, respectively.

- Load in CU center: in Void CU_R2, Void CU_C4 and Void CU_C6, stresses are characterized by the same order of magnitude (N/mm^2) of the ones registered when load is along CU edge. Stresses are close to the applied load, but their distributions change according to the e-road cross-sectional geometries. In Void CU_R1, compression stress concentration rises around CU cavity. Considering Void CU_R2, a stress intensification appears in concrete between the two rectangular cavities; the highest compression value is in the bottom right corner of the left cavity (4.88 N/mm^2). As regard the remaining CU geometries, stresses are concentrated between the two central circular cavities.

Considering "CU center" analysis section (Figure 3 and Figure 4), the highest compression stresses in CU concrete are equal to -1.37 N/mm^2 (Void CU_R2), -0.93 N/mm^2 (Void CU_C4) and -1.28 N/mm^2 (Void CU_C6). Comparing these results with the one obtained in Void CU_R1 (-0.61 N/mm^2), it is possible to note stress increase of 126%, 54% and 111%, respectively.

Similarly, considering the same analysis section, the highest stresses in compression into wearing course are equal to -1.05 N/mm^2 (Void CU_R2), -0.83 N/mm^2 (Void CU_C4), -0.82 N/mm2 (Void CU_C6). Comparing these results with the Void CU_R1 stress (-0.83 N/mm^2), a significant increment is registered for Void CU_R2 (26%).

Regarding the horizontal stresses σ_{xx}, the main results on the structural performance are listed below:

• Load centered on CU: pavement surfaces show significant compression stresses close to the applied load, in line with the results discussed in (Nodari *et al.* 2021). Moreover, some compressive stresses arise at the CU bottom, while tensile stresses appear at the bottom of cement stabilized subbase layer.

Stress distributions are similar in all the analyzed geometries; this means that CU cavity shape does not affect stresses. As regard values, the order of magnitude is 0.1 N/mm^2 for tensile stresses and N/mm^2 for compression ones.

• Load in CU edge: stresses increase is appreciable compared to the ones obtained when load is centered on CU. Regarding distributions into the entire pavement, stress concentrations appear around the cavity closest to the load (e.g., left cavity in Void CU_R2, first cavity in Void CU_C4, etc.). In particular, in Void CU_R2, the highest compression stress in CU concrete along "Cavity down" analysis section (Figure 3) is equal to -3.24 N/mm^2. Considering the same analysis section in Void CU_R1, the compression stress value is -2.39 N/mm^2. Therefore, the increase of compression stress in CU concrete of Void CU_R2 is 36% respect to Void CU_R1. Similarly, the highest tensile stress, in Void CU_R2 in concrete is 1.97 N/mm^2, while in CU with one rectangular cavity is equal to 0.07 N/mm^2. In this case, the stress in the CU with two rectangular cavities is 28 times higher than the one obtained in Void CU_R1.

In line with the outcomes of (Nodari *et al.* 2021), tensile stresses develop at the CU left bottom corner. Considering "Binder course-Cement stabilized subbase Top" analysis section (Figure 3 and Figure 4), the highest tensile stresses in CU concrete are equal to 1.75 N/mm^2 (Void CU_R2), 0.94 N/mm^2 (Void CU_C4) and 1.20 N/mm^2 (Void CU_C6). Comparing these results with the one obtained in Void CU_R1 (1.02 N/mm^2), it is possible to note stress increase of 72%, 8% and 18%, respectively.
Moreover, slight tensile stresses appear at the bottom of cement stabilized subbase layer.

• Load in CU center: in Void CU_R2, Void CU_C4 and Void CU_C6, stresses are characterized by the same order of magnitude (N/mm^2) of the ones registered when load is along CU edge.

Tensile stresses can be marked at the bottom of the CU. Considering "Binder course-Cement stabilized subbase Top" analysis section (Figure 3 and Figure 4), the highest tensile stresses in CU concrete are equal to 2.16 N/mm2 (Void CU_R2), 1.74 N/mm2 (Void CU_C4) and -1.89 N/mm2 (Void CU_C6). Comparing these results with the one obtained in Void CU_R1 (2.20 N/mm2), it is possible to observe stress increase of 2%, 21% and 14%, respectively.

As regard compression, the highest values in CU concrete are registered along "Wearing course-Binder course down" analysis section (Figure 3 and Figure 4) for Void CU_R1 (-2.83 N/mm^2), Void CU_C4 (-1.91 N/mm^2) and Void CU_C6 (-1.81 N/mm^2). The highest result (-6.89 N/mm^2) is obtained along "Cavity down" analysis section (Figure 3) of Void CU_R2. Comparing these results with the Void CU_R1 stress, it is possible to observe stress increase of 33% (C4), 36% (C6) and 143% (R2).

5 CONCLUSIONS

Dynamic inductive charging for Electric Vehicles (EV) is a promising technology to increase the environmental sustainability of road transport sector in the future years. This on-the-road charging system involves a new charging infrastructure network made of prefabricated concrete Charging Units (CUs) embedded into the bituminous road pavement. The present research focuses on the study of void CU shape and dimension, using a FEM approach, in order to optimize e-road cross-sectional geometries from a structural point of view. Therefore, the study is aimed at maximizing the CU cavity characteristics (to accomplish the electrical technology needs), preserving the pavement structural performance.

Based on the obtained outcomes, the main conclusions and further proposals of this research can be drawn as follow:

- Optimizing both shape and dimension of the CU cavity, stresses are lower than the tensile/compression failure value of the pavement materials.
- Among the analyzed e-road cross-sectional geometries, Void CU_R2 (CU characterized by two rectangular cavities) maximizes CU cavity area, bringing to advantages for electrical technologies accommodation. However, other CU cavities are proposed in the study with the aim of providing to the electronics different solutions for accomplishing electrical technology needs.
- From a structural point of view, the highest vertical/horizontal stresses in concrete are obtained in Void CU_R2, deviating from the results gained in Void CU_R1 (which is the target cross-sectional geometry). In particular, stress concentrations grow in the area close to the rectangular cavities' corners.
- Stress distribution into pavement is affected by both load positions and cavity characteristics (shape and dimension). Considering load centered on CU, slight difference can be observed according to cross-sectional geometries. When load is on CU edge or on CU center, significant differences grow in stress distribution as a result of e-road cross-sectional geometries modifications.

As further research steps, some aspects can be investigated concerning pavement design and maintenance of the proposed e-road cross-sectional geometries, such as fatigue resistance, rutting behavior, and surface distresses monitoring. Regarding CU, the existing studies evaluate the cavity shape and dimension; therefore, in future research the structural effects of the external CU geometry can be investigated (e.g., fillet-shaped exterior corner, etc.). Finally, full scale tests are needed in order to validate the achieved theoretical and numerical results.

REFERENCES

Ajanovic A. and Haas R., 2016. Dissemination of Electric Vehicles in Urban Areas: Major Factors for Success. *Energy* 115, 1451–1458.

Azad A.N., Echols A., Kulyukin V.A., Zane R. and Pantic Z., 2019. Analysis, Optimization, and Demonstration of a Vehicular Detection System Intended for Dynamic Wireless Charging Applications. *IEEE Trans. Transp. Electrif.* 5, 147–161.

Bi Z., Keoleian G.A., Lin Z., Moore M.R., Chen K., Song L. and Zhao Z., 2019. Life Cycle Assessment And Tempo-spatial Optimization of Deploying Dynamic Wireless Charging Technology for Electric Cars. *Transp. Res. Part C Emerg. Technol.* 100, 53–67.

Chabot A. and Deep P., 2019. 2D Multilayer Solution for an Electrified Road with a Built-in Charging Box. *Road Mater. Pavement Des.* 20, S590–S603.

Chen F., Balieu R., Córdoba E. and Kringos N., 2019. Towards an Understanding of the Structural Performance of Future Electrified Roads: A Finite Element Simulation Study. *Int. J. Pavement Eng.* 20, 204–215.

Cirimele V., La Ganga A., Colussi J., Gloria A., De Diana M., Bellotti F., Berta R., Sayed N. El Kobeissi A., Guglielmi P., Ruffo R. and Khalilian M., 2020. The Fabric ICT Platform for Managing Wireless Dynamic Charging Road Lanes. *IEEE Trans. Veh. Technol.* 69, 2501–2512.

European Commission, 2021. EU Transport in Figures - Statistical Pocketbook 2021, Notes.

Mahesh A., Chokkalingam B. and Mihet-Popa L., 2021. Inductive Wireless Power Transfer Charging for Electric Vehicles-A Review. *IEEE Access.* 9, 137667–137713.

Marghani A., Wilson D. and Larkin T., 2019. Performance of Inductive Power Transfer-based Pavements of Electrified Roads. *2019 IEEE PELS Work. Emerg. Technol. Wirel. Power Transf.* WoW 2019 196–201.

Nodari C., Crispino M., Pernetti M. and Toraldo E., 2021. Structural Analysis of Bituminous Road Pavements Embedding Charging Units for electric Vehicles. *Comput. Sci. Its Appl. – ICCSA 2021.* Lect. Notes Comput. Sci. vol. 12952 1, 149–162.

Soares L. and Wang H., 2022. A Study on Renewed Perspectives of Electrified Road for Wireless Power Transfer of Electric Vehicles. *Renew. Sustain. Energy Rev.* 158.

Trinko D., Horesh N., Zane R., Song Z., Kamineni A., KonstantinouT., Gkritza K., Quinn C., Bradley T.H. and Quinn J.C., 2022. Economic Feasibility of In-motion Wireless Power Transfer in a High-density Traffic Corridor. e-Transportation 11.

155

Data monitoring and performance assessment

Probe vehicle data as input source for road maintenance

Björn Zachrisson*, Johan Hägg, Håkan Frank, Johan Petersson & Olle Noren
NIRA Dynamics AB, Linköping, Sweden

ABSTRACT: Since 2014 NIRA Dynamics has been working with connected vehicles, bringing the previous 20+ years of experience of developing vehicle software products into the connected and aggregated world of today. The NIRA software is installed at the production facilities of the VW Group passenger vehicles. Each one of the vehicles (~2 million annually, starting late 2020) act as an individual road probe, gathering important data for the vehicle industry as well as the asset management and winter maintenance operations. NIRA creates virtual sensors through sensor fusion of multiple signals using software only. The output is data such as road state (roughness in terms of IRI), potholes & driving obstacles that is being collected and updated on, at least, a daily interval. Combining data from a large fleet of vehicles give an objective up-to-date information of the roads at a full network level. Besides having a full inventory, the daily updates enable the possibility to track road wear trends and planning for reactions on sudden road damages. How can traditional methods be improved or replaced by modern methods? There are many challenges in the field of vehicle data. One of them is to challenge the traditional methods and standardizations of the equipment used; the focus has been on improving the technology but never really looking into the added value from the increased precision. In comparison to a laser scanner, the output from vehicles will always be different but does that really matter? NIRA is also researching how well the "slippery when wet" phenomena can be detected by vehicles. The major difference in comparison to measurement equipment is that the vehicles report actual experienced friction given any weather situation instead of focusing on repeatability. Initial tests in the UK (Mira test track) and France (Nantes) show great promise for the technology.

Keywords: Probe Vehicle Data, Road Safety, Road Monitoring

1 INTRODUCTION

Damaged and deteriorated roads is costing society as a whole and motorists globally a great deal. A motorist in the U.S. is losing on average $599 annually – a total of $130 billion nationally – in additional vehicle operating costs due to driving on roads in need of repair.[1] These costs include additional repair costs, accelerated vehicle deterioration and depreciation, increased maintenance costs, and additional fuel consumption.

The environment is also affected a great deal[2] by the road surface quality. Pavement preservation lowers environmental impact by reducing CO_2 emission, even when adding emission generated at construction stage. Timing is critical here, applying a new thin layer of

---

*Corresponding Author: björn.zachrisson@niradynamics.se
[1]TRIP, October 2018. *Bumpy Road Ahead: America's Roughest Rides and Strategies to Make Our Roads Smoother*. https://tripnet.org/wp-content/uploads/2019/03/Urban_Roads_TRIP_Report_October_2018.pdf
[2]Wang H. et al., 2020. *International Journal of Sustainable Transportation*, Volume 14 - Issue 1 https://www.tandfonline.com/doi/abs/10.1080/15568318.2018.1519086?journalCode=ujst20

asphalt before the IRI threshold is reached means a major reconstruction of the road surface is prevented, lowering road repavement emission.

Digitalization is changing the way of working, connected vehicles continuously monitoring the road surface condition offering new opportunities for creation of services that improve sustainability and safety. New data types, applications and tools bring new possibilities for the operators to make better decisions based on objective measurements. Millions of vehicles collecting data means greater data availability than ever before.

Some challenges are big. Implementing modern methods for road surface monitoring in a traditional industry requires a lot of education, intuitive tools, measurable result, and patience.

2 BACKGROUND

Every country in Europe has their own unique organization for maintenance of the road network and their own standard and methods. How road surveys are conducted depends on the available resources. It ranges from periodical high-performance measurements with traditional methods, via manual inspection, to reports from the public. Independent of how the road surveys are conducted, some frequent occurring challenges are: coverage of the entire road network with objectively inspections, the development of the road network between surveys and to reliably detect fast occurring damages, for example potholes.

Today, the most common way to measure the state of the roads is to either perform manual ocular inspections or use special equipped vehicles with sensors as for examples lasers and cameras. From the measurements, different road parameters are derived, as for example rutting and roughness.[3,4] The traditional machine method generates high-quality measurements, but it requires special equipment that is expensive and that makes it hard to motivate measurements more frequently than on an annual basis.

Also, a wide range of different new methods for monitoring the roads exists, from the use of mobile phones,[5] to modern sensors on vehicles, such as cameras and ultra-sonic sensors.[6] These methods are difficult to scale to create a complete coverage, and for the mobile phone data, it is difficult to create a reliable calibration for mobile phones of different brand and make.

LIDAR equipped passenger cars, has the potential to measure road parameters.[7,8] However, the authors have failed to find a reference to system that is operating today.

NIRA got the first connected after-market fleet up and running in 2015, and since mid-2020 the road monitoring software is installed directly on passenger vehicles from the VW-group during production. A limited fleet equipped with dongles is still used for testing of new features.

---

[3] Highways England, 2020. *CS230 Pavement Maintenance Assessment Procedure*, https://www.standardsforhighways.co.uk/dmrb/search/df44f6b0-4bf4-4189-90a2-912c19cc247a

[4] Stryk J. et al., 2021. Pavement Surface Characteristics - Differences in Measured Parameters and their Evaluation and Use in Different Countries, *Presented at ERPUG 2021*.

[5] Roadroid product sheet. https://www.roadroid.com/common/References/Roadroid%204%20pager.pdf

[6] Mercedes-Benz Predictive Infrastructure Monitoring and evaluating infrastructure with high quality data from our Mercedes-Benz customer fleet. https://data.mercedes-benz.com/products/predictive-infrastructure

[7] SICK US Blog, 2020. Automating Road Maintenance with LiDAR Technology. https://sickusablog.com/automating-road-maintenance-lidar-technology/

[8] Yu Y. et al., 2020. "Road Manhole Cover Delineation Using Mobile Laser Scanning Point Cloud Data," in *IEEE Geoscience and Remote Sensing Letters*, vol. 17, no. 1, pp. 152-156, Jan. 2020, doi: 10.1109/LGRS.2019.2916156.

In compliance with GDPR, NIRA Dynamics is collecting Probe Vehicle Data (PVD) from about 2 000 000 vehicles in Europe. The data is quality assured and processed to be used within all road monitoring and maintenance industries. The data flow is presented in Figure 1.

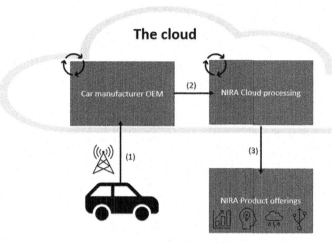

Figure 1.   Sketch of the data flow with (1) being the connection between the vehicle and the manufacturer cloud, (2) being the connection between the manufacturer cloud and the NIRA cloud and (3) being the generated NIRA product offerings based on the collected data.

3.1   *In-vehicle measurements*

The NIRA software Road Surface Monitoring (RSM) is installed in customer vehicles as a module. The module continuously monitors the road surface characteristics such as roughness, potholes and bumps etc. The input is signals from the common sensors already installed in the vehicle. This technology enables increased scalability for road surface measurements since every vehicle is a potential probe, no additional hardware is needed.

3.2   *Road roughness*

The RSM-roughness output is calibrated against laser scanner profiles conducted by the Swedish National Road and Transport Research Institute.[9] The roughness measurements are translated to the International Roughness Index (IRI) standard.[10] Roughness by NIRA Dynamics is always measured continuously above speeds of 7 km/h. The performance of the roughness has been evaluated on all types of roads, ranging from very smooth to very rough roads, with a mean absolute error of 0.51 [mm/m].

[9.]Road Surface Tester system of the Swedish National Road and Transport Research Institute. https://www.vti.se/en/services/highway-engineering-and-geotechnics/on-road-measurement/measurement-of-road-surface

[10.]International Roughness Index. https://en.wikipedia.org/wiki/International_roughness_index

3.3 Road obstacles

RMS's road obstacles include detection of potholes and bumps in the road. The output is calibrated against obstacles on a test track owned by NIRA Dynamics. The robustness is further evaluated against data collected on public roads with reference annotations.

3.4 Road friction measurements

NIRA Dynamics Tire Grip Indicator[11] (TGI) measures friction between the vehicles tires and road surface. The focus in the original development of the TGI was winter condition, in recent years, NIRA Dynamics have investigated the performance during summer conditions to detect slipperiness when wet. Result from the tests tracks Mira in UK and Nantes in France is showing that passenger vehicles with normal tires experience higher friction than traditional methods using tires without treads.

3.5 Data transfer to the cloud

The measurements from the different in-vehicle software modules are packed into data batches together with GPS positions and environmental data such as temperature and sun irradiance. These data batches are sent to the cloud of the car manufacturer and then forwarded to the NIRA cloud where the processing of the vehicle measurements starts.

3.6 Cloud processing

The NIRA cloud processing pipelines ensures the quality of the vehicle data and performs map matching of the vehicle measurements to a road segment map. Further on, the vehicles measurements are fed into different aggregation pipelines where data from multiple vehicles are combined to get one refined output per road segment and data type. The output is then packaged and stored to be able to supply the NIRA end-customer services.

3.7 Data aggregation - Road roughness

The roughness aggregation combines several measurements over time for a given road segment into a single representative road roughness value. The aggregation process runs daily and use historical data for increased precision and coverage. The spatial resolution of the road segments is 25 meters. A map over the current roughness state for the larger roads in Europe is shown Figure 2.

3.8 Data aggregation - Road obstacles

The obstacle aggregation combines the detected road obstacles from all vehicles to identify where and when obstacles such as potholes and bumps appears on the road surface. The spatial resolution of the aggregated obstacles are pointwise locations, where the locational accuracy increases with the number of vehicle measurements.

3.9 Data aggregation - Slippery when wet

By combining friction measurements, in-vehicle weather data, and external weather sources, this aggregation can over time identify and report which road segments that have a tendency of getting slippery when the road surface is wet. In Figure 3 detections of slippery when wet

---

[11.]Zachrisson B., et al., 2021. FCD Enabler for More Efficient WM Operations, S11-S12. Presented at XVI WORLD WINTER SERVICE AND ROAD RESILIENCE CONGRESS 2022.

Figure 2.   Current Road roughness map for the larger roads in parts of Europe.

Figure 3.   Slippery when wet hot spots in The Netherlands.

conditions in the Netherlands is shown. This service is in development and the result shown are from an early concept study. In 2022 the usability and performance will be evaluated in selected markets.

3.10 *Data verification and validation*

With a crowd sourced fleet with millions of vehicles there is a huge set of potential errors sources that might affect the measurements. There are diagnostic modules built into the vehicle software components that detects faulty sensor signals. If a failure is detected, no data are transmitted to the cloud. In addition to the in-vehicle validation there are quality assurance and statistical validations in the cloud environment.

4 NIRA SOLUTIONS

Connected vehicles as road monitoring probes generates tremendous amount of information. To leverage this information and create value, the way of working with road maintenance and asset management will need some adjustment. To face this challenge, NIRA has implemented a set of tools that can support the way road maintenance is performed today, but also allow the possibility to transform road operations. It allows a step-by-step approach towards implementing a fully data-driven road maintenance.

Some big upsides are that regular manual inspections to grade road state can be removed, and any sudden damage or rapid degeneration will immediately be detected and highlighted for the road owner/operator to perform actions to prevent more extensive damages.

Additionally, planning of maintenance can be based on the current status of the road network, low quality road segments can be pinpointed to find unknown problem areas and verify known problems, and it makes it possible to validate the success of performed maintenance actions.

4.1 *Current road state - Inventory*

Having a fleet source approach enables an always up to date snapshot of the road state. The data is updated daily and can be filtered to show only the worst segments of the roads. See an example of this in Figure 4.

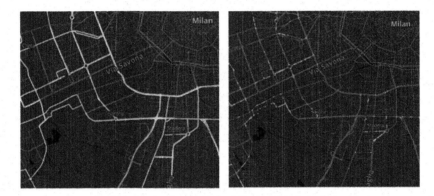

Figure 4.   Road Class 2 and 3, to the left all roads are displayed, to the right only the roads with an IRI level above 6.

For cities this is information that previously was inaccesible due to the high cost of having laser equipped machines out scanning the entire road network. Instead they tend to rely on manual and ocular inspection as well as public complaints. On a federal level, reoccuring scannings are usually performed as it is being part of the strategic national network. However, on both local and federal level the use-case of validating the maintenance plan post-winter is very valid. For Transport Scotland the worst places (from the PVD) per region were brought up with regional responsible; some areas were known and planned to be maintained, or there was a good reason for why not to maintain them. But some of the very worst segments, was not known and not planned, before identified by the inventory tool. Another example is to look at the Freeze thaw cycles as was done in Indiana by Purdue University. During the 7 cycles in February of 2022, a seemingly good road segment went into a very rough road in less than a month, see Figure 5.

4.2 *Road state trend – Deviation detection*

The A5 highway has a viaduct close to its connection to the A10 highway near Amsterdam, The Netherlands. The PVD data on this viaduct displayed elevated international roughness index (IRI) values, see Figure 6. The IRI values also indicate that the degeneration rate is

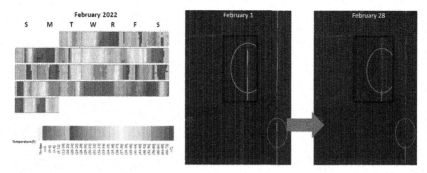

Figure 5. Showing the freeze thaw cycles, seven of them, in February and the impact on a specific road segment in the Lafayette city center.

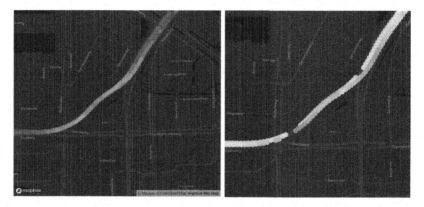

Figure 6. Zoomed in view of the A5 viaduct, to the left current IRI and to the right IRI degeneration.

higher on the viaduct than in its surroundings. According to highway experts at Rijkswaterstaat, this viaduct is a multiple span viaduct with several joints. The contractor miscalculated the height of the joints, leading to the joints being installed too low. The consequence is that the asphalt layer is not aligned vertically. The stress on the joints is therefore higher than normal and the steel expansion joint show significant more damage than what might be expected.

There are rough areas on some spots but not the entire viaduct, hence pinpointing the locations using real time roughness data and with the aid of an IR trend dashboard is possible, this in an early stage before more severe damage has occurred.

4.3 *Viaduct – Broken drainage*

Another example of road state trend is the viaduct at A28 near Assen. Due to the drainage construction being outdated and failed, the underlying sand layer got soaked during heavy rain falls in the end of February 2020. In consequence the materials around the foundation got transported away with the water. The foundation of the viaduct was also affected and started to move. This was transferred to the surface causing distress on the pavement, increasing the roughness values. This was discovered almost instantly, in March 2020, by the connected fleet, as seen in the graph of Figure 7. The drainage was repaired during the summer of 2020 and the problem was solved.

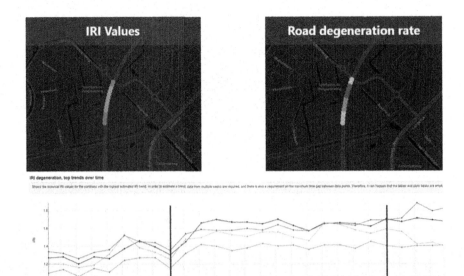

Figure 7. In the top left, actual IRI values are shown, and the top right shows the degeneration rate of the segments. In the bottom the four worst segments are displayed with their trend over 2020.

4.3 *Alerts – Highlight*

Alerts can be triggered from any generic source, the current ones are triggered by; segments exceeding threshold value, segments with high degeneration, or severe potholes where severe can be either high hit-ratio or high vehicle impact. The major strength with the Alerts function is that this can be easily incorporated with existing ways of working, without extensive educational efforts.

As the alerts are triggered, the user can quickly enter the map dashboard and automatically find the areas of interest that need immediate attention, allowing much quicker response time for correcting anomalies in the road network.

5 CONCLUSIONS

Digitalization and probe vehicle data is a game-changer. Roads can be maintained in a much more efficient way, making them safer, while lowering cost and reducing environmental impact. Probe vehicle data enables new ways of working, and objective, automatic measurements is the foundation.

Implementation of road surface monitoring in the daily operations allows road owners and operators to implement a pavement preservation program. It postpones the need for significant resurfacing or reconstructing sub-grade road layer by performing initial maintenance and preservation on road surfaces while they are still in good condition.

Monitoring road state trends and current road state allows resurfacing of roads in a timely fashion using trend analysis, and alerts functionality makes it possible to maintain an aggressive pothole patching program. Everything done without the need for manual inspections or expensive special equipment.

The NIRA roughness values and the IRI values based on laser measurements are very similar, but not identical. However, the great strength using probe vehicle data is the unprecedented coverage of the road network. This means that tracking the road conditions over time and ensuring correct trend analysis, is now possible. Allowing operators to set their

166

own warning thresholds makes it possible to tailor each individual tool for a specific region, allowing the right road maintenance performed at the right time.

With regards to detecting Slippery When Wet scenarios, the foundation is already there. Detection of slippery conditions and wetness state of the roads is monitored. In theory it works, but next step, the service will be tested on a full scale in production, to be able to set proper detection thresholds to bring value to road operators and safety to drivers, but also allow the possibility to transform road operations. It allows a step-by-step approach towards implementing a fully data-driven road maintenance.

REFERENCES

Highways England, 2020. CS230 *Pavement Maintenance Assessment Procedure*, https://www.standardsfor-highways.co.uk/dmrb/search/df44f6b0-4bf4-4189-90a2-912c19cc247a

Mercedes-Benz *Predictive Infrastructure Monitoring and Evaluating Infrastructure with High Quality Data From our Mercedes-Benz Customer Fleet*. https://data.mercedes-benz.com/products/predictive-infrastructure

Roadroid Product Sheet. https://www.roadroid.com/common/References/Roadroid%204%20pager.pdf

Road Surface Tester system of the Swedish National Road and Transport Research Institute. https://www.vti.se/en/services/highway-engineering-and-geotechnics/on-road-measurement/measurement-of-road-surface

SICK US Blog, 2020. *Automating Road Maintenance with LiDAR Technology*. https://sickusablog.com/automating-road-maintenance-lidar-technology/

Stryk J. et al. 2021. *Pavement Surface Characteristics - Differences in Measured Parameters and their Evaluation and use in Different Countries*, Presented at ERPUG 2021

TRIP, October 2018. Bumpy Road Ahead: *America's Roughest Rides and Strategies to Make our Roads Smoother*. https://tripnet.org/wp-content/uploads/2019/03/Urban_Roads_TRIP_Report_October_2018.pdf

Wang H. et al., 2020. *International Journal of Sustainable Transportation* Volume 14 - Issue 1 https://www.tandfonline.com/doi/abs/10.1080/15568318.2018.1519086?journalCode=ujst20

Yu Y. et al 2020. "Road Manhole Cover Delineation Using Mobile Laser Scanning Point Cloud Data," *In* IEEE Geoscience and Remote Sensing Letters, vol. 17, no. 1, pp. 152–156, Jan. 2020, doi:10.1109/LGRS.2019.2916156.

International Roughness Index. https://en.wikipedia.org/wiki/International_roughness_index

Zachrisson B. et al 2021. FCD Enabler for More Efficient WM Operations, S11–S12. *Presented at XVI World Winter Service and Road Resilience Congress 2022.*

Analysis and visualization for pavement condition assessment using network-level survey data

Nyunt Than Than*
Land Transport Authority, Serangoon, Singapore

Cheng Zhuoyuan
School of Science and Technology, Singapore University of Social Sciences, Singapore

Zulkati Anggraini, Koh Puay Ping & Chin Kian Keong
Land Transport Authority, Serangoon, Singapore

ABSTRACT: Road condition survey plays an important role in road pavement management and different types of survey are typically performed to monitor the conditions of road network for management. In this study, the data obtained from the network-level condition surveys using Sideway-Force Coefficient Routine Investigation Machine (SCRIM), Laser Crack Measurement System (LCMS), and Falling Weight Deflectometer (FWD) were analyzed, using different analytics tools to assist with maintenance decision-making. Using the data from the SCRIM survey, the map-based analyses were performed to evaluate the in-service skid resistance performance of the road sections paved with different asphalt mixes at different times. Using the data from the LCMS and FWD surveys, different parameters such as cracks, ruts and deflections were overlaid on the map to identify the hotspots where the undesirable values of these different parameters overlapped at the same road section. A dashboard has also been built to allow for a quick assessment of the hotspots in a lane-level accuracy. In addition, the deflection data from the FWD tests were compared with the outputs from the 3D numerical simulation to better understand the structural performance of the road network. In summary, this paper demonstrates and discusses the different analysis approaches using the network-level condition data to assess the project-level pavement performances.

Keywords: Road Network Condition Data, Map-based Analyses, Assessment of Pavement Performances.

1 BACKGROUND

Road inspection regime plays a critical role for management of road pavements and in Singapore, it comprises two parts: (i) visual inspections currently performed by human inspectors and (ii) specialised inspections using high-speed survey equipment. Visual inspections are performed more frequently (e.g. a weekly frequency for high-speed roads) so that defects that pose immediate safety concerns to road users can be promptly rectified. Specialised inspections are currently performed yearly for high-speed roads so that the network performance indicators, e.g. riding comfort in terms of international roughness index (IRI) and rut, can be monitored and used for planning of maintenance work. In this paper,

---

*Corresponding Author: than_than_nyunt@lta.gov.sg

DOI: 10.1201/9781003429258-17

the data obtained from the specialised surveys using the 3 equipment namely, (i) Laser Crack Measurement System (LCMS), (ii) Sideway-force Coefficient Routine Investigation Machine (SCRIM) and (iii) Falling Weight Deflectometer (FWD), were analysed to assess pavement surface and structural conditions to assist with maintenance decision-making. The paper covers the safety condition assessment performed using the data from the SCRIM survey for Singapore's expressways which have speed limits up to 90 km/h and the structural condition assessment performed using the data from the LCMS and FWD surveys for the selected stretches of expressways as well as selected roads in the industrial areas.

2 NETWORK REFERENCING SYSTEM AND DATA FORMATS

Similar to most pavement management systems, the road network condition data are collected in a standardised format so that data collected from various sources (e.g. different types of condition surveys, maintenance history and other related road pavement inventory info) can be integrated for pavement management purposes e.g. reporting, budgeting, prediction and etc. Figure 1 describes how the road network is standardised through a referencing system in which a road is represented by a centre line with a unique road code. The centre line is segmented into road sections, referencing to physical road features such as junctions and/or exit and entry of expressways. Road sections are also given a unique ID following one direction of the traffic flow. Road lanes are also labelled systematically by using different symbols, "L" and "R" to differentiate the traffic directions as well as the lane positions, e.g. CL1 indicating the fast lane travelling towards the traffic direction where the road section IDs are in the ascending order and CR1 indicating the fast lane travelling in the opposite direction of the road.

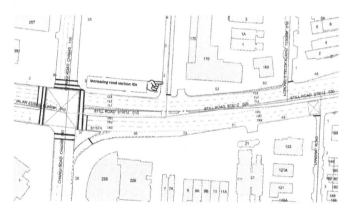

Figure 1.   Network referencing system.

The condition data from the LCSM and SCRIM surveys are available in every 10 m section of road length and that from the FWD survey are available in every 25 m. The LCMS and SCRIM surveys are typically performed for each lane of the roads except the FWD survey which is performed on the slow lanes of the selected roads. Besides the network referencing system, x and y coordinates are also available in a lane-level accuracy for every data point. The availability of geodata as well as the standardised network referencing system allows network data analysis to be more efficient and intuitive with the use of analytics tools, e.g. Tableau and QGIS.

169

3 SAFETY CONDITION ASSESSMENT

3.1 *Different types of asphalt mixes used in Singapore*

Skid resistance of road pavement is dependent on the 2 key parameters namely: micro and macro texture of asphalt mixes, in which the micro texture is related to the properties of aggregates usually expressed in terms of polishing stone values (PSV) and macro texture is related to the surface texture of asphalt mixes. The commonly used asphalt mixes in Singapore are a gap-graded asphalt mix for expressways and a dense-graded asphalt mix for arterial roads. An open-graded asphalt mix is also used for selected stretches of expressways for noise mitigation. Granite aggregates are commonly used for asphalt mixes and other types of aggregates such as calcine bauxite and steel slag aggregates are also used to improve skid resistance performance of road pavements e.g. slip roads and junction approaches.

3.2 *Data analysis and visualisation of skid resistance*

In 1999, the Land Transport Authority (Singapore), LTA, and the National University of Singapore, NUS, carried out a study to evaluate the skid resistance performance of different asphalt mixes used then. Grip Tester was used to collect the in-service skid resistance data of different asphalt mixes at different locations to investigate the skid resistance deterioration trends w.r.t the ages of asphalt mixes and traffic volumes. The study by Fwa, T. F and Tan, S. A (1999) postulated a typical deterioration trend as depicted in Figure 2, highlighting that the skid resistance decreases quickly at an initial phase and tends to level off after 40 months of service life, at a value close to the SCRIM coefficient (SC) of 0.26. Based on the data obtained from the previous studies, the initial skid resistance of asphalt mixes using different aggregates (e.g. granite or steel slag aggregates) was observed to be around 0.5 SC. In the past, Grip Tester was also used to obtain the skid resistance data and the grip numbers were then converted using the conversion equation (SC = 0.89 * GN) recommended in the TRL report (2009) for easy comparison.

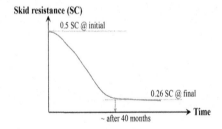

Figure 2.   Skid resistance deterioration trend reported in the Fwa T. F and Tan S. A (1999) study.

For management of road network, the skid resistance performance is monitored using the threshold limits namely, investigatory and intervention levels. The investigatory levels describe a range of skid resistance values at which agencies should start to perform necessary investigatory or detailed monitoring works, and the intervention levels describe a range of skid resistance values at which agencies must take corrective actions (e.g. maintenance works/restorative treatments and/or other improvement works) for safety of road users. Different road agencies adopt different threshold levels depending on the local practices, economic assessments, site conditions, accident statistics, etc. Highways England (2021) recommends the investigatory levels of 0.3 and 0.35 CSC, for low traffic motorway and heavy traffic motorway, respectively, in which CSC is the Characteristic SCRIM Coefficient computed from measured SCRIM coefficients corrected for the effect of seasonal variation.

Fwa (2017) presented the different approaches for setting of threshold levels. One of the methods suggested that the investigatory level can be set when the skid resistance starts to decrease significantly over time and the intervention level can be set at a certain skid resistance magnitude or percentage (e.g. 10 percent) below the investigatory level.

For network monitoring, a statistical analysis was performed in Tableau using datasets from 10 expressways and these data were obtained from the survey performed in February 2020. From the analysis, the first quartile, median and third quartile values were obtained as 0.30, 0.34 and 0.38 SC (Figure 3). Besides the data obtained in February 2020, similar analysis was performed for the historical data of the expressway network. It was observed that the skid resistance data distribution was similar to the one shown in Figure 3. Varela F., and Casero E.R., (2019) also discussed the important of performing statistical analyses for the network condition data before feeding them into the database of the management system.

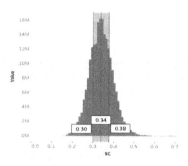

Figure 3. Skid resistance (SC) distribution of Singapore's expressways with the first quartile, the median and the third quartile values.

Based on the statistical analysis, the first quartile value of SC 0.30 (25% of the data are less than or equal to 0.30 SC) can be set as the investigatory level for the expressway network. The SC < 0.26 (the value less than the investigatory level of 0.30 SC, typically constitutes about 15% or less of the road network data, and the SC 0.26 is also the lower bound levelling value reported in the 1999 study) can also be set as the intervention level for the management of Singapore's expressways. For the dashboard visualisation, the following colour coded bins, 0-0.25 (red), 0.26-0.30 (pink), 0.31-0.40 (orange), 0.41-0.45 (blue) and >0.45 (green), are used for easy identification of pavement stretches with SC values in the investigatory and intervention levels. Using these colour bins, the network-level condition of individual expressways can be visualised as shown in Figure 4. The conditions of road pavement across different lanes can vary significantly due to different ages and traffic conditions and thus it is useful to present the data lane-by-lane for efficient planning of maintenance works. Figure 4 shows the lane-by-lane presentation of skid resistance data on the map using the same colour bins.

3.3 Case study on skid resistance performance of different asphalt mixes on expressways

In Singapore, asphalt mixes with high PSV aggregates (e.g. calcine bauxite) are used at the stretches of road pavement, where self-skidding accidents are evident. This is to improve the surface skid resistance performance and also to reduce self-skidding incidents. Steel slag aggregates have been reported to have good engineering properties e.g. PSV. They are also used for the dense-graded asphalt mix typically applied at high stress areas (e.g. junction

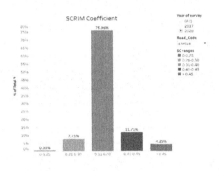

(a) The SC distribution in percent for the different ranges of values

(b) Map-based visualisation of the SC with colour coded bins for different lanes

Figure 4.   Different data visualisations for management of road network.

approaches) and for the open-graded asphalt mix at selected stretches of expressways for noise mitigation. Since the skid resistance is dependent on the micro and macro texture of surface materials, the effect of surface textures on the skid resistance performance of the 3 different asphalt mixes was evaluated. The 3 asphalt mixes are: (i) gap-graded asphalt mix with 50% calcine bauxite aggregates, denoted as M1, (ii) open-graded asphalt mix with 100% steel slag aggregates, denoted as M2 and (iii) gap-graded asphalt mix having the same gradation as M1 but with 100% granite aggregates, denoted as M3.

Table 1 summarises the locations of M1, M2 and M3 mixes with their respective ages and the average SC values. From this table, it can be observed that M1 mix showed the good and sustained SC values over a longer service life (~10 years). The average SC values of the 3 mixes at different locations are compared in Figure 5. The initial skid resistance values of M1 mix are not available, however it was assumed to be around 0.55 SC in this study. The hypothetical deterioration trend reported in the Fwa T. F and Tan S. A (1999) study was also added in Figure 5 for comparison. It was observed from this figure that the deterioration trends of M2 and M3 agreed well with the findings observed in the 1999 study. It was also noted that M1 showed a gentler deterioration trend compared with that of M1 and M2 mixes. Moreover, M1 mix has yet to reach the lower limit value of 0.26 SC even after 10 years of service life. For assessment of the scheme effectiveness, a 5-year before and after evaluation was performed at the 3 stretches paved with M1 mix using the accidents data.

Table 1.   Summary of asphalt mixes and their respective locations, ages and average values of SC as of February 2020.

	Location	(months)	Average SC
M1	Slip road from KJE into BKE towards PIE	119	0.32
	Slip road from BKE into SLE towards CTE	119	0.36
	Slip road from SLE into BKE towards Woodlands	100	0.43
M2	ECP bet Still Rd to Siglap Rd (both bounds)	30	0.36
	PIE Aljunied Flyover to Paya Lebar Flyover (both bounds)	24	0.34
	PIE bet Jln Eunos and Bedok North Rd (both bounds)	1	0.43
M3	AYE between Benoi Flyover and Tuas Flyover (towards Tuas)	30	0.31
	SLE,Ulu Sembawang Flyover and Marsiling Flyover (both bounds)	22	0.36
	TPE, Api Api Flyover and Tampines Flyover (towards SLE)	6	0.43

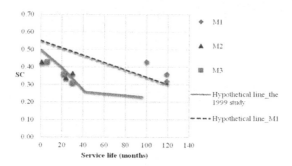

Figure 5.   Comparison of M1, M2 and M3 mixes w.r.t the hypothetical SC degradation trends.

Figure 6 summaries the numbers of before and after accidents at the locations and most of the accidents were self-skidding motorcycle accidents. This figure showed that M1 mix is relatively effective in reducing the number of self-skidding accidents over the years.

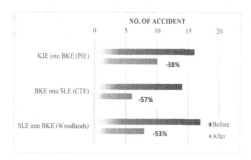

Figure 6.   Summary of road accidents, 5 years before and after periods with the reductions in percentage.

4   STRUCTURAL CONDITION ASSESSMENT

4.1   *Identification of hotspots using the LCMS and FWD data*

For structural condition assessments, the analyses were performed at the open-source GIS platform (QGIS) and Tableau using the data from the LCMS and FWD surveys. Table 2 summarises the various outputs from the 2 surveys. For the LCMS crack rating system, the approach presented in Tan J.Y. et al., (2021) was used and for the FWD deflection bowl parameters, the method described in the Horak E. and Emery S. (2009) study was referred. A statistical analysis was first performed to understand the data distribution at network-level and Figure 7 showed the typical data distributions of IRI and D_{max}. For identification of hotspots where pavements are likely to have structural issues, the parameters from the 2 surveys were overlaid on the map for assessment at the QGIS, following the steps shown in Figure 8.

The spatial analysis was performed using the structural, longitudinal, and transverse cracks defined in the 3 severity levels (low, medium and high), by ignoring the low severity cracking data. Similarly, the deflection bowl parameters (D_{max}, BLI, MLI, LLI except RoC) were analysed, by referring to the green, amber and red benchmarking values for a granular base pavement as a guide (Horak E. & Emery S. 2009). The analysis showed that the deflection bowl indices (BLI, MLI and LLI) in the green, amber and red ranges did not

Table 2. Summary of typical outputs from the LCMS and FWD surveys.

Equipment	Parameters		Unit	Remark
LCMS		IRI	mm/ m	Riding quality expressed in terms of International Roughness Index (IRI)
		Rut	mm	Depressions along wheel paths
	Cracking	Structural crack	% area	Fatigue cracks in both wheel paths, available in 3 severity levels, low, medium and high
		Longitudinal crack	% m	Cracks in longitudinal directions, available in 3 severity levels, low, medium and high
		Transverse crack	m	Cracks in transverse directions, available in 3 severity levels, low, medium and high
		Texture	mm	Macro texture expressed in terms of mean texture depth
FWD		D_{max} or D_1	μm	Maximum deflection
		RoC	μm	Radius of curvature $RoC = \frac{L^2}{2D_1}\left(1 - \frac{D_{300}}{D_1}\right)$, where, $L = 200$ mm for FWD
		BLI	μm	Base layer index, $BLI = D_1 - D_{300}$
		MLI	μm	Middle layer index, $MLI = D_{300} - D_{600}$
		LLI	μm	Lower later index calculated using, $LLI = D_{600} - D_{900}$

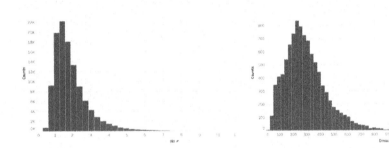

Figure 7. IRI and D_{max} distributions.

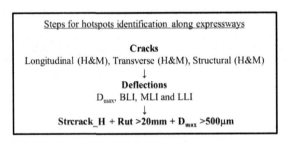

Figure 8. Data analysis flow used for identification of hotspots for structural assessments.

correspond well with the respective benchmarking values of D_{max}. Therefore, further analysis was performed by correlating D_{max} to the deflection bowl parameters obtained from the measured datasets (Figure 9).

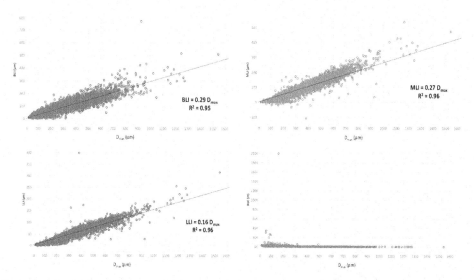

Figure 9.   Correlations of D_{max} with the deflection bowl parameters, BLI, MLI, LLI and RoC

Table 3 summarises the correlation equations with the respective R^2 values and it showed that the BLI, MLI and LLI correlated well with D_{max} in a linear manner, except RoC. It was also noted that the slope coefficients of BLI and MLI with D_{max} are very close. Thus it is unlikely to estimate the performance of asphalt base layer and aggregates middle layer

Table 3.   Relationships of D_{max} with deflection bowl parameters (BLI, MLI, LLI and RoC) and the benchmarking values from the measured datasets.

Correlation Equations	R^2	Benchmarking values		Calculated Benchmarking values			
		Range	D_{max} (µm)	Range	BLI (µm)	MLI (µm)	LLI (µm)
BLI = 0.29 D_{max}	0.95	Green	< 500% (<500)	Green	< 145 (<200)	< 135 (<100)	< 80 (<50)
MLI = 0.27 D_{max}	0.96	Amber	500-750 (500-750)	Amber	145-218 (200-400)	135-203 (100-200)	80-120 (50-100)
LLI = 0.16 D_{max}	0.96	Red	750-1000[#] (>750)	Red	218-290 (>400)	203-270 (>200)	120-160 (>100)

individually through these indices for Singapore's pavements. The ranges of D_{max} obtained from the measured datasets agreed well with that presented in the Horak E. and Emery S. (2009). However, the values of D_{max} in the amber and red bands are relatively narrow for the measured datasets and thus the respective calculated BLI, MLI and LLI in these bands become very narrow as well. As a result, the calculated values of BLI, MLI and LLI for the amber and red benchmarking bands have become ambiguous. Thus, for identification of hotspots to assess structural condition, only D_{max} was used as the critical parameter in this study. Figure 10 shows the map overlays for hotspots identification. From this, it was noted

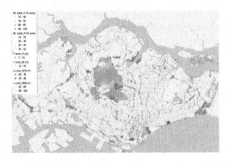

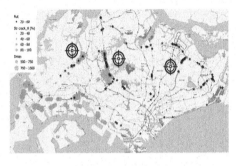

(a) Map overlay for different types of cracks in   (b) Map overlay for the 3 critical parameters,

different severity levels   Strcrack_H, Rut >20mm, D$_{max}$ >500µm

Figure 10.   Analysis for the LCMS and FWD surveys at the QGIS platform.

that the structural cracks or fatigue cracks along the wheel paths were more significant than the other types of cracks for Singapore's expressway network. Moreover, the hotspots for the structural assessment can be identified by using the following 3 critical limits/parameters, i.e. structural cracks in the high severity level (Strcrack_H), Rut >20 mm and D$_{max}$ > 500 µm.

As demonstrated, the map overlays of multiple parameters can be performed efficiently in the QGIS. However, it is difficult to pinpoint an exact location of the hotspot in a lane-level accuracy. This is because the critical parameters were filtered using numerical categories regardless of the lane position. For a more accurate positioning of hotspots, Tableau software was used and the same critical limits were used to build the dashboard (Figure 11). Using the x and y coordinates and the network referencing system, the hotspots can be identified in lane accuracy. Other useful condition parameters (e.g. IRI) at the same road section on the same lane can also be compared in the line graphs as shown. One key observation from this work is that data visualisation in Tableau is more interactive than the QGIS, however map overlays with a multiple filtering option is limited in Tableau. With these analytics tools and data in standardised formats, analysing network-level condition data for a project-level assessment becomes feasible for an effective pavement management.

% denotes a typical deflection for Singapore's pavement (mainly for expressways, major arterial/industrial roads) approximately 1 year after opening to traffic, # denotes maximum

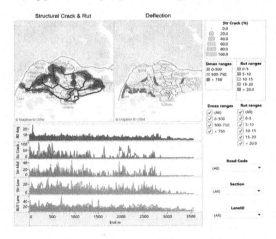

Figure 11.   Dashboard concept for pavement structural condition assessment at lane-level.

176

deflection obtained from the datasets, values in brackets are referred from Horak E. and Emery S. (2009) for granular base.

4.2 *Comparison of measured FWD deflections with the 3D numerical simulation*

A numerical modelling was performed in SAP2000 to better understand the pavement responses under a FWD test environment. The 3-layer rectangular model (2050×1020×600 mm) comprises the 170 mm asphalt layer that sits on the 550 mm aggregate layer with the 300 mm subgrade as a foundation layer. The layers are modelled as elastic materials and the bottom of the model is pinned in all directions. The model is excited at one end using a force pulse in the shape of a half-sine with a 20ms duration. The applied load is equivalent to 50kN in accordance to ASTM D4694-09 (2015). The end with excitation and the two longitudinal sides are restrained from the out-of-plane movements. The furthest end is restrained using a spring constant equivalent to a subgrade reaction modulus recommended in the FAA (2009).

The first run was performed using the elastic modulus values of 3 GPa, 0.3 GPa and 0.1 GPa, for Layer 1 (asphalt), Layer 2 (aggregates) and Layer 3 (subgrade), respectively. Poisson's ratio of 0.3 was used for all three layers. The elastic modulus values for the first run were estimated from the laboratory resilient modulus of standard dense asphalt mix used in Singapore and the minimum CBR requirements of aggregate layers and subgrade specified for construction. Figure 12 summaries the deflection outputs from the multiple runs using different ranges of material properties for the 3 layers. The numerical deflections were also compared with the typical upper and lower bound deflections obtained from the field data. It was noted from Figure 12 that the simulated deflections tend to converge beyond the fourth sensor regardless of the different Layer 1 and Layer 2 properties. In addition, the varying properties of Layer 1 showed a limited effect on the deflections.

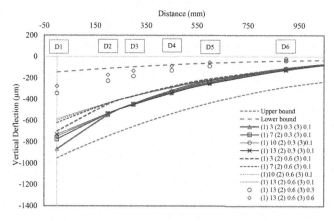

Figure 12.   Comparison of numerical deflections with the measured deflections.

To simulate the lower bound field deflection data, the Layer 3 modulus value of 0.1 GPa was increased to 0.3 GPa and 0.6 GPa for the 2 additional runs with the modulus values of 13 GPa and 0.6 GPa for Layer 1 and Layer 2, respectively. From these 2 runs, it was observed that the properties of subgrade have significant effects on the deflections at different sensor positions. This observation may suggest that the pavement structure with the layer thickness details modelled in this study can be stiffer/stronger ($D_{max} < 500$ mm) by having stiffer materials for Layer 2 and Layer 3, but not for Layer 1. It will also be interesting to explore the effect of layers' moduli for the pavements with different layers' thickness under the simulated FWD testing. This aspect of work can be studied under the future work.

5 SUMMARY

Network condition surveys provide a useful information to assess the safety, functional and structural condition of road pavements for maintenance planning. In this study, the data analysis and visualisation were performed using the QGIS and Tableau to draw useful insights from the large network datasets. The methodologies for assessment of safety and structural conditions using the 3 types of network condition survey are discussed. In addition, the numerical simulations were performed to better understand the pavement responses under the FWD testing.

In summary, from the safety condition assessment, it was observed that the asphalt mix with calcine bauxite aggregates provides a good and sustained skid resistance over the longer lifespan, compared with the other types of asphalt mixes with steel slag and granite aggregates. The calcine bauxite asphalt mix has shown to be one of the effective measures in reducing self-skidding incidents. From the structural condition assessment, the 3 critical limits (Strcrack_H, Rut >20 mm, D_{max} > 500 mm) can be used as a guide to identify the hotspots, where pavements may be subjected to structural related distresses. This can then be useful to locate the stretches of pavement that require further detailed investigation through an invasive field assessment and testing (e.g. coring and dynamic cone penetration test). In addition, the methodologies provide a quick assessment of the network-level condition data at the lane-level accuracy for a better pavement management. Lastly, the outputs from the numerical simulations assisted in making the convincing findings from the analysis of FWD field datasets.

ACKNOWLEDGEMENTS

The first author greatly appreciates the support given by the Management of Traffic and Road Operations Group, Land Transport Authority, Singapore, for a successful completion of the case studies presented in this paper.

REFERENCES

ASTM. 2015. *Standard Test Method for Deflections with a Falling-Weight-Type Impulse Load Device of ASTM D4694 - 09*, USA: ASTM International.
Federal Aviation Administration (2009), "*Airport Pavement Design and Evaluation.*" *Advisory Circular 150/5320-6E*, Office of Airport Safety and Standards, Washington DC.
Fwa T.F and Tan S.A (1999)., "Report Prepared for LTA-NUS Joint Research Project Titled "*Development of Performance Evaluation Procedure for Drainage Mix*".
Fwa T.F. (2017)., "Skid Resistance Determination for Pavement Management and Wet-weather Road Safety", *International Journal of Transportation Science and Technology* 6, 217–227.
Highways England (2021)., "UK Design Manual for Roads & Bridges (DMRB)", Highways England, Guildford, UK.
Tan J.Y. *et al.*, (2021)., "Automatic Pavement Crack Rating for Network-level Pavement Management System." *12th International Conference on Road and Airfield Pavement Technology (ICPT)*, Colombo, Sri Lanka.
TRL PPR497 (2009)., "Grip Tester Trial Including SCRIM Comparison", *Transport Research Laboratory*.
Varela F., and Casero E.R., (2019)., "Good Practices in the Analysis of Input Data for Asset Management: Control of Coherence", *The 26th World Road Congress*, paper no. 361.

Roads and Airports Pavement Surface Characteristics – Crispino & Toraldo (Eds)
© 2023 The Author(s), ISBN 978-1-032-55149-4

Simplified road pavement surface deterioration model in urban area

Monica Meocci* & Valentina Branzi
Department of Civil and Environmental Engineering, Università degli Studi di Firenze, Firenze, Italy

ABSTRACT: The Road Pavement Condition (RPC) represents one of the main indices that describe both the political stability and the economic level of a country. To date, maintaining the high efficiency of the road network is one of the greatest challenges for the Road Authorities (RAs).

Pavement Degradation Prediction Models (PDPMs) can be a valuable tool for assessing current and future RPC, helping RAs to keep roads in proper service condition. Currently, many PDPMs are already developed. However, these appear either too complicated or too specific and usable only within the boundary conditions on which were developed or only after proper recalibrations with local data. Furthermore, the constrained budgets and the consequent limited technical resources to collect data on the condition of pavements that RAs normally have available for urban roads, highlight their evident need to have specific and simple deterioration models for these roads based on a process that allows them to optimize the road pavement management as a function of the available budget and data.

This research aims to illustrate a simple PDPM which intends to support the RAS' making-decision tool to describe both the current road pavement condition and the road surface distresses evolution over time along with the urban road network.

The proposed PDMP was developed on the basis of data from a global index collected on 246 homogeneous road sections within the road network of the Municipality of Florence. The monitoring process for data collection was performed using a high-performance and low-cost methodology which is based on the vertical accelerations recorded by black boxes located inside the vehicles that routinely pass on the road network.

A simplified linear model was used to describe the distress trend of the road surface conditions. The results, evaluated by means of the coefficient of determination (R^2), show that the model predicts with good precision the pavement distress, thus supporting the usefulness of these tools in assisting the decisions of the RAs for the allocation of adequate and timely funds for the maintenance of a high efficiency also of the urban road network.

Keywords: Pavement management System, Deterioration Prediction Model, Vertical acceleration, Urban Road

1 INTRODUCTION

The road pavements condition (RPC) plays an essential and active role in the progress of cities and communities, being one of the key sectors that determine the socio-economic development of countries, as well as one of the main factors for estimating the welfare of people and the safety and comfort of all road users (Queiroz & Gautam 1992; Shtayat et al.

---

*Corresponding Author; monica.meocci@unifi.it

DOI: 10.1201/9781003429258-18

179

2022). Therefore, proper maintenance of road pavements is essential to preserve and improve social benefits.

Adopting effective road pavement maintenance programs is one of the current challenges for road authorities (RA) in the appropriate decision making, financing and management, to ensure that an acceptable level of efficiency is achieved for the entire road network. A poor or wrong maintenance strategy could cause a significant deterioration of this assets (Santero & Horvath 2009) or lead to a non-optimal use of the available budgets for emergency interventions that have proved, in the long term, to be less effective than preventive and corrective maintenance (Alberti et al. 2017).

The Pavement Management System (PMS) represents one of the most widespread sustainable tools to assist RAs in monitoring, planning, evaluating, managing, and implementing recommendations capable of efficiently maintaining the road network in a timely and economic manner, as well as to guarantee comfort and safety for all users. However, to be effective, it requires the availability of road pavement distress data, as well as the ability to constantly update them (Meocci et al. 2021). All of this can be too expensive for a local transport administration.

Many researchers have agreed that the best way to accurately monitor pavement condition performance is by the use of prediction models. Indeed, these models, having the ability to describe the minimum and maximum changes in road pavement performance using certain arbitrary or weighted values which vary within a certain range, play a crucial role in several aspects of an effective PMS (Elhadidy et al. 2021; Shtayat et al. 2022). Specifically, at the network level they make it possible to predict the pavement performance to plan maintenance and rehabilitation activities, while at the project level they are able to determine the most suitable maintenance and rehabilitation actions to be undertaken for a specific project, such as preventive maintenance, rehabilitation or reconstruction (Radwan *et al.* 2020).

In the literature there are several studies that have developed models to estimate the road pavement degradation, using different approaches and analytical methods for their development, such as statistical regression analyses based on historical field performance data (Dong et al. 2015; Elhadidy et al. 2021; Osorio et al. 2014; Pérez-Acebo et al. 2019, 2020; Piryonesi & El-Diraby 2020; Ziari *et al.* 2019), Markov chains and Monte Carlo simulations (García-Segura et al. 2020; Moreira et al. 2018; Osorio-Lird et al. 2018; Pérez-Acebo et al., 2018), machine learning algorithm (ML) models (Piryonesi *et al.* 2020; Wang et al. 2020; Ziari et al. 2016) and artificial neuronal networks (ANN) (Jalal et al. 2017; Mathew et al. 2008; Terzi 2007) and genetic algorithm (GN) (Fwa et al. 2000). A recent literature review (Shtayat et al. 2022), although it found that the performance models developed using ML algorithms and ANN modelling have accurate estimation results for pavement conditions while regression models showed high accuracy in detecting and classifying pavement damage, concluded that each model has specific features, strengths, and weaknesses and therefore selecting an appropriate prediction model is the first step towards a high-quality prediction performance system.

In general, the development of accurate Pavement Degradation Prediction Model (PDPM) mainly depends on the creation of an accurate database and the correct identification of variables that affect road pavement degradation. In fact, several variables (such as traffic load, environmental conditions, road functional classification, pavement age, Structural Number and layers thickness) could impact road pavement lifespan and, for this, they have been taken into consideration in several prediction models proposed. Therefore, different components of these variables could lead to specific and different decay trends depending on the observed situation. Obviously, it is very difficult to implement an effective and, at the same time, comprehensive model that includes all the factors.

Currently, several countries have already developed PDPMs for specific road types and using different performance indices. However, most of these models are developed for assessing the pavement degradation of highway and/or suburban roads. Therefore, their application may not be representative for urban road pavements due to the differences that are present in terms of road traffic (e.g., short trip length, high number of diversion, traffic intensity is heavy and uneven-no uniform), pavement structure, cross-section and the influence of suffering on functionality (Llopis-Castellò *et al.* 2018; Loprencipe et al. 2017; Osorio *et al.* 2014).

Furthermore, since pavement performance indices can be considered time-dependent variables, they have often been used as main variables in the development of pavement performance prediction models. However, most of the current models are based on single pavement performance indices (e.g., IRI, cracking, rutting or uniformity) and not on global performance indices (e.g., PCI), which as shown in the literature review conducted by Llopis-Castellò *et al.* (2018) appear to be more effective and reliable for assessing pavement conditions than a single index, especially for urban roads. It is in fact easy to understand how, for example, the IRI index that is calculated considering a simulated car traveling at 80 km/h cannot adequately represent the pavement conditions for urban pavements due to the difficulty of its collection.

In conclusion, existing PDPMS appear either too complicated in trying to include as many factors as possible, becoming difficult to implement due to data limitations, or too specific as they are developed specifically for a data set and therefore, usable only within the boundary conditions on which have been developed or only after proper recalibrations with local data. Furthermore, for urban roads, the constrained budgets and the consequent limited technical resources to collect the data on pavement conditions that the RA usually have available for these particular road environments make the analysis of the deterioration process on urban pavements even more complicated. All this highlights that there is a clear need for RAs to have specific deterioration models for urban road environments for decision-making for the different situations they have to manage based on simple and fast approaches that allow them to predict the temporary deterioration of road pavements and to update it over time with a minimum availability of data and economic resources. In recent years, some studies have proposed simplified approaches to the pavement deterioration prediction based on a few years of observations, using different approaches and analytical methods (Abaza 2015; Mohammadi *et al.* 2019; Pérez-Acebo *et al.* 2019). The first results of these studies have shown that this type of approach is applicable and useful to local and national RAs for the decision-making process of budget allocation for a targeted road pavement maintenance. Furthermore, it emerged that among the analytical methods used, RAs may prefer deterministic models, because they are easier to understand and to implement in the decision-making process (Mohammadi *et al.* 2019). This type of easy and fast approach, however, seems to be still unexplored for the development of PDPMs of the Italian urban roads. This research opportunity motivated our present study.

This paper aims to illustrate a simple and easy PDMP based on the fast and inexpensive monitoring process. This will make RAs independent and able to develop and update their specific deterioration curves, improving decision-making and budget distribution with minimum data availability and lower costs. Through the monitoring conducted in the 246 homogenous road sections within the road network of the Municipality of Florence, the approach used for the development of the prediction model was illustrated. It is based on the data of the Global Pavement Box (GPB) index collected in pavements characterized by the structural number, the traffic demand and the pavement age (measured in terms of the last overlay intervention). The data relating to the road surface condition were collected by the pave box methodology, a high-performance, low-cost methodologies and an immediately operational distress detection approach based on the exploitation of data collected by the black boxes located inside the vehicles that routinely pass on the road network.

2 DATA DESCRIPTION AND RESEARCH METHOD

2.1 *Overview of the research method*

The research method is summarized in Figure 1. The analysis was based on two datasets describing the pavement characteristics (in orange) and the traffic demand (in blue). Within the pavement datasets, information on pavement distresses and age was included. Additional information on the pavement structure was only used to select the road sections analysed, characterized by the same (or similar) layers thicknesses. Pavement distresses data are based on monitoring activities. Finally, traffic demand was evaluated by surveys along with the road network, during different peak hours during the week (working days).

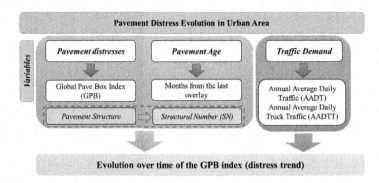

Figure 1. Research method overview.

The evolution over time of the Global Pave Box (GPB) index is conducted on different road segments characterized by different ages and different traffic demands. The distress trend is estimated using a regression procedure.

2.2 *Road section*

The road network of the Florence Municipality extends for over 1,000 km. The tested roads were selected from the municipality street classified as "local street – F type" in urban area), characterized by the same (or similar) pavement structure but with different ages. In terms of "age", the independent variable considered was represented by the number of months that have passed since the last overlay intervention performed.

To define a set of homogeneous road sections, only the carriageway consisting of one-lane one-way was considered. This is to describe the maximum distress evolution, due to the limited standard deviation of traffic wander.

Each road segment was characterized by:

- different length;
- absence of road intersection;
- same or similar cross section (section width).

The entire test roads selected were divided into 246 homogeneous segments characterized by different levels of distress. Each segment starts and ends at an intersection.

2.3 *Distresses evaluation*

The distress evaluation procedure was conducted according to the proposal described in (Meocci et al. 2021; Meocci & Branzi 2022), which consists of an intensive (and distributed) vibration-based monitoring campaign conducted along with the road urban road network of the Florence municipality.

Data (vertical acceleration) were recorded through a common Black-Boxes, with no specific performances, installed on a Taxi (Toyota Prius). Prior to the monitoring process, each tire was checked and adjusted to the recommended tire pressure for the car used. According to the pavement condition procedure proposed by the Pave Box methodology (Meocci et al. 2021), the data resulting from the monitoring performed were used for the evaluation of the Global Pave Box Index. The index offers a comprehensive damage assessment of the entire road segment considered. In (1) the equation used for the GPB index evaluation was shown.

$$GPB = \left(\int_0^L \widehat{\sigma(l)}^2 \, dl \right) \times \frac{100}{L} \tag{1}$$

where $\widehat{\sigma(l)}^2$ represents the moving variance of the signal measured by a black box (a_v) and L represents the total length of the homogeneous road segment. The moving variance $\widehat{\sigma^2}$ is given by equation (2).

$$\widehat{\sigma^2} = \frac{\sum_{i=1}^{n} \left(a_{v,i} - \overline{a_v} \right)^2}{n} \tag{2}$$

where $a_{v,i}$ represents the i-th vertical acceleration value and $\overline{a_v}$ represents the centred moving average across a range of twenty (20) values (0.2 s).

As a function of the GPB values obtained, the road section can be classified into different classes, characterizing the different distress severity: low, fair and high. These classes were then used to prioritize the maintenance intervention. When the GPB value was greater than 1.45, another class was defined that describes the need for an immediate repair.

The threshold values were based on a PCI survey conducted along with the road network in previous research. As described in Table 1, PCI and GPB correspondences are detailed in Meocci et al. (2021).

Table 1. GPB index rating scale.

GPB rating scale	GPB<0.40	0.40<GPB<0.65	0.65<GPB<1.45	GPB>1.45
Severity levels	good	fair	high	need to repair
PCI	100-70	70-55	55-25	≤25

Each lane was monitored one or more times at speeds between 40 km/h and 50 km/h. The car's transversal position was assumed in the middle of the lane (as far as possible).

2.4 Road pavement structure characterization

The road pavement structure, as previously mentioned, was considered only as a relevant parameter in the road section selection process. This parameter, although considered among the main ones for determining the distress evolution, it has not been considered in the distress trend.

2.5 Road pavement age

The road pavement age represents a key factor affecting the road surface quality and performance. The information used to develop the model refer to the number of months that have passed since the last overlay intervention. It is important to highlight that the dataset used doesn't contain any additional information about the pavement construction time and on the material the characteristics/performances used for construction and/or restoration.

Table 2 summarizes the available data concerning the "pavement age" for the 246 road sections considered.

Table 2.   Road pavement age section by section.

Number of sections	17	101	32	46	50
Months since the last overlay	20	30	40	50	80

2.6   *Traffic demand*

One of the most important variables that can affect road surface pavement deterioration is traffic demand. Two variables of traffic demand were considered in the present research:

- annual Average Daily Traffic (AADT) in vehicles per day;
- heavy vehicle (HGV) in percentage (%).

The variables' values were provided by in-field traffic analysis conducted in the morning and evening peak hours, over several working days. The AADT was evaluated considering a traditional traffic distribution during the day. In Table 3, four classes of traffic demand in terms of AADT have been defined. The mean heavy vehicle demand was shown as percentage of the total traffic demand based on the traffic analysis performed. Table 3 also classified the different section in terms of traffic demand.

Table 3.   Traffic demand classes.

Classes	AADT (veh/day)	HGV (%)	# of section included
1	<2500	<1	96
2	2500-5000	1-2	0
3	5000-7500	1-2	32
4	>7500	>2	118

Traffic demand was higher in the sections characterized by the lower severity level. On the other hand, the traffic demand on the sections characterized by higher distress levels is extremely lower. This is because the pavement age affects the severity level more than the traffic demand value, especially when a lot of time has passed since the last maintenance intervention.

2.7   *Data analysis*

Data collected during the monitoring process is shown in Figure 2. Data depicted represent the GPB index relating to each road segment analysed and they have been plotted as a function of the months that have passed since the last maintenance (overlay) intervention. No section had either by the same (or similar structure) and 60 and/or 70 months since the last overlay was found in the municipality database.

Each data group is statistically analysed in terms of mean value and standard deviation. Then, an outliers' detection procedure was conducted to understand whether there were outliers on the data set with reference to the first and third inner quartiles. The results

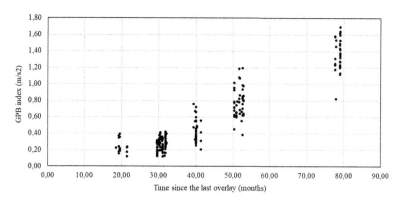

Figure 2. GPB data measured.

obtained are summarized in Table 4. Only one value was considered outlier according to the process and then deleted from the database to obtain the best fitting of the model.

Table 4. Statistical results for the outliers' analysis.

ID	Months since the last overlay intervention	Mean value (standard deviation)	Lower inner fence	Upper inner fence	Outliers presence
#1	20	0.23 (0.08)	0.11	4.59	no
#2	30	0.27 (0.07)	0.87	4.58	no
#3	40	0.44 (0.14)	-0.07	8.73	no
#4	50	0.77 (0.17)	2.95	11.94	no
#5	80	1.37 (0.20)	7.66	19.85	yes (1)

For each group of observations (characterized by the same time since the last maintenance intervention), data are represented in the following graph (Figure 3) and the percentage of the GPB index classified in the different distress levels was estimated.

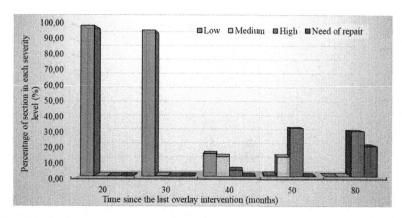

Figure 3. GPB distribution in the different severity classes.

185

The graph represented in Figure 3 allows to observe that for limited periods since the last maintenance, 100% of the values were classified in the low severity level. As the time elapsed since the last intervention increases, the percentage drops to zero. At the same time, the percentage of sections included in the high severity level increases (e.g., after 80 months from the last intervention, the percentage of sections classified in the "need of repair" class was about 20%). The percentage of the sections classified in medium and high severity levels may vary as a function of the overall road pavement condition monitored.

2.8 *Distress trend model fitting*

A simplified linear model was used to describe distress trend of the road surface conditions.

The proposed function allows describing the GPB trend evolution as a function of the pavement age. The 245 homogeneous sections considered for the model fitting were also selected according to the pavement structure and traffic demand.

The proposed linear equation is shown in (3).

$$GPB = \mathrm{A}age + B \tag{3}$$

Where the independent variable "age" represents the time (months) elapsed since the last overlay conducted by the RA on the section considered.

The goodness of fit was evaluated by the R-squared coefficient of determination R^2, which represents the traditional coefficient usually adopted to estimate the percentage of variation explained by the regression model. According to this criterion, the closer the coefficient R^2 is to one, the better the model explains the adopted dataset.

3   RESULTS AND DISCUSSION

Figure 4 shows the data trend obtained. The black data represent the GPB index evaluated for each of the 245 monitored sections (1 section was deleted as an outlier). The average values, about the maintenance period, are instead shown in blue. The different colours of the graph represent the different distress severity levels that characterized the values of the GPB indices evaluated by the monitoring process conducted.

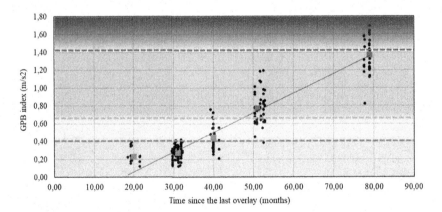

Figure 4.   Linear distress trend.

The equation obtained from the regression is shown in (4).

$$GPB = 0.221\,age - 0.3841 \qquad (4)$$

The function obtained was characterized by an R-squared coefficient of determination R^2 equal to 0.89, demonstrating a good ability of the model to describe the dataset.

The model shows a tendency to under-predict the value of the GPB index for those sections characterized by the shorter time since the last maintenance. The phenomena can be considered acceptable as these sections have just been maintained and did not require short-term maintenance intervention. However, for the short-term distress trend, the best regression model was represented by a power function, which instead did not fit the GPB index value when they had been characterized for a long time since the last overlay maintenance. For this reason, the authors preferred greater accuracy in estimating the GPB value when the pavement needs (with greater probability) to be repaired.

4 CONCLUSION

This paper describes the distress trend developed through a low-cost, reliable, user-friendly and crowdsourcing monitoring procedure.

To provide Italian municipalities and urban road authorities with a tool that allows them to deal with potential problems of the road surface and manage the road network maintenance plan, identifying the sites with the greatest need for maintenance.

The simplified model proposed was developed using a dataset containing information on 246 sections within the road network of the municipality of Florence monitored in the last period and characterized by different traffic demands and different ages since the last overlay intervention. The pavement structure, included in the database, was used to select the homogeneous section to develop the provided distress model. From these data, the model allowed to estimate the Global Pave Box index along with the urban road network as a function of the number of months that have passed since the last maintenance intervention. In these terms a linear function has been provided.

The goodness of fit of the model was studied by evaluating the coefficient of determination R^2, which was found to be equal to 0.89, demonstrating a good ability of the model to describe the analysed dataset.

Future research will focus on improving of the model based on the introduction of different variables that could affect the phenomena evaluated (e.g., specific traffic demand and pavement structure).

The reliable results obtained promoted the development and future evaluation of a tool that merges the importance of quickly detecting the road pavement distresses and the development of a pavement management system based on the specific index and integrated with distress trends to allow the RAs to optimize the type of intervention, times and maintenance costs within the urban road network.

REFERENCES

Alberti A., Crispino M., Giustozzi F. and Toraldo E., 2017. Deterioration Trends of Asphalt Pavement Friction and Roughness from Medium-term Surveys on Major Italian Roads, *Int. J. Pavement Res. Technol.*

Chan W.T., Fwa T.F. and Hoque K. Z., 2001.Constraint Handling Methods in *pavement maintenance programming. Transportation Research Part C: Emerging Technologies*, vol. 9, no. 3, pp. 175–190.

Dong Q.; Huang B. and Richards S.H., 2015. Calibration and Application of Treatment Performance Models in a Pavement Management System in Tennessee. *J. Transp. Eng.*, 141, 04014076.

Elhadidy A.A., El-Badawy S.M. and Elbeltagi E.E., 2021. A Simplified Pavement Condition Index Regression Model for Pavement Evaluation. *International Journal of Pavement Engineering*, vol. 22, no. 5, pp. 643–652.

Elhadidy A.A., El-Badawy S.M. and Elbeltagi E. E, 2021. A Simplified Pavement Condition Index Regression Model for Pavement Evaluation. *International Journal of Pavement Engineering*, vol. 22, n. 5, pp. 643–652, 2021.

Fwa T.F., Chan W.T. and Hoque K.Z., 2000. Multi-objective Optimization for Pavement Maintenance Programming," *Journal of Transportation Engineering*, vol. 126, no. 5, pp. 367–374.

García-Segura T.; Montalbán-Domingo L.; Llopis-Castelló D.; Lepech M.D.; Sanz M.A. and Pellicer E., 2020. Incorporating Pavement Deterioration Uncertainty into Pavement Management Optimization. *Int. J. Pavement Eng.*, 1–12.

Hassan R.; Lin O. and Thananjeyan A., 2017. A Comparison Between Three Approaches for Modelling Deterioration of Five Pavement Surfaces. *Int. J. Pavement Eng.*, 8, 26–35.

Jalal, M., Floris, I., and Quadrifoglio, L., 2017.*Computer-aided Prediction of Pavement Condition index (PCI) using ANN*," In Proceedings of the International Conference on Computers and Industrial Engineering, CIE, Portugal, 2017.

Lethanh N., Kaito K. and Kobayashi K., 2014. Infrastructure Deterioration Prediction with a Poisson Hidden Markov Model on Time Series Data *J. Infrastruct. Syst.* 21 (3).

Loprencipe G., Pantuso A. and Di Mascio P., 2017. Sustainable Pavement Management System in Urban Areas Considering the Vehicle Operating Costs. *Sustainability*, vol. 9, p. 453.

Mathew B.S., Reshmy D.S. and Isaac K.P., 2008.Performance Modeling of Rural Road Pavements Using Artificial Neural Network, *Indian Highways*," IRC, vol. 36, no. 1, pp. 31–39.

Meocci M., Branzi V. and Sangiovanni A., 2021. An Innovative Approach for High-performance Road Pavement Monitoring using Black Box. *Journal of Civil Structural Health Monitoring*, 11, 485–506.

Meocci M. and Branzi V., 2022. *Black Boxes Data for Road Pavement Condition Monitoring*: A Case Study in Florence, BCRRA proceedings.

Moreira A.V.; Tinoco J.; Oliveira J.R.M. and Santos A., 2018. An Application of Markov Chains to Predict the Evolution of Performance Indicators Based on Pavement Historical Data. *Int. J. Pavement Eng.*, 19, 937–948.

Osorio A., Chamorro A., Tighe S. and Videla C., 2014. Calibration and Validation of Condition Indicator for Managing Urban Pavement Networks. *Transp. Res. Rec.*, 2455, 28–36.

Osorio-Lird A.; Chamorro A.; Videla C.; Tighe S.; Torres-Machi C., 2018. Application of Markov Chains and Monte Carlo Simulations for Developing Pavement Performance Models for Urban Network Management. *Struct. Infrastruct. Eng.*, 14, 1169–1181.

Pérez-Acebo H.; Bejan S.; Gonzalo-Orden H., 2018. Transition Probability Matrices for Flexible Pavement Deterioration Models with Half-year Cycle Time. *Int. J. Civ. Eng.*, 16, 1045–1056.

Pérez-Acebo H.; Gonzalo-Orden H.; Findley D.J. and Rojí E., 2020. A skid resistance prediction model for an entire road network. *Constr. Build. Mater*, 262, 120041.

Pérez-Acebo H.; Gonzalo-Orden H. and Rojí E., 2019. Skid Resistance Prediction for New Two-lane Roads. *Proc. Inst. Civ. Eng. Transp.*, 172, 264–273.

Piryonesi S.M. and El-Diraby T.E., 2020. Data Analytics in Asset Management: Cost-effective Prediction of the Pavement Condition Index. *Journal of Infrastructure Systems*, vol. 26, no. 1.

Queiroz C.A. and Gautam S., 1992. Road Infrastructure and Economic Development*: Some Diagnostic Indicators; Policy Research Working Paper, 921*; World Bank: Washington, DC, USA

Santero N.J. and Horvath A., 2009. Global Warming Potential of Pavements. *Environ. Res. Lett.* 4, 034011.

Shtayat A., Moridpour S., Best B. and Rumi S., 2022. An Overview of Pavement Degradation Prediction Models". *Journal of Advanced Transportation*. https://doi.org/10.1155/2022/7783588.

Suman S.K. and Sinha S., 2018.Pavement Surface Condition Prediction by Markov Chains," *Highway Research Journal*, vol. 9, no. 1.

Terzi S., 2007. Modeling the Pavement Serviceability Ratio of Flexible Highway Pavements by Artificial Neural Networks. *Construction and Building Materials*, vol. 21, no. 3, pp. 590– 593.

Wang X., Zhao J., Li Q., Fang N., Wang P., Ding L. and Li S., 2020. A Hybrid Model for Prediction in Asphalt Pavement Performance based on Support Vector Machine and Grey Relation Analysis. *Journal of Advanced Transportation*, vol. 2020. https://doi.org/10.1155/2020/7534970

Ziari H., Sobhani J., Ayoubinejad J. and Hartmann T., 2016.Prediction of IRI in Short and Long Terms for Flexible Pavements: ANN and GMDH Methods," *International Journal of Pavement Engineering*, vol. 17, no. 9, pp. 776–788.

Ziari H.; Maghrebi M.; Ayoubinejad J. and Waller T., 2016. Prediction of Pavement Performance: Application of Support Vector Regression with Di_erent Kernels. *Transp. Res. Rec.*, 2589, 135–145.

Roads and Airports Pavement Surface Characteristics – Crispino & Toraldo (Eds)
© 2023 The Author(s), ISBN 978-1-032-55149-4

Road asset management with a mobile game and artificial intelligence

Markku Knuuti
AFRY, Helsinki, Finland

Konsta Sirvio*
SirWay, Helsinki, Finland

Vesa Männistö & Susanna Suomela
Finnish Transport Infrastructure Agency, Helsinki, Finland

ABSTRACT: July 2021, Finnish Transport Infrastructure Agency conducted a pilot project to produce data on the pavement condition of the pedestrian and cycling routes in the region of Pirkanmaa (421 km), through crowdsourcing video-data collection with a mobile game application to ordinary road users. Focus on this research project was to find new fast and affordable ways for pavement defect data collection. Crowdsourcing allows safe, fast and cost-effective video-data collection of the network. Although, the network was very fragmented and scattered around the region of over 15,000 km2, 90% of the network length was collected just over the weekend. When surveys were done in a single town, with 500 km of network, data have been collected in only 2 hours. A mobile game was used to collect GPS-tagged video data. In the game, virtual objects worth money, were placed on the routes. Using bicycles, people in the region collected these by recording videos in the application and were paid money for each collected object in the end of the day. Objects disappear in real time when somebody has driven over it, so routes will not be surveyed more than once. Videos which don't meet quality standards will be rejected and returned to the application map for somebody else to collect. Generally, feedback has been good from the participants. It was seen as entertaining game and participating event for the local citizens. Using these videos, Artificial Intelligence was then used to detect altogether 15 different types of pavement defects: Cracking (Alligator, Minor Longitudinal, Moderate Longitudinal, Severe Longitudinal, Moderate Transverse, Severe Transverse, Wheel Track), Potholes (Minor, Moderate, Severe), Fretting (Severe, Moderate), Settlements, Edge deterioration and Bleeding. The method uses machine learning application for automatic pavement defect detection. Data was saved in to Road Asset Management System to plan periodic and routine maintenance works and the needed budget for the following years. Instant data makes it possible to react promptly to the defects that are dangerous for the road users. Repeatable surveys, makes it possible to create deterioration models for predicting the future condition. This concept elevates the data collection and analysis in totally new level in terms of efficiency and reliability. If in Friday afternoon, you don't know in which condition your road asset is, on Monday you may have maintenance plans on your desk. Service was provided in consortium of consulting agency (AFRY-Finland), game platform (Crowdchupa), company (Vaisala) and provider of for Road Asset Management System (SirWay).

---

*Corresponding Author: konsta.sirvio@sirway.fi

DOI: 10.1201/9781003429258-19

Keywords: Crowdsourcing, Artificial Intelligence, Road Asset Management System, Maintenance planning, Bicycle routes

1 INTRODUCTION

The tools and methods to manage bicycle paths are often copied from the road pavement management sphere and the specific needs are not always fully addressed (El Said 2018). However, the importance of cycling is growing, and the length of the bicycle paths is already substantial to deserve dedicated measures to improve their management.

In 2021 FTIA (Finnish Transport Infrastructure Agency) launched a road asset management developing project in which the agency is adjusting the development to meet international standard ISO 55 000 framework and adapting it to agency's processes. The goal to asset management development is to manage the state-owned roads in a life-cycle economic manner, that is to ensure that the value of the assets is maintained, that funding is used efficiently, and that the performance of the assets matches the need for service levels.

An important part to be able to manage the assets is data. It should be known what type of assets exist, where they are located, in what condition they are and how their condition will develop. The data should be extensive, reliable and up to date. In road asset management development project, each asset type is considered one at a time to check the accuracy of the data, and to create condition and condition forecasting models for the strategic management.

Road asset management does not only include maintenance of road pavements, but also other assets such as pedestrian and bicycle paths to ensure the safety of the road network and day-to-day operation. The total length of pedestrians and cycling paths managed by the Finnish Transport Infrastructure Agency in 2021 was 6 062 km, of which 732 km in 2020 were in poor condition. An inventory of maintenance needs is currently carried out using an annual cycle of about 1/3 of the length of the network. Maintenance needs are assessed by visual examination of five different variables.

The aim of the study was to reflect on a new, more precise way of collecting condition information on pedestrians and cycling paths, and to establish a condition index for pedestrians and cycling paths serving maintenance purposes, from the point of view of both the road owner and the user. The project tried a new crowd-sourced data collection developed by Crowdchupa, a computer vision system and a road asset management system applied to bicycle paths provided by SirWay.

2 DATA COLLECTION AND PROCESSING

2.1 *Current methodology*

Currently, the damage inventory is done visually by driving 5-15 km/h along pedestrian and cycling routes. The inventory is carried out on a vehicle by visually assessing the severity of the different types of damage to the pavements of pedestrian and cycle paths on the basis of which either the need for repair or no need for repair is recorded for the period. The inventory is performed by a crew of two with one driving the vehicle and the other storing the periods in need of repair on the data acquisition device. The inventory period starts on the 1st of June and ends on the 30th of August.

There are criteria for different types of pavement damage to determine the need for maintenance for the section. Otherwise, the section is marked as "no repair". If a section has several different types of pavement damage, the need for repair is recorded for the section if one or more types of damage exceed the criteria for the need for repair.

In the pavement damage mapping of pedestrian and cycling lanes, only pavement damage in need of repair is inventoried, so minor damage is not inventoried as pavement damage. The results are stored in the Finnish Transport Agency's information system in 100-metre segments. Each segment contains information on the amount of repair needed (% or metres). If the result is zero, the target is perfectly fine. In other words, after corrective action, the results are reset. The data only tells you how much repair is needed, but does not include information about the different types of damage because it is not stored separately during inventory. The information system does not currently have a deterioration model for predicting the condition of pedestrian and cycling routes, unlike, for example, road rutting data. The current problems include having subjective opinions and not information on the actual condition that might not correctly drive the maintenance planning. The survey is slow and expensive and thus they are not conducted often.

Images of pedestrian and cycling routes are currently out of date and not comprehensive. This is a clear shortcoming, especially as the condition of the fairways can deteriorate rapidly during the year, for example due to damage caused by winter maintenance equipment. Mild and rainy winters caused by climate change also contribute to the rate of damage, due to the increasing number of freeze-thaw cycles.

The selection of maintenance sites is based in part on the collected condition data, but on the other hand, the prioritization favours routes known to be busy, although the network does not perform systematic traffic volume counting. The unity of the condition level in relation to the network managed by the municipalities is also taken into account. Thus, the selection of maintenance sites is not automated, but the selection of repair sites is started with pavement damage mapping data, but is completed manually.

In 2020, a total of 170 kilometres of pedestrian and cycle paths were rehabilitated, but in previous years the number has been lower, at around 100 kilometres per year. The number of poor roads has increased in recent years. In 2020, 732 kilometres were classified as in poor condition, which is about 12.1% of the total length of the network (Finnish Traffic Infrastructure Agency 2021). The aim is to invest more in pedestrian and cycling routes in the future as part of sustainable development.

There is currently a two-tier condition level requirement (K1, K2) that varies by region. If the need for refurbishment in category K1 is more than 50 metres, then that 100-metre segment is defined as in poor condition, otherwise the segment is in good condition. If there is a rehabilitation need in the K2 class of more than 65 metres, then the 100-metre segment in question is defined as in poor condition, otherwise the segment is defined as in good condition. Centres for Economic Development, Transport and the Environment (ELY) have been able to determine for themselves which routes belong to K1 and which to K2, and the policy is not uniform between ELY centres.

Typical maintenance work includes temporary manual and longer-lasting machine patching of potholes, edge damage and cracking. In new pavements, AB pavement is usually used, with a maximum grain size of 11 mm. In frost risk areas, fiberglass mesh is often used on the underside of the pavement. In new construction projects, structural lightening solutions are also possible on soft subsoil.

In many ELY centres, repairs are handled within maintenance contracts, but separate repair contracts are also used. With regard to resurfacing, there are paving contracts, which may also include structural improvements if necessary. Based on the data alone, a site plan is not made directly, but it also requires site visits and more detailed planning. In other words, the data only gives a general idea of where there might be a need for maintenance.

2.2 *Implementation of pilot measurements*

As part of the project, crowd-sourced data collection and artificial intelligence processing in the Pirkanmaa ELY centre was tested in the condition surveys of pedestrian and cycling routes with a network of about 412 km. The mapping was done by aggregating the collection of video data for road users using the Crowdchupa application. In the app, road users collect money-worthy objects by recording videos marked with GPS coordinates (Crowdchupa 2021).

The survey was marketed in a captivating manner, and prior to the surveys, Crowdchupa ran a marketing campaign to ensure that a sufficient number of people were encouraged to participate. The Transport Infrastructure Agency also supported the marketing campaign by sharing information about the capture on social media. The event was well publicized and covered in a total of more than 20 media outlets, including Finnish Broadcasting Company (YLE) and local media. Challenges to the mapping were caused by the geographical dispersion of the road sections to be mapped. Short sections and necessary transfer times were compensated by setting higher fees along them.

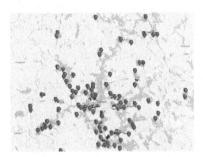 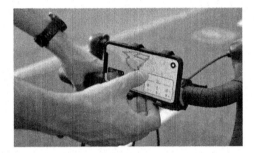

Figure 1.   Left: virtual collectible objects on the map. Right: Crowdchupa app on a phone attached to a bike.

There were 109 participants in Pirkanmaa. Approximately 82.5% of the target network was mapped during the first day, 90.0% was mapped after three days, and the entire network was mapped ten days after the event. The actual measurement length (video data) was 97.5% of the entire network. The survey was generally good and the quality of the data was mainly model, but there were also a few quality deficiencies.

The total length of the measured network was 412 km. Of this, 97.5% (401.2 km) was video-recorded, and usable data was obtained for approximately 95% (392 km) of the total network length. The missing sections are explained as follows:

1. Approximately 2% was due to individual 10-metre segments that did not get the measurement point. Such proportions are updated if the network is re-measured;
2. Approximately 2.5% was explained either by roadblocks in situations where the road was crossed and data bound to the wrong centreline, and by situations where the bike path was close on both sides of the road and objects from both bike paths could be collected by driving on one side only.

2.3   *Defect detection using Artificial Intelligence*

Collected videos were analysed in a service that enables the collection of videos combined with spatial data and the generation of automated pavement distress data based on video analysis. The technological solution consists of a data collection application running on a mobile phone, analysis of collected videos based on artificial intelligence and machine vision, and a browser-based interface where videos and generated data can be viewed.

Road videos collected by computer vision processing are used to identify and classify pavement damage and damage repairs, and to determine their area of influence at each point on the road. The information can be exported from the computer vision platform in a compatible format for use in an asset management system.

Road video can be collected while driving at normal driving speeds, and because the analysis with machine vision technology is fully automatic, the results of the analysis are available within a few hours of uploading the video data to the service. However, these factors make it possible to collect data from the entire network several times a year with lower costs and resources compared to traditional inventory methods.

Road condition data can be collected without special equipment. The data collector does not need to be a trained pavement expert when the machine vision produces the data systematically and objectively.

3 RESULTS

3.1 Survey results

The results were analysed using computer vision and 15 different types of defects were identified from the pavement. The model was created for car lanes, so many damage types do not serve to manage the condition of pedestrian and cycling lanes. A more detailed analysis of the measurement results showed at least the following problems in the data (Table 1).

Table 1. Problems and solutions of the data collection and processing.

Problem	Solution
Well covers are often interpreted as potholes.	The computer vision model has been improved in this respect. The data is rerun with an improved model. Misinterpretations are expected to decrease.
Tiled and cobblestoned segments interpreted as cracking or other defects.	The computer vision model was improved in this respect. The situation is different on cycling paths, where paving or tiling has been used. The aim is to develop a model so that the paved sections do not appear as damage.
At some points, the video image was out of focus, shaky, and the damage might not have been recorded in the best possible way.	The problem is solved by developing the app's auto-focus feature and thus improving the camera's focus on the road.
Centreline problems	Collecting objects are placed farther apart on parallel / both sides of the road) and adjacent lanes so that two lines cannot be collected in one run, leaving the other unmeasured. The problem of intersecting roads will remain, because at road crossings, maintenance belongs to the highway.
The computer vision model sometimes picks up damage from nearby driveways.	The collection application is installed as standard. a horizon line that unifies the angle of view of the data collection. This feature is already available in the video capture application. The possibility of automatically demarcating the area to be described and interpreted is being investigated. Improvements to the model are underway and will be completed shortly.

3.2 Control survey

Control measurements were performed on the results of the measurement, in which the results of computer vision were compared with a visual inventory, for a section of about 10 km. Five people (surveyors) made inventories from selected sites (Table 2) and the results were benchmarked.

Table 2.   Summary of the inventory results.

Road and section	Surveyor 1	Surveyor 2	Surveyor 3	Surveyor 4	Surveyor 5	People	Computer vision
70325_409	3.34	4.57	4.23	4.47	3.46	4.02	4.54
70325_461	2.17	2.89	2.40	3.05	2.61	2.63	2.47
73230_440	2.15	2.53	2.35	2.81	2.60	2.49	2.18
73230_491	2.76	3.56	3.17	3.88	2.50	3.17	3.89
70325_415	2.52	3.47	3.09	3.59	2.91	3.12	3.76
Mean	**3.02**	**4.08**	**3.74**	**4.09**	**3.19**	**3.62**	**4.08**

Control measurements were initially made for a general condition rating of 1-5. There were five human inventories and a computer vision model. The standard weighting of various condition variables was used in the computer vision model. Not all the people, either inadvertently or due to an inaccurate picture, fully inventoried all items. A total of 8,891 meters of comparable pedestrian and cycling routes remained. Table 2 summarizes the results of the different inventories. The results describe the average of the condition class of the different segments, where 5 is a segment in very good condition and 1 is a segment in very poor condition. The standard condition index calculation rule meant for roads gives a better result than the human eye. Differences in condition between different parts of the road become apparent both in the human eye and in computer vision. There were larger differences in inventory results between different people than between the mean of computer vision and human inventory results.

Table 3, in turn, shows the correlations between different inventories with 1-meter-modified condition data. The results show that the correlation between machine vision and human mean is reasonable, 0.69. On the other hand, the correlation between the results of different people and their mean is relatively high, between 0.81 and 0.94. The correlation between people ranges from 0.67 to 0.86. Based on the correlations, it can be said that the weighting of the different condition variables in the machine vision model does not fully correspond to the human eye's perception of the condition of pedestrian and cycling routes.

Table 3.   Correlations between the surveyors.

Surveyors	Surveyor 1	Surveyor 2	Surveyor 3	Surveyor 4	Surveyor 5	People	Computer Vision
Surveyor 1	1	**0.77**	0.82	0.79	0.67	0.89	0.60
Surveyor 2	0.77	1	0.85	0.86	0.69	0.93	0.65
Surveyor 3	0.82	0.85	1	0.83	0.69	0.94	0.66
Surveyor 4	0.79	0.86	0.83	1	0.68	0.93	0.66
Surveyor 5	0.67	0.69	0.69	0.68	1	0.81	0.52
People	0.89	0.93	0.94	0.93	0.81	1	**0.69**
Computer vision	0.60	0.65	0.66	0.66	0.52	0.69	1

3.3   *Road user needs*

The road user perceives the roughness and safety of the fairway as the most important factors, without necessarily recognizing the different types of damage from each other. The most important aspects of road safety are sudden discharges (frost damage, raised rocks etc.)

and depressions (ponding), wide longitudinal and transverse cracks, and moderate and large potholes. Ponding is not only an inconvenience factor for cyclists, but also a safety risk when a pond freezes. Potholes and wide longitudinal cracks are dangerous for the cyclist, especially if they are covered with snow or tree leaves and are not foreseeable by the cyclist.

In order to obtain the widest possible information from the road user's point of view regarding the level of service, a questionnaire was designed to find directly from users the main types of damage and other issues that affect usability of the fairway. Service level issues were divided into four categories; condition, safety, functionality and comfort as often used in the past studies (Arellana *et al.* 2020). There were six to fourteen questions in the categories, and each question had to be marked according to how important the topic was to the development of bike paths. Responses were given on a scale of 1 to 5, with 1 being "not relevant" and 5 "very important". 401 people responded to the survey. In addition to this, we had the opportunity to make free-form comments and we received a commendable number of these, almost 200. The most important of the condition variables were large potholes and wide longitudinal cracks. Nearly 85 percent of respondents felt that big potholes were a very important issue when it comes to the level of service on bike paths. Wide length cracks were considered very important by 247 respondents, well over half of all respondents.

All service-related service factors clearly have some relevance from the perspective of road users. "Not relevant" responses were highest for edge lesions, but even less than four percent of respondents. While a few of the damage emerges from the masses in importance, in reality all of these have an impact on the road user. No condition variables can be ignored when considering the maintenance of bike paths in the future.

3.4 *New condition index*

Both the common view of the experts and the results of the user survey have been used to weight the variables in the condition index. The emphasis of the condition variables is to highlight the most urgent sections requiring maintenance on pedestrian and cycling routes, both from the users' point of view and from the road owner's point of view. Figure 2 summarizes the responses from the survey on condition variables.

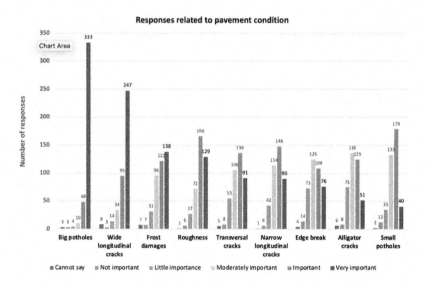

Figure 2.   Importance of the condition variable based on the cyclist survey.

Based on user survey and road maintenance experts, the weightings for the different condition variables were sketched. The scale is 1-5, where 5 represents the highest weight for that condition variable. Table 4 shows the results of the weightings used for maintenance planning.

Table 4.   Condition index weighing.

Condition variable	Weight	Upper limit
Big pothole	2.5	100
Medium pothole	2	100
Small pothole	1.5	100
Severe ravelling	0.5	20
Moderate ravelling	0.5	20
Left edge break	1.5	100
Right edge break	1.5	100
Alligator cracking	2	100
Severe longitudinal cracking	2.5	100
Moderate longitudinal cracking	1	30
Minor longitudinal cracking	1	30
Moderate transverse cracking	0.5	10
Severe transverse cracking	2	20
Area patching	0.5	20
Spot patching	1	50
Subsidence (ponding)	1.5	100

3.5   *Maintenance planning*

On the basis of the condition measurements, periodic and routine maintenance plans were drawn up for the area of Pirkanmaa ELY Centre for the year 2022. As a rule, periodic maintenance planning (paving) would be guided by a condition index weighted from the perspective of the road operator. In addition, treatment is guided by a weight index that is weighted from the perspective of the road user. The weightings of the condition index are based on a user survey of pavement damage and expert evaluation. The need and amount of maintenance can be assessed through three variables and alternative periodic maintenance plans were developed with SirWay's RAMS software (SirWay 2021) as in Table 5.

Table 5.   Condition index weighing.

Weighting of condition variables	Thresholds values for the condition index	Merging rules for segments
Road owner's perspective (periodic maintenance plan)	Condition index < 60% Condition index < 50% Condition index < 40%	Without merging rules Minimum length = 20 metres
Road user's perspective (routine maintenance plan)	Big potholes Medium potholes Severe longitudinal cracking Moderate longitudinal cracking	No merging rules

196

The resulted length of the periodic maintenance plan with different condition index threshold values with and without merging smaller segments is shown in Table 6 with the coloured line as the selected one.

Table 6. Effect of condition index on plan length (Unlimited budget, 392 km from measured data).

Condition index threshold	Without merging (% / km)		With 20 m merging (% / km)	
65	8.6	33.7	5	19.6
60	10	39.2	6	23.5
55	11.8	46.3	7.4	29.0
50	13.7	53.7	9	35.3
45	16.1	63.1	10.9	42.7
40	19	74.5	13.2	51.7
35	22.5	88.2	16.2	63.5
30	26.7	104.7	20.1	78.8

In the second phase, the paving program was formed by setting a higher threshold for the condition index and selecting only those objects that were at least 20 meters in length. The limit value of the condition index was chosen to be 60, giving a length of 13.2% (51.7 km) and a total cost estimate of EUR 2,065,535.

With these values, a limited-budget maintenance plan was created as a basic measure for balancing + fiberglass mesh + AB 11/120. The unit price used was last year's price (9.0 € / m2) plus a 15% cost increase and 10% additional costs due to wells, curbs, paving, etc. With a budget constraint of € 500,000, the length of the paving program was 12.5 kilometres (Figure 3). The sites were prioritized by taking the average of the segments in the unrestricted program for each road section separately and selecting the sections with the worst average condition.

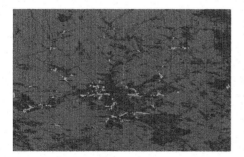 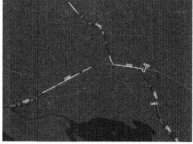

Figure 3. Left: periodic maintenance plan (unconstrained in blue and constrained in red). Right: Example of selected periodic maintenance marked as yellow and segments in poor condition as red, but not selected due to short length.

A treatment program requiring urgent patching was created by selecting from a network outside the constrained periodic maintenance plan segments that exceeded the limits in Table 5. Since the values given by the computer vision model used are the relative maximum percentages of damage to individual segments from the width of the road, the number of potholes and the length of the wide cracks were estimated. The width (diameter) of a single pothole was estimated to be 20 cm. The computer vision result was divided by one hundred, multiplied by the width of the pavement and the default width of the hole. The unit cost of patching the potholes is 30 euros. For wide cracks, the default was a maximum of 3 parallel on the same segment, with the crack being the entire length of the segment. Thus, the total length of the cracks in one segment is 30 meters. The unit price for the crack sealing is 3.5 euros per meter.

The resulted cost estimate is EUR 26,700 for patching potholes and a cost estimate of EUR 70,000 for sealing cracks. Outside the paving program, repairable 10-meter segments with severe damage are shown on the map with a red symbol (Figure 3).

Several discrepancies between visual and computer vision selection were identified and three examples are reported in Figure 4. At point 1, (Road 70325 / Road section 409 / Distance approx. 6640 m), the poor condition pavement changes to a slightly better junction, leaving only a moderate length crack at the edges, as well as occasional narrow cross-cracks. Based on the inventories, the need for maintenance has been identified, based on computer vision data, there is no need for periodic maintenance. The current merging rules as well as the inventory method would seem to take longer proportions to the pavement than necessary.

Figure 4. Comparison of selected segments for maintenance by computer and human vision.

At point 2 (Road 70230 / Road section 491 / Distance approx. 770 m), the site has a large number of narrow and moderate length cracks in some places and the inventory would justify going from a long distance to paving. Computer vision data would exclude portions

198

of the pavement that are considered satisfactory in terms of condition and for which, for example, a simple crack sealing will suffice.

At point 3 (Road 70325 / Road section 465 / Distance approx. 50 m), the maintenance inventory seems to have excluded short sections of road from the inventory, but according to computer vision, there is a clear need for resurfacing at this point. Elsewhere, it can be seen that even short sections, even in very poor condition, are ignored. This is due to the current merging rules and the minimum length (100 m), but on the other hand possibly the "generosity" of the current method, i.e. the shortest sites may not be inventoried. In this sense, the computer vision method is more sensitive to finding individual, even dangerous, damage sites.

4 CONCLUSIONS

Crowd-sourced data collection proved to be fast and cheap. Computer vision data provides more accurate positioning of pavement sites and the critical sections for road users and the systematic nature of planning is improved if new up-to-date condition data is available each year.

Improving aggregation rules will lead to a plan that may be more economically viable to implement. In the programming, a distinction was made between periodic and routine maintenance, from which periodic maintenance segments were selected first. Routine maintenance should not be applied on the sections that are about to be paved. For prioritization, there should be information on the number of potential users, as well as maintenance categories of the segments (K1, K2, L).

There is no standard methodology for optimizing the maintenance of pedestrian and cycle paths similar to the motorways, where the HDM-4 methodology seeks to minimize the sum of costs for both the road owner and road users. In this model, a correlation has been found between road roughness and road user costs (Onyango et al. 2015). For cyclists, similar models have not been found. Thus, it is not known to what extent the deterioration of the surface condition of cycle paths will increase accidents as well as the maintenance costs of bicycles and, on the other hand, slow down the pace of cycling, which could be converted into time costs.

Instead of total costs, the optimization could be targeted at the costs of the road owner alone, while maintaining a certain level of service for road users. If the long-term maintenance costs of an individual cycle path are minimized from the perspective of the road owner, it would be important to know the patterns of deterioration of cycle paths so that certain damage can be addressed at the right time. According to the hypothesis, cracks allow water to enter structures and the formation of potholes, which could be prevented by preventive maintenance by sealing cracks in their initial stage. However, it is not known how much this would slow down the damage and whether this would be cost effective maintenance.

The repair debt for the pavement of the cycle road network under investigation was about four times the budget constraint. As a result, a large number of very poorly maintained segments were excluded from the first year's paving program, for which maintenance was selected, such as patches for holes and wide cracks. However, these measures may not always be economically viable if the segment in question is included in the paving program for the following years. In this case, the options are to change the prioritization rules so that sites with hazardous damage are included in the first year's paving program, increase the budget for paving, or neglect care if the site is to be paved the following year.

ACKNOWLEDGEMENTS

We express our gratitude to Finnish Transport Infrastructure Agency for an important topic and forward-looking approach in development Road Asset Management.

REFERENCES

Arellana J.; Saltarín M.; Larrañaga A.M.; González V.I. & Henao C.A. Developing an Urban Bikeability Index for Different Types of Cyclists as a Tool to Prioritise Bicycle Infrastructure Investments. *Transportation Research Part A: Policy and Practice*. Volume 139, September 2020, Pages 310–334.

Crowdchupa 2021. https://crowdchupa.com.

Feras Mohammad El Said. *Road Management Systems to Support Bicycling*: A Case Study of Montreal's Bike Network. *A Thesis In the Department of Building, Civil and Environmental Engineering*. Concordia University. Montreal, Quebec, Canada. June, 2018.

*Finnish Traffic Infrastructure Agency, 2021.*Väyläviraston Tilinpäätös 2020. Väyläviraston Julkaisuja 10/2021.

Onyango M., Sen T., Fomunung I., Owino J. & Maxwell J. Evaluation of Treat- ment Choice, User Cost and Fuel Consumption of Two Roadways in Hamilton County Tennessee using HDM-4. *Athens Journal of Technology & Engineering* 2, 2015.

SirWay, 2021. https://solidstreet.eu.

Roads and Airports Pavement Surface Characteristics – Crispino & Toraldo (Eds)
© 2023 The Author(s), ISBN 978-1-032-55149-4

Comprehensive road infrastructure measurements – The solution for the next generation sustainable road infrastructure

Bjarne Schmidt*
ARRB Systems Europe, Sweden

Simon Tetley & Verushka Balaram
ARRB Systems South Africa, South Africa

Jerome Daleiden
ARRB Systems Americas, USA

ABSTRACT: It is important to maintain the financial and economic value of the road infrastructure to provide an operational, environmentally sustainable, and safe asset to society. Comprehensive measurements, as performed by ARRB Systems iPAVe (Intelligent Pavement Assessment Vehicle), is a robust method to ensure that the road infrastructure value is maintained, and a high serviceability level of mobility is provided. State-of-the-art comprehensive road monitoring equipment will support an efficient, safe, and sustainable road infrastructure, that meets next generation mobility needs and supports the UN agenda 2030, by providing vital data for key elements such as:

1. Accessibility and mobility, 2. Security, 3. Environmental impact, 4. Air quality, 5. Noise, 6. Traffic safety.

For the road infrastructure, it is not possible to separate the different subjects, as the impact from the roads are very complex. Hence, we must see how the condition of the road infrastructure, traffic flow and pattern, together with environmental and economic challenges affect the overall strategic goals. New methods of building and maintaining road infrastructure, and performance indicators quantifying the economic and environmental challenges, must be considered in the effort to produce optimal user- and environmentally friendly road infrastructure. Apart from delivering essential transport services to society, roads also pose adverse side-effects such as injuries and fatalities from traffic accidents and other health threats. Accidents, noise, pollution, and stress are a few of the negative road related issues that we must consider, which are derived, or more dominant, from poorly planned/designed or maintained roads. For road owners, it is of important to balance the positive and negative effects of new constructions and through robust maintenance strategies of existing roads. Number of fatal accidents, traffic safety accessibility, noise, CO_2 emissions etc. must be considered and balanced against economy. Comprehensive Road condition inventory performed by the iPAVe, provides vital digitised data for systematic analysis through ARRB Systems BIG-data handling processes providing rapid evaluation that supports and optimizes operating, maintaining, and upgrading the road infrastructure, in a cost-effective way. The paper will show how road inventory BIG data can be handled and analyzed systematically to support management of road infrastructure.

Keywords: Road Infrastructure, Measurements, Sustainable

---

*Corresponding Author: bjarne.schmidt@arrbsystems.com

DOI: 10.1201/9781003429258-20

Most road users don't think about the strip of asphalt or concrete that they travel on, or what impact the condition thereof may have on them or the country, so why is making roads "safe" so important and why do we spend time and resources in monitoring and developing standards for evaluating road condition?

Building and maintaining roads to a certain standard is based on decades of expertise and experiences. A road network becomes dangerous to use and loses its economic support to the society if this significant investment is not protected and maintained by provision of robust and reliable condition data, which, in turn, is evaluated to provide invaluable information that supports cost-effective road infrastructure strategies. According to the Global Plan 2021 – 2030, developed by the World Health Organization and other stakeholder's (WHO 2020) road traffic collisions causes nearly 1.3 million preventable deaths and an estimated 50 million injuries each year.

The World Road Association, (PIARC 2020) has stated that only a few studies into the socio-economic costs related to injury have been undertaken but shows that for the European Union (covering 27 countries) the annual costs range from 134 billion to 172 billion Euro representing a cost equivalent of around 2% of GDP over the last decade. According to The Organisation for Economic Co-operation and Development (OECD 2008) the socio-economic cost of road crashes using different methods of evaluation amounted to between 1.5–5% of GDP. McInerney, (2012) reports that in Africa, the International Road Assessment Programme (IRAP) estimates indicated annual costs of up to 7% of GDP.

Road distress such as patching, surface defects, larger longitudinal and transversal cracks, undulation, and rutting are associated with a higher incidence of accidents and personal injury. Of these road surface defects; rutting seems to be the cause of most injuries and fatal accidents according to (Hyldekær et al. 2019). Other types of pavement failure such as potholes can have a dramatic influence on traffic accidents as these can cause sudden loss of vehicle control. It is, however, important to state that causality between road condition and road accidents can be difficult to statistically correlate, due to vice versa causality, ie, a bad road, for example, causes road users to be more vigilant, slow down or undertake corresponding defensive driving behavioral adjustments.

The economic cost of traffic accidents, due to a degrading infrastructure is significant, but is not the only financial issue given that road infrastructure requires an increasing investment due to inflation and backlog situations. As an example, according to reports of the Asphalt Industry Alliance in the UK (AIA 2022) it is stated that although an increase in the highway maintenance budgets of up to 4% in 2020/21 has been implemented, the reported backlog of carriageway repairs has increased by 23% on last year's figure to approximately 15 billion Euro – a situation which, if not addressed in the short term, will become impossible to resolve.

Another example, noted in the United States, where TRIP, National Transportation Research Nonprofit, in their report (TRIP 2018) states that the U.S. needs 740 billion dollars of maintenance and rehabilitation investments for roads and bridges.

The European Court of Auditors (ECA 2018) states that maintenance budgets are often insufficient and have not kept up with the increasing scale of the ageing infrastructure resulting in significant maintenance backlogs.

1.1 *The challenges today for a sustainable road infrastructure*

The condition of many road infrastructure networks has, over time, gradually deteriorated. With insufficient funding in most countries, the rate of degradation will increase with an associated increase in backlog actions. This means that the road system will deteriorate faster, than what the road authorities can accomplish with maintenance.

The increased maintenance and growing need for reconstruction of roads, has resulted in many roads reaching their technical service life due to increased traffic volumes, a trend towards ever higher permissible maximum axle loads and climate change is a concerning factor that influences the pavements bearing capacity. Many road networks were built decades ago, using designs that were appropriate for the traffic volume, axle loading and environmental impact at that time, but as years progressed many road networks do not have the ability to carry the ever-increasing traffic load and environmental impacts currently experienced. Road pavements are generally designed and built based on strength characteristics or bearing capacity, but generally managed according to their functional condition. It has been noted that bearing capacity has historically been difficult and expensive to measure, on a routine basis. Until now, overall pavement condition has largely been determined using riding quality and rutting, often expressed in IRI and RUT depths. The assumption often made is, if a road is even, in the longitudinal and transverse direction, the pavement is likely not in a state of structural distress and has not exceeded its structural capacity. However, with modern technology for measuring bearing capacity, it has become easier and more cost effective to more accurately assess existing structural capacity. It must also be realized that a road on the surface may visually be in good condition, but the structural capacity could have reached its limits, though this is not evidently shown on the surface. Modern road infrastructures must be sustainable and provide an increasing number of functionalities for the benefit of the environment and the road user. A poorly maintained road will be costly for the road owner, and have an impact on road user costs, such as vehicle maintenance costs, fuel consumption etc. For the environmental impact road noise, particle and CO2 emissions are vital to reduce to secure the sustainability. Figure 1 shows the parameters influencing the road infrastructures, but also what is needed to be in focus to establish and provide a sustainable road infrastructure.

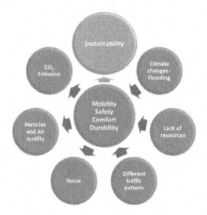

Figure 1. Impacts and challenges for a road infrastructure in balance contributing to sustainability (authors own figure).

PIARC's report (PIARC 2020) states that approximately 15% of the GNP is spent within road transport and infrastructure in industrialized countries, and that these huge investments must be preserved in a sustainable way. So, if the investment is lacking for maintenance, the road infrastructure contributes in a negative way to a sustainable society.

A critical and essential element in protecting road infrastructure investments are robust maintenance strategies and plans. With increasing challenges of providing sustainable roads, including requirements for them to be safe and environmentally friendly, we have until now been challenged by the capability of obtaining comprehensive, robust, and reliable condition data and use this information to develop complete performance models, to predict pavement performance with time. Seeing the need of comprehensive measurements to obtain complete evaluations of road conditions, two concepts have been developed, both being very similar. The German Federal Highway Research Institute (BASt) have developed their multi-functional assessment tool for structural evaluation and design of pavements (MESAS). ARRB Systems went on further and developed their Intelligent Pavement Assessment Vehicle (iPAVe). These systems collect both pavement strength and pavement functional characteristics simultaneously and, thereby, provides a complete picture of the condition of a road pavement, which greatly assists in the development of robust maintenance strategies and, hence, optimizing the use of the budgets available.

1.2 What are Comprehensive measurements

Figure 2 shows the principles of Comprehensive measurements of road infrastructure condition data collected by the intelligent Pavement Assessment Vehicle (iPAVe). The iPAVe is a fully integrated survey vehicle capable of collecting both structural and functional pavement condition data at traffic speeds, with the following information being acquired:

- Pavement strength and pavement layer thicknesses
- Cracking, type, extent, and severity
- Longitudinal and transverse road profile
- Pavement macro texture
- Road geometry
- Geospatial position
- Digital imaging
- Asset inventory and condition

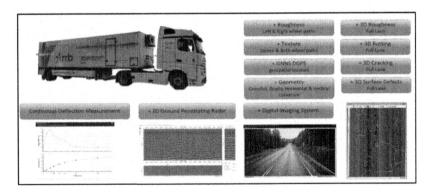

Figure 2.   Simultaneous collection of functional + pavement layer + structural data.

1.3 Measuring pavement structural condition

The component on the iPAVe that measures pavement strength, is the Traffic Speed Deflectometer (TSD) equipment, developed by Greenwood Eng. The TSD can measure the bearing capacity of a road pavement at traffic speeds, with a typical speed of up to 80 km/h. Performing structural measurements at such speeds introduces a tremendous benefit in terms of traffic safety, as it does not create any obstacle on the road, as opposed to "traditional" stationary or slow-moving equipment, such as an FWD or Deflectograph.

The sensors used for deflection measurements are Doppler lasers. These lasers measure the instantaneous deflection velocity of the pavement, through a load applied by the trailer tyres on the rear axle as it occurs in reality. Using Doppler sensors for deflection measurement provides time-based results taken in real time at traffic speed. This minimizes the effect of travel speed, surface texture, and roughness, on the resultant data output, which presents a significant advantage in terms of data repeatability and precision.

Performing traffic speed bearing capacity measurements creates a large data volume of approximately 6 megabytes per kilometre, depending on the driving speed. For every 0.02 m travelled the system provides deflection velocity information to an equivalent resolution of 5 μm. Approximately 1000 data samples are recorded per second, per sensor. Based on the measured velocities the deflection bowl is derived using the Area Under the Curve principle (Muller *et al.* 2013). When the deflections have been calculated, different parameters representing the structural capacity of the pavement can then be established. To transfer this large amount of data into something that is practical to handle, the iPAVe averages the individual 20 mm data points to provide deflection data outputs for every 10-meter interval for each sensor. These intervals can be reduced or increased as required. Figure 3 graphically indicates an example of how segmentation of the data can be applied, turning the 10-metre detailed information into a more maintenance related evaluation of the bearing capacity. The segmentation can be implemented to comply with the individual clients' requirements and maintenance strategies.

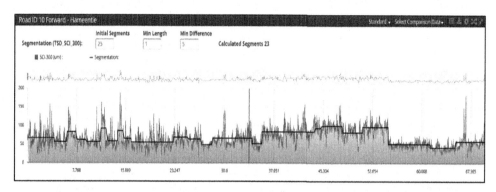

Figure 3.   Bearing capacity measurements, represented by SCI 300 and an applied segmentation.

2   MEASURING PAVEMENT SURFACE CONDITION

2.1   *Automatic Crack Detection (ACD)*

The pavement images are collected in conformance with AASHTO R 86-18 Standard Practice for Collecting Images of Pavement Surfaces for Distress Detection, (AASHTO 86 2018) with the processing of the images, separated into two distinct phases. Firstly, the raw data is processed into crack maps, which are produced through proprietary algorithms developed by the company Pavemetrics. The data for the "crack maps" is stored in an XML format.

These XML files can then be analysed using ARRB Systems own proprietary algorithms to classify, weight and aggregate the cracking information into multiple formats as required, such as (AASHTO 85 2018) AASHTO R 85-18 Standard Practice for Quantifying Cracks in Asphalt Pavement Surfaces from Collected Pavement Images Utilizing Automated Methods.

2.2   *ACD systems pavement profiling*

Additionally, the transverse profile of the pavement can be measured by the system, at a 28 kHz sampling rate. This equates to one transverse profile, nominally 4 metres in width and being recorded every 1 mm of longitudinal travel at survey speed.

205

Figure 4.   Example of cracked area and the image from automated crack detection.

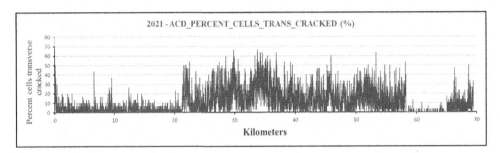

Figure 5.   Surveying a 70 km road – detecting the amount of transverse cracking.

2.3   *Digital laser profile measurement*

For measuring longitudinal pavement evenness, the iPAVe is mounted with the Hawkeye Digital Laser Profiler (DLP). These are Class 1 non-contact inertial laser profiler as per EN 1306-6:2008 and ASTM E950. The system uses a set of paired lasers and accelerometers to obtain the longitudinal profile of the pavement.

The DLP have a vertical accuracy of 0.01 mm, with profiles sampled every 1 mm of longitudinal travel to report longitudinal profile height at 25 mm intervals or greater in mm units to 1 decimal place. Based on the measured longitudinal profile typical indices for pavement evenness, such as the International Roughness Index, IRI, can be calculated.

For measuring the pavement texture the iPAVe uses the same three non-contact 32 kHz lasers utilized for longitudinal profile. The macrotexture is traditionally reported in terms of Mean Profile Depth (MPD) in accordance with EN ISO 13473-2019 (EN 2019), ASTM E1845 (ASTM 2016), in accordance with TRL Lab Report. 639 (Merrill 2005) for Sensor Measured Texture Depth. With the iPAVe pavement profiling are made every 1 mm or less at speeds of up to 80 km/h.

For more in-depth analysis of the pavement texture, the actual texture profile sampled at 1 mm interval can also be exported from the wheel path and the road lanes center position.

2.4   *Digital imagery*

The iPAVe is equipped with an Asset View imaging package which is an advanced video acquisition system that utilises the latest in digital camera technology developed by Basler and manufactured specifically for the road environment. The system is in combination with high grade Kona optics with auto iris lenses that are optimized for high-speed road collection.

All digital Images have a positional accuracy of less than 0.5 m and measurement accuracy of less than 0.25 m. The images are high-resolution, and the set-up of the cameras easily allows the reading of lettering down to + /- 100 mm high.

Using the Hawkeye Insight viewer systems also allows for measuring lane width etc in the images, as shown in Figure 6.

Figure 6.   Front view digital image including a measure of the lane width.

2.5   *Handling BIG data*

The comprehensive measurements performed by the iPAVe generates up to 3 Gb of data per kilometer. A key element is not necessarily the actual collection and storage of the various information sets but rather how to utilize the vast amount of comprehensive data and information collected. The goal is to ensure that the data collected is verified and validated in providing value to the road authorities and supports the decision makers in making better informed, appropriate, and cost-effective maintenance strategies, that comply with the challenges that influence road infrastructure condition, quality, and safety.

To provide an optimal contribution to the decision-making process and to utilize the BIG data collected, the iPAVe is supported by the Hawkeye system, which is the main operating system of the iPAVe including various integrated tracking, processing, and visualization software components, including:

- An interactive and real-time acquisition control interface to control and integrate all systems simultaneously.
- A quality control tool that can alert both operators in the field and office field supervisors of anomalous data in near real-time.
- A full data processing, extraction, analysis and management system for the collected images, profiles, surface and structural distress, and other pavement data. This software allows for full control and synchronization for all raw and processed data streams collected and enables analysis down to all incremental levels and up to all aggregative levels as required.
- A web-based application that provides users with a customizable workspace for creating data visualization and providing simultaneous viewing of synchronized pavement images.

2.6 Data processing

Data processing is a vital aspect in delivering correct data and the backbone for establishing useable data complying with national and international standards. The processing software used for the iPAVe measurements provides full data processing, analysis and management capabilities on all imaging, profile, distress, and pavement data collected.

The software allows for full control, synchronization and integrations of all raw and processed data collected and enables analysis to whatever incremental and aggregated levels is/are required by the client.

Figure 7.   Interface for Hawkeye processing toolkit.

Using data information from the road condition surveys to establish and maintain extensive inventories and risk assessments of road and roadside assets are also of critical importance in terms of road safety. This is possible as the geo-referencing system allows users to create inventory of roadsides assets such as road signs, guard rails, etc. The images are referenced against GPS coordinates by logging the latitude, longitude, and altitude of these objects and potential hazards.

2.7 Visual assessment of collected data

How to utilize the vast amount of collected data, BIG data, as a means for optimizing maintenance and rehabilitation budget is important. To assist and support interpretation of the data collected by the iPAVe a web-based platform has been developed that enables the visual assessment of collected data, in a simple, user-friendly format. The platform includes the functionality needed to view, search, QC, and edit the data and images collected by the iPAVe.

Utilizing the mapping functionality of Google Maps, the platform allows the user to navigate through the collected data and assess the condition of the road network. The platform allows the user to transfer from a network level approach and assessment to a project level investigation without the need for the usual additional project level evaluation.

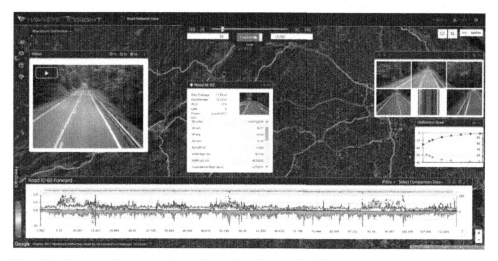

Figure 8.   Example of a road network displayed with the online data viewer.

2.8   *Advanced filtering of road network data*

The visual assessment platform provides the functionality to filter the network level data according to specifications related to accepted road condition or intervention levels. By doing this filtering, it is possible for the road authority to locate the roads sections where maintenance is needed. By using comprehensive measurements including both structural and functional pavements information, the selection or filtering process brings a new dimension into locating roads with present or future maintenance needs.

Figure 9 illustrates an example of a filtering process whereby pavements that might not indicate obvious surface or functional defects but do have a poor structural capacity can easily be combining individual measurements from the iPAVe.

When the road network has been filtered and the relevant road sections requiring maintenance are located it is relatively straightforward to start prioritizing the roads that are in highest need of maintenance by utilising user defined parameters of key factors such as general condition, traffic volumes, daily axle loading, road class, maintenance history and current need etc.

Figure 9.   Filtering network data to locate road sections with a poor asphalt surface and weak structure.

In addition to finding and optimizing the maintenance needs within the Insight platform, previous measurements can be compared directly, providing vital information on comparative condition change year to year and more importantly rate of deterioration. Figure 10 shows an example of three consecutive measurements of a roads structural capacity and how it changes with time.

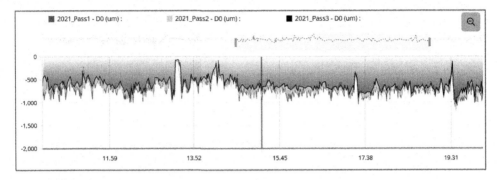

Figure 10. Consecutive measurements to compare pavement structural changes with time.

2.9 *How does Comprehensive measurements support selective UN 2030 goals?*

Comprehensive measurements provide information that can be used to proactively meet UN 2030 goals – better roads for safer, environmentally friendly, and accessible roads, by providing road authorities with crucially improved ability to:

- Proactively ascertain and refine treatment forecasts and strategies to mitigate more extensive areas of structural and functional concern, that currently are not identifiable from the surface condition alone i.e., better decision tools.
- Ability to conduct immediate local repairs at precisely defined locations (Potholes and crocodile (alligator) cracked areas) in advance of surfacing treatments to improve treatment performance and optimize required overlay type and thicknesses.

There is no doubt that the use of comprehensive full spectrum pavement measurement greatly assists in an optimal use of financial resources for maintenance and rehabilitation which in turn will provide a cost beneficial road network as described in (Tetley & Visser 2021) with a higher degree of mobility and safety which will contribute to the lowering of fuel consumption and, hence, support the goals of reducing road traffic CO_2 emissions.

2.10 *Address road safety whilst utilising the iPAVe*

Credible road condition and characteristic data is increasingly considered as critical in addressing road safety. The International Road Assessment Program (iRAP), which is endorsed by the United Nations, has certified the Hawkeye technology that is used in the iPAVe as a Class B Inspection System.

The iPAVe records road features that impact road safety to allow for a streamlined assessments and address high risk roads. The data is used to undertake a network level assessment and determine Star Ratings based on the existing attributes of the road. This is done by analysing:

- Geo-referenced imagery - used to code (post rate) the road. The 360-degree coverage through multiple camera lenses has the added benefit of being able to accurately measure road attributes within the toolkit application. This enables the users to determine the distance to roadside hazards such as trees, light poles, etc. from the road edge.

- Alignment Data - collected with the GypsiTrac system which measures road geometry is also integrated with the safety assessment to identify road sections with steep grades, incorrect cambers and crossfalls (transverse slopes), which leads to better understanding of drainage.
- LCMS and profile data- for quick pothole identification, areas with low macro-texture (more prone to skidding at high speeds) and rutting (leads to poor directional stability and areas where water can pond and lead to possible aquaplaning).

Once the road has been Star Rated, the data is presented spatially to provide the Road Agency with a holistic illustration of hazardous locations. The information is also used to determine appropriate countermeasures to improve the level-of-safety, prioritise investment, and benchmark performance for year-on-year comparison. This is a more proactive approach in reducing road traffic accidents and collisions as opposed to retroactively addressing the "accident black spots" which are typically determined by the number of historic accidents and are often incorrectly referenced. The iPAVe has successfully been used to determine IRAP Star Ratings of around 16000 km's of provincial roads in South Africa, and acts as a catalyst to implement road safety initiatives and ultimately save lives.

3 CONCLUSION

Greater emphasis is now being placed on road safety and environmental sustainability. It is recognized that the road infrastructure community can have a significant impact on these concerns. Obtaining the right information on road conditions and utilizing the data available correctly, are critical to successfully achieving these objectives. Using comprehensive measurements, all aspects of a network's performance can be assessed simultaneously, saving time and money. One must also have the proper tools to evaluate such data sets however to facilitate securing that the road infrastructure, its safety, and the resources needed are optimized.

REFERENCES

AASHTO 86 2018; *AASHTO R 86-18 Standard Practice for Collecting Images of Pavement Surfaces for Distress Detection*

AASHTO 85 2018; *AASHTO R 85-18 Standard Practice for Quantifying Cracks in Asphalt Pavement Surfaces from Collected Pavement Images Utilizing Automated Methods*

AIA 2022; *Asphalt Industry Alliance (AIA) 2022 - Annual Independent Survey*

ASTM 2016; ASTM E1845, *Standard Practice for Calculating Pavement Macrotexture Mean Profile Depth*, 2016

ECA 2018; European Court of Auditors 2018 "Towards a Successful Transport Sector in the EU: Challenges to be Addressed"

EN 2019; EN ISO 13473-2019, *Characterization of Pavement Texture by Use of Surface Profiles — Part 1: Determination of Mean Profile Depth.*

Janstrup H. K., Møller M. and Pilegaard N. 2019, *Vejens Omgivelser, Udformning og Tilstand - Betydningen for Trafiksikkerhed.* DTU Management, Institut for Teknologi, Ledelse og Økonomi.

McInerney R. 2012, *A World Free of High-Risk Roads, International Assessment Progamme, Presentation to Millennium Development Bank Training Programme*, World Bank, Tunis

Merrill D. 2005, Guidance on the Development, *Assessment and Maintenance of Long-life Flexible Pavements. TRL Lab Report.* 639

Muller W. and Roberts J. (2013). Revised Approach to Assessing Traffic Speed Deflectometer Data and Field Validation of Deflection Bowl Predictions. *International Journal of Pavement Engineering.* 14.

iRAP (the International Road Assessment Programme) https://irap.org/rap-tools/infrastructure-ratings/star-ratings/

OECD 2008; *Towards Zero: Achieving Ambitious Road Safety Targets Through a Safe System Approach.* OECD, Paris.

PIARC 2020; World Road Association, 2020, *The Contribution of Road Transport to Sustainable and Economic Development* https://roadsafety.piarc.org/en/strategic-global-perspective-scope-road-safety-problem/socio-economic-costs

Tetley S. and H. Visser 2021, *A Study into the Benefit and Cost-effectiveness of Using State-of-the art Technology for Road Network Level Condition Assessment IRF 2021 S.*

TRIP 2018; National Transportation Research Nonprofit 2018 *Bumpy Road Ahead: America's Roughest Rides and Strategies to make our Roads Smoother*

WHO 2020; World Health Organization *et al.* 2020 *Global Plan, Decade of action for Road Safety 2021 – 2030*

Comparison of bearing capacity measurements with the Traffic Speed Deflectometer (TSD) and surface characteristics on intra-urban roads in Germany

Barbara Esser*, Pahirangan Sivapatham & Stefan Koppers
University of Wuppertal, Wuppertal, Germany

ABSTRACT: Due to the various boundary conditions specific to intra-urban roads such as fixtures or excavation patches the surface characteristics of those roads might not correspond to their structural condition as well as those of country roads and state roads. Within the scope of this research project, bearing capacity measurements were carried out on selected municipal main roads in North Rhine Westphalia using the Pavement-Scanner of the University of Wuppertal, which is equipped with a TSD as well as multiple devices to record surface characteristics. The aim is to investigate the correlation between the surface characteristics and the bearing capacity of intra-urban roads and to evaluate the importance of a common consideration of both factors in pavement management.

In the present project, measurements were carried out with the TSD over the period of a year on main municipal roads in Wuppertal, bearing capacity data of a total of more than 700 km of roads (federal roads, state roads and district roads) were collected and evaluated. The bearing capacity characteristics presented provide a meaningful overview of the bearing capacity of the road structure and allow an estimate of the depth of possible damage or weak points. The test sections were driven at different speeds, temperatures and over the duration of one year. It can be seen that plausible and valid characteristic values result from speeds of 15 km/h and above. Dynamic axle loads were taken into account by means of a normalization factor. It can also be determined that even after the progression of several seasons and at different temperatures, there is a high repeatability of the measurements. In summary, it can be stated that with the help of the bearing capacity, visually undetectable weak points in the traffic surface pavement can be identified and, if the appropriate measures are initiated at an early stage, the structural road conditions can be improved in a targeted manner.

Keywords: Traffic Speed Deflectometer, Pavement Structural Performance, Bearing Capacity

1 INTRODUCTION

In Germany, measurement runs are carried out every four years to assess the condition of roads (ZEB). Various indicators are measured during these measurement runs: Evenness in the longitudinal and transversal profile, road grip as well as structural characteristics of the surface. So far, the measured characteristics only allow for an evaluation of the surface condition of roads, which can be a valid indicator of their structural condition. In order to assess the structural condition of the entire pavement structure reliably, the load-bearing capacity of the road should be taken into consideration. Generally, the bearing capacity can be determined on a point-by-point basis, for example with the Falling Weight Deflectometer

---

*Corresponding Author: b.esser@uni-wuppertal.de

DOI: 10.1201/9781003429258-21

(FWD). However, a network level determination of the bearing capacity is only possible and economically reasonable if the measurements can be carried out in flowing traffic. This can be achieved with the Traffic Speed Deflectometer (TSD), which is capable of measuring the bearing capacity of asphalt roads in flowing traffic at speeds of up to 80 km/h.

In Germany, network-wide the pavement condition survey and assessment (ZEB) are conducted every four years. Until now, the measured characteristics only allow an evaluation of the surface condition of roads, based on evenness (longitudinal unevenness and transverse unevenness) and structural characteristics (cracks, path holes, patch, ravelling etc.). After the survey of surface condition, an assessment is carried out to calculate pavement condition value formed from the above-mentioned measured condition variables. In a data-synthesis, a serviceability value is then calculated for the assessment of serviceability and a structural performance value (surface) for the assessment of the pavement structure performance, with the worst value being used as the critical value for maintenance planning. The structural performance value (surface) is composed of irregular network cracks, patches, and other surface damage (path holes, ravelling and binder accumulation), as well as rut depth or longitudinal evenness (max. value). (ZTV ZEB-StB, 2006/2018, E EMI, 2012).

The pavement condition characteristics currently measured only allow an assessment of the surface condition of the roads. Deep-located damage such as a defective layer bond and/or fatigue cracks that propagate in the asphalt from bottom to top and significantly affect the structural condition of the road pavement are not taken into account or are only taken into account when they appear on the road surface. Consequently, the structural performance value (surface) according to ZEB alone is not sufficient to describe the condition of the structural performance of the pavement (AP 9 Serial S 2019).

2 STRUCTURAL PERFORMANCE OF THE PAVEMENT

In order to assess the structural performance of the entire road structure, the bearing capacity of the road must be examined in addition to the surface condition. The bearing capacity is defined as the resistance of the road structure to short-term loads. It provides information on the structural condition of the pavement and the subgrade or sub-base (Beckedahl et al. 2019). Bearing capacity measurements have been carried out since the 1950s using stationary bearing capacity measurement systems such as the Benkelman beam or the Falling Weight Deflectometer (FWD), which are usually extended by investigations of cores. Neither destructive area-wide coring nor the use of non-destructive stationary bearing capacity measurement systems are suitable for a network-level survey due to the intensive time and manpower requirements and the need for traffic control.

The most efficient bearing capacity measurement, which is least disruptive to traffic and also advantageous in terms of occupational safety, is achieved by the Traffic Speed Deflectometer (TSD) in the Pavement-Scanner of the University of Wuppertal, which can measure continuously in flowing traffic at speeds of 5 to 80 km/h. The TSD has been used worldwide since 2004 to measure the bearing capacity of pavement materials. Since 2004, the TSD has been used worldwide in Australia, Europe, the USA and China, among other countries, in the network level recording of the condition of roads. Although it is already being used successfully internationally, the TSD is a relatively new procedure in Germany, so that there is still a need to gather experience - especially in the municipal sector (FGSV 433 C 5 2020).

In order to meet this need, municipal main roads in the area of the city of Wuppertal were measured with the pavement scanner for more than one year in this project. The aim of this project was to demonstrate the feasibility and practicability of bearing capacity measurements in flowing traffic on main municipal roads, taking into account the special boundary conditions (low speeds, unevenness due to road fixtures, etc.). Multiple measurements of the same route under different conditions make it possible to identify and potentially eliminate various influences of the vehicle dynamics on the measurement system. For the investigation of the change in load-bearing behavior over the course of several seasons and at different temperatures, the measurements were realized on several measurement days between August 2020 and June 2021 and added up to approximately 710 km of measurement distance.

3 MEASUREMENT SYSTEM AND DATA PROCESSING

The objective of the TSD measurement is to measure the deformation (deflection) that occurs in the pavement when a defined axle load passes over a measurement point. For this purpose, in the TSD, 11 Doppler-Lasers are used to measure the relative velocity of the road surface to the vehicle at specific distances from the load axle, and a function for the deflection bowl is derived from this (black line in Figure 1). The deflection bowl is then used to form various characteristic values that can be used to determine the load-bearing capacity.

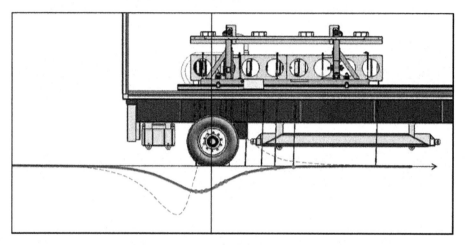

Figure 1.   Course of the vertical velocities (green dashed line) and the derived deflection bowl (black line) with points at the individual laser sensor positions (red dots) of TSD (Krarup 2016).

The most apparent information about the deflection bowl is provided by the maximum deformation. Since its position is variable, but by nature always close to the load application point, the deformation below the load axle is currently used as an approximation to the maximum. The so-called D_0-value already provides a good indication of the magnitude of the load-bearing capacity. (FGSV 433 C 5 2020)

However, according to the research results of Čičković et al. (2020), the D_0 value is not as reliable or stable as the so-called "Surface Curvature Index" (SCI), which denotes the difference between the deflection under the load axle and the deflection at a certain distance in front of it. The most commonly used value is SCI_{300}, which is the difference between D_0 and D_{300} (deflection 300 mm in front of the load axle). Through it, it is possible to conclude the load-bearing capacity of the asphalt layers. In addition to SCI_{300}, the values SCI_{200} (D_0 - D_{200}) and SCI_{SUB} (D_{900} - D_{1500}) are of interest. With the help of the SCI_{SUB}, the bearing capacity of the subgrade or the unbound layers can be determined, and the SCI_{200} provides information on the condition of the upper asphalt layers. In general, it can be stated that the further away the deflections are located from the load axle the deeper the layer about which they provide information (Beckedahl et al. 2019).

To investigate the unbound layers, the subgrade damage index BDI (D_{300} - D_{600}) and the subgrade curvature index BCI (D_{600} - D_{900}) can provide valuable additional information, especially in conjunction with the SCI_{SUB}. (Lukanen et al. 2000)

4 INVESTIGATION OF POSSIBLE INFLUENCES OF VEHICLE DYNAMICS ON THE MEASUREMENT RESULTS

Since the TSD does not measure statically, but in flowing traffic on existing (often curved and uneven) pavements, varying speeds and accelerations in different directions occur, which can influence the measurement results.

Due to the varying vertical accelerations of the measuring vehicle - induced by unevenness or curves - the load on the road surface caused by a rolling wheel does not correspond to the static wheel load. The right wheel load, which is decisive for the measurement, varies selectively by several tons in the individual measurements, although the static wheel load of 5 tons does not change. In order to be able to determine the influences of the driving dynamics on the load and eliminate them later (if necessary), strain gauges are installed to the TSD on both sides of the load axle. From their measurement data, the current dynamic wheel load can be determined, taking into account the vehicle-specific and climatic boundary conditions. If the influences of the dynamic load are to be limited, the measurement data can be adjusted to a standard load (FGSV 433 C 2.1 2014). All measurement data in the present project are adjusted after export before further evaluation. The standard load is the arithmetic mean of the right wheel load of all data sets.

The measurement speed can also have an influence on the measurement results. Low measurement speeds and standing times cannot be avoided, especially in municipal areas, due to intersections and disruptions in traffic flow. A change in the bearing capacity measurement data is to be expected at very low measurement speeds, because the duration of loading influences the response of the asphalt (Mollenhauer 2008) and influences from dynamic loading disappear at low speeds. In addition, the influences from acceleration and braking processes might gain in importance (Schäfer 2009).

By default, the processing software therefore excludes data sets with speeds of less than 5 km/h as not usable. The manufacturer recommends a constant speed of more than 40 km/h or 80 km/h on highways. In urban areas, constant speeds can only be achieved over short distances and speeds of less than 40 km/h are not unusual. Therefore, the following section examines the extent to which speeds below 40 km/h can be used. The manufacturer's processing software automatically excludes data from which implausible deflection result and outputs data gaps instead. Figure 2 can be derived from the measurement data of the present project, showing the percentage of valid data measured (without data gaps) in the total measured data at different measurement speeds. The percentage of valid data records in the total data records increases with increasing measurement speed, whereby it is already over 95% from speeds of 30 km/h.

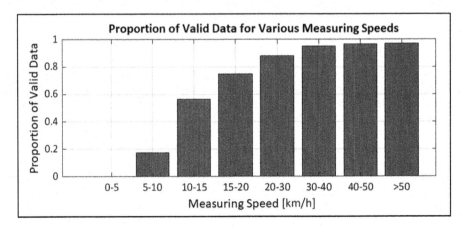

Figure 2.   Proportion of valid records in different speed classes, source: own diagram.

Overall, it can be stated that from 15 km/h, three quarters of the measured values can be used and thus a satisfactory number of valid data is already available at these measurement speeds. However, data recorded at speeds of less than 15 km/h account for only a small proportion of the total measurement data from Wuppertal; 2697 out of 70957 data records (3.8%).

5 REPEATABILITY OF THE MEASUREMENTS

The FGSV working paper of bearing capacity measurement states that the TSD is "very well suited for network-level monitoring" of structural pavement condition (FGSV 433 C 5 2020). In addition, a large number of studies, both in Germany (Beckedahl et al. 2019; Čičković et al. 2020), and internationally (FHWA 2016; Rasmussen et al. 2008; Wix et al. 2016) prove the reproducibility and repeatability of TSD measurements.

A visual comparison of the measurement results in this study confirms this, too. As an example, Figure 3 shows two measurements of the same section ("A") of a state road ("L") in the direction of stationing ("Direction R"), with the SCI values depicted in 10 m intervals in each case. The measurements took place nine months apart. The charts show that repeatability is given even after several seasons. The fact that the TSD also produces reliable measurement results at extreme temperatures is illustrated by an example in Figure 4. The figure shows the SCI values of a measurement of a section ("B") of a federal road in June 2021 at normal temperatures and the lower diagram shows a second measurement on the same section a few days later at an unusually high temperature. Despite different boundary conditions, the local maxima and minima of the section's bearing capacity can be clearly seen.

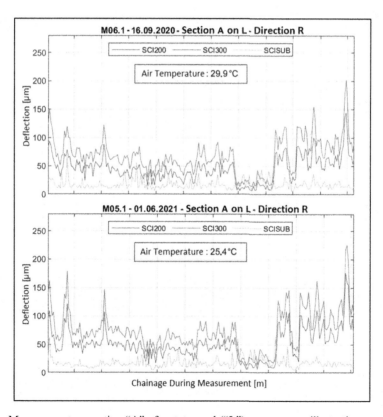

Figure 3.   Measurement on section "A" of a state road ("L"), source: own illustration.

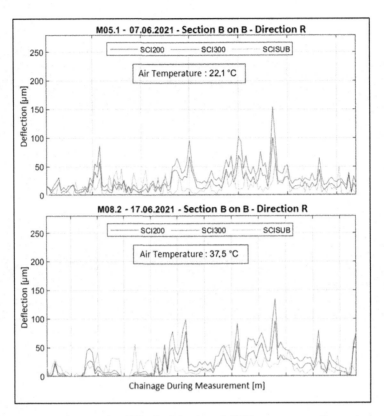

Figure 4.   Measurement on section "B" of a federal road ("B"), source: own illustration.

6   EVALUATION OF BEARING CAPACITY

There is currently no valid approach in Germany for evaluating the bearing capacity from the measurement data of the TSD (FGSV 433 C 5 2020). In other countries, various methods derived from empirical values and adapted to their own country are used. However, due to the differences in climate, pavement type, pavement material and loads in Germany from those in other countries, the assessment system of another country cannot be recommended for the assessment of the bearing capacity of pavements in Germany. A comparison of the evaluation methods from Great Britain, Poland, Ireland and the USA confirms this for the measured bearing capacity parameters in the context of this study.

In the working paper on the bearing capacity of traffic pavements, it is therefore recommended to divide the bearing capacity parameters into classes according to frequency distributions and to evaluate them based on these classes (FGSV 433 C 5 2020). For the measurement data in the present project, a local evaluation scheme is consequently used, which is formed based on the quartiles (25%, 50% and 75%) as well as the 5% and 95% quantiles of all valid data sets for the individual characteristic values (Tables 1 and 2).

The tables show, among other things, that the subdivision in terms of loading and structure is of central importance for the evaluation. With an SCI_{300} of 35 μm, one is in the worse half of the data for federal roads and in the better half for state roads. This evaluation of the bearing capacity measurements clearly shows that continuous empirical monitoring of the road quality by means of the road network level is possible and that an economically and technically reasonable maintenance strategy can be realized based on the objective aspects. In addition, the development of the road structural performance at the network level can be

218

Table 1. Evaluation of the measurement data based on the SCI_{300} separately per road class.

SCI_{300} [μm]	Federal road	State road	Country road
High (best) 5%	< 5.64	< 13.16	< 12.33
High (best) 25%	< 17.89	< 25.48	< 25.29
Median	30.69	41.79	45.05
Low (poor) 25%	> 54.46	> 70.41	> 80.17
Low (poor) 5%	> 114.80	> 139.64	> 145.36

Table 2. Evaluation of the measurement data based on the SCI_{SUB} separately per road class.

SCI_{SUB} [μm]	Federal road	State road	Country road
High (best) 5%	< 1.66	< 2.73	< 1.84
High (best) 25%	< 6.28	< 7.91	< 8.25
Median	13.32	15.16	15.90
Low (poor) 25%	> 24.24	> 26.13	> 26.28
Low (poor) 5%	> 44.49	> 46.34	> 46.52

determined comparatively by carrying out load-bearing capacity measurements on a regular basis and, based on this, maintenance sections that are to be rehabilitated with the same maintenance methods can be determined.

7 EXAMPLE FOR THE EVALUATION OF THE INDIVIDUAL ROAD CLASSES

In the following, one section with a heavy load-bearing capacity and one section with a lower load-bearing capacity are examined as examples for each type of the different road classes examined - federal roads, state roads and county roads.

A highly loaded federal road, which is considered the most important main road in the city, was frequently driven on with the TSD because this historically developed road is well suited for a more in-depth investigation. As an example, a selected section A_B of the federal road shows very satisfactory bearing capacity values (Table 3). The SCI_{200} value (5.53 [μm]) is below average for federal roads and indicates a high bearing capacity. Damage is visible in a few places based on the Pavement-Scanner images, but this is not reflected in the bearing capacity data, so it is probably only surface damage. In contrast, the bearing capacity parameters on a section B_B of the federal road are not satisfactory. The bearing capacity parameter SCI_{200} amounts to 56.40, which corresponds to the worst quarter of the values on federal roads (Figure 5).

As an example for state roads, two sections of different state roads are compared with each other. Section A_L shows satisfactory values in terms of bearing capacity, whereas section B_L performs poorly in all parameters (cf. Table 3). In the case of the latter, both a damaged asphalt package and a low bearing capacity of the unbound layers or an unsuitable subgrade can be suspected. While the bearing capacity number of section A_L meets the requirements for higher load classes, the bearing capacity of section B_L has to be allocated to a lower load class.

Figure 6 compares two photographs on the sections. It can be seen that at least the asphalt pavement on Section A_L is younger than that of Section B_L. The many visible excavation

Table 3. Load-bearing capacity values on federal roads, state roads and county roads; as an example, one section with a high load-bearing capacity and one section with a low load-bearing capacity respectively.

Section	SCI$_{200}$ [μm]	SCI$_{300}$ [μm]	SCI$_{SUB}$ [μm]	BCI [μm]	BDI [μm]	R0 [m]
A_B	5.53	8.93	7.59	5.31	7.90	51474.12
B_B	56.40	86.82	25.25	34.72	67.39	406.70
A_L	14.73	19.88	8.22	8.22	12.88	1860.36
B_L	80.95	125.86	31.02	44.36	93.42	309.41
A_K	17.86	28.21	16.92	18.57	28.24	1351.65
B_K	73.96	111.21	26.40	36.34	75.36	339.22

Figure 5. Example photos of the front camera on sections A_B (left) and B_B (right); source: own photos.

Figure 6. Road with good result (left, section A) and poor result (right, section B) in comparison, source: own photos.

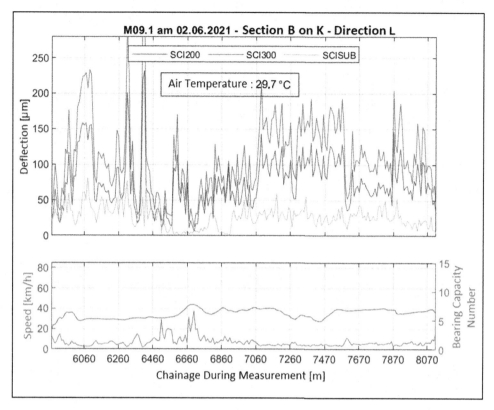

Figure 7. Load capacity characteristics SCI and measuring speed on section B_K, source: own illustration.

patches as well as the clear rutting on section B_L confirm the determined lower load capacity of this section.

As an example for district roads, two sections of different district roads in different directions of travel are compared. The selected section A_K has very satisfactory bearing capacity values (Table 3). With the measured high bearing capacity of 17.9, the section meets high bearing capacity requirements. Section B_K, on the other hand, has a very low bearing capacity of 74.0 (Figure 6).

8 SUMMARY AND OUTLOOK

In this study, bearing capacity data of a total of more than 700 km of roads (federal roads, state roads and district roads) were collected and evaluated over a period of one year. The bearing capacity characteristics presented provide a meaningful overview of the bearing capacity of the road structure and allow an estimate of the depth of possible damage or weak points.

The test sections were driven at different speeds, temperatures and over the duration of one year. It can be seen that plausible and valid characteristic values result from speeds of 15 km/h and above. Dynamic axle loads are taken into account by means of a normalization factor. It can also be determined that even after the progression of several seasons and at different temperatures, there is a high repeatability of the measurements.

As an example, this report gives one example of a good and one of a poor bearing capacity for federal roads, state roads and county roads. It should be noted that, in particular, locations with high SCI_{SUB} values, i.e. a lower bearing capacity of the unbound layers or the subgrade, are usually not detectable by visual condition measurement. Areas with a low bearing capacity of the bound layers are more often - but by far not always - noticeable by visual damage. In contrast, areas with varied and pronounced visual damage patterns often have satisfactory bearing capacity. From these observations, it can be concluded the great importance of a joint investigation of surface properties and bearing capacity when determining the condition of roads. Only if the structural condition is directly assessed by determining the bearing capacity can the most sensible maintenance measure be found in each case from the point of view of construction technology and economy, and thus successful maintenance management can be pursued.

REFERENCES

Beckedahl H.J., Koppers S., Schrödter T., Balck H.; Skakuj M., 2019, *Comparison of Different Continuous Measuring Systems for the Determination of the Bearing Capacity of Asphalt Pavement on Network Level* (original in German), FE_04-0276-2013-EGB, Final Report, University of Wuppertal, Germany.

Čičković M., Bald J., Middendorf M., 2020. *Analysis of Evaluation and Assessment Procedures for the Application of the Traffic Speed Deflectometer on Asphalt Pavements*, FE 04.0318/2018/MRB, Final Report, Darmstadt, Germany.

FGSV 433 C 2.1, 2014. Working Paper – Bearing Capacity of pavement, Part C 2.1: Falling Weight Deflectometer (FWD): Evaluation and Assessment – Asphalt pavement. *Research Society for Road and Traffic Engineering (original in German)*, Cologne, Germany.

FGSV 433 C 5, 2020. Working Paper, Bearing Capacity of Pavement, Part C 5: Traffic Speed Deflectometer (TSD), Evaluation and Assessment – Asphalt pavement. *Research Society for Road and Traffic Engineering (original in German)*, Cologne, Germany.

AP 9 Serial S, 2019. Maintanance Planing, - Serial S: Structural value, *Research Society for Road and Traffic Engineering (original in German)*, Cologne, Germany.

E EMI, 2012. Recommendations for the Maintenance Management of Urban Roads, *Research Society for Road and Traffic Engineering (original in German)*, Cologne, Germany.

ZTV ZEB-StB, 2006/2018. Additional Technical Contract Conditions and Guidelines for the Condition Measurement And Assessment Of Roads, *Research Society for Road and Traffic Engineering (original in German)*, Cologne, Germany.

Krarup J., 2016. *TSD Basics, and Latest Software Update*. Prag, Czech Republic.

Lukanen E.O.; Stubstad R. and Briggs R., 2000. *Temperature Predictions and Adjustment Factors for Asphalt Pavement*, HWA-RD-98-085, DBNX94822-D, C6B, Final Report, USDOT, USA.

Mollenhauer K, 2008, Design-relevant Prediction of the Fatigue Behavior of Asphalt by Means of Uniaxial Tension-swell Tests, *Dissertation* (original in German). URL https://d-nb.info/990703789/34 – checked on 11.01.2022, Braunschweig, Germany.

Soren R.; Lisbeth A.; Susanne B.; Jorgen K. 2008: *A Comparison of Two Years of Network Level Measurements with the Traffic Speed Deflectometer*, Denmark.

Schäfer F, 2009. Loading of Asphalt Pavement from Acceleration Processes, *Dissertation* (original in German) URL http://docplayer.org/207628925-Asphaltbeanspruchung-aus-beschleunigungsvorgaengen.html. – checked on 11.01.2022, Bochum, Germany.

Structural Evaluation, 2016. *Pavement Structural Evaluation at the Network Level: Final Report*, U.S. Department of Transportation, Federal Highway Administration, USA.

Wix R; Murnane C.and Moffatt M., 2016. *Experience Gained Investigating, Acquiring and Operating the First Traffic Speed Deflectometer in Australia*. Transport Research Arena, Australia.

Roads and Airports Pavement Surface Characteristics – Crispino & Toraldo (Eds)
© 2023 The Author(s), ISBN 978-1-032-55149-4

The fundamental characteristics of sideway-force skid resistance measurement devices

Peter D. Sanders*, Cormac Browne & Martin Greene
TRL, Crowthorne house, Wokingham, UK

Andrew Mumford
National Highways, The Cube, Birmingham, UK

ABSTRACT: The effective management of road surface skid resistance is critical in providing a safe means of travel to road users. Essential to this task is understanding the skid resistance supplied by road surfaces and how this aligns with the frictional demands of road users. Throughout Europe and Australasia, skid resistance is characterised by devices utilising the sideway-force measurement principle; Sideway-Force Devices (SFDs). On the English trunk road network, the Sideways-force Coefficient Routine Investigation Machine is used.

The measured skid resistance emerges from the interaction between the test device and the road surface. Hence it is dependent on the fundamental characteristics of the test device (such as its % wheel slip and operational velocity). An understanding of these fundamental characteristics is therefore essential to establish how the skid resistance reported by the device relates to the frictional demands of vehicles, and how the measurements made by a device can be compared with the measurements provided by other test devices (which is a key requirement in developing standards for harmonising skid resistance measurements).

The current hypothesis regarding the measurement properties of SFDs (utilising a 20 degree wheel angle and a 50 km/h vehicle speed) is that they characterise skid resistance at 34% wheel slip and therefore an operational velocity of 50 km/h. However, whilst theoretically sound, this hypothesis is yet to be supported by dedicated experimental data.

This paper presents a two-part study into the measurement properties of SFDs. Part one is a desk study deriving the theoretical characteristics of SFDs. Part two is an experiment comparing skid resistance measurements made with SFDs with a device that reports % wheel slip and operational velocity directly.

The desk study demonstrated that the fundamental characteristics of SFDs can be theoretically described in four ways, three of which are equally valid. The experimental results overwhelmingly supported one of the characterisations derived from the desk study; that SFDs characterise skid resistance at 100% wheel slip and 17 km/h operational velocity. This conclusion falsifies the current hypothesis and prompts questions about how the skid resistance reported by SFDs relates to the frictional demands of vehicles, and how these measurements compare with those provided by other test devices.

Keywords: Friction, Skid Resistance, Road Management.

---

*Corresponding Author: pdsanders@trl.co.uk

DOI: 10.1201/9781003429258-22

1 INTRODUCTION

The effective management of road surface skid resistance is critical in providing a safe means of travel for all road users. The characterisation of skid resistance through direct measurement is a key component in the effective management of skid resistance.

In Australia, Belgium, Czech Republic, England, France, Germany, Hungary, Ireland, New Zealand, Northern Ireland, Scotland, Slovenia, Spain, and Wales, the skid resistance of the road network is characterised by devices utilising the sideway-force measurement principle; Sideway-Force Devices (SFDs). On the English strategic road network this is carried out using the Sideways-force Coefficient Routine Investigation Machine; an SFD that utilises a 20 degree wheel angle and standard test speed of 50 km/h ($SFD_{(20,50)}$). Understanding the fundamental measurement properties of SFDs would therefore provide a major benefit to UK and international road authorities.

The current hypothesis regarding the measurement properties of $SFDs_{(20,50)}$ has been in place for some time (Henry 2000). However, this view is based on a theoretical analysis of the device and (to the Authors knowledge) is unsupported by practical assessment. It is the aim of this work to investigate the measurement properties of $SFDs_{(20,50)}$ through desk study and practical experiment, such that a well-supported conclusion regarding the measurement properties of $SFDs_{(20,50)}$ can be made.

2 THE SIDEWAY-FORCE COEFFICIENT ROUTE INVESTIGATION MACHINE

Sideway-force Coefficient Route Investigation Machines are used for routine monitoring of the skid resistance of the English strategic road network. Measurements from these devices provide the information used to compare the performance of surfacings with the requirements laid out in National Highways[1] (NH) (formerly Highways England) standard CS228 (Highways England *et al.* 2019).

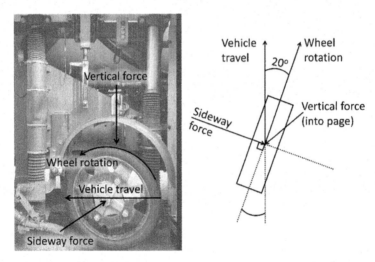

Figure 1.   Image of an $SFD_{(20,50)}$ measurement system with annotations (Left), diagram of the measurement system in plan view (Right).

---

[1]The body responsible for the maintenance of the English strategic road network.

These devices operate on the sideway-force friction measurement principle and use a smooth, freely rotating test tyre installed at an angle of 20 degrees to the direction of travel, which is mounted on an instrumented axle. The skid resistance (SR) value is the average ratio between the measured horizontal (along the rotational axis) and vertical (normal) force, multiplied by 100. This principle is represented in Figure 1.

3 NOMENCLATURE

The properties of skid resistance measurement devices are described in this paper using the following nomenclature:

- Vehicle speed (V_v) is the speed at which the test device traverses the test surface.
- Operational velocity (V_o) is the speed at which the rotational axis of the tyre moves with respect to the direction of friction measurement.
- Wheel speed (V_w) is as the tangential linear speed of the circumference of the test tyre about its rotational axis.
- Slip speed (V_s) is the speed differential between V_w and V_o with respect to the direction of friction measurement.
- The percentage wheel slip (% Wheel slip) is V_s expressed as a percentage of V_o.

4 GENERATION OF HYPOTHESES

A desk study was carried out which resulted in the generation of four hypotheses regarding the measurement characteristics of SFDs. The methodology used was to resolve the V_o and V_s of SFDs into the directions shown in Figure 2. The currently accepted hypothesis for the measurement characteristics of SFDs$_{(20,50)}$ is H$_1$A and the additional hypotheses generated as part of this work are H$_1$B, H$_1$C, and H$_1$D.

The derivations of the hypotheses were carried out under the following assumptions:

- V_v = 50 km/h.
- θ = 20 degrees.
- V_w = 46.98 km/h (V_v Cos (θ)).

The key parameters (% Wheel slip and operational velocity) resulting from these derivations are presented in Table 1.

5 EXPERIMENTALLY CHALLENGING THE HYPOTHESES

5.1 *Equipment used*

The Skid Resistance Development Platform (SkReDeP) (Figure 3) is a research tool used by National Highways for the development of their skid resistance management policy. SkReDeP incorporates sideway-force measurement equipment which conforms to the UK standards for SFD$_{(20,50)}$ (British Standards Institution 2006) (Highways England *et al.* 2019). For this work, SkReDeP was used in its standard configuration in accordance with BS 7941-1-2006 (British Standards Institution 2006).

The Pavement Friction Tester (PFT) (Figure 4) is a longitudinal variable slip (friction is measured in line with the direction of travel at a range of % Wheel slips) friction testing device used as a research tool by National Highways. The PFT is comprised of a tow vehicle and trailer. The trailer holds the test wheel, which is mounted on an instrumented axle. The test wheel can be independently braked and the forces acting upon it measured to determine

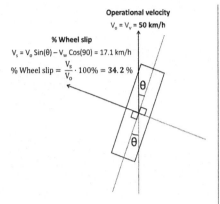

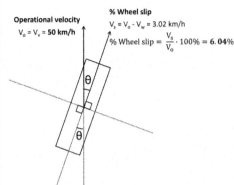

In H_1A friction is considered to be measured in the direction of vehicle travel and so V_o is resolved in this direction, and the V_s is considered to act along the rotational axis of the test tyre.

In H_1B friction is considered to be measured in the direction of vehicle travel and so V_o is resolved in this direction, and the V_s is considered to act about the rotational axis of the test tyre.

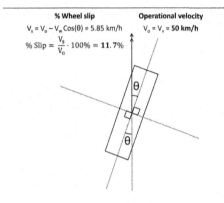

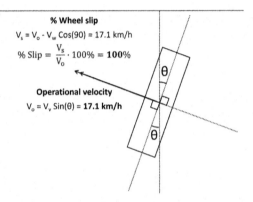

In H_1C friction is considered to be measured in the direction of vehicle travel and so V_o is resolved in this direction, and the V_s is considered to act in the direction of vehicle travel.

In H_1D friction is considered to be measured along the rotational axis of the test tyre and so V_o is resolved in this direction, and the V_s is considered to act along the rotational axis of the test tyre.

Figure 2. Derivation of H_1A (Upper Left), H_1B (Upper Right), H_1C (Lower Left), and H_1D (Lower Right).

Table 1. Key parameters resulting from desk study.

Hypothesis	% Wheel slip	Operational velocity
H_1A	34.2	50 km/h
H_1B	6.04	
H_1C	11.7	
H_1D	100	17 km/h

Figure 3. The National Highways Skid Resistance Development Platform (SkReDeP).

Figure 4. The National Highways Pavement Friction Tester (PFT).

the friction between the test tyre and road surface. Critically for this work, the PFT reports operational velocity and % Wheel slip directly.

Modifications were made to the PFT trailer and measurements were therefore made outside of normal testing standards. These modifications were carried out in order to make

227

the PFT "look" as much like the SkReDeP as possible such that a direct comparison between the measurements made by both devices could be made:

- SkReDeP wheels and tyres were added to the PFT using bespoke interference plates. These plates allowed two tyres to be installed on each side of the PFT trailer.
- The PFT trailer weight was amended to be representative of that of the SkReDeP. It was only possible to reduce the trailer weight of the PFT to 400 kg per wheel without affecting the structural stability of the PFT trailer. This is twice the load applied to a SkReDeP wheel, hence the use of two wheels per side.
- The angle between the PFT test tyres and direction of travel remained unchanged (zero degrees) as this was the dependant variable being assessed.

To ensure that this setup would not introduce error into the results, a preliminary study was carried out comparing the friction measurements made with the PFT using one SkReDeP tyre at 300 kg and a vertical load of 300 kg (per side), with measurements made using two SkReDeP tyres at 600 kg load (per side). This study is detailed by Sanders (2021) and concluded that both setups produced similar results (as determined using Student's t-test).

5.2 *Materials assessed and measurements made*

The experimental work required the determination of the properties of materials with different nominal friction levels, at a wide range of vehicle speeds, and with multiple measurements being made at each vehicle speed. To this end measurements were made on the straight line wet grip area at the HORIBA-MIRA proving ground. Table 2 summarises the measurements made.

Measurements were made in such a way as to minimise the effects of track conditioning[2]. This was achieved by using an alternating test programme where measurements were made in rotation such that track conditioning effects would be averaged out in the data analysis.

Table 2.   Summary of measurements made.

Material	Material description	SkReDeP passes at speeds (km/h):			PFT measurements at speeds (km/h):		
		30	50	80	30	50	80
Bridport Pebble	An asphalt material with bridport pebble aggregate	8	8	8	16	13	4
Basalt Tiles	A smooth surface constructed from basalt tiles	8	10	8	24	28	43
DeluGrip 1	An asphalt surface designed to have	8	8	7	27	26	22
DeluGrip 2	high friction values	6	5	6	39	46	42
ISO braking asphalt 1	An asphalt surface used in standardised	8	8	8	20	26	25
ISO braking asphalt 2	braking tests	8	8	8	27	27	18

---

[2]The act of making a friction measurement alters the friction of the surface.

5.3 *Data assessment and results*

The data collected were assessed by plotting friction values against the operational velocities for measurements made using the PFT (Fn), with measurements made using SkReDeP (SR), for each of the hypotheses being challenged and on each of the surfaces assessed. This yielded 24 plots (4 plots for each of the 6 materials tested) which follow the format of that shown in the example in Figure 5 where:

- solid series markers show the average Fn or SR at each nominal operational velocity,
- shaded series markers represent the individual Fn or SR values,
- solid lines represent lines of best fit through the average Fn or SR values,
- the broken black line represents the line of best fit through all average values.

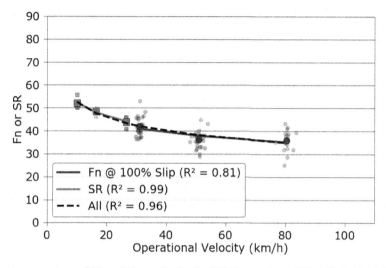

Figure 5. A comparison of SR and Fn results for the ISO 2 material and H_1D. Individual data points are shown in the translucent series markers. Average values at each operational velocity are shown in the solid series markers. R^2 values are shown in the legend.

The agreement between measurements made using the PFT and SkReDeP was quantified using the R^2 value for the "All" series. This methodology was used in preference to traditional statistical methods (Student's t-test / Cohen's d-test) because such tests could not be applied to data assessed under H_1D.

For brevity Figures 6 to 11 present results relating to the currently accepted hypothesis (H_1A), and the hypothesis resulting in the best agreement between PFT and SkReDeP

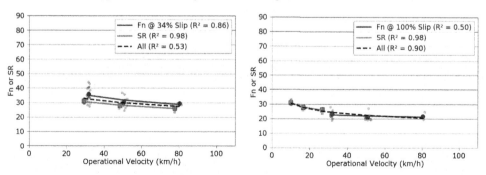

Figure 6. Measurements made on the **bridport pebble** material for H_1A (Left), and the hypothesis resulting in the largest R^2 value for the "All" series, H_1D (Right).

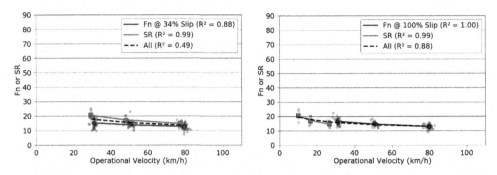

Figure 7. Measurements made on the **basalt tiles** material for H_1A (Left), and the hypothesis resulting in the largest R^2 value for the "All" series, H_1D (Right).

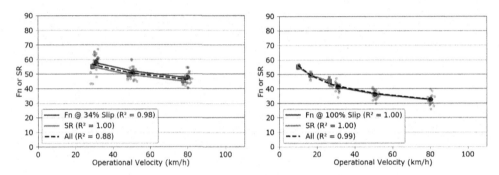

Figure 8. Measurements made on the **DeluGrip 1** material for H_1A (Left), and the hypothesis resulting in the largest R^2 value for the "All" series, H_1D (Right).

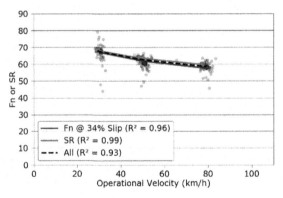

Figure 9. Measurements made on the **DeluGrip 2** material for H_1A which was the hypothesis resulting in the largest R^2 value for the "All" series.

measurements. Summary results for each material and hypotheses are presented in Table 3. When viewing these figures it should be kept in mind that friction as measured with the PFT is dependant upon % Wheel slip and therefore different friction values are measured under each hypothesis.

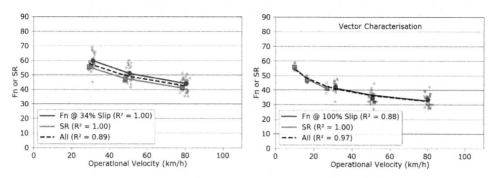

Figure 10. Measurements made on the **ISO braking asphalt 1** material for H_1A (Left), and the hypothesis resulting in the largest R^2 value for the "All" series, H_1D (Right).

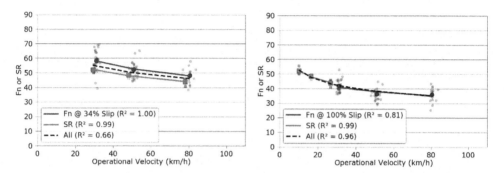

Figure 11. Measurements made on the **ISO braking asphalt 2** material for H_1A (Left), and the hypothesis resulting in the largest R^2 value for the "All" series, H_1D (Right).

Table 3. Summary of testing results.

Material	R^2 value for all average values relating to hypothesis:			
	H_1A	H_1B	H_1C	H_1D
Bridport Pebble	0.53	0.03	0.09	0.90
Basalt Tiles	0.49	0.14	0.27	0.88
DeluGrip 1	0.88	0.65	0.53	0.99
DeluGrip 2	0.93	0.72	0.57	0.87
ISO braking asphalt 1	0.89	0.69	0.60	0.97
ISO braking asphalt 2	0.66	0.80	0.29	0.96

6 CONCLUSIONS, DISCUSSION AND RECOMMENDATIONS FOR FUTURE WORK

From the results presented, the following observations can be made:

- The results of measurements made on the; Bridport Pebble, Basalt Tiles, DeluGrip 1, ISO braking asphalt 1, and ISO braking asphalt 2 materials support a single hypothesis; H_1D.
- The results of measurements made using the PFT and SkReDeP on the DeluGrip 2 material supported H_1A; a finding outlying from the overall results of this study.
- The DeluGrip 2 material provided the largest friction values of all the materials assessed.

231

These observations led to the following overall conclusions:

1. H_1A should be rejected as the working hypothesis for the measurement characteristics of $SFDs_{(20,50)}$ in favour of H_1D.
2. $SFDs_{(20,50)}$ characterise friction at 100% Wheel slip and 17 km/h operational velocity ($V_v Sin(20)$).

These conclusions, demonstrate that the mechanism of skid resistance characterised using $SFDs_{(20,50)}$ differs from that previously accepted. It would be prudent to investigate the implications of this finding on the management of road surface skid resistance when considering the frictional demands that vehicles place on the network. Such an investigation could consider the following research questions:

- How do the frictional demands of vehicles relate to % Wheel slip and operational velocity?
- How should this update in knowledge (gained from the work presented in this paper, and that provided by the answer(s) to the above research question) affect the use or collection of skid resistance data for the management of friction on the network?

A second point of discussion relates to the outlying results observed on the DeluGrip 2 material. Sanders (2021) explores this in some detail in Chapter 10 of that work and submits the hypothesis that the results are due to SkReDeP making measurements below the critical wheel angle; a fundamental property of all SFDs that it is essential to make measurements above. It should be noted that as yet, this hypothesis is not supported by dedicated experimental data. Experimental works designed to challenge this hypothesis should be carried out before it is accepted.

REFERENCES

British Standards Institution. (2006), *BS 7941-1-2006 Methods for Measuring the Skid Resistance of Pavement Surfaces – Part 1 SCRIM*, BSi, London, England.
Henry J.J. (2000), *Evaluation of Pavement Friction Characteristics – A Synthesis of Highway Practice*, Transportation Research Board and National Research Council, Washington DC, USA.
Highways England, Transport Scotland, Welsh Government, Department for Infrastructure, (2019), *CS 228 Pavement Assessment and Assessment – Skidding Resistance*, National Highways, London, England.
Sanders P.D. (2021), *PPR980 Characterising The Measurements Made by Sideways-force Skid Resistance Devices – An Experimental Study*, TRL, Wokingham, England.

Roads and Airports Pavement Surface Characteristics – Crispino & Toraldo (Eds)
© 2023 The Author(s), ISBN 978-1-032-55149-4

Alternative methods of road condition monitoring

Helena Angerer* & Berthold Best
Faculty of Civil Engineering, Nuremberg Institute of Technology, Germany

ABSTRACT: Municipal road condition monitoring relies both on measurement campaigns and visual inspections. This is often unsuitable, as conditions in urban areas change rapidly and recording requires enormous human and financial resources. Due to this periodic recording, there is a lack of a continuous overview of the road condition's development. The aim of this research project is to develop a cost-effective measurement system for continuous monitoring road conditions in urban areas. The use of new data recording methods allows data monitoring and condition assessment to be carried out continuously and with minimum human resources. This innovation supports the road asset management of municipalities and has an impact on economic and political issues. Through such continuous monitoring, damage detection is possible at an early stage, and the condition data updates permanently. The continuous road condition monitoring leads to optimized maintenance strategies instead of expensive early renovation or renewal. The reduction of the employment of human resources needed to record the damages and to take action at an early stage to intervene has cost-saving effects. In addition, the period of use of a road can be extended and traffic safety increases. Overall, this leads to an improvement in sustainability as well as an extension of the road availability. Through this innovation in asset management the mobility requirements of the next generations can be met.

Keywords: Road Condition Monitoring, Road Asset Management, Maintenance

1 RESEARCH GAP AND MOTIVATION

Municipal road condition monitoring relies both on measurement campaigns and visual inspections. This is often unsuitable, as conditions in urban areas change rapidly and recording requires enormous human and financial resources (FGSV 2012). Due to this periodic recording, there is a lack of a continuous overview of the road condition's development. Considering that municipal roads account for about three quarters of the total road network in Germany in terms of distance (BMVI 2020), for this reason it is mandatory to develop a system focused on municipal issues (FGSV 2012).

Therefore, the main aim of the research is to develop a cost-effective measurement system for continuous monitoring road conditions in urban areas. This paper presents a proof of concept how the acceleration sensor technology of a GoPro can be used for road condition monitoring.

2 CONDITION CHARACTERISTICS OF AN ASPHALT ROAD

The road surface can consist of three different materials: asphalt, concrete, stone paving. However, this article is limited to the road condition monitoring of roads with asphalt as road surface. The condition of asphalt roads in urban areas is divided into different condition

---

*Corresponding Author: helena.angerer@th-nuernberg.de

DOI: 10.1201/9781003429258-23

characteristics. These condition characteristics are combined into two groups of characteristics. The evenness characteristic includes the condition characteristics of longitudinal and transverse unevenness. The substance characteristic for the asphalt surface includes the following condition characteristics: cracks, open seams and joints, applied patches, inlaid patches as well as other surface damage such as break-outs, leaching and binder accumulation (FGSV 2016).

As soon as a condition characteristic describes a damage, it is a damage characteristic. As an alternative method, these damage characteristics shall be recorded by measuring the vertical acceleration of a vehicle as it passes over the damage. The damage characteristics that can be recorded with acceleration sensor technology were already determined in the first thesis about this topic at the TH Nuremberg by Michael Knüpfer (see Table 1) (Knüpfer 2019).

Table 1. Detectability of damage characteristics with acceleration sensor technology.

Damage features are ... ... with acceleration sensor technology	... detectable ...	... not detectable ...
longitudinal unevenness: short-wave	X	
longitudinal unevenness: long-wave		X
transverse unevenness		X
crocodile cracks	X	
longitudinal cracks		X
transverse cracks	X	
open seams and joints	X	
applied patches	X	
inlaid patches	X	
break-outs	X	
leaching		X
binder accumulation		X

Due to the fact that the acceleration sensor technology is limited to the contact area of the wheels to the road, the surface next to the road, the edge areas as well as the drainage facilities cannot be recorded by the acceleration sensor. In order to include these areas in the recording, an additional visual recording is necessary.

3 QUALIFICATION OF THE GOPRO AS A MEASURING SYSTEM

At the research project "Alternative methods for road condition monitoring" of the Institute of Technology in Nuremberg, the GoPro was used to visually record the condition characteristics via video and to record the location data (GPS). While doing this, the GoPro also records the internal time. The accelerations that affect the vehicle while driving on the road were previously recorded with an additional acceleration sensor. This external sensor stores the system time of the computer that controls the sensor. Synchronising the data over time is therefore not possible with sufficient accuracy.

When acceleration data was detected in the metadata of the GoPro, the question came up whether the GoPro could be used as a measurement system on its own. If the GoPro can record the video and GPS data as before and additionally the acceleration data, further steps such as synchronising the various data could be made easier, or even unnecessary.

A study at the Institute of Technology in Nuremberg (Angerer 2021) investigated whether the GoPro records all the required data (video, GPS, acceleration) with a sufficient accuracy. Furthermore, it was analysed if the data are already synchronised or if the synchronisation can be done afterwards. The evaluation showed if it is possible to identify condition characteristics from the acceleration data. All in all, the study researched whether the GoPro is suitable as an alternative measurement system for recording road conditions.

4 PERFORMING MEASUREMENTS WITH THE GOPRO

The GoPro was mounted on the bonnet of the VW T5 using the supplied curved adhesive mount and the mounting clip. While the GoPro is not only used for visual recording, but also for recording acceleration data, it is important to take care that the orientation of the GoPro also corresponds to the axes for acceleration measurement. This means that it is not possible to angle the GoPro forwards to get a different image frame. The GoPro has to be mounted vertically on the bonnet.

In the settings, a resolution of 4 K at 30 frames per second with a linear lens was selected. The HyperSmooth function, which stabilises the image, was intentionally deactivated to make it easier to link the visual recording and the deflections in the course of the acceleration data later. The GoPro was connected to a smartphone to start and stop the recording from inside the vehicle.

The measurements were carried out on sections with few intersections and no roads with right of way. Intersections with traffic signals as well as intersections and intersecting roads with right before left force drivers to reduce their speed and pay attention to the traffic. Since the measurements are speed-dependent, the recordings should be made at an approximately constant speed. Considering the points mentioned above, a section in Nuremberg harbour was chosen for the measurements. The measurements started at the Duisburger Straße corner Hamburger Straße, extended over the Duisburger Straße, left onto the Bochumer Straße until it meets the Hamburger Straße again. The route has a length of 1.6 km and was driven three times.

The measurements were carried out at a speed of approximately 30 km/h. In order to reduce the natural oscillation of the driver around the planned speed, cruise control was activated after the speed of 30 km/h was reached.

5 OUTPUT DATA FROM THE GOPRO

The video is saved as an MP4 file. A web-based application called Telemetry Extractor for GoPro (Lite) is used to extract more data from the MP4 file. A CSV file with the acceleration data and a CSV file with the GPS data are created. The web-based application only has the option of outputting both data packages in two individual tables. The acceleration data is therefore not yet synchronised with the GPS data. Figure 1 below shows a section of the acceleration data as raw data:

```
"cts","date","Accelerometer [m/s2]","1","2","temperature [°C]"
"116.009","2021-11-25T12:42:30.116Z","10.24220623501199","0.4244604316546763","0.0407673860911271","16.927734375"
"121.08482741116751","2021-11-25T12:42:30.121Z","10.045563549160672","0.1223021582733813","0.21103117505995203","16.927734375"
"126.16065482233502","2021-11-25T12:42:30.126Z","9.94484412470024","-0.3165467625899280","0.2494004796163069","16.927734375"
"131.23648223350253","2021-11-25T12:42:30.131Z","8.973621103117505","-0.0575539568345323","0.2014388489208633","16.927734375"
"136.31230964467005","2021-11-25T12:42:30.136Z","8.815347721822542","-0.019184652278177457","0.2565947242206235","16.927734375"
```

Figure 1.   Section of the acceleration data as raw data.

"cts" is a time stamp in milliseconds based on the time stamp of the MP4 data. This must be selected in the web-based application for both the acceleration data and the GPS data in order to match the cts timestamp with the video footage. If this is taken into account when extracting both data sets, both files will have the same time stamp. Under this condition it is possible to synchronise the acceleration data and the GPS data via the timestamp "cts" later on. This is advantageous because the format as decimal number is more suitable for calculations than the format of the column "date". The column "date" contains the date and time with hours, minutes, seconds and milliseconds.

The next three columns show acceleration values in m/s^2. Because the acceleration values in the first column oscillate around a rest value of 9.81 m/s^2 and the gravity of 9.81 m/s^2 acts permanently on the z axis, it is obvious that this is the z axis of the GoPro. Through brief measurements, it could be determined that the x-axis and then the y-axis are recorded. The arrangement of the axes is shown in Figure 2. And the last column shows the temperature and is irrelevant in the context of the measurement so far.

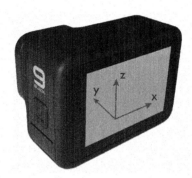

Figure 2.   Arrangement of the axes on the GoPro.

Several measurements show that the acceleration values are recorded by the GoPro with a frequency of up to 200 Hz.

The GPS data can also be exported as a CSV file by the web-based application. Figure 3 shows the beginning of the data package:

```
"cts","date","GPS (Lat.) [deg]","GPS (Long.) [deg]","GPS (Alt.) [m]","GPS (2D speed) [m/s]","GPS (3D speed) [m/s]","fix","precision","altitude system"
"74.905","2021-11-25T12:42:30.074Z","49.4037696","11.053111","318.09","3.761","3.93","3","129","MSLV"
"129.7647894736842","2021-11-25T12:42:30.129Z","49.4037699","11.0531081","318.092","3.778","3.76","3","129","MSLV"
"184.6245789473684","2021-11-25T12:42:30.184Z","49.4037702","11.0531052","318.086","3.802","3.78","3","129","MSLV"
"239.4843684210526","2021-11-25T12:42:30.239Z","49.4037706","11.0531024","318.053","3.819","3.8","3","129","MSLV"
"294.34415789473684","2021-11-25T12:42:30.294Z","49.403771","11.0530996","318.022","3.78","3.82","3","129","MSLV"
```

Figure 3.   Section of the GPS data as raw data.

The first two columns "cts" and "date" are identical to the acceleration file. However, it is noticeable that the GPS data records have a greater time interval than the data records with the acceleration data. A closer look shows that the GPS data are recorded with a frequency of up to 18 Hz.

The next two columns contain the geographical coordinates. The first of the two columns states the latitude and the second the longitude as decimal degrees. They are based on the geodetic reference system called the World Geodetic System of 1984 (short: WGS84). The coordinates are recorded by the GoPro with seven decimal places.

The measurement technology of the GoPro relating to the video quality, the decimal places and the frequencies of the acceleration and GPS data would be sufficient based on the characteristics determined so far. A negative aspect is that the user cannot make any settings on the acceleration sensor and does not receive any other information about the built-in measurement technology.

Due to the fact that the two data packages, as already shown in Figures 1 and 3, are output in two separate tables, they have to be merged in a further processing step. This is necessary in order to be able to display the data in a geographic information system such as QGIS. Several Excel macros were written to read in the two CSV files in Excel, prepare them and then link them in a table. These macros can be variably applied to all CSV files, as long as they were generated according to the procedure described above using the web-based application.

At the beginning of the first macro the acceleration file and then the GPS file needs to be selected. The macro reads the acceleration data in raw format into the first Excel worksheet called "Tabelle1" and the GPS data into "Tabelle2". In "Tabelle3" and " Tabelle4" are then the processed data in a table format of Excel. In "Tabelle5" the macro lists all data records, sorted by time (cts). The resulting table is limited to the relevant columns which makes the intermediate step look like this (see Figure 4):

cts	Date	Time	Accelerometer z [m/s2]	Accelerometer x [m/s2]	Accelerometer y [m/s2]	GPS (Lat.) [deg]	GPS (Long.) [deg]	GPS (2D speed) [m/s]
116,009	2021-11-25	12-42-30,116	10,24220624	0,424460432	0,040767386	49,40376982	11,05310883	3,773737344
121,0848274	2021-11-25	12-42-30,121	10,04556355	0,122302158	0,211031175	49,40376985	11,05310856	3,775310246
126,1606548	2021-11-25	12-42-30,126	9,944844125	-0,316546763	0,24940048	49,40376988	11,05310829	3,776883148
131,2364822	2021-11-25	12-42-30,131	8,973621103	-0,057553957	0,201438849	49,40376991	11,05310802	3,778643835
136,3123096	2021-11-25	12-42-30,136	8,815347722	-0,019184652	0,256594724	49,40376994	11,05310775	3,780864402
141,3881371	2021-11-25	12-42-30,141	9,887290168	-0,215827338	-0,175059952	49,40376996	11,05310749	3,783084969
146,4639645	2021-11-25	12-42-30,146	10,24940048	0,12470024	0,302158273	49,40376999	11,05310722	3,783305537
151,5397919	2021-11-25	12-42-30,151	10,27338129	0,136690647	0,035971223	49,40377002	11,05310695	3,787526104
156,6156193	2021-11-25	12-42-30,156	10,13908873	0,18705036	0,302158273	49,40377005	11,05310668	3,789746671
161,6914467	2021-11-25	12-42-30,161	9,585131894	-0,194244604	0,189448441	49,40377007	11,05310641	3,791967239
166,7672741	2021-11-25	12-42-30,166	8,472422062	0,033573141	0,395683453	49,4037701	11,05310614	3,794187806
171,8431015	2021-11-25	12-42-30,171	9,052757794	-0,143884892	0	49,40377013	11,05310588	3,796408373
176,9189289	2021-11-25	12-42-30,176	10,34292566	-0,318944844	0,035971223	49,40377016	11,05310561	3,798628941
181,9947563	2021-11-25	12-42-30,181	10,25899281	0,438848921	0,237410072	49,40377019	11,05310534	3,800849508
187,0705838	2021-11-25	12-42-30,187	10,20143885	0,18705036	0,038369305	49,40377022	11,05310508	3,80275797
192,1464112	2021-11-25	12-42-30,192	9,853717026	0,086330935	0,513189448	49,40377025	11,05310482	3,804330872

| Tabelle5 | Tabelle4 | Tabelle2 | Tabelle3 | Tabelle1 |

Figure 4.   Acceleration and GPS data in one table sorted by cts.

Figure 4 shows that the worksheet "Tabelle5" consists of both the acceleration and the GPS data. After sorting the data by time, the data could be declared as synchronised. However, it is obvious that the frequency of the acceleration data is higher than that of the GPS data. In order to display the acceleration data in a geographic information system, such as QGIS, concrete coordinates must be assigned to the acceleration values. This is done by the second macro, which completes the missing GPS data by linear interpolation over time (cts).

To be able to carry out this interpolation, the table has to be prepared first. It is quite possible that an acceleration data set and a GPS data set were recorded at the same time, for this reason it is necessary to identify these cases and combine them in one row. The other row is deleted. Additionally, it is possible that first an acceleration value is recorded and then a GPS value or at the end of the video the acceleration value is recorded longer because of the frequency. Therefore, acceleration values must be deleted before the first and after the last GPS value. Afterwards, the GPS values are linear interpolated, which means that each acceleration data set is given a specific GPS data. Due to the fact that the primary GPS data sets do not contain any acceleration data, these are deleted after the interpolation. Figure 5 shows the final table:

With the macro "Save" the finished file is then automatically saved using the path of the raw data as an Excel file with macros. The name of the file results from the name of the video, the recording date of the data and the addition "_ACC_und_GPS". In addition, the finished table, worksheet "Tabelle5", is saved separately as a CSV file in UTF-8 coding. The name of the CSV file is added the suffix "_CSV". The file with the macros is then closed automatically without saving. This means that the source file is never changed and can be used again each time as a kind of programme for synchronisation. If all the macros mentioned so far wish to be executed directly one after the other, this is possible with the macro "Execute".

cts	Date	Time	Accelerometer z [m/s2]	Accelerometer x [m/s2]	Accelerometer y [m/s2]	GPS (Lat.) [deg]	GPS (Long.) [deg]	GPS (2D speed) [m/s]
116,009	2021-11-25	12-42-30,116	10,24220624	0,424460432	0,040767386	49,40376982	11,05310883	3,773737344
121,0848274	2021-11-25	12-42-30,121	10,04556355	0,122302158	0,211031175	49,40376985	11,05310856	3,775310246
126,1606548	2021-11-25	12-42-30,126	9,944844125	-0,316546763	0,24940048	49,40376988	11,05310829	3,776883148
131,2364822	2021-11-25	12-42-30,131	8,973621103	-0,057553957	0,201438849	49,40376991	11,05310802	3,778643835
136,3123096	2021-11-25	12-42-30,136	8,815347722	-0,019184652	0,256594724	49,40376994	11,05310775	3,780864402
141,3881371	2021-11-25	12-42-30,141	9,887290168	-0,215827338	-0,175059952	49,40376996	11,05310749	3,783084969
146,4639645	2021-11-25	12-42-30,146	10,24940048	0,12470024	0,302158273	49,40376999	11,05310722	3,785305537
151,5397919	2021-11-25	12-42-30,151	10,27338129	0,136690647	0,035971223	49,40377002	11,05310695	3,787526104
156,6156193	2021-11-25	12-42-30,156	10,13908873	0,18705036	0,302158273	49,40377005	11,05310668	3,789746671
161,6914467	2021-11-25	12-42-30,161	9,585131894	-0,194244604	0,189448441	49,40377007	11,05310641	3,791967239
166,7672741	2021-11-25	12-42-30,166	8,472422062	0,033573141	0,395683453	49,40377701	11,05310614	3,794187806
171,8431015	2021-11-25	12-42-30,171	9,052757794	-0,143884892	0	49,40377013	11,05310588	3,796408373
176,9189289	2021-11-25	12-42-30,176	10,34292566	-0,318944844	0,035971223	49,40377016	11,05310561	3,798628941
181,9947563	2021-11-25	12-42-30,181	10,25899281	0,438848921	0,237410072	49,40377019	11,05310534	3,800849508
187,0705838	2021-11-25	12-42-30,187	10,20143885	0,18705036	0,038369305	49,40377022	11,05310508	3,80275797
192,1464112	2021-11-25	12-42-30,192	9,853717026	0,086330935	0,513189448	49,40377025	11,05310482	3,804330872

Tabelle5 | Tabelle4 | Tabelle2 | Tabelle3 | Tabelle1 | (+)

Figure 5. Table with acceleration and GPS values for each data set.

It can be noted that the data from the GoPro are not synchronised at the beginning, but they can still be synchronised afterwards with some macros. With the macros, the time required per measurement is only half a minute.

7 EVALUATION OF THE DATA

For evaluation, the final table with the suffix "_CSV" is imported into QGIS. This is illustrated in Figure 6.

Figure 6 can be used to identify the road where the measurements took place. Accordingly, the GPS data are sufficient for this. Figure 7 shows the inaccuracy of the GPS data due to the fact that the position of the measurement vehicle was recorded at different coordinates while standing. The inaccuracies in the location recording led to a maximum variation of 0.57 m for video GH010073 (see Figure 7).

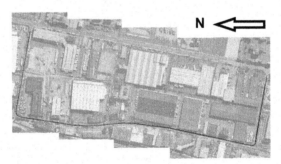

Figure 6. Data imported into QGIS.

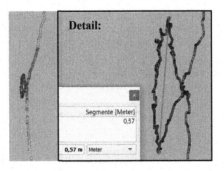

Figure 7. Inaccuracy of GPS data while standing.

Unfortunately, the GPS is not accurate enough to reliably output the driven lane. Figure 8 shows the damage on the opposite lane.

Figure 8.   Road damage located in the wrong lane.

In QGIS, the measurement points were displayed in different colours depending on the measured acceleration value. Measuring points that were within $+/- 0.2$ m/s^2 of the rest value of 9.81 m/s^2 were displayed in dark green. Measuring points with a greater difference from the rest value were displayed in a colour range from light green to yellow to red.

In order to be able to display the measured values in a diagram, the data have to be prepared further. Because speed changes occur during the measurements and the measuring vehicle has to stop to give way to another vehicle, it does not make sense to display the acceleration values in a diagram over the time. To compensate for the speed differences, the travelled distance s must first be calculated from the speed v and the time t using the following equation:

$$s = v * t \tag{1}$$

This was also done with a macro. When the vehicle is standing still, the speed is zero, which means that the travelled distance is also zero. If the acceleration data is displayed in a diagram over the distance, any acceleration values from standing are assigned to only one value on the axis of the distance.

As this leads to an uneven distribution of the acceleration values on the axis of the distance, an xy diagram has to be chosen for visualisation in Excel. The dot plot allows an uneven distribution of the values over both axes. This happens because the data points, consisting of the acceleration value and a value for the travelled distance, are plotted at the intersection of both values in the diagram. In comparison, a line diagram would always mark the acceleration values one unit further. A second value axis with the distance or time is not available.

Before the three measurement runs can be entered in the same diagram, a common kilometre classification has to be selected. Assuming that the maximum acceleration value was generated for all three measurement runs when passing over the same road damage, the kilometre classification can be made on the basis of the maximum measurement value. The measurement run that has the greatest travelled distance to the maximum measurement value was started first. In order to bring the other measurement series to the same kilometre value, the distance value of the other measurement series is added to each distance value. The distance values are then equal at the maximum acceleration value.

239

The following diagram (Figure 9) shows the acceleration values of the three measurements on the y-axis and the travelled distance in metres on the x-axis. In addition, the rest value of 9.81 m/s² and the threshold value at a tolerance of + /- 2,5 m/s² from the rest value are marked. This threshold value was determined by Christoph Raab as part of his bachelor's thesis at the Institute of Technology in Nuremberg (Raab 2020).

Based on the distance and personal impression, the maximum deflection between 1200 m and 1300 m is caused by the following damage (see Figure 10).

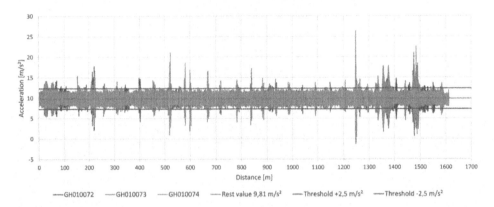

Figure 9.   Diagram with the acceleration data of the three measurement runs.

Figure 10.   Road damage at maximum acceleration deflection.

A closer look at the accelerations in the range of the maximum deflection shows the following diagram (see Figure 11):

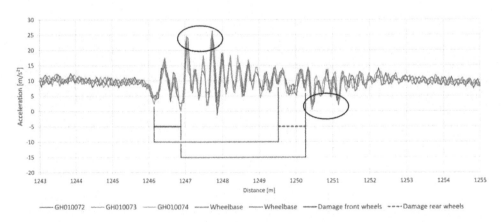

Figure 11.   Acceleration in the range of the maximum deflection.

240

The diagram shows the acceleration values on the y-axis and the distance in metres on the x-axis. It shows the acceleration data for all three measurement runs. At the beginning, the acceleration values oscillate around the rest value of 9.81 m/s². Then, the profile is influenced by the road damage and reaches a low point at the distance 1246.15 m in data series GH010073. The next, similarly significant, low point is located at the distance 1246.88 m. The distance between the two points is 0.73 m. This corresponds approximately to the spread of the damage in the longitudinal direction. The former vibrations were only initiated by the passage of the two front tyres over the damage. Consequently, there are two more extrema. One low point is at the distance 1249.52 m and another at 1250.26 m. The distance between these two points is 0.74 m.

The distance of the first low point of the front wheels to the first low point of the rear wheels is 3.37 m and the distance of the second low point of the front wheels to the second low point of the rear wheels is 3.38 m. This reflects the measuring vehicle's wheelbase of 3.40 m.

It can be critically noted that the effects of the rear wheels are not obvious in the diagram without further knowledge. But it is interesting to note that the accelerations overlap in the later part of the diagram in such a way that more obvious extremes are obtained. These extremes are circled in black in the diagram in Figure 11. The distances between these extrema also represent the extent of the damage in the longitudinal direction and the wheelbase of the measuring vehicle. These superimposed extrema are significantly visible in the diagram and form an initial basis for identifying damage from the acceleration data and determining its extension in the direction of travel.

8 CONCLUSION AND OUTLOOK

When investigating whether the GoPro is suitable as an alternative measurement system for road condition monitoring, it was determined that all the required data (video, GPS, acceleration) are recorded sufficiently accurately by the GoPro, only the GPS could sometimes be more accurate. Even though the data are not synchronised with each other at the beginning, this is not a problem because the data can be synchronised afterwards with a macro. In the evaluation, it was possible to identify damage features from the acceleration data. Furthermore, it was even possible to determine the spread of a damage feature in the direction of travel. The wheelbase of the measuring vehicle could also be derived from the acceleration data. Nevertheless, the visual image material is currently still needed to support the evaluation.

Because the measurements are speed-dependent and the influence of the speed has not been investigated so far, the measurements must be recorded at a constant speed. In the further steps, on the one hand, the dependence on the speed and the longitudinal acceleration of the measuring vehicle should be investigated, and on the other hand, acceleration data from various types of road damage must be recorded and evaluated. If these investigations confirm the findings of this paper, an artificial intelligence can be trained on the basis of the data.

For further investigations, the GoPro can certainly be used. However, within the framework of the research project "Alternative methods for recording road conditions", it has become apparent that it is useful if there are more information and setting options for the acceleration sensor. The disadvantage of the GoPro is that there is no information about the installed components. Even the measuring frequency was only found by evaluating the measurement logs.

REFERENCES

Angerer H., 2021. *Eignung der GoPro als Alternatives Messsystem zur Straßenzustandserfassung.*
BMVI (Bundesministerium für Verkehr und digitale Infrastruktur), 2020. *Verkehr in Zahlen 2020/2021.*

FGSV (Forschungsgesellschaft für Straßen- und Verkehrswesen), 2012. *Empfehlungen für das Erhaltungsmanagement von Innerortsstraßen (E EMI 2012)*.

FGSV (Forschungsgesellschaft für Straßen- und Verkehrswesen), 2016. Arbeitspapier Nr. 9/K 2.1 zur Systematik der Straßenerhaltung; Reihe K: Kommunale Straßen; Abschnitt K 2: Zustandserfassung; Unterabschnitt K 2.1: *Vorbereitung und Durchführung der messtechnischen Zustandserfassung für innerörtliche Verkehrsflächen*.

Knüpfer M., 2019. *Bildreferenzierung bei der Straßenzustandserfassung mittels Beschleunigungssensorik*.

Raab C., 2020. *Ermittlung von Warn- und Schwellenwerten für die Straßenzustandsbewertung mittels Erschütterungssensorik*.

The multi-speed deflectometer: New technology developed for traffic-speed non-destructive structural testing of pavements

Alessandro Marradi*
Department of Civil and Industrial Engineering, University of Pisa, Italy

Lily Grimshaw & Graham Salt
Pavement Analytics Group, GeoSolve Ltd, New Zealand

ABSTRACT: The Traffic Speed Deflectometer has transformed pavement structural data collection on highways, where network testing was formerly carried out with Falling Weight Deflectometer, Deflectograph or Beam. However, the Multi-Speed Deflectometer (MSD) is now also available, which can test highways but more significantly, fills a gap for an efficient device for structural testing of urban roads. In these locations, issues that are often overlooked include the frequent slowing or stopping at intersections, cornering, access, the extreme variability of structural stiffness due to pavement subservices and the collection of quality structural data over a wide range of speeds while still ensuring the unimpeded flow of traffic at all times. The Multi-Speed Deflectometer is an economical non-destructive traffic speed pavement testing device used to benchmark the structural capacity of large networks of roads. Data are collected at 1 m intervals, usually in both wheelpaths and averaged to 10 or 20 m intervals in each lane. MSD structural data have been collected over the last 4 years in multiple regions throughout New Zealand and Italy. When paired with traditional surface profiling from the high-speed data (HSD), reliable traffic records and maintenance history, a comprehensive understanding of the mechanisms of pavement performance can be achieved including both the surfacing and the structural layers. Examples are provided to demonstrate application. Pavements with a poor surface condition can be cross checked against the structural condition to verify whether there is an underlying structural issue. If so, these sites can then be flagged for project level testing and renewal. Sites with poor surfacing condition and no structural issues can be flagged for maintenance or re-surfacing treatment. The right solution for the right problem at the right time and over the right extents can now be economically identified, providing authorities with the capability of assessing the optimum Net Present Value expenditure for any large roading network.

Keywords: Multi-Speed Deflectometer, pavement deflections, structural evaluation, non-destructive testing

1 INTRODUCTION

The Traffic Speed Deflectometer (Xiao et al. 2021; Zofka & Sudyka 2015) has transformed pavement structural data collection particularly because standard reporting at 10 m intervals or less addresses the extreme variability of structural stiffness inherent in many

---

*Corresponding Author: alessandro.marradi@unipi.it

DOI: 10.1201/9781003429258-24

pavements. However, its cost and the limited number of units worldwide means it is not always readily available for pavement screening. Traditionally, the Falling Weight Deflectometer (Ullidtz 1998) has been used for both network and project level surveying in many countries worldwide. While FWD testing has proven extremely useful to confirm the distress mode and most effective type of rehabilitation design at project level, it is much less effective for network level surveying because it is slow and hence often is used with low test density (points per road area coverage). Furthermore, the FWD requires costly traffic management to minimise health and safety risks to the operators and road users. Similar limitations are associated with other traditional devices, such as the Deflectograph and Benkelman Beam.

The Multi-Speed Deflectometer (MSD) is now also available, which can test highways but also fills a gap for an efficient device for structural testing of urban roads where access, cornering, frequent reductions in speed with stopping at intersections, and the collection of quality structural data over a wide range of customary traffic speeds, are important considerations. The Multi-Speed Deflectometer is ideal for economical non-destructive traffic speed pavement structural testing in these conditions to benchmark the structural capacity of a large network of roads. Data are recorded at 1 m intervals, usually in both wheel tracks (300,000 test points per day) and averaged to 10 or 20 m, providing near continuous structural data useful for defining structurally homogenous sections and to indicate the location of reduced capacity within the pavement cross section i.e., which pavement layer will first develop distress and hence become critical.

Network level pavement management based solely on surface condition observations relies on identifying distress only once it manifests. Additional structural testing is required to identify the cause of distress, because assessment from surface parameters enables only short-term Forward Works Programming (1 to 2 years), hence inhibiting the planned intervention prior to the initiation of distress reaching a terminal condition. Most of the traditional surface condition parameters (rutting, roughness, cracking and visual imaging) can be collected simultaneously with the same MSD vehicle, greatly reducing the overall cost and carbon emissions for provision of comprehensive state-of-the-art network management.

2 COMPARISON OF TSD, FWD AND MSD

The science underlying FWD and TSD is limited to recording of vertical velocity of the pavement surface at unloaded points near a heavy uniaxial load on a plate (FWD) or between moving wheels (TSD), whereas the science underlying the MSD involves capturing all forms of 3-dimensional deformation of the pavement surface using multiple sensors and images recording data both from beneath and around the contact patches of heavily loaded moving wheels. Differences between the FWD and MSD are compared in detail in Table 1.

The measures are fundamentally different, but it is important to note that all of the differences are such that the MSD deformations are more representative of the actual in situ deformations that occur under a heavy vehicle. Therefore, the deformations from the MSD should be more suitable for predicting pavement performance particularly where there are multiple distress modes, or where models that acknowledge only uniform layers with vertical loading are less appropriate. ASTM D5858 (2020) highlights the issues involved for calculating layer moduli from FWD test results, particularly for cracked pavements or locations without pavement layering information.

The use of lasers on the TSD limits surveys to drier conditions which in the case of New Zealand surveys and limited TSD availability has led to avoidance of testing in wet seasons

Table 1. Key Differences between the MSD and FWD.

Multi Speed Deflectometer	Falling Weight Deflectometer
Pneumatic tyre (deformable) with 30mm rubber and steel mesh/ply	Steel/fibre circular plate (stiff) covered with 3mm of ribbed rubber
Rolling load creating a mini "bow wave" at traffic speed	Stationary position and weights dropped to mimic vertical load at traffic speeds
Rotation of principal stresses	Fixed orientation of principal stresses
Measurement of 3D longitudinal, transverse and vertical deformations characterising the asymmetric deflection bowl	Measurement of vertical deformations only, characterising a symmetric deflection bowl
Transverse accelerations affect wheel load to match those of actual heavy vehicles	No consideration of any transverse (radial) accelerations on corners or due to camber or superelevation
Using a rolling wheel inherently acknowledges that the longitudinal profile (at all wavelengths) induces changes in dynamic vertical loads which have a consequent impact on pavement life prediction.	Static location provides a reading which relates only to loading from a smooth road (IRI=0). *This leads to both under and over prediction of remaining structural life, and substantially so for mature roads*
Near continuous spatial coverage at about 1m centres optionally presented as median each 10 or 20m	Spatially separated individual test points every 20 or 50m centres staggered across lanes –no indication of variation on the vast majority of the pavement
Both wheeltracks tested simultaneously at minimal additional cost.	Normally only one wheel track is tested, otherwise costs are double. Data collection and traffic management can be difficult when surveying the offside wheel path
Response is always from loading within each wheeltrack as no additional edge clearance is required.	As the FWD load plate is centrally located, the wheelpath cannot always be tested if there is inadequate clearance (eg from parked vehicles)

when pavements are in their most susceptible condition. The MSD can survey in both wet and dry conditions and because a dedicated vehicle is not required (installation of the various devices takes only a few hours), multiple MSDs can be readily mobilised and available, including in remote locations.

ASTM D4695-03 (2020) General Pavement Deflection Measurements also includes FWD testing intervals according to the different goals, ranging from an upper limit of 500 m for network level, reducing to 10 m where necessary for detailed project level. These limitations do not apply to the MSD given the continuous nature of testing.

3 MSD DESIGN OBJECTIVES

The MSD has been developed by installing and exploring the recordings of all types of high performance sensors and continually upgrading their configuration as available specifications for these are progressively enhanced. The prime objective is to extend beyond the traditional limitation (recording only vertical deformation) to more realistically characterise the "myriad ways" (Dawson 2002) in which pavements respond when experiencing different modes of distress. Effectively recording their multi-dimensional dynamic behaviour provides the basis of a more mechanistic approach for performance prediction.

MSD vehicles can be supplemented with other sensors (such as GPR & TDR), but these substantially increase the cost/km, yet the consequential effects of their parameters are already incorporated in the primary deformations beneath and around the tyre contact patch as recorded by standard MSD.

4 MSD DATA COLLECTION AND RATIONALE FOR INTERPRETATION

State-of-the-art pavement condition data collection and its structural evaluation requires:

• Collection of data to be non-destructive at traffic speed (no impediment to road users).

- Coverage of both the surface of existing roads and where practical, each layer of any road under construction, recording all data, near-continuously from both wheel paths of all appropriate lanes.
- Processing that determines all parameters relevant to pavement performance in a manner that also enables mechanistic characterisation.
- Identification of all modes of distress in all layers.
- Characterisation of spatial and temporal maintenance or renewal needs (extents, depths, and optimum timing) for each test point.
- Sub-sectioning all test points into homogenous Structural Treatment Lengths (STL), with ongoing re-sectioning (dynamic incremental-recursive model).
- Design of the most economic form of maintenance and timing for sub-intervals within each STL, and categorise each for local maintenance versus full length renewal
- Prediction of Remaining Structural Life, with a usefully reliable "Hit Rate" for each STL
- Determination of the optimum Forward Work Programmes for both Maintenance and Renewals (with due recognition of their interdependence) and determination of their respective costs.

Historically, such evaluations with FWD have been slow, costly and of variable reliability (Arnold et al 2009). Speed has been greatly increased with the advent of the Traffic Speed Deflectometer, although the length of the TSD makes it impractical on many local authority roads. Now with the Multi-Speed Deflectometer as well, all roads (under construction or completed, surfaced or unsurfaced, dry or wet in any condition) can be tested at traffic speed. MSD provides the additional advantages of measurements where the rubber meets the road (beneath the contact patch not just in the unloaded gap between dual wheels) as well as providing mechanistic insight into 3-dimensional deformations, testing continuously in both wheelpaths. The instrumentation is readily transportable to remote sites and can be installed or adapted to fit most heavy vehicles (including trailers or forklifts). Calibration is carried out using FWD, TSD, (or even Deflectograph or Beam if necessary), initially for seamless transition by their practitioners but ultimately for the more comprehensive characterisation of pavement properties and performance obtainable from the new technology.

Since the introduction of non-destructive testing of pavements by A C Benkelman in 1952 (Highway Research Board 1955) until now, the focus has been almost exclusively on one parameter: vertical deflection.

The science underlying FWD or TSD is somewhat limited in view of the above. Both devices record only vertical velocity of the pavement surface at unloaded points near a heavy uniaxial load on a plate (FWD) or between moving wheels (TSD). Widely recognised analytical models are then used for quantification of moduli, stresses and strains for known as-built layering.

The science underlying the MSD is somewhat different in that it focuses on capturing all forms of 3-dimensional deformation of the pavement surface. The relevant stress/strain tensor field throughout the deflection bowl (with each point having 9 components), and its observed asymmetry beneath a moving wheel precludes using just a simplistic measure (vertical deformation) if pavement life for a network is to be predicted with any reliability (particularly where there is minimal as-built information). Technology now provides a practical option with the capability for much more relevant, more comprehensive and more extensive data collection at traffic speed and at much lesser cost. MSD uses multiple sensors and images recording data both from beneath and around the contact patches of heavily loaded moving wheels then applying primarily machine learning to correlate the large volumes of data with equivalent simple data from an FWD or TSD recording of the same interval of road. Machine learning is then extended to associate other forms of 3-dimensional deformation recorded, using calibrations to sites that have known precedent performance in that region, including those observed to be experiencing specific distress modes or are in a terminal condition. This approach is taken because often there is little or no as-built information and so far, there appears to be no existing analytical model that will:

1. interrogate all of the recorded 3-dimensional dynamic characteristics of the deformations induced by a moving wheel and
2. output relevant parameters for an asymmetric layered visco-elastic model in a practical timeframe for network structural analysis and
3. evaluate them using any existing recognised criteria (fatigue limits).

Machine learning provides pavement engineers using MSD with a particularly effective tool to advance this new discipline mechanistically, beyond the limitations of the traditional scientific method, paraphrasing Anderson (2003):

"This is a world where massive amounts of data can, to a large degree at least, replace every other tool or test that might be brought to bear. Numbers give us not only immediate lessons from relevant history (regional precedent performance), but also unlimited potential for ongoing improvement.

Who knows the full theory of why roads perform the way they do? The point is they do, and for every region's permutation of terrain, sources, practices, loadings and climate, machine learning can now track and quantify their precedent performance with unprecedented fidelity.

With enough data, the numbers speak for themselves."

Pavements are highly variable structures that are not often amenable to simplistic analysis yet many of the traditional models are uni-variate (sometimes bi-variate). Experience with MSD data from large networks has demonstrated that multi-variate models that give due recognition to the myriad ways in which pavements become distressed, provide more reliable solutions. Many pavement models are based on results from laboratory testing or Accelerated Pavement Test facilities located at great distance from the relevant region. Few practitioners use relevant calibrated models that take into account all of the local conditions; subgrades, aggregate sources, construction methods, maintenance practices, environment etc. Until recently there was little choice. Such regionally-specific, calibrated mechanistic models based on historic observations of all relevant distress modes and precedent performance were often too costly or time-consuming to establish. However, high-speed collection of both structural and surface condition data together with the recent advances in big-data machine learning technology has effectively transformed the industry and provided a choice. Informed pavement management, more reliable performance prediction and optimised planning of forward work have become practical and economic realities for both categories of pavement networks, (highways and local roads).

Software has been developed, e.g., Regional Precedent Performance (RPP) which uses multi-variate analysis to analyse these huge data sets providing informed understanding of pavement deterioration and modelling of future performance. The cost is typically orders less than the cost of one kilometre of pavement rehabilitation, and benefits continue for many years.

Traditional methodology with visual inspections provides some information on pavement life predictions for up to 1-2 years ahead at best. The MSD provides the potential for a significant step forward that addresses Transport Agency focus on improving longer term predictions i.e. from 30 months out to 30 years. While reliability has been very low to at least until 2010, the potential for better reliability on highways with FWD supplemented by TSD data was indicated more recently by Stevens & Schmitz (2018), and with appropriate MSD output as well this is now being successfully extended to wider networks, including for the first time, local authority roads. Regional Precedent Performance longer term prediction of pavement life (RPP 30-30) is now being targeted with the latest MSD upgrades in hardware, firmware and software.

Outputs are now able to be delivered in close to real time, (the same day if necessary) enabling much more cost-effective and timely decision making for construction projects.

5 MSD OUTPUTS

MSD data output comes in three forms with varying detail in their characterisation: Basic, Empirical or Developmental.

5.1 *Basic MSD outputs*

Basic output is generated simply by correlation to the widely recognised FWD parameters, i.e. central deflection and curvature, standardised to 40kN load by default (50kN if required). Curvature for thick structural surfacings is commonly required as Surface Curvature Index, although where thin surfacings predominate, Curvature Function may be preferred.

5.2 *Empirical MSD outputs*

Empirical outputs include the HDM IV parameter, Adjusted Structural Number (SNP). In addition, more pertinent indices are available, similar to those promoted in Italy by ANAS (2021) since 2009 and in South Africa by Horak (2008), that focus on which layer is of interest and are determined from vertical deflection bowl offsets (at unloaded locations). Horak uses indices (with units of distance) and suffix of I for Index. To distinguish from these, MSD uses the prefix SN as the range of values is tied to SNP range for the network (normally 0 to 8). The corresponding MSD layer parameters are generated at or near loaded locations and are:

- Structural Number for Rutting (SNR) reflecting the stiffness of the whole pavement. It is similar to structural number (SNP) and relates inversely to central deflection. SNR relates to the resistance to rutting from the combination of movement in all layers resulting from both vertical and longitudinal deformations, scaled to the same range as SNP. The Structural Number for Vertical deformation (SNV) is also generated, relating to the vertical component of rutting deformation only.
- Structural Number for Base (SNB) a measure of the strength of the main structural layer and relates inversely to surface curvature index.

The above are the principal indices that may be provided for those familiar with FWD, TSD, Deflectograph or Beam, and calibration may be to whichever form of data is most readily available for any individual network.

5.3 *Developmental MSD outputs*

The MSD processing also outputs "Developmental" indices which relate to more specific characteristics which are at present recorded only by the MSD or are newly developed or under development (because they can be collected at minimal additional cost with the same vehicle). MSD research began in 2015 and the "signatures" of the multi-dimensional tensor field deformations present an enigma of which about 10% has been able to be deciphered each year, using principally, machine learning calibrations to observed performance. Many of the recorded features are not yet fully understood in relation to the progression of specific distress modes. Note not all of the following developmental indices have yet been advanced to the stage they can be used for production, but are documented here so that longer term goals can be indicated, and others may elect to use them for research (eg by applying them on sites where the reasons for premature distress are unknown but can then be explored by observing whether the extents of distress severity correspond consistently with extreme values). Feedback of this type of information and re-analysis greatly accelerates understanding of the relevant distress mechanisms, and ongoing feedback loops become successively more useful each year especially on heavily trafficked roads, as the significance of the MSD deformations becomes more evident from distress progression on each network. Re-processing to incorporate any changes in distress severity that are observed is fully automated. On most local roads where the traffic loading is reasonably well known or recorded, the structural testing should remain current and not need to be re-tested for several years.

Some of the developmental indices can be utilised in lieu of traditional HSD parameters. If HSD data are already available or become available in due course, they should be used in preference, otherwise the interim MSD equivalents may be adopted for network evaluation to refine or guide remaining life algorithms using MSD deformations.

- Structural Number for the Surface (SNS) a measure of the resistance to near surface instability along the wheelpath. It is significant only occasionally and is relevant to distress in unbound aggregates or thin surfacings.
- Modular Ratio Index (MRI) is a measure of the ratio of the moduli of successive layers above the subgrade, calibrated to the Normalised Modular Ratio parameter for FWD. A value of 1.0 indicates compaction is likely to be satisfactory and conforming with the Austroads modular ratios expected from good quality unbound granular aggregates. Values less than 1.0 may indicate under-compaction. Significantly higher values indicate bound layers may be present.
- Structural Number for Transverse Shear. (SNT) is a measure of the resistance to trans-verse shear. Low values are expected to be relatively rare in full width pavements but occasionally experienced in narrow (rural) thin surfaced unbound granular pavements on low strength shallow subgrade where the outer wheelpath is too close to a soft shoulder, and as a result may be accompanied by deep-seated shear or possibly edge break. There is no closely equivalent parameter in traditional tests using vertical deflection. Interim calibration uses the ratio of the FWD shear strain at the top of the subgrade to the equivalent thickness (as far as the transition only with truncation of values). Beyond the transition, an interim mirror calibration could be attempted, to see what can be learnt. Very low values will suggest subgrade deformation is likely. The intermediate values around the transition are all expected to indicate soundly compacted unbound granular pavements or thick bound layers, that may also relate to high mod-ular ratios. Further trials to find suitable correlations are needed.
- Bound Cracking Index (BCI) is a new parameter that quantifies the potential for cracking of a near surface bound layer because it is underlain by a significantly more flexible layer. It is correlated to FWD data using pavements that have known construction (usually those with thick AC or cement stabilised basecourses) and known current condition.
- Apparent Cracking Index (ACI) is generated by MSD as a simplistic measure of cracking from JPeg images, 300 mm square, taken in the wheeltrack at 1 m intervals. Machine learning is used to quantify in real time, just the number of cracks which are essentially continuous ie pass fully from one side to another, returning numbers of 0, 1, 2, 3, or 4 with counting truncated at 4. Shorter cracks are ignored.
- Estimated International Roughness Index (eIRI) and Estimated Mean Texture Depth (eMTD). The estimated descriptions are used to distinguish the parameters from those collected using traditional equipment, as the MSD uses laser imagery to provide localised measures that approximate the traditional International Roughness Index and Mean Texture Depth, both correlated to existing data typically measured by HSD in roading databases such as RAMM (New Zealand).
- Apparent Rolling Resistance (ARR) is the ratio of the dynamic shear resistance (acting longitudinally on the pavement surface at the tyre contact patch) that is generated against the direction of motion of a free rolling wheel, to the normal force on the pavement, expressed as a percentage. The shear force is the resultant of the forces contributed by tyre deformation (including contact patch hysteresis losses around the patch perimeter as well as internally from texture indentation) and pavement layer deformations (that impose

energy losses as the wheel continually attempts to "climb out" of the deflection bowl). The bowl becomes progressively more asymmetric with speed. Because Rolling Resistance has been found to be strongly speed dependent (Cenek 1996), it is standardised to a reference speed (currently 50 km/hr) as well as other aspects, particularly tyre temperature and pressure. It has associated parameters that allow correction to other vehicle speeds, tyre types and pressures where required. In recent years, Rolling Resistance has been a feature of detailed research in Europe (for identification of pavement types which result in reduction of carbon emissions) using more costly traditional test procedures. However, it was recently discovered that the same parameter was generated incidentally (an unexpected "by product" of the machine learning technology) in the MSD interpretation. For that reason, it may also be outputted when required by interested researchers.

The advantage of this extended form of data collection available via MSD is that users may elect either to use simply one or two parameters such as SNP or central deflection, along with traditional HSD data collected separately, or they may elect to encompass the dozen or so supplementary parameters that can now be readily generated in a single MSD pass. In either case, basic interpretation can be limited to dTIMS or Austroads, or extended to include the more versatile tools of a Regional Precedent Performance evaluation and hence Remaining Structural Life and a Forward Work Programme, generated from calibrations to terminal sites in the network – the ultimate reality checks.

6 MSD CASE HISTORIES

6.1 *Auckland transport, Auckland, New Zealand*

Over two months in May and June 2021, 4,460 lane km in both left (outer) and right (inner) wheelpaths were collected using MSD data technology on behalf of Auckland Transport. Readings were typically collected at 1 to 3 m intervals and reported as the median value of the readings within each 10 m road segment. Left and right wheelpath data were staggered. Roads tested comprised mainly arterials and primary collectors.

The final outputs are as per the MSD outputs outlined earlier in the report. Structural Treatment Lengths (section lines in lieu of points) have yet to be determined and reported at time of writing this paper, however their characterisation can at present be readily inferred on inspection as shown in Figure 1 for Meola Rd and will in due course be computed algorithmically.

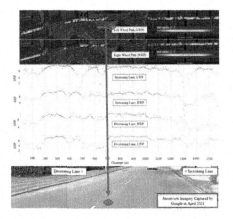

Figure 1.   Meola Rd, Auckland example of MSD data in all lanes and wheelpaths, well supported by visual reality checks.

6.2 *Rome municipality, Rome, Italy*

Over three days in April 2021, 300 lane km in the right (outer) wheel path were collected using MSD technology. Roads tested mainly comprised arterials and primary collectors of the municipality network as shown in Figure 2.

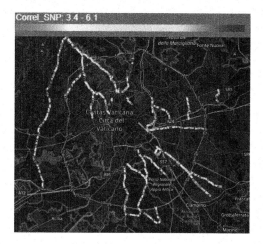

Figure 2.   MSD Test coverage for Rome.

Via Prenestina in the vicinity of Villa Gordiani was selected for closer inspection as shown in Figure 3. Sub-sections of sustained low and high SNP were reality checked with Google Street View Imagery captured in January 2022, just a few months after MSD testing. Review of historical imagery indicates that the pavement had been resurfaced or rehabilitated circa 2015. Within 2-3 years distress manifested at the surface in the form of fine alligator cracks and pumping. Distress is more severe in the left rather than right wheelpath highlighting the potential benefit of dual wheelpath MSD surveys particularly for mature roading networks such as Rome.

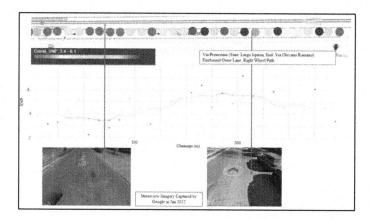

Figure 3.   Reality checks on sub-sectioning of Via Prenestina.

6.3 Florence municipality, Florence, Italy

Over three days in December 2021, 185 lane km in the right (outer) wheelpath were collected using MSD technology. Roads tested comprised arterials and primary collectors. The scale of the data collected over the entire network is best appreciated geospatially as shown in Figure 4.

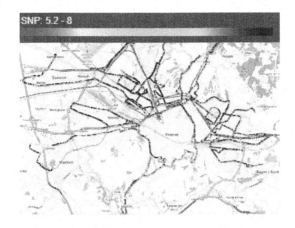

Figure 4. MSD Test coverage for Florence.

Viale Francesco Talenti was selected for closer inspection as shown in Figure 5. Sub-sections of sustained low and high SNP were reality checked with Google Street View Imagery captured in January 2022, just a few weeks after MSD testing. Once again the MSD appears to have correlated well with identified sections of weak and strong pavements.

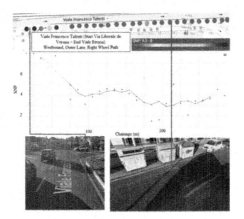

Figure 5. Viale Francesco Talenti reality checks.

7 CONCLUSIONS

The Multi-Speed Deflectometer, fills a gap for an efficient device for rapid low-cost testing and structural evaluation of a large network of urban roads. The above recent case histories demonstrate its effectiveness using Google Streetview. Management of pavement deterioration can now be expedited by development of an optimised Forward Works Programme which can be readily validated with traditional methods (visual inspection, destructive tests or minimal Falling Weight Deflectometer testing).

REFERENCES

ANAS - Gruppo FS Italiane - Coordinamento Territoriale/Direzione - Capitolato Speciale di Applato - Norme Tecniche per l'esecuzione del contratto - Parte 2 - IT.PRL.05.21 - Rev. 3.0 - Pavimentazioni Stradali, Giugno 2021.

Anderson C., 2003. *The End of Theory: The Data Deluge Makes the Scientific Method Obsolete*. Wired Magazine. June 2003.

Arnold G., Salt G., Stevens D., WerkmeisterS., Alabaster D. and van Blerk G., 2009. NZTA Research Report 381. Compliance Testing Using the Falling Weight Deflectometer for Pavement Construction, Rehabilitation and Area-Wide Treatments.

ASTM D4695-03, 2020. General Pavement Deflection Measurements.

ASTM D5858, 2020 Reapproval Notice to 1996. Standard Guide for Calculating In-Situ Equivalent Elastic Moduli of Pavement Materials Using Layered Elastic Theory. Note 5.

Cenek P. D., 1996. Rolling Resistance Characteristics of New Zealand Roads. Transit NZ Research Report 61.

Dawson A. 2002. The Mechanistic Design and Evaluation of Unsealed & Chip-Sealed Pavements. University of Canterbury Workshop, Briefing Paper.

Horak E., 2008. Benchmarking the Structural Condition of Flexible Pavements with Deflection Bowl Parameters. Journal of the South African Institution of Civil Engineers 50 (2).

Stevens D. and Schmitz G., 2018. *Traffic Speed Deflection Data Applied to Network Asset Management. The Kaikoura Bypass Reality Check*. Road Infrastructure Management Forum, *Auckland*.

Ullidtz, P., 1998. Modelling Flexible Pavement Response and Performance. Polyteknisk Forlag. ISBN, 8750208055, 9788750208051.

Xiao F., XiangW., Hou X. and Amirkhanian S., 2021. Utilization of Traffic Speed Deflectometer for Pavement Structural Evaluations. Measurement 178 (2021) 109326. Elsevier.

Zofka A and Sudyka J. 2015. *Traffic Speed Deflectometer Measurements for Pavement Evaluation. International Symposium Non-Destructive Testing in Civil Engineering. Berlin*.

Road condition monitoring and assessment in Germany – how to use the data for pavement analysis

Ferdinand Farwick zum Hagen*
Federal Highway Research Institute Germany (BASt), Germany

Felix Lau
Federal Ministry for Digital and Transport, Germany

ABSTRACT: Efficient maintenance management is essential to ensure high availability of the German federal trunk road network. It is the basis for carrying out a precise life cycle analysis, which in turn have a major influence on sustainability assessments of the transport infrastructure. In this regard, an objective road condition monitoring and a subsequent assessment are fundamental. As a result, basic information about the road surface condition as well as sections of prior maintenance needs are provided. In Germany, the road condition monitoring and assessment (in German: Zustandserfassung und -bewertung, ZEB) is an established process and has been carried out regularly for more than 25 years on behalf of the Federal Ministry for Digital and Transport. Within this process, certain requirements for measuring techniques are defined and standardized data processing methodologies as well as quality control mechanisms are applied. The ZEB is performed on the entire federal trunk road network within a four-year cycle. Thereby, the focus is on a reliable and robust recording of the road surface condition while using fast-moving measuring vehicles to evaluate relevant road condition indicators, such as evenness, skid resistance, or cracks. The pavement condition data is subsequently added to a comprehensive database keeping the information of the road infrastructure up-to-date. The database allows comparisons between actual recorded data and historical data, thus conclusions about condition development of the road can be drawn. A deeper analysis can be performed by taking road layer information, such as material types, layer age or layer thickness, into account. The paper gives insights into the entire ZEB process by brief descriptions of the measuring principles, the basic steps of data processing, and the relevant condition indicators. An analysis of the ZEB data shows the correlation between different types of top layer materials and surface properties as well as their development.

Keywords: Road Surface Condition Assessment, Condition Development, Long-Term Pavement Performance

1 INTRODUCTION

The federal trunk road network in Germany has grown historically and therefore has very different structural designs and age distributions. Major parts of the carriageways are about to reach their originally scheduled technical service life that is approximated to be 30 years in accordance to the "Guidelines for the Standardization of Pavement Structures of Traffic Areas" (RStO 2012). In addition, the analyses of the pavement condition show that

---

*Corresponding Author: farwickzumhagen@bast.de

DOI: 10.1201/9781003429258-25

maintenance, rehabilitation and reconstruction (MRR) measures are required for a large part of the existing network in the coming years. Despite today's traffic loads, which are often significantly higher than traffic forecasts (Stöckert et al. 2019), the quality of the federal trunk roads has been maintained on a constant high level (Zander 2017). The strategic orientation of an appropriate and efficient maintenance is essentially considered within the framework of the 2030 Federal Transport Infrastructure Plan by the Federal Ministry for Digital and Transport.

The federal trunk roads are under a permanent development. As for example, in period from 1970 to 1985, the length of the federal motorway network (in German: Autobahn) has almost doubled to 8,200 km. Another significant increase to 11,000 km was recorded after the reunification of Germany in 1990. Afterwards, the development of the infrastructure in Germany was focused on the expansion of an all-German motorway network. Today, the federal motorway network has a length of about 13,000 km (BMDV 2022). The proportion of concrete pavements is about 25% and that of asphalt is 75%. The federal highway network (in German: Bundesstraße) covers a length of approx. 38,000 km and is almost exclusively made of asphalt pavements.

Despite the significant expansion of the road network length in recent years, the network is reaching its capacity limits today, especially in regions with high traffic volumes. In order to nevertheless ensure a trouble-free mobility, resilient road constructions with high availability are required. This can be achieved by an efficient maintenance management that uses information about the current road surface conditions as a basis. The road condition monitoring and assessment (in German: Zustandserfassung und -bewertung, ZEB) process provides the necessary data basis, as it evaluates the lane-based surface condition within an objective and defined framework. This allows maintenance measures to be planned in the long term and implemented in a demand-oriented manner.

This paper gives insights into the entire ZEB process by brief descriptions of the measuring principles, the basic steps of data processing, and the relevant condition indicators. It shows the current road surface condition of federal trunk roads in Germany. Using the example of federal highways, the condition indicators for cracks and rut depths were analysed regarding their historical development and their influence on comprehensive condition values. Finally, the results were aggregated with road construction data, thus a more precise analysis of the development of condition indicators could be done.

2 ROAD MONITORING AND ASSESSMENT

The ZEB process is a nationwide standardized procedure for recording and assessing the road surface condition of federal trunk roads. As already mentioned, the federal trunk road network covers approx. 51,000 km, of which 13,000 km are federal motorways and 38,000 km are federal highways. Expressed in lane-kilometers, the entire network length adds up to more than 140,000 km. Against the background of efficient and economic road maintenance, the ZEB guidelines consider it necessary to record every single lane in both directions on federal motorways. On federal highways, however, it is intended to record only one direction of travel on a two-lane road that is usual in Germany. As a result, approx. 60,000 lane-kilometers on federal motorways and approx. 40,000 lane-kilometers on federal highways need to be assessed and monitored regularly. Since this would be an enormous effort for an annual road monitoring and assessment, the measurement volume is distributed over four years within the ZEB process. Thus, around a quarter of the federal trunk road network is recorded every year and divided into different federal states as exemplarily shown in Figure 1.

The ZEB measurement campaigns have been carried out regularly since 1992. In 2015 a revised calculation of the condition indicators was introduced that will be described later. Previously recorded data was retrospectively converted back to 2005. In this way, road condition developments under the same conditions can be analysed over the last 16 years.

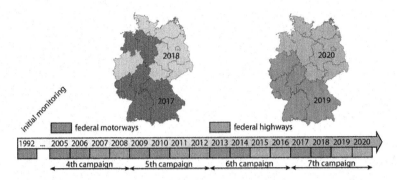

Figure 1. ZEB measurements in Germany. The measurements on federal motorways and federal highways are depicted in blue and orange, respectively. For each campaign the measuring sequence and areas are exemplarily shown in maps.

2.1 *Measurement techniques and condition parameters*

The focus of the ZEB is on determining the structural condition as well as safety relevant properties of the road surface. These two aspects are expressed through road surface properties such as longitudinal and transverse evenness, skid resistance, and substance characteristics of the road surface (e.g. cracks, patches). Table 1 provides an overview of relevant attributes and their associated physical state variables, sometimes differentiated according to measurement velocity or road construction type. For a more detailed description the reader is referred to the "Additional Technical Terms of Contract and Guidelines for Road

Table 1. Relevant properties for pavements according to ZTV ZEB-StB (2018).

Characteristic Group	Attribute	Description	Physical state variable
Evenness, longitudinal profile	General unevenness	Measure for unevenness, spectral density of unevenness heights Φ_h [cm^3]	AUN
Evenness, transverse profile	Rut depth	Maximum of the mean value of the right and left rut depth [mm]	MSPT
		Maximum of the mean values of the right and left notional water depth [mm]	MSPH
Roughness	Skid resistance	Side friction coefficient temperature and speed corrected to 40, 60, or 80 km/h, mean value [a.u.]	GRI_40GRI_60GRI_80
Substance characteristics (surface) for asphalt pavements	Cracking	Partial areas of a section affected by cracks [%]	RISS
	Residual damage	Partial areas of a section that are not affected by cracks but affected by patches or material losses [%]	RSFA
Substance characteristics (surface) for concrete pavements	Longitudinal and transverse cracking	Longitudinal and transverse cracking, medium length [m]	LQRL
		Longitudinal and transverse cracking, affected slab area [%]	LQRP
	Residual damage	Partial slab areas of a section that are not affected by cracks but affected by corner breaks, edge damages, material losses, or asphalt patches [%]	RSFB

Monitoring and Assessment" (ZTV ZEB-StB 2018), which describes the entire ZEB process regarding metrological recording, evaluation, and quality assurance.

The road condition measurements are carried out by non-destructive measuring systems recording the road surface properties with a typically cruising velocity between 40 km/h and 80 km/h depending on the road type and local circumstances. A brief description of the measuring principles is given below and can be seen in Figure 2. Additionally, the measurement vehicles of the Federal Highway Research Institute (BASt) are shown.

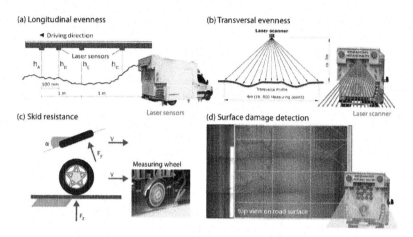

Figure 2.   Measuring principles and vehicles from BASt. The devices for longitudinal (a) and transversal evenness (b) and surface damage detection (d) are installed on the same vehicle ('MEFA') (BASt 2022a and 2022b).

2.1.1   *Longitudinal and transverse evenness*

Road evenness is a very important surface property and it is distinguished in longitudinal and transverse evenness. In case of longitudinal evenness, the measurements take place in the right wheel track parallel to the axis of the road. In the case of transverse evenness, ruts are measured symmetrically by scanning the road surface perpendicular to the road axis. Both, either longitudinal or transversal evenness have an impact on the driving comfort, influence the traffic safety, and affect road stress of the entire road superstructure. The evenness (more common: unevenness) characterizes the road in terms of wavelengths in regions above 0.5 m to 50 m. For that matter, unevenness is considered as geometric irregularities of the road surface that cannot be attributed to the texture of the surface or the gradient of the roadway. (Un)evenness is therefore generally understood to mean height deviations from the planned road surface geometry.

The most common parameter for longitudinal profile measurements is the International Roughness Index (IRI). The IRI is a physical state variable expressed in terms of slope, e.g. [m/km], that is calculated through a simulated rolling of an imaginary small wheel of a vehicle model along a two-dimensional road profile. Within the model, vehicle specific parameters such as mass and spring constant (damper) are defined (Sayers 1995). The method by which the road profile is obtained is subsidiary and may vary. Nevertheless, the IRI includes contributions not only of the roadway but also from the vehicle (i.e. the vehicle model).

The AUN, however, is a measure of the evenness of the road pavement that is independent from any vehicle model assumptions. The road profile is obtained by relative distance readings taken from four laser sensors mounted at specified positions on a rigid beam. The rigid beam is fixed at the measuring vehicle on the right-hand side and in such way that all laser sensors are exactly aligned in driving direction. The state variable AUN is calculated through a spectral density analysis of the measured height profiles (Becker 1995). As the calculation of the AUN is based on a Fast Fourier Transform (FFT) at 1024 sample points and the sample points are equally spaced at an interval of 100 mm, the AUN represents the evenness over sections of approx. 100 m length (exact 102,4 m). Since the AUN represents all necessary wavelengths in an equal manner while avoiding any dependence on a vehicle model, it is the preferred method for determining the evenness in Germany.

The transverse evenness is measured by up to 41 single laser sensors that are aligned at a lateral distance of 10 cm from each other. Analogously to this arrangement, rotating laser scanner can be used to generate the same number of data points along the transverse road axis. The measured height values are collected and averaged over a travel distance of 100 mm representing a single transverse profile. The single profile is recorded in an interval of one meter in the direction of travel. For each transverse profile the rut depth is obtained by the state variable MSPT, which is the maximum of the mean values of the right and left rut depth. It is calculated by a simulated two-meter sliding beam and separated for left and right parts of the lane. Similarly, the maximum notional water depth (MSPH) is calculated considering the cross fall of the whole measured transverse profile as regression line (TP Eben 2009).

2.1.2 *Skid-resistance*

In addition to the evenness, the skid resistance (grip) of road surfaces is also of crucial importance for road safety in the context of friction between tire and road. In the ZEB process, the skid resistance is described by the coefficient of friction measured according to the sideway-force principle (μSFC) (FEHRL 2006). The coefficient of friction is determined on an oblique measuring wheel of $20°$, which is loaded with a normal force of 1960 N under static conditions. It is defined as the ratio of measured lateral force (F_y) and given normal force (F_z) between the road surface and the tire (TP Griff-StB(SKM) 2007). A speed dependent water amount is applied to the road surface for measuring so that a theoretically constant water film with a thickness of 0.5 mm is ensured. The state variable for skid resistance is expressed in terms of the nominal measuring velocity (see Table 1) (ZTV ZEB-StB 2018). The Sideway-Force measuring method is the most widely used in Europe, i.e. SKM and SCRIM method.

2.1.3 *Surface damages*

Besides the evenness and skid resistance, the substance characteristics (surface) are the third technical measure within the ZEB. The surface properties skid resistance and evenness are functional parameters that relate to the utility value of the road and therefore influence road safety. However, the substance characteristics (surface) allow indicating the destruction of the road substance by judging the road surface condition. Therefore, the road surface is pictured in high resolution images that are used later to manually examine the road surface regarding cracks, patches, and other kind of surface damages in certain sections. According to ZTV ZEB-StB (2018), a distinction is made between concrete and asphalt pavements: entire slabs are evaluated for concrete pavements and for asphalt pavements a continuous grid is used. Based on the damage detection of the road surface, the corresponding state variables (RISS, RSFA, LQRL, LQRP, and RSFB) are very important for maintenance planning on federal trunk roads.

The MEFA vehicle in Figure 2 records the road surface with two-line scan cameras in addition to the measurement of longitudinal and transverse evenness. Each camera has 2,048 pixels and records one-line per approx. 1.25 mm travel distance. This corresponds to approx. 18,000-line captures per seconds at a measuring speed of 80 km/h. The necessary lighting is provided by pulsed red LED lights. The image width covers up to 4.5 m in cross section of the road.

2.2 Quality assurance

The road condition data are the basis for maintenance management and therefore high demands are placed on the data quality. In order to obtain high-precision data, the BASt carries out quality assurance measures throughout the entire ZEB process on behalf of the Federal Ministry for Digital and Transport. The quality assurance measures are applied at three points of the process, namely before the measurement process, during the measurements, and after the measurements in the course of data post-processing.

The acquisition of the measured data is carried out by external contractors, who use versatile equipped measurement systems. Although the measuring systems might be based on different measuring techniques, the requirements for the systems are standardized by regulations (TP Eben 2009; TP Griff-StB(SKM) 2007; TP OF-StB 2020). In order to ensure the performance of the measuring systems according to the ZEB standards, the measurement systems are approved by BASt. Therefore, the vehicles undergo an annual technical inspection at BASt, which is the only institution in Germany that is allowed to perform such tests independently of the measuring technique of the vehicle. For approved vehicles BASt issues time-limited licences. Additionally, the measuring vehicles of the BASt are used as a reference within the quality assurance process and must undergo self-monitoring at regular intervals.

During the measurement period, which is usually set from May to September due to weather conditions, BASt carries out control measurements in intervals of about 2,000 km. Before the measurement data are used, technical experts are reviewing the data on behalf of BASt. Thereby the focus is on the completeness of the data as well as the technical correctness of the data. If any deficiencies are identified, corrections or remeasurements are initiated.

2.3 Method of assessment and data management

The assignment of the measurement data to the road network is essential for an analysis of the road condition. As a prerequisite for this, discrete sections of a certain length are created for each lane of the road network. Each section is precisely located either within a node-edge model. In the context of road maintenance management, an usual section length of 100 m turned out to be suitable (ZTV ZEB-StB 2018). However, shorter sections are also considered in order to cover the entire road network, for example at the end of a network node section or within built-up areas. The contractors use this basic data for route planning during the measurement period.

After the measurements, as part of the data post-processing, the raw data is allocated to the given sections using special software. The software not only assigns the raw data to the road network, which are given for every travelled meter, but also calculates a single physical state variable for every single section based on the raw data (see Table 1). Afterwards, the physical state variables are normalized in order to obtain condition indicators. Thus, for each section and for each attribute (e.g. evenness or skid resistance) a condition indicator is issued as single value. Additionally, partial values and overall values are generated according to the ZTV ZEB-StB (2018) as it is shown in Figure 3.

All condition indicators, partial values, and overall values are subdivided into different categories from 1 to 5. The categories are performance measures, which are similar to a grade. They enable a high comparability of road sections among each other, especially when they are visualized in maps with defined colour scheme (Stöckert & Schüller 2018). Figure 4 gives an example of a lane-based visualization of a road surface indicator. Additionally, the categorisation can be seen.

The most common values for the assignment of the road surface condition in general are the utility value and the substance value (surface). The utility value is a measure of the road condition related to safety relevant aspects. A poor rating of a road in this category indicates short-term actions that can be implemented, for example, by speed limits or an extra roughening of the road surface. In contrast, the substance value (surface), indicates road

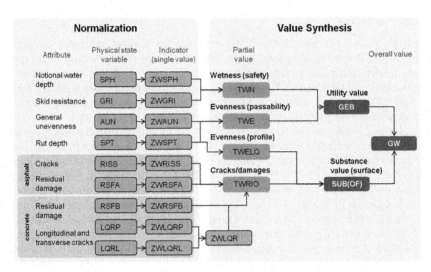

Figure 3. Method of assessment. On the left the normalization of the physical state variables to condition indicators is shown. The further determination of the overall values can be seen in the right. Detailed information on the generation of the values can be found in the ZTV ZEB-StB (2018) (BASt 2022a).

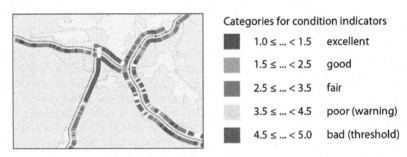

Figure 4. Categories of road surface condition indicators and an example for a visualisation in a map. Adopted from (Stöckert & Schüller 2018), ©GeoBasis-DE / BKG2018.

conditions with regard to the need for maintenance. Road sections that are in a poor or bad condition are preferentially considered. The overall value is the combination of utility and substance value (surface), whereas the worst condition values are considered.

For a standardized deployment and graphical representation of the results, an online information system was developed on behalf of BASt 10 years ago. The platform is called IT-ZEB Server (Heller & Wasser 2013). Every year, approx. 12 to 15 TB of road condition data (incl. images) are added. The server offers the possibility to visualize the condition data, show statistics and has a download area. It also acts as an archive that contains the relevant data for the federal trunk road network since the road monitoring in 1997.

3 DEVELOPMENT OF THE ROAD SURFACE CONDITION AS A DECISION-MAKING AID FOR NETWORK MANAGERS

The latest ZEB results are from 2017/18 for federal highways and 2019/20 for federal roads. In case of federal motorways, the nationwide road network condition can be seen in Figure 5

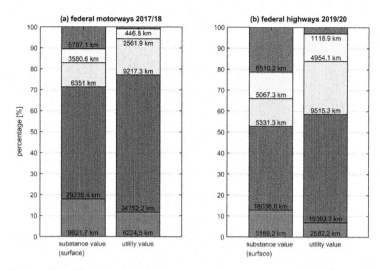

Figure 5. Nationwide ZEB results for (a) federal motorway 2017/18 and (b) federal highways 2019/20. Only valid results are shown according to the five condition categories (i.e. excellent (blue), good (dark green), fair (light green), poor (yellow), and bad (red)) for substance value (surface) and utility value, respectively.

(a) expressed through the condition indicators substance value (surface) and utility value. In case of the substance value (surface), about 18% of the road network is in an excellent condition, whereas 10.5% are in a bad condition. These sections are given priority in the course of maintenance. In case of the utility value, the overall road condition is much better. Only 0.9% are in a bad condition, which are about 447 lane-kilometers that need to be monitored from a safety perspective. But still, the majority of road sections, which are in a fair, good, or excellent condition, is notably high. Thus, the overall condition of the federal motorways can be rated as good.

In comparison, the federal highways are not ranked as good as federal motorways. As it can be seen in Figure 5 (b), the amount of road sections that are either in a poor or bad category is much higher. However, within a holistic view, the road surface condition is also better ranked by the utility value than by the substance value (surface). In case of the utility value, the bad category is below 3%, but in case of the substance value (surface) it is around 21%. Without showing here, a comparison with the results of the last few years has even shown that the bad category of the substance value (surface) has increased slightly. The reasons for such a high contribution of the bad category may be found through a more detailed analysis of the condition indicators for rut depth (ZWSPT) and cracks (ZWRISS) that are relevant input parameters of the substance value (surface). The actual condition indicators ZWSPT and ZWRISS can be seen in Figure 6 and 7, respectively. Additionally, their development over the last 12 years is shown.

In case of rut depth, the condition indicator ZWSPT shows an exceptional good condition as there are only narrow contributions of the bad category. The development is improving almost continuously. This can be seen within the frequency distribution of Figure 6 (b) in more detail. In contrast, the condition indicator ZWRISS is deteriorating. Although there is a huge contribution of the excellent category representing nearly crack-free road sections, the contribution of the bad category has increased from about 11% to 16%. Since the ZWRISS has a strong influence on the substance value (surface) (ZTV ZEB-StB 2018), the bad as well as the poor conditions states are nearly directly transferred to the substance value (surface). Again, the development of cracks within the road network can be seen in the frequency distribution. The different onsets arise out of the huge contribution of the excellent category.

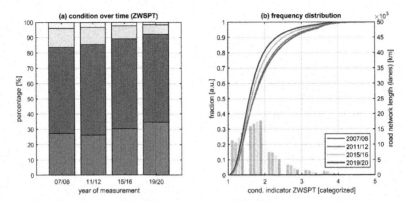

Figure 6. Development of the condition indicator ZWSPT (rut depth) over time for federal highways is shown in (a). The corresponding frequency distributions are shown in (b). The solid lines correspond to the left axis (fraction) and the bars correspond to the right axis (binned network lengths in lane-kilometers).

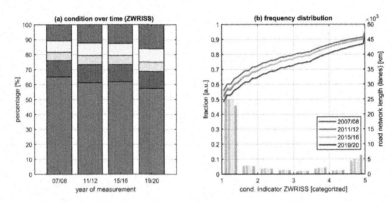

Figure 7. Development of the condition indicator ZWRISS (cracks) over time for federal highways is shown in (a). The corresponding frequency distributions are shown in (b). The solid lines correspond to the left axis (fraction) and the bars correspond to the right axis (binned network lengths in lane-kilometers).

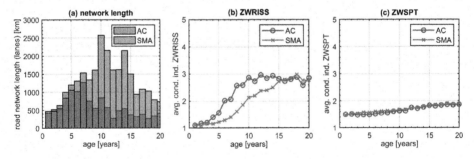

Figure 8. In (a) the network length for AC and SMA constructions vs. age is shown for federal highways. The averaged condition indicators ZWRISS and ZWSPT depending on the construction age are shown in (b) and (c), respectively.[1]

---

[1]*The Analysis is Based on Aggregated Actual ZEB Data and Structural Data From 12/31/2021.* Note That Not All Recent Maintenance Measures Could be Considered.

The contrary development of ZWSPT and ZWRISS shows on the one hand a reduction of rut depths within the road network and on the other hand an increased occurrence of cracks. This effect is most presumably related to the types of road construction materials for which tougher binders were used in the last decades (Arand & von der Decken 1996; Hüning 1999; Woltereck 1997). The crack formation may also has been favoured by hot weather periods of recent years (Brienen et al. 2020). A more precise explanation can only be provided by material-specific investigations taking into account further information on the road structure.

[1]First results of a material-specific analysis of asphalt concrete (AC) and stone mastic asphalt (SMA) constructions, which make up about 43% and 51% of all federal highways, respectively, are shown in Figure 8. As it can be seen, the development of the averaged condition indicators is quite different: In case of ZWRISS, new road constructions are in an excellent condition nearly independent of the material type within the first five years. At advanced age, however, the road conditions become worse approaching an average condition value of 3 after 15 years. Comparing the two types of materials, the SMA construction appears to be more resilient to cracking as the curve has a less steep slope. This observation is consistent with the length distribution by construction age as SMA construction have an apparently longer service life (Figure 8 a). In contrast, in case of ZWSPT, the development of the road condition regarding rut depth is almost stable as there is only a slight increase over 20 years. Any material-specific differences cannot be identified here. Note, the averaged condition indicators do not serve as specific maintenance indicators but rather than giving supporting explanations of the road condition development of federal highways in Germany. However, further investigations are necessary in order to gain a more precise examination of the condition development of federal highways and motorways.

4 SUMMARY AND CONCLUSIONS

The ZEB in Germany is an essential element of modern road maintenance management, as it provides the necessary data on road conditions to support network operators in their decision-making. Witthin the ZEB process, measurement and data post-processing is carried out according to defined standardisations ensuring a constant and robust data quality.

In the present work, the ZEB process and the most relevant condition indicators, such as the utility value and the substance value (surface), were introduced. The latest results were presented for both the federal motorways and the federal highways and show that the overall road condition is quite good. In a direct comparison, however, the federal motorways performed better. An analysis of the condition development regarding rut depth and cracks was made for federal highways consisting entirely of asphalt. While in general, a network-wide reduction of rut depths was observed, the amount of road sections affected by cracking slightly increased. This development was presumably related to the road construction materials, for which increasingly tougher binders have been used in recent years. To investigate this assumption in more detail, further analyses are required, which can be obtained, for example, by intersecting ZEB data with information about material properties. Such further investigations have already been initiated and are to be carried out more deeply in the near future.

REFERENCES

Arand W. and von der Decken S., 1996. *Quality Planning in Asphalt Road Construction – Using the Example of Deformation Resistance.* Final Report, Braunschweig Pavement Engineering Centre, Technische Universität, Braunschweig.

Becker W., 1995. The Spectral Density - a Measure for Longitudinal Road Evenness, *National Journal of Roads and Motorways*, 10, 583–592.

Brienen S. Walter A., Brendel C., Fleischer C., Ganske A., Haller M., Helms M., Höpp S., Jensen C., Jochumsen K., Möller J., Krähenmann S., Nilson E., Rauthe M., Razafimaharo C., Rudolph E., Rybka H., Schade N., and Stanley K., 2020. *Climate Change-induced Changes in the Atmosphere and Hydrosphere: Final Report of the Topic Scenario-Building (SP-101) in Topic Area 1 of the BMVI-Expertennetzwerk.* 56–68. DOI: 10.5675/ExpNBS2020.2020.02.

Forum of European National Highway Research Laboratories (FEHRL), Harmonisation of European Routine and Research Measuring Equipment for Skid Resistance. FEHRL rapport 2006/01. Brussel, 2006.

Federal Highway Research Institute (BASt), IT-ZEB Server. https://itzeb.bast.de, 2022. [online; accessed 05/18/2022]

Federal Highway Research Institute (BASt), Sideway-Force Measuring Method. https://www.bast.de/DE/Strassenbau/Technik/SKM.html, 2022. [online; accessed 05/18/2022]

Federal Ministry for Digital and Transport (BMDV), Length Statistics for Supralocal National Roads. https://www.bmvi.de/SharedDocs/DE/Artikel/StB/bestandsaufnahme-strassen-ueberoertlich.html, 2022. [online; accessed 05/18/2022]

Heller S. and Wasser B., 2013. User Application of ZEB-data – The IT-ZEB Server. *In: Conference Transcript Infrastructure Management Workgroup, Cologne.*

Hüning P. and Rode F., 1999. *Additional Technical Terms of Contract and Guidelines for the Construction of Road Surfacing from Asphalt, edition 1998 (ZTV Asphalt-StB 94), Kirschbaum, Bonn.*

Road and Transportation Research Association (FGSV), 2018. Additional Technical Terms of Contract and Guidelines for Road Monitoring and Assessment (ZTV ZEB-StB), FGSV 489, Cologne.

Road and Transportation Research Association (FGSV), 2012. Guidelines for the Standardization of Pavement Structures of Traffic Areas (RStO), FGSV 499, Cologne.

Road and Transportation Research Association (FGSV), 2009. Technical Test Specification for Measuring the Evenness of Road Surfaces in Longitudinal and Transversal Direction (non-contact) (TP Eben (non-contact)), FGSV 404/2, Cologne.

Road and Transportation Research Association (FGSV), 2007. Technical Test Specification for Measuring the Skid-Resistance on Road Constructions (SKM) (TP Griff-StB(SKM)), FGSV 408/1, Cologne.

Road and Transportation Research Association (FGSV), 2020. Technical Test Specification for the Recording of Substance Characteristics (Surface) with Fast-moving Measuring Systems (TP OF-StB), FGSV 434/1, Cologne.

Sayers M.W. 1995. *On the Calculation of International Roughness Index from Longitudinal Road Profile, Transportation Research Record* 1501, 1–12.

Stöckert U., Schmerbeck R. and Lau F., 2019. Maintenance Management for Federal Roads – Current Status and Outlook, National Journal of Roads and Motorways, 5, 399–409.

Stöckert U. and Schüller S., 2018. The New Assessment Procedure for the Road Monitoring and Assessment Process (ZEB) – Explanatory Notes, National Journal of Roads and Motorways, 7, 574–580.

Woltereck, G., 1997. *Experiences with Stone Mastic Asphalt on Bavarian Motorways, Bitumen,* 2, 50–53.

Zander U., 2017. The Challenge of Infrastructure Maintenance. National Journal of Roads and Motorways, 9, 710–717.

Human-centered evaluation with biosignals for localized surface roughness on expressway pavements

Marei Inagi
Graduate School of Engineering, Kitami Institute of Technology, Kitami, Japan

Kazuya Tomiyama*
Division of Civil and Environmental Engineering, Kitami Institute of Technology, Kitami, Japan

Ryo Kohama
East Nippon Expressway Company Co., Ltd., Omiya-ku, Saitama, Japan

Masayuki Eguchi
East Nippon Expressway Company Co., Ltd., Iwamurada, Saku, Nagano, Japan

Masakazu Sato
East Nippon Expressway Company Co., Ltd., Omiya-ku, Saitama, Japan

ABSTRACT: The International Roughness Index (IRI) has been applied to the roughness evaluation of pavement surfaces on expressways in Japan with two specific criteria: a fixed interval of 200 meters for the quality in average of a pavement and a fixed interval of 10 meters for the localized roughness treatment. However, the gap between the maintenance criteria in terms of the IRI and the road users rating has been reported for the ride quality. This is mainly because the types of surface distress have been localized severely since porous asphalt pavements made with polymer modified bitumen have been introduced for the surface layer. Against this background, this study conducted field surveys on in-service expressway sections where the poor ride quality was reported by road users even though the surface conditions were less than the maintenance threshold. At the locations, the longitudinal surface profiles including different roughness features as well as the distress types were acquired in the survey. After that, the data were virtually reproduced and examined by a motion base driving simulator which involved road users as participants. The simulator experiment introduced two measurers of biosignals, the electrocardiogram (ECG) and the electrodermal activity (EDA), to comprehend unconscious mental stress of the participants toward vehicle vibrations induced by the surface roughness. This study examines the relationship between road surface characteristics in terms of the profile wavelength and the mental stress quantified by the biosignals. The result shows that the mental stress increases with increasing amplitudes of wavelength ranging from 4 to 8 meters of a profile even if the IRI indicates an acceptable level. Consequently, the IRI brings underestimation of ride quality because it less responds to this waveband. In contrast, the Ride Number which emphasizes these wavelengths well corresponds to the mental stress. This study describes how the localized surface roughness corresponds to the mental stress of road users regarding ride quality from biosignal viewpoints. The outcomes of this study contribute to the development of sustainable road transportation on the basis of human-centered design concept.

Keywords: Road Surface Roughness, IRI, Heart Rate Variability, Skin Conductance Response, Human-centered Design

---

*Corresponding Author: tomiyama@mail.kitami-it.ac.jp

DOI: 10.1201/9781003429258-26

1 INTRODUCTION

The derivative transportation demand such as activities at a destination has been decreasing because people "no longer make non-essential trips" due to COVID-19 for the past few years. On the other hand, essential transportation demand such as work that requires visiting the actual sites, education that involves practical training and skills, and travel and driving that is the purpose of transportation itself, is possibly kept in the foreseeable future, and their relative importance is expected to increase (Kashima *et al.* 2021). Therefore, transportation infrastructures including paved roads need to be controlled in the quality according to the essential demand in order to improve the travelling safety and comfort of users.

Road roughness is an important factor of road infrastructures that directly affects the ride quality of road users. For this reason, the International Index (IRI) has been applied as an index of pavement surfaces expressways in Japan with two specific criteria: a fixed interval of 200 meters for the quality in average and a fixed interval of 10 meters for the localized roughness treatment. However, the IRI assumes an average evaluation of road surface roughness condition, and then underestimates road surface conditions including localized deformations. Thus, the IRI is potentially difficult to provide comprehensive maintenance and repair for all road surface deformations under the current management standards. Localized deformation is damage especially specific to porous asphalt (Kawamura *et al.* 2007) and tends to deepen over time, therefore early repair is important (Kamiya 2021). In addition, over one hundred of complaints regarding ride quality have annually been reported to the customer center of an expressway company. This fact indicates a gap between the road surface rating based on the specific criteria and the expectation of road users for the ride quality. Consequently, a new road surface evaluation method is required to correspond with the characteristics of road surface deformations that are difficult to control with the current management standards.

As alternative approaches, physical (Chanjun & Seung-ki 2019; Fujita *et al.* 2013), psychological (Ishida *et al.* 2004) and physiological (Jennifer & Rosalind 2005; Tomiyama *et al.* 2015) have been studied as methods of road surface evaluation that relates pavement surface characteristics to road users. Fujita *et al.* (2013) showed a response-type roughness measuring system and Geographic Information System (GIS) contribute to the efficient identification of deteriorated surface locations in city planning areas. Chanjun and Seung-ki (2019) proposed a road surface damage detection technique using fully convolutional neural networks. For the psychological approach, Ishida *et al.*(2013) evaluated ride comfort using a questionnaire and biosignals, and showed that biosignals captured potential stress transitions that are difficult to obtained by subjective evaluation. The physiological index relates physiological information such as occupant heart rate and emotional hidrosis to road surface roughness. Tomiyama *et al.* (2015) showed that heart rate variability analysis is applicable to the evaluation of surface smoothness based on mental fatigue caused by vehicle vibration from road surface deformations. Jennifer and Rosalind (2005) showed physiological monitoring, by measuring physiological signals, allowed indicating driver stress and also showed the potential to measure continuously the effects of different road and traffic conditions on drivers. Many studies (Eguchi *et al.* 2015; Fukuda *et al.* 2012; Kawamura *et al.* 2000) applied psychological indices for the surface evaluation of expressways, whereas there were only a few studies (Tomiyama *et al.* 2015; Ishida *et al.* 2007) using physiological indices. Studies involving actual complaints of road users regarding ride quality are particularly rare.

In this study, in order to fill the gap between the maintenance criteria and the road users demand, a field survey was conducted on expressway sections where users often inquire about road surface damage. Then, the measurement data obtained were loaded into a motion base driving simulator (DS) for driving tests. In addition, focusing on road surface evaluation using physiological indicators on expressways, pulse wave and skin potential activity measurements were conducted on people cooperating in the DS experiment, and analyses of heart rate variability and electrodermal activity were conducted to study the wavelength characteristics of stressed road surfaces necessary for developing road surface management indicators for human-centered evaluation.

2 OVERVIEW OF COMPLAINT AND ROAD SURFACE CONDITION

2.1 *Distribution of road surface-related complaints*

The complaint data were organized and visualized using a GIS. The following are criteria for mapping a user's complaint regarding road surface conditions in the service area of Kanto Branch of East Nippon Expressway Company Limited (NEXCO) for this study.

1. Complaints from other than driver or passengers such as neighbors are excluded.
2. Complaints identified within one Kilometre-post (kp) are involved. The longitude and the latitude are then recognized for mapping on GIS.
3. The secondary partition mesh of the Geospatial Information Authority of Japan of Ministry of Land, Infrastructure, Transport and Tourism (MLIT) as shown in Figure 1 is used for the mapping.
4. The number of complaints included within a mesh is counted and color-coded according to the numbers.

Figure 1 shows a mapping result of the complaints regarding road surface conditions during the one-year period from April 2020 to March 2021. According to this result, a road survey for driving experiment mentioned later was conducted on three road surfaces with the highest number of complaints.

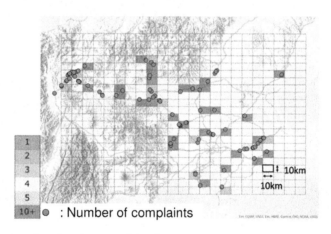

Figure 1.   Mapping of road surface complaint data.

2.2 *Target road surface*

Three road surfaces to be evaluated in driving experiments mentioned later are selected on the basis of the following criteria:

- Under the maintenance criterion of which IRI for a fixed interval of 200 m (hereafter IRI$_{fix}$ (200)) and 10 m (hereafter IRI$_{fix}$ (10)) are less than 3.5 mm/m and 8 mm/m, respectively.
- No emergency repair is required, that is, no pothole is appeared.
- An extension of the road profile should be 1600 m in order to obtain sufficient biometric information and to avoid excessive strain on the subject.

267

Table 1 shows the properties of the road surfaces dedicated to the driving experiment. Table 2 indicates the profiles of the road surfaces obtained in the field survey. In the table, fixed interval and continuous IRI denoted by IRI_{Fix} (segment length) and IRI_{Cont} (segment length), respectively are also depicted. Here, the surface profiles measured at a sampling interval of 0.1 m. The surface conditions are dry.

Table 1. The Characteristics of experimental road surfaces.

Surface	Characteristics
A	Localized deteriorations causing the lack of smoothness are dominated on the surface. Surface condition is potentially complained in terms of the ride quality.
B	Severe localized deteriorations of surface are successively observed. Complaints were reported.
C	Sinusoidal and cyclic profile deformations cause the decrease of ride quality.

Table 2. Profiles and IRI for the DS experiment.

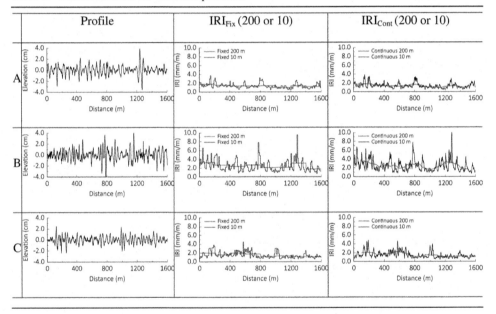

3 DRIVING SIMULATOR EXPERIMENT

3.1 *Apparatus*

The Kitami Institute of Technology Driving Simulator (KITDS) shown in Figure 2 (a) was used to conduct the experiment. For the KITDS, a six-axis motion base system consists of a cockpit with six cylinders activated by electric motors and reproduces the position and slant of the cockpit (6 degrees of freedom) by expanding and contracting the cylinders. It has the ability to generate the vehicle vibration induced by road profiles as well as the behaviour of the driver on the brake, the accelerator, and the steering wheel. The function of the KITDS allows loading actual road surface data directly into the simulator to perform a full vehicle

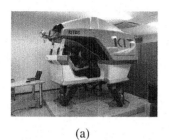
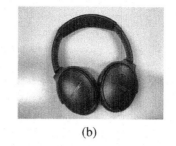
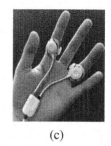

(a)   (b)   (c)

Figure 2.   Apparatus used in DS experiment: (a) Driving simulator, (b) Noise-canceling headphones, (c) biosignal sensor.

simulation controlled by the CarSim software of Mechanical Simulation Corporation with the front view of computer graphics provided by the screen. The details of the KITDS can be shown elsewhere (Kawamura A. *et al.* 2004).

In this experiment, there was a big difference between the sound inside the car when driving on the highway and the environmental sound heard when seated in the driver's seat during the DS test due to the loud motion operation sound of the KITDS. Accordingly, a headphone with noise-canceling function (BOSE QuietComfort 35 wireless headphones I) shown in Figure 2 (b) was employed.

For biosignals, heart rate variability was measured by an electrocardiogram (ECG) using the chest dielectric method, referring to previous studies (Tomiyama *et al.* 2015, 2017), and electrodermal activity (EDA), which is related to human emotional states, cognitive activities, and information processing processes, was measured using the skin conductance response. The biosignals were measured using Creact biosignalsplux, which transmits biosignals to a PC via a wearable hub for real-time monitoring. Figure 2 (c) shows an image of the EDA sensor employed. The ECG sensor is also worn on the upper body.

3.2   *Biosignals interpretation*

Heart rate variability (HRV: Heart Rate Variability) is the fluctuation of the heart rate and is a time series of the interval between R waves in the electrocardiogram (ECG) waveform shown in Figure 3 (a) and (b), and reflects the activity of the autonomic nervous system, which reflects mental stress. It is a physiological response that indicates the physiological state of the body. In the frequency domain, the low frequency (LF: 0.04-0.15 Hz) component represents sympathetic and parasympathetic nervous system activity, and the high frequency (HF: 0.15-0.5 Hz) component represents high frequency activity. Here, the HF component represents the parasympathetic nervous system activity. As shown in Figure 3(c), the HF component increases in the resting state, i.e., under stress-free conditions, and LF/HF, which is the LF component divided by the HF component as an index of the sympathetic nervous system, has the property of increasing under stressful conditions (Nakagawa 2016). The HF component decreases under the influence of road surface deformations that cause vehicle vibration (Tomiyama *et al.* 2015). This study focuses on the 5th percentile variation (lnHF5th) by obtaining percentile values for log-transformed HF (lnHF) taking into account the sensitivity to vibration among subjects.

Perspiration secretion alters the electrical properties of the skin, which is referred as electrodermal activity (EDA). By applying constant voltage, the change in skin conductance (SC) can be measured non-invasively (Fowles *et al.* 1981). It is measured as Skin Conductance Response (SCR) and has an unit of micro siemens (μS). Examples of characteristic waveforms of EDA signals and SCR are shown in Figure 4. Since the aim of this study is to understand subconsciousness toward localized road surface deformations, the

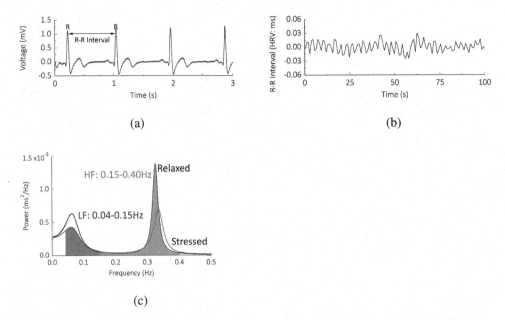

Figure 3. An Example of HRV in the time- and frequency-domain: (a) ECG, (b) HRV, (c) Changes in HRV in the frequency domain.

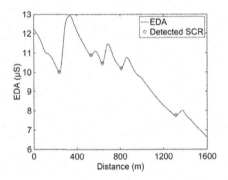

Figure 4. Examples of EDA signals and SCRs.

SCR which is a transient response of SC is focused on. The frequency of SCRs of the subjects acquired in order to extract the road surfaces that most the subjects find stressful is analyzed for the purpose of this study. Hereafter, the frequency of SCR occurrence is referred as SCRpp. When SCRpp increases, it can be interpreted that people are experiencing mental stress.

3.3 Test scenario

In order to evaluate physiologically the reaction of road users, a driving experiment by use of KITDS was conducted with a constant driving speed of 80 km/h. The road surface conditions were the combination of the properties listed in Table 1 and Table 2. In the experiment, an additional extension of which the IRI_{Fix} (200) is 1.7 mm/m which is the average IRI on the in-service expressways in Japan is set in the front of experiment section. Figure 5 shows

270

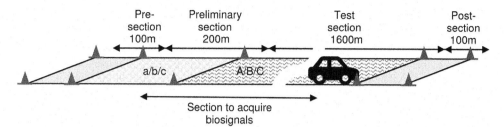

Figure 5.   DS Test scenario.

Table 3.   IRI of the preliminary section.

Surface	$IRI_{Fix}(200)$	Features
a	1.0 mm/m	Average $IRI_{Fix}(200)$ of construction quality control on expressway (Eguchi *et al.* 2015)
b	2.7 mm/m	Values rated as "good" or "very good" by more than half of the experiment participants (Kumada, *et al.* 2005)
c	3.5 mm/m	Rehabilitation threshold of expressway

an overview of the driving scenario in the DS experiment. A preliminary section of 200 m in length is provided in order to examine the impact of road surface conditions on the experience of road users. The road surfaces in the preliminary section include no localized deformations, and the IRIs are shown in Table 3 referring to the previous studies (Eguchi *et al.*, 2015; Kumada *et al.* 2005).

Each trial takes about 2 minutes and thus the total time is about 22 minutes. Seventeen participants in their 20s to 50s voluntary participated in the experiment. Two participants were tested in a different day due to error of the measurement. Finally, a total of 18 ECG data and 17 EDA data were involved in the analysis.

4   ROAD SURFACE EVALUATION METHOD

4.1   *Analysis of two gaps*

This section describes the analysis method for two gaps assumed. The first gap is assumed between the road surface evaluation based on specific criteria and the road surface evaluation by users. To find the gap location, the changes in SCRpp obtained for each subject road surface are shown together with the lnHF5th component and $IRI_{Cont.}(10)$. The second gap is assumed between the road surface evaluation indices and the road surface evaluation by users. The relationship between IRI and RN, which are road surface evaluation indices, and SCRpp is linearly interpolated and visualized to analyze the correlation of mental stress with each road surface evaluation index.

4.2   *Analysis of road surface characteristics associated with mental stress*

Wavy characteristics are analyzed for surface profiles associated with mental stress of participants. For this purpose, the road surface profiles are decomposed into a series of sinusoids by the wavelet transform and are examined for the wavelength characteristics at locations where mental stress is observed.

271

5 RESULTS AND DISCUSSION

5.1 *Gaps between maintenance criteria and rating of users*

Focusing on the stress-induced SCRpp, Figure 6 shows the following results. At a distance of about 200 m, the mental stress is detected even though the IRI below the current maintenance criteria. On the other hand, at a distance of about 500 m, no mental stress was observed, despite the similar IRI values with a distance of about 200 ms.

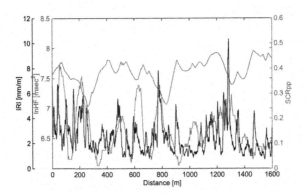

Figure 6.   Changes in mental stress against IRI.

Figure 7 illustrate the relationship of SCRpp with IRI and RN. As shown in the figure, high mental stress was observed at the points where IRI was 4 to 6 mm/m below the criteria and RN was 2.4 to 2.6. Currently, the maintenance criteria are set at 3.5 mm/m for IRI_{Fix} (200) and 8 mm/m for IRI_{Fix}(10) on expressways in Japan. Thus, we can say that IRI underestimates the surfaces that people find stressfull. In addition, RN was suggested to be consistent with mental stress. However, since RN is a nonlinear indicator, care must be taken in setting the evaluation interval length and threshold in road surface maintenance.

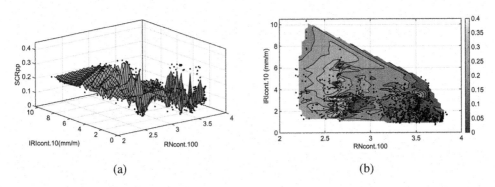

(a)	(b)

Figure 7.   Relationship of SCRpp with IRI and RN: (a) IRI and RN and SCRpp 3D contour map, (b) IRI and RN and SCRpp 2D contour map

5.2 *Road surface characteristics associated with mental stress*

Figure 8 shows (a) the decomposition result of a surface profile and (b) changes in the HF of 5th percentile against the IRI. As shown in the figure, the mental stress sensitively

272

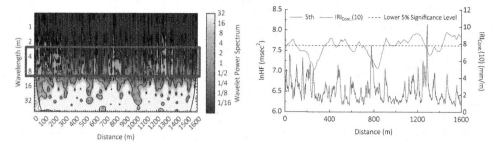

Figure 8.   Relationship between the Road Surface Characteristics and Mental Stress: (a) Decomposition result of a Road Surface Profile, (b) Changes in lnHF5th against IRI.

responds to the wavelength ranging from 4 to 8 m even if the IRI reveals an acceptable level. This is because the IRI is less sensitive to the wavelength ranging from 4 to 8 m. In contrast, the sensitivity range of the human body for vertical vibrations is 4~8 Hz (ISO1997) corresponding to the wavelength ranging from 2.8 to 5.6 m at a speed of 80 km/h, which overlaps with the insensitive wavelength of the IRI. Accordingly, the maintenance criteria with IRI are inconsistent with the mental stress corresponding to complaints of road users.

6   CONCLUSIONS

The findings obtained in this study are as follows:

- According to the relationship between biosignal and road surface wavelength, surface roughness inducing mental stress is potentially underestimated by the IRI.
- The mental stress increases with increasing amplitudes of wavelength ranging from 4 to 8 m of a profile even if the IRI reveals an acceptable level.
- The insensitivity of IRI to wavelengths between 4 and 6 m generates the gaps between the maintenance criteria in terms of the IRI and the road user rating.

These results suggest a need to develop a new road surface evaluation idea with human-cantered concept which contributes to the reduction of the number of complaints. According to the results of this study, the development of a maintenance criteria consistent with user rating leads to the human-cantered pavement management system.

REFERENCES

Chanjun C. and Seung-Ki R., 2019. Road Surface Damage Detection Using Fully Convolutional Neural Networks and Semi-Supervised Learning., *Sensors Vol.* 19, No.24.

Eguchi M., Tanaka Y., Kawamura A. and Tomiyama K., 2015. A New IRI-based Assessment for Detecting Spotted Surface Defects., *Journal of Japan Society of Civil Engineers, Ser. E1 (Pavement Engineering)*, Vol. 71, No. 3, pp. I_17–I_23.

Fowles D.C, Christie M.J., Edelberg R, Grings W.W., Lykken D.T. and Venables P.H., 1981. Publication Recommendations for Electrodermal Measurements. *Psychophysiology*, 18(3): pp.232–239.

Fujita S., Tomiyama K., Abliz N. and Kawamura A., 2013. Development of a Roughness Data Collection System for Urban Roads by Use of a Mobile Profilometer and GIS. *Journal of Japan Society of Civil Engineers, Ser. F3 (Civil Engineering Informatics)*. Volume 69, No. 2, pp. I_90–I_97.

Fukuda S., Matsumoto T., Okada H. and Momiyama Y., 2012. Estimation of Unacceptable Expressway Road Roughness With Short Wavelength Detected by Truck Drivers. *Journal of Japan Society of Civil Engineers, Ser. E1 (Pavement Engineering)*, Vol. 68, No. 3, pp. I_45–I_53.

Ishida T., Kawamura A., Alimujiang Y. and Tomiyama K., 2007. A Basic Study on Ride Comfort Evaluation Based on Biosignals. *Journal of Pavement Engineering, JSCE*, Vol. 12, pp. 197–204.

Ishida T., Takeomoto H., Kawamura A. and Shirakawa T., 2004. Impact of Pavement Roughness on Riding Comfort and Sense of Safety Evaluated Using a Driving Simulator., *Journal of Pavement Engineering, JSCE*, Vol. 9, pp. 49–56.

ISO Mechanical Vibration and Shock – Evaluation of Human Exposure to Whole-body Vibration - Part1: General Requirements, ISO2631–1:1997, 1997.

Jennifer A. and Rosalind P., 2005. Detecting Stress During Real-world Driving Tasks Using Physiological Sensors. *IEEE Transactions on Intelligent Transportation Systems*, Vol. 6, No. 2.

Kamiya K., 2021. Life Cycle Analysis of Road Pavement utilizing NEXCO-PMS. PDRG Meeting JRPUG2021.

Kashima H., Ito M., Ono S., Hiraoka T., Nakano K., Oguchi T. and Suda Y., 2021. Future Prospects for Mobility Triggered by Change of the Era. *Seisankenkyu.*, Vol. 73, No. 2, pp. 87–92.

Kawamura A., Sakakimoto T., Oono S., Sato M. and Suzuki K., 2000. Investigation on Road Profile of Expressways From the Point of View of Road Users. *Journal of Pavement Engineering, JSCE*, Vol. 5, pp. 102–111.

Kawamura A., Shirakawa T. and Maeda C., 2004. KIT Driving Simulator for Road Surface *Evaluation. Proceedings of the SURF 2004*, Pages 1–10.

Kawamura K. and Kamiya K., 2007. Current Status and Issues of Pavement Rehabilitation in NEXCO. *Asphalt*, Vol.50, No.222.

Kumada K., Sato M. and Kamiya K., 2005. Study on Road Surface Properties and Ride Quality Evaluation, *Nihon Doro Kodan Research Institute Report*, Vol. 42, pp. 197–208.

Nakagawa C., 2016. Special Issues No.3: Measurement Technique for Ergonomics, *The Japanese Journal of Ergonomics, Section 4: Measurements and Analyses of Bioelectric Phenomena and Others (5) Measurement and Analysis of Autonomic Indices*.

Tomiyama K., Kawamura A., Rossi R., Gastaldi M. and Mulatti C, 2015. Evaluation of Surface Unevenness Considering Mental Fatigue Based on Heart Rate Variability Analysis., *Journal of Japan Society of Civil Engineers, Ser. E1 (Pavement Engineering)*, Vol. 71, No. 3, pp. I_1–I_8.

Tomiyama K., Kawamura A., Rossi R., Gastaldi M. and Mulatti C., 2017. Physiopsychological Response of Vehicle Passengers to Surface Roughness and the Acceptable Limit., *Journal of Japan Society of Civil Engineers, Ser. E1 (Pavement Engineering)*, Vol. 73, No. 3,pp. I_89–I_96.

.

Surface features and performances

Medium to long term stockpiling performance of bitumen stabilized materials

Niel Claassens[*]
Department of Civil Engineering, Stellenbosch University (Hatch Africa (Pty) Ltd), Stellenbosch, South Africa

Kim Jenkins
Department of Civil Engineering, Stellenbosch University, Stellenbosch, South Africa

ABSTRACT: Worldwide, road rehabilitation exceeds the demand for constructing new roads. This is not only due to the significant cost of constructing new roads but also a shift in focus towards sustainability principles in our modern society. Cost and sustainability have indeed become the major considerations of roads projects. Bitumen Stabilised Materials (BSMs) are well suited for road rehabilitation projects and provide a more environmentally friendly and sustainable solution compared to hot mix asphalt (HMA).

Unlike HMA, BSMs are manufactured at ambient temperatures providing savings in energy consumption and emissions. Furthermore, BSMs are typically manufactured using reclaimed asphalt (RA) or recycled pavement layers (granular or cemented). Many road authorities around the world are overwhelmed by reclaimed asphalt stockpiles. Bulk manufacturing and stockpiling of BSMs can provide a durable and sustainable pavement material at a fraction of the cost of HMA. Stockpiled BSMs can be used as and when required, reducing the cost of road rehabilitation projects considerably.

Current guidelines regarding stockpiling of BSMs are empirical and not based on experimental data. This has created a grey area on construction projects where it is widely accepted that BSMs can be stockpiled for a period of between 7 and 10 days. The medium to long term stockpile performance of two BSM mixtures at varying moisture conditions were investigated over a period of 6 months. In particular the performance characteristics described by cohesion, angle of friction, ITS and dissipated energy (flexibility) were analysed over the 6-month period.

Based on a recent research project, the detrimental effect of increased moisture content on the performance characteristics of the BSM mixes were evident. In general, the test results showed a decrease in cohesion and dissipated energy with an increase in stockpile life. No correlation between the ITS results and the cohesion results were observed and therefore, the ITS test does thus not give a good indication of the shear properties of BSMs. It is hypothesized that stockpile size has a direct impact on the stockpile performance of BSMs. Air can move more freely through smaller stockpiles leading to premature ageing of the bitumen and a reduction in shear strength.

Keywords: BSM, Sustainability, Recycling

1 INTRODUCTION

With the ever-increasing focus on sustainability in our modern society, the roads industry faces a challenge of balancing the use of virgin materials with recycling of materials.

[*]Corresponding Author: niel.claassens@hatch.com

DOI: 10.1201/9781003429258-27

Worldwide, and especially in developed countries, the need for rehabilitating existing roads far exceeds the need for constructing new roads. Road rehabilitation typically entails the replacement of existing pavement layers by milling and resurfacing. Consequently, large volumes of reclaimed asphalt (RA) can be generated on road rehabilitation projects. Many road authorities around the world are overwhelmed by RA stockpiles which continue to grow (Williams et al. 2019).

Reclaimed asphalt consists of high-quality aggregates and residual bitumen making it a valuable resource. There is a big drive in the roads industry to develop Hot Mix Asphalt (HMA) mixes incorporating large amounts of RA. The use of RA in asphalt mixtures is however typically limited to roughly 20% by weight. Until specifications are widely accepted for asphalt mixes containing higher percentages of RA, RA stockpiles will continue to grow (Williams et al. 2019). To reduce the size of RA stockpiles many road authorities use high quality RA material as fill material, gravel wearing course material or gravel shoulder surfacing.

High quality RA is however also well suited for bitumen stabilization using foamed bitumen or bitumen emulsion to manufacture Bitumen Stabilised Materials (BSMs). BSMs are well suited for road rehabilitation projects and are mainly utilised as base layers. BSMs are manufactured by mixing foamed bitumen or bitumen emulsion and a small amount of active filler (usually 1%) with the recycled pavement materials. BSMs are considered to be more sustainable than HMA as they are manufactured at ambient temperatures and offer savings in energy consumption and emissions.

Collings & Jenkins (2011) have previously reported on the successful use of BSM on various road rehabilitation projects around the world. Bulk manufacturing and stockpiling of BSMs manufactured from 100% RA can provide a durable and sustainable pavement material at a fraction of the cost of HMA. Stockpiled BSMs can be used as and when required, reducing the cost of road rehabilitation projects considerably.

Current guidelines regarding stockpiling of BSMs are empirical and not based on experimental data. It is widely accepted that BSMs manufactured from foamed bitumen can be stockpiled for a period of between 7 and 10 days. Stockpiling of BSMs manufactured from bitumen emulsion is not recommended. Claassens (2021) recently completed a research study in Cape Town, South Africa which investigated the medium to long term stockpiling performance of two BSM mixtures at varying moisture conditions over a period of 6 months. In particular the performance characteristics described by cohesion, angle of friction, ITS and dissipated energy (flexibility) were analysed over the 6-month period. This paper will present the findings of the research study and give recommendations for future studies which aim to investigate the stockpiling performance of BSMs.

2 CURRENT BSM SPECIFICATIONS

BSMs have been around for over 30 years however major improvements have been made to this technology over the last decade. In South Africa, over 20 years' BSM experience has culminated in the publication of the revised *Technical Guideline: Bitumen Stabilised Materials – A Guideline for the Design and Construction of Bitumen Emulsion and Foamed Bitumen Stabilised Materials (TG2 2020)* by the Southern African Bitumen Association (Sabita) in 2020. TG2 does not distinguish between BSM manufactured using foamed bitumen (BSM-foam) and BSM manufactured using bitumen emulsion (BSM-emulsion).

The design of BSMs is based on a mix design procedure. Typically, reclaimed pavement materials (from which the BSM is to be manufactured) are sampled and various test specimens are manufactured. The target grading requirements for BSM-foam and BSM-emulsion is shown in Figure 1 below.

The test specimens are manufactured at varying mix proportions but limiting the bitumen to less than 3% and the active filler to a maximum of 1%. Indirect tensile strength (ITS) and

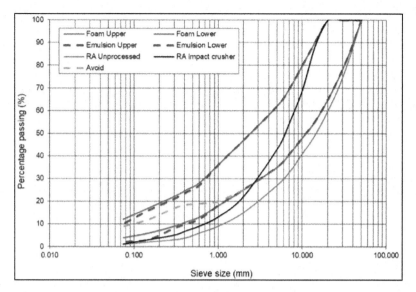

Figure 1. Target gradings for BSM-foam and BSM-emulsion (Sabita 2020).

triaxial testing is carried out on the specimens from where the indirect tensile strength and shear properties of the mix are determined. The mix proportions are then specified based on the mix which yields the highest ITS and triaxial test results.

The shear properties (angle of friction and cohesion) are the two main design variables used in the BSM design method. Generally, a higher cohesion and angle of friction provides for increased material strength and thus a more resilient pavement material. Two different BSM classes are specified in TG2 2020 as shown in Figure 2 below.

BSM 1 is typically specified for roads with high volumes of traffic whereas BSM 2 is more suitable for roads with lower traffic volumes. Due to the expensive and specialised nature of triaxial testing equipment, BSM quality control during construction is based on the ITS specifications. Current guidelines in TG2 2020 states that BSM-foam may be stockpiled for several days provided that:

- hydrated lime is used as the active filler;
- moisture content of the stockpile is maintained close to optimum moisture content;
- the stockpiled material remains in an uncompacted state;
- regular ITS testing is carried out on BSM samples taken from the stockpile to monitor the loss of strength.

The influence of stockpiling BSMs on the performance characteristics such as cohesion, angle of friction and ITS is largely unknown.

Class	RA (%)	ITS (kPa)[1]		Triaxial		
		ITS$_{DRY}$	ITS$_{WET}$	Cohesion (kPa)	Friction Angle (°)	Retained Cohesion (%)
BSM 1	< 50%	225	125	250	40	75
	50 – 100%	225	125	265	38	75
BSM 2	< 50%	175	100	200[2]	38	65
	50 – 100%	175	100	225	35	75

Note:
1. 152 mm diameter specimen geometry used for ITS tests and 150 mm diameter for Triaxial tests

Figure 2. BSM Specification Limits (Sabita 2020).

3 EXPERIMENTAL SETUP

A simple experiment was carried out whereby two different BSM mixtures were produced and stockpiled under varying moisture conditions over a period of 6 months in Cape Town, South Africa. The material used in the experiment was procured from a BSM project on National Route 7 (N7) in Cape Town. One BSM mix consisted of 100% RA (RA-BSM). The other mix consisted of 100% reclaimed BSM (BSM-BSM). The reclaimed pavement materials were crushed before being processed. Both BSM mixtures were stabilised with 2% foamed bitumen and 1% hydrated lime using a Wirtgen KMA 200 mobile mixing plant. The mix designs are summarised in Table 1 below:

Table 1. Summary of BSM mix designs.

Mix	MDD (kg/m^3)	OMC (%)	Active Filler (%)	BC (%)	ITS Dry (kPa)	ITS Wet (kPa)	Cohesion (kPa)	Retained Cohesion (%)	Angle of Friction (°)
RA-BSM	2021	3.5	1	2	299	225	276	72	42.1
BSM-BSM	2180	5	1	2	460	221	386	76	42.4

For each BSM mix, three separate stockpiles were established each with a different moisture condition: mixing moisture (MMC), optimum (OMC) and Wet. The moisture conditions were maintained for the duration of the experiment. Six different stockpiles were thus established. To ensure consistency with the experimental results, the BSMs were manufactured on the same day and taken to stockpile immediately. Each separate stockpile consisted of roughly 4m^3 of material. The size of each stockpile was limited due to the following reasons:

- limited materials and stabilising agents were made available for the study;
- limited space was available to establish the stockpiles. Furthermore, the stockpiles had to be located close to a water source which was only available at the project site camp;

Each stockpile was covered with a thick plastic impervious blanket for protection against the elements and to control the moisture conditions as far as possible. The height of each stockpile was limited to a maximum of 1.5 m.

For the OMC and wet stockpiles, a water application system was developed using agricultural drip irrigation pipes connected to a standpipe to ensure constant water flow. The flow rate of the water was controlled with a ball valve. For the OMC stockpiles the valve was opened by a quarter to limit the flow rate of the water. Initially, this application rate of water was found to keep the material close to OMC. For the wet stockpiles, the valve was left open completely to ensure that the stockpiles were completely soaked. For the OMC and wet stockpiles, water was allowed to continuously flow into the material for the duration of the experiment. No water was added to the MMC stockpiles after processing in the Wirtgen KMA 200 mobile mixing plant.

Periodic sampling of each stockpile was conducted over a 6-month period and the performance of the BSMs in stockpile were assessed based on the following material properties: angle of friction, cohesion, indirect tensile strength (ITS) and dissipated energy (flexibility). Bulk samples were sent to BSM Laboratories (a specialist BSM testing laboratory in Durban, South Africa) who carried out the testing on the samples. A summary of the sampling regime is shown in Table 2 below.

Table 2. Summary of BSM sampling regime.

Sample No.	1	2	3	4	5	6	7	8	9
Stockpile Age (days)	0	3	7	14	28	84	112	139	168

4 EXPERIMENTAL RESULTS & DISCUSSION

It should be noted that over the course of the experiment it became apparent that the water application system was not accurate enough to maintain the OMC stockpiles at Optimum Moisture Content. As a result, both the OMC and Wet stockpiles became fully saturated and yielded similar test results. For this reason, the results from the OMC stockpiles will not be presented.

4.1 *Shear properties*

The angle of friction and cohesion results for the RA-BSM and BSM-BSM stockpiles are shown in Figures 3 and 4 below respectively. The angle of friction results of all the stockpiles remained relatively constant over the duration of the experiment. Angle of friction is a shear property mainly influenced by aggregate shape and texture. This result can be expected since stockpiling does not alter these properties. In general, increased moisture content did not have any effect on the angle of friction – this result can also be expected.

Generally, there was a decrease in cohesion with increasing stockpile age. The almost immediate decrease in cohesion of the BSMs at a short stockpile age is significant. The detrimental effect of increased moisture content on cohesion is also evident from the test

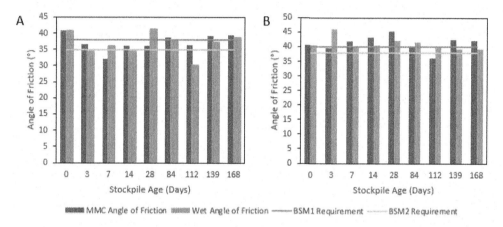

Figure 3.   Angle of Friction Results for RA-BSM (A) stockpiles and BSM-BSM (B) stockpiles.

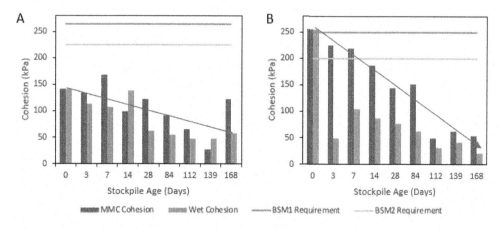

Figure 4.   Cohesion Results for RA-BSM stockpiles (A) and BSM-BSM stockpiles (B).

results. It is also evident that the cohesion results obtained from the RA-BSM stockpiles were well below the cohesion requirements for either a BSM1 or a BSM2.

Increased cohesion due to stabilisation with small amounts of bitumen and active filler is a general characteristic of BSMs (Sabita 2020). Further investigation has revealed that the RA-BSM mix used in the research study was manufactured as backup material for the N7 project (in case of a shortfall) from finer fractions of RA. The finer RA material generally yielded low cohesion results and was never used in construction.

Generally, a decrease in the cohesion results with increasing stockpile age is evident for all the stockpiles. The exact reason for the general decrease in cohesion, and especially for the materials in the RA-BSM MMC and BSM-BSM MMC stockpiles, are unknown. Further research is required to understand the mechanisms causing this decrease in cohesion of the stockpiled materials. The following factors or combinations thereof could provide insight into the decrease in cohesion:

- Due to the size of the stockpiles, air can potentially move more freely through the material and cause premature aging of the miniscule bitumen "spot welds" between aggregate particles. As a result, the miniscule bitumen "spot welds" become brittle which in turn leads to a reduction in cohesion;
- Time delays between sampling in Cape Town and preparing the samples in Durban were roughly 3 days and could have affected the cohesion results;
- Material samples were placed in thick plastic bags which were zip-tied to control the moisture content of the samples however, the environment in which the samples were transported may have affected the cohesion results.
- Active filler generates a bond between the aggregate surface and surface of the bitumen droplet. This bond increases whilst curing in the stockpile. As the material is disturbed due to loading, placing and compacting the bonds can be broken resulting in a decrease in cohesion;
- The longer the delay in stockpile, the more the efficiency of the cohesive bond deteriorates i.e. a combination of point 1 and point 4 described above.

The detrimental effect of moisture content on cohesion is also evident from the cohesion results. The cohesion results of the wet stockpiles were generally lower than the cohesion results of the MMC stockpiles (for both mixes). Twagira (2010) identified various moisture damage mechanisms applicable to BSMs and their effect on cohesion and adhesion. Based on the research carried out by Twagira (2010), the following factors or combinations thereof can provide reasons for the decrease in cohesion due to increased moisture content:

- Excess moisture in the mix could affect the adhesion between the aggregate and bitumen. This can result in poor bond strength at the aggregate interface and thus lead to decreased cohesion;
- Excess moisture in the mix can flush out some of the binder rich mastic which in turn will also cause a decrease in cohesion;
- It is possible that the constant addition of water to the wet stockpiles changes the micrograding of the materials which in turn has a negative effect on cohesion.

4.2 *Indirect Tensile Strength (ITS)*

The ITS results of the RA-BSM and BSM-BSM stockpiles are shown in Figure 5 below. In general, the ITS test results of the wet stockpiles were lower than the ITS test results of the MMC stockpiles. The detrimental effect of increased moisture content on ITS is thus evident from the results. The moisture damage mechanisms discussed above are also valid for the ITS. It is known that the ITS test does not provide highly repeatable results (Wirtgen 2012). The variability of the ITS test is also evident from the results shown below. Due to this variability, no clear trend in ITS was evident from the analysis.

When comparing the ITS test results with the cohesion test results, the same trend is not observed. In general, with an increase in stockpile age, a downward trend in the cohesion of the

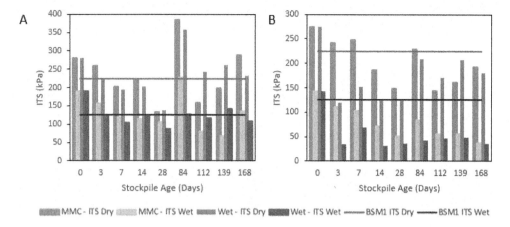

Figure 5.   ITS Results for RA-BSM stockpiles (A) and BSM-BSM stockpiles (B).

stockpiled BSM was observed. Cohesion results obtained from the RA-BSM stockpiles were well below the cohesion requirements for a BSM1. The ITS results however indicate that the RA-BSM met the requirements for a BSM1 up to a stockpile age of 14 days. The ITS test does thus not give a good indication of the shear properties of BSMs. It can thus also be concluded that ITS does not give a good indication of the loss of strength of the BSM in stockpile.

ITS testing is used for acceptance control on BSM projects however, the ITS did not give a good indication of angle of friction or cohesion. This is concerning as cohesion and angle of friction are the two main parameters used for the design of BSMs. The test results indicate that a significant decrease in cohesion does not necessarily result in a significant decrease in ITS. A significant decrease in cohesion would, however, have a greater influence on the long-term performance of BSMs. A different test method, which gives a better indication of the shear properties of BSMs, is required for acceptance control testing.

4.3   Dissipated energy (Flexibility)

The dissipated energy results of the RA-BSM and BSM-BSM stockpiles are shown in Figures 6 and 7 respectively. It is evident that dissipated energy increases with increasing confining pressure. The stress dependent nature of BSMs is evident from the results. At increasing confining pressure, the resilient modulus of the BSM increases resulting in an increase in the dissipated energy. More energy is dissipated to cause failure of the BSM and thus at increased confinement a higher strain at break is achieved. Llewellyn (2015) reported similar findings.

Generally, the dissipated energy of the BSMs decrease as stockpile age increases. BSMs acquire flexibility due to stabilisation with small amounts of bitumen and active filler (Sabita 2020). The decreasing dissipated energy can be attributed to the general decrease in cohesion over time. Possible mechanisms contributing to the decrease in cohesion have been discussed and are also valid for dissipated energy. The general decrease in cohesion over time is indicative that the positive effects of bitumen stabilisation diminish with increasing stockpile age. This in turn leads to a decrease in dissipated energy.

It is hypothesized that due to aging, the bitumen "spot welds" become brittle in stockpile over time. The brittle bitumen "spot welds" have significant surface area considering their diameter, however, they do not contribute significantly to the flexibility of the BSM and as a result, the dissipated energy decreases.

The detrimental effect of increased moisture content on BSM flexibility is also evident from the results. Generally, the dissipated energy results of the Wet stockpiles were lower

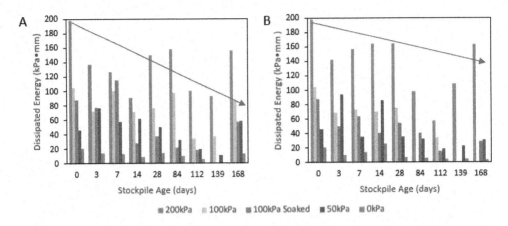

Figure 6. Flexibility Results for RA-BSM MMC stockpile (A) and RA-BSM Wet stockpile (B).

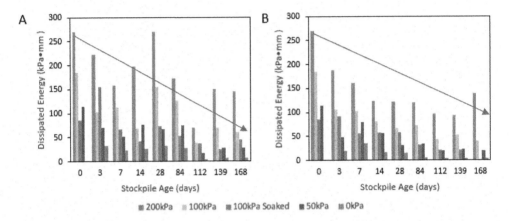

Figure 7. Flexibility Results for BSM-BSM MMC stockpile (A) and BSM-BSM Wet stockpile (B).

than that the dissipated energy results of the MMC stockpiles. The following factors or combinations thereof can provide reasons for the decrease in dissipated energy over time:

• Excess moisture in the mix could affect the adhesion between the aggregate and bitumen. This can result in poor bond strength at the aggregate interface with impaired efficacy of the active filler and thus lead to decreased dissipated energy;
• As the material becomes more saturated, the voids are being filled with water which leads to the build-up of pore pressure. As the confinement increases after compaction and during loading, pore water pressure is exerted on the individual aggregate particles and leads to the breakdown of the integrity of the BSM and thus a decrease in dissipated energy.

5   CONCLUSIONS & RECOMMENDATIONS

The results of the study have indicated that stockpiling BSMs for extended periods of time can have a detrimental effect on the performance characteristics of the material in stockpile. The drastic deterioration of the shear properties within three days of being placed in stockpile is significant. The detrimental effect of stockpiling BSMs at increased moisture content is also evident. Based on the outcomes of this research study, it is recommended that BSMs are not kept in stockpile. It must however be noted that the results obtained from this study may not hold true for all BSMs. Further research is required to ascertain whether these trends hold true for all BSMs.

Further recommendations of this research study are given below:

- Similar studies should be conducted to verify the results of this research study;
- A similar study should be carried out on BSM-emulsions to investigate the stockpiling performance of BSM-emulsions;
- The effect of increased quantities of bitumen and active filler on the stockpiling performance of BSMs should be investigated;
- The influence of climate on the stockpiling performance of BSMs should be investigated in future studies;
- A new test, which gives a better indication of the shear properties of BSMs, should be developed for acceptance control testing;
- Larger stockpiles are more representative of the stockpiling conditions on construction sites and should be considered in future studies;
- A more sophisticated water application system using a solenoid valve (or similar) is required to control the moisture condition of stockpiles;
- Future research studies should be located closer to a testing laboratory to prevent the possible influence of material transportation (including the environment under which samples are transported) on test results;
- The performance of BSMs should initially be monitored daily over a period of 7 days in future studies to monitor the daily change in BSM performance. Thereafter testing can be conducted at 14 days and then monthly (as required).

ACKNOWLEDGEMENTS

- Western Cape Government, Department of Transport and Public Works;
- Wynand van Niekerk from BSM Laboratories;
- Hatch Africa (Pty) Ltd.

REFERENCES

Claassens N., 2021. *Medium to Long-Term Stockpiling Performance of Bitumen Stabilised Materials*, Master's Thesis, Stellenbosch University.

Collings D. and Jenkins K., 2011. The Long-Term Behaviour of Bitumen Stabilised Materials, *10th Conference on Asphalt Pavements for Southern Africa*.

Llewellyn G.B., 2015. *Flexibility Behaviour of Bitumen Stabilised Materials*, Master's Thesis, Stellenbosch University.

Southern African Bitumen Association (Sabita), 2020. *Technical Guideline: Bitumen Stabilised Materials – A Guideline for the Design and Construction of Bitumen Emulsion and Foamed Bitumen Stabilised Materials.*

Twagira E., 2010. *Influence of Durability Properties on Performance of Bitumen Stabilised*, PhD Dissertation. University of Stellenbosch.

Williams B.A., Willis J.R. and Shacat J., 2020. Asphalt Pavement Industry Survey on Recycled Materials and Warm-mix Asphalt Usage: 2019 (No. IS 138 (10e)).

Wirtgen Group, 2012. *Cold Recycling: Wirtgen Cold Recycling Technology*, 1st edition, Windhagen, Germany.

A correlation between laser-based device and close range photogrammetry technique for road pavement macrotexture estimation

Mauro D'Apuzzo & Azzurra Evangelisti*
Department of Civil University of Cassino and Southern Lazio, Cassino, Italy

Giuseppe Cappelli
Department of Civil University of Cassino and Southern Lazio, Cassino, Italy
University School for Advanced Studies, IUSS, Pavia, Italy

Vittorio Nicolosi
University of Rome "Tor Vergata", Rome, Italy

ABSTRACT: Both public and private roads Agencies face the problem of having to guarantee high levels of service and safety on their managed roads, having limited economic resources available. In this context, the necessity to accelerate and improve the monitoring techniques for road pavement quality, promotes the research towards the development of easily accessible and reliable tools.

For this purpose, the present work shows the preliminary results obtained from the application of photogrammetric techniques to the evaluation of the macrotexture of road pavements. In fact, it is widely recognized that macrotexture is a fundamental indicator for indirectly defining the grip level of a road pavement, especially in the wet conditions.

In order to investigate the correlation between laser scanning and photogrammetric technique, in terms of Mean Profile Depth (MPD), three different road pavements have been selected and two test points for each of them, have been studied.

The three-dimensional models of the six test points have been developed for both laser scanner and photogrammetry acquisition methods and six hundred profiles have been analyzed. Preliminary results have shown an acceptable correlation ($R^2 = 0.956$ and Pearson = 0.98) between MPD values obtained from both three-dimensional reconstructions, confirming the potential of the photogrammetry also for the road pavement surface acquisition and macrotexture characterization.

Keywords: Macrotexture, Mean Profile Depth, Photogrammetry, Laser-based Device, Profile Analysis, Road Pavement

1 INTRODUCTION

The effectiveness of using photogrammetry to pavement engineering has been recognized very early and first applications have been performed since the end of the sixties. Without claiming to provide an exhaustive overview of the photogrammetry applications in the engineering field, main research within road pavements analysis have been summarized below.

As fa as pavement detection activities, within road inspections are concerned, with the aim to map the actual condition of the road pavement, several applications of the

---

*Corresponding Author: aevangelisti.ing@gmail.com

DOI: 10.1201/9781003429258-28

photogrammetric analysis have been introduced: road pavement measurement systems, primarily designed for pothole and crack detection, but also for surface deformation measurements, mounted on vehicles or Unmanned Aviation Vehicle (UAV), have been developed (Kertész et al. 2008; Knyaz & Chibunichev 2016; Mustaffara et al. 2008; Zhang 2010).

Some relevant applications have been provided for microtexture and skid resistance investigations as Schonfeld (1970) who, defining seven texture-geometrical parameters, proposed a method for estimating skid resistance from stereophotographs of the pavement surface. Other studies showed comparisons between Close Range Photogrammetry (CRF) and different texture or skid measurements, in term of surface roughness (Slimane A.B. et al. 2006), of Abbott-Firestone curve's parameters (McQuaid G. et al. 2014) or in term of Skid Number (Sarsam S.I. et al. 2015).

Good correlations have been found between macrotexture, expresses in terms of MTD Sand Patch and Mean Profile Depth (MPD) evaluated with photogrammetry applications (Medeiros S. et al. 2016; Sarsam S.I. et al. 2016). Some researches have been performed for investigating correlations between photogrammetry and laser applications using the Fast Fourier Transform (FFT) (Elunai R. et al. 2010), or describing macrotexture in terms of areal surface texture parameters (Edmondson V. et al. 2019), or in terms of Mean Profile Depth (MPD) (El Gendy A. et al. 2011).

The aim of the present work is to enrich this research field, presenting preliminary findings to establish correlation between MPD values evaluated on profiles extracted on 3D reconstructions of both photogrammetry and Laser applications.

2 BRIEF OVERVIEW ON MACROTEXTURE ESTIMATION

According to the Standard ASTM E965-15 (2019), the pavement macrotexture can be defined as *"the deviations of a pavement surface from a true planar surface with the characteristic dimensions of wavelength and amplitude from 0.5 mm up to those that no longer affect tire-pavement interaction"*.

A direct measure of macrotexture, expressed by the Mean Texture Depth (MTD) coefficient, can be provided with the Volumetric Technique (ASTM E965-15 2019) historically known as Sand Patch Method, due to the Ottawa natural silica sand used for this test method.

Currently indirect macrotexture evaluations are worldwide used and the most widespread is the calculation of the Mean Profile Depth (MPD) from a profile of pavement macrotexture, which is regulated by (ASTM E1845-15 2015; UNI EN ISO 13473-1 2019).

In particular, the MPD is calculated as the average value of the Mean Segment Depths (MSD) for all the segments of the measured profiles. In the Figure 1 the procedure for the

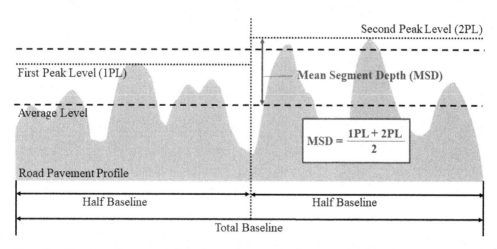

Figure 1. Procedure for computation of mean segment depth.

computation of the MSD, has been provided. Furthermore, linear transformations from the MPD to the Estimated Texture Depth (ETD) have been found (ASTM E1845-15 2015; ASTM E2157-15 2019).

Due to its usefulness in predicting the speed constant (gradient) of wet pavement friction (D'Apuzzo M. et al. 2020a; Wambold J.C. et al. 1995), the MPD can be used as stand-alone macrotexture value (D'Apuzzo M. et al. 2020b; Evangelisti A. et al. 2015) or as indirect descriptor of road skid degradation models (Nicolosi V. et al. 2020).

3 BASIC PRINCIPLES OF PHOTOGRAMMETRY

The photogrammetry can be defined as "the "science of measuring in photos", and is traditional a part of geodesy, belonging to the field of remote sensing (RS)" (Linder W. 2016).

Primordial studies on the principles of photogrammetry seem to be attributed to the engineer M.A. Cappeler in 1726. In 1759 thanks to the Perspectiva liber by J. H. Lambert the mathematical laws on which photogrammetry is based, have been defined. However only in the second half of the nineteenth century, the first applications have been performed and the development of new technologies and application fields do not seem to stop nowadays, such as in: architecture, cultural heritage, topographic mapping, geology and engineering.

The principle of Triangulation is the foundation of the photogrammetry. By taking at least two photos from different positions of the same object, three-dimensional coordinates of any point which is represented in both photographs, could be calculated (see Figure 2).

However, as matter of fact that the accuracy of a photogrammetric measurements depends on different related factors, among which:

1. camera's features as resolution, lens, optical or digital zoom, etc.;
2. number of analyzed photographs;
3. relative position of the photos, also respect to the object;
4. the size of the object.

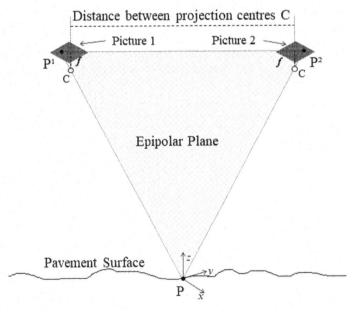

Figure 2.   Photogrammetry principle where: f, focal length; P1 and P2, P representation in the left and right pictures respectively.

Several processing software are available to perform photogrammetry applications, some of them require preliminary calibration of the camera and relative distances, others provide self-calibration algorithms (but a larger number of pictures are needed).

4   EXPERIMENTAL CAMPAIGN

In order to investigate the effectiveness of photogrammetry to the macrotexture evaluation, by comparing the MPD results to the values obtained on 3d reconstructions based on laser measurements, two different equipment have been selected.

In particular, the acquisition of digital photographs has been performed with a Canon Digital Ixus 75 camera and its main feature have been summarized in the Table 1. On the other hand, the laser measurements have been carried out by the Nikon MCAx Handheld scanner.

Table 1.   Camera s' features.

Canon Digital Ixus 75		
F-stop	f/2.8	
Exposure time	[1/sec]	1/60
Sensitivity	[ISO]	500
Focal distance	[mm]	6
Focal length	[mm]	35

Three road pavements, characterized by different textures and aggregates' sizes and types have been selected: Pavement A with large and rough aggregates (high macrotexture), Pavement B with small and smooth aggregates and Pavement C, closed and compact (low macrotexture). A representation of each pavement has been reported in the following figure (Figure 3).

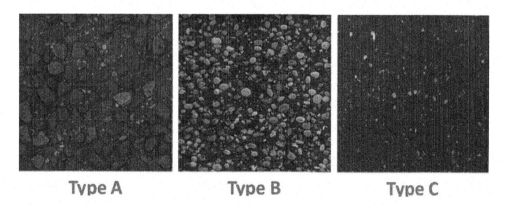

Figure 3.   Particular of selected road pavements.

For each type of pavement, two points, at least three meters apart, have been identified and subjected to measurements.

Several pictures, rotating at 360° with different angles and respecting the images overlap between 50% and 70%, have been taken. The features summarized in the following table (Table 2), characterized each photograph:

Table 2. Photographs' features.

Dimension	[pixel]	3072 ×2304
Vertical resolution	[dpi]	180
Horizontal resolution	[dpi]	180
Depth	[bit]	24
Flash		NO
ISO		80

In order to provide the dense point cloud from the pictures, an open-source software has been used and the highest level of detail has been selected. To make visible the different reconstruction quality of the point clouds obtained by both photogrammetric method and laser scanning, the three dimensional models, one for each pavement, have been reported in the following figure (Figure 4).

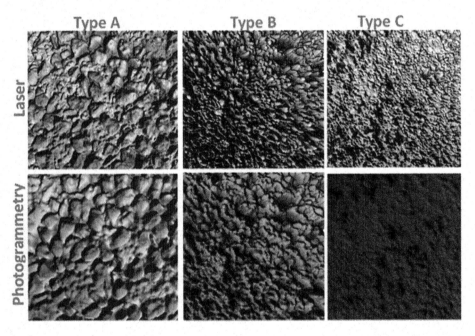

Figure 4. 3D reconstruction of road pavements by dense point cloud of both laser measurements and photogrammetry applications.

Finally, basing on the 3D reconstructions obtained on point clouds of both elaborations, for each sample, 50 profile (25 along X and 25 along Y direction) have been extracted and the MSP and MPD values have been evaluated.

5 ANALYSIS OF THE RESULTS

As previously introduced, two separate point on three different road pavements have been selected as tests. Two different devices for the point cloud evaluation, have been used and, for each 3D model, 50 profiles have been extracted. In total 600 profiles have been analysed and the related MSD values have been evaluated, according to (ASTM E1845-15 2015).

Defining an acceptance range of $\mu \pm 2\sigma$ (where μ is the average value and σ is the standard deviation of MSDs of each selected test) the MSD values of the six tested points have been summarized in the following figure (Figure 5).

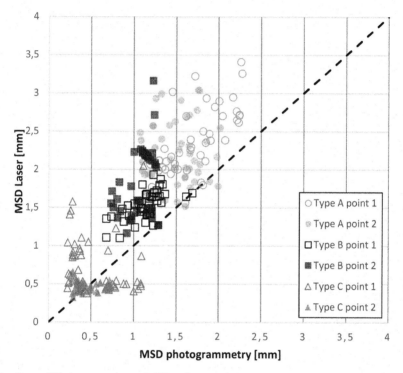

Figure 5. Laser VS photogrammetry MSD values.

As it is possible to see, both laser and photogrammetry applications are enabled to identify with a good agreement, the different pavement types from coarse (Type A) to fine (Type C) texture.

Finally, according to the standard (ASTM E1845-15 2015), the MPD values, one of each tested point, have been evaluated and a correlation between the two different applications have been defined (see Figure 6). Observing the Figure 6 it is possible to state that a good correlation ($R^2 = 0.956$ and Pearson = 0.98) between laser and photogrammetry applications for the MPD evaluation, have been found.

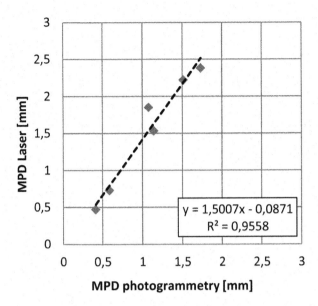

Figure 6. Laser - photogrammetry MPD correlation.

6 CONCLUSIONS

The application of photogrammetry for the analysis of road pavement macrotexture has shown satisfactory perspectives, thanks to the relatively low cost of the devices and the availability of free software for images post-processing.

In this work a satisfactory correlation (R^2 = 0.956 and Pearson = 0.98) between MPD values evaluated on three dimensional models of three different road pavements has been found. Two different techniques for the point cloud construction have been applied and compared: the laser scanner and the photogrammetric procedures.

Six hundred profiles, extracted on both 3d models have been analysed and the comparison between MSD values has shown that photogrammetry applications, as well as the more consolidated laser scanning, are enabled to characterize the macrotexture of the road pavements.

The results obtained so far, strengthen the intent to deepen the knowledge and open the way for new applications.

REFERENCES

ASTM E1845-15, (2015). *Standard Practice for Calculating Pavement Macrotexture Mean Profile Depth, ASTM International*, West Conshohocken, PA, www.astm.org.

ASTM E2157-15, (2019). *Standard Test Method for Measuring Pavement Macrotexture Properties Using the Circular Track Meter*, ASTM International, West Conshohocken, PA, www.astm.org.

ASTM E965-15, (2019). *Standard Test Method for Measuring Pavement Macrotexture Depth Using a Volumetric Technique*, ASTM International, West Conshohocken, PA, www.astm.org.

D'Apuzzo M., Evangelisti A. and Nicolosi V., (2020). An exploratory step for a general unified approach to labelling of road surface and tire wet friction. *Elsevier, Accident Analysis & Prevention*. Vol. 138, 105462. https://doi.org/10.1016/j.aap.2020.105462.

D'Apuzzo M., Evangelisti A., Santilli D. and Nicolosi V., (2020). Theoretical Development and Validation of a New 3D Macrotexture Index Evaluated from Laser Based Profile Measurements. In: Gervasi O. et al.

(eds) *Computational Science and Its Applications*. ICCSA 2020. Lecture Notes in Computer Science, vol 12251. Springer, Cham. https://doi.org/10.1007/978-3-030-58808-3_27.

Edmondson V., Woodward J., Lim M., Kane M., Martin J. and Shyha I., (2019). Improved Non-contact 3D Field and Processing Techniques to Achieve Macrotexture Characterisation of Pavements. In: Construction and Building Materials, Vol. 227, 10.12.2019, 116693.

El Gendy A., Shalaby A., Saleh M. and Flintsch G.W., (2011). Stereo-vision Applications to Reconstruct the 3D Texture of Pavement Surface, *International Journal of Pavement Engineering*, 12:03, 263–273, DOI: 10.1080/10298436.2010.546858.

Elunai R., Chandran V., & Mabukwa P. (2010). Digital Image Processing Techniques for Pavement Macro-texture Analysis. In Doyle, Neil (Ed.) *Proceedings of the 24th ARRB Conference: Building on 50 Years of Road Transport Research*, ARRB Group Ltd., Sebel Hotel, Melbourne, Vic, pp. 1–5.

Evangelisti A., D'Apuzzo M., Nicolosi V., Flintsch G., De Leon Izeppi E., Katicha S. and Mogrovejo D., (2015). Evaluation of Variability of Macrotexture Measurement with different Laser-based Devices. *Airfield & Highway Pavement Conference*.

Kertész I., Lovas T. and Barsi A.. (2008). Photogrammetric Pavement Detection System, in: The International Archives of the Photogrammetry, Remote Sensing and Spatial Information Sciences. Vol. XXXVII. Part B5. Beijing.

Knyaz V.A.; Chibunichev A.G., (2016). Photogrammetric Techniques For Road Surface Analysis. ISPRS International Archives of the Photogrammetry, Remote Sensing and Spatial Information Sciences, Volume XLI-B5, 2016, pp.515–520. 2016. DOI: 10.5194/isprs-archives-XLI-B5-515-2016.

Linder W., (2016). Introduction. In: *Digital Photogrammetry*. Springer, Berlin, Heidelberg. https://doi.org/10.1007/978-3-662-50463-5_1.

McQuaid G., Millar P. and Woodward D. (2014). A Comparison of Techniques to Determine Surface Texture Data. Conference paper for Civil Engineering Research in Ireland, 2014. DOI: 10.13140/2.1.2138.5927.

Medeiros S., Underwood S., Castorena C., Rupnow T., and Rawls M., (2016). 3D Measurement of Pavement Macrotexture Using Digital Stereoscopic Vision. *Presented at 95th Annual Meeting of the Transportation Research Board*, Washington, D.C.

Mustaffara M., Lingb T. C. and Puanb O. C., (2008). Automated Pavement Imaging Program (APIP) for pavement cracks classification and quantification—a photogrammetric approach, in *Proceedings of the Congress of the International Society for Photogrammetry and Remote Sensing*, vol. WG IV/3.

Nicolosi V., D'Apuzzo M. and Evangelisti A., (2020). Cumulated Frictional Dissipated Energy and Pavement Skid Deterioration: Evaluation and Correlation, Construction and Building Materials, Volume 263, 120020, ISSN 0950-0618, https://doi.org/10.1016/j.conbuildmat.2020.120020.

Sarsam S.I., Al Shareef H.N., (2015). Assessment of Texture and Skid Variables at Pavement Surface, *Applied Research Journal*. Vol.1, Issue, 8, pp.422–432, October, 2015. ISSN: 2423-4796.

Sarsam S.I., Daham A.M. and Ali A.M., (2016). Assessing Close Range Photogrammetric Approach to Evaluate Pavement Surface Condition, J. Eng. 2016; 22(1).

Schonfeld R., (1970). Photo-Interpretation of Skid Resistance, *Highway Research Record*: No.311.

Slimane A.B., Khoudeir M., Brochard J., Do M-T., (2006). Characterization of Road Microtexture by Means of Image Analysis, Wear, Volume 264, Issues 5–6, 2008, Pp. 464-468, ISSN 0043-1648, https://doi.org/10.1016/j.wear.2006.08.045.

UNI EN ISO 13473-1, (2019). Characterization of Pavement Texture by Use of Surface Profiles - Part 1: Determination of Mean Profile Depth. www.iso.org.

Wambold J.C., Antle C.E., Henry J.J., and Rado Z., (1995). *International PIARC Experiment to Compare and Harmonize Texture and Skid Resistance Measurements, Final report*, Permanent International Association of Road Congresses (PIARC), Paris.

Zhang C., (2010). 3D Reconstruction From UAV-acquired Imagery for Road Surface Distress Assessment.

Roads and Airports Pavement Surface Characteristics – Crispino & Toraldo (Eds)
© 2023 The Author(s), ISBN 978-1-032-55149-4

Skid resistance and braking distance revisited – The case of grinding

Roland Spielhofer* & Matthias Hahn
Center for Low-Emission Transport, AIT Austrian Institute of Technology GmbH, Vienna, Austria

ABSTRACT: Correlation between RoadSTAR skid resistance coefficient and braking distance has been investigated and well documented in various research projects. In these projects, the relationship between skid resistance (expressed by μRoadSTAR) and braking distance was investigated on different asphalt and concrete pavements that had isotropic textures. Anisotropic textures, as represented by the longitudinal grinding texture, have not been investigated so far. In Austria, several sections of concrete surfaces have been textured using (diamond) grinding. This gave rise to the motivation to investigate to what extent the findings obtained on isotropic textures can be transferred or confirmed to this new type of concrete pavement texturing. Skid resistance measurements and braking tests for different (initial) speeds have been carried out on two surfaces, on an exposed aggregate concrete and on the same, but this time grinded concrete with PIARC and commercial passenger car tyres. Measurements were done on the all-new surface and half a year later, after the first winter period. In the paper, the measurement setup and the braking tests are described and the relation of measured skid resistance and achieved decelerations/braking distances are discussed.

Keywords: Skid Resistance, Braking Distance, Grinding, Exposed Aggregate Concrete, Brake Test

1 MOTIVATION

Grinding or Next Generation Concrete Surface (NGCS) were developed with regard to reduction of rolling noise, improvement of evenness and improvement of skid resistance (Dare et. al. 2009; Scofield 2016). The mechanism to improve skid resistance is described in the cutting of the coarse aggregate fraction and the associated exposure of the aggregate surface. The additional space created for the absorption of water in the grooves is deemed to have a positive effect on the drainage capacity and thus reduce the risk of aquaplaning. This effect is more pronounced with grooving textures.

In the USA several projects have been carried out on the construction and optimization of grinding and NGCS. The skid resistance values reported in these projects, measured with the blocked wheel method, can be described as high or very high. However, the effect on braking distances of passenger cars has not been investigated. In Germany, several projects have already been carried out on the production and optimization of grinding and grooving. The skid resistance values reported in these projects, measured with the sideway-force measuring method (SKM), can be described as 'high' or 'very high' - they are clearly above the requirement values of newly produced road pavements (which is $\mu_{SKM} = 0.45$ [-] at 80 km/h according to (FGSV 2007)). Values between $\mu_{SKM} \sim 0.6$ to $\mu_{SKM} \sim 0.9$ were reported.

---

*Corresponding Author: roland.spielhofer@ait.ac.at

DOI: 10.1201/9781003429258-29

Since 2016, various concrete motorway sections in Austria were produced with (diamond) grinding and grooving textures. In Austria, the RoadSTAR skid resistance measurement system represents the state of the art in the field of network-wide condition monitoring. The measurement principle is described in (CEN 2009). In contrast to SKM, the RoadSTAR measures the skid resistance by means of 18% longitudinal slip and thus simulates the (full) braking of a passenger car with antilock braking system. The correlation of $\mu_{RoadSTAR}$ and braking distance has been investigated and well documented in various research (Maurer 2007, 2009). In this research, the relationship between skid resistance (expressed by $\mu_{RoadSTAR}$) and braking distance was investigated on different asphalt and concrete pavements that had isotropic textures. Anisotropic textures, as represented by the longitudinal grinding texture, have not been investigated so far. This gave rise to the motivation to investigate to what extent the findings obtained on isotropic textures can be transferred or confirmed to this new type of concrete pavement texturing. A research project provided the opportunity to monitor skid resistance development over time, both measuring the skid resistance and doing brake tests at the same time.

2 EXPERIMENTAL DESIGN

To investigate the relationship between skid resistance and braking distance, brake tests to a standstill are indispensable. These brake tests must be full braking leading to the maximum possible deceleration ("emergency braking"). On motorways, they require at least one lane to be closed. This lane closure is difficult to implement and requires the support and cooperation of the motorway operator.

The newly built motorway A5, direction Vienna near exit Schrick was chosen as the subject of the study. The motorway was opened to traffic on December 8, 2017, and in the period prior to this it was possible to carry out extensive investigations regarding skid resistance, texture, and braking distance. While the whole contract section (about 8 km in length) was built as exposed aggregate concrete (EAC) pavement, two grinding textures were applied over a length of 500 m before opening for traffic. The northern grinding section with the texture 2.8/2.2 mm (segment width/segment spacing) was investigated by means of brake tests before grinding, immediately after grinding and 1.5 years after grinding. In parallel, brake tests were carried out on the unchanged exposed aggregate concrete located immediately before the grinding section. Skid resistance measurements were made on all these sections with the RoadSTAR for comparison, using PIARC ribbed standard tire and a commercial passenger car tire.

3 SETUP BRAKE TESTS

3.1 *Planning*

To plan the brake tests, literature (Roos et al. 2005; Van der Sluis 2002), was reviewed, from which the following conclusions were drawn for the test setup:

- Use of a passenger car representative of the vehicle collective on motorways: a Ford Focus, a typical representative of the compact class with front-wheel drive, was used.
- Use of summer tires: A Bridgestone Turanza T001, 195/65 R15 was used. This tire was repeatedly ranked in the top positions in various tire tests of automobile magazines. It is also available in a dimension that allows it to be used on the RoadSTAR skid resistance measurement system.
- Braking must take place on a wet road surface: This was ensured by means of tank trucks, provided by motorway company ASFINAG. The exact height of the water film was not decisive for the braking tests; the aim was to produce "typical conditions" as they occur

during rain. From the literature it was evident that the height of the water film only has an influence on the braking distance in the peripheral areas (very low water film, shortly before drying or very high water film over 2 mm, beginning of aquaplaning). During the third round of testing in May 2019, wetting by tanker was not necessary due to rain.
- Braking was performed at initial speeds of 40, 60, 80, and 100 km/h with ABS on. A minimum of three trials per speed were conducted.

3.2 *Hardware*

The commercial system "VBOX" from the company RaceLogic was used to determine the braking distance. This system is designed for automotive applications and allows easy data acquisition and evaluation of various typical questions in the field of vehicle dynamics - including the determination of the braking distance.

- The system consists of several components
- a central data acquisition unit (cf. Figure 1a)
- Pressure switch on the brake pedal to determine the start of braking (cf. Figure 1b)
- GPS receiver (cf. Figure 1c)
- Inertial measurement unit (cf. Figure 1c)
- Base station with GPS receiver for determining the correction signal to improve the GPS position accuracy (cf. Figure 1d)
- Radio system for communication between vehicle system and base station
- Connected laptop for control and monitoring during the measurement

Furthermore, the system includes an evaluation software to handle various typical questions in the field of vehicle dynamics. In our application, the module *Brake Test from Brake*

(a)   (b)

(c)   (d)

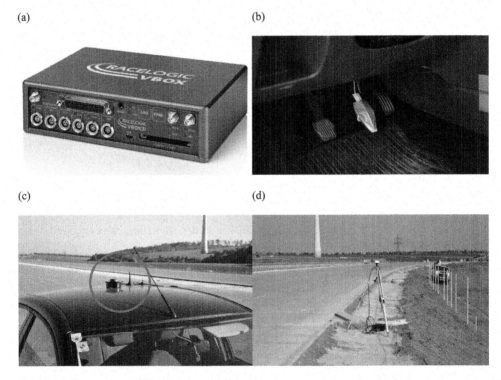

Figure 1.   Components of the brake test hardware setup.

Trigger was used, which on the one hand automates the determination of the different characteristic values of the braking, and on the other hand enables the control of the braking itself. The use of GPS, IMU and brake pedal switch also means that the triggering of braking is not tied to a specific location ("starting line" or similar), which facilitates the execution of the tests. During the braking tests, the target starting area was signalled by two marking cones, which served to orient the driver but did not have to be hit exactly.

In summary, it can be stated that this system eliminates the need for a specially trained test driver. The braking tests can be carried out by experienced drivers after some practice. A certain amount of concentration is necessary to safely reach and maintain the approximate range of the initial speed on the one hand, and to actually initiate and sustain an emergency braking manoeuvre on the other. Because emergency braking rarely occurs in normal everyday driving, this was practiced in advance with the test vehicle.

4 SETUP SKID RESISTANCE MEASUREMENTS

The skid resistance measurements were carried out immediately after each of the brake tests in the same right wheel path using the RoadSTAR measuring system:

- The skid resistance measurements were carried out at 40, 60 and 80 km/h.
- The skid resistance measurements were carried out both with the standard PIARC ribbed tire and with the Bridgestone Turanza T001, 165/50 R16 tire (the same type of tire that was also mounted on the braking vehicle) (cf. Figure 2).
- Two skid resistance measurements were carried out for each speed. As the measured sections appeared to be homogenous, skid resistance was averaged over the whole section of braking.

Figure 2.   Profiles of the tires used on the brake vehicle (left) and on the skid resistance device (right) - Bridgestone Turanza T001.

5 REALISATION OF THE BRAKE TESTS AND SKID RESISTANCE MEASUREMENTS

The tests and measurements were performed on the following days:

Date	Surface
2017-10-12	EAC new condition (produced in September 2017)
2019-05-28	EAC age approx. 1.5 years, after 2 winters
2017-11-16	Grinding new condition (produced on 11-10-2017)
2019-05-28	Grinding age approx. 1.5 years, after 2 winters

Measurements taken in 2017 were taken prior to traffic opening. Except for construction traffic, there was no traffic load at that time. Brake tests in 2017 on EAC and grinding were done on the same location, before and after grinding. All measurements were taken on the right lane. The AADT at that section is approx. 15,400 vehicles, the truck share approx. 11% (data from the ASFINAG permanent counting station Schrick 2018).

Before the start of the brake tests, the tire pressure on the test vehicle was checked and set to 2.0 bar. The sections on which the brake tests were carried out have hardly any longitudinal inclination. During the brake tests, the pavement was wetted by means of a tanker wagon. At least after each speed series, the wetting was repeated so that a uniform water film was constantly present. The brake tests in May 2019 were carried out during rain, and no additional watering was required here. The skid resistance measurements were carried out immediately after the tests to ensure uniform environmental conditions.

6 ANALYSIS

After all measurements, the following characteristic values were available for further analysis per pavement and year:

- braking distance at initial speeds of 40, 60, 80 and 100 km/h.
- mean deceleration for the four speeds mentioned
- mean fully developed deceleration (MFDD) for the speeds mentioned
- $\mu_{RoadSTAR}$ with PIARC ribbed tire at a measurement speed of 40, 60 and 80 km/h on the braking sections
- $\mu_{RoadSTAR}$ with Bridgestone Turanza tire at a measurement speed of 40, 60 and 80 km/h on the braking sections

The MFDD is defined as the deceleration averaged with respect to distance over the interval of v_b to v_e according to the following formula (from UN-ECE 2016):

$$MFDD\ d_m = \frac{vb^2 - v_e^2}{25.92(S_e - S_b)} m/s^2$$

where: v_o = initial vehicle speed in km/h,
v_b = vehicle speed at 0.8 v_o in km/h,
v_e = vehicle speed at 0.1 v_o in km/h,
S_b = distance travelled between v_o and v_b in metres,
S_e = distance travelled between v_o and v_e in metres

6.1 *Brake tests*

First, the collected or calculated parameters were visualized. Figure 3 show the braking distance, the mean deceleration and the MFDD, each per pavement and year. A comparison

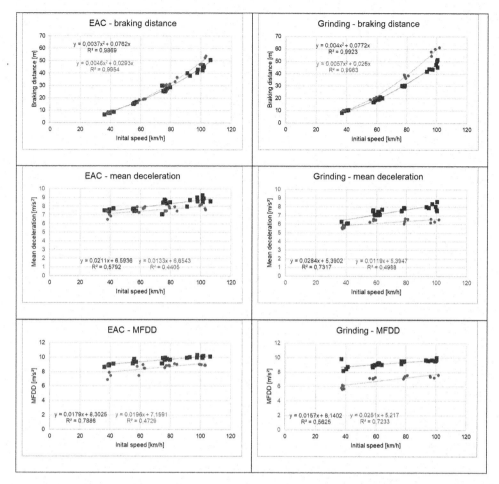

Figure 3. Plot of stopping distance, mean deceleration and MFDD on exposed aggregate concrete (left column) and grinding (right column). Red = measurements 2017, grey = measurements 2019).

of the 2017 parameters (red) shows similar values on exposed aggregate concrete and grinding. Larger differences can be seen with the 2019 characteristics (grey).

As can be seen in the formulas of the regression equations, the braking distance increases quadratically with the initial speed. The regression analysis shows very high R^2, which on the one hand allows the braking distance to be calculated well for other speeds. On the other hand, the high R^2 shows that the braking distance measurement is consistent and provides plausible results. The strong "concentration" of the points at the individual speeds also shows the good repeatability of the braking tests, so that effects such as fading or changing tires obviously do not play a role. Likewise, no outliers due to measurement errors can be detected. The slight increase in mean decelerations is also expected, because initial and final effects play less of a role with increasing braking durations due to higher initial speeds. The MFDD should actually be independent of the initial speed and shows a very small increase compared to the mean deceleration.

6.2 Skid resistance

First, the averaged skid resistance was plotted against speed, separated by tire and pavement. This is shown in Figure 4 below. The following can be seen:

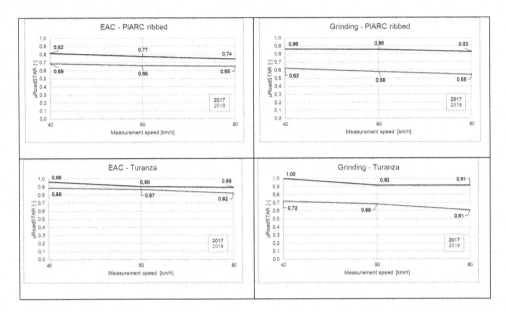

Figure 4.　Skid resistance at different speeds. Left: Exposed aggregate concrete, right: Grinding. Top: PIARC ribbed, bottom: Bridgestone Turanza T001. (Red = measurements 2017, grey = measurements 2019).

- The passenger car tire Bridgestone Turanza delivers higher skid resistance values than the PIARC ribbed on all pavements as expected.
- The skid resistance decreases with increasing speed as expected.
- The skid resistance changes between 2017 and 2019 is significantly lower on exposed aggregate than on the grinding section although the material is the same.

6.3　*Relationship between skid resistance and braking distance*

A comparison of the changes in skid resistance and the changes in the characteristic values of the braking distances over time were made. For this purpose, braking distance, mean deceleration and MFDD were calculated for speeds of 40, 60, 80 and 100 km/h using the regression formulas given in the diagrams above (cf. Figure 3). The designators used in the Table 2 below are described in Table 1.

Table 1.　Parameters used for comparison of skid resistance measurements and braking distance.

Pavement	EAC (exposed aggregate concrete) or GR (grinding)
$\mu_{x\text{km/h}}$	average skid resistance at x km/h measured speed, measured with RoadSTAR
$\text{MFDD}_{x\text{km/h}}$	Mean fully developed deceleration at x km/h initial speed
$a_{AVGx\text{km/h}}$	Mean deceleration at x km/h initial speed
$\text{BDist}_{x\text{km/h}}$	Length of braking distance at x km/h initial speed
$\Delta\mu_{x\text{km/h}}$ (abs)	absolute change in averaged skid resistance at the braking section from 2017 to 2019
$\Delta\mu_{x\text{km/h}}$ (rel)	Relative change of averaged skid resistance at the braking section from 2017 to 2019 in percent
$\Delta\text{MFDD}_{x\text{km/h}}$ (abs)	Absolute change in MFDD at x km/h initial speed from 2017 to 2019.

(continued)

Table 1. Continued

Pavement	EAC (exposed aggregate concrete) or GR (grinding)
$\Delta MFDD_{x km/h}$ (rel)	Relative change of MFDD at x km/h initial speed from 2017 to 2019 in percent
$\Delta a_{AVG x km/h}$ (abs)	Absolute change in mean deceleration at x km/h initial speed from 2017 to 2019
$\Delta a_{AVG x km/h}$ (rel)	Relative change in mean deceleration at x km/h initial speed from 2017 to 2019 in percent
$\Delta BDist_{x km/h}$ (abs)	Absolute change in braking distance at x km/h initial speed from 2017 to 2019
$\Delta BDist_{x km/h}$ (rel)	Relative change in braking distance at x km/h initial speed from 2017 to 2019 in percent.

Table 2. Comparison of skid resistance measurements at 40, 60 and 80 km/h and braking at 40, 60 and 80 km/h. Absolute and relative changes in the characteristic values after 1.5 years on exposed aggregate concrete (EAC) and grinding (GR). The tire column refers to the tire used in the skid resistance measurements.

40 km/h		2017	2019	2017	2019	2017	2019	2017	2019
Tyre	Surface	$\mu_{40km/h}$	$\mu_{40km/h}$	$MFDD_{40km/h}$	$MFDD_{40km/h}$	$a_{Avg40km/h}$	$a_{Avg40km/h}$	$BDist_{40km/h}$	$BDist_{40km/h}$
Turanza	EAC	0,96	0,89	9,02	7,92	7,44	7,19	8,97	8,53
Turanza	GR	1,00	0,72	8,77	6,22	6,53	5,87	9,49	10,16
PIARC	EAC	0,82	0,69	9,02	7,92	7,44	7,19	8,97	8,53
PIARC	GR	0,86	0,63	8,77	6,22	6,53	5,87	9,49	10,16

change after 1.5 years		$\Delta\mu_{40km/h}$ (abs)	$\Delta\mu_{40km/h}$ (rel)	$\Delta MFDD_{40km/h}$ (abs)	$\Delta MFDD_{40km/h}$ (rel)	$\Delta a_{Avg40km/h}$ (abs)	$\Delta a_{Avg40km/h}$ (rel)	$\Delta BDist_{40km/h}$ (abs)	$\Delta BDist_{40km/h}$ (rel)
Turanza	EAC	-0,07	-7%	-1,10	-12%	-0,25	-3%	-0,44	-5%
Turanza	GR	-0,28	-28%	-2,55	-29%	-0,66	-10%	0,67	7%
PIARC	EAC	-0,13	-16%	-1,10	-12%	-0,25	-3%	-0,44	-5%
PIARC	GR	-0,23	-27%	-2,55	-29%	-0,66	-10%	0,67	7%

60 km/h		2017	2019	2017	2019	2017	2019	2017	2019
Tyre	Surface	$\mu_{60km/h}$	$\mu_{60km/h}$	$MFDD_{60km/h}$	$MFDD_{60km/h}$	$a_{Avg60km/h}$	$a_{Avg60km/h}$	$BDist_{60km/h}$	$BDist_{60km/h}$
Turanza	EAC	0,90	0,87	9,38	8,30	7,86	7,45	17,89	18,32
Turanza	GR	0,92	0,68	9,08	6,72	7,09	6,11	19,03	22,08
PIARC	EAC	0,77	0,66	9,38	8,30	7,86	7,45	17,89	18,32
PIARC	GR	0,86	0,58	9,08	6,72	7,09	6,11	19,03	22,08

change after 1.5 years		$\Delta\mu_{60km/h}$ (abs)	$\Delta\mu_{60km/h}$ (rel)	$\Delta MFDD_{60km/h}$ (abs)	$\Delta MFDD_{60km/h}$ (rel)	$\Delta a_{Avg60km/h}$ (abs)	$\Delta a_{Avg60km/h}$ (rel)	$\Delta BDist_{60km/h}$ (abs)	$\Delta BDist_{60km/h}$ (rel)
Turanza	EAC	-0,03	-3%	-1,08	-11%	-0,41	-5%	0,43	2%
Turanza	GR	-0,24	-26%	-2,36	-26%	-0,99	-14%	3,05	16%
PIARC	EAC	-0,11	-14%	-1,08	-11%	-0,41	-5%	0,43	2%
PIARC	GR	-0,28	-33%	-2,36	-26%	-0,99	-14%	3,05	16%

80 km/h		2017	2019	2017	2019	2017	2019	2017	2019
Tyre	Surface	$\mu_{80km/h}$	$\mu_{80km/h}$	$MFDD_{80km/h}$	$MFDD_{80km/h}$	$a_{Avg80km/h}$	$a_{Avg80km/h}$	$BDist_{80km/h}$	$BDist_{80km/h}$
Turanza	EAC	0,89	0,82	9,73	8,68	8,28	7,72	29,78	31,78
Turanza	GR	0,91	0,61	9,40	7,23	7,66	6,35	31,78	38,56
PIARC	EAC	0,74	0,65	9,73	8,68	8,28	7,72	29,78	31,78
PIARC	GR	0,83	0,55	9,40	7,23	7,66	6,35	31,78	38,56

change after 1.5 years		$\Delta\mu_{80km/h}$ (abs)	$\Delta\mu_{80km/h}$ (rel)	$\Delta MFDD_{80km/h}$ (abs)	$\Delta MFDD_{80km/h}$ (rel)	$\Delta a_{Avg80km/h}$ (abs)	$\Delta a_{Avg80km/h}$ (rel)	$\Delta BDist_{80km/h}$ (abs)	$\Delta BDist_{80km/h}$ (rel)
Turanza	EAC	-0,07	-8%	-1,06	-11%	-0,56	-7%	2,01	7%
Turanza	GR	-0,30	-33%	-2,17	-23%	-1,32	-17%	6,78	21%
PIARC	EAC	-0,09	-12%	-1,06	-11%	-0,56	-7%	2,01	7%
PIARC	GR	-0,28	-34%	-2,17	-23%	-1,32	-17%	6,78	21%

The differences in the development of the skid resistance of the two surfaces over time are evident. At 60 km/h (standard measuring speed), the skid resistance with the PIARC tire decreases by 14% for the exposed aggregate concrete, while it decreases by 32% on the Grinding surface. For the passenger car tire (Turanza), the overall reductions are smaller than for the PIARC tire, but the differences between the surfaces are just as significant (-4% on exposed aggregate, -26% on Grinding). These differences continue with the braking parameters. Here, too, the losses (in MFDD and a_{AVG}) are greater on Grinding than on EAC, as are the increases (in braking distance). With an increase in the initial speeds, the different behaviour becomes even clearer; the braking distance at 100 km/h (relevant speed level on the highway) increase by 10% on the EAC, and by 25% on the grinding section.

7 CONCLUSIONS

Using the same passenger car tire on the RoadSTAR and on the braking vehicle (in this case Bridgestone Turanza T001) clearly shows the good transferability of the RoadSTAR measurement results to the braking distances or mean decelerations and confirms the conception of the RoadSTAR skid resistance measurement system as a simulation of passenger car full braking, also on anisotropic textures like grinding. The observed reduction of skid resistance on the grinding surface is fully reflected in the increased brake distances and therefore relevant to the road user.

The PIARC tire shows the same tendencies as the passenger car tire and is on a lower level in absolute terms, which fits the assessment of the Bridgestone Turanza as a premium tire and conforms to results of prior research. The PIARC tire is thus representative of tires with lower tread depths and generally poorer quality found in the vehicle collective.

Viewed over time, both the grinding surface and the EAC lost grip over the first winter, but to a much greater extent in the case of the grinding. What could not yet be answered with this test program is the question why the same concrete material shows such different skid resistance development depending on the texture designed as exposed aggregate concrete and as grinding. One possible explanation would be the different contact area resulting from the grinding grooves and the associated smaller contact area between tire and road. In new condition, the cut grooves could possibly compensate for this effect due to their sharp edges. Due to the smaller contact area, the polishing effect could be greater and would then lead to a decrease in grip. Long term monitoring of the skid resistance is recommended to trace the skid resistance development of the grinding section Further research is needed on the causes of the different development of skid resistance of the two sections.

ACKNOWLEDGEMENTS

We would like to express our thanks for the comprehensive support provided by ASFINAG employees. In addition to the lane closure, this also included the provision of equipment and personnel for watering the right lane in order to be able to carry out braking operations on wet pavement.

REFERENCES

CEN 2009. *CEN/TS 15901:2009, „Road and Airfield Surface Characteristics ⊠ Part 1: Procedure for Determining the Skid Resistance of a Pavement Surface Using a Device with Longitudinal Fixed Slip Ratio (LFCS)*: RoadSTAR", Brussels, 2009

Dare T., Thornton W., Wulf T. and Bernhard R., Purdue Final Report Acoustical Effects of Grinding and Grooving on Portland Cement Concrete Pavements HL 2009-1, Purdue University's Institute of Safe, Quiet, and Durable Highways, 2009

FGSV 2007. *ZTV Beton-StB 07: Zusätzliche Technische Vertragsbedingungen und Richtlinien für den Bau von Tragschichten mit hydraulischen Bindemitteln und Fahrbahndecken aus Beton, Ausgabe* 2007, FGSV, Cologne (in German)

Mauer P. 2007, Aspects of Road Skid Resistance and its Influence on Achievable Car Braking Decelerations., Österreichische Straßenforschung, Heft 964, Wien, 2007 (in German)

Maurer P. 2008, The New Austrian Skid Resistance Evaluation Background based on the Correlation of Skid Resistance Values μRoadSTAR and Braking Deceleration of Passenger cars, 6th Symposium on Pavement Surface Characteristics - SURF 2008, Portoroz, Slovenia

Roos R., Zimmermann M. and von Loeben W. 2005. Possible Braking Deceleration as a Function of Road Skid Resistance, *Forschung Straßenbau und Straßenverkehrstechnik*, Heft 912, Bonn, 2005 (in German)

Scofield L. 2016. *Development and Implementation of the Next Generation Concrete Surface 2016 Report-Living Document*, ACPA, 2016

UN-ECE 2016. Regulation No 13 of the Economic Commission for Europe of the United Nations (UN/ECE) — Uniform Provisions Concerning the Approval of Vehicles of Categories M, N and O with Regard to Braking [2016/194] (OJ L 42 18.02.2016, p. 1, CELEX: https://eur-lex.europa.eu/legal-content/EN/TXT/?uri=CELEX:42016X0218(01))

Van der Sluis S. 2002. Derivation of a Correlation Between Skid Resistance, Speed and Stopping Visibility Distance based on Real Brakings, Dissertation, RWTH Aachen, 2002 (in German)

Roads and Airports Pavement Surface Characteristics – Crispino & Toraldo (Eds)
© *2023 The Author(s), ISBN 978-1-032-55149-4*

A methodology to assist in the management of friction supply on the Irish national road network

Emmanouil Kakouris*, Enda Burton, Mark Tucker & Ilaria Bernardini
Roughan & O'Donovan Consulting Engineers, Dublin, Ireland

Tom Casey
Transport Infrastructure Ireland, Dublin, Ireland

ABSTRACT: It is well established that there are a number of factors that play a role in the occurrence of skid related incidents on roads, including, but not limited to, geometric alignment factors, operating speed, driver workload and skid resistance. The interplay between the various parameters is complex and there is also potential for significant variation in these parameters along different road sections and over the operational life of a road. Management of materials is central to the objectives of a circular economy approach, and any methodology to balance the demand for high grade materials with a balanced response will be of significant benefit. Therefore, appropriate assessment of these factors can facilitate improvement of the management of skid resistance, whilst meeting environmental constraints.

This paper introduces a methodology, developed by ROD on behalf of Transport Infrastructure Ireland (TII), to assist in the investigation of the risk of a vehicle sliding on asphalt pavements by utilizing a combination of site testing and a friction model. The friction model has been used in conjunction with an operating speed model to compare friction supply and demand. The magnitude of the friction demand is estimated by using the aforementioned friction model while in-situ pavement condition surveys are used as a proxy for the friction supply. Two measures of road friction are calculated, the friction utilization and the friction residual. The methodology has been applied to a representative case study on the Irish National Road network.

Keywords: Pavement Skid Resistance, Pavement Management, Asset Management

1   INTRODUCTION

This paper describes outcomes of work undertaken by ROD on behalf of Transport Infrastructure Ireland (TII) to provide information about the level of friction demand required by vehicles on the Irish National Road network and to compare this demand against the underlying friction supply. The methodology has been developed to assist infrastructure managers in investigating and deciding the most appropriate mitigation measures when considering demand and supply issues at specific locations on the network which have already been identified by other asset management procedures.

---

*Corresponding Author: Emmanouil.Kakouris@rod.ie

DOI: 10.1201/9781003429258-30

2 MODELS AND DATA USED

2.1 *The point mass model*

The point mass model is the simplest vehicle model that represents the entire vehicle as a point at the centre of gravity of the vehicle, with the entire mass of the vehicle acting at this point. The point mass model does not consider any vehicle characteristics except its total mass, m, and it does not distinguish between rear and front wheels. According to the point mass model, the braking and cornering equations that describe the vehicle response are expressed by Equations (1) to (6) (Varunjikar 2011):

Braking equation:

$$F_b = m \cdot a_x - m \cdot g \cdot \frac{G}{100} \tag{1}$$

Cornering equation:

$$F_c = m \cdot \frac{V^2}{R} - m \cdot g \cdot \frac{e}{100} \tag{2}$$

Weight balance equation:

$$N = m \cdot g \tag{3}$$

where F_b and F_c are the longitudinal and lateral tyre forces, respectively. The vehicle mass is represented by the variable m, g is the gravitational acceleration, a_x is its longitudinal acceleration, V the longitudinal velocity of the vehicle, G the horizontal grade (negative is downgrade) and e is the superelevation of the road. Considering those equations, the friction demand of the vehicle in the longitudinal and lateral direction is given by:

$$\mu_{x',demand} = \frac{F_b}{N} = \frac{a_x}{g} - \frac{G}{100} \tag{4}$$

and

$$\mu_{y',demand} = \frac{F_c}{N} = \frac{V^2}{g \cdot R} - \frac{e}{100} \tag{5}$$

For the vehicle to maintain contact with the road without skidding, the longitudinal and lateral forces should not excess the friction supply:

$$\sqrt{\left(\frac{\mu_{x',demand}}{\mu_{x',supply}}\right)^2 + \left(\frac{\mu_{y',demand}}{\mu_{y',supply}}\right)^2} \leq 1 \tag{6}$$

Equation (6) defines what is known as the "friction ellipse" or "friction circle", as shown in Figure 1.

From this, a value for the friction utilisation (n) can be defined, which is the proportion of the available friction supply which is used by the friction demand:

$$n = \sqrt{\left(\frac{\mu_{x',demand}}{\mu_{x',supply}}\right)^2 + \left(\frac{\mu_{y',demand}}{\mu_{y',supply}}\right)^2} \tag{7}$$

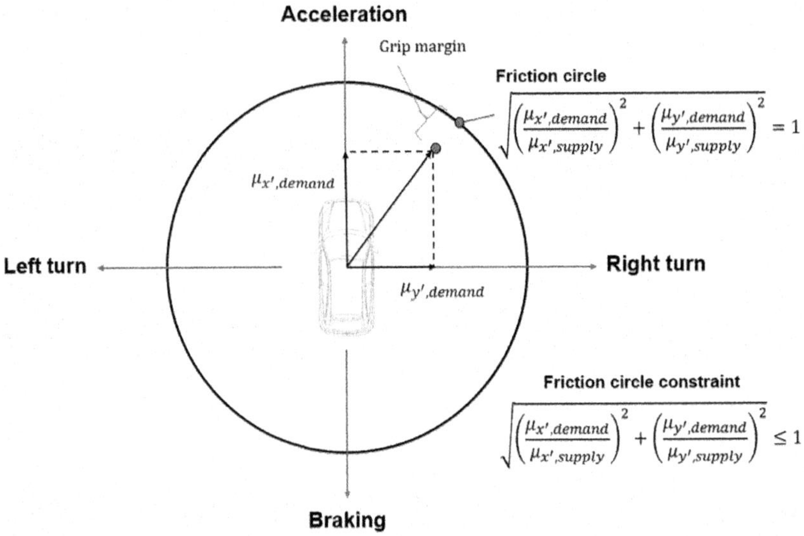

Figure 1. Point Mass Model. Friction circle.

2.2 *Variables included in the model*

The variables included in the point mass model are given in Tables 1 and 2.

The vehicle parameters used are those for an E-Class sedan car as provided in Appendix B of (Torbic et al. 2014). Values for grade and superelevation were taken from annual pavement surveys conducted annually by Pavement Management Services Ltd. (PMS) on behalf of TII. Bend radii were taken from geometric data developed as part of a related project by ROD for TII, the RibGeom project. (Details of this project will be published in a forthcoming report.) This geometric data was ultimately derived from Global Positioning System (GPS) data recorded as part of the pavement surveys. Values for speed and acceleration

Table 1. Road variables used in the point mass model.

Variable	Description	Value
G	Road horizontal grade [%]	Varies
e	Road superelevation [%]	Varies
r	Road curve radius [m]	Varies
$\mu_{x,supply}$	Longitudinal friction supply [-]	Varies
$\mu_{y,supply}$	Lateral friction supply [-]	Varies

Table 2. Vehicle variables used in the point mass model.

Variable	Description	Value
m	Vehicle mass [kg]	1830.7
T	Vehicle width [m]	1.60
h_{cg}	Height of vehicle sprung mass [m]	0.60
h_r	Height of vehicle roll centre [m]	0.20
V	Vehicle velocity [m/s]	Varies
a_x	Vehicle acceleration/deceleration [m/s^2]	Varies

came from the Operating Speed model developed as part of the RibGeom project. CSC data (see below) from the pavement surveys was used as a proxy for the friction supply.

2.3 Pavement data

The true values of the friction supply along the National Road network in all situations are unknown, as measurements of friction are necessarily a "friction pairing" between the road surface and the specific measurement device used. TII undertakes a programme of annual pavement condition assessment on the entire National Road network. Measurements for monitoring the in-service microtexture skid resistance are made with a Sideway-force Coefficient Routine Investigation Machine in accordance with TII publication AM-PAV-06045 (Transport Infrastructure Ireland 2020). In this context, the term "skid resistance" refers to the characterisation of the friction of a road surface when measured using a specific device in accordance with a standardised method as set out in AM-PAV-06045. These measurements are used to characterise the road surface and assess the need for maintenance but cannot be related directly to the friction available to a road user making a particular manoeuvre at a particular time or to specific collision situations (Transport Infrastructure Ireland 2020).

Values for the Characteristic Skid Coefficient (CSC), as described in TII Standard AM-PAV-06045, are recorded as part of pavement surveys conducted annually by PMS on behalf of TII. The entire National Road network is surveyed annually for one side of each road, with the other side of the road being surveyed the following year, such that each lane is surveyed once every two years. The CSC is an estimate of the underlying skid resistance once the effect of seasonal variation has been taken into account. In this research work, the CSC values have been used as a proxy for the friction supply in order to calculate values for the friction utilisation, as per equation (7).

2.4 Investigatory level and the friction residual

TII Standard AM-PAV-06045 defines Site Categories and Investigatory Level (IL) values that have been developed for the Irish national road network. Site Categories are assigned to each part of the network based on the road type and geometry. The IL values are defined for each Site Category based on the level of traffic and according to the perceived level of risk within each site category. AM-PAV-06045 defines the concept of an IL, as "*a level above which the skid resistance is considered to be satisfactory and at or below which the road is judged to require further assessment of the site specific risks in more detail*". The Standard defines how a specific IL can be assigned to each segment of the road network (based on road type, curvature, proximity to junctions, etc.); these ILs are then used for network management purposes.

The current work proposes a new parameter, the friction residual ($R_{sliding}$), which is defined as the Investigatory Level minus the resultant friction demand:

$$R_{sliding} = IL - \sqrt{\mu^2_{x',demand} + \mu^2_{y',demand}} \qquad (8)$$

This parameter allows the appropriateness of the IL values to be examined by comparing them with the estimated friction demand.

Geometric data from the RibGeom project and Traffic data was used to determine the Investigatory Level for each point on the roads in question, according to Table 5.1 of AM-PAV-06045.

2.5 Operating speed model

The speed data used for the friction models was taken from an Operating Speed model developed in the RibGeom project. This model predicts the 85th percentile speed of passenger cars at 5 m or 10 m intervals as they travel along a given route. The model provides a prediction for how vehicles decelerate at the approaches to bends and accelerate as they exit the bends. Equations are used to characterise:

- the Curve Speed, the speed that vehicles will travel when travelling around a bend
- the deceleration rate at the approach to the bend
- the acceleration rate at the exit from the bend
- the Desired Speed, which is the speed that vehicles will accelerate to on sufficiently long tangents.

The speed profile is then developed by combining these elements in ways similar to those illustrated in Figure 5 of (Fitzpatrick et al. 2000). It should be noted that the Operating Speed model assumes that all acceleration and deceleration takes place between the bends, with the speed along bends being constant. This assumption is reasonable for the assessment of the Alignment Consistency of a given route; Operating Speed models have traditionally been developed for this purpose (Lamm et al. 1986). However, it does mean that there is potential for the modelled friction demand to be somewhat different on bends where the true speed profile deviates significantly from this assumption.

It should also be noted that an Operating Speed model can only account for vehicle deceleration related to bends and cannot model vehicle deceleration that might occur for other reasons, such as at the approaches to junctions.

3 PRACTICAL APPLICATION

The current approach to the management of skid resistance in Ireland is described in TII Standard AM-PAV-06045. This Standard considers bends according to whether their radii are greater than or less than 250 m radius, but it does not take account of the specific values of their radii. As can be seen from Equation (6), vehicle speed is a key component of the lateral friction demand on bends. The presented methodology is therefore intended to provide assistance to the asset manager in the assessment of individual bends, in a more detailed manner than is possible with the current 250 m threshold. Use of the operating speed model also allows the analysis to take account of locations where the vehicle speed on a given bend is reduced by the presence of another bend immediately prior to it, which again is something that the current Standard cannot take account of.

3.1 *Case study*

The model was applied to a site on the Irish National Road network. Figures 1 and 2 show the CSC and the friction utilisation at this site, for the forward and reverse directions respectively.

The CSC values were used as proxy friction supply values for both the longitudinal and lateral directions. There was a significant difference between the values for the forward and reverse directions. In the forward direction, the CSC values vary from about 0.45 to about 0.6. In the reverse direction, most of the CSC values are between 0.27 and 0.5, with the lowest values occurring on the middle bend. The difference may be associated with a resurfacing of the road which occurred between the dates of the surveys in the forward and reverse direction which occurred in alternate years.

This has a significant effect on the values of the friction utilisation. In the forward direction, the maximum values are close to 0.9 on the middle bend. In the reverse direction, it can be seen that the utilisation values are higher, with values of close to 1.2 on the middle bend.

Next, the friction residual of the site is examined, which is the difference between the investigatory level, shown in Figure 3(a) and (b) for the forward and reverse direction, respectively, and the friction demand magnitude, shown in Figure 4(a) and (b). The

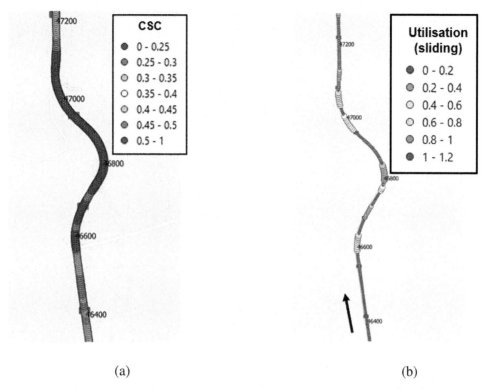

(a)   (b)

Figure 2.   (a) CSC and (b) friction utilisation for the case study site (forward direction).

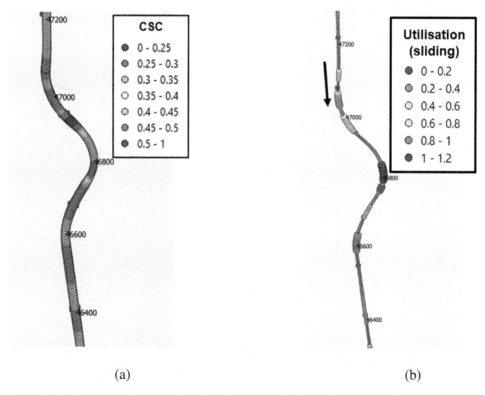

(a)   (b)

Figure 3.   (a) CSC and (b) friction utilisation for the case study site (reverse direction).

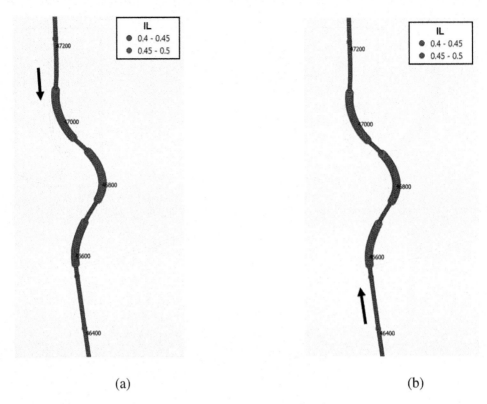

| (a) | (b) |

Figure 4.   Investigatory level for (a) forward and (b) reverse direction.

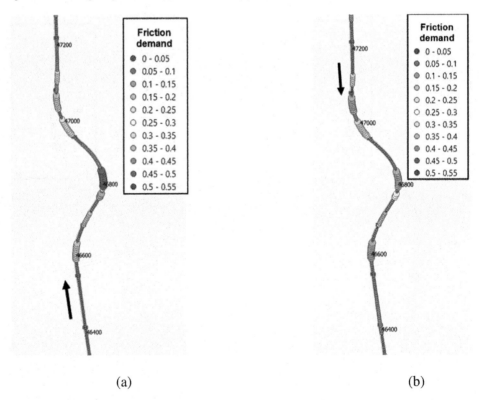

| (a) | (b) |

Figure 5.   Friction demand for (a) forward and (b) reverse direction.

310

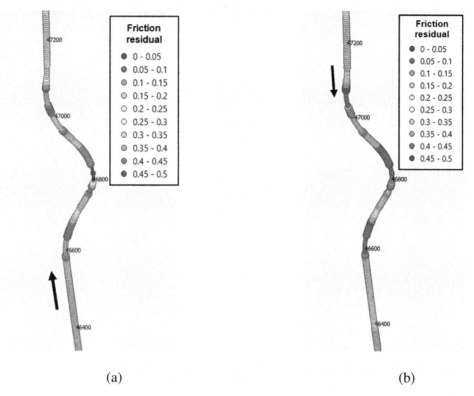

(a)   (b)

Figure 6.   Friction residual for (a) forward and (b) reverse direction.

corresponding friction residual plots are shown in Figure 5(a) and (b). In Figure 6 it can be seen that while the friction residual is above 0.0 at all points, the values are nonetheless quite low on the tightest bends.

4   CONCLUSIONS

A methodology has been developed which allows sites (on bends) to be individually characterised and estimates of friction supply and demand to be assessed on a site specific basis (where the sites themselves have been identified by other asset management procedures).

A case study undertaken on a site on the Irish National Road network was used to examine the use of the model at a localised scale. The modelling was found to provide useful insight into the relationship between operational speed, estimated friction supply and demand at the scale of individual bends on the road.

It is intended that the friction utilisation and friction residual will be of assistance to asset managers, allowing them to manage skid resistance more effectively, and investigating locations that are not performing as expected. As demonstrated in the case study, these tools can aid in the characterisation of the nature of specific bends when examining friction issues at a local scale. This should enable treatments such as pavement resurfacing works to be targeted and designed more effectively. In locations where the friction demand can be shown to be low, this method can provide support for a decision to maintain an appropriate level of supply while avoiding the unnecessary use of high friction aggregates, which are a scarce resource.

Conversely, in locations where the friction demand is likely to be high, appropriate remedial actions can be considered. For example, more durable materials could be specified if appropriate, or the driver's behaviour could be improved by the addition of signage or road marking features to induce the driver to reduce speed and thereby to reduce the friction demand.

In some locations it may be the case that the existing surface is already providing sufficient friction supply without the need for any resurfacing works at all, reducing both monetary and environmental costs. Reducing the unnecessary use of scarce construction materials and maximising the life of existing assets are key aspects of the "circular economy".

ACKNOWLEDGEMENTS

This work was carried out on behalf of, and funded by, Transport Infrastructure Ireland, the Irish National Roads Authority.

REFERENCES

Fitzpatrick *et al.* 2000. *Evaluation of Design Consistency Methods for Two-Lane Rural Highways: Executive Summary.* Vol. No. FHWA-RD-99-173. McLean, Virginia: Federal Highway Administration.

Lamm R., Hayward J.C. and Cargin J.G. 1986. "Comparison of Different Procedures for Evaluating Speed Consistency." Transportation Research Record 1100.

Torbic D.J., O'Laughlin M.K., Harwood D.W., Bauer K.M., Bokenkroger C.D., Lucas L.M. and Ronchetto J.R. *et al.* 2014. *NCHRP Report 774: Superelevation Criteria for Sharp Horizontal Curves on Steep Grades.* Washington, D.C.: Transportation Research Board.

Transport Infrastructure Ireland. 2020. *AM-PAV-04045 Skid Resistance Assessment.* June 2020. Dublin: Transport Infrastructure Ireland. https://www.tiipublications.ie/.

Varunjikar T. 2011. *Design of Horizontal Curves with Downgrades Using Low-Order Vehicle Dynamics Models.* M.S. Mechanical Engineering Thesis, University Park, Pennsylvania: Pennsylvania State University.

Reference calculations and implementation of mean profile depth

Thomas Lundberg* & Ulf Sandberg
Swedish National Road and Transport Research Institute (VTI), Linköping, Sweden

Luc Goubert
Belgian Road Research Centre (BRRC), Brussels, Belgium

Wout Schwanen
M + P Raadgevende Ingenieurs, Vught, The Netherlands

Robert Rasmussen
The Transtec Group, Austin, Texas, USA

ABSTRACT: ISO/TC 43/SC 1/WG 39 is a working group under the Acoustics/Noise committee working with standardization of the subject, "Measurement of pavement surface texture using a profiling method". The standardization is focusing on road surface properties in the texture range, i.e., wavelengths under 500 mm. Currently, the main topic is to characterize road surface texture by use of surface profiles and providing standardized measures to be used for noise and road condition investigations. One of the standards developed by WG 39 is ISO 13473-1, Characterization of pavement texture by use of surface profiles — Part 1: Determination of mean profile depth (MPD) (International Organization for Standardization (ISO) 2019). The standard was revised 2019 and a correction was published in June 2021.

WG 39 has proposed an unofficial procedure to make a common interpretation of the standard to help researchers, system providers and developers of macrotexture calculations with a "reference" program code. Four members of WG 39 made calculations of MPD from eight digital reference raw texture profiles. During the work some ambiguities were detected in the standard that led to the corrected version in June 2021. The main reasons for the correction are related to how to process dropouts and some filtering details. It was decided how to address the ambiguities before the work was finalized. The calculations of the four members organizations of WG 39 provided their interpretation of the calculations and the results were compared. The comparison between calculated MPD values from the reference profiles shows very good results with only small differences that can be explained. To spread the work done by WG 39, the reference data, reference program code and the calculated MPD values are publicly available via https://www.erpug.org/ (European Road Profile Users' Group). This work shows the importance of making a reference implementation of a standard when the standard is under development to ensure avoiding parts which can be interpreted in different ways.

Keywords: Macrotexture, Mean Profile Depth, Reference Calculations and Implementation of a Standard

*Corresponding Author: thomas.lundberg@vti.se

DOI: 10.1201/9781003429258-31

1 INTRODUCTION

ISO celebrates 75 years as a worldwide standardization organization in 2022. ISO has 167 members, of which 124 are full member bodies, the category that is part of the standardization work. Standardization is a democratic process in which different member states agree on a solution. A standard can be used in many areas; procurement, in research, to increase quality, in trade and export and to increase credibility in a specific subject. A standard also makes it possible to compare a property in different countries under the same conditions and to create uniform and transparent routines that provide a solution to an issue. Representatives from companies, organizations, authorities, municipalities, non-profit associations, and sole proprietorships are examples of participants in the ISO committees.

ISO ISO/TC 43/SC 1/WG 39 is a working group under Technical Committee 43, "Acoustics" and the Subcommittee 1, "Noise". Working Group 39 (WG 39) specializes in standardization of the subject, "Measurement of pavement surface texture using a profiling method". The standardization is focusing on road surface properties in the texture range, i.e., texture wavelengths under 500 mm. Currently, the main topic is to characterize road surface texture by use of surface profiles and providing standardized measures to be used for noise and road condition investigations.

One of the standards developed by WG 39 is ISO 13473-1, Characterization of pavement texture by use of surface profiles — Part 1: Determination of mean profile depth (MPD). The standard was revised in 2019 and a correction was published in June 2021. Other standards or technical specifications produced by WG 39 describe procedures for calculating megatexture (wavelengths between 50 mm and 500 mm, ISO 13473-5:2009) and how to perform spectral analysis (ISO/TS 13473-4:2008) in the texture wavelength range. Additional standards in the 13473 series describe terminology, specifications, and verification procedures.

To help the end user, WG 39 has proposed an unofficial procedure to make a common interpretation of 13473-1:2019. This will help researchers, system providers and developers of macrotexture calculations by providing "reference" programming code for verification of their in-house developed software. The study presented in this paper is based on the work done by four members of WG 39.

2 PURPOSE OF THE STUDY

The main purpose of this study is to support the end users with an unofficial (i.e., not sanctioned by ISO) interpretation of the standard and make it available as a free downloadable software. However, the benefit of the study was greater than first anticipated.

Most standards include a calculation routine that is often coded in a programming language. When a prudent programmer reads a standard, every sentence in the text is scrutinized and every possible interpretation considered as the text is converted to program code. Possible errors or incorrect input data must be taken into consideration and handled by the software. This is an optimal method to determine if a standard can be interpreted in more than one way. During this process for the ISO 13473-1 standard, ambiguities were found and documented including those related to spike removal and missing data procedures. After discussing different interpretations at a working group meeting, it was agreed how to solve these problems and make corrections to the MPD standard.

3 THE MPD STANDARD – A HISTORICAL VIEW

The first major method to describe the macrotexture of pavement surfaces was the Sand Patch Method which presented the measure named Mean Texture Depth (MTD). It was developed in the United Kingdom in the 1950's and the first official publication seems to be in a Road Note 27 in 1960 (Road Research Laboratory (RRL) 1960) although a second edition in 1969 has been more widely referenced. Although this method has been and still is frequently used, it was not standardized internationally until a European Standard was published in 2001, then with the name changed to Volumetric Patch Method and using glass spheres instead of sand (European standards (EN) 2001).

A competing method was developed in the UK in the 1970's based on a laser sensor method, establishing a parameter called Sensor-Measured Texture Depth (SMTD). This recorded samples taken from the road surface profile, processed using methods similar to Root Mean Square (RMS). But SMTD did not have the vertical-directional features of the MTD, i.e., extra sensitivity to peaks in the profiles, and become common only in equipment used in the U.K., Australia and New Zealand.

The sand patch or volumetric patch methods are crude spot measurements and depend largely on the operator and can be used only when roads are partly or fully closed to traffic. Along with developments in contactless surface profiling techniques–first stationary and later mobile–it become possible to replace patch measurements with those derived from profile measurements. With a suitable approach, profile processing could imitate the patch method, thus ensuring that the peaks in the profile would have more influence on the measured value than the lower parts. When lasers could function well in mobile highway speed operations, large road networks could be surveyed safely with the new procedure, named Mean Profile Depth (MPD).

In the 1980's, based on an original idea by Dr. Descornet (BRRC), a cooperation between Drs. Guy Descornet (BRRC) and Dr. Ulf Sandberg (VTI) resulted in a first draft for a Mean Profile Depth, intended to imitate the sand patch method, by calculating the profile depth (area) below a defined top reference on a two-dimensional profile curve over a 100 mm long profile segment. The draft method was tested extensively the first time within the International PIARC Experiment in the lower 1990's (Wambold et al. 1995), and presented at the PIARC conference in 1995 (Sandberg 1995). Most notably, the measure showed its potential to estimate the speed coefficient of skid resistance measurements in the large PIARC experiment. Sandberg proposed this method for standardization in ISO, which resulted in the first publication of the MPD standard in 1997 (International Organization for Standardization (ISO) 1997).

But the MPD standard is published by the ISO Acoustics Committee. How come? In the late 1980's, work in the Noise subcommittee of the Acoustics Committee was initiated to produce a reference test surface for vehicle noise testing, a standard which later became ISO 10844:1994 (International Organization for Standardization (ISO) 1994). The working group responsible for this was convened by Dr. Sandberg but with Dr. Descornet being instrumental in developing the technical specifications. This new noise standard needed a reference to a standard for measurement of road surface texture. At that time the MPD method, although drafted by Descornet/Sandberg, was not published and experience of it was still lacking. As a temporary solution, the volumetric patch method published by ASTM (American Society for Testing And Materials (ASTM) 1987) was used as a model and largely copied as an Annex A into ISO 10844:1994. It thus became a predecessor to EN 13036-1, but only as an Annex in ISO 10844. To provide a better texture measurement standard for the next edition of ISO 10844, work on the MPD standard was initiated by Dr Sandberg in the Noise subcommittee. After this, although the MPD measure is used in numerous road surface applications, the standard still stays in the ISO noise subcommittee, although also CEN and ASTM International have approved it in their series as well.

4 THE MPD CALCULATION PRINCIPLES

MPD is a geometrical description of the shape of a single 2D-profile of the road surface and is used to quantify the macrotexture level of the surface. Macrotexture describes irregularities in the surface in the wavelength range 0.5 mm up to 50 mm. The size of the aggregate in a pavement is normally 8 to 16 mm, in rare cases smaller or bigger. Hence, the aggregate size commonly influences MPD, as does the type of pavement, positive/negative texture, and the age and accumulated traffic. The macrotexture level has an influence on numerous properties that are important for the road user as well as the road owner: internal and external noise, surface drainage, splash and spray, friction, tire wear, and rolling resistance (Figure 1).

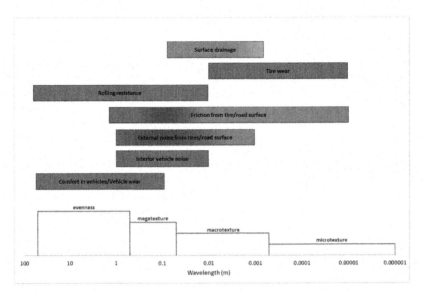

Figure 1. Properties affected by macrotexture. Red colour indicates negative effects and green positive.

The source or input to the MPD calculation is a high-resolution texture profile. The longitudinal resolution should be at most 1 mm. To measure the texture profile a point laser or a line scan laser can be used if the specification of the sensor fulfils the requirements in the standard. The macrotexture profile can contain inaccuracies (spikes) and missing data that affect the result of an MPD calculation. To compensate for this, the MPD standard (International Organization for Standardization (ISO) 2019) has standardized routines to take care of these imperfections. Spikes are identified and replaced with linear interpolation between the nearest valid data points at each side of the spike. After the "spike cleaning" action, the texture profile is resampled to either 0.5 mm or 1 mm. To illustrate the technique used when resampling to 1 mm, all samples that fall within the distance > 0 mm and ≤ 1 mm are averaged for the first resampled 1 mm value, and so on for the rest of the profile, see Figure 2.

The third step in the pre-processing is to handle missing data. Missing data are replaced with linear interpolation between the nearest valid data points at each side of the missing

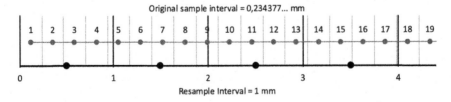

Figure 2. Resampling the spike cleaned texture profile to 1 mm samples.

316

data, the same technique as for spike removal. The next step is to filter the profile to only the wavelength range of interest. After cleaning, resampling and filtering, a Mean Segment Depth (MSD) is calculated for every 100 mm along the section (Figure 3).

The principles for calculating MSD can be described by the following steps. The texture profile is divided into 100 mm segments. The 100 mm segment is in turn divided into two 50 mm parts ("left" and "right"). MSD is determined as the as the average of the peak values at the left and right part of the segment. The MSD data is averaged to an MPD value representing a part of a road, normally 1 meter up to 20 meters or as an average for the entire section. To give a brief introduction to normal MPD levels, Table 1 gives statistics of state roads in Sweden measured at network level between 2015 and 2021.

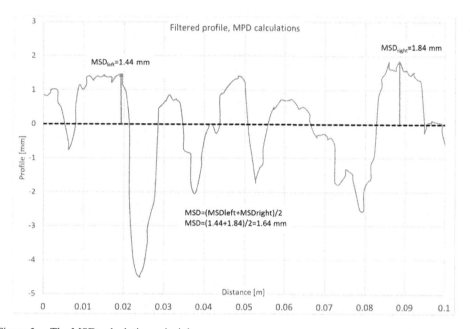

Figure 3.   The MSD calculation principles.

Table 1.   Median 20 m MPD values in Sweden. Pavements in the age span of 0 to 5 years.

Pavement, max. aggregate size	MPD Left (mm)	MPD middle (mm)	MPD right (mm)	Average age (year)	Observations (#20 m values)
SMA11	0.96	0.93	0.95	2.8	300305
SMA16	1.11	1.20	1.12	2.6	2050689
DAC11	0.75	0.74	0.78	2.3	381222
DAC16	0.83	0.79	0.83	2.4	1326832
Thin overlay11	1.08	1.06	1.12	2.8	61638
Thin overlay16	1.18	1.44	1.22	2.7	705409
PAC11	1.48	1.59	1.49	2.6	10760
PAC16	1.52	1.79	1.54	3.3	952
Chip seal11	1.01	1.04	1.06	2.5	61756
Chip seal16	0.91	0.93	0.93	2.7	789578
Chip seal22	1.17	1.33	1.30	2.5	107775
Surface dressing11	1.37	2.01	1.57	2.4	1193479
Surface dressing16	1.10	1.71	1.19	2.9	13930

317

5 RESULTS

The calculations of MSD and MPD from the eight raw texture profiles were performed by the four members organizations of WG 39. Each organization did their own implementation of the standard. The analysis and evaluation were compiled by VTI. The comparison of the MPD average for the entire sections can be seen in Figure 4.

The results from the comparison showed a good agreement between the four organizations with only small differences (reproducibility). The average difference of MPD at a section level was 0.003 mm and the greatest difference between MPD at a single section was 0.02 mm (3.6%). Furthermore, the average standard deviation between the average values per section was 0.005 mm. Overall, a very satisfactory result.

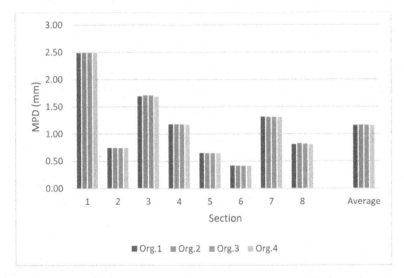

Figure 4.   MPD calculated from raw data from eight different sections by four organizations. Calculations according to ISO 13473-1:2019.

The emergence of the differences was discussed and pinpointed. The main contributor to the differences was how the programmer handled missing data and also resampling of the texture profile to either 0.5 mm sample interval or 1 mm. Organizations numbered 1 and 4 had the closest results and almost the same interpretation of the standard. A comparison between these organizations revealed reproducibility measured to the fourth decimal. However, one section had small differences which appeared as a noise around zero. It was the section with the sampling distance 0.234337 mm. The MSD was compared to find out the cause, see Figure 5.

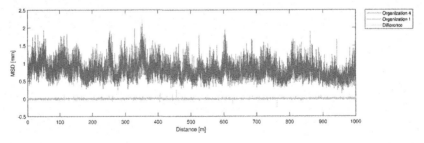

Figure 5.   Comparison of MSD between organizations 1 and 4 at section eight. (Note, data from organization 4 is covered by data from organization 1).

Even with the average almost the same, differences can be observed in Figure 5 indicating that the programmers had different interpretations of the standard. In this case, the resampling technique and how to interpretate the raw texture profile was the reason for the differences. There are three ways to interpret the raw data and to do the resampling of the raw texture profile to a texture profile that can be used for the MPD calculation, see Figure 6.

The three different assumptions can be described by the following points:

1. If assumed that the first sample represents the position dx/2, the first 1 mm resampled value will be represented by the first four raw data values.
2. If assumed that the first sample represents the position dx, the first 1 mm resampled value will be represented by the first four raw data values.
3. If assumed that the first sample represents the position 0, the first 1 mm resampled value will be represented by the first five raw data values.

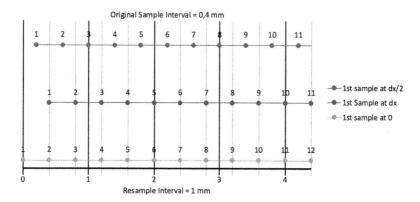

Figure 6.   Different options how to interpret the raw data and how to resample the raw texture profile to an equidistant texture profile.

This led to small differences in resampled texture profiles and thus inputs to the MSD calculation, which consequently led to slightly different MSD and MPD values. The programmers of organization #1 did the assumption according to point (2) above and organization #4 used point (1). We can see from Figure 5 that the maximum difference for a single MSD value is somewhere between 0.3 mm and 0.4 mm due to the different input to the resampling.

A similar comparison was done in 2013 when four organizations calculated MPD from almost the same raw data profiles as used for the present study. The first seven out of eight raw data texture profiles were the same. At that occasion an earlier version of the standard was valid, ISO 13473-1:1997. This version had more options and not as good specifications, particularly how to filter the profile and the routine to identify and remove spikes was not yet introduced. The test was presented at a CEN meeting 2013 (The European Committee for Standardization). The comparison led to the following results, see Figure 7. The differences from this test can be summarized and compared with the later test from 2020 (Table 2).

This shows the importance of making a standard that is easy to follow and has no alternative ways to do the calculations.

When the programmers translated the text in 13473-1:2019 to program code, several ambiguities were also detected. They had minor impacts on the results but were still considered to be important enough to be corrected. These ambiguities were documented and discussed at an ISO WG 39 meeting. The ambiguities found in ISO 13473:2019 can be summarized by the following points:

• The segment length was corrected to 100 mm throughout the document, earlier 100 ± 10 mm;

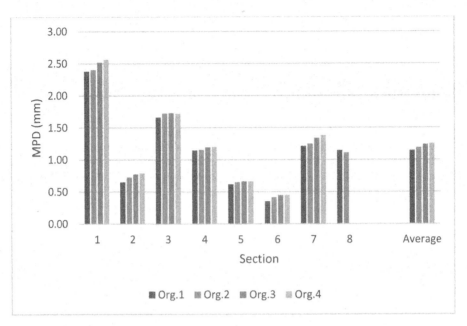

Figure 7.   MPD calculated from raw data from eight different sections by four organizations. Calculations according to ISO 13473-1:1997.

Table 2.   Results from calculations of MPD from raw data profiles according to ISO 13473-1:2019 and ISO 13473-1:1997.

	Results 2020 (mm)	Results 2013 (mm)
Maximum difference at a single section	0.02	0.14
Average difference	0.003	0.09
Average standard deviation	0.005	0.17

- A better description of how to handle invalid values in the beginning of a raw texture profile;
- A better routine how to handle the pre-processing if there are no data available before and after the section to be computed. In those cases, one should extend the raw texture profile by mirroring the first and the last segments before filtering;
- A more distinguish text of how to handle spikes, to avoid being misinterpreted;
- Some rephrased sentences to improve the readability;
- Compulsory reporting information was corrected as well as one figure.

This led to a correction of the standard (13473-1:2019) that was published in June 2021. To spread the implementation and reference program code done by WG 39, the reference data, reference program code and the calculated MPD values were published via https:// www.erpug.org/ (European Road Profile Users' Group[1] homepage). Hopefully, this gives the end user inspiration and ideas for programming solutions and to verify the results.

---

[1]ERPUG is a none profit organization that gives an annual conference in the area of road surface characteristics and associated measurement techniques.

6 CONCLUSIONS

It is difficult to write a standard that is flawless. It is ideal to avoid giving options to a routine or calculation description in order to achieve a uniform result. Writing a sentence that can be interpreted in more than one way could easily slip through all control points and reviews of the standard developers (working group) and the multiple ballots before final publication. In the development of a standard involving calculation routines, it is highly recommended to let at least two programmers scrutinize the standard in a coding process that goes on in parallel with the development of the standard or as a last control point before the first ballot. Unfortunately, the commonly occurring tight timetable seems to be a hindrance to this.

The MPD standard, 13473:2019 corrected in June 2021, has routines and specifications that gives a result with only small differences when different implementations are used. The test described in this paper shows that the four organizations that did their own implementation of the standard got an average deviation of the overall MPD close to zero, for the eight 1 000-meter sections. Compared with a similar test done according to the previous version of the MPD standard, 13473-1:1997 the uncertainties and deviations have been reduced considerably.

The work performed in this study led to a reference implementation guide for calculating MPD. The Guide includes an example of an MPD program written for Matlab (MathWorks®). Together with the program code, reference raw texture data and MPD results are downloadable. This will support the end user with the implementation and verification in the development of an MPD software.

7 DISCUSSION

The MPD standard, ISO 13473-1, has from the first publication in 1997 been updated and rewritten to a standard that is easier to follow and that gives very few alternative ways to do the calculations. The single most important improvement since the first version is the specification of the filtering method and filter parameters that ensures a standard way of handling the texture profile. The second most important improvement is a description of how to clean the raw texture profile from spikes. The lack of this operation could lead to single big deviations when doing repeated measurements.

The MPD standard is spread and used world-wide for several purposes, for example research, complement to friction measurements, management of road network and as an indicator of distress in the road surface. The reference implementation of a standard done by WG 39 is a good way to guide and ease the interpretation for the end users. With the knowledge we have today, we would probably have the program code as an informative annex in the standard instead of distributing the program at the ERPUG homepage.

REFERENCES

American Society for Testing And Materials (ASTM). (1987). *ASTM E 965-87, Standard Test Method for Measuring Pavement Macrotexture Depth Using a Volumetric Technique*. West Conshohocken: ASTM, (Latest edition from 2019).

European standards (EN). (2001). *EN 13036-1:2001, Road and Airfield Surface Characteristics – Test Methods – Part 1: Measurement of Pavement Surface Macrotexture Depth Using a Volumetric Patch Technique*. Brussels: EN, (Later replaced by a version from 2010).

International Organization for Standardization (ISO). (1994). *ISO 10844:1994, "Acoustics – Specification of Test Tracks for the Purpose of Measuring Noise Emitted by Road Vehicles. Annex A: Measurement of Pavement Surface Macrotexture Depth Using a Volumetric Patch Technique"*. Geneva: ISO, (Latest edition is from 2021).

International Organization for Standardization (ISO). (1997). *ISO 13473-1:1997, Characterization of Pavement Texture Utilizing Surface Profiles - Part 1: Determination of Mean Profile Depth.* Geneva: ISO (Latest edition is from 2021).

International Organization for Standardization (ISO). (februari 2019). *ISO 13473-1:2019, Characterization of Pavement Texture Utilizing Surface Profiles - Part 1: Determination of Mean Profile Depth.* Geneva: ISO. doi:ISO 13473-1:2019(E)

Road Research Laboratory (RRL). (1960). *Instructions for Using the Portable Skid Resistance Tester. Road Note 27, Road Research Laboratory.* United Kingdom Ministry of Transport: RRL (the sand patch method is described in Appendix 1 to this Road Note).

Sandberg U. (1995). Development of an international standard for measurement of pavement macrotexture - ISO/DIS 13473-1, PIARC World Congress. Montreal (3-9 September 1995): Piarc.

Wambold J., Antle C., Henry J., & Rado, Z. (1995). International PIARC Experiment to Compare and Harmonize Texture and Skid Resistance Measurements. Paris: The World Road Association (PIARC).

Roads and Airports Pavement Surface Characteristics – Crispino & Toraldo (Eds)
© 2023 The Author(s), ISBN 978-1-032-55149-4

Pavement friction characterization with the Wehner & Schulze (W&S) machine

Donatien de Lesquen du Plessis Casso*, Julien Waligora & Simon Pouget
EIFFAGE Infrastructure Research and Innovation, Corbas, France

ABSTRACT: Skid resistance is a major characteristic of road surfaces. It reduces braking distance, limits the risk of skidding and therefore considerably improves driver's safety. Several parameters control skid resistance of road surfaces, as aggregates properties (granularity, mineralogy, hardness, Polished Stone Value …) and the type of bituminous mixture (dense-graded, open-graded, porous, …). Thus, each aggregate will lead to different skid resistance associated with a specific evolution in time, and it is not possible to establish a reliable correlation between all previously mentioned parameters.

The aim of this research is to study the incidence of friction after polishing (FAP) using the Wehner & Schulze (W&S) machine on roads skid resistance. FAP test is the only standardized test method able to predict skid resistance of bituminous mixtures. It allows to measure a friction coefficient according to polishing cycles, simulating the evolution in time of the skid resistance of a road surface. However, it is difficult to correlate FAP results with other parameters, such as aggregates PSV.

In this paper are exposed results of a graduation work that aimed to identify and correlate different parameters having an incidence on FAP values. Thus, PSV effect was studied using different aggregates nature, and showed that it is not sufficient to explain skid resistance evolution in time. Thus, it is necessary to identify other parameters to explain the obtained results. The average depth of texture was studied and appears to be an interesting parameter. It could be added to an existing logarithmic model for skid resistance of bituminous mixtures to improve its prediction.

Keywords: Road skid resistance, Friction After Polishing, Wehner & Schulze machine

1 INTRODUCTION

As part of the pavement structure, road's surface layers have two major purposes: ensure comfort and safety for users. In order to provide safety it is necessary to maintain a good level of adhesion through time regardless of the weather conditions and the traffic's aggression.

In France, the level of skid resistance of a road is measured by the Sideways force Coefficient Routine Investigation Machine (SCRIM), a machine that can measure the transverse coefficient of friction (CFT) thanks to a smooth-tyred wheel loaded and oriented at an angle of 20 degrees from the track of the measuring vehicle. The CFT translates the effectiveness of a road to prevent skidding by simulating a skid situation (NF P 98 200-4 1996). There are other measures such as the longitudinal coefficient of friction (CFL) but those measures are less important in the French classification of road safety.

---

*Corresponding Author: donatien.delesquen@eiffage.com

DOI: 10.1201/9781003429258-32

The skid resistance of a vehicle is conditioned by many factors such as the transverse evenness of the pavement, the presence of pollutants on the surface or water. These parameters are variable and cannot be controlled. However, there are two parameters that heavily influence the skid resistance of an asphalt concrete and that can be analyzed in the laboratory: the macro-texture and the micro-texture:

Macrotexture characterizes surface irregularities ranging from 0.2 to 10 mm vertically, and from 0.5 to 50 mm horizontally. It is related to the size of the aggregates, to the mix design (grading, binder content), to the implementation of the bituminous mix (compaction) and to possible surface treatments. A strong macrotexture makes a better evacuation of the water slides on the pavement, reduces the projections and removes the mirror effect on the surface of the coating. It also allows a lower decrease in adhesion as a function of speed. However, it can lead to an increase in rolling noise and rolling resistance. On the other hand, a low macrotexture leads to a risk of hydroplaning in case of rain due to the persistence of a water film between the tire and the pavement (Cerema 2015).

The microtexture corresponds to the micro-roughness at the top of the surface layer. It allows to break the water film on the surface of the aggregates and therefore the contact with the tire. The size range of these micro-rough spots is between 0 and 0.2 mm vertically and 0 and 0.5 mm horizontally. The micro texture of an aggregate is due to its ability to retain sharp edges, which largely depends on the nature of the rock from which the aggregate comes, its mineral composition and the way used to produce it (crushing, ...). The micro texture of a bituminous asphalt can be influenced by more parameters. First, it is effective only after stripping the binder film covering the surface aggregates. It is then related to the NMAS (Nominal Maximum Aggregate Size), the percentage of sand and the polishing resistance of the aggregates.

To get an asphalt concrete with a good skid resistance, it is necessary to optimize macrotexture and preserve microtexture. Both of these parameters need to be optimized through laboratory testing. But since it is not possible to test an asphalt concrete's skid resistance with the SCRIM in a laboratory, road industry has searched ways to predict the skid resistance with several tests performed in laboratory.

One of the first test used, which is still the reference in France nowadays, is the Polish Stone Value (PSV) test that consists in polishing aggregate specimens and measuring their pendulum friction values (NF EN 1097-8 2009).

However, even though this test can be used to characterize the polishing strength of aggregates and the preservation of aggregates' microtexture, it cannot be used to characterize the adhesion of bituminous concrete and explain neither the CFT measured on roads nor the importance of macrotexture of an asphalt concrete. This explains why road industry looked for a machine that can measure skid resistance on an asphalt concrete in a laboratory.

1.1 *W&S machine and skid resistance*

In the 60's the Technical University of Berlin built a machine that is able to measure the skid resistance of an asphalt concrete: the Wehner & Schulze (W&S) machine. More importantly, this machine can simulate the polishing carried out by road traffic on 225 mm diameter test specimen. By repeating this measure after several stages of polishing it became possible to observe the evolution of the skid resistance through polishing time. The name of this test is Friction After Polishing (FAP). A new standard has recently been published for this test (NF EN 12697-49 2022).

Currently, we do not know to what extent the FAP test allows an accurate prediction of the evolution of the CFT on real roads, but this machine offers the possibility to gain a better understanding of skid resistance and of all the parameters of the bituminous mix design that have an influence on it.

W&S machine is composed of two major parts:

- The polishing station: it simulates the polishing due to traffic thanks to a set of three rubber polishing cones and a mixture of water and abrasive
- The measuring station: it is able to measure skid resistance of the sample thanks to three measuring skates. These measuring pads are rotated at a speed of 100 km/h and then applied with a vertical static force of 253 N. The torque of the measuring head is recorded according to the speed of the skids. The torque value at 60 km/h is retained and then translated into a coefficient of friction μ

$$\mu = M/(F_V \times R)$$

With M the torque value, F_V the vertical strength applied on the skates and R the radius of the circle that goes through the center of the skates.

In contrast to the PSV test, the W&S machine allows automatic measurements at different polishing stages. The adhesion is then plotted as function of the number of polishing cycles. An estimation of the evolution of skid resistance over time is obtained for the asphalt mix studied with curves similar to the one presented in Figure 2.

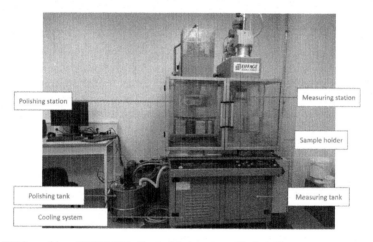

Figure 1.   W&S machine, EIFFAGE's Central Laboratory, Corbas, France.

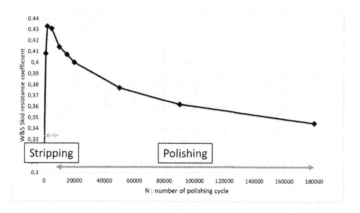

Figure 2.   Example of curve obtained with the W&S machine: BBSG 0/10, 06/2021.

On Figure 2, we can observe two phases: a phase of increasing adhesion, which is linked to the progressive stripping of the binder, and a phase of decreasing adhesion which is linked to the polishing of the aggregates. Those two phases also appears while measuring road's skid resistance with the SCRIM at different maturity of pavement.

1.2 *The main studies carried out with the W&S machine*

Since the W&S machine is installed in France, many studies were conducted in order to improve our knowledge of skid resistance. Within the major results that were obtained, we can mention the thesis of Zhao D. (2011) who worked on the ageing of an asphalt concrete and its effect over skid resistance. Zhao also proposed a logarithmic prediction model of the skid resistance measured by W&S machine by identifying the PSV as a major parameter of skid resistance and using linear regressions to calculate the values of formulas 'constants.

In the study conducted by Lafon et al. in 2012 different asphalt formulations with different characteristics were analyzed It was found that:

- Higher PSV results in higher coefficients of friction.
- The skid resistance of 0/6 asphalt is systematically higher than that of 0/10 asphalt.
- Some kind of asphalt mixes have a better skid resistance:

$$\mu_{W\&S}(dense\ gradded\ asphalt) > \mu_{W\&S}(open\ graded\ asphalt) > \mu_{W\&S}(porous\ asphalt)$$

Another important thesis was presented by Senga Y. (2012) about the effects of mixing different types of aggregates on W&S skid resistance. Senga showed that a linear mixing law appears while mixing aggregates with different PSV values, but this law cannot be perfectly transposed when using W&S machine on asphalt concrete. However there is still an improvement of skid resistance by adding aggregates with a high PSV.

In order to continue the work of understanding the parameters of a mix design that influence the adhesion of a bituminous concrete, this study was based on three research axes:

- The main role of the 6/10 aggregates
  The aim was to find correlations between the characteristics of these aggregates and the results of the W&S machine. The classic characterization tests for wearing course aggregates have been carried out and many asphalt mix formulas have been created with the care to get the same grading curves and the same sand, but with different 6/10 aggregates.
- The main role of the 0/6 aggregates
  In France, the characterization of the adhesion of the aggregates is traditionally done on the 6/10 aggregates via the PSV test. So The influence of the aggregates of lower dimension have been observed by testing asphalt mixes having the same grading curve and the same void content but with different 0/6 aggregates nature.
- Verification of a prediction model for skid resistance

In this section, the skid resistance model of Zhao for W&S machine (Zhao 2011) have been applied with our own results, then it has been found that this model could be completed with the consideration of the macrotexture.

2 MATERIALS AND PROCEDURES

Our study is based on 8 types of aggregates with grain size between 6 and 10 mm coming from different quarries that are currently used in France. Those aggregates had various petrographic nature and characteristics in order to see if we could correlate some of them to the skid resistance level and/or evolution.

All of those aggregates were firstly characterized using the traditional French character-ization for aggregates (NF P 18-545 2011).

Table 1. Original quarries of the aggregates used and characteristics.

Quarries	petrographical nature	PSV	LA	MDE	MVR	Absorption	FL
QA	sillico-limestone	50	18	8	2,61	0,3	11
QB	Porphyric Microgranite	50	16	5	2,57	1,1	10
QC	Eruptive	47	12	6	2,84	0,6	19
QD	porphyry	50	14	3	2,6	0,7	8
QE	Eruptive	54	15	5	2,61	0,7	5
QF	Amphibolite Microgranite	55	16	16	2,72	0,9	10
QG	Rhyolith	56	19	7	2,62	0,7	14
QH	Rhyolith	52	13	7	2,66	0,7	8

LA: Los Angeles coefficient, characterizes the resistance to fragmentation of aggregates.
MDE: Micro-Deval coefficient, characterizes the attrition resistance of aggregates.
MVR: Density of the aggregate.
FL: flattening coefficient, percentage of flat aggregates.
Aggregates with grain size between 0 and 6 mm were provided from the quarries QA and QB.
During the whole experiment, the same bitumen has been used. It was a neat bitumen of class 35/50.
The study started with as first aim to look at the role of 6/10 aggregates within a classic mix design.
For this purpose, an asphalt concrete formula has been created using a particle size curve commonly used in France (figure X). This dense-graded asphalt concrete name is BBSG, it is generally used as a surface layer.
This kind of mix has the particularity to have the best levels of skid resistance compared to the other type of mix (with the same size of bigger aggregates) according to the bibliography previously mentioned. It also includes a sufficiently large 0/6 element dosage to allow their observation in contrast to other types of formulas used in France.
The first formula was made only with aggregates from QA's quarry. Then this formula have been modified by replacing the 6/10 aggregates from QA with 6/10 aggregates from another quarry, and it have been declined with each of the seven others quarry.

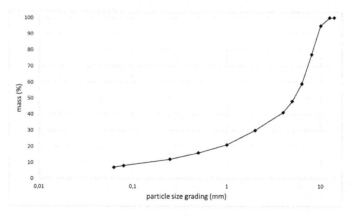

Figure 3. BBSG's granulometric curve used for the study.

For each formula, the dosages were adapted in order to obtain particle size curves as close as possible to each other. The compactness has also been controlled for each sample that was put to the test in order to assure the same macrotexture for every formula.

The second aim of the study was to observe the effect of the lower size aggregates.

For this part, the grading curve presented at the beginning of the study, have been reworked in order to replace the 0/6 QA aggregates by 0/6 aggregates from the QB quarry.

So it has been obtained a BBSG formula with 0/6 QB aggregates and 6/10 QA aggregates. Then, its grading curve has been reproduced by changing the 6/10 QA aggregates by 6/10 QG aggregates for the second formula and by 6/10 QB aggregates for the third.

Thus it was possible to compare six asphalt concrete formulas having the same granular skeleton, the same binder content and the same compaction (Table 2).

Table 2. Formulas used to study the impact of 0/6 aggregates.

6/6 \ 6/10	QA	QB	QG
QA	BBSG 100% QA	BBSG 0/6 QA 6/10 QB	BBSG 0/6 QA 6/10 QG
QB	BBSG 0/6 QB 6/10 QA	BBSG 100% QB	BBSG 0/6 QB 6/10 QG

Moreover, the first two columns of the table were intended to find out if it would be possible to obtain a law of mixtures in the continuity of the work of Y.SENGA mentioned in the bibliography.

3 THE MAIN ROLE OF THE 6/10 AGGREGATES

First of all, the effect of 6/10 aggregates on the skid resistance of an asphalt concrete has been studied by realizing eight FAP tests with the same mix, while changing only the 6/10 aggregate. We started by comparing the aggregates' characteristics to the value of skid resistance after 180 000 polishing cycles. Some of those comparisons are presented on Figure 4:

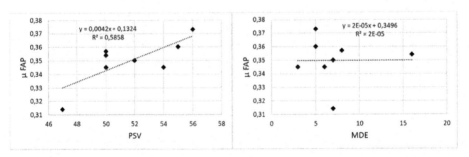

Figure 4. Correlation between PSV measurement and skid resistance after 180,000 polishing cycles (left) and correlation between MDE measurement and skid resistance after 180,000 polishing cycles (right), Lyon 2021.

Those comparison has shown that only one parameter could be correlated to μ with a R^2 superior to 0.5: the PSV. This level of correlation with PSV do not surprise because of the meaning of the tests and the results of previous studies, but we can notice that $R^2 = 0.58$ is not as high as we could expect.

LA coefficient correlation is 0,48 which is higher than MDE's one. But this result is difficult to explain because LA test is representative of the resistance to crushing while MDE give an indicator of the resistance to abrasion.

A few linear multiple regression were tried using as parameters LA and PSV but those correlations led to nothing.

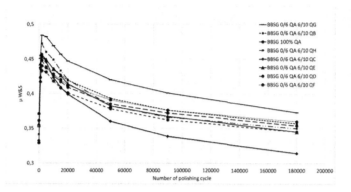

Figure 5.   Results obtained with the W&S machine for bituminous concrete with variation of aggregates 6/10.

The curves obtained with the W&S machine are presented in Figure 5.

Having the evolution of the skid resistance gives more information on the evolution of an aggregates' polishing property. As we can see on Figure 5, QG have the best values and QC the worst after 20000 cycles, which is coherent with their PSV's respective values: 56 and 47. However, for the other curves the skid resistance value at 90 000 and 180 000 cycles is not related to the PSV and the curves do not have the same evolution: for some of them, adhesion drops faster which leads to intersecting curves.

This shows that PSV can describe a difference of global adhesion for two type of aggregates if the values are very different. However, for two close values of PSV, it is impossible to asset that the higher one will provide better skid resistance results.

Thus, the W&S machine is partially correlated to the PSV result on asphalt, but it seems necessary to find other aggregate characteristics to add to a potential prediction model.

4   THE MAIN ROLE OF THE 0/6 AGGREGATES

In 2012, SENGA studied the possibility to use a proportional mixing law for the evolution of skid resistance considering the mastic made from bitumen and sand would be the same for each mix, and thus considering that it did not affect its results. In order to check this assumption, it has been decided to observe what happens to the evolution of skid resistance if the 0/6 aggregates are changed.

Therefore, six formulas of asphalt mixes have been tested using three different types of 6/10 aggregates and two different types of 0/6 aggregate. The grading curve was similar to those obtained with the 0/6 QA aggregate and the samples were compacted at the same void content.

The results given by the W&S machine appear in the Figure 6:

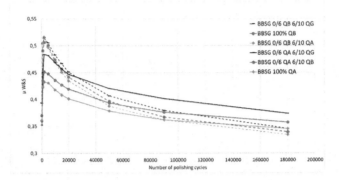

Figure 6.   W&S machine measurement results after change of aggregates 0/6.

Figure 6 shows that there is a major difference between curves with 0/6 QB (dotted lines) and with 0/6 QA (continuous lines). The first ones have a higher maximum of skid resistance but then it drops faster than the other group of curves. Within each group of curves, the shapes of the curves are similar but translated in relation to each other.

This result suggests that the lower fraction of a grading curve has a major impact over the evolution of skid resistance, however another parameter has been introduced: the surface aspect of the specimens. Indeed, 0/6 QB aggregates are more rough than QA aggregates, which results in a more difficult compaction. In order to fit the material into the mould, it was necessary to press the compactor further, which resulted in a closure of the asphalt surface that can be seen on Figure 7.

Figure 7.   Difference in appearance between a bituminous concrete test piece with 0/6 QA aggregates and 0/6 QB aggregates.

The difference of surface can be characterized by the average Mean Texture Depth which is about 0,975 mm for the pieces with 0/6 QA and 0,802 mm for pieces with 0/6 QB. This difference is high considering usual MTD values for BBSG are between 0,4 and 1,4 mm.

Thus, it is not possible to conclude from these results whether the difference of evolution is due to the origin of the 0/6 aggregate or to the surface aspect. It will therefore be necessary in the following studies to test both possibilities, taking care to isolate each phenomenon.

Nevertheless, this part of the study showed the existence of at least one, if not two, parameters having a very important influence on the value of the adhesion and its behavior: the origin of the 0/6 aggregates and the surface aspect.

5 STUDY OF D. ZHAO ADHESION PREDICTION MODEL: LIMITATIONS AND POTENTIAL FOR IMPROVEMENT

Among ZHAO's research (Zhao 2011), there is an improvement of TANG's adhesion prediction model for W&S machine which aims to analyze the curve with two parameters. This model is the following one:

(1) $\mu = (1 - d)\mu_B + d\mu_G$

With μ the skid resistance coefficient of the W&S machine, μ_B a function of bitumen aging, μ_G a polishing function of aggregates and d, a function of bitumen stripping. She has obtained the following function for μ_G:

(2) $\mu_G(N) = -A\log(N) + B$ for N > 0

And $\mu_G(0) = \mu_0$

A and B are obtained by linear regressions of W&S tests on 6/10 aggregates samples. According to Zhao, A is a function of the PSV of the aggregate and B is a function of the load exerted on the measuring head.

(3) $A = -0,0025.PSV + 0,1787$

(4) $B = -0,0167.P + 0,6445$

These formula only takes 6/10 aggregates into account as a reference for the evolution of adhesion and that the texture of the surface is not taken into account in this model. Therefore these formula have been compared with the results obtained in the laboratory.

First, the ZHAO model has been confronted to the curves obtained on asphalt concrete (Figure 8).

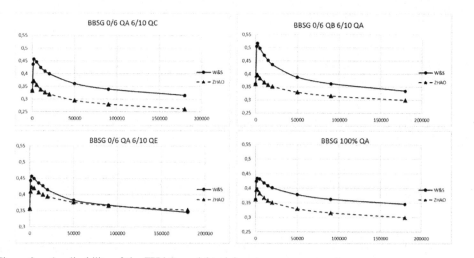

Figure 8.   Applicability of the ZHAO model to laboratory measurements.

It is possible to observe important differences between the results despite attempts to recalculate the coefficient A which depends on the PSV, the P value imposed by the measuring head, and the duration of stripping which is different from that observed by ZHAO (likely due to differences in bitumen and bitumen-aggregate adhesion).

Since the measured curve could not be approached by simply modifying the value of A, another method has been tested. It consisted in modifying the values of both A and B in formula (2) in order to minimize the difference between the ZHAO curve and the measured curve with seven measures. An example of modified formulas is illustrated on Figure 9.

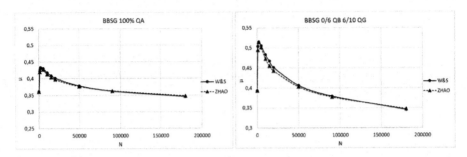

Figure 9.   Graph after correction of ZHAO model parameters A and B.

Then a multiple linear regression have been applied, taking into account two parameters on the value of A:

- The PSV because its value has a proven influence over the curve
- The MTD which is the Mean Texture Depth: it is a measure used to characterize the macro-texture of a road surface.

A linear regression has also been applied on the value of B taking as parameter, PSV, MTD and both. The best correlation was obtained with a simple linear regression using MTD. In order to conserve the work of ZHAO over the role of the parameter P, it has been decided to apply the regression over a factor of the ZHAO's expression.
The following formulas were obtained:

$$A' = -0,1409.MTD - 0,0003PSV + 0,2260$$

$$B' = (-0,0167.P + 0,645).(-1,1652.MTD + 2,3820)$$

With $R^2 = 0{,}62$ which is quite low, but considering the regressions were made with only seven measures, this correlation coefficient remains satisfying.
Then those modified formulas were applied on two new measures of the W&S machine on BBSG with aggregates 6/10 that have not been used for linear regressions and we obtained the graphs on Figure 10:

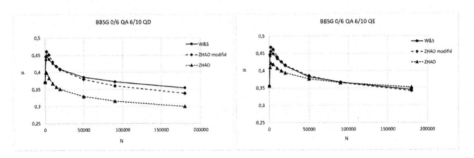

Figure 10.   Comparison of results with ZHAO model, modified ZHAO model and laboratory measurements.

As it can be seen, the modified model is much closer to the measure than the original one, which is satisfying considering the small number of measurements used to correlate.

According to these results, the surface aspect plays a major role over the road adhesions, almost as important as the PSV.

Further studies need to confirm these results with more measures using a larger number of aggregates, MTD values, and asphalt mixes types.

At this point, the model cannot explain why a porous asphalt which has a greater MTD and a lower value of skid resistance.

6 CONCLUSION

The purpose of this study was to deepen our knowledge of the evolution of adhesion for the W&S machine.

In this context, it is possible to confirm the importance of the polishing resistance of aggregates characterized by the PSV value. Two parameters that could have a significant effect on the adhesion behavior of a bituminous concrete to the W&S machine have been identified. However, it remains to determine the gains that could potentially be achieved with those parameters. Finally, a method have been proposed to improve the logarithmic prediction model of the evolution of the W&S coefficient that was proposed by Zhao D. in 2011. Although the proposed model was based on a small number of samples, the first comparisons are encouraging. This report therefore proposes an improvement in the understanding of the results of the Wehner & Schulze machine with the addition of the MTD parameter. By continuing this work of identifying the parameters influencing the result of the machine, it might eventually be possible to propose a model that works for all types of mixes.

However, it is necessary to bear in mind that there is not yet a direct correlation established between the SCRIM and W&S machine measurements. Studies are currently underway to establish this potential correlation. If they succeed, they should make it possible to optimize the adhesion of the formulas created in the laboratory and thus to guarantee the safety of road infrastructures.

Two protocols will be tested in the near future to verify the impact of sand and surface appearance:

- The impact of the surface will be tested by using the same formula with different levels of compactness that should lead to differences in the surface aspect. It can be expected that there is an optimum of surface aspect since an increase in density will increase the surface of contact with the measuring skates, but if the surface is too closed, an aquaplaning effect will occur.
- The impact of the sand 0/2 will be observed by using another type of asphalt mixes such as BBTM class, which has a 2/6 discontinuity. Thus, it would be possible to obtain materials that are compacted similarly without significantly changing the particle size distribution curve. The workability of the mixtures will be checked with the gyratory compactor test and it will be adjusted by changing the bitumen and fine particulate content.

REFERENCES

Cerema, l'adhérence des chaussées, Etat de l'art et recommandations. Octobre 2015.
Lafon R., Drouadaine I., Le Vagueresse A., Giacobi C. and Chaix C., *Evaluation Prédictive de l'adhérence des Matériaux de Chaussées Avec la Machine de Polissage Wehner et Schulze.* RGRA N°907-908, Décembre 2012 Janvier 2013
NF P 98 200-4, *Essais liés à l'adhérence, Partie 4: Essai Permettant d'obtenir un Coefficient de Frottement Transversal Avec un Appareil* SCRIM, 1996

NF EN 1097-8, *Essais pour Déterminer les Caractéristiques Mécaniques et Physiques des Granulats, Partie 8: Détermination du Coefficient de Polissage Accéléré*, 2009

NF EN 12697-49, *Mélanges Bitumineux: Méthodes d'essai, Partie 49: Détermination du Frottement après Polissage*, 2022

Senga Moundele Y., *Influence de la variabilité des Propriétés des Granulats sur la Résistance au Polissage des Enrobés Bitumineux*. Université Pierre et Marie Curie, France, 2012.

Zhao D., *Evolution de l'Adhérence des Chaussées : Influence des Matériaux, du Vieillissement et du Trafic, Variations Saisonnières*. IFFSTAR Nantes, France, 2011.

Roads and Airports Pavement Surface Characteristics – Crispino & Toraldo (Eds)
© 2023 The Author(s), ISBN 978-1-032-55149-4

Improving the understanding of user experience of ride quality

Craig Thomas*, Alex Wright, Kamal Nesnas & Nathan Dhillon
TRL Limited, Wokingham, UK

Neng Mbah & Stuart McRobbie
National Highways, Guildford, UK

ABSTRACT: On the UK Strategic Road Network, the maintenance standards for ride quality implement parameters derived from laser measurements of surface profile to identify lengths for maintenance investigation. Feedback from road users has suggested that there would be benefit in establishing a better understanding of the levels of ride quality experienced by road users, to optimise the identification of lengths adversely affecting user experience. Therefore, a study has been carried out to understand how the experience of road users relates to measurements of road surface profile.

A tablet-based Application has been developed that enables road users to rate their experience of roughness and bumpiness at regular intervals during a journey. Test routes were established containing a wide range of road roughness and a set of users engaged to carry out trials on these routes. Significant effort was made to select participants that had a broad range of demographics, and the trials were carried out to either minimise, or to understand, the potential effects of systematic issues.

The user's ratings of roughness and bumpiness recorded by the App (with location) were compared with road shape data collected using a laser profilometer. A number of ride quality parameters (obtained from the measured profile, including IRI, LPV and waveband measures) have been investigated to assess their ability to predict user experience. It was found that 3 m and 10 m eLPV (the current standard UK parameters) displayed marginally better capability in predicting user experience than other parameters, although broad agreement was demonstrated between parameters. However, comparisons between user ratings of bumpiness and these roughness parameters identified gaps in the ability of current parameters to predict user experience of bumpiness. Although the use of shorter wavelength roughness parameters showed some capability in predicting bumpiness, additional parameters developed for the specific identification of bumps are likely to show higher capability to predict user experience of these types of features.

This paper will describe the development process for this novel App-based approach to understand user experience, the trials undertaken and the application of the data to identify parameters to better predict user experience of ride quality and bumpiness.

Keywords: Ride Quality, IRI, Profile, User Survey, Laser, LPV, Bumps, Roughness, Road, Variance, Pavement

1 INTRODUCTION

Studies of road user satisfaction on the UK Strategic Road Network, carried out by Transport Focus (TF) and the Office of Rail and Road (ORR), have suggested that the

---

*Corresponding Author: cthomas@trl.co.uk

DOI: 10.1201/9781003429258-33

335

current approach taken to assess and quantify road surface condition does not fully reflect the experience of condition reported by road users (Transport Focus & Highways England 2017; Transport Focus & Office of Rail and Road 2017). In the light of these recommendations National Highways commissioned research to determine how the ability to relate the experience of road users with objective measurements of road pavement surface condition could be improved.

This paper discusses the development of an App based tool to quantify the experience of ride quality of users on the network, and its application in a practical study to update our understanding of the relationship between users' experience of ride quality and measurement of surface profile, encompassing both general roughness and localised roughness, which is experienced as bumps.

2 A METHODOLOGY FOR QUANTIFYING USER EXPERIENCE OF RIDE QUALITY

User perception of roughness is a complex area that is influenced by the visual, auditory and dynamic (vibration/roughness) experience of the user. As the dynamic experience is assumed to be the most strongly linked to ride comfort, road administrations typically undertake network surveys of the shape of the road surface and use this data to estimate the experience of the road user, in relation to comfort. On the English Strategic Road Network this data is provided by the TRACS (TRAffic speed Condition Surveys) survey, which provides a measure of the 3D profile of the surface. However, to understand how the profile affects user comfort the profile data must be processed through an algorithm to deliver parameters with intensities that can be used to quantify the ride quality. The most common parameter applied for this is the International Roughness Index (IRI), with Longitudinal Profile Variance (eLPV) being applied in the UK (Wright & Benbow 2016).

In this work, questions have been asked in relation to the robustness of the parameters that are used to understand ride quality in the UK, and the intensities of these parameters that are considered to define poor/good ride quality. The first objective of this project was therefore to develop a robust methodology to quantify users' experience of ride quality, to provide data that could be compared with the parameters provided by network surveys of road profile. Whilst there have been previous user studies such as questionnaires and focus groups (TRL 2011), and the use of probe vehicle data (Van Geem et al. 2014) these have not provided spatially located data that can be directly compared with profile measurements. Therefore, the chosen methodology for this work was to collect information on user experience directly by asking users to report their ratings of ride quality when travelling over the network.

An App was therefore developed that operated on a touch screen tablet. Its design was developed through experimentation, trialling the use of buttons and sliders and a number of different physical layouts, roughness rating terminology and scales. However, it was identified early in the work that road roughness could be viewed in two distinct ways, as either the overall general smoothness/roughness of a section of road, or the individual bumps that are encountered. Therefore, the App was developed to seek views on both general roughness and bumpiness. Trials undertaken to test the App resulted in the following approach:

- The App deploys a 5-point scale for the assessment of general road roughness. Users are prompted to enter a rating by a chime and visual indication every 30 seconds.
- An initial trial with the App to report the presence of bumps (yes/no) showed that this would not provide enough information to assist in the development of a profile parameter for the assessment of bumps. Therefore, further versions were trialled. The trials sought to determine whether asking users to rate bump defects would adversely affect the quality

(repeatability) of the surveys. There was significant difference in performance and hence the App uses a 3-point scale to report bump severity.

- As it was not considered safe for drivers to operate the App. Controlled trials investigated the difference between passenger and driver ratings. Tests of repeatability were also carried out using the same person in the front and rear of the vehicle. The tests suggested that ratings are not strongly affected by being the driver or passenger. However, the trials highlighted large differences between users, as might be expected.
- An external GPS antenna was used to improve location referencing of the data provided, and to improve the ability to align user experience data with the measured profile.

The final App interface taken forward for user experience surveys is as shown in Figure 1.

Figure 1.   The final design of the App front end – Roughness App left, Bump App right.

3   USER TRIAL DESIGN

The outcomes of the pilot trials, combined with a desk study (e.g., Turtschy *et al.* 2003) suggested that a number of factors could introduce uncertainty into the results of user trials. Each factor was considered, whether it would have a significant effect on the outcomes, whether it would be feasible to control this factor, and how. A summary of the factors is given in Table 1. Whilst a number of factors can be controlled through the trial design (weather, route selection etc.), it was necessary to undertake 'Focussed' trials to investigate the significance of a number of factors – these included vehicle type and speed. For the speed trials, repeat user surveys were carried out over a range of speeds between 80 and 110 km/h (the expected speed range for larger trials) to identify any change in the overall ratings reported by users. The results suggested that speed has only a small impact on user ratings within the range of speeds that would be targeted within user trials. Therefore, it would not be necessary to undertake user trials using a range of set speeds - users would be asked to maintain a general speed within the range 80–110 km/h.

Focussed trials were also undertaken using a range of vehicles (a small car through to HGV). The results were not quite as expected. Participants travelling in the larger vehicles (SUV and HGV) reported higher (poorer) road roughness, with participants in the mid-range vehicle reporting the smoothest ride, and the budget/van falling between these - although the overall ratings of users travelling in different vehicle types broadly overlapped (Figure 2). The results suggested a complex relationship that breaks down into make and model of vehicle. As formal controls on vehicle type at this level (i.e., using a set fleet of makes and models) would not be practical in full user trials it was decided that the user trials would use an "informal" range of vehicle types, allowing participants to use their own vehicles. This was desirable as the trials were undertaken during the Covid pandemic and this simplified social distancing.

Table 1. Confounding factors considered in the design of the trials, the colours group the factors.

Factor	Resolution – approach	Factor	Resolution - approach
Suspension	Focussed trial - car types	Traffic	Trial design (off peak)
Seat	Focussed trial - car types	Vegetation	Low influence
Vehicle stability	Not practical to address	Lane width	Low influence
Vehicle type	Focussed trial – on types	Street lighting	Not practical to address
Weather	Control via trial design (dry)	Flooding	Low influence
Temperature	Control via trial design (no extremes)	Visibility	Not practical to address
Noise	Focussed trial - how noise affects roughness ratings	Road category / type	Control via trial design (routes on different road categories)
Day/night	Not practical to address	Driving speed	Focussed trials - assess user ratings at different speeds
Distraction	Control via trial design (App)	Journey time	Control via trial design
Driver attitude	Control via trial design (selection)	Journey purpose	Not considered significant
Elderly and disabled users	Incorporate into sample for public trials	Previous driving experience	Control via trial design (user selection)
Other drivers	Not practical to address	Incidents /roadworks	Control via trial design (avoid roadworks)
Road markings	Visual appearance - not practical to address. Geometry – avoid extremes in the trials	Debris and skid marks	Low influence
Surface texture (splash & spray)		Skid resistance	
Visual condition		Stiffness	
Surface colour and reflectance		Surface type	Control via trial design (range of surfaces)
Road geometry			

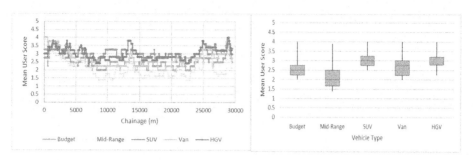

Figure 2.   Focussed trial on vehicle type – mean experience values along site (left) average (right).

4   THE USER TRIALS

Building on the above, user trials were undertaken with the App to collect data on the experience of users and provide a better understanding of the relationship between user experience and road profile. The outline trial design is summarised in Table 2. 174 people took part in the trials, carried out on Lane 1 of each of 6 routes.

Table 2.   Outline of user trials objectives.

Component	Approach
Routes	6 locations distributed over the strategic road Network. Each route located near a suitable centre to allow for meeting and greeting. The routes were between 20 minutes and an hour maximum to maintain participant concentration and were easy to navigate by trial participants. A route with a concrete surface was included to allow sufficient user opinions to be collected on this surface type.
Participants and Recruitment	A minimum of 30 people were recruited in each location. This number of participants was insufficient to allow the differences between types of road user to be ascertained. The trial therefore incorporated a wide range of road users replicating a demographic spread similar to that of the national norm. Pairs of participants were recruited, to drive and operate, enabling two sets of data to be collected per pair (assuming that both could drive). Participants used their own vehicle, the type of which was recorded for later analysis.
Shape data	The HARRIS3 profilometer was used to provide surface profile data
The user trials	Tablets (and GPS) were provided for data collection. Two Apps were used - recording Roughness with a 5-point scale, and Bump with a 3-point scale. Each participant completed two journeys, once for Roughness and once for Bump. The trials randomised which App was used first. The journey was completed in Lane 1 (or whichever is most suitable and legal for uninterrupted travel) at a speed of approximately 80 km/h to ensure minimal variation between trials. After the journey the participants completed a questionnaire collecting demographic data and general opinions on road condition.

5   FINDINGS

5.1   *Effect of demographic / group*

An initial analysis of the ratings of roughness considered all the user ratings by demographic to identify systematic bias or trends. Although there were small differences between the datasets, there was no strong evidence of systematic bias by demographic (e.g. Figure 3),

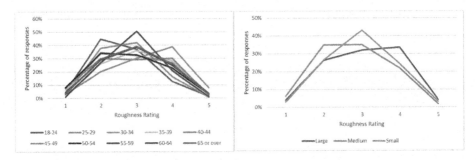

Figure 3.   Distribution of Roughness responses by participant age (left) and vehicle type/size (right).

given the level of repeatability or range of ratings reported within a particular demographic. As a result, it was concluded that the development of roughness criteria could be based on the combined dataset, with the result applying across demographics. Therefore, the remainder of the analysis combines data from all users for comparison with the profile measurements.

5.2   A representative user rating of roughness for each length

The ratings reported by the users were each fitted to the network of HAPMS sections using the GPS data. This gave ratings for each user, reported by section and distance. To compare this with the profile data it was necessary to combine the user data (for which there could be up to 30 user ratings per length) to obtain what we refer to as the "representative user rating" for each length. However, as users recorded ratings every 30 seconds using the App, the user rating was reported over varying lengths (up to a few hundred metres each) for each user. We assumed that each rating represented the overall user rating for every segment in that length. Therefore, each user value was resampled into 10 m lengths so that all user ratings could be aligned. Various approaches to obtain a representative user rating for the length were then investigated, for example the statistical mode, weighted average etc. However, it was found that a simple mean of all the user ratings provided a reasonably representative user experience rating of the roughness of each length of each route.

5.3   Comparison between user ratings of roughness and profile

The parameters currently deployed to quantify ride quality from profile can be divided into 1) indices derived from directly measured and processed road surface geometry data, or 2) calculated indices based on the processed effects of road surface undulations (vehicle or its components dynamic response, vibrations, driving comfort, etc.) (Ueckermann & Steinauer 2015). The four most common and acknowledged longitudinal unevenness indices in Europe are presented in prEN 13036-5 (CEN 2017)

(1) International Roughness Index (IRI) (normative), used in international comparison and benchmark tests
(2) Wave band analysis. Calculated evenness-energy in three different wavelength bands using the bi-octave processing (RMS) (informative)
(3) Wave band analysis. Calculated evenness-energy in three wave band bands using profile variance (LPV) (informative)
(4) Calculated range of evenness-variation and deviation using the weighted longitudinal profile (WLP) (informative)

Although there are other indices used across the world they are, in most respects, comparable to those presented in prEN 13036-5. To understand the relationship between the parameters and the user ratings, several ride quality parameters were calculated and compared with the representative user ratings, including IRI and the current UK parameters 3 m and 10 m Longitudinal Profile Variance (eLPV). It was shown that there is a relationship between each parameter and the user rating. This was generally logarithmic. Initial direct comparisons between the profile parameters and user ratings suggested that the correlation was strongest when comparing user ratings with 3 m eLPV. However, when all the routes were compared, two relationships were identified between the eLPV and the user data, depending on the test route under consideration. This is shown in Figure 4. The left graph shows two distinct clouds (which are colour coded). The lower orange cloud primarily contains data from test routes on two motorway sites in the south of England. For these sites the users rated lengths with moderately low 3 m ELPV to have higher roughness ratings. The right-hand graph plots 10 m eLPV against the user ratings, with the points that were colour coded as blue/orange on the left plotted in the same colour on the right. It can be seen that these locations contained generally higher levels of 10 m eLPV. This confirms that parameters to estimate user experience do need to consider multiple wavelengths, and this is the case for IRI and where both 3 m and 10 m eLPV are considered. It complicates the development of a single relationship between user ratings and profile. Given the long-term experience in the UK of 3 m and 10 m eLPV there is benefit in maintaining a link with these existing parameters.

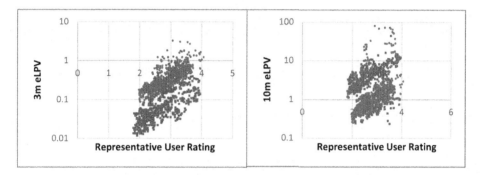

Figure 4. Comparing 3 m eLPV (left) and 10 m eLPV (right) with representative user ratings.

It is possible to combine 3 m and 10 m eLPV to obtain a general proxy for IRI, referred to as the Roughness Index (RI), using the equation

$$Roughness\ Index = Max\left(\sqrt{\frac{10}{3} \times eLPV3} + \sqrt{eLPV10} - 0.1,\quad 0 \right) \qquad (1)$$

This is compared with the user experience data in Figure 5. The combined index delivers a reasonably strong relationship with the user ratings. Indeed, a slightly higher correlation was found with the Roughness Index than with IRI. However, this relationship is not the same for all road types or surfaces. There was strong evidence for separate relationships on motorways and single carriageway roads, and on concrete roads. The reason for the different relationship for concrete roads is unclear. However, we suspect that this could be a result of the additional noise on these surfaces influencing the users' ratings.

341

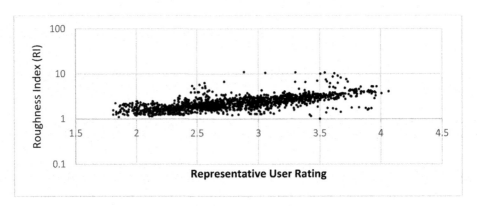

Figure 5. Comparing roughness index with representative user ratings.

5.4 *A representative user rating of bumpiness for each length*

As for the general roughness data, before comparing the user data with the profile it was necessary to establish a "representative user bump" reference dataset. The App recorded where each user reported individual bumps on each route, and the severity of each bump. However, to compensate for the lag in recording bumps (which is different for each user) and the errors in the location data, each user's bump data over the 50 m *before* the user recorded the bump was extrapolated and it was assumed that each 10 m in this 50 m length contained that bump. The user datasets were then aligned and each 10 m length considered in terms of the number the number of users that agreed there was a bump in that 10 m length ("user agreement") and the severity they reported.

A user Bump Index combines "user agreement" with the severity ratings. The parameter A_i expresses the percentage of users (j from N) that agree there is a bump at location i:

$$A_i = \sum\nolimits_{j=0}^{N} \frac{A_{ij}}{N} \qquad (2)$$

Bump intensity parameters S_i expresses the severity of the bump reported by the user at location i (modal value). The bump index parameter BI_i at location i is defined as:

$$BI_i = A_i S_i \qquad (3)$$

The Bump Index values were then aggregated over 100 m lengths.

5.5 *Comparison between user ratings of bumpiness and profile*

There are no published parameters to quantify the intensity of bumps using profile. Therefore, we considered two potential parameters. These draw on the 3D profile measurements, rather than the single line of longitudinal profile data typically applied for the measurement of roughness (e.g. to calculate IRI). The first method, called the "Area" method, calculates lines of best fit to the 3D profile and the area between the measured data and the line of best fit. The "CDM" method calculates the central second derivative of profile to highlight the presence of stepping in the data which may arise at rises or depressions. As the vehicle travels in the wheelpaths, these measures are calculated from the 3D profile collected in the wheelpaths. Figure 6 shows a heat map of individual CDM values when this is calculated from the 3D profile, highlighting a bump in the Nearside wheelpath (NSWP). To report an overall value to compare with the user data the parameter is calculated as the 95th percentile in each wheelpath in each 1 m, which is then smoothed, and the maximum reported over each 100 m length.

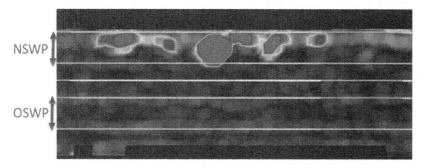

Figure 6. Heat map of lane roughness obtained from the CDM method (high values are shown as red) in the Nearside and Offside Wheel Path (NSWP and OSWP).

A bump (Area or CDM) value was obtained for each 100 m length on the test routes and compared with the user ratings (Bump Index). Whilst the relationship between the parameter and the user data is not as strong as for the combined roughness parameter above (Figure 7), the bump parameter does broadly indicate the lengths that contain bumps that would have been considered severe by users (Figure 8). We have found that the Area approach provides the strongest relationship with users on asphalt pavements.

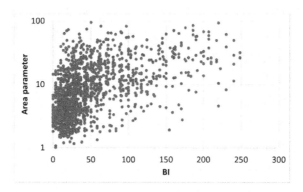

Figure 7. Comparing the bump parameter (here the Area method) with the user Bump index.

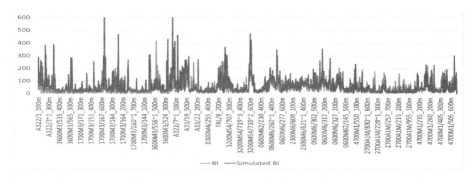

Figure 8. Comparing the simulated bump index with the user Bump Index, along the test route.

343

The approach taken to obtaining objective user ratings using the App described in this work can support the development of thresholds to help road administrations predict how users would rate lengths of the network. This could be achieved by considering the relationship between the user rating of roughness or bumpiness and the proportion of users who would agree that a length has a higher level of roughness or bumpiness. This is shown graphically for our test route in Figure 9, for roughness. Figure 9 relates the value of roughness reported by the users (the representative user rating for a length) with the proportion of users that reported that length to have a roughness of 4 or 5 (i,e., rough). To illustrate this – a length with a representative user value of 5 would need to have 100% of users rating it as 5, whilst a length with a representative user value of 3 is likely to have had different proportions of users rating it across all rating levels. The graph suggests that lengths with a representative user rating above ~3.5 would be rated as "rough" by more than 60% of users. By using the relationship established between representative user rating and the profile parameters (Figure 5), thresholds could be established for the profile parameter based on the proportion of users that would consider a length to be rough. A similar approach can also be applied to understand the proportion of users that would consider a length to be bumpy. Although both parameters would appear to measure similar quantities (road unevenness) there was found to be little overlap between the two for the data collected on the trial sites. Depending on the thresholds chosen the overlap could be as little as 3%. This reinforces the need to use both measures to understand the unevenness.

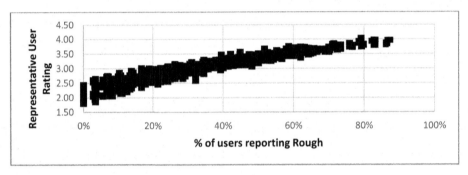

Figure 9. Relationship between the representative user roughness value and the percentage of users who would report that length to have a rating of 4 or 5 (rough).

6   CONCLUSIONS

This work aimed to determine how the ability to relate the experience of road users with objective measurements of road pavement surface condition could be improved/updated. An approach has been developed that uses an App to obtain users ratings of roughness and bumpiness on the network in a way that enables spatial analysis and direct comparison with data provided by profile measurement systems. The App provides robust data on user experience. The trials found that the variation between users (i.e. users do not tend agree on the rating for each length) is comparable to the variation between demographics (user type, vehicle type etc.). This supported the combination of the user data into a single group for the analysis.

When comparing the user data with parameters for general roughness we have found that parameters comparable with the commonly applied IRI provide a reasonable estimation of user experience. The observation of different responses by users on sites containing different wavelength components supports the use of parameters that include a range of wavelengths. However, roughness alone does not rate users experience of ride quality and hence we have developed a separate approach to seek users experience of bumps and have developed a bump parameter that attempts to estimate the user ratings. The lower level of agreement with the user data reflects the more complex nature of this parameter but it is felt that, with some refinement, the Area method developed in this work should provide a reasonably robust tool to rate bumpiness. The next stage in this work will be to consider how these tools can be applied to better understand the experience of users on the network.

ACKNOWLEDGEMENTS

This study was funded and supported by National Highways (SPaTS contract 1-1095).

REFERENCES

CEN/TC 227. prEN 13036-5:2017 *Road and Airfield Surface Characteristics - Test Methods - Part 5: Determination of Longitudinal Unevenness Indices* (2017)

Transport Focus, & Highways England. (2017). *Road Surface Quality: What Road Users Want From Highways England.*

Transport Focus, & Office of Rail and Road. (2017). *Measuring Performance of England's Strategic Roads: What Users Want Transport Focus.*

TRL, AIT, BRRC, TNO, & VTI. (2011). EXPECT - Stakeholders' Expectations and Perceptions of the future Road Transport System. *State of the Art and Current Practice in Asset Management.*

Turtschy J.C., van Ooijen W., Fontul S., Simonin J.M., Ferne B., Kokot D., ... Teng P. (2003). FORMAT - Fully Optimised Road Maintenance. *D6 Optimised Pavement Condition Data Collection Procedures.*

Ueckermann A., & Steinauer B. (2008). The Weighted Longitudinal Profile. A New Method to Evaluate the Longitudinal Evenness of Roads. *Road Materials and Pavements Design,* 9(2), 135–157. https://doi.org/10.3166/RMPD.9.135-157

Van Geem C., Mocanu I., Nitsche P., & Sjogren L. (2014). *Monitoring Road Functionality in Real-Time with Probe Vehicle Data.* TRIMM project deliverable D4.4.

Wright A., & Benbow E. (2016). *HiSPEQ - Hi-speed Survey Specifications, Explanation and Quality. Final Report.*

Are there new ways to improve the asphalt mixtures' surface functions? For sure! By functionalization process

Iran Rocha Segundo
University of Minho, ISISE

Cátia Afonso
University of Minho, CF-UM-UP

Orlando Lima Jr
University of Minho, ISISE

Salmon Landi Jr
Federal Institute Goiano

Elisabete Freitas*
University of Minho, ISISE

Joaquim O. Carneiro
University of Minho, CF-UM-UP

ABSTRACT: Titanium Dioxide (TiO_2) semiconductor material and Polytetrafluorethylene (PTFE) have been applied in the form of nano/microparticles over asphalt road pavements to provide them with new surface functionalities. This is a FUNCTIONALIZATION process that aims to improve the sustainable characteristics of the asphalt mixtures through the photocatalytic, superhydrophobic, and self-cleaning capabilities, which are related to the degradation of hazardous pollutants from the atmosphere (NOx, SO2, among others) for environmental remediation and the cleaning of the road surface (dust, greases, and oils) for the mitigation of the decrease of friction. Thus, in this research work, dispersions containing TiO_2 nanoparticles and PTFE microparticles were sprayed over an AC 10 asphalt mixture to coat and functionalize it. To confirm the photocatalysis, super hydrophobicity, and self-cleaning capacities, this smart material performance was evaluated under the degradation of the organic compound Rhodamine B (RhB) as a pollutant model and Water Contact Angle (WCA). The results indicate the photodegradation of the pollutant, confirming the proper functioning of the functionalization process, and a WCA of 150°, proving that the nanoparticles were well dispersed just over the surface of the asphalt mixture. In general, this multidisciplinary research contributes to social and environmental enhancement and enlarges the opportunities for applications of nanomaterials in a very large-scale field such as Civil Engineering. Moreover, it showed that the surface characteristics of the asphalt pavements could be studied and improved not only with a conventional approach based on noise, friction, texture, and rolling resistance measurements but also through a physicochemical approach, as the functionalization processes, using knowledge of Materials Science.

Keywords: Functionalized Asphalt Mixtures, Photocatalysis and Self-Cleaning, Surface Characteristics, Smart Coatings, Sustainable Road Pavements

---

*Corresponding Author: efreitas@civil.uminho.pt

DOI: 10.1201/9781003429258-34

1 INTRODUCTION

Recently, new capabilities have been applied to several materials, including Civil Engineering materials, such as asphalt mixtures (Segundo et al. 2021). New capabilities such as photocatalytic (Carneiro et al. 2013; Hassan et al. 2010; Hassan et al. 2012; Leng & Yu 2016; Osborn et al. 2014), superhydrophobic (Arabzadeh et al. 2016a; Nascimento et al. 2012), self-cleaning (Carneiro et al. 2013; Nascimento et al. 2012), deicing/anti-ice (Aydin et al. 2015; Liu et al. 2014; Ma et al. 2016), self-healing (Agzenai et al. 2015; Hou et al. 2015; Lv et al. 2017; Pamulapati et al. 2017; Tabakovic & Schlangen 2015), thermochromic (Cardoso 2012; Hu et al. 2015; Zhang et al. 2018), and Latent Heat Thermal Energy Storage (LHTS) (Guan 2011; Ma et al. 2014; Manning et al. 2015; MeiZhu et al. 2011) are being investigated in road pavements. They intend to contribute positively by improving environmental, safety, and economic issues (reducing costs).

Heterogeneous semiconductor-mediated photocatalysis attracts significant interest due to its ability to efficiently convert ultraviolet (UV) light from solar radiation into chemical energy that can photodegrade harmful air pollutants. Due to the huge surface area of photocatalytic asphalt pavements and their proximity to car exhaust gases, they are cited as promising surfaces, with growing interest in the literature, for reducing the concentration of atmospheric pollutants SO_2 and NO_x (Cao et al. 2017; Ma et al. 2021; Segundo et al. 2021). In addition, some semiconductors act as catalysts when exposed to the action of UV radiation, promoting the photodecomposition of organic molecules adsorbed to the surface, such as oils and greases, cleaning it and mitigating the reduction of friction caused by these compounds, reducing the number of road accidents (Carneiro et al. 2013).

Several functionalization methods of asphalt mixtures for photocatalytic purposes are reported in the literature, namely asphalt binder modification, bulk incorporation, spreading, and spray coating, which is the most important. The spray coating is a surface treatment consisting of the deposition of particles dispersed into a solvent and spread with a paint gun. For the evaluation of this functionalization process, there are several methods; in which the most important ones are the gas degradation, using pollutants such as NO_x and SO_2, and the utilization of a dye as a model of pollutant (Segundo et al. 2019).

Asphalt pavement friction at safe levels is one of the primary concerns of Transportation Engineering. The presence of water, ice, and snow on the surface significantly decreases friction, negatively contributing to accidents' frequency and gravity (Baheri et al. 2021; Dalhat 2021; Lee & Kim 2021). Thus, the water, ice, and snow must be rapidly removed from the pavement's surface. An alternative to that is providing the superhydrophobic capability to the pavement surface with the application of particles, such as TiO_2 (Wu C et al. 2021), and PTFE (Arabzadeh et al. 2016a), usually by spray coating (Arabzadeh et al. 2016b; Peng et al. 2018; Segundo et al. 2021). It is achieved when its water contact angle is higher than 150° (Arabzadeh et al. 2016b; Lee & Kim 2021).

The main objective of this research is to functionalize an asphalt mixture using micro-PTFE and nano-TiO_2 to provide it with photocatalytic, superhydrophobic, and self-cleaning properties. This smart material will be evaluated concerning wettability, photocatalysis, and chemical and morphological properties to ensure these functions are achieved.

2. MATERIALS AND METHODS

2.1 *Asphalt mixture*

The asphalt mixture selected for the functionalization process was an AC 10, designated by R. The grading is composed of 68% of 4/10 and 28% of 0/4 aggregate fractions, 4% of filler., Its main volumetric properties are maximum bulk density of 2.428 g/cm^3, bulk density of 2.305 g/cm^3, and air voids of 5.1% and 5.5% of asphalt binder content.

The particles used in the functionalization process were nano-TiO_2 and micro-PTFE. The characteristics of these particles were checked through Fourier Transform Infrared Spectroscopy (FTIR) to analyse the chemical composition and the presence of chemical bonds (Figure 1), Scanning Electron Microscopy (SEM) to analyse their surfaces (homogenization, dispersion, and particle size), and Energy Dispersive Spectroscopy (EDS) for chemical element analysis (Figures 2 and 3).

Regarding the FTIR, for TiO_2, which is a simple structure of Ti and O, note the strong bands attributed to a stretching vibration of Ti–O bound (near to 401 cm-1). Also, two other bands are observed, one at 3309 cm−1, corresponding to the stretching vibration of the hydroxyl group O-H of the TiO_2 and another around 1643 cm−1, corresponding to bending modes of water Ti-OH (León et al. 2017; Rocha Segundo et al. 2018). For PTFE, which is composed of a chemical structure of the polymer macromolecules of (-CF2-)n, symmetric

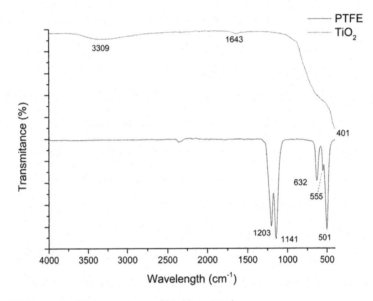

Figure 1.   FTIR results of the particles used in this research.

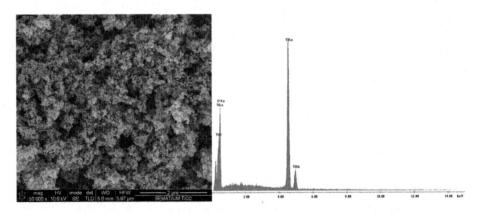

Figure 2.   SEM and EDS results of the nano-TiO_2.

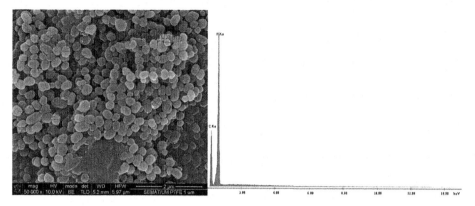

Figure 3.   SEM and EDS results of the micro-PTFE.

and asymmetric stretching vibrations of CF2 are observed at 1141 and 1203 cm-1. The peaks at 501-632 cm−1 are ascribed to CF2 rocking, wagging, and bending vibrations (oscillations of the pendulum and the deformation CF2-groups) (Fazullin et al. 2015; Piwowarczyk et al. 2019; Wang et al. 2018).

For SEM, the particle size of TiO_2 and PTFE was about 25 nm and 260 nm, respectively. Concerning EDS, the results indicate peaks of Ti and O for TiO_2 and C and F for PTFE, being in accordance with FTIR.

2.3   Sampling

The asphalt samples prepared for testing were extracted from slabs compacted in the laboratory, later cut with the dimension of 25 ×25 x 15 mm^3.

2.4   Functionalization procedure

For the functionalization process, two successive spraying coatings were carried out: i) the first one with a diluted epoxy resin (1:1 butyl acetate and resin) and the second one, immediately after the first one, with an alcoholic solution with nano-TiO_2 (4 g/L) and micro-PTFE (4 g/L), previously studied as the best solution (BS) (Segundo et al. 2022). The epoxy resin with two components was selected, as it is usually used in road pavements. This material was already used as a binder in the spreading method (Wang et al. 2016) and spraying coating method for the functionalization process (Arabzadeh et al. 2016b).

2.5   Analysis procedure

To analyze the photocatalytic efficiency, a dye (Rhodamine B - RhB) was used as a pollutant model, and its degradation was evaluated under solar simulation irradiation (Madhukar et al. 2016; Zhang et al. 2016). The cut asphalt samples were immersed in a 30 mL RhB aqueous solution with a concentration of 5 ppm. They were placed inside a box, 25 cm below a lamp (11 W/m^2 of power intensity). The samples were kept for 3 h and 8 h under dark and light conditions, respectively. To quantify the efficiency, the maximum absorption of the dye was observed over time. Using Equation 1, the photocatalytic efficiency was calculated (Carneiro et al. 2013).

$$\Phi \quad (\%) = \quad \left(\frac{A_o - A}{A_o}\right) \times 100 \qquad (1)$$

Where Φ is the photocatalytic efficiency, A0 and A represent respectively the RhB maximum absorbance for time 0 and "t" hour after irradiation.

To analyze the wettability, the samples were submitted to the Water Contact Angle test (WCA) (Arabzadeh et al. 2016a; Rocha Segundo et al. 2021). Three 5 μL water drop readings were performed in 2 samples for 2 minutes at room temperature and relative humidity.

Also, a SEM image and an EDS spectrum of the surface of the functionalized asphalt mixture were obtained to evaluate the particles' distribution and fixation.

3 RESULTS AND DISCUSSIONS

The results of the photocatalytic efficiency are presented in Figure 4 for samples with the epoxy coating (R-0.25 g) and both epoxy and BS spraying coating (R-0.25g-BS). The efficiency of the sample with the BS coating was higher, achieving 43%.

Figure 5 shows the results of the WCA for the same samples. The results indicate a much higher WCA for the sample with the BS coating (WCA > 150°), achieving the super-hydrophobic capability.

The results of SEM of the functionalized asphalt mixture are shown in Figure 6. It can be seen that the particles are dispersed on the resin, but some of them sank into the adhesive layer. On the one hand, this may improve the fixation of the particles. On the other hand, it may decrease the efficiency of the surface treatment since the particles are not exactly over

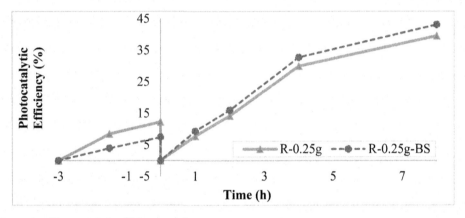

Figure 4.   Photocatalytic efficiency results.

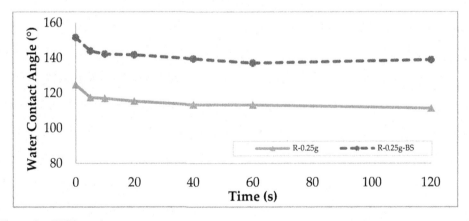

Figure 5.   WCA results

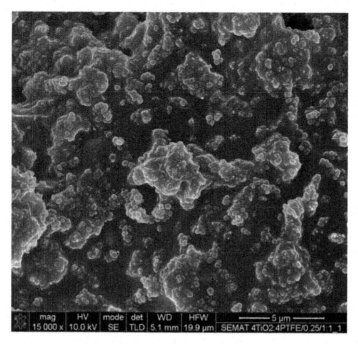

Figure 6.   FTIR results of the particles used in this research.

the surface, reducing the availability of the particles for the irradiation, the contact with water (to repel it), and the contact with the pollutant (to photodegrade it).

The sinking effect of the particles in the resin layer can be avoided by using a more diluted resin or by reducing the mass of this layer to the lowest possible, as long as the immobilization of the particles is still satisfactorily guaranteed.

The results of the EDS are shown in Figure 7. It can be seen that C, F, Ti, and O are presented in the spectrum, indicating the existence of the particles TiO_2 and PTFE on this

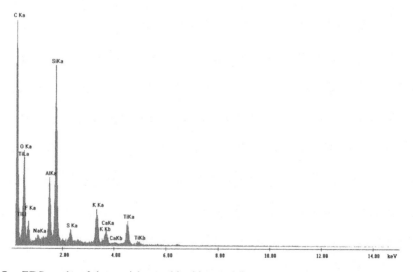

Figure 7.   EDS results of the particles used in this research.

surface treatment. Other peaks are present, namely F, Na, Al, Si, K, S, and Ca, which probably are due to the resin or even to the asphalt mixture composition.

4 CONCLUSIONS

The main objective of this research was to provide photocatalytic, superhydrophobic, and self-cleaning properties to an asphalt mixture using micro-PTFE and nano-TiO$_2$ by spraying coating. The main conclusions are that the new capabilities were achieved using two successive spraying coatings: the first one with an adhesive coating and the second one with a particle's solution.

The performance assessment considered the photocatalytic efficiency, water contact angle, SEM, and EDS tests. The SEM showed that the particles are fixed over the surface of the asphalt mixture, but the sinking of some particles in the resin can decrease the efficiency. Thus, changes in the amount of resin (adhesive coating), like its reduction, would be beneficial.

Overall, the achievements of this multidisciplinary research provide social and environmental benefits and contribute to the large-scale application of nanomaterials in the field of Civil Engineering, specifically to asphalt mixtures. Moreover, the surface characteristics of the asphalt pavements can be evaluated by essential characteristics such as friction, texture, noise, and enhanced physicochemical characteristics such as wettability and photocatalysis.

ACKNOWLEDGEMENTS

This research was funded by Portuguese Foundation for Science and Technology (FCT) under the framework of the projects NanoAir PTDC/FIS-MAC/6606/2020, UIDB/04650/ 2020, and UIDB/04029/2020.

REFERENCES

Agzenai Y., Pozuelo J., Sanz J., Perez I. and Baselga J (2015) Advanced Self-healing Asphalt Composites in the Pavement Performance Field: Mechanisms at the Nano Level and New Repairing Methodologies. *Recent Pat Nanotechnol* 9(1):43–50. https://doi.org/10.2174/1872208309666141205125017

Arabzadeh A., Ceylan H., Kim S., Gopalakrishnan K and Sassani A (2016a) Superhydrophobic Coatings on Asphalt Concrete Surfaces. *Transp Res Rec J Transp Res Board* 2551(April):10–17. https://doi.org/10.3141/2551-02

Arabzadeh A., Ceylan H., Kim S., Gopalakrishnan K. and Sassani A. (2016b) Superhydrophobic Coatings on Asphalt Concrete Surfaces: Toward Smart Solutions for Winter Pavement Maintenance. *Transp Res Rec* 2551(June):10–17. https://doi.org/10.3141/2551-02

Aydin D., Kizilel R., Caniaz R.O. and Kizilel S. (2015) Gelation-Stabilized Functional Composite-Modified Bitumen for Anti-icing Purposes. https://doi.org/10.1021/acs.iecr.5b03028

Baheri F.T. Poulikakos L.D. Poulikakos D. and Schutzius T.M. (2021) Ice Adhesion Behavior of Heterogeneous Bituminous Surfaces. *Cold Reg Sci Technol* 192(April):103405. https://doi.org/10.1016/j.coldregions.2021.103405

Cao X., Yang X., Li H., Huang W. and Liu X. (2017) Investigation of Ce-TiO$_2$ Photocatalyst and its Application in Asphalt- based Specimens for NO Degradation. *Constr Build Mater* 148:824–832. https://doi.org/10.1016/j.conbuildmat.2017.05.095

Cardoso P.D. (2012) Development of Microcapsule-Based Thermosensitive Coatings for Application to Road Pavements (Translated from Portuguese Language). University of Minho

Carneiro J.O.O., Azevedo S., Teixeira V., Fernandes F., Freitas E., Silva H. and Oliveira J. (2013) Development of Photocatalytic Asphalt Mixtures by the Deposition and Volumetric Incorporation of TiO2 Nanoparticles. *Constr Build Mater* 38:594–601. https://doi.org/10.1016/j.conbuildmat.2012.09.005

Dalhat M.A. (2021) Water Resistance and Characteristics of Asphalt Surfaces Treated with Micronized-recycled-polypropylene Waste: Super-hydrophobicity. *Constr Build Mater* 285:122870. https://doi.org/10.1016/j.conbuildmat.2021.122870

Fazullin D.D., Mavrin G.V., Sokolov M.P. and Shaikhiev I.G. (2015) Infrared Spectroscopic Studies of the PTFE and Nylon Membranes Modified Polyaniline. *Mod Appl Sci* 9(1):242–249. https://doi.org/10.5539/mas.v9n1p242

Guan B. (2011) Application of Asphalt Pavement with Phase Change Materials to Mitigate Urban Heat Island Effect. *2011 Int Symp Water Resour Environ Prot*:2389–2392. https://doi.org/10.1109/ISWREP.2011.5893749

Hassan M.M., Dylla H., Asadi S., Mohammad L.N. and Cooper S. (2012) Laboratory Evaluation of Environmental Performance of Photocatalytic Titanium Dioxide Warm-Mix Asphalt Pavements. *J Mater Civ Eng* 24(5):599–605. https://doi.org/10.1061/(ASCE)MT.1943-5533.0000408

Hassan M.M., Dylla H., Mohammad L.N. and Rupnow T. (2010) Evaluation of the Durability of Titanium Dioxide Photocatalyst Coating for Concrete Pavement. *Constr Build Mater* 24(8):1456–1461. https://doi.org/10.1016/j.conbuildmat.2010.01.009

Hou Y., Wang L., Pauli T. and Sun W. (2015) Investigation of the Asphalt Self-Healing Mechanism Using a Phase-Field Model. *J Mater Civ Eng* 27(3):04014118. https://doi.org/10.1061/(ASCE)MT.1943-5533.0001047

Hu J., Asce S.M., Yu X.B. and Asce M (2015) Reflectance Spectra of Thermochromic Asphalt Binder: Characterization and Optical Mixing Model. *J Mater Civ Eng* 28(2):1–10. https://doi.org/10.1061/(ASCE)MT.1943-5533.0001387.

Lee E. and Kim D.H. (2021) Simple Fabrication of Asphalt-based Superhydrophobic Surface with Controllable Wetting Transition From Cassie-Baxter to Wenzel Wetting State. *Colloids Surfaces A Physicochem Eng Asp* 625(March):126927. https://doi.org/10.1016/j.colsurfa.2021.126927

Leng Z. and Yu H. (2016) Novel Method of Coating Titanium Dioxide on to Asphalt Mixture Based on the Breath Figure Process for Air-purifying Purpose. *J Mater Civ Eng* 28(5):1–7. https://doi.org/10.1061/(ASCE)MT.1943-5533.0001478

León A., Reuquen P., Garín C., Segura R., Vargas P., Zapata P. and Orihuela P.A. (2017) FTIR and Raman Characterization of TiO2 Nanoparticles Coated With Polyethylene Glycol as Carrier for 2-Methoxyestradiol. *Appl Sci* 7(1):1–9. https://doi.org/10.3390/app7010049

Liu Z., Xing M., Chen S., He R. and Cong P. (2014) Influence of the Chloride-based Anti-freeze Filler on the Properties of Asphalt Mixtures. *Constr Build Mater* 51:133–140. https://doi.org/10.1016/j.conbuildmat.2013.09.057

Lv Q., Huang W. and Xiao F. (2017) Laboratory Evaluation of Self-healing Properties of Various Modified Asphalt. *Constr Build Mater* 136:192–201. https://doi.org/10.1016/j.conbuildmat.2017.01.045

Ma B., Si W., Ren J., Wang H.N., Liu F.W. and Li J. (2014) Exploration of Road Temperature-Adjustment Material in Asphalt Mixture. *Road Mater Pavement Des* 15(3):659–673. https://doi.org/10.1080/14680629.2014.885462

Ma T., Geng L., Ding X., Zhang D. and Huang X. (2016) Experimental Study of Deicing Asphalt Mixture with Anti-icing Additives. *Constr Build Mater* 127:653–662. https://doi.org/10.1016/j.conbuildmat.2016.10.018

Ma Y., Li L., Wang H., Wang W. and Zheng K. (2021) Laboratory Study on Performance Evaluation and Automobile Exhaust Degradation of Nano-TiO2Particles-Modified Asphalt Materials. *Adv Mater Sci Eng* 2021. https://doi.org/10.1155/2021/5574013

Madhukar B., Wiener J., Militky J., Rwawiire S., Mishra R., Jacob K.I., Wang Y., Kale B.M., Wiener J., Militky J., Rwawiire S., Mishra R., Jacob K.I. and Wang Y. (2016) Coating of Cellulose-TiO2 Nanoparticles on Cotton Fabric for Durable Photocatalytic Self-cleaning and Stiffness. *Carbohydr Polym* 150:107–113. https://doi.org/10.1016/j.carbpol.2016.05.006

Manning B.J., Bender P.R., Cote S.A., Lewis R.A., Sakulich A.R. and Mallick R.B. (2015) Assessing the Feasibility of Incorporating Phase Change Material in Hot Mix Asphalt. *Sustain Cities Soc* 19:11–16. https://doi.org/10.1016/j.scs.2015.06.005

MeiZhu C., Jing H., Wu S., Lu W. and Xu G. (2011) Optimization of Phase Change Materials Used in Asphalt Pavement to Prevent Rutting. *Adv Mater Res* 219–220 (May):1375–1378. https://doi.org/10.4028/www.scientific.net/AMR.219-220.1375

Nascimento J.H.O., Pereira P., Freitas E. and Fernandes F. (2012) Development and Characterization of a Superhydrophobic and Anti-ice Asphaltic Nanostructured Material for Road Pavements. *In: 7th International Conference on Maintenance and Rehabilitation of Pavements and Technological Control.* At Auckland, New Zealand

Osborn D., Hassan M., Asadi S. and White John R. (2014) Durability Quantification of TiO2 Surface Coating on Concrete and Asphalt Pavements. *J. Mater. Civ. Eng.* 26:331–337

Pamulapati Y., Elseifi M.A., Cooper S.B., Mohammad L.N. and Elbagalati O. (2017) Evaluation of Self-Healing of Asphalt Concrete Through Induction Heating and Metallic Fibers. *Constr Build Mater* 146:66–75. https://doi.org/10.1016/j.conbuildmat.2017.04.064

Peng C., Zhang H., You Z., Xu F., Jiang G., Lv S. and Zhang R. (2018) Preparation and Anti-icing Properties of a Superhydrophobic Silicone Coating on Asphalt Mixture. *Constr Build Mater* 189:227–235. https://doi.org/10.1016/j.conbuildmat.2018.08.211

Piwowarczyk J., Jedrzejewski R., Moszyński D., Kwiatkowski K., Niemczyk A. and Baranowska J. (2019) XPS and FTIR Studies of Polytetrafluoroethylene Thin Films Obtained by Physical Methods. *Polymers (Basel)*. 11(10):1–13. https://doi.org/10.3390/polym11101629

Rocha Segundo I., Freitas E., Branco V.T.F.C., Landi S., Costa MF. and Carneiro J.O. (2021) Review and Analysis of Advances in Functionalized, Smart, and Multifunctional Asphalt Mixtures. *Renew Sustain Energy Rev 151.*(June 2020):111552. https://doi.org/10.1016/j.rser.2021.111552

Rocha Segundo I.G. da, Dias E.A.L., Fernandes F.D.P., Freitas E.F. de, Costa M.F. and Carneiro J.O.. (2018) Photocatalytic Asphalt Pavement: the Physicochemical and Rheological Impact of TiO$_2$ Nano/microparticles and ZnO Microparticles Onto the Bitumen. *Road Mater Pavement Des.* https://doi.org/10.1080/14680629.2018.1453371

Segundo I.R., Freitas E., Branco V.T.F.C., Landi S., Costa M.F. and Carneiro J.O. (2021) Review and Analysis of Advances in Functionalized, Smart, and Multifunctional Asphalt Mixtures. *Renew Sustain Energy Rev* 151(June 2020):111552. https://doi.org/10.1016/j.rser.2021.111552

Segundo I.R., Freitas E., Landi S., Costa M.F.M., Carneiro J.O. (2019) Smart, Photocatalytic and Self-Cleaning Asphalt Mixtures: A Literature Review. *Coatings* 9(11). https://doi.org/10.3390/coatings9110696

Segundo I.R., Zahabizadeh B., Landi S., Lima O., Afonso C., Borinelli J., Freitas E., Cunha V.M.C.F., Teixeira V., Costa M.F.M. and Carneiro J.O. (2022) Functionalization of Smart Recycled Asphalt Mixtures: *A Sustainability Scientific and Pedagogical Approach*. *Sustain.* 14

Tabakovic A., Schlangen E. (2015) Self-Healing Technology for Asphalt Pavements. *Adv Polym Sci.* https://doi.org/10.1007/12

Wang D., Leng Z., Hüben M., Oeser M., Steinauer B., H??ben M., Oeser M. and Steinauer B. (2016) Photocatalytic Pavements with Epoxy-bonded TiO2-containing Spreading Material. *Constr Build Mater* 107:44–51. https://doi.org/10.1016/j.conbuildmat.2015.12.164

Wang H., Wen Y., Peng H., Zheng C., Li Y., Wang S., Sun S., Xie X. and Zhou X. (2018) Grafting Polytetrafluoroethylene Micropowder via in Situ Electron Beam Irradiation-induced Polymerization. *Polymers (Basel)* 10(5). https://doi.org/10.3390/polym10050503

Wu C.; Li L.;Wang W.; Gu Z.; Li H.; Lin X.and Wang H. (2021) Coating on Asphalt Pavement. *Nanomaterials* 11:1–19

Zhang H., Chen Z., Xu G. and Shi C. (2018) Physical, Rheological and Chemical Characterization of Aging Behaviors of Thermochromic Asphalt Binder. *Fuel 211.*(September 2017):850–858. https://doi.org/10.1016/j.fuel.2017.09.111.

Zhang W., Xiao X., Zeng X., Li Y., Zheng L. and Wan C. (2016) Enhanced Photocatalytic Activity of TiO$_2$ Nanoparticles using SnS$_2$/RGO hybrid as co-catalyst: DFT study and Photocatalytic Mechanism. *J Alloys Compd.* 685:774–783. https://doi.org/10.1016/j.jallcom.2016.06.199

Characterization of surface roughness on the basis of vibration response corresponding to the ride quality of micromobilities

Yuki Kotani
Graduate School of Engineering, Kitami Institute of Technology, Kitami, Japan

Kazuya Tomiyama*
Division of Civil and Environmental Engineering, Kitami Institute of Technology, Kitami, Japan

Hayato Nishigai & Kenichiro Sasaki
Graduate School of Engineering, Kitami Institute of Technology, Kitami, Japan

Yuki Yamaguchi & Kazushi Moriishi
Obayashi Road Corporation, Sarugaku-cho, Chiyoda-ku, Japan

ABSTRACT: Micromobilities including electric standing scooters and electric four-wheel mobility scooters are spreading rapidly in many cities as a popular mode of transportation. This relatively new type of transportation is battery-operated personal small vehicles driven by higher speed than conventional personal mobilities such as bicycles and wheelchairs operated by human power. In return to the convenience, users of the micromobilities are affected by the surrounding environment including road and transportation conditions more directly than motor vehicles. Road surface roughness is one of the potential factors inducing negative effects on the safety and comfort of the users. However, the term "roughness" and its definition in terms of road surface characteristics developed has so far been intended to be applied to the roadways for motor vehicles. Although the definition of roughness, more specifically, is a surface characteristic dimensions that affect vehicle dynamics and ride quality, it depends on the specifications of a vehicle of interest. In the light of this background, this study examines surface characteristics associated with the vibration response of micromobilities in terms of ride quality. For purpose, the driving experiments employing an electric standing scooter, electric four-wheel mobility scooters and a human-powered bicycle were conducted on various surfaces in terms of the roughness and texture as well as the materials including dense-graded and porous asphalt concrete pavements. In the experiments, the vibration responses of the micromobilities were measured with accelerometers for various locations of the employed micromobilities to identify the interaction with surface characteristics. This study also employed valid measuring devices such as a low-speed profiler and a circular track meter so that surface properties can be characterized in terms of roughness and texture. The analysis for the data obtained in the experiment focuses on the wavelength of a surface profile and the excited frequency response of the micromobilities. It identified the range of surface wavelengths corresponding to the ride quality for mobility vehicles that is the "roughness" from the viewpoints of micromobilities. This study finally redefines the roughness of the road surface that affects micromobilities enabling the quantitative evaluation of surface condition for diverse road spaces other than roadways.

Keywords: Micromobility, Vehicle Vibration, Ride Quality, Roughness, Texture

---

*Corresponding Author: tomiyama@mail.kitami-it.ac.jp

DOI: 10.1201/9781003429258-35

1 INTRODUCTION

Micromobilities including electric standing scooters and electric four-wheel mobility scooters are rapidly spread in many cities as a popular transportation tool. This relatively new type transportation is battery-operated personal small vehicles driven at higher speeds than conventional personalmobilities such as bicycles and wheelchairs operated by human power and thus are also becoming popular in many countrries. In particular, electric standing scooters are being offered as a new mobility service in metropolitan areas, and four-wheel mobility scooters are becoming popular as an alternative to automobiles for elderly people who have given up their automobile licenses. Micromobility users are affected by the surrounding environment including road and transportation conditions more directly than motor vehicles. Road surface roughness is one of the potential factors inducing negative effects on the safety and comfort of the users. ISO 2631-1 (ISO 1997) declares that the evaluation criteria for vibration safety and comfort, along with vibration evaluation during the ride is important. However, the term "roughness" and its definition in terms of road surface characteristics developed has so far been applied mostly to the roadways for motor vehicles. Although the definition of roughness, more specifically, is a surface characteristic dimensions that affect vehicle dynamics and ride quality, it depends on the vehicle of interest.

In human-powered bicycles, a new road surface evaluation index was developed and proposed for bicycles that was different from the International Roughness Index (IRI) (Tomiyama et al. 2019). Human-powered bicycles were also studied for texture (Li et al. 2015; Thigpen et al. 2015). A study (Li et al. 2015) reported a correlation between Mean Profile Depth (MPD) and vibration compared to the correlation between IRI and vibration, with a greater influence of texture. For wheelchairs, a research simulated vibration transmission to a wheelchair-bound passenger riding in a wheelchair-accessible vehicle (Matsuoka et al. 2003). In the handle-type four-wheel mobility scooters, the distinctive response frequencies were reported at 8 Hz and 30 Hz (Tomiyama & Moriishi 2020). A study evaluated design and safety standards and improved product design of electric standing scooters (Garman et al. 2020). Few researches regarding the electric standing scooters have not so far been studied conducted in relation to the road surface. According to the above-mentioned literatures, although vibration verification in terms of safety measures for electric standing scooters and wheelchairs have been studied, researches considering the relationship between micromobilities and road surface have so far been limited to human-powered bicycles and handle-type four-wheel mobility scooters. The literatures especially focused on specific micromobilities, but not on the relationship between various micromobilities and road surfaces. However, since electric standing scooters are driven in a standing position, a study is required to consider a method different from that of a vehicle that moves in a seated position.

Against this background, driving experiments employing an electric standing scooter, electric four-wheel mobility scooters and a human-powered bicycle are conducted on various surfaces in terms of the roughness and texture as well as the surface materials including dense-graded and porous asphalt concrete pavements. In the experiments, the vibration responses of the micromobilities are measured with accelerometers for various positions of the vehicles to examine the interaction with surface characteristics. This study also employs a low-speed profiler and a circular track meter so that surface properties can be characterized in terms of roughness and texture. The analysis for the data obtained in the experiment focuses on the wavelength of surface profiles and the excited vibration frequency response of the micromobilities.

2 TESTED ROAD SURFACES

Road surfaces measurements were taken on the pavements as shown in Figure 1: (a) Soil Pavement 1, (b) Soil Pavement 2, (c) a dense-graded asphalt concrete pavement commonly used in Japan, (d) porous asphalt concrete pavement with voids and tile-like pavement cut joints filled with cement milk for water retention called water retention asphalt concrete pavement, and (e) a porous asphalt concrete pavement with high porosity and drainage properties. In order to measure various road surfaces, a total of five measuring lines were prepared in two test yards A and B as shown in Figure 2. Figure 2 (a) shows No.1, No.2, and No.3 of the three measurement lines in Test Yard A. Figure 2 (b) shows the pavements content of test yard A. The measuring line of No.1 and No.2 consists of a total length of 26 m, with the pavement of Figure 1 (a) being 14 m, the pavement of Figure 1 (b) being 4 m, and the pavement of Figure 1 (c) being 8 m. The measuring line of No.3 consist of a total length of 24 m, with the pavement of Figure 1 (d). Figure 2 (c) and (d) shows No.4 and No.5 of the two measurement lines in Test Yard B. Figure 2 (e) shows the pavements content of test yard B. The measuring line of No.4 is composed of the same pavement as Figure 1 (c) and has a total length of 18 m. The measuring line of No.5 is composed of the same pavement as Figure 1 (e) and has a total length of 18 m.

The road surfaces were measured with two devices such as shown in Figure 3. First, the Multi Road Profiler (MRP), which is a low-speed profiler shown in Figure 3 (a), was used to measure road surface profiles on the measuring lines. The sampling interval was set as 10 mm. Second, the texture depth of the surface was measured with a Circular Track Meter (CTM) as shown in Figure 3 (b). One representative point for pavements on the measurement line was selected and measured with CTM. Table 1 shows the values of the MPD for

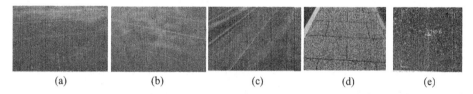

Figure 1. Pavements: (a) Soil Pavement 1, (b) Soil Pavement 2, (c) Dense-graded asphalt concrete pavement, (d) Water retention asphalt concrete pavement, (e) Porous asphalt concrete pavement.

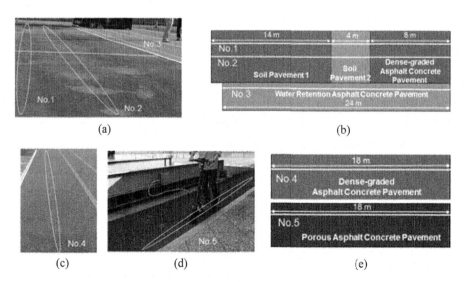

Figure 2. Test yards: (a) No.1, No.2 and No.3, (b) Test yard a of pavements content, (c) No.4, (d) No.5, (e) Test yard b of pavements content.

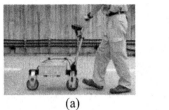
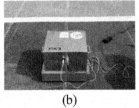

(a)   (b)

Figure 3.   Pavement surface measurement equipment: (a) Multi road profiler (MRP), (b) Circular Track Meter (CTM)

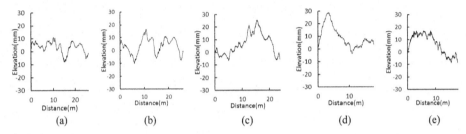

(a)   (b)   (c)   (d)   (e)

Figure 4.   Road surface profiles: (a) No.1, (b) No.2, (c) No.3, (d) No.4, (e) No.5.

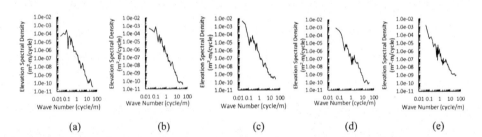

(a)   (b)   (c)   (d)   (e)

Figure 5.   Profiles power spectral density: (a) No.1, (b) No.2, (c) No.3, (d) No.4, (e) No.5.

each measurement line. Figure 4 shows the Road Surface Profile measured by MRP on each measurement lines. The profile in Figure 4 was processed with a 50 m high-pass filter. Figure 5 shows the Power Spectral Density (PSD) of each road surface profile.

3   APPARATUS

Figure 6 shows the four micromobilities employed in the experiment. (a) electric standing scooters, (b) handle-type four-wheel mobility scooters, (c) joystick-type four-wheel mobility scooters, and (d) human-powered bicycles. The running position of the micromobilities was measured with the wheel for electric standing scooters and rickshaws, and with the left wheel for four-wheel mobility scooters. Figure 6 shows the accelerometers installed at the measurement in test yard A. The measurement experiment of No.5 is only for electric standing scooters, handle-type four-wheel mobility scooters, and human-powered bicycles.

| (a) | (b) | (c) | (d) |

Figure 6.   Location of accelerometers in test yard A: (a) Electric standing scooters, (b) Handle-type four-wheel mobility scooters, (c) Joystick-type four-wheel mobility scooters, (d) Human-powered bicycles.

Table 1.   Measurement Results.

Measurment Line	Pavement Classification	MPD
No.1	Soil Pavement 1	0.18 mm
	Soil Pavement 2	0.90 mm
	Dense-graded Asphalt Concrete	0.23 mm
No.2	Soil Pavement 1	0.20 mm
	Soil Pavement 2	0.90 mm
	Dense-graded Asphalt Concrete	0.39 mm
No.3	Water Retention Concrete	1.30 mm
No.4	Dense-graded Asphalt Concrete	0.39 mm
No.5	Porous Asphalt Concrete	2.65 mm

The positions where the accelerometers were installed on each mobility scooter are shown below. The Electric Standing Scooters and Human-powered Bicycles ran at 10 km/h, and the Four-wheel Mobility Scooters ran at 4 km/h.

- Figure 6 (a) show the locations of the accelerometers for the electric standing scooters. The accelerometers were installed in four locations: on the handle, unsprung mass and sprung mass of the front wheel, and near the rear wheel. An accelerometer with a sampling interval of 500 Hz was installed in test yard A. A wireless accelerometer with a sampling interval of 200 Hz was installed in test yard B.
- Figure 6 (b) show the locations of the accelerometers for the handle-type four-wheel mobility scooters. The accelerometers were installed at three locations: under the seat, sprung mass and unsprung mass of the front wheel. An accelerometer with a sampling interval of 100 Hz was installed under the seat. An accelerometer with a sampling interval of 500 Hz was installed on the sprung mass and unsprung mass.
Figure 6 (c) show the locations of the accelerometers for the joystick-type four-wheel mobility scooters. The accelerometers were installed at three locations: under the seat, sprung mass and unsprung mass of the front wheel. An accelerometer with a sampling interval of 100 Hz was installed under the seat. An accelerometer with a sampling interval of 500 Hz was installed on the sprung mass and unsprung mass.
- Figure 6 (d) show the locations of the accelerometers for the human-powered bicycles. The accelerometers were installed in four locations: near the handle, near the seat, front wheel, and rear carrier. The under seat of the human-powered bicycles was installed near the pedals due to the difference in the size of the accelerometer used. An accelerometer with a sampling interval of 500 Hz was installed in test yard A. A wireless accelerometer with a sampling interval of 200 Hz was installed in test yard B.

Figure 7 shows an example of longitudinal acceleration directly linked to the ride quality measured by micromobilities. Figure 7 shows the waveform of unsprung mass of the electric standing scooters measured on the No.1. Figure 8 shows the correlation between MPD of texture and RMS (Root Mean Square) acceleration.

As shown in Figure 7, soil pavement 2 tends to have a larger accelerometer waveform than soil pavement 1 and dense-graded asphalt concrete pavement. As shown in Table 1, the No.1 MPD value is 0.18 mm for soil pavement 1, 0.90 mm for soil pavement 2, and 0.23 mm for dense-graded asphalt concrete pavement. Figure 7 and MPD results show that micromobilities is texture sensitive.

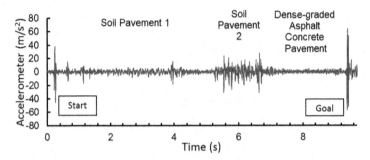

Figure 7.   An example of acceleration in the longitudinal direction of micromobilities.

As shown in Figure 8, the results of comparing the MPD of the texture and the RMS of the acceleration show that the electric standing scooters are correlated with sprung mass and handle, handle-type four-wheel mobility scooters are correlated with unsprung mass, and the

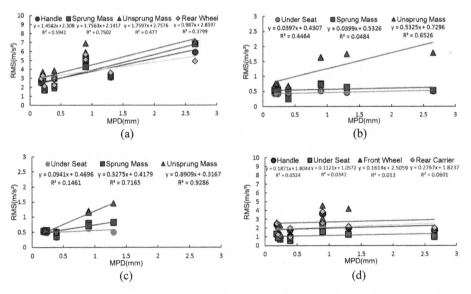

Figure 8.   Relationship between MPD and acceleration RMS: (a) Electric standing scooters, (b) Handle-type four-wheel mobility scooters, (c) Joystick-type four-wheel mobility scooters, (d) Human-powered bicycles.

four-wheel mobility scooters are correlated with the unsprung mass and sprung mass. Human-powered Bicycles showed no correlation. As shown in Figure 8, the results that cannot be said to have a correlation between micromobilities RMS and road surface texture are due to the fact that there are few measurement points for MPD and that the RMS of acceleration is easily affected by the values of Roughness above texture.

From the results of vertical acceleration such as Figure 7 and previous studies (Li et al. 2015) (Tomiyama & Moriishi 2020), it is considered that there is a relationship between the vertical acceleration of micromobilities and the texture of the road surface.

5 FREQUENCY RESPONSE FUNCTION

This section deals with the frequency response function for the ride quality of micro-mobilities. The frequency response function of micromobilities was calculated with reference to The Little Book of Profiling (Sayers & Karamihas 1998). The analysis procedure calculated the Response Gain (Gain) after calculating the PSD from the acceleration obtained from the road surface profile and the accelerometers installed in each micromobilities. These calculations were computed using MATLAB R2022a (MathWorks 2022). Detrending of acceleration data was performed as a preprocessing step for PSD calculation. The PSD was calculated using the Yule-Walker method of parametric spectral estimation. The gain results tended to increase at frequencies below 1 Hz, but this is considered to be due to road gradient when micromobilities ratings are taken into account. In this study, frequencies below 1 Hz were excluded from the verification.

Figure 9 shows the calculated response gain of electric standing scooters, Figure 10 shows the calculated response gain of handle-type four-wheel mobility scooters, Figure 11 shows

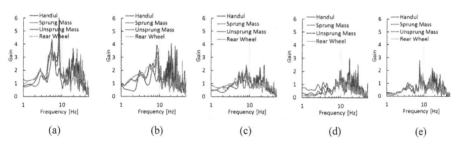

Figure 9.   Electric standing scooters frequency response function: (a) No.1, (b) No.2, (c) No.3, (d) No.4, (e) No.5.

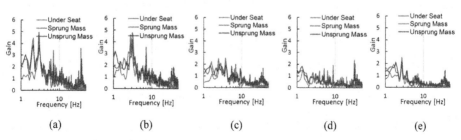

Figure 10.   Handle-type four-wheel mobility scooters frequency response function: (a) No.1, (b) No.2, (c) No.3, (d) No.4, (e) No.5.

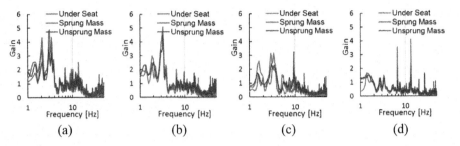

Figure 11. Joystick-type four-wheel mobility scooters frequency response function: (a) No.1, (b) No.2, (c) No.3, (d) No.4.

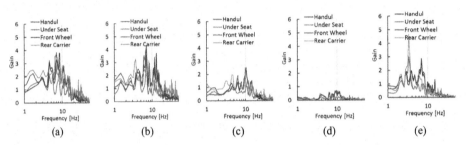

Figure 12. Human-powered bicycles frequency response function: (a) No.1, (b) No.2, (c) No.3, (d) No.4, (e) No.5.

the calculated response gain of joystick-type four-wheel mobility scooters, and Figure 12 shows the calculated response gain of human-powered bicycles.

5.1 *Electric standing scooters*

The results in Figure 9 shows the frequency response function of the electric standing scooters. The electric standing scooters shows that the characteristic response frequencies of the peak are around 9 Hz and 18 Hz. The results for the area near the rear wheels show a tendency for the gain to be higher at 30 Hz to 50 Hz compared to other installation locations. This is due to the motor of the electric standing scooter near the rear wheel. The natural frequency of the electric standing scooters was influenced by the megatexture with a wavelength of 50 to 500 mm in the road surface characteristics classification (PIARC 1987).

5.2 *Handle-type and joystick-type four-wheel mobility scooters*

The results in Figure 10 and 11, for the handle-type and joystick-type were not significantly different. This is because the vehicle dimensions and wheel sizes are similar despite the different driving controls for handle-type and joystick types.

The results in Figure 10 and 11, for the four-wheel mobility scooters shows that the characteristic response frequencies of the peak are around 3 Hz. On the axle side, a peak is seen around 30 Hz in addition to 3 Hz, and it can be seen that it is also affected by the macrotexture with a wavelength of 0.5 to 50 mm according to the road surface characteristics classification. In previous study (Tomiyama & Moriishi 2020) that investigated the frequency response of handle-type four-wheel mobility scooters, the resonance frequencies of the natural frequencies of the sprung mass and under seat are around 8 Hz. The natural frequency of the unsprung mass is around 30 Hz. From the results in Figure 10 and 11, The graph shows the peaks at around 8 Hz and 30 Hz, it will be necessary to manufacture and verify a vibration model for the four-wheel electric scooters in the future.

The natural frequencies of the body side of four-wheel mobility scooters was influenced by the megatexture with a wavelength of 50 to 500 mm in the road surface characteristics classification. The axle side the body side of four-wheel mobility scooters can be seen that it is also affected by the megatexture with a wavelength of 50 to 500 mm and the macrotexture with a wavelength of 0.5 to 50 mm according to the road surface characteristics classification.

5.3 Human-powered bicycles

The results in Figure 12 shows the frequency response function of the human-powered bicycles. The human-powered bicycles show that the characteristic response frequencies of the peak are around 8 Hz. The natural frequencies of the human-powered bicycles were influenced by the megatexture with a wavelength of 50 to 500 mm in the road surface characteristics classification.

6 DISCUSSION

This study investigated the characterization of surface roughness based on the vibrational response corresponding to the ride quality of micromobilities using vertical accelerations. The findings are summarized below.

- The micromobilities are sensitive to texture.
- The natural frequencies of electric standing scooters when traveling at 10 km/h are around 9 Hz and 18 Hz. In the road surface characteristics classification, they correspond to the megatexture region with a wavelength of 50 to 500 mm.
- The natural frequencies of four-wheel mobility scooters when traveling at 4 km/h is around 3 Hz, which corresponds to the 50 to 500 mm wavelength megatexture. On the axle side, a peak is seen around 30 Hz in addition to 3 Hz, and it can be seen that it is also affected by the macrotexture with a wavelength of 0.5 to 50 mm according to the road surface characteristics classification. The distinctive response frequencies of the different types of four-wheel mobility scooters prepared for this measurement did not change significantly.
- The natural frequencies of human-powered bicycles when traveling at 10 km/h are around 8 Hz. In the road surface characteristics classification, they correspond to the mega-texture region with a wavelength of 50 to 500 mm.

This study clarified that micromobilities corresponds to the mega-texture range of 50 to 500 mm in the road surface property classification. It is generally said that the IRI index is related to the ride quality of the car. Figure 13 shows the difference in wavelength that corresponds to the ride quality of the cars and micromobilities.

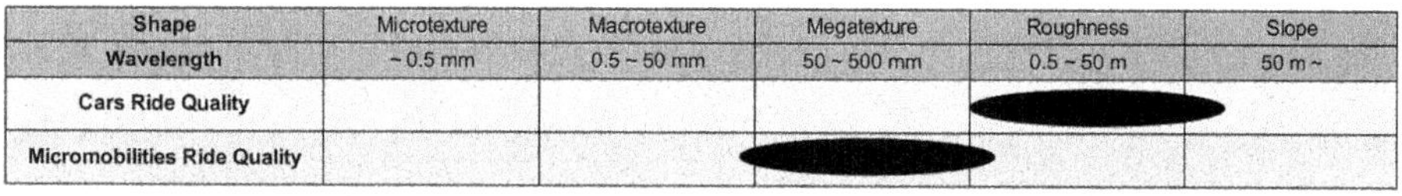

Shape	Microtexture	Macrotexture	Megatexture	Roughness	Slope
Wavelength	~ 0.5 mm	0.5 ~ 50 mm	50 ~ 500 mm	0.5 ~ 50 m	50 m ~
Cars Ride Quality				⬤	
Micromobilities Ride Quality			⬤		

Figure 13. Difference in characteristic wavelength between micromobilities and cars.

7 CONCLUSIONS

This study clarified that micromobilities corresponds to the mega-texture range of 50 to 500 mm in the road surface property classification. Since IRI corresponds to the roughness wavelength range of 0.5 to 50 m in the road surface characteristics classification, it is necessary to demonstrate the necessity of developing a new index that is different from the IRI index in order to create a micromobilities index that corresponds to ride quality.

In the future, we would like to vibration model and study more specific vibration characteristics of micromobilities. In particular, electric standing scooters require a different perspective than conventional vehicles because they are driven in a standing position. The results of this research are expected to contribute to the development of new road surface management indicators for micromobilities related to ride comfort, and to improve the road environment such as micromobilities lanes.

REFERENCES

Garman C.MR, Como S.G., Campbell I.C., Wishart J., O'Brien K., and McLean S., 2020, Micro-Mobility Vehicle Dynamics and Rider Kinematics during Electric Scooter Riding, *SAE Technical Papers*, April.

International Organization for Standardization, 1997, *Mechanical Vibration and Shock -Evaluation of Human Exposure to Whole-body Vibration. Part 1: General Requirements.* ISO 2631-1.

Li H., Harvey J., Chen Z., He Y., Holland T.J., Prince S. and McClain K., 2015, Measurement of Pavement Treatment Macrotexture and Its Effect on Bicycle Ride Quality: Transportation Research Record: *Journal of the Transportation Research Board*, No.2525, pp. 43–53.

MathWorks, 2022, https://jp.mathworks.com/products/matlab.html

Matsuoka Y., Kawai K. and Sato R., 2003, Vibration Simulation Model of Passenger-Wheelchair System in Wheelchair-Accessible Vehicle, *Journal of Mechanical Design, Transactions of the ASME*, Volume 125, Issue 4, Pages 779–785, December.

PIARC, 1987, *Optimization of Surface Characteristics, Technical Committee Report on Surface Characteristics–PIARC Xviii World Road Congress*, Brussels, Belgium.

Sayers M.W. and Karamihas S.M., 1998, The Little Book of Profiling Basic Information about Measuring and Interpreting Road Profiles, University of Michigan, Ann Arbor, Transportation Research Institute, Engineering Research Division.

Thigpen C.G., Li H., Handy S.L., and Harvey J., 2015, Modeling the Impact of Pavement Roughness on Bicycle Ride Quality. Transportation Research Record: *Journal of the Transportation Research Board*, No.2520, pp. 67–77.

Tomiyama K. and Moriishi K., 2020, Pavement Surface Evaluation Interacting Vibration Characteristics of an Electric Mobility Scooter, *9th International Conference on Maintenance and Rehabilitation of Pavements (Mairepav9)*, Dübendorf, July.

Tomiyama K., Takahashi K., Sasaki Y., Hagiwara T., Watanabe K., and Moriishi K., 2019, Development of a Road Roughness Index based on the Bicycle Vibration Response, *Transportation Research Board*, Vol. 98, No.19-04363.

Roads and Airports Pavement Surface Characteristics – Crispino & Toraldo (Eds)
© 2023 The Author(s), ISBN 978-1-032-55149-4

How modern petrographic analysis of roadstone aggregate can provide a better understanding of their frictional properties

Richard P. Unitt* & Patrick A. Meere
School of Biological, Earth and Environmental Sciences, University College Cork, Ireland

Tom Casey
Transport Infrastructure Ireland, Dublin, Ireland

Brian Mulry
Pavement Management Services Limited, Athenry, County Galway, Ireland

ABSTRACT: The overall frictional performance of road wearing courses is directly related to the microtexture of the included stone particles. This study develops a detailed understanding of how mineralogy, petrography, and microstructure in different rock types, used as high polished stone value (hPSV) surfacing in Ireland, is related to the provision of frictional resistance.

Whilst the surface microtexture of any stone particle is predominantly a function of grain-size, it is mostly the distribution and shape of surface topographic peaks and dales (asperities) that determines its frictional properties. These features are investigated using traditional microscopy in tandem with digital metrology and 3D Raman spectroscopy.

The study follows the development of microtexture during artificial wear using standard PSV test specimens on an Accelerated Polishing Machine (APM). Measurements and analysis are conducted on raw aggregate (before polishing) and then at regular intervals in the polishing process at 3, 4.5, 6, 7.5 and 9 hours. The measurements follow the evolution of the microtexture, its relation to a selected set of metrological parameters and the petrography of the stone particle.

In addition, road cores extracted from the oil track and wheel track of variably trafficked sites provide details on the wear of in-service stone particles for direct comparison with the laboratory worn samples.

This work demonstrates that by combining traditional engineering tests, such as and PSV and AAV (Aggregate Abrasion Value), with novel modern methodologies, such as digital metrology and Raman spectroscopy, we can develop a better understanding of how the petrography of stone particles affects surface microtexture and by extension its potential frictional properties. As a result, this data can be utilised by road engineers to design sustainable surfaces by matching frictional demand with available local resources.

Keywords: Microtexture, Metrology, Raman Spectroscopy

1 INTRODUCTION

Many studies have examined the correlation between the mineralogy and petrography of stone aggregates in road surface courses and their frictional properties (Dokic 2015; Kane

---

*Corresponding Author: r.unitt@ucc.ie

DOI: 10.1201/9781003429258-36

et al. 2013; Krutilová & Přikryl 2017; Perry 2014; Shabani et al. 2013; Unitt & Meere 2018; Wang et al. 2019, 2015). Each study is often based around stone aggregates commonly used in a particular country and in-service conditions, such as trafficking and climate, which can play a major role in how those aggregates perform.

This study has undertaken comprehensive testing of stone aggregates utilised in the Republic of Ireland as high polished stone value (hPSV) surfacing materials. Sources were selected based on their geographic distribution, petrographic variance, range of PSV values, and their common usage in Ireland's National Road Network (Figure 1).

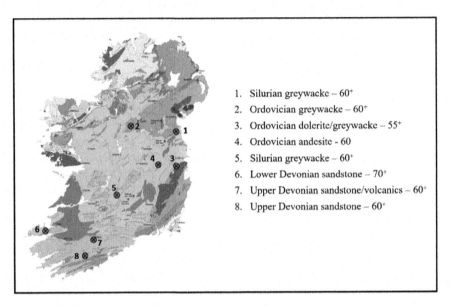

1. Silurian greywacke – 60[+]
2. Ordovician greywacke – 60[+]
3. Ordovician dolerite/greywacke – 55[+]
4. Ordovician andesite - 60
5. Silurian greywacke – 60[+]
6. Lower Devonian sandstone – 70[+]
7. Upper Devonian sandstone/volcanics – 60[+]
8. Upper Devonian sandstone – 60[+]

Figure 1. Geological map of Ireland (gsi.ie) showing the location of quarry sources selected for this study. The key identifies the geological age of the source plus the expected maximum PSV.

Previous studies, based in Ireland (Woodside et al. 1998; Woodward et al. 2005) have often identified greywacke lithologies as the ideal high friction surfacing material although recent studies undertaken by Transport Infrastructure Ireland (TII126) have shown that heterogeneous sources, such as those that produce greywacke and other siliciclastic rocks, need careful management to ensure the end-product meets the required specification.

Additional problems have been identified with other lithologies which together with heterogeneous sources has led to overspecification in Irish surfacing Standards (TII_DN-PAV-03023). Overspecification results in the depletion of valuable high friction surfacing materials and the omission of other sources that could provide sufficient frictional properties over large areas of the road network.

One of the goals of this study is to develop a framework for the characterisation of stone aggregates using a combination of traditional mechanical testing and modern, state-of-the-art methodologies.

2 METHODOLOGIES

Each of the quarry sources were examined and bulk samples taken from the working face at the time of visit. In addition, the working face was mapped to understand the current distribution of the different rock lithologies. In total, several tonnes of aggregate (dust, 6 mm, 10 mm, and 14 mm) were extracted each year, over a 3-year period, for petrographic examination and mechanical testing (PSV, AAV, Micro Deval, FAP). The products collected represent general aggregates rather than specific stockpiles designated as high-performance aggregate. This was undertaken to ensure that the complete range of lithologies available at the source were recovered.

For the current study, only six of the quarries were selected: 1, 2, 4, 5, 6 and 8. These are all quartzo-feldspathic rock sources ranging from sedimentary to meta-sedimentary and meta-igneous. The sources can be highly heterogenous, especially those described as greywacke, with considerable variation in grain-size.

2.1 *Mechanical testing*

Four PSV test specimens were prepared from each quarry source, according to IS EN 1097-8: 2020 and measurements taken at specific stages during the artificial wear process at 0, 3, 4.5, 6, 7.5 and 9 hours. Standard PSV measurements are normally taken after 6 hours of wear.

To test the overall durability of the aggregates both Aggregate Abrasion Value (AAV) and Micro-Deval (M_{DE}) tests were conducted on the selected aggregates.

2.2 *Digital metrology (microtexture measurements)*

The stone particles on each PSV test specimen were numbered and described petrographically. Stacked images (up to 40 per particle) were acquired using a digital microscope to generate a 3D topographical representation of the surfaces. These topographic models can be examined in metrological software and relevant parameters extracted to characterise the microtexture.

The methodology, developed during this project, utilises a 3×3 mm area to produce a series of standard measurements. This areal measurement can be used on aggregate particles ≥ 6 mm and has proven to provide accurate measurements on lithology grain-sizes ≥ 2 mm.

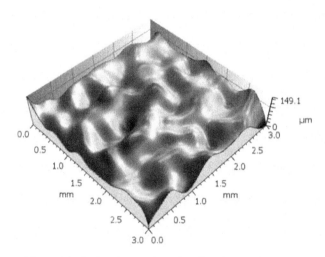

Figure 2.   Topographic model of a 3×3 mm area taken from a stone particle, generated by Mountains Map v8.2.

Measurements are taken on selected particles at the same stages of artificial wear described for the measurement of PSV (0, 3, 4.5, 6, 7.5 and 9 hours).

Specific areal parameters, defined by ISO 25178, have been chosen to characterise the microtexture of stone particles based on detailed analysis of numerous datasets. These are:

• Sq – Root Mean Square Height – This gives the overall roughness of the measured surface and considered the strongest measurement of microtexture.

- Sku – (kurtosis) Measures the sharpness of the roughness profile.
- Ssk – (skewness) Degree of bias in asperities. Ssk = 0 means height distribution is symmetrical around plane.
- Sdq – Root mean square gradient across defined surface - a level surface = 0. Increases with number of components with a strong gradient.
- Sdr – Developed interfacial area ratio. The percentage of the area's additional surface contributed by the texture over the planar surface (Sdr = 0). The ratio of Sdq to Sdr for various rough surfaces, including stone aggregates, should fall on a curvilinear trend.
- Vmp – This measures the amount of material making up the topmost points of asperities (peaks) on a given surface. If those asperities are the resistant minerals present at the tyre/stone aggregate interface, then this parameter gives us a clear indication of how much material is present at that contact and may have a bearing on adhesion and hysteresis. As this parameter generates very small areas, the data is presented x100.
- Spd – Density of peaks. The number of peaks per unit area. The density of peaks is another parameter to determine the amount of material potentially present at the tyre/stone aggregate interface and therefore can be used when investigating adhesion and hysteresis.

2.3 Raman spectroscopy

Hyperspectral Raman maps of selected stone particles are generated from both the PSV test specimens and rock samples taken directly from quarry faces. Raman spectroscopy uses monochromatic light (laser) to interact with the molecular bonds of a material. Different materials provide unique spectra which can be correlated with a known spectrum to aid identification. Both 2D and 3D maps can be generated of a stone particle surface. This can reveal detailed mineralogy as well as the ability of specific minerals to form asperities.

Hyperspectral maps are generated with a Renishaw inVia Qontor Confocal Raman Microscope using LiveTrack realtime focus modulation and can be displayed as simple topographic projections, 2D maps or combined 3D mineralogical maps.

3. RESULTS

3.1 Mechanical testing

The results of the mechanical testing provided a range of values for PSV, AAV, and Micro-Deval (Table 1). A control stone (PSV 50) was used to obtain corrected values. Many of the results were close to the predicted maximum PSV reported for each source. It is interesting to note that the three greywacke sources provided a PSV of 59, despite differences in AAV and M_{DE}.

Table 1. Results of the mechanical testing of aggregates from selected quarries.

Quarry Number	PSV (6 hr)	Aggregate Abrasion Value (AAV)	Micro-Deval (M_{DE})	Rock-Type
1	59	7.3	27	Greywacke
2	59	4.7	16	Greywacke
4	53	5	13	Andesite
5	59	6	20	Greywacke
6	72	11.2	42	Sandstone
8	62	6.3	23	Sandstone

3.2   *Digital metrology*

Digital metrology results reveal the close association between Sq (microtexture) and grain-size in sedimentary and metasedimentary rock types (Figure 3). Except for specific aggregates containing strong microstructural fabrics (Unitt & Meere 2018), the size of individual mineral grains corresponds with the height of asperities i.e., a fine-grained sandstone will provide a lower Sq than a medium-grained sandstone.

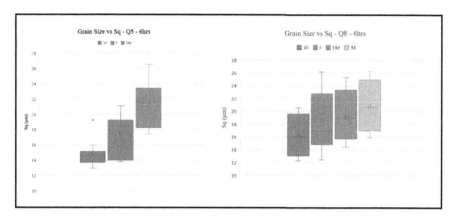

Figure 3.   Box whisker plots displaying variation in Sq in relation to grain-size in greywacke (Q5) and a sandstone (Q8) taken following 6 hours artificial polishing. VF – very fine sand, F – fine sand, FM – fine to medium sand, M – medium sand.

A comparison of PSV and Sq during the accelerated polishing experiment (hours 4.5 to 9) demonstrates that they do not necessarily follow the same trajectory. Both PSV and Sq display an approximate maximum value at 6 hours with PSV remaining relatively constant in 4 of the samples from 6 to 9 hours but Sq values drop in all 6 samples. Interestingly the PSV does drop during hours 6 to 9 for samples from sources displaying the lowest AAV and M_{DE} values (Q2 and Q4).

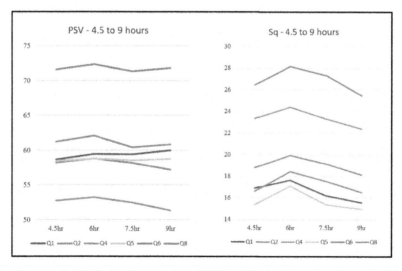

Figure 4.   Line graphs displaying the evolution of PSV and Sq (μm) over time for each of the six quarry sources.

When examining the correlation between PSV and overall microtexture (Sq) there is a relatively clear correlation amongst the sedimentary/meta-sedimentary rock types but a large disparity for the igneous source Q4 which has a large Sq value at 6 hr (24.42 μm) but the lowest PSV value of 53.

This indicates that there are additional aspects of microtexture that may influence the frictional characteristics. One such characteristic can be measured as the amount of material occurring at the tyre-aggregate interface. This can be assessed using the parameters Spd (peak density) and Vmp (volume of peak material over a specified height).

The strongest correlation in the data sets from this study is between Sq and Vmp (Figure 5). This suggests that the average grain-size of any given source has a strong influence on the amount of material present at the tyre aggregate interface.

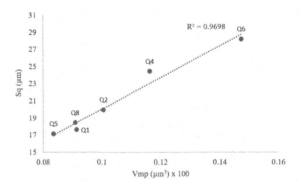

Figure 5. Average measurements for peak material volume (Vmp) versus Sq, taken after 6 hours of artificial polishing.

The skewness (Ssk) measurement provides an insight into the type of microtexture present. A positive Ssk can be equated to a surface topography represented by numerous peaks surrounded by a relatively flat plain. A negative Ssk represents a relatively flat plain punctuated by enclosed valleys (pits). Figure 6 displays the variation in Ssk across the different quarries with values ranging from highly positive to slightly negative. Q4 (metaigneous) and Q6 (sedimentary) represent sources with the highest Sq measurements.

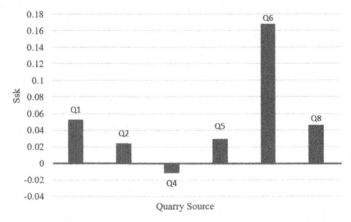

Figure 6. Average skewness data from each of the quarry sources, measured after 6 hours artificial wear.

3.3   *Petrography/Raman spectroscopy*

The quarry sources were chosen to represent a broad cross-section of the most widely used high PSV roadstone aggregates in the Republic of Ireland. The predominant mineral components of these rocks are quartz and feldspar. All examples display significant hardness differences between grains/crystals and matrix. This differential hardness is essential for the development of a strong microtexture. However, in the sedimentary and meta-sedimentary lithologies quartz and feldspar occur as individual grains surrounded by weaker, finer-grained matrix minerals whereas in the meta-igneous rock, recrystallised quartz and feldspar form the finer-grained matrix surrounding large, altered feldspar crystals and amygdaloidal cavities. These two scenarios result in either a positive (hard grains, soft matrix) or negative (hard matrix, softer crystals) microtexture (Figures 7 and 8).

It should also be noted that the peaks generating a positive microtexture are predominantly formed from single mineral grains (Figure 8a), whereas the peaks generating a negative microtexture often comprise fine-grained aggregations.

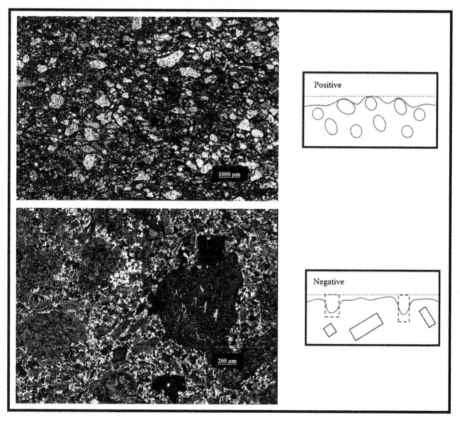

Figure 7.   Hyperspectral Raman maps of samples from Q2 (top) and Q4 (bottom) displaying the distribution of minerals in each lithology and a sketch of how this will generate either a positive or negative microtexture. Dominant minerals present are coloured white – quartz, shades of pink – feldspar, dark blue – anatase, red – rutile, yellow – titanite, orange – muscovite.

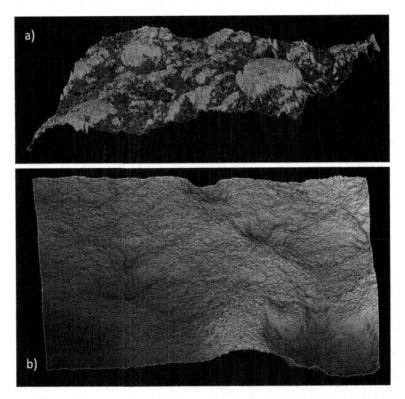

Figure 8. a) 3D hyperspectral Raman map of greywacke from Q2 displaying large peaks composed from quartz (grey) and b) laser topographic map of andesite from Q4 displaying prominent depressions characteristic of a negative microtexture. Both measured after 6 hours artificial wear.

4 DISCUSSION

The results of this study describe how the petrographic features of different aggregate sources provide fundamental controls on their frictional properties.

The initial property for any lithology utilised as a road surfacing material is the presence of both hard and soft minerals to produce differential wear and therefore a microtexture. Monomineralic sources, regardless of their resistance to erosion will polish under trafficking e.g., marble or quartzite.

In the Republic of Ireland, the dominant high friction surfacing products are quarried from rocks containing the minerals quartz and feldspar (quartzo-feldspathic). These range from sedimentary to meta-sedimentary through to meta-igneous rocks.

Greywacke has been described as the ideal lithology (sandy sediment with < 5% clay minerals) for generating and maintaining a good microtexture and therefore in-service skid resistance. However, greywacke deposits, by their nature, are heterogeneous, with large variations in grain-size, from mud-size ($<3.9\ \mu$m) to very coarse sand (>1 mm). This study recognises that lithologies with grain-size $< 125\ \mu$m (fine sand) do not produce a significant microtexture and therefore any lithologies classified as very fine sandstone or below (silt-stone, mudstone) are often considered deleterious in any aggregate used as high-friction surfacing and should form a small percentage of any bulk aggregate. There are, however, exceptions whereby pervasive microstructural fabrics can yield a sustainable microtexture in finer grained lithologies (Unitt & Meere 2018). This is demonstrated in Figure 3 where there is gradual increase in microtexture (Sq) from very fine sandstone to medium sandstone in rocks from Q8 which contain a strong microstructural fabric but there is a sharp transition in the scale of microtexture in rocks from Q5 that contain little or no internal fabric.

As Sq (microtexture) is controlled predominantly by grain-size it is unsurprising that there is a strong correlation with Vmp which represents the volume of peak material at the tyre/stone interface (Figure 5). What may be less understood is the potential difference if the peak is composed from a single mineral or an aggregation of minerals. Although quartz has reportedly poor adhesion properties it may prove to provide better adhesion than fine-grained aggregations of minerals.

One of the major findings of this project is the recognition that aggregate microtexture can either be described as positive or negative. The only source to produce a consistently negative texture in this study is Q3. Despite having the second highest Sq measurements this source provided the lowest PSV readings. Negative microtexture has been described in industrial publications as a plateau honed surface (Podulka et al. 2014) and has been recognised to reduce friction between two mechanical parts with the addition of a lubricant. It is considered that a similar mechanism could explain the lower frictional properties displayed by negatively textured aggregate in the presence of rainwater.

One of the questions that this study hoped to address was the hypothesis that road aggregate microtexture is periodically refreshed by in-service conditions. This would involve the 'plucking' of individual minerals from the stone surface to allow a further stage of differential wear to begin. The data obtained from the extended APM experiment suggests that this process is not prevalent in typical Irish aggregate lithologies with only one source (Q6) displaying the loss of mineral grains during the extended period of polishing (6 to 9 hours). All the lithologies appear to continue to polish during extended polishing, represented in the data by the gradual decrease in Sq over time although it was noted in the Raman results that some large grains had been 'plucked' from Q6. This contrasts with the PSV results which suggest very little change during extended polishing. A study is underway to see if similar results can be obtained using extended FAP (Friction After Polishing) cycles, which utilises grinding mediums more comparable to those found in typical road conditions. The reason why polishing prevails over plucking in most of the lithologies can be identified in the AAV tests. The relatively low AAV (5 to 7.3) and M_{DE} (13 to 27) of five of the six sources represents the degree of crystallisation present. The lithologies are all Lower Palaeozoic in age and have been subjected to various levels of heat and pressure (metamorphism). The lowest degree of recrystallisation can be found in Q6 and this source is known for its high frictional properties but also a lower ability to withstand wear over time.

5 SUMMARY

This study demonstrates how modern techniques such as digital metrology and 3D Raman spectroscopy can be combined with more established mechanical testing to provide detailed insights into the specific properties that control friction supply in stone aggregates.

At present, the most widely utilised high friction stone aggregates in the Republic of Ireland are derived from relatively crystalline Lower Palaeozoic sources. As a result, most are highly durable but appear to show a reduction in microtexture over extended periods of wear. Research is underway examining road core from a range of in-service sites across the Irish Network to determine whether there is a correlation between artificial wear and in-situ trafficking.

It is expected that the methodologies generated by this study and others will be incorporated into new testing requirements for stone aggregate sources. This will enable road engineers to have greater confidence in materials, reducing over specification, and allowing the design of sustainable surfaces by matching frictional demand with available local resources.

REFERENCES

Dokic O., Matovic V., Eric S. and Saric K., 2015. Influence of Engineering Properties on Polished Stone Value (PSV): A Case Study on Basic Igneous Rocks from Serbia. Construction and Building Materials, 101, 1088–1096.

DN-PAV-03023 (Oct 2020). Surfacing Materials for New and Maintenance Construction for Use in Ireland. *TII Publications.*

Dunford A.M., Parry A.R., Shipway P.H. and Viner H.E., 2012. Three-dimensional Characterisation of Surface Texture for Road Stones Undergoing Simulated Traffic Wear. Wear, 292-293, 188–196.

International Standards Institution. *Tests for Mechanical and Physical Properties of Aggregate – Part 8: Determination of the Polished Stone Value.* IS EN 1097-8, 2020.

International Standards Institution. IS EN ISO 25178-2 Geometric Product Specifications (GPS) – Surface Texture: Areal Terms, Definitions, and Surface Texture Parameters, 2012.

Kane M., Artamendi I. and Scarpas T., 2013. Long-term Skid Resistance of Asphalt Surfacings: Correlation Between Wehner-Schulze Friction Values and the Mineralogical Composition of the Aggregates. Wear, 303, 235–243.

Krutilová K. and Přikryl R. 2017. Relationship Between Polished Stone Value (PSV) and Nordic Abrasion Value (A N) of Volcanic Rocks. Bulletin of Engineering Geology and the Environment. 76 (1), 85–99.

Perry M.J., 2014. Role of Aggregate Petrography in Micro-texture Retention of Greywacke Surfacing Aggregate. Road Materials and Pavement Design, 15:4

Podulka P., Dobrzański P., Pawlus P. and Lenart A. 2014. The Effect of Reference Plane on Values of Areal Surface Topography Parameters from Cylindrical Elements. Metrology and Measurement Systems, 21 (2), 247–256.

Shabani S., Ahmadinejad M. and Ameri M., 2013. Developing a Model for Estimation of Polished Stone Value (PSV) of Road Surface Aggregates Based on Petrographic Parameters. International Journal of Pavement Engineering, 14:3, 242–255.

Unitt R.P. and Meere P.A. 2018. Mineralogical and Microstructural Controls on the Surface Texture of High Polished Stone Value Aggregates. Wear 408, 13–21.

Wang D., Chen X., Xie X., Stanjek H., Oeser M. and Steinauer B., 2015. A Study of the Laboratory Polishing Behaviour of Granite as Road Surfacing Aggregate. Construction and Building Materials, 89, 25–35.

Wang D., Liu P., Oeser M., Stanjek H. and Kollmann J. 2019. Multi-scale Study of the Polishing Behaviour of Quartz and Feldspar on Road Surfacing Aggregate. International Journal of Pavement Engineering, 20 (1), 79–88.

Woodside A.R., Lyle P., Woodward W.D.H. and Perry M.J., 1998. Possible Problems with High PSV Aggregate of the Gritstone Trade Group. In: Latham, J.-P. (ed.) *1998.* Advances in Aggregates and Armourstone Evaluation. London, Geological Society, Engineering Geology Special Publications, 13, 159–167.

Woodward W.D.H., Woodside A.R. and Jellie J.H., 2005. Higher PSV and Other Aggregate Properties. *First International Surface Friction Conference*, Christchurch, New Zealand, Transit New Zealand, Wellington, NZ 12p.

Evaluation of the application of tailing from magnesite extraction in pavement by adhesion properties

Mateus Brito* & Suelly Barroso
Department of Transportation Engineering, Federal University of Ceara, Fortaleza, Brazil

Lilian Gondim
Science and Technology Center, Federal University of Cariri, Juazeiro do Norte, Brazil

ABSTRACT: Mining activity is one of the biggest generators of environmental liabilities today. Brazil is one of the largest producers of magnesite in the world, which makes it also one of the largest generators of material tailings. The demand for more sustainable materials for paving that reduce the environmental impact caused by this activity has been growing in recent decades at the same time as the tendency to use recycled materials giving them a new destination is increasing. In this sense, this work aimed to analyze the compatibility between the magnesite tailings and asphalt binders used in chip seal layers, through the analysis of the adhesion between them. The materials studied in this research were the tailings of magnesite extraction, an aggregate of granitic origin (control material), traditional asphalt emulsion and polymer-modified asphalt emulsion. The adhesive test was carried out by means of an adaptation of the Binder Bond Strength (BBS) test, using the mortar pullout equipment, in order to measure the pullout strength and verify the failure conditions in the bond between the materials. The magnesite tailings showed similar behavior when compared to the granitic rock, presenting even better resistance to moisture damage, indicating a good compatibility in terms of adhesion between the mineral substrate and the asphalt binder, enhancing its use in chip seals and other paving solutions.

Keywords: Recycled Materials, Chip Seal, Paving, BBS Test, Adhesivity.

1 CONTEXTUALIZATION

Brazil is the one of largest magnesium compounds productors in the world, only second to China (USGS 2022). The magnesium compounds are extracted by mining of magnesite, which generates a substantial amount of waste, mainly consisting of particles of smaller size that cannot be used in the calcination due the risk of blockage of the ovens. These tailings are stored outdoors in large piles generating the possibility of soil and water resources contamination.

The materials traditionally used in paving are from non-renewable sources, such as the petroleum asphalt emulsion and the stone aggregates, as the gravel of granitic origin. In order to minimize environmental impacts generated by the disposal of waste and to obtain materials for construction in places where adequate natural sources materials are scarce, several studies have been developed, some of which aim at replacing conventional aggregates with mining waste (Castelo Branco 2004; Loiola 2009; Parente *et al.* 2003; Pereira 2010, 2013; Santos Neto 2007).

---

*Corresponding Author: mateus.brito@det.ufc.br

DOI: 10.1201/9781003429258-37

The chip seal is a type of pavement technique commonly used in Brazil due to its lower cost compared to asphalt mixtures, in addition to the ease of execution. It is a surface course widely used in low traffic roads consisting of the successive application of asphalt emulsion and aggregates layers (Larsen 1985). One of the main distresses of this type of surface course is the loss of aggregates (Adams 2014; Lee & Kim 2008, 2009, 2010; Loiola 2009; Pereira 2010, 2013).

One of the possible causes for the loss of aggregates may be the incompatibility between the aggregate and the asphalt emulsion, generating a low adhesivity that can lead to detachment of the aggregate from the pavement when exposed to traffic abrasion. Thus, the objective of this work is to evaluate the compatibility of the tailings of magnesite mining in seal chip by analyzing the compatibility between this alternative aggregate and two asphalt emulsions, comparing its performance with the traditionally used granite aggregate.

2 MATERIALS AND METHODS

Gondim (2017) and Bezerra (2018) methods were used to perform this test with the use of mortar pullout equipment for the BBS test. The test was performed according to AASHTO T 361 (2016), except for the equipment used, which in this case was the mortar pullout equipment. Two types of rapid setting cationic emulsions were used, a conventional (RR-2C) and a polymer modified one (RR2C-E), in addition to two aggregate substrates, the granitic and the magnesite.

The test was performed using the residue on rock (ROR) method and the residues were obtained by subjecting the emulsions to the oven at a temperature of 60°C for a period of 24 hours, according to the Figure 1.

The rocks were cut and sanded, then cleaned with a pressure washer and brought to the oven at a temperature of 80°C for drying for 24 hours. For each combination of aggregate and emulsion residue, 6 specimens were prepared, 3 of them were immersed in water for 24 hours (wet conditioning) and the other 3 were dry conditioned. Figure 2 illustrates the combinations and samples prepared for testing.

The temperature of 80°C was maintained for the heating of the aggregates and metal stubs for the test. The residue of the asphalt emulsions was molded into small spheres of 0.45 g each, placed in the heated stub and then the stub was pressed onto the substrate for 10 seconds. The samples were conditioned in two ways, dry and wet. In dry conditioning the samples were left at room temperature for 24 hours after applying the residue, while in wet

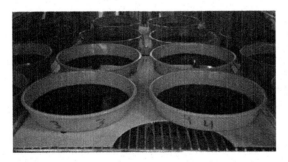

(a) Emulsions in oven   (b) Residue after drying
in an oven

Figure 1. Obtaining the residue in an oven. (a) Emulsions in oven (b) Residue after drying in an oven.

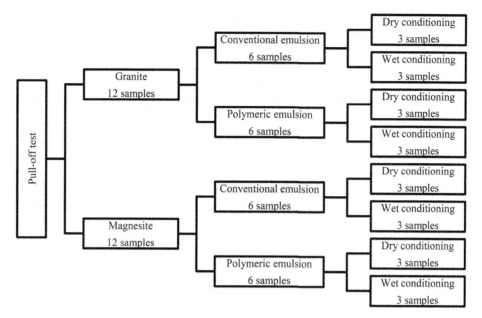

Figure 2.   Experimental plan of the pull-off test.

conditioning the samples were left for 1 h at room temperature for the consolidation of the bond of the binder with the aggregate and the stub and then submerged in water at 40°C for 24 h. The wet samples were then taken from the bath and left for 1 hour at room temperature to stabilize before the pull-off test. Figure 3 the samples in its two conditioning states for the test.

(a) Dry conditioning   (b) Wet conditioning

Figure 3.   Conditioning of samples for pull-off test. (a) Dry conditioning (b) Wet conditioning.

The test is carried out by placing the specific screw for the equipment in the stub that is attached to the rock with the residual binder. The equipment and the test-ready sample are illustrated in the Figure 4.

The equipment is then fitted to the stub, which is pulled out by the rotation of the crank, however with the limiting of 25 turns per test. The test is interrupted when the stub is detached from the substrate or the 25 turns are reached, whichever is to occur first. At the end of the test, the maximum force reached in kilograms-force is read on the display. The calculation of the pull-off tensile strength (POTS) is obtained by dividing the force by the

(a) Mortar pull-out equipment   (b) Sample with stub for the test

Figure 4.   Test equipment and sample. (a) Mortar pull-out equipment (b) Sample with stub for the test.

contact area between the stub, the residual binder and the substrate, according to Equation 1.

$$POTS = \frac{F}{A} \tag{1}$$

Where:
     POTS = pull-off tensile strenght, in MPa;
     F = pull-out force of the stub, in N;
     A = contact area between the stub, the aggregate and the residual binder, in mm^2.
     It is noteworthy that POTS is the expression used here for the pull-off, however it does not translate into the POTS of the AASHTO T 361 standard, due to the difference in the equipment used. After the stub was removed from the rock, a visual analysis was performed to determine the cause of the failure, whether it was adhesive, cohesive or a combination of both, according to Figure 5.

 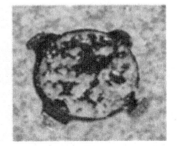

(a) Cohesive failure   (b) Adhesive failure

Figure 5.   Failure mechanisms in the BBS test (AASHTO, 2016). (a) Cohesive failure (b) Adhesive failure.

A ratio was also calculated between POTS in wet condition and dry condition according to Equation 2, which should present values greater than 70% (AGUIAR-MOYA 2015).

$$RPOTS(\%) = \frac{POTS_{wet}}{POTS_{dry}} \qquad (2)$$

Where:
POTS$_{wet}$ = POTS obtained in the wet test condition (MPa), and
POTS$_{dry}$ = POTS obtained in dry test condition (Mpa).

3 RESULTS AND DISCUSSIONS

Figure 6 shows the mean of the POTS results comparing the aggregates while the binders vary. It was verified that the granite obtained an average POTS very similar to the alternative aggregate in all combinations and conditionings used. The POTS results of this test may not be the maximum of each combination, since in all cases the equipment reached 25 crank turns before the total rupture, causing the sample test to be finished with only a partial rupture.

In general, the values found by Gondim (2017) and Bezerra (2018) were higher than those found in this research. It is noteworthy that both studies were carried out with the neat asphalt binder, which alters the application temperatures of the binder, aggregate and stub. In addition, the samples of Gondim (2017) and Bezerra (2018) suffered total disruption before reaching the 25 crank turns.

On the present paper, the type of binder did not cause a significant change in POTS values, causing them to remain at the same level in each test combination and conditioning, as shown in Figure 6. Possibly this lack of variation may have occurred due to equipment limitation, since at the end of the test it was found that the partial rupture of samples with the polymeric emulsion was much lower than the samples with conventional emulsion. The failure mechanism of the samples was the cohesive type, which means that the rupture happened in the binder, as shown in Figure 7. The absence of adhesive failure even in the wet conditioning could be explained also by the equipment limitation.

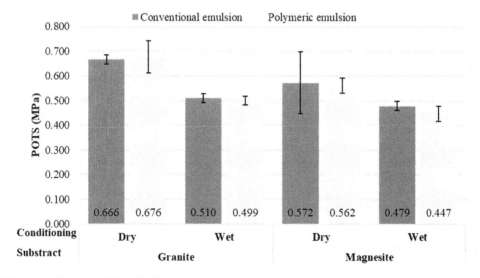

Figure 6.   Results of the pull-off test.

Figure 7.   Cohesive failure in a sample.

The RPOTS were then calculated according to Equation 2. The results are presented in Table 1. It was verified that all combinations presented RPOTS higher than 70%, indicated as minimum limit in Aguiar-Moya *et al.* (2015), which indicates that aggregates do not suffer excessive influences on their adhesivity when exposed to water. It is also observed that the highest values were found for magnesite, in both combinations with the two binders used, indicating that this aggregate has good resistance to moisture damage.

Table 1.   RPOTS for the analysed combinations.

Emulsion	Substract	RPOTS (%)
Conventional	Granite	76.6
	Magnesite	83.6
Polymeric	Granite	73.8
	Magnesite	79.6

4   CONCLUSIONS

Regarding the adhesivity, the type of emulsion did not cause a significant change in the pull-off stress values in any combination and conditioning for the test. It is possible that this lack of variation is due to the limitation of the equipment, since the samples with polymeric emulsion presented less partial disruption. All combinations obtained a ratio between the pull-off stress in wet condition and that in dry condition greater than 70%, as recommended by the literature, showing the low susceptibility to detachment of these samples to the presence of water.

It is important to emphasize that the adaptation of the BBS test using mortar pullout was not satisfactory in this research, contrary to what was observed by Gondim (2017) and Bezerra (2018), because it was not able to generate total sample disruption. Nevertheless, it was possible to verify that the basic character of magnesite was manifested in favor of improving the adhesivity, presenting superior RPOTS.

The results of the magnesite material were similar to the granite, which indicates the possibility of the use in chip seals. More comprehensive experimental testing should be performed, such as the aggregate loss test, to verify the performance of the alternative material. A broader characterization of the particles is also necessary, as the form of the aggregate impact the performance in the field.

ACKNOWLEDGMENTS

The authors would like to express gratitude to the Ceará Foundation for Support to Scientific and Technological Development (FUNCAP) and the Coordination for the Improvement of Higher Education Personnel (CAPES) for financing this research.

REFERENCES

AASHTO. Standard Method of Test for Determining Asphalt Binder Bond Strength by Means of the Binder Bond Strength (BBS) Test. *AASHTO T 361*, Washington, DC. 2016.

ADAMS J.M. (2014) *Development of a Performance-Based Mix Design and Performance-Related Specification for Chip Seal Surface Treatments. Doctoral's dissertation, North Carolina State University, Raleigh.*

Aguiar-Moya J., Salazar-Delgado J., Baldi-Sevilla A., Leiva-Villacorta F. and Loria-Salazar L. (2015) Effect of Aging on Adhesion Properties of Asphalt Mixtures with the Use of Bitumen Bond Strength and Surface Energy Measurement Tests. Transportation Research Record, v. 2505, p. 57–65.

Bezerra S.L.O. (2018) Análise Comparativa Entre os Ensaios BBS e Arrancamento de Argamassa Utilizados Avaliação da Adesividade e do Dano por Umidade na Interface Ligante Agregado. Undergraduate Thesis. Universidade Federal do Cariri. Juazeiro do Norte.

Castelo Branco V.T.F. (2004) *Caracterização de Misturas Asfálticas Com o Uso de Escória de Aciaria Como Agregado. Master's thesis.* COPPE, Universidade Federal do Rio de Janeiro. Rio de Janeiro.

Gondim L.M. (2017) Investigação Sobre a Formulação de um bio-ligante a Base da Seiva da Euphorbia Tirucalli Para Emprego em Pavimentação. Doctoral's Dissertation. Departamento de Engenharia de Transportes, Universidade Federal do Ceará. Fortaleza.

Larsen J. (1985). Tratamento superficial na conservação e construção de rodovias. ABEDA. Rio de Janeiro.

Lee J.S. and Kim Y.R (2008). Understanding the Effects of Aggregate and Emulsion Application Rates on Performance of Asphalt Surface Treatments. Transportation Research Record. V. 2044, p. 71–78.

Lee J. S. and Kim Y. R. (2009). Performance-Based Uniformity Coefficient of Chip Seal Aggregate. Transportation Research Record, v. 2108(1), p. 53–60.

Lee J. and Kim Y.R. (2010). Optimal Distribution of Rolling Coverage in Multiple Chip Seals. Transportation Research Record, v. 2150(1), p. 70–78.

Loiola P.R.R. (2009) *Utilização de resíduos de construção e demolição para pavimentos urbanos da regîao metropolitana de Fortaleza.* Master's thesis. Departamento de Engenharia de Transportes, Universidade Federal do Ceará. Fortaleza.

Parente E.B., Boavista A.H. and Soares J.B. (2003) *Estudo do comportamento mecânico de misturas de solo e escória de aciaria para aplicação na construção rodoviária na região metropolitana de Fortaleza. Anais do XVII Congresso de Pesquisa e Ensino em Transportes, ANPET, Rio de Janeiro, v. 1, p. 215–222.*

Pereira S.L.O. (2010) Avaliação de Tratamentos Superficiais de Rodovias através de Análise de Laboratório. Undergraduate Thesis. Universidade Federal do Ceará. Fortaleza.

Pereira S.L.O. (2013) Avaliação dos Tratamentos Superficiais Simples, Duplo e Triplo de Rodovias Através do Emprego de Diferentes Agregados da Região Metropolitana de Fortaleza. Master's Thesis. Departamento de Engenharia de Transportes, Universidade Federal do Ceará. Fortaleza.

Santos Neto P.F. (2007) *Estudo do Uso de Escória de Aciaria em Camadas de Pavimentos na Região Metropolitana de Fortaleza. Undergraduate Thesis.* Universidade Federal do Ceará. Fortaleza.

USGS (2022). *Mineral Commodity Summaries 2022. Mineral Commodity Summaries.* U.S. Geological Survey, https://pubs.er.usgs.g gov/publication/mcs2022.

Roads and Airports Pavement Surface Characteristics – Crispino & Toraldo (Eds)
© 2023 The Author(s), ISBN 978-1-032-55149-4

Raveling algorithm for PA and SMA on 3D data

W. van Aalst* & P. Piscaer
TNO, The Hague, The Netherlands

B. Vreugdenhil & F. Bouman
Rijkswaterstaat, Dutch Ministry of Infrastructure and Water Management, Utrecht, The Netherlands

M. van Antwerpen, G. van Antwerpen & J. Baan
TNO, The Hague, The Netherlands

ABSTRACT: A method to quantify raveling on Porous Asphalt (PA) and Stone Mastic Asphalt (SMA) based on high resolution 3D laser triangulation data is develop. The method consists of two steps: correcting the 3D data and raveling computation. Correcting the 3D data is firstly done in two steps. First, the data is cleaned by removing outliers and filling in missing values. The cleaned data is then flattened to remove influences like motion of the measurement system and deformations like rutting. The flattening is done by removing the background estimated by a 2D percentile filter. The size of the filter is slightly dependent on the used 3D measurement setup and resolution. Raveling quantification is performed by the so-called 'coin' algorithm. A virtual coin with a certain radius is used to estimate the surface area percentage of areas big enough to fit the coin at a certain depth. The percentage of raveling per square meter can be used as indicator for raveling. The radius and depth can be varied for various pavement types with different aggregate sizes and distributions. The developed method is successfully applied to 3D data from different triangulation systems: two lab setups and multiple LCMS systems at various resolutions and measurement accuracies. The method is in use since 2012 for yearly measurements on Dutch highway network and experimentally on other Dutch and European roads. Examples of use on various PA (16, 8 mm) and SMA (11, 16 mm) pavements will be shown.

Keywords: Raveling, PA, SMA.

1   INTRODUCTION

A coarse description of the way raveling is computed for the Highway network in the Netherlands has been published and presented before [van Aalst *et al.* 2015, 2016, 2021], this paper however describes the used algorithm in more detail and in such a way it should be possible for others to use this algorithm on similar 3D data.

1.1   *Dutch highway network*

As described in [van Aalst *et al.* 2016]: accurate and reliable information on road surface condition is essential for effective pavement management. The use of in-traffic measurement systems for this purpose is booming. Visual inspection performed at low speed is

---

*Corresponding Author: willem.vanaalst@tno.nl

DOI: 10.1201/9781003429258-38

however still often used on many roads in Europe. Visual inspections are time consuming, expensive, subjective and dangerous to conduct as more and more hard shoulders are being used during rush hours. To assess the road surface conditions with minimal impact on road users Rijkswaterstaat (RWS), part of the Dutch Ministry of Infrastructure and the Environment, decided to automate the survey processes of raveling and cracking. Together with the automated survey of skid resistance, rutting and unevenness this delivers the required input for pavement planning on all main roads in the Netherlands. Up-to-date condition information allows predictive maintenance, so that (small) maintenance activities can be planned and undertaken the moment they are needed and least impactful on traffic, also avoiding more expensive last-minute maintenance can be avoided.

As described in [van Aalst et al. 2016]: More than 90% of Dutch main roads are applied with porous asphalt. Raveling is the most common form of damage that occurs on these roads. Raveling is the process of separation of aggregate particles (rocks) from the road surface. Eventually, larger aggregate particles can get loose as shown in Figure 1. In such case the road has to be resurfaced. A road with a serious level of raveling reduces ride quality, increases noise and the risk of damage to vehicles.

Figure 1. Left: Severe raveling on porous asphalt [IALCCE16]. Right: RWS road survey vehicle with the developed road surface measurement system for raveling on porous asphalt [photo: TNO].

1.2 Raveling measurement systems

As described in [van Aalst et al. 2016]: since 2007 RWS and TNO are collaborating in several research & development projects to fully automate the survey processes of raveling, cracking and other types of road surface deterioration. This has led to a road surface measurement system (DOS system, Figure 1) and algorithms that automatically determine the road surface type, the amount of raveling, cracking and the estimated remaining service life of each 100 meter road section based on the damage assessment and intervention levels in the Netherlands.

1.3 3D data

The DOS system uses a LCMS-1 or LCMS-2 sensors [Laurant 2008] to capture 3D profiles of the road surface with a transversal resolution of 1 mm and a longitudinal resolution of typically 1, 2 or 5 mm. During feasibility studies also two prototype laser triangulation setup were used to acquire 3D road surface data, as shown in Figure 2. All these systems create transversal profiles which can be combined into larger 3D range or height images. In Figures 3 and 4 some examples of these 3D image data is shown.

Figure 2. Prototype laser triangulation setups used (in 2007 (left) and 2016 (right)) to capture 3D surface of small sections of pavement.

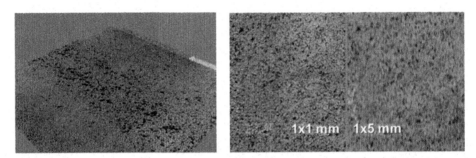

Figure 3. Left: 3D visualisation of a 2×2 meter PA-16 showing raveling (darker spots). Color indicates the height. Right: Two similar positioned measurements of PA-8, but measured at a different resolution (using a LCMS-1 system).

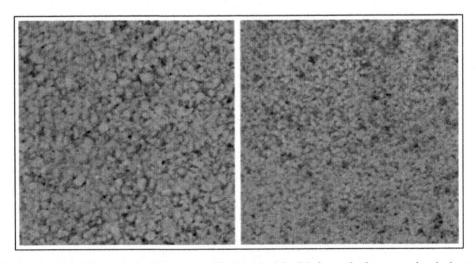

Figure 4. Two different types of pavement, left: PA-16, right: PA-8 mm, both measured at 1×1 mm resolution using an LCMS-1 system.

2 RAVELING ALGORITHM

A method to quantify raveling on Porous Asphalt (PA) and Stone Mastic Asphalt (SMA) based on high resolution 3D laser triangulation data is develop. The method consists of two steps:

1) Correcting the 3D data
2) Raveling quantification

These two steps are described in the next sections and shown in Figure 5.

Figure 5.   Example of the raveling algorithm on small section of PA-16 (LCMS-1, 1×5 mm). The dark areas in the flattened and cleaned image are deeper areas in the input image. The white areas in the output raveling image are areas indicating raveling.

2.1 *Correcting the 3D data*

Correcting the 3D data consist of the following consecutive steps:

- Fill missing data points
- Flattening
- Removing outliers
- Zero levelling
- Resampling

The above steps are described in the following paragraphs and should lead to a flattened, cleaned and uniform image of the 3D data.

2.1.1 *Fill missing data points*
In 3D laser triangulation it sometimes occurs that data points are missing, for example deep voids or shiny surface prevents sufficient laser light returning to the camera. Missing data point are filled in using a nearest neighbour approach. The performance of this approach does of course depend on the total number of points missing and is only accurate when only a few points are missing. If locally many data points are missing, for example when measuring wet or moist pavement, the 3D data is considered invalid.

2.1.2 *Flattening*
The measured 3D data consists of high frequent and low frequent structures. The profile must be corrected and flattened for some low frequent structures like:

- Rutting
- Crossfall
- Movements of the measuring profile/system

The flattening is performed by means of estimating the low frequency structures by a percentile filter and subtracting this 'background' from the data. The size of the filter is slightly dependent on the used 3D measurement setup and resolution. By default, a filter of 150 mm long and 90 mm wide is used on Dutch highway LCMS data. For computational reasons this background is computed on a lower resolution such that the percentile filter has a size of 30×30 pixels.

2.1.3 *Removing outliers*
All points which deviate more than x times the standard deviation (default: x = 5) of the flattened data are considered outliers and are substituted by the nearest value. This step removes possible spikes in the 3D data.

2.1.4 *Zero levelling*
All data should be oriented in such a way that lower values represent deeper points. First, the sign of the LCMS data is changed, followed by an artificial zero level is defined as the 85 percentile point of each local section (typically a section of 1×1 or 1×2 meters), see also Figure 7, which is subtracted from the height(image). This ensures that data captured using two different sensors (for example the right and left LCMS sensor) have no offset to each other and can be merged, more or less without a seem, as shown in Figure 6.

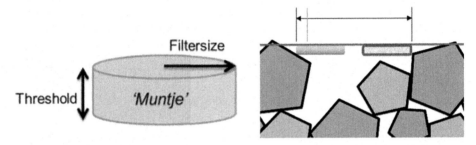

Figure 6.   Example of the input and output of the correction of a square meter of data. Note that the input image(left) consist of the data of two different 3D sensors, with a offset in height, and that the resulting image is continues.

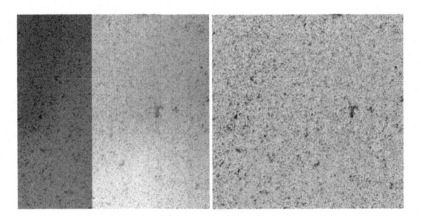

Figure 7.   Left: graphical representation of the 'Coin' (in Dutch: 'Muntje') used to locate areas big enough to be missing stones [van Aalst 2021], right: graphical representation of the virtual coin and raveled area on a cross-section of porous pavement. In red the zero-level (85 percentile) [source: J. Lucas, RWS].

2.1.5 *Resampling*

The height data is then resampled to a uniform grid to simplify the raveling computation. Default the resampling is performed to a 1×1 mm grid using bilinear interpolation.

2.1.6 *Remarks*

Other structures, like roadmarkings, give rise to artifacts when flattening the data. Preferably, the described methods should be applied on clean pavement that does not include any roadmarking.

2.2 RAVELING COMPUTATION – COIN ALGORITHM

Raveling computation is performed by the so-called 'coin' algorithm. A virtual coin with a certain radius is used to estimate the surface area percentage of areas big enough to fit the coin at a certain depth, as shown in Figure 7. The percentage of raveling per square meter can be used as indicator for raveling. Radius and depth can be varied for various pavement types with different aggregate sizes and distributions. This develop method is inspired on the 2D method described by [van Ooijen 2004].

The used implementation consists of two steps:

1) thresholding the flattened 3D image (using the 'Threshold' value), resulting in a binary image.
2) a morphological closing operation using a round structuring element (see Figure 7)

The remaining true pixels in the binary image correspond to raveling.

Note that the threshold (depth/height of the virtual coin) is relative to the zero level earlier defined. An example of the round structuring element corresponding to two coins of different filter sizes is shown in Figure 8.

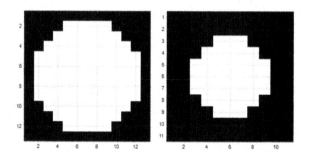

Figure 8.   Two examples of round structuring elements, coins with radius of 6 and 4 mm (on a 1×1 mm grid).

3 RESULTS

The algorithm has been extensively used and tested on the Dutch Highway network, consisting mainly of PA-16 and PA-8, as well as various PA and SMA pavements with different aggregate sizes. The default 'coin' setting for PA-16 are 6 mm radius and -2.0 mm depth and for PA-8 the default 'coin' parameters are 3 mm radius and -3.0 mm depth. This result section will show the performance of the discussed algorithm on different types of pavement.

3.1 Porous asphalt

In the following figures various examples of raveling on PA are shown. Figure 9 shows sections of PA-16 and the corresponding raveling for three different levels of raveling severity. Figure 10 shows an example for German OPA-6.8 with various 'coin' settings, highlighting the dependency of the computed raveling on the depth and radius of the coin.

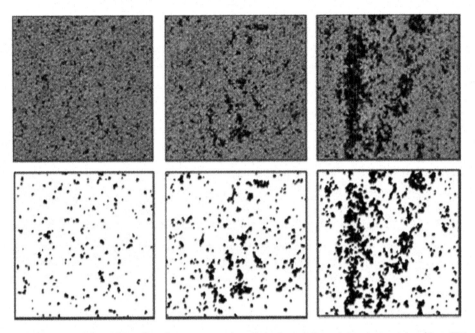

Figure 9.   3D height and raveling data on porous asphalt, from left to right an example of low (6% stone loss), moderate (12%) and high (22,5%) severity class [van Aalst *et al.* 2016]. Data is measured with a LCMS-1 system.

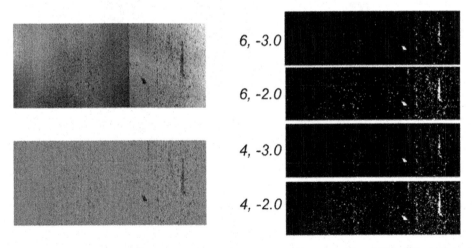

Figure 10.   Section of OPA-6.8 (left, top to bottom: raw range data, cleaned and flattened range data) with raveling image output using different parameters sets (radius, depth). Measured with a LCMS-1 system.

3.2 Stone mastic asphalt

In Figures 11 and 12 two examples of the algorithm's performance on SMA are shown.

Figure 11.   SMA-6, N209, The Netherlands, L: photo, M: flattened and cleaned data (1×1 mm ARMS prototype), R: raveling depicted as red contours.

Figure 12.   Raveling on SMA-11 (Sweden), L: raw LCMS-1 data (resolution: 1×5 mm), M: flattened and cleaned data, R: raveling image (coin settings, radius: 6.0 mm, depth: -2mm).

3.3 Correlation to visual inspections

Several experiments have been performed to correlate the output of the described algorithm to human visual inspection. Figure 13 shows an example for a patch of PA-16 from such an experiment. Pavement experts have judged several sections to obtain consensus on a set of parameters for PA-16 on 1x1mm data (radius: 5.9 mm and -3.125 mm depth).

In Figure 14, the raveling percentage given by the algorithm is compared to the raveling percentages given by several visual inspection experts. This experiments shows a clear

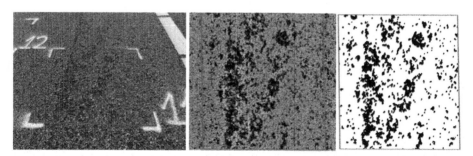

Figure 13.   example of a square meter of PA-16, with left: photo, middle: 3D range data and right: raveling image.

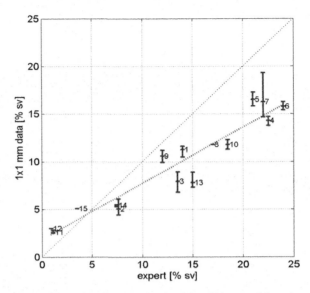

Figure 14.   raveling percentages given by the coin algorithm (6.0 mm, -2.0 mm) compared to average percentages estimated by a group of visual inspection experts. The bars indicate the minimum, maximum and mean percentages determined by the algorithm in three different measurements. The numbers indicate the different patches of pavement.

correlation between the two forms of raveling inspection. For most asphalt patches, the raveling percentage given by the algorithm falls within the minimum-maximum-range given by the inspectors. Moreover, the large variance between the different inspectors implies that visual inspection can be a complicated tasks.

3.4   *Repeatability and reproducibility*

To show the repeatability of the results, Figure 14 shows an example of repeated DOS measurements of the same section of highway. The results show only minor deviations when comparing the measurements, even when using two different LCMS-2 systems.

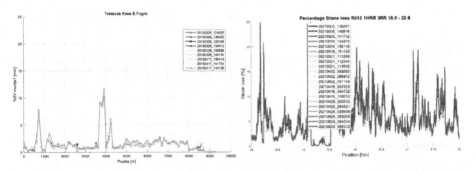

Figure 15.   Some examples of repeated raveling measurements on the Dutch highway network. Left: 9 measurements with two different LCMS-2 systems in 2019. Right: several measurements with the same DOS (LCMS-2) measurement vehicle in 2021, as presented at the [ERPUG 2021].

A quantitative analysis of the reproducibility is realised by computing the following measure:

$$\mu(\sigma) = \frac{1}{N}\sum_{m=1}^{N}\sigma_m$$

Where σ_m is the standard deviation of a set of measurements performed at the same location and N is the number of locations. On a test covering 2 km of asphalt measured for 14 times, the results obtained are given in Table 1.

Table 1. Repeatability results. Coin 1 has a radius of 6 mm and depth of -2 mm. For Coin 2 this is 3 mm and -3 mm respectively. The indications Left and Right correspond to the left and right wheel paths in the measured lanes. The first row shows the average raveling percentage, the second row the corresponding value of $\mu(\sigma)$.

	Coin 1 Left	Coin 1 Right	Coin 2 Left	Coin 2 Right
Avg. Raveling [%]	1.8%	1.1%	4.1%	3.1%
$\mu(\sigma)$	0.112	0.067	0.221	0.166

4 DISCUSSION

Aggregate sizes close to or smaller than the resolution (or accuracy) of the 3D data are more difficult to measure properly. For example, raveling on PA-5 or SMA-5 is still difficult to measure with the current accuracy of the highest resolution systems tested (1×1 mm LCMS-2).

The proposed algorithm estimates raveling based on a single measurement, regardless of the initial condition of the pavement. As raveling is defined as the loss of aggregates from the pavement over time, ideally a time series of the pavement, or the initial situation, should be considered to more accurately determine the real stone loss, raveling.

5 CONCLUSIONS

A method has been presented that is able to quantify raveling on Porous Asphalt (PA) and Stone Mastic Asphalt (SMA) based on high resolution 3D laser triangulation data. The method consist of two steps: 1) Correcting the 3D data and 2) Raveling computation. The first step consists of cleaning the data by removing outliers and filling in missing values, followed by a flattening step, which is done by removing the background estimated by a 2D percentile filter. For the second step, the so-called 'coin' algorithm was presented. A virtual coin with a certain radius and depth is used to estimate the raveling in terms of a surface area percentage. The radius and depth of the coin can be varied for various pavement types with different aggregate sizes and distributions.

The method was demonstrated on several examples of various PA and SMA pavement types using various types of 3D data with different resolutions. When compared with visual inspection, the method has been shown to be a suitable indicator of raveling. Furthermore, it is shown to be a robust measure of raveling due to its high repeatability and reproducibility results.

REFERENCES

Aalst W.L.C., Derksen G.B., Schackmann P.P.M., Bouman F.G.M. & Paffen P. 2015. Automated Raveling Inspection and Maintenance Planning on Porous Asphalt in the Netherlands, *International Symposium Non-Destructive Testing in Civil Engineering*. 15-17 September 2015. Berlin.

Aalst W.L.C., Derksen G.B., Schackmann P.P.M., Bouman F.G.M. & Paffen P. 2016, Automated Road Survey and Pavement Management on Porous Asphalt, *Fifth International Symposium on Life-Cycle Civil Engineering*, October 16-19, 2016 / The Netherlands, Delft.

Aalst W.L.C, 2021, *The Current Dutch Highway Pavement Monitoring System*, European Road Profiler User Group, 10-12 November 2021, Vienna.

Laurent J., Lefebvre D. and Samson E. 2008, Development of a New 3D Transverse Profiling System for the Automatic Measurement of Road Cracks. *Proceedings of the 6th Symposium on Pavement Surface Characteristics*, Portoroz, Slovenia.

Van Ooijen W., Van den Bol M. & Bouman F. 2004. High-speed Measurement of Ravelling on Porous Asphalt, *5th Symposium on Pavement Surface Characteristics*, Roads and Airports. 6-10 June. Toronto.

Analysis of road damage with computer-vision based drive-by system

Sergio Caccamo*, Rafael Mrden & Amit Dekel
Univrses, Stockholm, Sweden

ABSTRACT: Different measurement approaches can be used to assess the quality of the driveable road surface in cities. Mobilizing fleets of vehicles already operating in the city, such as taxis and buses, by retro-fitting sensors (such as a camera) offers a new paradigm in monitoring the road surface. This sensing approach often called drive-by has the potential to deliver a step change improvement in terms of data quality, update frequency and cost. This, in turn, enhances road safety and leads to reduced traffic accidents and loss of life. In this paper we study road damage detections and city coverage from data collected using a drive-by sensing approach using smartphones and computer-vision based perception. We compare harvested road damage data and coverage patterns for different fleet types in several European cities. We show how the data harvesting effort is influenced by the choice of vehicle and that different cities, despite their diverse geography and population, have similar road damage distributions. A key finding is that data collected on roads with lower speed limits show higher counts of road damage. We also show how a small number of vehicles can give valuable insights about the road quality in a short time scale. This work aims to illustrate how drive-by sensing approaches can offer road management authorities precious insights for maintaining their current road network.

Keywords: Road Damage, Drive-By Sensing, Road Coverage

1 INTRODUCTION

Assessing the status of the road network quality is a challenging and laborious task for road management authorities. It requires costly and time consuming data collection efforts as well as a great amount of tedious supervision work to process and identify damages in the pavement. The importance of understanding the quality of the road network is linked to the social and economical impact that road damages have on the population.

Taking as an example the potholes problem, studies have shown how crucial it is to identify and fix potholes as soon as possible (Eaton 1989). A survey of the American Automobile Association estimated that approximately 16 million vehicles were damaged between 2011 and 2016 in the United States due to potholes, with several billions of dollars in costs per year (AAA 2016, 2022). Considering the impact on human lives, the Indian ministry of road transportation assessed that in 2019, 2% of all road accidents and 2.1% of all road accident deaths were attributed to the presence of potholes (Government of India 2019). The total number of accidents caused by potholes in India was 4775 in 2019 and 3564 in 2020 (The Hindustan Times 2021). More generally, also other kinds of pavement distress may affect accident rates as was shown in a study on several U.S. states (Vinayakamurthy M *et al.* 2016, Baskare s. *et al.* 2016). Looking at the environmental

---

*Corresponding Author: sergio.caccamo@univrses.com

DOI: 10.1201/9781003429258-39

impact, studies have also shown that delaying road pavement maintenance increases gas emissions, with drivers forced to consume more fuel than otherwise needed (Wang H *et al.* 2019).

From a high level perspective, one can distinguish at least 5 phases connected to the abovementioned task: firstly, data about the road pavement (e.g. road damage, friction coefficients, roughness, etc.) is collected and geographically or temporarily ordered. Secondly, the data is processed and inspected. In a third phase, road management technicians or urban engineers plan and prioritize the work needed to either fix the roads or prevent issues connected to poorly designed urban structures. In the last two phases, the work is carried out by the construction workers and on a later stage verified through inspections or new reports. In most cases, the entire process can take anywhere from a few months to several years and be extremely costly. In the UK, for instance, it is estimated that $17 billion is needed just to fix potholes (The Economist 2016).

Data regarding road quality is collected and processed following different approaches. A traditional one is to let municipalities' employees or road maintenance authorities drive and survey predefined sections of road and collect data manually. This process is done by having the user identify road features like the status of pavement markings, presence of cracks, potholes, water accumulation, etc. Usually, the driver uses a tablet or smartphone to take pictures of the damage and to roughly annotate its geographic position. A different and very diffuse method is to let citizens report faulty road sections. Special service phone numbers or smartphone apps are made available for citizens to submit reports. Unfortunately this approach relies on the good will, unskilled and potentially biased judgment of non-experts. This makes the data collected inconsistent between damage of a similar nature, too sparse and often outdated. A third approach is the so-called drive-by sensing. In this case, several detection methods can be used depending on the kind of damages (Mohan P. *et al.* 2008; Taehyeong K *et al.* 2014; Wang. P *et al.* 2012). Among these, the most flexible is the use of modern computer vision algorithms to automatically detect road features and annotate them. In this case, vehicles are equipped with cameras and GPS antennas. The data is either processed onboard (edge computing), streamed to a server, or collected and post processed. Static cameras represent yet another widely diffuse method of road network assessment. These devices are however most commonly used to monitor traffic or weather conditions and are used to assess the road damage indirectly (e.g. assuming that more traffic leads to faster road wear). This is due to the fact that in order to cover a sufficient amount of road surface one would need a prohibitively high number of costly high resolution cameras capable of detecting road imperfections from high view points.

In this paper we do not focus on the pros and cons of drive-by sensing methodologies but rather on a common issue shared among them: optimize data coverage to capture the distribution of road damage. More specifically we will share some of the insights gathered from the deployment of computer-vision aided data harvesting devices mounted on different kinds of fleets in several European cities having different dimensions and topology. We show how different kinds of road damage are distributed in the cities and discuss the data coverage achieved by the different types of fleets involved.

2 DATA HARVESTING

Several studies have taken into consideration the importance of coverage in drive-by sensing paradigms (Agarwal et al. 2019; O'Keeffea et al. 2019). The drive-by approach enables a large coverage at a low cost, to maintain an up-to-date road-condition database of the city, and to monitor changes and trends in the city. In this study the system is deployed in different cities and is carried by different fleet types, and so different coverage patterns are expected to emerge.

Table 1. Summary of the fleet's activity and coverage. In the "# of km covered inside the city boundaries" column we indicate in parenthesis the percentage of the total city graph length. The graph length considers lanes and directions, so a two way street contributes for both directions.

City name	Fleet type	# of active days	# of km driven inside the city boundaries	# of km covered inside the city boundaries	City graph length km	Total # of km driven (also outside the city boundaries)	Total # of km covered (also outside the city boundaries)
Stockholm	Taxi	120	35,829	2,054 (59%)	3,470	81,621	8,606
Zurich	Taxi	79	6,279	732 (59%)	1,235	7,953	1,343
Turin	Law enforcement vehicles	70	2,405	698 (30%)	2,331	3,758	830
Bern	Store delivery trucks	56	1,765	124 (18%)	676	7,322	859
Helsingborg	Waste collection trucks	30	1,975	884 (44%)	2,001	1,994	900

The drive-by agents in our study are passive; i.e. they were not instructed to cover specific locations in the cities, but rather to install the system and continue with their usual daily tasks. As a result we get the native patterns for each fleet. In our analysis we focus on the city coverage within the city boundaries. However, the agents were not restricted to stay within such boundaries, and so some of the trips were dropped or truncated. The data collection activity varies between the different fleets and cities as summarized in the table below (see Table 1.). All the data was collected between January and April 2022.

All the agents involved rely on a mobile phone mounted on the front windshield of a vehicle to detect road features. The phones are equipped with an app[1] that collects data about the position, and detects and processes road damages "at the edge" using a deep neural network.

The network yields the bounding boxes on the image plane (see Figure 1), as well as the class of the road damage which could be: crack, pothole, patch and an iron-plate. Iron-plates are not strictly road damages, but their proximity with other types of road damages are known to increase the likelihood that the road will deteriorate more quickly. Moreover, for

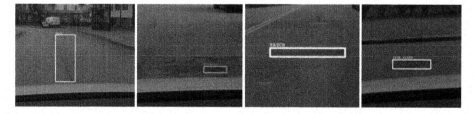

Figure 1. Examples of road damages detected by the deep network used by the agents.

---

[1]In This Study, We Make Use of a Commercially Available Data Collection Tool Called 3DAI City, Developed by Univrses.

the study of the potential coverage and detections they are relevant candidates as any other common object on the road. The data was also geographically binned in road segments of fixed lengths to further minimize the effect of outliers and then normalized over the number of times a vehicle traversed a specific bin, as we will discuss in the next section.

3 DATA COVERAGE

Road-coverage is an important aspect of any data collection pipeline that relies on the drive-by sensing paradigm. Ideally, one wishes to cover (and therefore sense and inspect) as many roads as possible in the shortest time, especially if the data to be harvested is of timely importance or decay quickly, becoming outdated. As expected, the more agents operating on a road network, the more road segments are covered in a short period of time. Slightly less obvious is the importance of the type of agent motions involved and the city topology. In Figure 2 we provide an example of the road coverage evolution in Turin.

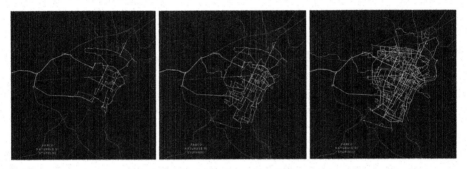

Figure 2.   Coverage of 5 agents in Turin, Italy. The images are 2 weeks apart. Yellow roads are the ones visited at least once.

We indicate by G a bidirectional graph representing the road network, where each node is an intersection or a road junction and each edge a segment of road of a certain length. The coverage of the city graph can be defined in different ways and with respect to different variables. In this work we consider an edge to be covered if it was visited at least one time, and the total coverage as the ratio of visited edges with all the edges of the graph. Then we consider the coverage as a function of the total distance traveled by all the agents. Later we normalize the traveled distance by the city graph length, which renders all the parameters to be dimensionless. This last transformation allows us to study and compare the coverage in different cities and for different agent types directly.

The coverage function $C(x)$ (where x is the total distance traveled in city length units) has some universal properties. It is a monotonic function by construction, starts at zero, and saturates a number less or equal to one (or 100%) in the limit $x \to \infty$. It is also expected that the first derivative $C'(0) = 1$, though this last condition is of lesser interest when looking at intermediate and late times.

In a recent study O'Keeffea (2019) suggested a model-based way to parametrize and semi-analytically describe the street-covering factor $C(x)$ of taxis depending on a city parameter. This parameter however changes for different cities and needs to be computed from costly simulations and fitted to empirical data and is not guaranteed to fit well for other fleet types or fleets and cities of different sizes. For example, their model will always saturate 100% coverage of the entire city graph, while in practice some areas will never be visited by some fleets. In this study we apply a practical approach to fit and extrapolate the data (see next section).

When looking at the coverages of different cities one might want to observe the rate at which the curve reaches a certain coverage ratio and the asymptotic value. This is equivalent to asking how many kilometers a specific vehicle or fleet needs to travel in a city in order to cover a percentage of the total roads or what is the maximum expected percentage or road covered after a certain time.

3.1 *Discussions about the road coverage*

In our experiments, we had different amounts of vehicles operating for different periods of time in various cities. To homogenize the comparison between cities we would like to predict the coverage based on the collected vehicles positions (trips). The total distance traveled in the different cities ranges roughly between 1–10 times the city size. In general, such a prediction is a difficult task, and any extrapolation is not guaranteed to properly estimate the coverage. Nonetheless, in the following we apply and motivate a naive approach to allow such a comparison. Our attempt to predict the coverage is based on the assumption that there are common characteristics between the coverage curves that could be captured by some simple empirical parametric form.

In order to justify our approach, we divided our datasets into training and test sets for each city, containing the 25–75 and 75 percentiles of the data respectively, where the data is ordered by the total distance traveled. We omitted the first 25% of data points as they appear to be less informative for the extrapolation, and occupy a short period of the traveled time as the data is denser at the beginning.

The coverage for every city was computed with working day increments, namely accumulating over a day the total distance traveled by the fleet, and the length of the new traveled edges. As such a curve could depend on particular behaviors and events, we regenerated the coverage curves several times by reshuffling the dates, and thus augmented the data and made it more robust. Finally, we trained several empirical, sigmoid like, family of function on the test set, and found the following 4-parameter family function to yield a good fit on the test set

$$C(x;a,b,c,d) = \frac{100e^{-c^2}x^d}{(a + x^{bd})^{\frac{1}{b}}},$$ (1)

where x is the traveled distance (in city graph length units) and $a, b, d \geq 0$ are the fitted parameters. In Figure 3 we show the (augmented) coverage data and a prediction for the coverage. We also provide the comparison between the test set and the fitted data based on the training set, for the three cities where at least twice the city graph size was travelled.

We show both the coverage calculated from the data and the one from fitting equation (1) with the thicker and dotted part of the curve respectively. We do not study the specific behavior of a single agent but look at the cumulative driving effort of the fleet in a given period of time. The number of kilometres driven by each fleet and the time window in which the data was collected varied between cities. This explains the different lengths of the thicker part of the curves. We start by observing that none of the fleet seem to converge to 100% after driving a distance of at least 10 times the size of the city. The agents were not given the goal to explore the whole city but were acting according to their purposes. A taxi would always prioritize certain roads to go from one side of the city to the other.

Another observation is that similar fleets acted in a similar way, as one can see in the curves of Turin, Stockholm and Zurich. Waste collection units did the best job in covering the city compared to other fleet types. This is because these operators need to access many residential and less trafficked roads and because they follow a more structured routing strategy. Taxis and law enforcement vehicles were the second most efficient fleet type. The destinations or picking points of these fleets tend to be focused on strategic and repetitive

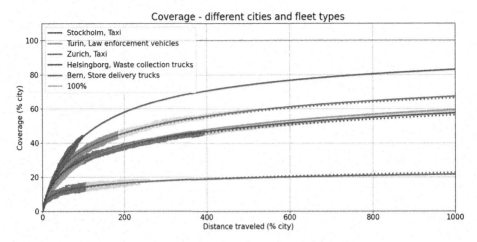

Figure 3.   Coverage curve fitting. Street coverage of different fleets operating in 5 European cities. The scattered points represent the coverage coming from the data. The solid lines represent the function in equation (1) fit to the whole data. In the 3 cases where the distance travelled was at least twice the city size, we divide the data into training and test set (in darker and lighter colours respectively) and further fit a curve (dotted) to the training set and observe the fit with respect to the test set in order to motivate the prediction capabilities of the fitting function. The horizontal axis indicates the distance covered by the vehicles as a percentage of the total size of the city (see Table 1. for specific the city size numbers). For example, a value of 200 in Stockholm indicates that almost 7000 km were driven by the taxis.

landmarks like stations, airports or restaurants or more densely populated areas. Also, these vehicles drive over faster predefined roads to traverse the city. The least efficient fleet was the store delivery trucks. In this case the trucks were driving in repetitive routes from the warehouses to distribution centres and between the centres. Saturation was indeed fast and assessed at around 20% of the city size.

4   ROAD DAMAGE DISTRIBUTION

We now look at the road damage data collected in the covered area from the same fleets discussed above. In this case we focus on the four largest cities where more data is present. To describe the road damage distribution, we split the road graph of the city into bins of size 20 meters, which is approximately half of the relevant longitudinal field of view of the camera. Then, for each bin we consider the average number of detections, that is, the number of detections divided by the number of visits to the bin. These average numbers of detections suggest the density level of the road damage on each bin. Averaging again for each city and each road type (where the road type can be given by different speed limits, or by highway vs. urban road), we obtain the values presented in Figures 4 and 5.

By looking at the density distribution of the damages in the city, (see example in Figure 6), it is possible to identify neighbourhoods or specific roads that would benefit from maintenance work.

Another way to assess the distribution of the road damage is to associate a binary variable indicating the presence of damage to each bin, and each damage type. This approach does not consider the damage density of a bin but mitigates the effect of severely damaged roads on the overall statistics. We declare that a bin has damage of a certain type if detections of that type were seen every time a vehicle traversed the bin. The percentage of observed bins with damage, for each city and each damage type, is given in Figure 7, and spatial distribution of such bins for Stockholm is given in Figure 8.

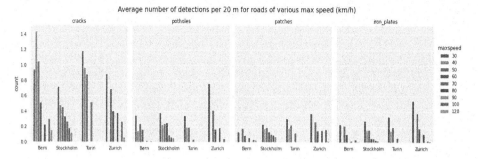

Figure 4. Average number of detections on each 20 m bin marginalized by roads of different speed limits. High speed roads usually carry less road damage than low speed roads.

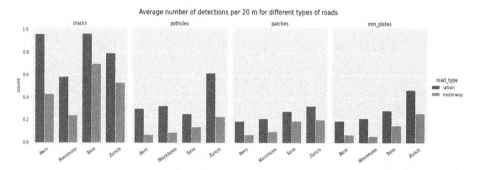

Figure 5. Average number of detections on each 20 m bin is marginalized by two categories of road type: motorways (including highways and primary roads) and urban roads (including secondary roads). Stockholm has the least crack-affected roads, followed by Zurich. We also notice that motorways usually carry a less amount of damage compared to urban roads.

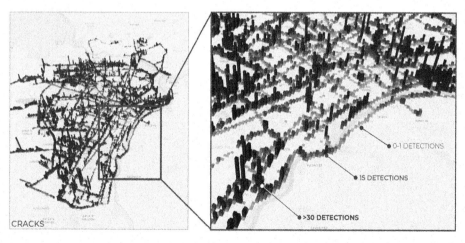

Figure 6. 3D close-up of the binned cracks detections in the visited areas of a neighbourhood in Turin between January and February 2022. The height of the blocks indicates the concentration of cracks in that specific bin (30 m2).

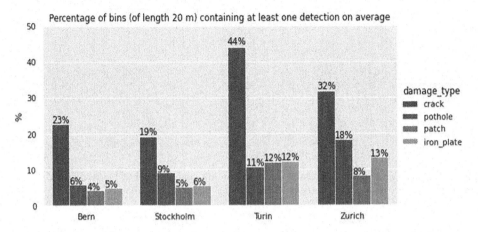

Figure 7. The percentage of bins (of 20 m length) containing at least one damage, out of all visited bins.

Figure 8. Detected road features distribution in Stockholm with a bin size 20 m. The visited roads with no detection are colored in gray, while the roads with at least one detection are colored in the indicated color. The percentage of bins with at least one detection is also summarized in Figure 7.

5 CONCLUSIONS

In this paper we have looked into the results of deploying small fleets in urban areas for the purpose of gathering road damage information following the drive-by sensing paradigm. We have shared insights collected from vehicles of different kinds, namely taxis, law enforcement vehicles, delivery trucks and waste disposal trucks, deployed for a period of 4 months in 5 cities in Europe. We showed how the different types of driving patterns affect the amount of data collected and therefore the outcomes of a hypothetical data collection campaign. Garbage trucks were shown to be the agents that covered the largest proportion of the city in a short time scale, followed by taxis and law enforcement vehicles and lastly, delivery trucks.

We proposed the use of a specific fitting function to mimic the road coverage trend when limited data is available, for instance because few vehicles are available or because the fleet has been deployed for a short period of time. Extrapolating such trend is useful to properly design an effective data acquisition campaign and answer questions like what amount of km of roads the fleet has to drive in order to properly cover a certain percentage of the city network, or equivalently, how long it takes for a certain fleet to sufficiently explore a portion of the target area. In general, our findings show that few agents have the power to cover large portions of the city in a short amount of time/traveled distance. This agrees with other previous findings.

We have also looked at the road damage data collected by the agents and shown how the damage types are distributed in different cities. We have marginalized the data collected by type of road, namely urban and motorways, and noticed that on average most of the damages are present in urban roads. We have also noticed that cracks are by far the most diffuse kind of pavement imperfection in all the urban areas and that roads with low speed limits carry the most damage. This paper provides precious hints to municipalities or road maintenance authorities on how to plan a successful data collection campaign aimed at understanding the status of the roads if a drive-by data harvesting paradigm is used.

REFERENCES

AAA American Automobile Association, Inc., 2016, Pothole Damage Costs U.S. Drivers $3 Billion Annually.

AAA American Automobile Association, Inc., 2022, Potholes Pack a Punch as Drivers Pay $26.5 Billion in Related Vehicle Repairs.

Agarwal D., Iyengar S. and Swaminathan M. 2019. *Modulo: Drive-by Sensing at City-scale on the Cheap. COMPASS '20*: Proceedings of the 3rd ACM SIGCAS Conference on Computing and Sustainable Societies.

Baskara S. Yaacob H., Hainin M. and Hassan S., 2016. Accident Due to Pavement Condition – A Review. *Jurnal Teknologi* 78(7-2)

Eaton R.A., Joubert R.H. and Wright E.A., 1989, Pothole Primer. *A Public Administrator's Guide to Understanding and Managing the Pothole Problem.*

Government of India, 2019, Report: *Road Accidents in India.* P. 39–58.

Kim T. and Ryu S. 2014. Review and Analysis of Pothole Detection Methods. *Journal of Emerging Trends in Computing and Information Sciences.*

Mohan P., Padmanabhan V. and Ramjee R. 2008. Nericell: Rich Monitoring of Road and Traffic Conditions using Mobile Smartphones. SenSys '08: *Proceedings of the 6th ACM Conference on Embedded Network Sensor Systems.*

O'Keeffea K., Anjomshoaaa A., Strogatzb S., Santia P. and Rattia C. 2019. *Quantifying the Sensing Power of Vehicle Fleets.* PNAS, 12752–1275.

The Economist, 2016. *Article*: The Hole Story. *Researchers are Inventing New Ways to Prevent a Motoring Curse.*

The Hindustan Times, 2021, Article: 3,500 Road Accidents Last Year Due to Potholes: *Govt Tells Parliament.*

Vinayakamurthy M, Mamlouk M., Underwood S. and Kaloush K., 2016. Report of the National Transportation Center at Maryland: *Effect of Pavement Condition on Accident Rate.*

Wang H., Al-Saadi I., Lu P. and Jasim A., 2019. Quantifying Greenhouse gas Emission of Asphalt Pavement Preservation at Construction and use Stages Using Life-cycle Assessment. *International Journal of Sustainable Transportation.*

Wang P., Hunter T., Bayen A., Schechtner K. and González M. 2012. Understanding Road Usage Patterns in Urban Areas. *Scientific Reports.*

The relationship between surface defects and structural distress on a 50-year-old Japanese toll road

Keizo Kamiya*

Advanced Engineering Development Department, NEXCO Central Headquarters, Nagoya, Japan

ABSTRACT: The stock of Japanese toll expressways aged over 50 years is on the increase. In addition, porous asphalt, now 80 percent of the entire road surface of NEXCO, has generated another problem: the water-retaining binder layer is faster damaged than sticky-bonded porous surface, resulting in a two-layer rehabilitation to be a repair standard. Toward achieving a best management on already aging and faster deteriorating porous pavements, it is crucial to properly monitor and evaluate their existing conditions and prevent from fatigue cracking or other irreversible damages. As a first step to the development of an effective screening method, the relationship between fatigue cracking and road surface defects was investigated. This paper introduces two years cut-open research on the 50-year-old Chuo Expressway in central Japan, as well as analysis of road surface data of the section, and associated FWD deflection survey. As a result, at partially 30 mm faulting spots on the road surface, despite having 30 centimeter AC thickness, almost associated size of permanent deformation was confirmed on the spot's underlying subbase layer. Moreover, it was observed that underground water weakens bearing support of lower base, resulting in higher FWD deflection and causes of road surface defects.

Keywords: Surface Defects, Pavement Structure, Cut-Open Survey.

1   INTRODUCTION

Since Japan's first expressway opened to traffic in 1963, toll road stock aged 50 years or older has been on the increase. It is high time for nationwide toll road operators, namely NEXCO East, Central and West, to rebuild a new rehabilitation strategy so that road pavements can be more sustainable in the longer term. In addition, thanks to its greatly reducing wet accidents, porous asphalt is now covered with 80 percent of the entire road surface at NEXCO. However, as the pavement welcomes rainwater from its porosity and runs it down on the binder layer, the layer is faster damaged rather than the porous surface, which can be strongly bonded with high SBS-content modified bitumen. Now, a two-layer rehabilitation has unfortunately become a standard repair style of porous asphalt at NEXCO. With the aging road structure on the increase and continued adoption of the pavement, some areas must be rehabilitated substantially with surface, binder and asphalt-treated base layers with structural problems (Figure 1). Figure 2 illustrates a tendency of damages going deeper. To prevent thicker rehabilitations, an effective screening method of detecting a signal of bottom-up cracking from the road surface is urgently needed.

---

*Corresponding Author: k.kamiya.ab@c-nexco.co.jp

DOI: 10.1201/9781003429258-40

To prevent from further aggravating existing damages into deeper layers, understanding of the mechanism of structural damage and how to effectively and promptly screen problematic sites are important. The fracture mechanism of asphalt pavement has long been investigated at many institutes and road agencies. Hopefully for the purpose of finding a screening method, here are literature reviews on pavement damages, each followed by the author's comments.

Toward long-life cycle of the pavement on an airport runway, its cracking data were used to determine the stress and displacement of the pavement. By using FEM based analysis, it was found that stress intensify factor decreases as layer thickness increases. Finally, it was concluded that 20 cm is best for surface layer thickness (Solanski et al. 2019). NEXCO already has a design policy of minimum AC thickness of 18 cm.

Theoretical investigations in the UK design method, taking into consideration traditional pavement failure modes, methods for calculating base strain in flexible pavements, and types of deformation and looking at how pressure dissipates in typical asphalt pavements were conducted. It was concluded that bottom-up fatigue cracking does not occur in well-constructed asphalt pavements of even moderate thickness (Hunter 2018). This contradicts the above evidence, as shown in Figure 1.

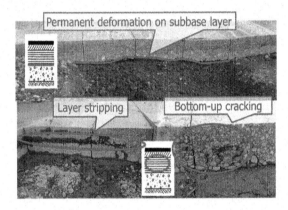

Figure 1.   Porous asphalt with good visibility on a rainy day.

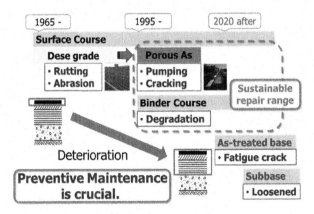

Figure 2.   Pavement damages going deeper.

The numerical analyses carried out with the use of FEM allowed to assess the strain and stress changes occurring in the process of cracking road pavement. It has been shown that low thickness pavements are susceptible to fatigue cracks arising "bottom to top" (Mackiewicz 2019). This substantially matches observations in the field at NEXCO.

To identify sources of moisture and other conditions that led to the early rutting problems observed, five projects were targeted. Overall, improper tack coat or failure, permeable dense-graded layers, inadequate drainage, and, possibly, inadequate compaction of AC materials were identified as the likely root causes of the problem (Scholz et al. 2009). This has revealed the importance of strengthening binder layer beneath porous surface at NEXCO.

The use of deflection-basin data to determine the in-place structural condition of AC layers for rehabilitation design in accordance with the Mechanistic-Empirical Pavement Design Guide—A Manual of Practice was evaluated, as agencies have started to use for rehabilitation objectives (Ayyala et al. 2018; Kim et al. 2021). For easier application for field practitioners, NEXCO has adopted a method of difference in deflection, but FWD cannot be a network-based screening method.

Ground Penetrating Radar (GPR) is used to evaluate the road damage. Attenuation of damaged and undamaged road sites has made it possible to trace the causes of the degradation of pavements (Colagrande et al. 2019). The author considers that the accuracy of the non-destructive approach is highly dependent on localized attenuation technique, which required high cost.

Likewise, currently available methods cannot easily screen bottom-up or fatigue cracking from the road surface, the author introduces a more informative cut-open survey at problematic spots on an existing NEXCO toll expressway. By focusing on the 50-year-old Chuo Expressway, this paper mainly tries to find the relationship between road surface defects and structural damage of pavement through cut-open research, time-series analysis of road surface characteristics and FWD deflection survey.

3 CUT-OPEN RESEARCH

Figure 3 are photos of road surface defects which were shot in September 2020 by field engineers of the Iida Operation Office which controls 93 km part of the Chuo Expressway, which connects Tokyo and Nagoya in central Japan. About 30 mm level of faulting or deformation was observed on the outer wheel path (OWP) of the leftmost lane intermittently within a certain 200-meter section. The phenomenon cannot be dismissed, because the defects took place after a 10 cm AC cut-and-overlay in 2009. To prevent such incidents from happening again in the future, the cause of 30 mm faulting in 10 years had to be investigated. For this reason, the Iida Office decided to conduct cut-open research on the section.

Figure 3. Road surface defects on the outer wheel path of the leftmost lane on northbound of the Chuo Expressway controlled by the Iida Operation Office (KP 284.450 to 284.650).

The purpose of cut-open research is to help understand the inner condition of existing pavements by removing its cut bulk component. Because of high cost and very time-consuming approach, it cannot be a standard research. Moreover, since the jobs must be finished during permitted several hours for lane closure, the research cannot always be done hopefully as planned. For example, in cutting AC materials, dry cutting rather than wet cutting is desirable to understand the actual condition of the pavement as it is. However, the dry cutting takes several times longer than the usual wet cutting. Due to time limitation, the Iida section had to rely on wet cutting.

Figure 4 shows the procedure of cutting road materials. Over the course of the procedure, technically important observations were recorded: cracking or splitting of AC layers, segregation of AC materials, bottom-up or top-down cracking direction, lower base layer's permanent deformation and so on.

Figure 4.   Procedure of cutting road materials.

As a result of the cut-open surveys at seven spots that had had 30 mm faulting, Figure 5 shows two of the prominent photos. The left photo reveals permanent deformation on the lower base layer. Its deformation was as large as 40 mm on the outer wheel path (OWP). The OWP, which is heavily trafficked, contained permanent deformation, while in the less trafficked area between wheel paths (BWP) there was no damage. Therefore, traffic accumulation can be a contributor to permanent deformation. The wet condition on the lower base was the remained water after the wet cutting. Meanwhile, from the right photo of the sampled block core, fatigue cracking that had originated from the bottom of the asphalt treated base layer was confirmed. The upward cracking almost reached the bottom of the top 10 cm surface and binder layers which had been rehabilitated in 2009. Though the two layers were not much damaged, it is deduced that displacement caused by the fatigue cracking led to the 30 mm faulting on the OWP.

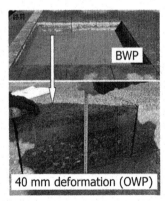 

Figure 5.   Permanent deformation on the lower base (left) and fatigue cracking (right).

Cut-open structural research has so far been conducted by NEXCO Research Institute on 20 nationwide problematic sections. Since fatigue cracking has mostly been reported on AC thickness levels of 18 to 25 cm, the finding of permanent deformation and fatigue cracking on the subbase layer capping 30 cm AC thickness was a big surprise. This sort of heavy damage should be avoided from all trunk roadways. As a first step to developing an effective screening method, the relationship between surface defects and structural damage was put into further investigation.

4   TIME-HISTORY OF SURFACE CHARACTERISTICS

Ordinary road surface data, such as IRI, cracking and rutting, were first compared between the latest past two years on the whole Iida section (KP 195.900 to KP 288.600). To identify spots of small defects, 10-meter based average and its standard deviation for each surface category were calculated.

Figure 6 refers to the leftmost lane on the north bound of Chuo Expressway. Because of network evaluation, there is almost no difference in each index between 2015 and 2018. Note that the macro data contains all road structures like earthwork, bridge and tunnel sections. To know more about impacts of small defects, further localization is to be introduced below.

Figure 7 focuses on the one-kilometre length (KP 284 to 285) that includes the above cut-open research sites. The average IRI-10 of the 1 km road surface was decreased from 3.0

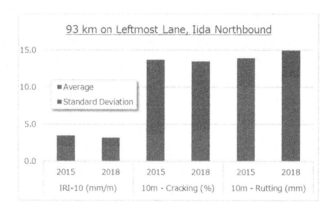

Figure 6.   Comparison of road surface data on the Iida section (KP 195.900 to KP 288.600).

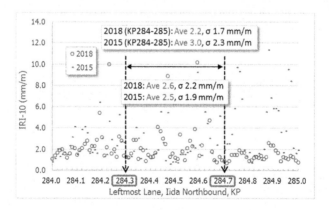

Figure 7.   Change in IRI-10 on one kilometre length (KP 284 to 285).

407

mm/m in 2015 to 2.2 mm/m in 2018. Why the reduction occurred? According to our NEXCO Pavement Management System, a section of 357 meter of the pavement was rehabilitated in the autumn of 2015. Therefore, this contributed to the reduction in IRI-10.

Meanwhile, on the inside 400-meter section (KP 284.3 to KP 284.7), the average value increased from 2.5 mm/m to 2.6 mm/m. This sensitivity happens depending on the ratio of safe and problematic sites of a focused section. If another selected section has many more defects, the value will be naturally higher.

Figure 8 is about rutting data on the same one kilometre. Average rutting values both on 1-kilometre and 400-meter sections in 2018 happen to be 9.8 mm. However, while the 1-km value decreased, 400-m value increased from 2015 to 2018. This tendency was similarly confirmed on the previous IRI-10 over the tree years. Comparison of cracking data could not be made, since most of the data was zero percent.

Figure 9 further narrows down into a 200-meter length of the section with seven specific spots to investigate the relation between defects and nearby road surface data. Photos here were taken in 2019. Despite a year later of the road survey, photos of the defects generally indicate an increase in IRI-10. If photos and surface data had been taken at closer time, more matching would have been confirmed.

Figure 10 focuses rutting data on the 200-meter section. The tendency of increase in rutting is confirmed similarly as with IRI-10. This means that a spotted defect may possibly relate both longitudinally and transversely. In other words, it is speculated that IRI-10,

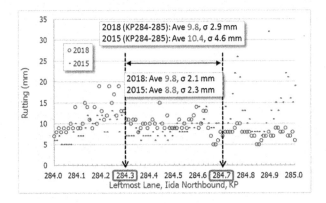

Figure 8.   Change in rutting on one kilometre length (KP 284 to 285).

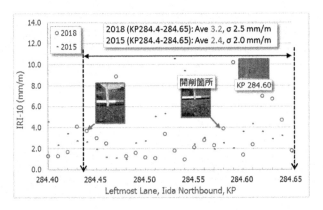

Figure 9.   Change in IRI-10 on 200-metre length (KP 284.450 to 284.650).

408

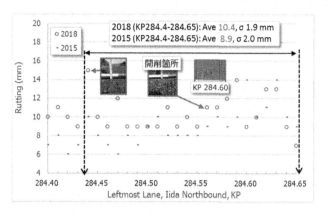

Figure 10. Change in rutting on 200-metre length (KP 284.450 to 284.650).

which accounts for longitudinal part of surface roughness, goes with increase in transverse rutting at some heavily problematic spots.

In 2021, other defected spots were also subjected to the cut-open survey and put into investigating the relation with road surface data. On site selection, faulting of 30 mm was primarily used to screen for road safety as expressway. Then based on the 2018 road surface data, sites for cut-open were conducted in the summer of 2021.

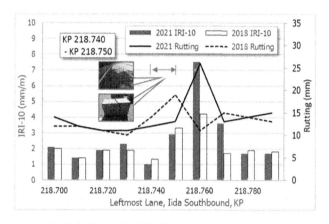

Figure 11. Cut-open at KP 218.740 to KP 218.750 and road surface data.

Figure 11 summarizes results of a cut-open site, where the rutting level was 19 mm and IRI-10 exceeded 3 m/mm, according to the 2018 data rather than the later obtained 2021 data. Though rutting was a high level, IRI-10 was lower. As shown in the photos inside the figure, top-down cracking was confirmed rather than bottom-up type as defined as "fatigue." If the 2021 data had been at our hands at the time of site selection, fatigue cracking might have been confirmed at the next KP 218.760 where both IRI-10 and rutting were higher. Importance of using the most updated road surface data is confirmed here again.

Table 1 shows levels of FWD deflection on the site. Both deflections at outer wheel path (OWP) and between wheel paths (BWP) were judged not to be a structurally problematic level, according to the local data record in the area.

Table 1.　FWD deflection results at KP 218.748 on southbound.

Loading site	D0 (μm)	D90 (μm)	D150 (μm)	D200 (μm)	AC thickness	Surface (°C)	Survey	Surface type	Lower base
OWP	247	76	40	28	22 cm	23.4	2021/8/27	Porous	Cement
BWP	132	61	32	22	22 cm	24.1	2021/8/27		treated

Figure 12 shows another site which was selected according to the 2018 surface data with 12.5 mm rutting and nearly 2 mm/m IRI-10. While rutting is the same level as the previous site, IRI-10 is much lower. As shown in the inside photos, fatigue cracking was confirmed. Table 2 of FWD deflections reveals that lower base layer is problematic, as D150 which relates to subgrade support shows higher levels (OWP 56 μm, BWP 68 μm) than those of nearby less problematic sites. Moreover, one notable thing is that the deflection on less trafficked BWP is higher than that on much more trafficked OWP. This is quite uncommon but there is a persuasive reason for that.

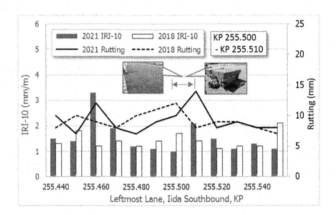

Figure 12.　Cut-open at KP 255.500 to KP 255.510 and road surface data.

Table 2.　FWD deflection results at KP 255.513 on southbound.

Loading site	D0 (μm)	D90 (μm)	D150 (μm)	D200 (μm)	AC thickness	Surface (°C)	Survey	Surface type	Lower base
OWP	188	142	56	30	24 cm	21.2	2021/8/23	Porous	Cement
BWP	281	143	68	36	24 cm	21.5	2021/8/23		treated

Figure 13 can clearly explain the phenomenon. First, the left photo shows 30 mm rutting on inner wheel path (IWP) at this site. Second, when lower base materials of the site were removed, as shown in the right photo, underground water beneath the IWP was leaking toward the BWP. The underground water must have loosened the subgrade materials beneath the IWP and BWP, resulting in a higher deflection on the BWP than on the OWP.

Figure 13. Photos of road surface faulting in immersion (left) and its underground water (right) both on inner wheel path (IWP) at KP 255.500 to KP 255.510.

Figure 14 summarizes another site where 15 mm rutting and a little over 4 mm/m IRI-10 were measured in 2018. IRI-10 is not much a high level. However, as shown in the inside photos, fatigue cracking was also confirmed here. According to Table 3, a higher deflection on OWP was confirmed. The maximum deflection level of 357 μm accounts for AC layers fracture in this area, which matches the photo of sampled block cores.

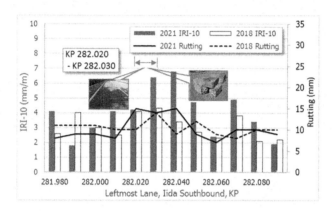

Figure 14. Cut-open at KP 288.020 to KP 288.030 and road surface data.

Table 3. FWD deflection results at KP 288.030 on southbound.

Loading site	D0 (μm)	D90 (μm)	D150 (μm)	D200 (μm)	AC thickness	Surface (°C)	Survey	Surface type	Lower base
BWP	138	63	33	23	28 cm	23.7	2021/8/24	Porous	Cement
OWP	357	66	33	24	28 cm	23.5	2021/8/24		treated

5 SUMMARY AND CONCLUSION

To avoid deeply rooted structural problems such as fatigue cracking, permanent deformation, or other irreversible damages, it is crucial to promptly screen pre-condition of such problematic sites. As a step toward the development of such an effective screening method, the relationship between fatigue cracking and associated road monitoring data was investigated. Based on cut-open survey on the Chuo Expressway and its road surface and deflection data, the followings were observed.

(1) At a site with 30 mm faulting on the road surface, permanent deformation and fatigue cracking were observed on the subbase layer beneath 30 cm AC thickness.

(2) Photos of road surface defects generally matched higher 10-meter based both IRI and rutting. It is speculated that increase in IRI-10, which accounts for longitudinal part of surface roughness, goes with increase in transverse rutting at problematic spots.

(3) In developing a screening method, it is suggested that all data collection including photos of road surface defects, surface characteristics and FWD monitoring data be done possibly within a closer time range.

(4) Fatigue cracking was confirmed at sites where FWD deflection was higher range that accounts for AC layers fracture.

(5) Underground water weakens bearing support of lower base, resulting in higher FWD deflection, and causing road surface defects.

Since the findings are short of getting a pre-condition of fatigue cracking, this sort of field research and study still needs to be continued.

Finally, the author of this paper would stress the importance of protecting base layers from being further aggravated, as is a principle in the Pavement Inspection Manual (Japanese government 2017). Since the concept is not just limited in japan, but also universally applicable all over the world. An effective monitoring approach will substantiate in the future.

REFERENCES

Ayyala D., Lee H., and Von Quintus H., 2018. *Characterizing Existing Asphalt Concrete Layer Damage for Mechanistic Pavement Rehabilitation Design*. FHWA-HRT-17-059. 2018.

Colagrande S., Ranalli D., and Tallini M., 2019. GPR Research on Damaged Road Pavements Built in cut and Fill Sections. AIIT 2nd International Congress on Transport Infrastructure and Systems in a changing world.

Hasni H., Alavi A., Chatti K., and Lajnef N., 2017. A Self-powered Surface Sensing Approach for Detection of Bottom-up Cracking in Asphalt Concrete Pavements: Theoretical/numerical Modelling. *Construction and Building Materials* 144, 728–746.

Hunter R., 2018. Disproving Bottom-up Fatigue Cracking in Well-constructed Asphalt Pavements. Proceedings of the Institution of Civil Engineering. *Construction Materials* 00 2016 Issue CM0, 1–8. 2018.

Kim R., Zeng Z., and Lee K., 2021. *Backcalculation of Dynamic Modulus from Falling Weight Deflectometer Data*. Research Project No. HWY-2017-03.

Mackiewicz P., 2018. Fatigue Cracking in Road Pavement. *IOP Conference Series: Materials Science and Engineering*.

Ministry of Land, Infrastructure, Transport and Tourism, 2017. *Pavement Inspection Manual (in Japanese)*.

Scholz T., and Rajendran S., 2009. *Investigating Premature Pavement Failure Due to Moisture*. FHWA-OR-RD-10-02. 2009.

Solanski H., and Nair K., 2019. Study on Fracture Analysis of Flexible Airport Pavement. *International Journal of Engineering Research & Technology*. Vol. 8 Issue 09, September-2019.

Zhang R., Khan A., Teng Y., and Zheng J., 2019. Back-calculation of Soil Modulus from PFWD based on a Viscoelastic Model. *Hindawi Advances in Civil Engineering* Volume 2019.

Roads and Airports Pavement Surface Characteristics – Crispino & Toraldo (Eds)
© 2023 The Author(s), ISBN 978-1-032-55149-4

Evaluation of porous asphalt hydrological performances through rainfall-runoff modelling by EPA SWMM

Diego Ciriminna, Giovanni Battista Ferreri, Leonardo Valerio Noto & Clara Celauro*
Department Engineering, University of Palermo, Palermo, Italy

ABSTRACT: Over the last few decades, the increase in extreme weather events has caused urban floods to become more and more frequent, with increased risks for road users. Among the different types of pavement structures, permeable pavements (PPs) can effectively increase the hydraulic resilience of the cities since they are specifically designed for reducing runoff by allowing water to infiltrate and be discharged into the soil and/or by drains. This study aims at evaluating the hydrological performance of a Porous Asphalt (PA) using numerical simulations for a case study carried out using EPA SWMM (Storm Water Management Model) software. Several simulations have been carried out, considering different rainfall events, covered areas, and spatial layouts of the permeable pavement itself. The results provide useful information on the use of PA.

Keywords: Permeable Pavement, Porous Asphalt, Hydroplaning, EPA SWMM

1 INTRODUCTION

Waterproofing of soil due to urbanization prevents rainwater from infiltrating the soil itself. Therefore, rainwater makes on the road surface a water film whose depth, usually indicated as Water Film Depth (WFD), depends on rainfall intensity and duration and flow path length as well as on the slope, material, texture, and permeability of the pavement (Gallaway *et al.* 1972). WFD can produce the hydroplaning phenomenon (Horne 1968), a hazardous phenomenon that occurs when a film of water introduces between the rolling tyres and the road pavement. The consequences are a loss of friction, possible vehicle instability, and a reduction of visibility through the splash and spray caused by the tyre-water film interaction (Anderson *et al.* 1998; Gallaway *et al.* 1971). The factors that contribute to triggering hydroplaning are the WFD, the vehicle speed, the characteristics of the tyre (such as the tyre tread depth and the tyre inflation pressure), and the geometry of the road (i.e., cross and longitudinal slope and pavement roughness) (Sitek *et al.* 2020).

An efficient way to reduce hydroplaning risk is the use of a Porous Friction Course (PFC), made in Open Graded Asphalt Mixture (OGAM), on the traditional pavement in bituminous layers (Eisenberg *et al.* 2011; Ferguson 2005). OGAM is a bituminous mixture with coarse, almost monogranular aggregates and characterized by high porosity that allows water to infiltrate the layer. Water storage in PFC allows for reduction of the WFD that provides improved tyre-pavement contact and vehicle stability (Do *et al.* 2013; Ong *et al.* 2005; Widyatmoko 2015).

The effectiveness in reducing runoff of four types of permeable pavements, namely Porous Asphalt (PA), Pervious Concrete (PC), Permeable Interlocking Concrete Pavement (PICP) and Grid Pavement (GP), has been examined by Ciriminna *et al.* (2022) for a case study by

---

*Corresponding Author: clara.celauro@unipa.it

DOI: 10.1201/9781003429258-41

numerical simulations, considering several scenarios each characterized by a different per-centage of area covered by PP and a different layout of permeable-impermeable areas. The results showed that all the PPs reduce the runoff, although in different percentages from one scenario to another. Indeed, PA proved to be less effective than the other PPs, however in a few cases, such as, for instance, covering of highways, where high traffic loads occur, and the short flow paths allow only thin WFD to form, the use of PA can be preferable to that of the other PPs. Therefore, it is useful to further investigate the performance of PA. In this paper, the latter topic has been stressed by the analysis of the results of further numerical simula-tions carried out using EPA SWMM software for the same case study used by Ciriminna *et al.* (2022). The case study was a car parking area located in the University Campus of Palermo (Italy) having an area of about 1400 m2. Seven scenarios were considered, differing from each other for the areal percentage of PA over the whole area and the PA-impermeable pavement layout.

In what follows, after the main characteristics of PA are reminded, the car parking area is described along with the seven scenarios. Then, the numerical simulations and their analysis are presented and commented on.

2 POROUS ASPHALT (PA)

PA can be considered as an alternative to the traditional pavement (Anderson *et al.* 1998) concerning which it provides important functional advantages such as reduction of the splash and spray risk with consequent better visibility, improved friction, reduction of the hydroplaning risk, low noise emission and reduction of the peak runoff velocity (Berbee *et al.* 1999; Stotz & Krauth 1994).

PA (Figure 1a) is composed of a moderate quantity of bitumen and a mix of homogeneous aggregates without fines that make the void space fall in the range of 18 – 25% (Eisenberg *et al.* 2011; Mullaney & Lucke 2014). This composition allows a very high permeability, usually ranging between about 4300 and 12500 mm/h, since the bitumen covering the grains does not fill the voids, that remain interconnected and allows water to follow not too winding paths (Chen *et al.* 2021; Kuang *et al.* 2011; Meng *et al.* 2020; Nooruddin & Hossain 2011; Zhong *et al.* 2016).

PA can be used both as the upper layer of permeable pavement and as a covering layer of a pre-existing traditional pavement, to improve its performance as reminded above. In the former case (Figure 1b), the layer of PA is usually from 75 mm to 180 mm thick (Mullaney & Lucke 2014). The rainfall water infiltrates the PA layer and then the underlying layers down to the native soil, to be finally drained through the soil itself and/or by a specific pipe system (Chaddock & Nunn 2010; Ballard *et al.* 2015; Ciriminna *et al.* 2022; Drake *et al.* 2013;

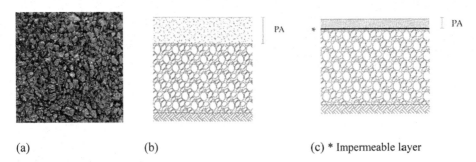

(a)   (b)   (c) * Impermeable layer

Figure 1.   Porous Asphalt: (a) Texture; (b) PA as the upper layer of a PP; (c) PA used as Porous Friction Course.

Eisenberg *et al.* 2011). In the latter case (Figure 1c), the PA layer is called Porous Friction Course (PFC). Its thickness ranges from about 25 to 50 mm. and the aggregates are smaller than in a PP (Anderson *et al.* 1998; Charbeneau & Barrett 2008; Drake *et al.* 2013; Eisenberg *et al.* 2011; Putman & Kline 2012). Water rained on the road drains vertically through PFC down to the bottom (i.e., the pre-existing impervious road surface, which in Figure 1c is indicated as the impermeable layer), and then flows through the voids toward the shoulders (Charbeneau & Barrett 2008) following the bottom slope.

3 NUMERICAL MODEL

3.1 *EPA SWMM model*

EPA SWMM is a dynamic model used to analyse stormwater runoff produced by rainfall events and to design and/or verify urban drainage systems. The runoff component of SWMM operates on a set of subcatchments that receive precipitation and generate runoff and pollutant loads (L. a. Rossman 2015).

SWMM is able to model LIDs, which are practices distributed in the urban context. Each LID is considered as a fraction of the whole subcatchment object (L. A. Rossman & Huber 2016). LIDs receive direct rainfall and runon from nearby areas as well as produce outflows towards the adjacent subcatchments. PPs can be modelled in SWMM through three layers, indicated as pavement, soil (an optional layer) and storage. In this pattern, water moves through each layer for evapotranspiration and/or infiltration. In the bottom layer (the storage) the accumulated water infiltrates into the subgrade and/or moves away by a drainage pipe system. In accordance with this pattern, a PP unit is modelled by classical continuity equations relating to, respectively, surface, pavement, soil and storage.

3.2 *The case study*

The case study adopted for examination of the hydrological performances of PA is a car parking area under construction located within the University Campus of Palermo (Sicily, Italy. Figure 2a). The car parking area is located along an inside lane of the campus and the relating boundary line was considered in the study as a watershed (no runon to the parking area). The S-W boundary line is a wall that hydraulically isolates the parking area. The total

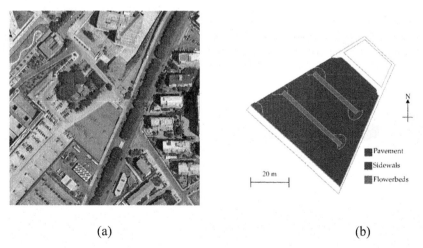

(a)   (b)

Figure 2.   The car parking area taken as a case study: (a) localization into the University Campus; (b) partition of the surface; (c) transit lanes and main lane; (d) Maximum slope direction and borders.

415

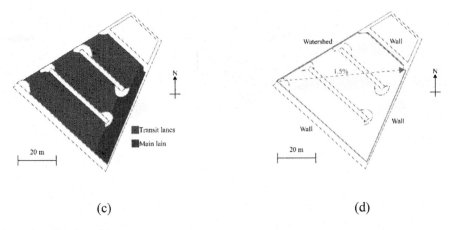

(c)   (d)

Figure 2.   (Continued)

area is 1387 m2, including sidewalks and flowerbeds, but the parking pavement is 1275 m2 only (Figure 2b). The two flowerbeds were not considered in the simulations as they were totally permeable and did not yield runoff. The car parking area is divided into three transit lanes, with parking spaces on both sides, and the main lane of connection along the S-E side (Figure 2c). Overall, the area can be considered an inclined plane with an average slope of about 1.5% (Figure 2d).

For the simulations, the car parking area was divided into 20 subcatchments (including 5 specific subcatchments for the sidewalks) connected to each other with the slope of the area. Eight scenarios named from 0 to 7 were considered (Figure 3), each relating to a different areal percentage of PA and a different PA – impermeable pavement layout. Scenario 0 relates to no-PA and was used as a benchmark to assess the hydrological performance of the scenarios with the presence of PA (scenarios 1 – 7). In the panels of Figure 3 relating to scenarios from 1 to 7, it is possible to distinguish three types of sub-areas (in scenario 1 coincident with the whole area): I) sub-area covered by PA; II) sub-area not covered by PA but it contributes to the runoff toward a sub-area covered by PA (we will refer to this type as *served* by PA); III) sub-area not covered by PA that receives a runon from sub-areas of the other two types that cumulates with the own rainfall. In the same panels of Figure 3, there are bound by a red line, in accordance with the runoff path, the subcatchments (differing from those used for the numerical modelling by SWMM) where a sub-area of type I processes the own rainfall and the runon from sub-areas of type II. In other words, these subcatchments benefit directly from the presence of PA. Table 1 summarizes, for each scenario from 1 to 7, the percentage of sub-areas of type I, the percentage of sub-areas of type I plus type II and the percentage of sub-areas of type III. In Figure 3 it is possible to notice how each scenario considered in this study has different percentages of the area covered by PA and area served by PA, over the total area of the parking lot.

In the numerical simulations carried out, the PA layer was modelled as PFC (Figure 1c), therefore no storage layer was considered (Zhu *et al.* 2019). The layer thickness was set to 50 mm and the void ratio to 18% (Table 2).

3.3   *Rainfall events*

For each scenario, 4 simulated rainfall events were considered and 2 real ones. In total, 48 simulations were carried out. The simulated rainfalls were obtained through TCEV (Two Component Extreme Value) method applied in Sicily (Forestieri *et al.*, 2018). These rainfall events referred to return periods of 5, 10, 50 and 100 years and were named, respectively, T5,

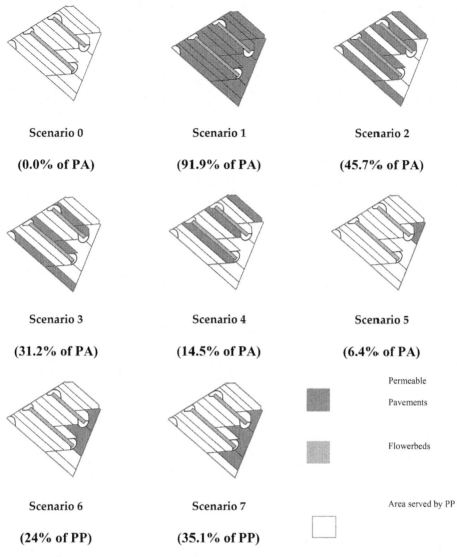

Figure 3. Scenarios analysed: Layout and percentage of PP considered.

Table 1. Percentages of the area covered by PA (Type I) and percentage of the area served (Type II) and not served by PA (Type III).

	Scenario 1	Scenario 2	Scenario 3	Scenario 4	Scenario 5	Scenario 6	Scenario 7
Type I (%)	91.9	45.7	31.2	14.5	6.4	9.6	35.1
Type II (%)	100.0	61.6	36.0	44.4	100.0	85.7	100.0
Type III (%)	0.0	38.4	64.0	55.6	0.0	14.3	0.0

Table 2. Parameters of PA assumed in SWMM.

Surface		Pavement	
Berm height (mm)	0	Thickness (mm)	50
Vegetation volume fraction	0	Void ratio (voids/solids)	0.18
Roughness (Manning's n)	0.01	Impervious Surface fraction	0
Surface slope (%)	1.5	Permeability (mm/h)	7000
		Clogging factor	0

T10, T50 and T100. The rainfall duration was 1 hour and, because of the short concentration time expected, it was divided into time intervals of 1 minute. To assess the rain depths minute by minute, the empirical relationship for short-duration rainfalls in Sicily (i.e., duration d < 60 min) by Ferreri and Ferro (1990) was used. The hyetographs were produced following Keifer and Chu (1957). As for the real events, they were recorded in Palermo on July 15th 2020, and July 18th 2021, and were named, respectively, R20 (Figure 4a) and R21 (Figure 4b). The return time of R20 was assessed by Francipane et al. (2021) to be of about 90 years, whereas that of R21 was assessed to be of about 5 years, by comparison to the depth duration frequency curves.

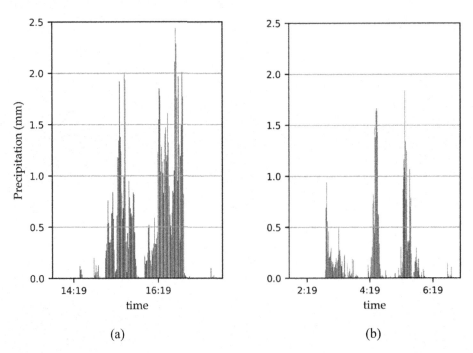

(a)   (b)

Figure 4. Hyetographs of rainfall events occurred in Palermo: (a) R20-July 15th 2020; (b) R21-July 18th 2021.

4 RESULTS AND DISCUSSION

The hydrological performance of a PA can be evaluated by the hydrographs in selected sections obtained by numerical simulations. For our purposes, the hydrograph in the downstream subcatchment was chosen, as the latter catches the runoff from the whole area. The results of the simulations were compared in terms of hydrograph peak and runoff volume. Therefore, for each scenario from 1 to 7, the hydrograph peak and the runoff volume relating to the 6 rainfall events were compared to the corresponding benchmark values of scenario 0.

Figure 5a shows, for the eight scenarios, the hydrographs relating to the simulated events T5, T10, T50, and T10, grouped by scenario, whereas Figure 5b the hydrographs for the real events R20 and R21.

The comparison between each scenario with the presence of PA in the Scenario 0 shows a decrease and a short delay of the peak concerning the latter. Both results are attributable to

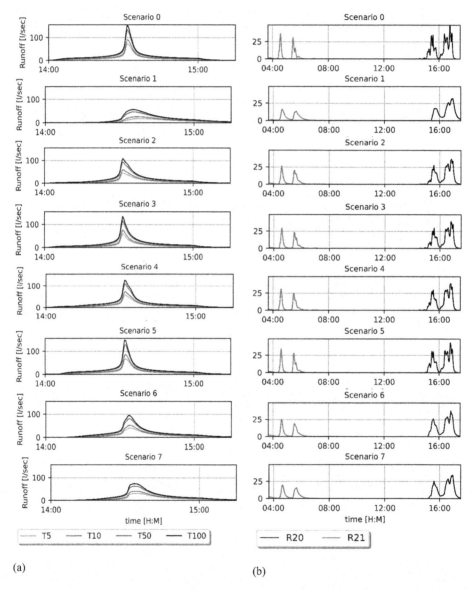

(a)   (b)

Figure 5. Hydrograph obtained by the simulations, grouped by Scenarios: (a) Simulated rainfall events T5, T10, T50 and T100; (b) Real rainfall events R20 (July 15th 2020) and R21(July 18th 2021).

419

the temporary storage of a part of the rainfall within the porous layer that lowers the runoff depth and then the runoff velocity. Both effects increase along with the areal percentage of PA. Larger peak attenuation and delay would likely occur for the higher thickness of the PA layer, that in the present simulations is of 50 mm only.

Further comparison is shown in the histograms of Figure 6a (simulated events) and Figure 6b (real events), where the ratios between the runoff volume of each scenario with PA and the volume of Scenario 0 are reported. For each scenario, identified by the related PA areal percentage, the ratios relating to the different rainfall events are grouped around the areal percentage itself. This concise representation allows the effects on runoff reduction due to PA areal percentage to be easily compared to each other for the different events considered. This figure confirms that the higher the PA areal percentage is the larger the runoff reduction is. Moreover, for each scenario, the reduction decreases as the rainfall intensity increases.

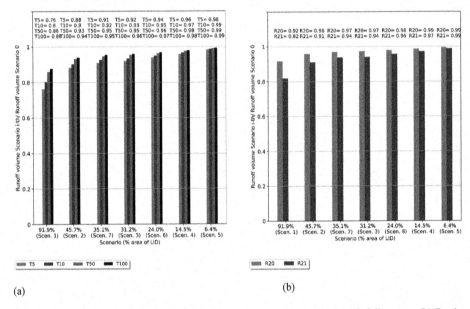

(a)   (b)

Figure 6. Comparison of the reductions of runoff volume: (a) Simulated rainfall events; (b) Real rainfall events.

It can be noted that, for the lower return times, the reductions in peak and volume can even be 20 – 30%, with the consequent reduction in the WFD producing benefits in hydroplaning risk and vehicle safety.

5 CONCLUSION

In the last years, extreme rainfall events have been more frequent than in the past, causing floods in urban areas where a large part of the surface is waterproof. Permeable pavements can be a good solution to increase the hydraulic resilience of paved surfaces. Porous asphalt is a specific type of permeable pavement, also known as Porous Friction Course (PFC), frequently used to cover existing impermeable pavements, to mitigate hydroplaning risk.

In this study, evaluation of the hydrological performance of PA has been carried out by numerical simulations using EPA SWMM (Storm Water Management Model) software.

A car parking area has been used as a case study, considering different scenarios, each relating to different areal percentages and planimetric layout of porous asphalt, and six rainfall events with return periods ranging between 5 and 100 years.

The results of analysis have shown attenuation of runoff peak and volume that, for the lower rainfall intensities considered, could be of a few dozens of percent. Reductions in peak and volume lower the runoff depth as well as the runoff velocity, which can be ascribed to the temporary storage of rainfall into the porous layer. The reductions in peak and volume increase along with the areal percentage of porous asphalt as well as they increase as the rainfall intensity decreases. A slight delay in the runoff peak was also observed in the presence of porous asphalt. In conclusion, porous asphalt seems to be a good solution for mitigating the peak and volume of the runoff and, consequently, for reducing the WFD and the related hydroplaning risk. At the same time, the study has proved that simulations carried out by effective software such as EPA SWMM provide useful information on the use of PA for mitigating the runoff, at the design level.

Further investigation will have to deal with the effects of the porous asphalt thickness and the slope of the considered area on which runoff propagation depends, as well as with the size of the area itself.

FUNDING

This research was financially supported by both the Italian Ministry of University and Research with the research grant PRIN 2017 USR342 Urban Safety, Sustainability and Resilience and the ERDF Regional Operational Programme Sicily 2014–2020 (POR FESR Sicilia 2014–2020), Action 1.1.5—"SMARTEP".

REFERENCES

Anderson D.A., Huebner R.S., Reed J.R. & Warner J.C., 1998. *Improved Surface Drainage of Pavements*. No. Project 1-29. NCHRP, Washington, DC, US.

Ballard B.W., Wilson S., Udale-Clarke H., Illman S., Scott T., Ashley R. & Kellagher R., 2015. *The SUDS manual*. CIRIA Publication: London, UK.

Berbee R., Rijs G., de Brouwer R. & van Velzen L., 1999. Characterization and Treatment of Runoff from Highways in the Netherlands Paved with Impervious and Pervious Asphalt. *Water Environment Research*, 71(2), 183–190.

Chaddock B. & Nunn M., 2011. *A Pilot-scale Trial of Reservoir Pavements for Drainage Attenuation*. Published Project Report n. 482. Transport Research Laboratory Wokingham, UK.

Charbeneau R.J. & Barrett M.E., 2008. Drainage Hydraulics of Permeable Friction Courses. *Water Resources Research*, 44(4), 1–10.

Chen S., You Z., Yang S. L., Garcia A. & Rose L., 2021. Influence of Air Void Structures on the Coefficient of Permeability of Asphalt Mixtures. *Powder Technology*, 377, 1–9.

Ciriminna D., Ferreri G.B., Noto L.V. & Celauro C., 2022. Numerical Comparison of the Hydrological Response of Different Permeable Pavements in Urban Area. *Sustainability*, 14(9), 5704. https://doi.org/10.3390/su14095704

Cossale G., Elliot R. & Widyatmoko I., 2013. The Importance of Road Surface Texture in Active Safety Design and The Importance of Road Surface Texture in Active Road Safety Design and Assessment. *In: International Conference Road Safety and Simulation*. 22–25, October. Rome. Italy

Do M.T., Cerezo V., Beautru Y., & Kane M., 2013. Modeling of the Connection Road Surface Microtexture/ Water Depth/Friction. *Wear*, 302(1–2), 1426–1435.

Drake J.A.P., Bradford A. & Marsalek J., 2013. Review of Environmental Performance of Permeable Pavement Systems: State of the Knowledge. *Water Quality Research Journal of Canada*, 48(3), 203–222.

Eisenberg B., Lindow K. C., & Smith D. R., 2015. *Permeable pavements*. ASCE. Reston, USA.

Ferreri G.B., & Ferro V. 1990. Short-Duration Rainfalls in Sicily. *Journal of Hydraulic Engineering*, 116(3), 430–435.

Ferguson B.K., 2005. *Porous Pavements*. Boca Raton, FL: Taylor & Francis.

Forestieri A., Conti F. lo, Blenkinsop S., Cannarozzo M., Fowler H.J., & Noto L.V., 2018. Regional Frequency Analysis of Extreme Rainfall in Sicily (Italy). *International Journal of Climatology*, 38, e698–e716.

Francipane A., Pumo D., Sinagra M., La Loggia G., Noto L.V. 2021. A Paradigm of Extreme Rainfall Pluvial Floods in Complex Urban Areas: The Flood Event of 15 July 2020 in Palermo (Italy). *Natural Hazards and Earth System Sciences*, 21(8), 2563–2580.

Gallaway B.M., Rose J.G., & Schiller Jr, R.E., 1972. The Relative Effects of Several Factors Affecting Rainwater Depths on Pavement Surfaces. *Highway Research Record*, 396, 59–71.

Gallaway B.M., Schiller R.E. Jr., & Rose J.G., 1971. *The Effects of Rainfall Intensity, Pavement Cross Slope, Surface Texture, and Drainage Length on Pavement Water Depths*. Research Report 138-5.

Horne W.B., 1968. Tire Hydroplaning and its Effects on Tire Traction. *Highway Research Record*, 214, 24–33.

Keifer C.J., & Chu H.H., 1957. Synthetic Storm Pattern for Drainage Design. *Journal of the Hydraulics Division*, 83(4), 1–25.

Kuang X., Sansalone J., Ying G., & Ranieri V., 2011. Pore-structure Models of Hydraulic Conductivity for Permeable Pavement. *Journal of Hydrology*, 399(3–4), 148–157.

Meng A., Tan Y., Xing C., Lv H., & Xiao S., 2020. Investigation on Preferential Path of Fluid Flow by Using Topological Network Model of Permeable Asphalt Mixture. *Construction and Building Materials*, 242, 1–12.

Mullaney J., & Lucke T., 2014. Practical Review of Pervious Pavement Designs. *Clean – Soil, Air, Water*, 42(2), 111–124.

Nooruddin H.A., & Hossain M.E., 2011. Modified Kozeny-Carmen Correlation for Enhanced Hydraulic Flow Unit Characterization. *Journal of Petroleum Science and Engineering*, 80(1), 107–115.

Ong G.P., Fwa T.F., & Guo J., 2005. Modeling Hydroplaning and Effects of Pavement Microtexture. *Transportation Research Record*, 1905(1), 166–176.

Putman B.J., & Kline L.C., 2012. Comparison of Mix Design Methods for Porous Asphalt Mixtures. *Journal of Materials in Civil Engineering*, 24(11), 1359–1367.

Rossman L.a., 2015. *Storm Water Management Model User's Manual Version 5.1*. Cincinnati: National Risk Management Research Laboratory, Office of Research and Development, US Environmental Protection Agency. Available online: https://nepis.epa.gov/Exe/ZyPDF.cgi?Dockey=P100N3J6.TXT (accessed on 16 October 2020).

Rossman L.A., & Huber W.C., 2016. *Storm Water Management Model Reference Manual. Volume III-Water Quality*. Cincinnati: National Risk Management Research Laboratory, Office of Research and Development, US Environmental Protection Agency. Available online: www2.epa.gov/water-research (accessed on 16 October 2020).

Sitek M.A., Lottes S.A., & Sinha N., 2020. *Computational Analysis of Water Film Thickness During Rain Events for Assessing Hydroplaning Risk Part 1: Nearly Smooth Road Surfaces (No. ANL-20/36)*. Argonne National Lab. (ANL), Argonne, IL (Unite States).

Stotz G., & Krauth K., 1994. The Pollution of Effluents From Pervious Pavements of an Experimental Highway Section: First Results. *Science of the Total Environment*, 146, 465–470.

Zhong R., Xu M., Vieira Netto R., & Wille K., 2016. Influence of Pore Tortuosity on Hydraulic Conductivity of Pervious Concrete: Characterization and Modeling. *Construction and Building Materials*, 125, 1158–1168.

Zhu H., Yu M., Zhu J., Lu H., & Cao R., 2019. Simulation Study on Effect of Permeable Pavement on Reducing Flood Risk of Urban Runoff. *International Journal of Transportation Science and Technology*, 8(4), 373–382.

Maintenance and preservation treatments

Roads and Airports Pavement Surface Characteristics – Crispino & Toraldo (Eds)
© 2023 The Author(s), ISBN 978-1-032-55149-4

Thermal performance of ultrathin multi-functional overlay

Mauro Coni*, Silvia Portas & Francesco Pinna
Department of Civil Engineering, Environment and Architecture, University of Cagliari, Cagliari, Italy

Francesca Maltinti
SOGAER – Cagliari International Airport, Cagliari, Italy

ABSTRACT: Ultrathin overlay (UTO) is a very thin layer used in the pavement to restore the surface quality and to seal and protect pavement, thus reducing the pavement's rutting and fatigue cracking and increasing its skid resistance. UTO is a non-structural layer that can improve the pavement's mechanical behavior by reducing the thermal field on the underlying asphalt materials. This paper presents an ultrathin multi-functional overlay (UMO) for this purpose. The research investigated the particular properties of mortars with different types of cement and aggregates and with a patented mixture of acrylic resins and additives. The technology, well known in concrete repair, can be extended to road asphalt pavement restoration. Experiments were performed in both the laboratory and the field. The tests demonstrate the remarkable ability of the UMO to maintain the stiffness of materials subjected to thermal stress and a high ability to support adhesion but a low skid resistance. The introduction of different grit sizes allows increasing the skid. The overlay could potentially affect many aspects of pavement maintenance, such as restoring degraded surfaces, increasing scratch resistance and hardness, sealing cracks and, in the case of airfield pavements, protecting the pavement from solvent fuels and very high temperatures.

Keywords: Paving ultrathin overlay, pavements maintenance, thermal behavior, asphalt preservation, bituminous surface, cement seal coat

1 INTRODUCTION

Safety requirements and pavement preservation impose a more efficient use of the available budget. As a result, management agencies have focused on maintaining and preserving existing pavement surfaces (Mamlouk & Zaniewski 2001). Preventive maintenance strategies allow extending the pavement service life, retarding progressive deficiency and distress. Earlier maintenance can be performed at lower costs and a high cost/benefit rate while maintaining constant safety conditions (Chenevière & Ramdas 2006]. Each highway agency has a maintenance manual that covers pavement defects, costs, alternative strategies and the lifespan of each strategy (Peterson 1991).

Thin bituminous surface (BTS) treatments are typically protective interventions in pavement maintenance with a minimum thickness of approximately 12–15 mm. In wearing course maintenance, a BTS is considered the most economical choice for low-traffic pavements (Liu *et al.* 2010). Ultrathin hot mix asphalt (HMA) comprises an HMA (12–20 mm) covering old wearing course with a thick asphalt emulsion layer or membrane. The system

---

*Corresponding Author: mconi@unica.it

DOI: 10.1201/9781003429258-42

can restore minor amounts of surface distress, thus providing a durable, friction resistant surface on existing pavement (Rahaman 2012; Ruranika & Geib 2007).

New materials with similar performances as ultrathin HMA are currently under investigation. These materials have thicknesses of less than 3–4 mm, and unlike previous materials, they permit the extension of durability with a simple laying technique. In 2007, the research investigated the particular proprieties of mortars with different types of cement, aggregates and others components with a nominal thickness of 2 mm (see Figure 1). Ultrathin multifunctional overlay (UMO) denotes the exceptionally low thickness of the new application investigated in the research and its capability to improve many different aspects of the pavement performance (Coni 2013).

Figure 1.   Bituminous sample surface with UMO.

UMO consists of high-strength concrete, micronized quartz powder, a mixture of acrylic resins and additives emulsified in water. The moisture content and proportion of components allow one to control the workability and consistency of the mixture. The application is well known in concrete repair and for protection of canals, dams, seawalls, docks, and airport apron slabs (Perkins 2003). The treatment allows for the sealing of cracks, restoration of a degraded surface, waterproofing, and protection from solvent fuels and can accommodate photocatalytic agents.

Previous studies determined that UMO has excellent scratch resistance and durability. Therefore, the application has the potential for use in port areas or in areas of bulk goods handling, where the action of abrasion attacks the pavement surface. Moreover, UMO can reduce a thermal field in the underlying asphalt materials with mechanical benefits for the pavement, deformations, and rutting. Thermal effects and oxidative aging might play a relevant role in the crack initiation and propagation mechanism. The expansion and contraction during thermal cycles can activate cracking that cannot fully recover in aged HMA (Croll 2009). Research performed in Kenya (Wamburga *et al.* 1999) indicated that solar radiation could be a most critical factor in accelerating surface age hardening and top-down cracking. In 2015, Alavi developed a compressive model for predicting thermal cracking in asphalt pavements considering the continuous evolution of the asphalt mixture properties with oxidative aging over time.

To assess the intrinsic parameters (thermal conductivity and heat capacity) of a bituminous mix, Houel in 2010 developed a procedure to determine the thermophysical characteristics of bituminous concrete and its influence on fatigue properties. Moreover, a fresh surface can control the absorbed energy and temperature, improving air quality and reducing heat pollution (Akbari 2001; Fang 2010; Golden & Kaloush 2006).

A recent study (Zachry 2017) demonstrated the importance of temperature reduction by controlling the thermal properties of HMA by selecting highly conductive aggregates and graphite powder. From this perspective, UMO has remarkable potential for controlling the urban heat island (UHI) phenomenon, and thermal distress in asphalt pavement.

2 RESEARCH OBJECTIVE AND APPROACH

Table 1. summarized the mechanical tests conducted.

Table 1. Tests conducted in this research.

Laboratory tests	No. of samples	In situ tests	No. of samples
Adhesion capability	8	Skid tests	28
Mechanical performance	24	Tenacity	4
Hot/cold cycles resistance	12	Laying technique	3
Skid resistance	16	Thermal	1 test section
Thermal	2×3 tests		

The experimental program involved laboratory and field tests. The tests focused on the adhesion capability, skid resistance, tenacity and mechanical performance of specimens with and without overlay and under hot/cold cycle aging. Furthermore, different laying techniques were analyzed in the in situ tests to adjust the laying technique and to optimize the total thickness and number of laying passes. After the application of the UMO at the experimental sites, the monitoring phase under real traffic conditions began.

The current paper summarizes the outcomes of the mechanical tests and investigates how the treatment could reduce the temperature inside bituminous pavements.

2.1 Summary of the experimental tests

2.1.1 Adhesion capability
The surface of eight asphalt cores, pull out from a road pavement (AADT: 13,000 vehicles/day with 32% heavy vehicles) assessed the bond of UMO on the surface layer of asphalt.

Four cores taken from pavements side, not subjected to traffic, and four taken from the wheel paths, ware used to make four specimens obtained by connecting UMO with two cores, matching the exterior face of the cores.

A comparison of the test results obtained by joining specimens with UMO (between 0.78 MPa and 1.03 MPa) and those joined by bitumen emulsion (between 0.010 MPa and 0.016 MPa) illustrates that the shear strength obtained using UMO is two orders of magnitude higher than that using bitumen emulsion (Figure 2).

When using bitumen emulsion, the failure surface occurs at the interface between the asphalt concrete and emulsion. In contrast, either the rupture of the bond between the cement mortar and asphalt sample or cohesive failure in the asphalt may occur in the specimens connected with UMO.

2.1.2 Mechanical performance
Indirect tensile tests were conducted to characterize the mechanical behavior of the specimens. Three sets of 4 samples (100 mm in diameter and 40 mm high) tested with Nottingham

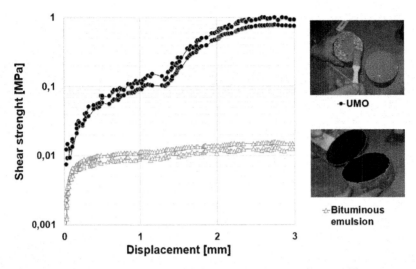

Figure 2.   Shear strength in samples connected with UMO and bituminous emulsion.

Asphalt Test (NAT) without a coating (0%), partially covered (only on one base face 28%), and entirely dressed over the surface (100%).

UMO has a stiffness modulus of approximately 31,000 MPa, and when applied to a bituminous surface, UMO increases the elasticity and surface modulus of a thin layer. The improvements derived by adding a layer on the surface; filling the superficial porosity of the HMA and strengthening the tie between superficial grains (Figure 3).

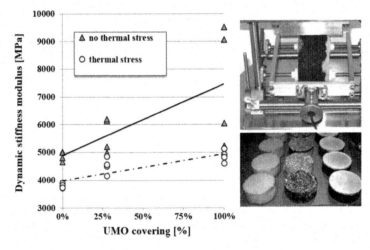

Figure 3.   Dynamic stiffness moduli obtained from NAT tests and samples dressed fully, partially and without UMO for indirect tensile tests.

The mean value of the dynamic stiffness modulus without UMO was 4,864 MPa. The dynamic stiffness modulus values of the second and third sets were 5,621 and 7,433 MPa, respectively.

The mean value increased by 15.56% for the samples partially covered and 51.81% for the fully covered specimens compared to the uncovered specimens. The data dispersion increased sharply with an increase in the amount of surface covered. This result confirms the positive effect of the UMO; however, the results are highly dependent on the laying method, texture characteristics and air void of the samples surface, in addition to various other factors. Therefore, the results are not sufficiently reliable and thus remain qualitative.

2.1.3 Stiffness under hot/cold cycles

Hot/cold cycle resistance tests were conducted by subjecting 24 specimens to cycling temperature between $-10°C$ and $+70°C$ for five days. Before the test, the samples were subjected to a temperature of $70°C$ in the thermostatic chamber for five hours and then immediately immersed in water at a temperature of $20°C$ for 4 hours. After a quick drip, the wet specimens were frozen for 8 hours, followed by a resting period at ambient temperature for 7 hours. Finally, before the NAT test, the materials were held in a thermostatic chamber at $20°C$ for 4 hours. Hot/cold cycles and water immersion before freezing affect HMA, as the water fills the pores and micro-cracks.

The fluid expansion during the cold phase opens the voids and decompresses the material. The specimens without thermal stress exhibit a mean stiffness of 5,972 MPa, whereas those subjected to hot/cold cycles, exhibit a mean value of 4,868 MPa (-18.49%). Compared to the uncovered specimens, the partially coated samples exhibit a 19.74% reduction, whereas the wholly covered specimens exhibit a 34.76% reduction in the dynamic stiffness modulus. Figure 3 summarizes the results.

2.1.4 Skid resistance

Friction tests were conducted in a laboratory using a British Pendulum under wet conditions. The preliminary results demonstrated the poor skid resistance of the pavement, ranging from 51 to 58. Four specimens (50 × 50 cm) with quartz grains (size 0.5 – 0.7 – 1.0 – 1.5 mm) over the UMO show acceptable average values (73 BPN).

In 3 in situ sections, the friction tests exhibited different results. The skid value obtained in the first tested site (fuel tanker parking area, in Figure 4a, with two-layer UMO, laid under the optimal conditions) exhibited a mean value of 74 BPN (British Pendulum Number). In the second section, located along a provincial road (Figure 4b), vehicles reach speeds of up to 70 km/h. The site was selected to establish the friction loss under different UMO thicknesses and laying techniques to limit the saturation of the existing pavement macro-texture.

Figure 4.   Specimens used for the skid tests (top and the left) and three different experimental sites: a) a parking area and transit path of fuel tankers; b) a road in an industrial area; and c) parking area in an urban park.

The initial mean value of 50 BPN obtained represents a significant decay compared to the existing bituminous pavement, which exhibited a value of 65 BPN. The problem can be mitigated using different size grains inside the UMO. Two different grain sizes (0.5–1.0 mm, 2.0–3.0 mm) introduced in the third section (parking area in an urban park, Figure 4c) improve the skid. The 25 skid tests yielded a mean value of 76 BPN for the 0.5/1.0 mm grain size and a mean value of 81 BPN for the 2.0/3.0 mm grain size.

2.1.5 *Tenacity*

Two sections of bituminous pavement, covered with UMO and without, subjected to 10 minutes of the repeated action of a caterpillar track excavator (Figure 5) were very stressed under loads in compression, shear, and torsion. This unconventional test is a qualitative and not a standard procedure.

Figure 5.   Tenacity test results. No damage with UMO treatment (on the left) and abrasion and cracks on the untreated pavement (on the right).

The UMO showed excellent adhesion and toughness on the asphalt surface. A comparative analysis of the two section shows that the aggressive action of the steel caterpillar track produced more extensive and severe damage in the absence of UMO. In the pavement covered with UMO no cracks, tears or degradations occur.

The only visible effect is related to the polishing; the coating abrades below the track due to the action of rubbing with steel. This remarkable performance represents a potential application for UMO in goods handling and loading areas, such as ports and warehouses.

2.1.6 *Laying technique*

The surface is first cleaned with high-pressure water jets to remove coarse dirt, debris, and any possible loose surface aggregates. The same jets are used with suitable nozzles to apply fluid mortar at a density of 3–4 kg/m^2. The dosage of components allows one to regulate the conditions of traffic limitations, the norm of which is 24 hours after laying. The laying technique, texture characteristics and surface conditions (e.g., moisture, temperature, aging of wheel course, raveling) are the main factors affecting the skid resistance.

The UMO surface treatment to improve adherence is sprayed in a double layer introducing high-quality grains, with sizes of 0.7–1.0 mm, after the second layer. Under wet conditions, the shedding of modified sodium silicate allowed for rapid hardening in approximately 2 min.

3   THERMAL ANALYSIS

The different thermal analysis investigated the response of the asphalt pavements covered with UMO. In the laboratory, thermal tests on specimens irradiated with an infrared source simulated solar heating.

The progression of the thermal field at the front and rear samples, via thermographic analysis, was recorded during blowtorch exposing to simulate jet blast from an aeronautic engine.

The first test subjected two rectangular samples (40 cm × 60 cm × 5 cm each) side by side, one covered and one not, to a heat source (see Figure 6).

Figure 6.   Test 1. Back and front of the rectangular specimens.

Two thermo-cameras located on the front (irradiated) and back of the samples measure the thermal field (Figure 7).

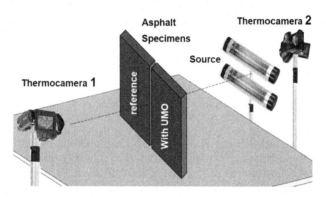

Figure 7.   Test 1. The configuration of the thermal analysis conducted in the laboratory.

The untreated sample showed the first signs of failure after approximately 60 min of irradiation. Figure 8 shows the evolution of the temperature of the back surface after 60 min of irradiation.

The maximum temperature in the covered sample was approximately 45°C, whereas that in the untreated sample was 57°C. Furthermore, the untreated sample collapses due to irradiation after about 75 min. Before the collapse, the temperature difference was 12°C (see Figure 9).

The second thermal tests took place in a parking area of a fuel transportation company, where stationary and heavy transit vehicles operate. An extended period of loads, high summer temperatures and the steel contact area from the landing gear of the fuel tank semi-trailer trucks, emphasize the rutting. A large circular area of the bituminous pavement at the site was covered with UMO, and a series of thermographs were taken from the aerial view

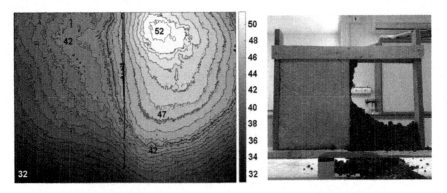

Figure 8.   Test 1. Temperature evolution of the back surface after 60 min. The untreated sample collapses after approximately 75 min.

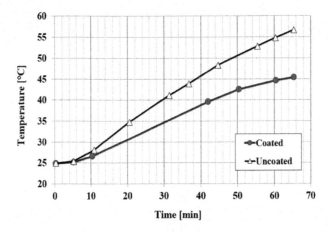

Figure 9.   Test 1. Evolution of the maximum temperature on the back surface.

over 24 hours. Early in the morning, the surfaces with and without UMO registered the same temperature, reaching a balanced condition during the night. During the day, approaching noon, the Sun heats the surfaces differently. The surface with UMO maintains a lower temperature compared to the adjacent asphalt pavement not covered with UMO. Figure 10 shows the temperature distribution 8 hours after sunrise (6:00 am to 2:00 pm).

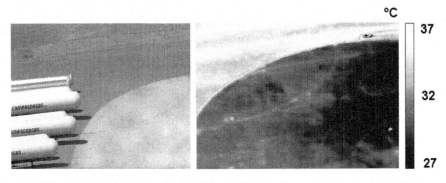

Figure 10.   Test 2. Temperature distribution 8 hours after sunrise.

Five points from the covered area and 5 points from the uncovered zone were monitored over 24 hours. Figure 11 shows the evolution of temperature and the mean temperature value of the two areas. The analysis shows a peak difference of up to 7.5°C.

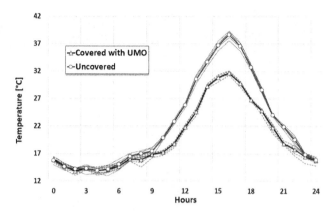

Figure 11. Test 2. Temperature evolution at 5 points in the treated zone and 5 points in the area without UMO and their mean values.

Figure 12 shows the temperature distribution at 2:00 pm in the corner of the concrete slab, asphalt pavement, grass, and steel, at the same experimental site. The thermograph indicates a lower temperature in the concrete slabs and grass and highlights the cracks in the corner. These properties of UMO overlay are particularly interesting for the durability of the

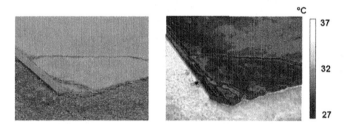

Figure 12. Test 2. Temperature distributions on the corner of the concrete, asphalt pavement, grass, and steel.

pavement, reducing viscous-elastic deformation and cracking. Among the available approaches, two were used to estimate the variation in the permanent deformation of HMA due to temperature reduction using UMO.

In the layered vertical permanent strain approach, the total deformation is the sum of each deformation as a function of repeated load applications. The Asphalt Institute (May and Witczak 1992) has developed a model that considers the effects of mix design properties:

$$\log \varepsilon_p = -14.97 + 0.408 \log N + 6.865 \log T + 1.107 \log \sigma_d - 0.117 \log \eta + 1.908 \log V_{eff}$$

$$+ 0.971 \log V_a$$

(1)

where:

ε_p: Permanent axial strain,
N: Number of load repetitions to failure,
T: Temperature, °F,
σ_d: Deviator stress, psi,
η: Viscosity at 70°F, Ps x 10^6,
V_{beff}: Percent by volume of effective asphalt content, and
V_a: Percent by volume of air voids.

By fixing σ_d, η, V_{beff}, V_a at typical values (80, 40, 5 and 5, respectively), the equation becomes:

$$\log \varepsilon_p = -11.04 + 0.408 \log N + 6.865 \log T \tag{2}$$

T has the most considerable contribution to the permanent deformation. The equivalent number of cycles reduction by considering the temperature reduction obtained with UMO (maximum $\Delta T \approx 7°C$ after 12 hours of daily solar irradiation). For 5,000 load cycles during a day, the $\Delta \varepsilon_p = -49.2\%$ is estimated, for each hour, related to a different temperature observed during Test 2. The same ε_p occurs by increasing the number of cycles in the area with UMO by 5.68 times.

A different approach uses constitutive relationship derived from the statistical analysis of a repeated load in laboratory tests, assuming a ratio of the permanent strain ε_p to the elastic (resilient) strain ε_r. Kaloush and Witczak in 1999 developed a simplified model with two main parameters (T and N):

$$\log(\varepsilon_p/\varepsilon_r) = -3.1555 + 0.3994 \log N + 1.734 \log T \tag{3}$$

where

ε_p: Accumulated permanent strain after N cycles,
ε_r: Resilient strain as a function of the asphalt mix properties,
N: Number of load repetitions,
T: Temperature, °F.

In this case, with the same hypothesis assumed above, $\Delta \varepsilon_p/\varepsilon_r = 88.6\%$ and the same $\Delta \varepsilon_p/\varepsilon_r$ are achieved, increasing the number of cycles by 1.35 times. Usually, the cracking phenomenon of asphalt pavements is an effect of the stress concentrations. Weak subgrade support (Hoque 2006), age strain hardening, low temperature and fatigue stress under repeated loading are believed to be the critical processes in asphalt cracking (Mates 1988; Harvey et al. 1999). However, many studies have noted that this cracking starts in the upper surface, and some authors (Lytton et al. 1993; Myers et al. 1998.; Wamburga et al. 1999.) suggest that solar radiation and thermal stress could be a principal factor in top-down cracking. Matsuno and Nishizawa in 1992 reported that in pavements affected by top-down cracking, cracks were not present when in the pavement shaded by overpasses. The expansion and contraction of pavement during alternations in heating and cooling often determine the amount of cracking and horizontal deformation. Therefore, although reliable predictive models are not available, thermal fatigue distress has significant implications for improving pavement design and maintenance strategies (Wang et al. 2013).

These properties of UMO are particularly promising for the durability of pavements and the mitigation of urban heat. Several studies are currently investigating how to increase the albedo of the coating materials, such as for road paving and roofing applications. Urban surfaces are the most significant cause of the UHI phenomenon and thermal concentration in urban areas (Figure 13). UHIs are significantly warmer than surrounding areas. Aside from their effect on temperature, UHIs may also produce significant effects on local

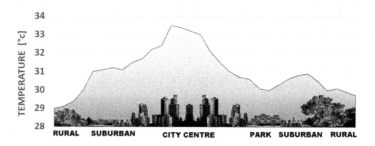

Figure 13.   Urban heat island (UHI) profile.

meteorology and ozone pollution and can increase the energy demands for refrigeration and air conditioning systems in urban areas (Santamouris & Kolokotsa 2016).

The solar reflectance index (SRI) is a measure (according to ASTM E 1980) of the constructed surface's ability to stay cool in the sun by reflecting solar radiation and emitting thermal radiation. A standard black surface has an SRI of 0, and a standard white area has an SRI of 100. UMO as a white cement topping is characterized by a high SRI (approximately 60), ten times more than that of an asphalt surface.

Table 2.   Typical SRI values.

Material	SRI	Material	SRI
New drain asphalt	1–6	White cement topping (UMO)	58–63
Aged asphalt	5–15	Light stone	55–62
Cement concrete gray	27–33	Dark stone	20–26
Cement concrete white	60–65	Forest	10–20
Snow white	79–80	Grass	20–25
Brick, stone	20–40	White paint road sign	70–90

Another promising application of UMO is for highly thermal stressed pavements, such as in airport pavement subject to engine blast, where temperatures higher than 200°C can hit the pavements (see Figure 14). Extreme situations occur in pavements subject to a vertical jet take-off, where temperatures of over 500°C can hit the pavements.

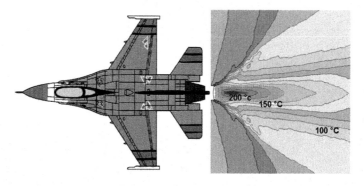

Figure 14.   Typical temperature distribution on the pavement due to jet blast.

435

The third (Figure 15 and Figure 16) and fourth (Figure 17 and Figure 18) tests were conducted with two cylindrical specimens with and without UMO (diameter = 100 mm, thickness = 20 mm) under a blowtorch jet, modifying the exposure time and source distance. During the laboratory tests, temperatures below 11–13°C occur in UMO-coated specimens and a lower value of 7.5°C in situ.

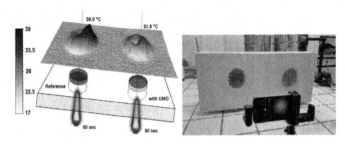

Figure 15.   Test 3. Temperature distributions in two samples with and without UMO under a blowtorch jet at a distance of 10 cm for 50 seconds.

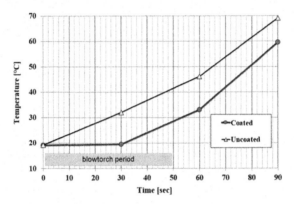

Figure 16.   Test 3. Temperature evolution on the back of two samples with and without UMO under a blowtorch jet at a distance of 10 cm for 50 seconds.

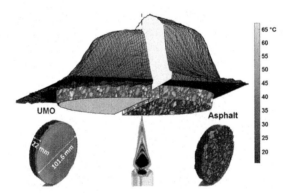

Figure 17.   Test 4. Temperature distributions in two samples with and without UMO under a blowtorch jet at a distance of 40 cm for 180 seconds.

436

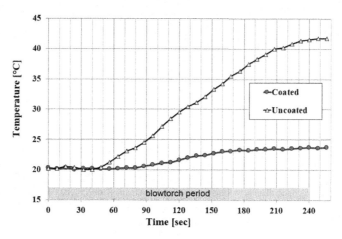

Figure 18. Test 4. Temperature evolution on the back of two samples with and without UMO under a blowtorch jet at distance of 40 cm for 180 seconds.

4 CONCLUSION

The test results demonstrate the remarkable capacity of UMO to protect asphalt pavement surfaces, seal cracks, and improve the scratch resistance while maintaining the stiffness of the materials subjected to thermal stress. The coating has shown a remarkable ability to support adhesion to asphalt and good toughness when subjected to the action of a tracked vehicle under different operating conditions. The low skid resistance has overcome by introducing different grit sizes, which has also made it possible to obtain good aesthetic qualities. The treatment could affect many aspects of pavement maintenance, including restoring degraded surfaces, improving scratch resistance and hardness, waterproofing and sealing cracks, and, in the case of airfield pavements, protecting the pavement from solvent fuels or extremely high temperatures.

In urban areas, an ultrathin cement layer can help to reduce the thermal concentration in urban areas, or UHIs. A temperature reduction of 7.5°C was observed in the full-scale test of the parking area.

UMO also reduces the service temperature of the asphalt due to its higher reflectivity. This characteristic is particularly important in warmer climates or areas subjected to strong irradiation, where the pavement can reach 50–60°C. Such high temperatures increase material deformation, resulting in the accumulation of plastic deformations and irregularities in the road pavement.

REFERENCES

Akbari H. et al., 2001. Cool Surfaces and Shade Trees to Reduce Energy use and Improve Air Quality in Urban Areas. *Solar Energy* 70 (3), 295–310. 8

Alavi M. et al. 2015. A Comprehensive Model for Predicting Thermal Cracking Events in Asphalt Pavements, *International Journal of Pavement Engineering*, DOI: 10.1080/10298436.2015.1066010

Chenevière P., Ramdas V. 2006. Cost-benefit Analysis Aspects Related to Long-life Pavements, *International Journal of Pavement Engineering*, 7:2, 145–152, DOI: 10.1080/10298430600627037-

Coni M. 2013. Ultrathin Multi-Functional Overlay, *Proceedings ASCE Airfield and Highway Pavements Conference* 2013, Los Angeles

Croll J.G.A. (2009). Possible Role of Thermal Ratcheting in Alligator Cracking of Asphalt Pavements, *International Journal of Pavement Engineering*, 10:6, 447–453, DOI: 10.1080/10298430902730547

Fang K. et al. 2010). *Reductions in Ground-Level Ozone Pollution through Urban Heat Island Mitigation Strategies Including Rehabbing Land Occupied for Transportation-Related Uses*: Case Study of Fresno, CA, 90th Annual Meeting of the Transportation Research Board – Transportation and Air Quality

Golden J.S. and Kaloush K.E. 2006. Mesoscale and Microscale Evaluation of Surface Pavement Impacts on the Urban Heat Island Effects, *International Journal of Pavement Engineering*, 7:1, 37–52, DOI: 10.1080/1029843 500505325

Harvey J.T. et al. 1999. CAL/APT Program: Test Results from Accelerated Pavement Test on Pavement Structure Containing Untreated Aggregate Base (AB) Section 501RF. Report Prepared for the California Depart ment of Transportation, Pavement Research Centre, University of California.

Hoque Z. 2006. Chapter 19. In: T.F. Twa ed. *Handbook of Highway Engineering*. Boca Raton, FL: Taylor & Francis Publication.

Houel A. et al. 2010. Thermomechanical Characterization of Asphalt Pavements in Laboratory Conditions, *International Journal of Pavement Engineering*, 11:6, 441–447, DOI: 10.1080/1029843090325411

Kaloush K. and Witczak M.W. *Development of Permanent to Elastic Strain Ratio Model for Asphalt Mixtures*. Development of the 2002 Guide for the Design of New and Rehabilitated Pavement Structure, University of Maryland, CollegePark, Maryland, 1999

Liu L. et al. 2010. *Costs and Benefits of Thin Surface Treatments on Bituminous Pavements in Kansas*. 89th Annual Meeting of the Transportation Research Board, National Research Council, Washington D.C.

Lytton B. et al. 1993, *Design for Asphalt Pavements for Thermal Fatigue Cracking, Research Report* 284-4, Texas Transportation Institute.

Mamlouk M.S. and Zaniewski J.P., 2001. Optimizing Pavement Preservation: An Urgent Demand for Every Highway Agency, *International Journal of Pavement Engineering*, 2:2, 135–148, DOI: 10.1080/10298430108901722

Mates 1988. *Michigan Department of Transportation*. Flexible Pavement Distress – part 1, July 21.

Matsuno S. and Nishizawa T., 1992. Mechanism of Longitudinal Surface Cracking in Asphalt Pavement. 7° International Conference on Asphalt Pavements. University of Nottingham.

May R.W. and Witczak M.W. 1992. *An Automated Asphalt Concrete Mix Analysis System*. Proceeding of the Association of Asphalt Paving Technologists, Volume 61. South Carolina.

Myers L.A. et al. 1998. Mechanisms of Surface Initiated Longitudinal Wheel Path Cracks in High-type Bituminous Pavements. *Asphalt Paving Technol.*, 67, 401–432.

Perkins P.H. 2003. *Repair, Protection and Waterproofing of Concrete Structure*, Third Edition F&FN Spon, London, ISBN 0 419 202 803

Peterson D.E. 1991. Evaluation of Pavement Maintenance Strategies, NCHRP Synthesis of Highway Practice, *Transportation Research Board*, ISSN: 0547–5570.

Rahaman F. et al. 2012. Evaluation of Recycled Asphalt Pavement Materials from Ultra-Thin Bonded Bituminous Surface, Proceedings of GeoCongress 2012 (#Sustainable Pavements) Oakland, California.

Ruranika, M.M., Geib J., 2007. Performance of Ultra-Thin Bounded Wearing Course (UTBWC), Report Research, *Minnesota Department of Transportation*

Santamouris M. and Kolokotsa D. 2016. *Urban Climate Mitigation Techniques Edited by Routledge*, NY

Wamburga J.H.G. et al. 1999. *Kenya Asphaltic Materials Study*, Transportation Research Board 78th Annual Meeting, Washington DC.

Wang H. Al-Qadi I Portas S. and Coni M. 2013. *Three-Dimensional Finite Element Modelling of Instrumented Airport Runway Pavement Responses*, TRB Transportation Research Board, 92th Annual Meeting, Washington DC.

Xijun Shi Zachry Department of Civil Engineering, Texas A&M University, College Station, TX, USA, et al., 2017. Effects of Thermally Modified Asphalt Concrete on Pavement Temperature, *International Journal of Pavement Engineering*, DOI:10.1080/10298436.2017.1326234

Roads and Airports Pavement Surface Characteristics – Crispino & Toraldo (Eds)
© 2023 The Author(s), ISBN 978-1-032-55149-4

Shrinkage crack induction for the study of repair agents in rigid pavements

Sebastián López S.
Facultad de Ingeniería, Universidad Nacional de Colombia, Bogotá, Colombia

Gloria I. Beltran C.*
Departamento de Ingeniería Civil y Agrícola, Facultad de Ingeniería, Universidad Nacional de Colombia, Bogotá, Colombia

Henry O. Meneses M.
Clínica de Pequeños Animales, Facultad de Medicina Veterinaria y Zootecnia, Universidad Nacional de Colombia, Bogotá, Colombia

Pedro F. B. Brandão
Departamento de Química, Facultad de Ciencias, Universidad Nacional de Colombia, Bogotá, Colombia

ABSTRACT: In the construction of rigid pavements, shrinkage cracking may occur at early stages, affecting the durability, strength, and aesthetic appearance of concrete surface layers. When studying the effectiveness of treatments to repair this type of cracking, resorting to reduced-scale physical models in the laboratory is quite relevant. However, simulating the dimensions and characteristics of shrinkage cracks is not an easy task. In this work, a successful protocol is proposed, based on the design and construction of an experimental assembly in a wind tunnel, to induce plastic shrinkage cracks in cement-based materials. Thus, ideal experimental conditions for crack generation were established: temperature, wind speed and relative humidity of the environment. Also, the dimensions and characteristics of moulds for sample preparation were defined. As a result, the system allowed to generate typical plastic shrinkage of 2 cm depth and 1 mm opening, whose geometry was defined by non-destructive tomographic images technique. Induced cracks are suitable for testing any remedial agent, including biological or chemical ones.

Keywords: Rigid Pavements, Shrinkage Cracks, Physical Modeling, Wind Tunnel, Computed Tomography

1 INTRODUCTION

The performance of hydraulic concretes is closely related to the materials used and to the construction processes of placement, compaction, and curing, as determining factors in the final quality (Orozco *et al.* 2018). Eventually, cracking problems can occur due to complex chemical reactions during the concrete curing process, causing adverse effects such as changes in strength, permeability, stiffness as well as its aesthetic appearance. It has also been mentioned that plastic shrinkage cracks may not significantly affect the strength of the concrete, but they do affect its durability (National Ready Mixed Concrete Association 2014). Considering these effects, several studies have addressed cracks in concrete in fresh

---

*Corresponding Author: gibeltranc@unal.edu.co

DOI: 10.1201/9781003429258-43

and hardened state (Becker 2015; Toirac 2004). Likewise, proposals have been made to classify cracks according to their depth as shallow or deep cracks, according to their movement as active or inactive, and according to their moisture condition as dry or wet (with or without water pressure). Depending on these, repair methods are recommended (Sika Colombia 2014).

The study of the effectiveness and efficiency of treatments to repair this type of cracks, usually requires physical modelling on a reduced scale in the laboratory, under controlled conditions. However, simulating the characteristics of shrinkage cracks can be an intensive trial-and-error task. In this work, a successful protocol, and an experimental assembly to simulate the dimensions and characteristics of plastic shrinkage cracking in the laboratory were developed. These contributions, complemented with non-destructive analysis by tomographic images are very useful to study and characterize shrinkage cracks resorting to reduced-scale physical models.

This paper initially presents some relevant fundamentals. Then materials and methods to plastic shrinkage cracking generation in cement-based materials are defined, including the design and construction of a wind tunnel to simulate environmental conditions, as well as the moulds for sample preparation. Then, experimental results obtained are discussed, establishing those optimal conditions for crack generation at laboratory scale. The analyses are supported with tomographic images illustrating the geometric characteristics of the induced cracks. Finally, some concluding remarks and recommendations are presented.

2 BACKGROUND – FUNDAMENTALS

2.1 Plastic shrinkage cracks

Plastic shrinkage cracks in concrete are caused by rapid evaporation of exudation water, generating tensile forces at the surface that exceed the early strength of the material. Typically, these cracks exhibit depths between 1 and 2 cm and openings of 0.5 to 1 mm. Becker (2015) mentioned the importance of the volumetric changes that the material undergoes during the setting process, mainly due to the hydration and evaporation processes of the water present in the mixture; this allows us to understand from another perspective the mechanism by which cracks appear in concrete. Once the concrete is placed, it begins to set in its fresh state, where part of the water hydrates the cement, another part is exuded, and still another part evaporates with time. At the end of the process, the concrete is hardened, and its volume is reduced due to shrinkage. The time of occurrence of plastic shrinkage cracking depends on the rate of evaporation relative to the exudation of concrete; cracking is more likely to occur when evaporation exceeds exudation.

The cracking mechanism is described in detail in several reports (Construcción y Tecnología en Concreto 2019; Grupo Polpaico 2000; Londoño 2012; National Ready Mixed Concrete Association 2014; Toxement 2010; Topçu et al 2017; Uno 1998; Zhang et al. 2022), which can be summarized in the following steps: a) exudation water appears on the surface, b) water evaporation is higher than exudation water velocity, c) concrete surface dries, d) concrete surface shrinks, e) wet concrete resists shrinkage, f) plastic concrete stresses are developed, and g) plastic shrinkage crack is formed.

Among the most influential factors in the generation of plastic shrinkage cracks, the following have been identified: environmental conditions (wind speeds above 5 mph, low relative humidity and high temperatures), material characteristics (high cement fineness, high percentage of filler, excessive mixing water), construction aspects that restrict volume changes (Londoño 2012; National Ready Mixed Concrete Association 2014; Toxement 2010). Other variables such as water/cement ratio, thickness of the concrete section, fines content, additives, and fibres are described by the phenomenon of plastic shrinkage cracking in concrete (Sayahi et al. 2014).

2.2 Experimental methods for crack induction

Field experimentation tends to represent more realistically the cracking phenomenon (Sayahi *et al.* 2022), but it is not always possible to carry it out due to costs and difficulty to control the incident variables. Hence, the study of plastic shrinkage cracking in cementitious materials usually requires laboratory experimentation. For this purpose, concrete or mortar mixtures are designed to handle small-scale samples, and environmental conditions promoting the induction of cracking are simulated. Sayahi *et al.* (2017) refer to two methods for crack formation: the ring test method NT Build 433, proposed by Johansen and Dahl (1993) and the ASTM C1579-21 method. Although the ASTM method was developed to study the plastic shrinkage cracking behaviour of concrete mixtures reinforced with fibres, it could be applied to the cementitious materials studied here, with appropriate modifications if the expected cracking does not occur.

The ASTM method indicates certain reproducible conditions in a wind tunnel to generate plastic shrinkage cracks: Temperature $36 \pm 3°C$, wind speed 4.7 m/s (17 km/h) and relative humidity of $30 \pm 10\%$. As for the characteristics of the samples, a width of 35.5 cm by 56 cm in length and 10 cm in height is established. A profiled sheet with three triangular notches is placed at the bottom of the moulds. The central notch is 5 mm at the base and 6.35 cm high, to have a smaller thickness of material in the middle of the sample, and to promote cracking by this section; the other two notches are half that size and work as internal restraint.

2.3 Non-destructive analysis with computerized tomography

Computed tomography (CT) is a non-invasive diagnostic imaging technique that uses a combination of X-rays (XR) and computer systems to obtain a series of cross-sectional images, conventionally used in health sciences (Agencia Valenciana de Salud 2015). In this technique, both the X-ray sources and receptors are mobile and by means of data analysis and computational algorithms, cross-sections of the observed object can be reconstructed.

The absorption of X-rays is directly proportional to the thickness of the study object and the linear attenuation coefficient of the tissues or structure through which the beam passes. Due to spatial resolution and image reconstructions, CT is an excellent tool that provides a detailed assessment of the internal structure in high dense tissues like bone, allowing the identification of fissures, fractures, and small irregular lesions (Ohlerth & Scharf 2007). In structural engineering, non-destructive experiments have been performed for high-precision measurements in pathological processes in concrete or mortar (Dong *et al.* 2018). Also, X-ray tomography is reported as an effective non-destructive methodology to monitor calcite deposition and the progress of crack repair during microbiologically induced carbonate precipitation (MICP) (Tamayo-Figueroa *et al.* submitted).

3 MATERIALS AND METHODS

3.1 Preliminary tests

To identify sensitive variables in the shrinkage crack generation process, some preliminary tests were carried out in a small-scale wind tunnel (0.35 m wide and 1 m long) available at the structural engineering laboratory of the National University of Colombia. Wooden moulds 15 cm long, 12 cm wide and 5 cm thick were used for preliminary samples preparation.

Mortar mixtures were tested with different cement/sand ratios (1/2.75 and 1/2.0), water/cement ratios (0.53 and 0.55), wind speeds (1.2 and 1.4 m/s), wind exposure times (6 and 24 h), sample thicknesses (3, 4 and 5 cm), as well as the insertion of metallic elements to promote cracking (nails and bent sheets). Two types of 0.5 mm thick aluminium sheets were used: one with the dimensions established in ASTM C1579-21 and the other modified according to the cutting and bending shown in Figure 1 (the central notch is 2 cm at the base and 2.5 cm high).

Figure 1.   Aluminium sheet bending diagrams at the base of moulds.

3.2   *Design and construction of wind tunnel for environmental conditions simulation*

Given the limitations found in the preliminary tests, it was necessary to design and build a wind tunnel that would allow greater control of variables. The dimensions of the tunnel were 60 cm long × 27.5 cm wide; the area for sample placement was 30.5 cm by 21.5 cm with a height of 12.5 cm; the remaining area was used for the electrical, electronic, temperature control and vent components (Figure 2). In this regard, 110V fans capable of producing wind speeds close to 5 m/s were installed. Electronic circuits were also implemented to regulate the speed.

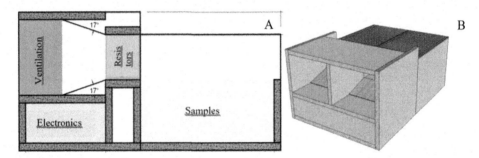

Figure 2.   General diagrams of designed wind tunnel. A) transversal side view; B) tridimensional view with ventilation ports at the front.

3.3   *Design and construction of moulds for sample preparation*

The moulds were designed to produce samples of 15 cm long, 12 cm wide and 5 cm thick. To facilitate sample extraction, a disassembled model was built with 4 mm thick rectangular acrylic sheets, joined by means of angles with nuts and bolts, as shown in Figure 3.

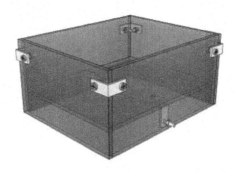

Figure 3.   General diagrams of the designed moulds.

3.4 *Mix design*

For this work, a mortar prepared with Portland cement for general use, water and two types of quartz sand in equal proportions was used, whose granulometries are shown in Figure 4 and the main properties are included in Table 1. The combined gradation of medium sand (0.4 to 2.0 mm) and fine sand (0.1 to 0.4 mm) promoted proper workability of the mortar.

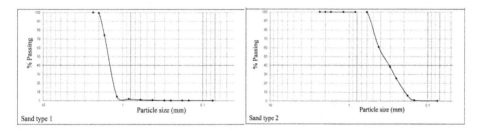

Figure 4.   Grain-size curves for the two types of quartz sands used.

Table 1.   Characteristics of the sands used for mortar preparation.

	Sand type 1	Sand type 2
Description	Light yellow to light gray sand; rounded to subrounded shape	
Curvature coefficient	0,96	0,88
Uniformity coefficient	1,30	2,16
Effective diameter (mm)	1,22	0,19

3.5 *Cracking characterization by means of CT scans*

Computed tomography (CT) was used as a non-destructive observation method for the characterization of the induced cracks in mortar samples. Images were obtained with a 1-detector row scanner (Tomoscan AV Philips) (Figure 5A). Whit axial scan mode, CT images were acquired with the following parameters: 170 mA, 120 kVp, 1.3 s rotation time and 2.0 mm slide thickness. Between 7 to 10 images were obtained from each mortar sample. To mark a physical reference point, a metal segment was placed on the upper right surface of each mortar sample (Figure 5B). After acquisition, the images were archived and organized digitally in DICOM format. Subsequent analysis of the selected images was carried out using the free software Weasis Version 1.2.5 (https://nroduit.github.io/en/), which allowed the morphological characterization of the shrinkage induced cracks.

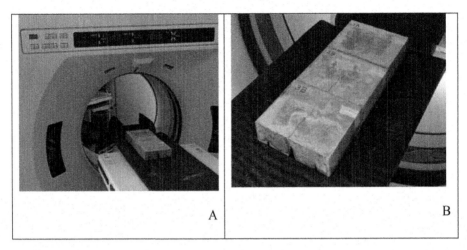

Figure 5.   Tomograph (A) used for imaging of cracks in mortar samples(B).

4   RESULTS AND DISCUSSIONS

4.1   *Preliminary test analysis*

In the preliminary tests, mortar samples were prepared with different conditions (Table 2). Most of the specimens did not exhibit cracks but some were completely vertically fractured, dividing the sample in two and did not meet the requirements for the purposes of this study. Additionally, the cracking times of 6 h and 24 h obtained, were excessive considering that cracking must be achieved in short times to perform efficient experimental processes (Becker 2015).

Table 2.   Variables considered in the preliminary tests for the preparation of cracked mortar samples.

| Test No. | Metal sheet | Mix mortar | | Nails | Time in tunnel (h) | Sample thickness (cm) | Wind speed (m/s) |
		cement/ sand	water/cement ratio				
1	No	1/2.75	0.53	No	6	5	1.2 ± 0.2
2	Yes	1/2.75	0.53	No	6	5	1.2 ± 0.2
3	Yes	1/2	0.55	Yes No	6	5	1.2 ± 0.2
4	Yes	1/2	0.55	No	24	5	1.2 ± 0.2
5	Yes (2 samples) No (2 samples)	1/2	0.55	No	24	3 (no sheet) 4 (with sheet)	1.2 ± 0.2
6	Yes (original and modified sheet)	1/2	0.55	No	24	3 4	1.4 ± 0.2
7	Yes (original and modified sheet)	1/2	0.55	No	24	5 4	1.4 ± 0.2
8	Yes (original and modified sheet)	1/2	0.55	No	24	3 4	1.2 ± 0.2

4.2 *Defining the optimal cracking conditions in the laboratory*

Test results and technical literature indicate that wind speed, exposure time of the mortar in the wind tunnel and the placement of metal restraining sheets are highly sensitive variables in the crack generation process. It was also possible to identify the optimum cement/sand and water/cement proportions that would allow obtaining cracks in the mortar in short times, with the typical geometric characteristics of plastic shrinkage.

As a result of the sensitivity analysis, the following ideal conditions were identified to adequately induce shrinkage cracks: Temperature: $36 \pm 2°C$; wind speed: 4.0 ± 0.2 m/s; testing time: 3 h maximum; optimum cement/sand proportion: 1 / 2; optimum water/cement ratio: 0.55; samples: rectangular prismatic with modified aluminum sheeting.

Figure 6 show the evolution of the cracks induced under controlled conditions in the established wind tunnel. The progression in extension and opening of a surface crack can be observed until appropriate geometric characteristics for its study are achieved.

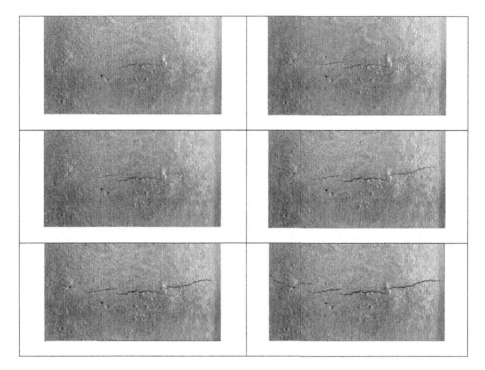

Figure 6.   Monitoring of shrinkage cracking process during curing.

4.3 *Proposed protocol for the preparation of shrinkage-cracked mortar specimens*

Once the ideal conditions for the simulation of plastic shrinkage cracks in mortar were identified, the following protocol was developed to produce samples under controlled laboratory conditions, whose repeatability was validated by the preparation of 14 satisfactory samples:

- Weigh 1000 g of each sand, on a balance with an accuracy of $\pm$ 0.1 g and metal weighing pans previously calibrated and zeroed.
- Use a third pan to weigh 1000 g of Portland cement for general purpose.
- Pour 500 mL of water into a graduated container.

- Integrate the components and mix until homogeneity is achieved.
- Pour the mixture into the assembled moulds previously lubricated
- Proceed to mortar compaction, following applicable local or international standards.
- Turn on the wind tunnel, to reach a speed higher than 4.0 m/s at the exit, and a temperature of 36°C
- Gently place the moulds inside the wind tunnel and leave them for the time required for cracking.
- While the specimens are cracking, all the elements used in the preparation of samples must be cleaned; traces of the remaining mixture must not be poured directly into the drain.
- After 1 hour and 45 minutes of placing the specimens in the wind tunnel, constant attention should be paid to the surface of the mortar. Normally, cracking starts in a window of 20 min.
- Once cracking starts in a specimen, it should be counted for 2 minutes and then removed from the wind tunnel; immediately place a damp textile on the surface to stop the cracking process.
- After 24 h of crack inducing, the samples should be removed by disassembling the moulds.
- Finally, each of the moulds should be cleaned and reassembled for later use.

4.4 Characterization of shrinkage cracks by computed tomography

By exposing the shrinkage-cracked mortar samples to tomographic tests, images were obtained that revealed the geometric characteristics and configuration of the induced cracks. Figure 7A illustrates the image obtained in a cross section of a sample; a top-down crack induced in the zone of influence of the notched metal sheet is clearly visible. Through initial inspection of all the images taken, those that were useful for the morphological characterization of cracking were selected. Conversely, those that did not correspond to plastic shrinkage cracking (Figure 7B), or where there was no cracking (Figure 7C), were discarded.

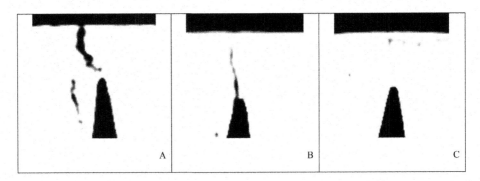

Figure 7. Examples of tomographic images obtained of cracks in mortar samples. A. Properly top-down plastic shrinkage induced crack; B. Incorrect induced crack; C. Unformed crack.

The software analysis of the selected images allowed direct measurements to be made on the tomograms. Figure 8 shows some determinations that can be made in crack characterization: openings (upper, middle, and lower), heights and areas. Measurements on the images are given in pixels, but it is possible to convert them into millimetres by defining the tomograms scale. For this study, it was established that three pixels (3 pix) correspond to one millimeter (1 mm). Moreover, Figure 9 shows several shapes and configurations of shrinkage cracks obtained by applying both the protocol and the experimental setup developed. Measurements indicate cracks ranging from 7 to 20 mm deep and 0.7 to 2 mm wide.

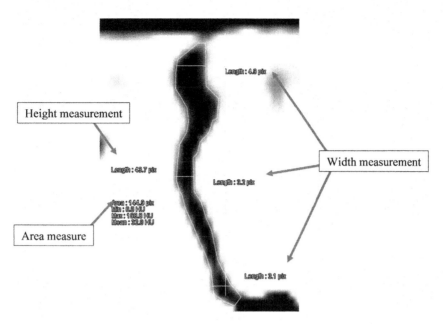

Figure 8.   Measurements of a mortar crack performed on a tomogram image with software Weasis v1.2.5.

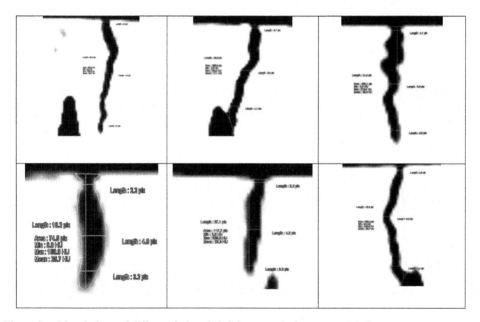

Figure 9.   Morphology of different induced shrinkage cracks in mortar samples.

5   CONCLUDING REMARKS

According to the results obtained in the tests, the following conclusions are established:

The aluminium foil imposes a restriction on the overall shrinkage of the mortar.

447

Increasing the water/cement ratio from 0.53 to 0.55 provided a higher probability of cracking.

The inclusion of nails at the ends of the specimens had no effect on physical restraint.

The exposure time of the mortar in the wind tunnel, the thickness of the samples and the wind speed, were identified as very sensitive factors in the crack generation process.

The intense sensitivity analysis carried out, allowed to establish the ideal conditions to adequately induce shrinkage cracks in the mortar in the experimental set-up developed in the laboratory.

The proposed methodology includes the development of an experimental set-up in a wind tunnel, with the corresponding protocol for the preparation of samples cracked by plastic shrinkage. Likewise, non-destructive testing with computed tomography for crack monitoring was applied.

This procedure constitutes a contribution to the study of plastic shrinkage cracks, overcoming the limitations of trial-and-error practices in the laboratory.

This protocol is being used in research on the application of calcite precipitating microorganisms MICP for sealing shrinkage cracks in mortars. The successful results obtained allow us to recommend it for future effectiveness studies of multiple repair agents for rigid pavements.

ACKNOWLEDGMENTS

The authors are grateful for funding received from Colciencias/Minciencias (Project No. 110180863795 CT-190-2019) and the Dirección de Investigación y Extensión sede Bogotá, Universidad Nacional de Colombia (Project No. 41833).

REFERENCES

Agencia Valenciana de Salud. 2015. *"Tomografía Computarizada (TAC)." Generalitat Valenciana*, 1–2. http://www.san.gva.es/documents/151744/512072/Tomografia+computarizada.pdf?version=1.0.

ASTM C1579-21 2021. *"Standard Test Method for Evaluating Plastic Shrinkage Cracking of Restrained Fiber Reinforced Concrete* (Using a Steel Form Insert)" 06 (Reapproved 2012): 1–7. https://doi.org/10.1520/C1579-06R12.2.

Becker Edgardo. 2015. *"Patrones de Fisuración En Pavimentos de Concreto: Algunos Conceptos Básicos."* Construcción y Tecnología En Concreto, octubre 2015. http://www.revistacyt.com.mx/pdf/octubre2015/tecnologia.pdf.

Construcción y Tecnología en Concreto. 2019. *"Fisuras Por Contracción Plástica Del Concreto."* Construcción y Tecnología En Concreto, 2019. http://www.imcyc.com/revistacyt/dic11/arttecnologia.html.

Dong B., Weijian D., Qin S., Han N., Fang G., Liu Y., Xing F. and Hong S. 2018. "Chemical Self-Healing System with Novel Microcapsules for Corrosion Inhibition of Rebar in Concrete." *Cement and Concrete Composites* 85: 83–91. https://doi.org/10.1016/j.cemconcomp.2017.09.012.

Grupo Polpaico. 2000. *"Agrietamiento Por Retraccion Plastica,"* 2000. https://www.nrmca.org/aboutconcrete/cips/CIP5es.pdf.

Johansen R. and Dahl P. 1993 *Control of Plastic Shrinkage of Cement.* Proceedings of the 18th Conference on Our World in Concrete and Structures, Singapore.Jonkers, Henk M. 2011. "Bacteria-Infused Self-Healing Concrete." HERON 56 (1/2): 1–12.

Londoño, Cipriano. 2012. *"3 Cosas Que Debes Saber Sobre La Retracción Del Concreto."* Argos 360. https://www.360enconcreto.com/blog/detalle/categoryid/157/categoryname/buenas-practicas/cosas-que-debes-saber-sobre-la-retraccion-del-concreto.

National Ready Mixed Concrete Association. 2014. *"Agrietamiento Por Contracción Plástica."* El Concreto En La Práctica, 2014. https://www.nrmca.org/aboutconcrete/cips/CIP5es.pdf.

Ohlerth S. and Scharf G. (2007). "Computed Tomography in Small Animals-basic Principles and State of the Art Applications." *The Veterinary Journal*, 173(2), 254–271. https://doi.org/10.1016/j.tvjl.2005.12.014.

Orozco M., Avila Y., Restrepo S. and Parody A. 2018. "Factores Influyentes En La Calidad Del Concreto: *Una Encuesta a Los Actores Relevantes de La Industria Del Hormigón.*" *Revista Ingeniería de Construcción* 33 (2): 161–72. https://doi.org/10.4067/s0718-50732018000200161.

Sayahi F., Emborg M., Hedlund H. and Ghasemi Y. 2022. "Experimental Validation of a Novel Method for Estimating the Severity of Plastic Shrinkage Cracking in Concrete." *Construction and Building Materials.* 339: 127794. https://doi.org/10.1016/j.conbuildmat.2022.127794

Sayahi F., Emborg M. and Hedlunda H. 2017. *"Plastic Shrinkage Cracking in Concrete – Influence of Test Methods"*. 2nd International RILEM/COST Conference on Early Age Cracking and Serviceability in Cement-based Materials and Structures – EAC2. ULB-VUB, Brussels, Belgium.

Sayahi F., Emborg M. and Hedlund H. (2014). "Plastic Shrinkage Cracking in Concrete: State of the Art." *Nordic Concrete Research.* 51: 95–110.

Sika Colombia 2014. *"Rehabilitación – fisuras en el Concreto Reforzado."* Departamento Técnico de Sika Colombia S.A. DCT-VO-182-07/2014, 1–19.

Tamayo-Figueroa D.P., Meneses-Martínez H. O., Darghan-Contreras A. E., Lizarazo-Marriaga J. and Brandão P. F. B. (Submitted). "A Comparison Index in Mortar Repair Treatments by Microbiologically Induced Carbonate Precipitation and its Evaluation by a Non-destructive Technique." *European Journal of Environmenal and Civil Engineering.*

Toirac J. 2004. "Patología de La Construcción. Grietas y Fisuras En Obras de Hormigón. Origen y Prevención." Ciencia y Sociedad Volumen XX. https://www.redalyc.org/articulo.oa?id=87029104.

Topçu İ. and Işıkdağ B. 2017. *"A Review on The Effect of Environmental Conditions on Concrete Evaporation and Bleeding."* 3rd International Sustainable Buildings Symposium At: Dubai, UAE.

Toxement. 2010. *"Fisuras Por Retracción Plástica."* 2010. http://www.toxement.com.co/media/3443/retraccio-n_pla-stica.pdf.

Uno P. (1998). "Plastic Shrinkage Cracking and Evaporation Formulas." *ACI Materials Journal.* 95-M34. 365–375.

Zhang L., Ruan S., Qian K., Gu X. and Qian X. (2022). "Will Wind Always Boost Early-age Shrinkage of Cement Paste?." *Journal of the American Ceramic Society.* 105: 2832–2846. https://doi.org/10.1111/jace.18245.

Roads and Airports Pavement Surface Characteristics – Crispino & Toraldo (Eds)
© 2023 The Author(s), ISBN 978-1-032-55149-4

Review and analysis of prediction models for as-built roughness in asphalt overlay rehabilitations

Rodrigo Díaz-Torrealba*, José Ramón Marcobal & Juan Gallego
Department of Civil Engineering, Transport, Technical University of Madrid (UPM), Madrid, Spain

ABSTRACT: The surface smoothness achieved in pavement construction, or as-built roughness, has great importance in road engineering; it's an indicator of both the level of service offered to users and the standard of construction quality in general. Thus, in pavement design and management, knowledge of as-built roughness is relevant for supporting pavement maintenance and rehabilitation studies decision-making. Pavements with increased initial smoothness will have a longer service life, will provide a better quality of service to the user over time, and the highway agency will spend less time and resources on their maintenance.

Traditionally, as-built roughness has been studied in the literature from an empirical approach, based on the statistical correlation between the roughness IRI observed before and after the execution of paving works. For the case of asphalt overlays, some of these investigations have focused exclusively on the factors whose influence predominates on the initial roughness achieved, without developing any prediction models. Other research has also proposed regression models to estimate as-built IRI, either by means of equations for direct prediction of post-construction IRI, or to estimate the magnitude in IRI reduction because of the works (called IRI-drop), leading to models of diverse natures and predictive capability.

In this study a review of the available prediction models for as-built roughness in asphalt pavement rehabilitation is presented; detailing their explanatory variables and analyzing models formulation. The results obtained allow us to establish important conclusions about the predictive capacity of the models and the potential limitations in their practical application.

Keywords: As-built Roughness, Pavement Rehabilitation, Asphalt Overlay, Surface Characteristics

1 INTRODUCTION

Surface roughness achieved in pavement construction, or as-built roughness, has proven to be of great importance for both future roughness progression and overall life cycle pavement performance. A study of the effects of initial smoothness on pavement life (Smith *et al.* 1997) showed that a 25% reduction in as-built roughness has a resulting 9% increase in pavement life. Another study (Janoff 1991 as cited in Stroup-Gardiner *et al.* 2004) indicates that pavements that are initially smoother have lower maintenance costs; estimating average annual savings of nearly $1,200 USD/mile-lane when as-built roughness (measured in Profile Index, PI) is reduced by 0.15 m/km. Therefore, pavements with reduced initial roughness will have a longer service life, will provide a better quality of service to the user over time, and the highway agency will spend less time and resources on their maintenance.

---

*Corresponding Author: rodrigo.diazt@alumnos.upm.es

DOI: 10.1201/9781003429258-44

When selecting pavements maintenance and rehabilitation strategies, one critical factor is to determine the effectiveness of different treatments, therefore, estimation of as-built roughness is an important issue for designers, contractors or any highway agency. As-built roughness prediction can be used by designers to evaluate the effect on smoothness of their design alternatives or by contractors to estimate the as-built roughness that will be achieved under a particular combination of construction factors. For highway agencies, correct estimation of as-built roughness will influence decision-making about whether or not to intervene, and about the type of intervention required.

Despite its importance, the estimation of as-built roughness has received little attention in the literature, and its study has traditionally been carried out from an empirical approach based on the statistical correlation between the International Roughness Index (IRI) observed before and after the execution of rehabilitation works. For asphalt pavements, some of the available research has focused exclusively on study the influence of various factors and its statistical significance on as-built roughness. To a lesser extent, some research have also proposed regression models to estimate as-built IRI, either by means of equations for direct prediction of post-construction IRI, or to estimate the magnitude in IRI reduction because of the works (called IRI-drop).

In this study, a review and analysis of available prediction models for as-built roughness in asphalt pavement rehabilitation is presented. The prediction capabilities of four regression models from the literature review where tested against real as-built IRI data from overlay sections of the North American Long Term Pavement Performance Program (LTPP) SPS-5 experiment. The results obtained allow us to establish important conclusions about the predictive capability of the models and the potential limitations in their practical application.

2 FACTORS AFFECTING AS-BUILT ROUGHNESS

The results available in the state of the art are diverse, and depending on the study, a series of explanatory variables have been proposed to influence the initial IRI achieved in asphalt pavement rehabilitation. The relative importance of these variables varies between authors, and for some of them, even contradictory conclusions can be found.

Perera et al. (1998) studied, among other research objectives, the IRI characteristics of flexible pavements for various rehabilitation strategies. The authors performed a before-after IRI analysis of asphalt overlays using available data from the SPS-5 test sections of the LTPP program. The results indicated that regardless of the IRI before the rehabilitation, the IRI after overlay would fall within a relatively narrow band. However, the limits of this range varied from project to project, and the factors that could determine this range include the profile of the pavement prior to overlay, the predominant wavelengths in the section that contribute to the IRI, and the capability of the contractor placing the overlay. Also, it was shown that even thin asphalt overlays (50 mm) reduce roughness substantially. The authors did not find any statistically significant relationship between the IRI before and after an overlay.

Based on the aforementioned study, Perera and Kohn (2001) carried out a series of complementary analyses for the SPS-5 experiment sections, including additional LTPP sections from the GPS-6 experiment (GPS-6B&C). One of the additional conclusions was that milling before overlay (intensive surface preparation) was statistically significant only in those sections with a pre-overlay IRI greater than 1.5 m/km. According to the authors, this effect of milling can be attributed to two factors. (1) Milling prior to rehabilitation provides a uniform surface for overlay placement, which translates into a lower IRI post-overlay. (2) Since the thickness of the milled pavement is replaced before overlay, the number of asphalt layers (paver lifts) required for placement of the total thickness in milled pavement sections could be greater than in the non-milled sections, creating an opportunity to improve smoothness on each layer.

McGhee (2000) conducted an investigation to identify the primary factors that affects the initial roughness achieved in asphalt overlays construction. The database used was 4,270 km-lane of IRI survey data from different rehabilitation works carried out over a two-year period, totaling 854 samples distributed throughout the state of Virginia, USA. The author concluded that the main factors that influence as-built IRI of an asphalt overlay are: (1) Existing roughness prior to overlay construction. (2) Roadway functional classification; it is likely that higher-classification roads, which are built to handle larger volumes of traffic, usually consist of roadways with more modern, less active geometries, which in turn contribute to smoother pavements. (3) Special provision for smoothness; in those construction sites in which a pilot plan of special provision for as-built roughness was implemented (payment adjustments), overlays reached a lower construction IRI compared to similar works without the special provision. The rest of the factors analyzed in the study such as: predominant surface distresses of the original pavement, traffic volume, original surface age, overlay thickness and mix type, milling of the original surface before overlay placement, laydown of additional structural layers (intermediate layer or milling and replacement) and time-of-day restrictions for paving works were not statistically significant or there were insufficient data samples to draw generalized conclusions.

Wen (2011) investigated the effect of various design factors on the initial roughness of asphalt-surfaced pavements. The author analyzed as-built IRI data of 442 asphalt overlay pavements constructed from years 2000 to 2004 in the state of Wisconsin, USA. The results indicate that: (1) Hot mix asphalt (HMA) layer thickness significantly affects initial roughness; thicker HMA layer shows lower initial roughness. The author explains that thicker asphalt layers usually needs more than one lift, which reduces the effects of base unevenness on the initial roughness of HMA surface. (2) The urban or rural location of the project was also significant; urban projects have a higher initial roughness than rural projects, probably due to utilities in urban pavements and/or geometric consideration for intersecting road and drainage. Rural pavement construction generally has fewer interruptions, resulting in a smoother surface. (3) Mixtures designed by the Marshall, Superpave and Stone Mastic Asphalt (SMA) methods were used in the studied projects. Statistical analysis indicated that SMA mixtures presented a higher construction IRI compared to the other asphalt mixtures. (4) The effect of base type underneath the new HMA layer was statistically significant only for pavement reconstruction cases with base type OGBC2 (open-graded base course #2), which resulted in a higher as-built IRI than other types of bases. (5) The effect of project length was also significant; the analysis showed that longer projects tend to have lower initial roughness.

Hung et al. (2014) analyzed Pavement Condition Survey (PCS) data from the California Department of Transportation (CALTRANS) to investigate the effect of various factors on initial overlay smoothness. Data from 193 maintenance and rehabilitation contracts between the years 2000 and 2009 where used for the analysis, totaling 228 projects and 4,475 analysis sections. The authors presented the following conclusions: (1) Pre-overlay condition; overlays applied on pavements with a lower pre-overlay IRI could achieve a lower initial IRI regardless of other factors. (2) Layer thickness; increase in the thickness of an overlay had no additional benefit when the pre-overlay IRI was less than 1.90 m/km. With open-graded mixes, thickness did not appear to have any significant effect on the initial IRI. (3) Pre-overlay surface preparation and mix type; milling before the overlay or use of rubberized binder alone did not provide any additional benefits toward the achievement of a lower initial IRI. Milling before overlay on pavement with an existing IRI of less than 1.90 m/km was disadvantageous and might result in higher roughness. Digouts before overlay provided a benefit in the reduction of as-built IRI when the pre-overlay IRI value was greater than 1.90 m/km. Projects with digouts performed better than those that were milled before overlay. Surface type (open-graded versus dense and gap-graded) had no effect in general on as-built IRI.

3 AS-BUILT ROUGHNESS PREDICTION MODELS

Similar research, in addition to study the factors affecting as-built roughness, have proposed a series of regression models to estimate as-built IRI, either by means of equations for direct prediction of post-construction IRI, or to estimate the magnitude in IRI reduction because of the works (Dong & Huang 2012; Djärf *et al.* 1995; Kwon *et al.* 2015; Lu & Xin 2018; Morosiuk *et al.* 2004; NDLI 1995a, 1991, 1995b; Qiao *et al.* 2016; Raymond *et al.* 2003; Watanatada *et al.* 1987). The basic principles of four main regression models from the literature are presented below, which will be used in the next section for as-built IRI prediction and analysis.

(i) One of the first studies related to immediate effects of different roadworks operations on the performance of bituminous pavements was developed by the World Bank within the HDM program (Highway Design and Maintenance Standards Study), leading to the fundamental models used in the HDM-III software. HDM-III models the reduction in roughness after overlay as a function of pre-overlay roughness and overlay thickness: two models were provided, one for a "regular" paver and one for an "automatic-levelling" long-base paver (Watanatada *et al.* 1987). The models developed for HDM-III were updated through successive revisions for the ISOHDM study (International Study of Highway Development and Management Tools); establishing a linear relationship between post-overlay roughness and pre-overlay roughness, overlay thickness and construction quality to be included in the first version of HDM-4 software (NDLI 1995a).

Subsequent revisions of the model suggested that the simplified linear relationship in HDM-4 (version 1) does not adequately take into account a dual effect of overlays; the short wavelength roughness corrected by thin overlays and the medium wavelength roughness corrected by thick overlays. A new model was proposed for HDM-4 version 2 software; establishing a bilinear overlay-roughness relationship for an specified overlay thickness, overlay execution technique and pavement type (Morosiuk *et al.* 2004).

$$\Delta RI = \max\{0, a_0[\min(a_1, RI_{bw}) - a_2] + a_3 \times \max[0, (RI_{bw} - a_1)]\} \tag{1}$$

$$RI_{bw} = \max(1.0, RI_{ap}) \tag{2}$$

$$RI_{aw} = RI_{bw} - \Delta RI \tag{3}$$

Where

ΔRI: Reduction in roughness after overlay (IRI m/km)
RI_{bw}: Roughness before overlay (IRI m/km)
RI_{aw}: Roughness after overlay (IRI m/km)
RI_{ap}: IRI after preparatory works (IRI m/km)

Coefficient a_0 default value is 0.9. Coefficient a_2 represents the minimum roughness after overlay. Coefficients a_1 to a_3 can be computed as a function of the overlay thickness H (mm) as follow.

$$a_1 = \max\left[4.0, 2.1 \times e^{(0.019 \times H)}\right] \tag{4}$$

$$a_2 = 1 + 0.018 \times \max[0, (100 - H)] \tag{5}$$

$$a_3 = \min\{a_0, \max[0, (0.01 \times H - 0.15)]\} \tag{6}$$

From Equation 2 it can be seen that the model establishes a minimum value of 1.0 m/km for pre-overlay roughness (RI_{bw}), that is, lower values of IRI will be considered as

1.0 m/km in the calculation. On the other hand, according to Equation 5, the minimum post-overlay roughness (a_2) is set by default to 1.0 m/km for any overlay thickness above 100 mm. From the analysis of the equations it can be concluded that thin overlays (less than 50 mm) will only produce a roughness reduction (ΔRI) when the pre-overlay IRI is greater than 2 m/km.

(ii) Using the SPS-5 experiment test sections from the LTPP database, Raymond *et al.* (2003) studied the influence of various design factors on as-built roughness in asphalt overlays. The analysis indicated four statistically significant factors influencing as-built roughness: (1) surface preparation prior to resurfacing (basic or intensive), (2) pavement roughness prior to resurfacing, (3) the interactive effect of surface preparation and pavement roughness prior to resurfacing and (4) the interactive effect of overlay thickness and pavement roughness prior to resurfacing. The type of overlay material (new or recycled) was determined to have no effect on the as-built roughness of a pavement. As a result, regression analyses were performed by the authors to provide prediction equations for as-built roughness that account for the influence of these factors.

3.1 Basic surface preparation (no milling)

$$\text{Thin Overlay}: \quad IRI_{as-built} = 0.44 + 0.31 \times IRI_{prior} \tag{7}$$

$$\text{Thick Overlay}: \quad IRI_{as-built} = 0.52 + 0.25 \times IRI_{prior} \tag{8}$$

3.2 Intensive surface preparation

$$\text{Thin Overlay}: \quad IRI_{as-built} = 0.40 + 0.29 \times IRI_{prior} \tag{9}$$

$$\text{Thick Overlay}: \quad IRI_{as-built} = 0.63 + 0.17 \times IRI_{prior} \tag{10}$$

The intensive surface preparation consists milling prior to rehabilitation, sealing of cracks and patching deteriorated areas, and replacement of the milled pavement before overlay placement. Overlay thickness were 50 mm and 125 mm for thin and thick overlays respectively.

The equations proposed by the authors establish that post-overlay IRI ($IRI_{as-built}$) is a function of pavement roughness prior to resurfacing (IRI_{prior}); and depending on the surface preparation and overlay thickness, it is equal to a minimum IRI value that varies from 0.40 m/km to 0.63 m/km plus a reduction in roughness respect to IRI_{prior} that varies from 69% to 83%. In all cases, a low R^2 coefficient was obtained (0.13 to 0.29), i.e. pre-overlay IRI explains only between 13% and 29% of the variation observed in the post-overlay IRI.

(iii) By investigating the LTPP database Dong and Huang (2012) analyzed the influence of overlay thickness, pavement thickness, traffic volume, and pre-overlay pavement conditions on the effectiveness and cost-effectiveness of different asphalt pavement rehabilitations. Multiple regression models were established for two types of treatment effectiveness measures: the initial effects, including the post-rehabilitation IRI value and the IRI drop because of the rehabilitation; and the long-term effects, including the IRI trend after the rehabilitation and the "benefit". Only IRI drop results are presented herein.

The cited authors obtained three significant predictors for roughness drop; IRI before rehabilitation (Pre_IRI), overlay thickness (inch) and the inclusion or not of milling before works (Mill). The following equation shows the multiple regression results for IRI drop.

$$IRI_{Drop} = 0.47 \times Pre_IRI + 0.03 \times OverlayThickness + 0.04 \times Mill \qquad (11)$$

From the analysis of the results presented by the authors it can be observed that roughness before overlay (Pre_IRI) is the most significant variable (t = 32.68). For each increase of 1 m/km in Pre_IRI, an increase of 0.47 m/km is obtained in roughness reduction (IRI_Drop) because of the rehabilitation. Similarly, although with less influence, a 1-in increase in overlay thickness results into 0.03 m/km increase in IRI_Drop, and milling execution before overlay produces a 0.04 m/km increase in roughness drop.

(iv) Qiao et al. (2016) studied the immediate maintenance effects on roughness and rutting of three types of interventions and provided a method to validate some immediate maintenance effect models. The authors introduce a data mining process to enhance the calibration of the models and used a linear regression to relate the reduction in IRI after maintenance (ΔIRI_n) to the existing roughness before the intervention (IRI_{n0}). After the regression analysis, the authors provide the calibration factors for the three maintenance alternatives and presented the validated maintenance effect models as a result of applying their data selection process.

Intervention Option 1: Surface course overlay (thickness 1.5 – 2.0 in)

$$\Delta IRI_n = 0.6307 \cdot IRI_{n0} - 22.491 \qquad (12)$$

Intervention Option 2: Surface + base course overlay (thickness 1.5 + 2.0 in)

$$\Delta IRI_n = 0.5234 \cdot IRI_{n0} - 8.0962 \qquad (13)$$

Intervention Option 3: Surface course mill and replacement (2.0 in)

$$\Delta IRI_n = 0.811 \cdot IRI_{n0} - 39.71 \qquad (14)$$

The interpretation of the models indicates that after the rehabilitation there is a roughness reduction that, depending on the intervention alternative, varies from 52% to 81% of the existing IRI prior to resurfacing plus a roughness value given by the constants 22.491 (0.36 m/km), 8.096 (0.13 m/km) and 39.71 (0.63 m/km) for interventions options 1, 2 and 3 respectively.

3.3 *As-built prediction models analysis*

The prediction capabilities of the regression models from the literature review where tested against as-built IRI data from overlay sections of the LTPP program; specifically the SPS-5 experiment, designed to evaluate the effects of different overlay alternatives on the performance of flexible pavements after rehabilitation.

Figures 1 and 2 presents the results of predicted as-built roughness (IRI-sim) against real as-built roughness data (IRI-post) obtained for sections corresponding to minimal surface preparation prior to overlay construction (no milling) and intensive surface preparation (includes milling) respectively. Two overlay thicknesses were constructed: thin overlays of 2 in approx. (SHRP ID code 502, 505, 506 and 509) and thick overlays of 5.0 in approx. (SHRP ID code 503, 504, 507 and 508). The HDM-4 v2 model was applied only for the cases of minimal surface preparation (Figure 1); assuming that roughness after surface preparation (RI_{ap}) is equivalent to IRI before works (IRI_{bw}), since no substantial changes in roughness are expected after minimal surface preparation.

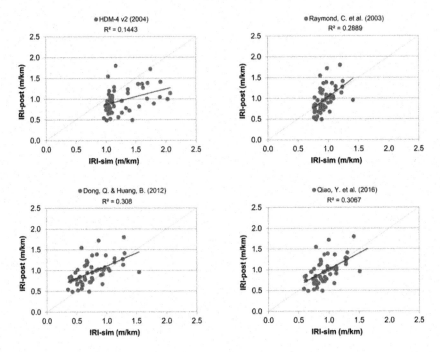

Figure 1.   Model results comparison – Minimal surface preparation.

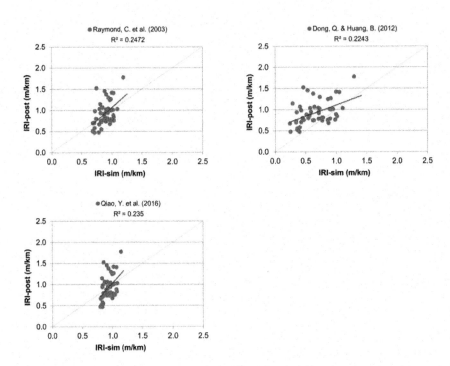

Figure 2.   Model results comparison – Intensive surface preparation.

The results shows that as-built IRI predicted by the models are fairly scattered (low precision) and to different degrees shifted respect to the equality line (bias), as indicated by the relatively low R^2 coefficients. Although in every case there are some as-built IRI values that show good prediction, in overall, the prediction capability seems to be insufficient, with some errors up to 1 m/km in the expected as-built IRI. 49% of the simulations showed a prediction error greater than 0.2 m/km IRI, being 37% in the range of 0.2 – 0.5 m/km IRI prediction error. 12% of the simulations have an error greater than 0.5 m/km.

3.4 *Practical application and potential limitations*

Although the regression analysis of these and other similar data series could show some correlation and statistically significant variables may be found; in practice, for pavement construction and management studies, a potential error of 0.5 m/km in the expected as-built IRI can lead to inaccurate decision-making and have significant impact both on works acceptance forecast and life cycle project analysis.

The use of IRI data measured before and after overlays to develop as-built prediction models seems to have a limitation: the fact that for the same pre-overlay IRI value and rehabilitation design (overlay thickness, asphalt mix type, previous repair, etc.), paving works that reach significantly different post-rehabilitation IRI can be observed, so adjusting a before-after IRI correlation seems to be insufficient to fully explain the observed variability. This issue was advised, but not addressed, in the regression model proposed by Kwon *et al.* (2015), who incorporated in their calculation the probability that the predicted IRI is equal to the real value; in the case of the cited study, for an initial IRI of 0.79 m/km there was only a 6.5% chance that the real IRI measured after works would be equal to the predicted IRI value of 0.62 m/km. The problem stated is illustrated in Figure 3, which shows the roughness IRI plot before and after overlay for the same pavement data set used for model testing in the previous section.

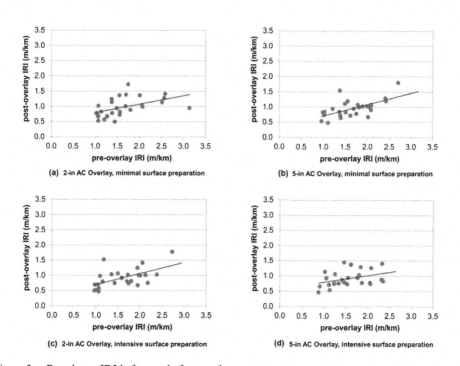

Figure 3.   Roughness IRI before and after overlay.

According to Figure 3(a), as an example, pavement sections with pre-overlay IRI values close to 1.5 m/km achieved post-overlay IRI values in the range of 0.5 m/km to 1.5 m/km, all of them for the same type of previous repair work (minimal) and overlay thickness (2 in). The same can be observed for thick overlays (5 in) in Figure 3(b). In a similar way, Figure 3(c) shows sections with pre-overlay IRI close to 2.0 m/km that ranged in as-built IRI values from 0.5 m/km to 1.5 m/km, for the same type of previous repair work (intensive) and overlay thickness (2 in).

The review and analysis presented shows that the most used variable in the literature to explain as-built roughness IRI is the existing IRI prior to overlay construction; leading to the development of several regression models although with insufficient capability to explain the observed variability in as-built roughness. In this regard, some studies have suggested that the nature of the existing roughness (i.e. predominant wavelengths that contribute to the IRI) must also be taken into account (Morosiuk et al. 2004; NDLI 1995a; Perera et al. 1998). However, the reviewed models have not effectively considered this influence, since IRI as an explanatory variable does not provide detailed information on roughness nature, that is, on the different wavelengths that make up profile defects.

The aforementioned relationship between the wavelength content of the existing profile and the smoothness achieved during construction can be explained by the asphalt paver working principle. The screed is the mechanical tool of the paver responsible for the placement, levelling and pre-compaction of the asphalt mix and is connected to the tractor unit at the 'tow point' by means of two lateral tow arms. As the tractor tows the screed forward, the paving material in front of the screed flows under it causing the screed to lift or 'float' over the mix, thus establishing the mat thickness and profile. Because of the 'free floating' principle the screed can pivot around the tow points and average out the irregularities of the surface being paved. By operating in this manner, the irregularities under the screed are levelled out through the paving material. The screed will compensate the bottom layer roughness by placing more paving material in the depressions and less over the humps, thus building a new longitudinal profile as a result of smoothing the underlying surface irregularities.

In consequence, the levelling response of the asphalt paver is directly related to the nature of the profile defects. Moreover, the levelling effectiveness of the screed is inversely proportional to the length of the humps or hollows. That is, the longer the undulation on the underlying surface, more closely the screed will reproduce these undulations (Barber-Greene 1972; Ulrich 1991); this is the reason why pavers use automatic control systems through levelling beams (long reference averaging) in order to regulate the tow point position of the screed and compensate the lowest levelling efficiency at longer wavelength defects.

According to the above, understanding of the asphalt laydown construction process can help to explain the observed variability in as-built IRI; since paving works with the same initial roughness IRI and overlay design may achieve different surface finishes, and therefore as-built IRI, depending on the wavelength content that make up the profile of the surface being paved (which will determine the paver levelling response). Thus, in order to improve as-built roughness prediction, it seems appropriate to consider into the analysis the nature of the existing roughness and its interaction with the paver during laydown operation.

4 CONCLUSIONS

A review of prediction models for as-built roughness in asphalt overlay rehabilitations has been presented, detailing their explanatory variables and analyzing their predictive capacity. The immediate effect of rehabilitation works on roughness correction has been studied in the literature from an empirical approach, based on the statistical correlation between the roughness IRI observed before and after the execution of rehabilitation works, leading to models of diverse nature and predictive capability.

The analysis of the prediction models suggest that adjusting a before-after IRI correlation seems to be insufficient to fully explain the observed variability. Even though the reviewed models can provide a method to quantify expected or typical as-built IRI values, the considerable amount of variability observed on the results can lead to information that lacks the precision required for decision-making in pavement construction and management practice. To account for the observed variability in as-built IRI data, an explanation based on the asphalt paver working principle is proposed.

REFERENCES

Barber-Greene 1972. *Principles of the Asphalt Finisher*. Illinois, USA.: Barber-Grenee Company, Manual Page 6900.

Djärf L., Magnusson R., Lang J. & Andersson O. 1995. *Road Deterioration and Maintenance Effects for Paved Roads in Cold Climates*. Stockholm: Swedish Road And Transport Research Institute.

Dong Q. & Huang B. 2012. Evaluation of Effectiveness and Cost-Effectiveness of Asphalt Pavement Rehabilitations Utilizing Ltpp Data. *Journal of Transportation Engineering*, 138, 681–689.

Hung S.S., Rezaei A. & Harvey J.T. 2014. Effects of Milling and Other Repairs on Smoothness of Overlays on Asphalt Pavements. *Transportation Research Record*, 2408, 86–94.

Kwon O., Lee H.S., Chun S., Holzschuher C. & Choubane B. Probabilistic Relationship Between Smoothness of Final Asphalt Surface and Underlying Layer. *94th Annual Meeting of The Transportation Research Board*, 2015 Washington, Dc.

Lu Q. & Xin C. 2018. *Pavement Rehabilitation Policy for Reduced Life-Cycle Cost and Environmental Impact Based on Multiple Pavement Performance Measures*. Final Report. Tampa, Florida. Ee.Uu.: *Center For Transportation, Environment, and Community Health Ctech*. University of South Florida.

Mcghee K.K. 2000. Factors Affecting Overlay Ride Quality. *Transportation Research Record*, 1712, 58–65.

Morosiuk G., Riley M.J. & Odoki J.B. 2004. *Modelling Road Deterioration and Works Effects*, Version 2. UK.: Hdm-4 Highway Development And Management Series. Volume Six.

Ndli 1991. *Nepal Road Maintenance Project – Final Report*. Vancouver. Canada.: N.D. Lea International.

Ndli 1995a. *Modelling Road Deterioration and Maintenance Effects in Hdm-4*. Final Report Asian Development Bank Project Reta 5549. Vancouver. Canada.: N.D. Lea International.

Ndli 1995b. *Thailand Pavement Management System – Ptmps Design Report*. Vancouver. Canada.: N.D. Lea International.

Perera R.W., C., B. & Kohn S. D. 1998. *Investigation of Development of Pavement Roughness*. Ee.Uu.: Federal Highway Administration. Publication No. Fhwa-Rd-97-147.

Perera R.W. & Kohn S.D. 2001. *LTPP Data Analysis: Factors Affecting Pavement Smoothness*. Washington, Dc, USA: Transportation Research Board, National Research Council.

Qiao Y., Dawson A., Parry T. & Flintsch G.W. 2016. Immediate Effects of Some Corrective Maintenance Interventions on Flexible Pavements. *International Journal of Pavement Engineering*, 19, 502–508.

Raymond C.M., Tighe S.L., Haas R. & Rothenburg L. 2003. Analysis of Influences on As-built Pavement Roughness in Asphalt Overlays. *International Journal of Pavement Engineering*, 4, 181–192.

Smith K.L., Smith K.D., Evans L.D., Hoerner T.E. & Darter M.I. 1997. *Smoothness Specifications For Pavements*. Washington, D.C. Ee.Uu.: Nchrp Web Document 1 (Project 1-31). Transportation Research Board, National Research Council.

Stroup-Gardiner M., Parker F., Burns Harris K. & Williams B. 2004. *The Effect of Material Transfer Devices on Hma Material Uniformity and Ride Quality*. Final Report. Alabama Department Of Transportation: Project Number 930-471.

Ulrich A. Automatic Levelling In Road Construction. *Proceedings Of The 8th International Symposium On Automation And Robotics In Construction* (Isarc), Pp. 467–478, 1991 Stuttgart, Germany.

Watanatada T., Harral C.G., Paterson W.D., Dhareshwar A.M., Bhandari A. & Tsunokawa K. 1987. *The Highway Design And Maintenance Standards Model*. Volume 1, Description of the Hdm-Iii Model. London, UK.: Published For The World Bank. Johns Hopkins University Press.

Wen H. 2011. Design Factors Affecting the Initial Roughness of Asphalt Pavements. *International Journal of Pavement Research And Technology*, 4, 268.

Laboratory and field evaluation of modified micro-surfacing to improve durability and lowering destructive environmental impacts

Hamidreza Sahebzamani* & Seyed Rasool Fazeli
School of Civil Engineering, University of Tehran, Tehran, Iran

Nader Mahmoodinia
Technical and Soil Mechanic Laboratory, Tehran, Iran

Masoud Mahmoodinia
Kandovan Pars Company, Tehran, Iran

ABSTRACT: Production and using aggregates and asphalt binder as the main components of flexible pavements have destructive effects on the environment. The consequences are oil and natural resources consumption, pollution in the production phase, high energy usage, and noise pollution. Production and implementation of asphalt mixtures are the other causes of pollution. As a result, researchers and artisans were always trying to find solutions for increasing the durability of roads and lowering the usage of raw materials. Using preserving methods such as micro-surfacing can increase the lifetime of roads. Additionally, using this technic would significantly lower the usage of minerals and oil resources and decrease greenhouse gases and noise pollution because of its high friction surface, cold implementation and very thin thickness. Research reveals that using this method compared to conventional hot mix asphalt will increase the existing pavement life by approximately 2 to 5 years. In this paper, the authors evaluated a new micro-surfacing mixture with better performance characteristics and less destructive environmental effects, called Eco-HP micro-surfacing. This mixture uses Reclaimed Asphalt Pavement (RAP) material and a chemical additive to increase the durability and eco-efficiency of conventional micro-surfacing as a maintenance layer. Both Laboratory and field investigations were performed to evaluate the effectiveness of these materials in improving the durability of the micro-surfacing layer and, as a result, increasing the sustainability. Cohesion, Loaded wheel Tracking (LWT) and wet track abrasion tests were conducted on the mixtures. Based on laboratory results, three mixtures passed the ISSA requirements. Based on laboratory results, the conventional micro-surfacing mixture and the mixture made with RAP material and chemical additive (Eco-HP micro-surfacing) were chosen for field evaluation. Evaluation of rutting performance and crack reflection were done between conventional and Eco-HP micro-surfacing. The Eco-HP micro-surfacing was almost intact after Four years of implementation compared to the conventional micro-surfacing overlay. Comparing Eco-HP micro-surfacing and conventional micro-surfacing showed a better riding quality for Eco-HP micro-surfacing with its Eco-friendly advantages and cost effectiveness.

Keywords: Micro-surfacing, Reclaimed Asphalt Pavement (RAP), Laboratory and Field Investigation

---

*Corresponding Author: hsahebzamani@ut.ac.ir

DOI: 10.1201/9781003429258-45

1 INTRODUCTION

Micro-surfacing is a maintenance method that typically contains aggregate, asphalt emulsion, water, polymer additive, and mineral fillers. Road repair and maintenance agencies mostly classify it as a surface treatment to correct rutting, improve surface friction and extend pavement life (Pederson *et al.* 1988).

Micro-surfacing is a maintenance layer suitable for economic and ecological advantages among different preservative treatment methods (Takamura *et al.* 2001). The National Cooperative Highway Research Program (NCHRP) reveals that by using the preservative maintenance method at the right time, for every dollar spent on the preventive method such as micro-surfacing, 3–4 dollars will be saved for future rehabilitation (Geoffroy 1996). Additionally, the Michigan Department of Transportation reported the amount of saving as six to ten dollars for future reconstruction costs (www.slurry.org).

Micro-surfacing effectiveness and pavement life increment can be varied for each project. Climate, traffic load, highway class, and underlaying layer conditions are the factors that affect this effectiveness. Labi *et al.* (2007) also concluded that micro-surfacing is more cost-effective than HMA thin overlays. Watson and Jared (1998) demonstrated that micro-surfacing would last 3–4 years in good condition under heavy traffic loads and Wood and Geib (2001) declared that micro-surfacing can be effective for at least seven years in medium to high traffic volume.

An eco-efficiency analysis was performed by BASF corporation to evaluate the micro-surfacing environmental impact compared to hot-mix asphalt. The results showed that the photochemical ozone creation potential, global warming potential, and acidification potential of micro-surfacing were four times lower than hot-mix asphalt. Results prove that micro-surfacing is more eco-efficient than hot-mix asphalt (www.slurry.org).

Robati (2014) tried to use RAP and RAS in micro-surfacing with different percentages. The author concluded that RAP could be added to the micro-surfacing up to 100 percent. This means that RAP can be used instead of virgin aggregates in micro-surfacing. This mixture will also pass ISSA criteria.

Saghafi *et al.* (2019) evaluated the feasibility of using RAP in slurry seal mixture. The findings express that slurry seal mixture with 87.5% RAP material is more sustainable and provides better laboratory performance than a conventional slurry seal mixture at a lower cost.

Wang *et al.* (2019) found that the optimum binder content will decrease with an increase in RAP content, and freeze-thaw resistance and mixing conditions of the micro-surfacing mixture will improve. This will help to save virgin materials.

Garfa *et al.* (2016) concluded that by using RAP in micro-surfacing, the minimum ISSA time specification for the cohesion test could not be achieved, and it needs more time to open to the traffic. The other significant result of this study was that emulsion affects the behavior of micro-surfacing made with recovered asphalt. This effect is related to the interaction between aged bitumen and emulsion binders.

2 OBJECTIVE

The purpose of this study is to use RAP material in micro-surfacing mixtures. Two different contents of RAP are used as an alternative for virgin aggregates in order to make this specific maintenance method more eco-efficiency. The research also used a chemical additive (PRIPOL 1017) to increase the acidity of bitumen and, therefore, have a better performance on a micro-surfacing mixture made with RAP. A field investigation was also conducted to achieve the best attitude for this laboratory evaluation. Two pavement sections, one with conventional micro-surfacing and one micro-surfacing mixture containing 100 percent RAP and acidifier, were constructed.

3 MATERIAL SELECTION AND MIX DESIGN METHOD

3.1 *Material*

3.1.1 *Aggregate*

The aggregates were taken from the dolomite quarry near Tehran. RAP materials were also taken from an asphalt layer that used the same aggregate from this quarry. Figure 1 shows the deposits of virgin aggregate and RAP material for preparing the mixtures and field implementation.

RAP material   Virgin aggregate

Figure 1.   RAP material and Virgin aggregate deposits.

Aggregate gradation was similar for laboratory mix and field sections. Although, some minor changes were needed for field implementation at the job site. The aggregate gradation is presented in Table 1, and aggregates properties are shown in Table 2. According to

Table 1.   Aggregate gradation.

Sieve Size	Type III specification	Passing %
9.5 mm	100	100
4.75 mm (# 4)	70–90	85
2.36 mm (# 8)	45–70	55
1.18 mm (# 16)	28–50	34
600 μm (# 30)	19–34	25
300 μm (# 50)	12–25	16
150 μm (# 100)	7–18	10
75 μm (# 200)	5–15	7

Table 2.   Aggregate properties.

Tests	Method	Results
L.A. abrasion, %	AASHTO T96	23
Flat and elongated particles, %	ASTM D4791	0.3
Fractured Particles, %	ASTM D5821	100
Soundness (Na2SO4), %	AASHTIO T104	1.0
Sand Equivalent, %	AASHTO T176	84
Fine Aggregates Angularity, %	ASTM C1252	46

deficiency of fine aggregates in the deposit of RAP material, virgin aggregates passing No. 100 sieve were also used for preparing the RAP mixture. This percentage was lower than 5 percent of the total mass RAP aggregates and the mixture with more than 95 percent of RAP aggregate is assumed as 100 percent RAP aggregate.

3.1.2 *Emulsion*
Pasargad oil company, PG 64-22 asphalt binder, was used for emulsion preparation. Water, Hydrochloric acid, Asfier N-480L KAO emulsifier, and SBR (Styrene-Butadiene-Styrene) latex, as a polymer, were the other components of the emulsion.

3.1.3 *Additives*
In order to obtain improvements in the adhesion of the asphalt binder, acidity modifier PRIPOL 1017 was used as an additive to enhance binder characteristics. This product is a dimer acid building block with great compatibility with bitumen emulsion, which prevents premature stripping. The other advantage of using PRIPOL 1017 in asphalt emulsion is that the breaking time would improve and the curing time would decrease.

As Iran's asphalt binder is paraffinic and paraffinic asphalt binders contain relatively few acidic components, it causes poor adhesion of the asphalt binder to the aggregate and lower emulsion properties. Consequently, the authors decided to use this product to improve the asphalt binder and emulsion properties in order to use RAP in a micro-surfacing mixture.

For determining the exact effect of using PRIPOL 1017 in the acid value of asphalt binder, Potentiometric titration equipment was used (Figure 2). This test was performed according to the ASTM D664 method. In this test method, the asphalt binder should be dissolved in a mixture of toluene and propan-2-ol with a small amount of water and titrated potentiometrically with alcoholic potassium hydroxide using a glass indicating electrode and a reference electrode. The device readings automatically measure the acid value of the sample.

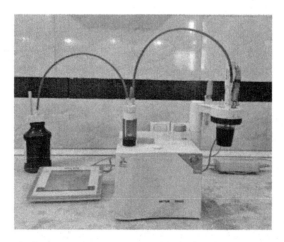

Figure 2.   Potentiometric titrator apparatus.

Based on the producer's instructions and practical experiences, 1.5 percent of PRIPOL 1017, by weight of asphalt binder, was used. The acid value of neat bitumen is almost zero for asphalt binders in Iran. 1.5 percent of acidifier, based on asphalt binder weight, increased the acid value of asphalt binder from zero to 3.264 (mg KOH/g). The changes in asphalt binder properties before and after using the acidifier are represented in Table 3.

Table 3. Comparing the acid value of neat and modified binder.

		Acid value	
Property	Method	Neat asphalt binder	Asphalt binder with 1.5% PRIPOL 1017
Softening Point, °C	ASTM D36	49.2	46.2
Penetration, 0.1 mm	ASTM D5	63	73
Specific Gravity, gr/cm3	ASTM D70	1.014	1.026
Ductility, cm	ASTM D113	138	143
Viscosity at 135°C, cst	ASTM D2170	326	307

3.1.4 *Mineral filler*

In order to have better consistency and regulate the setting time of the mixture, Portland cement is used both in the laboratory and field. As mentioned in ISSA A143, a one percent increase or decrease in the mineral filler of mixture design is permitted in the field when applying micro-surfacing.

3.2 *Mix design*

For assessing the feasibility of using RAP materials in micro-surfacing and the effect of acidifier on promoting the mixture characteristics, four types of micro-surfacing mixture were prepared.

- Conventional micro-surfacing
- Micro-surfacing with 50 percent of RAP as an aggregate
- Micro-surfacing with 100 percent of RAP as an aggregate
- Micro-surfacing with 100 percent of RAP and acidifier in the asphalt binder

Conventional micro-surfacing tests were performed in the laboratory in order to have a mixed design based on ISSA requirements and to check the compatibility of materials for field construction. The other mixtures were also prepared in the laboratory and tested by Cohesion, LWT Sand Adhesion, and Wet-Track Abrasion tests. The reports show test results to compare the effectiveness of using RAP and acidifier in the mixture.

4 TEST METHODS

Three laboratory tests were performed for all mixtures to evaluate the effect of RAP and PRIPOL 1017 on the micro-surfacing mixture. The authors intended to appraise the mixtures based on ISSA criteria, and for a more precise conclusion, a field investigation has been performed. The tests are as below:

- Cohesion test (ISSA TB-139)
- LWT Sand Adhesion test (ISSA TB-109)
- Wet-Track Abrasion test (ISSA TB-100)

4.1 *Cohesion*

This test is used to determine the time that a micro-surfacing layer needs to be settled before subjecting to straight rolling traffic. Six micro-surfacing samples should be cast in the mold with a height of 10 mm and 60 mm in diameter. When the sample is firm enough, the mold will remove and subjected to the apparatus. The apparatus in Figure 2 is set to impose 200 kPa pressure on the sample. After imposing the pressure, the torque reading will record.

4.2 *Loaded Wheel Test (LWT)*

This test is used to assess the mixture against severe asphalt flushing under heavy traffic loads. Three specimens with three different emulsion contents should be prepared for this test. The wheel is placed on the specimen, and the total weight of 56.7 kg should be fitted strongly on the wheel. After 1000 cycles, the wheel will unload, and the specimen's loose particles should be removed from the surface. Then the sample is weighted. Two hundred grams of standard sand were heated to 82°C and poured on the specimen, and placed on the load of 100 cycles. After removing the excess sand gently, the specimen should be weighted. The subtraction of these two weights is reported as sand adhesion value. Figure 3 shows the LWT apparatus.

Figure 3.   Cohesion test apparatus.

4.3 *Wet-Track Abrasion Loss test (WTAT)*

This test is developed to measure the quality of the micro-surfacing mixture under wet abrasion conditions. The specimen should be weighed after being placed in the oven for 24 hours at 60°C, then submerged in the 25°C-water bath for one hour. Then the specimen should be placed in the metal pan of the apparatus, and the abrasion head floats on the specimen. After washing the specimen and removing the debris, the specimen would be placed in the 60°C oven and dry to constant weight. The dried specimen should be kept at room temperature and weighted. The difference between this weight and the initial weight would be the grams of abrasion loss. Figure 4 shows the Wet-Track abrasion loss test device.

Figure 4.   LWT apparatus.

5   LABORATORY TEST RESULTS

As mentioned before, laboratory tests are conducted to verify the mixtures based on ISSA requirements. The materials were the same for all mixtures except the amount of RAP aggregates and the mixture made with PRIPOL 1017 as an asphalt binder additive. Three different emulsion contents (9, 11, and 13 percent) were tested to find the best composition for each mixture. All mixtures were passed the minimum requirements of ISSA with 11 percent emulsion except the mixture with 100 percent of RAP materials without additive. The results show that using RAP materials will decrease the consistency of micro-surfacing and has a notable effect on the cohesion of the mixtures, which means more time is needed to open the traffic. The effect of the PRIPOL 1017 additive on the final results was also significant. It promotes the mixture characteristics, as shown in Table 4.

Table 4.   Laboratory test results.

Mix Designation	Sample number	RAP content	Asphalt Additive	Emulsion Content	cohesion (kg-cm) 30 minutes (set)	60 minutes (traffic)	LWT (g/m^2)	WTAT (g/m^2)	Rejection or acceptance based on ISSA criteria
1	1	Without	–	9	12	24	230	642	Reject
	2	RAP		11			415	257	Accept
	3			13			650	98	Reject
2	4	50%	–	9	12	22	310	770	Reject
	5			11			475	342	Accept
	6			13			690	180	Reject
3	7	100%	–	9	10	19	452	990	Reject
	8			11			586	512	Reject
	9			13			737	418	Reject
4	10	100%	PRIPOL 1017	9	12	21	300	748	Reject
	11			11			448	329	Accept
	12			13			670	150	Reject

In order to have the most appropriate Eco-HP micro-surfacing and better evaluation of the mixtures on the job site, mixtures number 2 and 11 were constructed at the field. This will show a better aspect of the advantages of using RAP and PRIPOL 1017 in the real world.

6 FIELD EVALUATION

Two sections nearby in a site were chosen for field investigation of conventional micro-surfacing and Eco-HP micro-surfacing. These sites were located in the Qazvin province of Iran. For this evaluation, cracking, rutting and smoothness were considered. The previous condition of these two sections and the traffic load were almost similar, as they were located on the same freeway. Figure 5 shows the condition of these sections after four years of implementation. There were no significant cracks in both layers and reflective cracks don't reach out to the surface. But considering the smoothness, the conventional micro-surfacing almost lost its smoothness and the ride quality of this section decreased significantly. In some points, rutting deformation can be seen in conventional micro-surfacing. As a result, it could be concluded that conventional micro-surfacing was almost at the end of its maintenance life, and new rehabilitation was needed. However, the Eco-HP micro-surfacing was almost intact after four years of implementation and no notable ruptures were seen.

Eco-HP micro-surfacing   Conventional micro-surfacing

Figure 5.   Wet-Track abrasion apparatus.

Figure 6.   Field comparison of Eco-HP micro-surfacing and conventional micro-surfacing.

7 CONCLUSION

In this paper, the authors tried to develop a new micro-surfacing mixture with higher durability and omitting the use of virgin aggregates. Laboratory tests, including cohesion, LWT sand adhesion, and wet-track abrasion loss tests, are performed to assure that the mixtures could pass ISSA requirements. The selected mixtures were chosen for field execution. The evaluation of these mixtures based on deformations and riding quality were performed, and the main findings are as below:

(1) Laboratory tests revealed that using RAP material would decrease the consistency and cohesion of the micro-surfacing mixture.
(2) Using an acidity modifier can be a game-changer when using RAP material in micro-surfacing. PRIOPOL 1017 as an additive could have a significant effect on the acid value of the asphalt binder and, consequently, on the mixture characteristics.
(3) Using RAP aggregates in the micro-surfacing mixture could be up to 100 percent as an aggregate in case of using additives to enhance the mixture's characteristics.
(4) Although laboratory tests revealed that using conventional micro-surfacing has slightly better results, field investigation showed that Eco-HP micro-surfacing is more durable. Using this method increases the riding quality and smoothness of the micro-surfacing layer.
(5) Micro-surfacing mixture properties can be more than four years in case of using asphalt binder modifier even with RAP materials.

The findings of this study prove that RAP materials could be used in micro-surfacing mixtures in case of using additives to promote the mixture characteristics. Using this Eco-HP micro-surfacing not only extends this maintenance method's life-time but also eliminates using virgin materials. As a result, this will help save energy, virgin materials and oil resources and consequently lower greenhouse gases and air pollution.

REFERENCES

Garfa A., Dony A. and Carter A., 2016. Performance Evaluation And Behavior of Microsurfacing with Recycled Materials.
Geoffroy D.N., 1996. *Cost-effective Preventive Pavement Maintenance*. International Slurry Surfacing Association, https://www.slurry.org
Labi S., Mahmodi M.I., Fang C., and Nunoo C., 2007. *Cost-effectiveness of Microsurfacing and Thin Hot-mix Asphalt Overlays: Comparative Analysis*.
Pederson C.M., Schuller W.J., and Hixon C.D., 1988. *Microsurfacing with Natural Latex-modified Asphalt Emulsion: A Field Evaluation*.
Robati M., 2014. Evaluation and Improvement of Micro-surfacing Mix Design Method and Modelling of Asphalt Emulsion Mastic in Terms of Filler-emulsion Interaction.
Saghafi M., Tabatabaee N., and Nazarian S., 2019. Performance Evaluation of Slurry Seals Containing Reclaimed Asphalt Pavement. *Transportation Research Record* 2673(1): 358–368.
Takamura K., Lok K.P., Wittlinger R., and Aktiengesellschaft B., 2001. *Microsurfacing for Preventive Maintenance: Eco-efficient Strategy*. International Slurry Seal Association Annual Meeting, Maui, Hawaii.
Wang A., Shen S., Li B., and Song B., 2019. Micro-surfacing Mixtures with Reclaimed Asphalt Pavement: Mix Design and Performance Evaluation. *Construction and Building Materials* 201: 303–313.
Watson D., and Jared D., 1998. Georgia Department of Transportation's Experience with Microsurfacing. *Transportation Research Record* 1616(1): 42–46.
Wood T.J., and Geib G., 2001. 1999 *Statewide Micro-surfacing Project*.

Roads and Airports Pavement Surface Characteristics – Crispino & Toraldo (Eds)
© 2023 The Author(s), ISBN 978-1-032-55149-4

Use of traffic-speed ravelling data for the resurfacing intervention on open graded asphalt

Chloe Ip
Main Roads Western Australia, Perth, Australia

Richard Wix*
Australian Road Research Board, Melbourne, Australia

Lalinda Karunaratne
Main Roads Western Australia, Perth, Australia

ABSTRACT: Ravelling, defined as the progressive disintegration of the road surface by loss of binder and aggregate, is a prominent failure mode of open graded asphalt (OGA) surfaces. Once initiated, ravelling can progress rapidly, posing a hazard for road users, as loose aggregate can cause damage to vehicles. Additionally, as OGA surfaces are commonly used on high-speed roads (90km/h or faster) such as freeways and highways, field inspections can pose a safety risk to inspecting staff.

Currently, ravelling is identified by conducting visual surveys from a vehicle moving at traffic speed. It is a subjective assessment and is only effective in identifying high-severity ravelling. Moreover, while surface age is a common surrogate measure for determining ravelling progression, it is not consistent with age across all road sections, due to various factors.

The main objective of this study is to explore using a traffic-speed laser crack measurement system (LCMS) in an iPAVe survey vehicle to measure ravelling throughout the life of OGA surfaces objectively. It also aims to develop a ravelling progression model that determines the optimum investigatory and intervention levels for resurfacing works, thus reducing the risk of rapid progression by undertaking timely maintenance.

Ravelling data from the survey is reported in terms of a Ravelling Index (RI), which measures the volume of lost aggregate per road area. A trial survey was initially conducted on a lane section length of 50km, and the trial provided a well-correlated measure with visual inspections and surface age. Subsequently, RI data was obtained for the entire freeway and access-controlled highway network (approximately 830km) within the Perth Metropolitan Region.

This paper outlines the ravelling data measurement technology and the use of RI to develop an age-based ravelling progression model. The study identified an exponential relationship between surface age and RI. Further, it recommends that an RI of 38 or a surface age of nine years be used as an investigatory level to prevent the rapid deterioration of OGA surfaces. Data from a second survey of the same road sections conducted one year later is used to validate the developed model.

Keywords: Ravelling, Open Graded Asphalt, Laser Crack Measurement System.

---

*Corresponding Author: richard.wix@arrb.com.au

DOI: 10.1201/9781003429258-46

1   INTRODUCTION

1.1   *Background*

Main Roads Western Australia (MRWA) is responsible for operating and maintaining the Western Australian state roads, one of the world's most expansive road networks. The MRWA road network consists of approximately 830km of Open Graded Asphalt (OGA) surfaces on high-speed (90km/h or faster) and heavily trafficked roads in the Perth Metropolitan Region. Currently, the Perth Metropolitan Region's OGA Network is valued at $360 million AUD. The roads used for this analysis are in Table 1.

Table 1.   Summary of OGA surfaced roads in Perth Metropolitan Region.

Road Number	Length (Lane km)	Maximum AADT	Maximum % Heavy Vehicles
H015	360	89,279	24.8
H016	220	83,728	12.0
H017	120	56,135	23.0
H018	100	40,228	28.8
H021	30	46,377	14.3

OGA's particle size distribution has a large percentage of coarse aggregate, small amounts of fine aggregate and filler, and has 18–25% air voids, relying on the mechanical interlock of aggregates for stability (Austroads 2014). OGA is also known under different names around the world, one of the most common being porous asphalt (AAPA 1997).

The prominent failure mode of OGA is ravelling, which is defined as the progressive disintegration of the road surface by loss of binder and aggregate (Austroads 2019). Once initiated, ravelling can cause the road surface condition to deteriorate rapidly, degenerating into potholes, causing higher tyre noise and rough rides for road users. Loose aggregate can also damage windshields.

Ravelling condition assessments on the OGA network are currently undertaken from the cabin of a moving vehicle, making it easy to overlook emerging defects, due to OGA surfaced roads' higher traffic speed and multi-lane nature. Visual inspections also pose a safety risk to inspectors because of the high-speed environment. This has created the need to consider using automated traffic-speed data-collection technologies to measure ravelling.

Given the significant length of OGA network in the Perth Metropolitan Region, which is also continually expanding due to population growth, there is an increasing need to consider using technology to assist and potentially replace visual condition assessment to prioritise planned maintenance works.

MRWA maintains a pavement performance prediction and maintenance optimisation model using the Deighton Total Infrastructure Management System (dTIMS), which currently uses surface age as the parameter to predict ravelling progression. Visual inspections were conducted to validate the dTIMS model, identifying 94 road sections with ravelling. However, the dTIMS model, which currently uses 15 years as the surface life for OGA, failed to identify a vast majority of the 94 sections.

The decision-making process for identifying OGA sections in poor condition was reverse engineered using machine learning based on the condition data available to MRWA at the time. However, it was found that ravelling-related surface failures could not be predicted using roughness, rutting, surface age or texture. Traffic-speed laser-measured texture depth and low rutting were tested as surrogate measures for ravelling. They were deemed unreliable alternatives for ravelling as they increased the number of false positives and triggered treatment for sections where it was not required. The lack of an objective measure emphasised the need to develop a ravelling progression model to establish targeted inspection programs and prioritise funding for OGA resurfacing.

1.2 *Objectives*

The main objectives of the current study are to:

- compare ravelling data measured by the Laser Crack Measurement System (LCMS) with the surface age of road sections to determine if there is good correlation between ravelling and binder oxidisation of the surfacing
- select an appropriate indicator to measure the extent and severity of the ravelling, to be used for investigating and triggering resurfacing treatments and to model the future progression of ravelling
- develop a ravelling progression model that can be incorporated into MRWA's existing dTIMS pavement performance modelling system
- use the results of the data analysis to assist in estimating future funding needs, prioritising resurfacing sites and optimising treatment selection.

2 LITERATURE REVIEW

2.1 *Available laser technologies*

A review of traffic-speed laser systems from around the world was conducted to understand their capabilities in identifying ravelling.

Pavemetrics, the supplier of LCMS, has developed an algorithm to assess the severity of ravelling, which has been used to identify ravelling on porous asphalt in the Netherlands since 2012. The Rijkswaterstaat, which maintains roads and infrastructure in the Netherlands, reported a coefficient of determination (R^2) value of 93% between LCMS data and visual evaluations. The LCMS has gradually been replacing inspectors for ravelling detection since 2012 (Laurent *et al.* 2017). Pavemetrics have also indicated the development of a Ravelling Index, defined as the volume of aggregate lost per surface area, which also measures the volume of aggregate loss per surface area (Pavemetrics 2016).

In 2015, the Georgia Institute of Technology prepared a report for US Transportation Research Board to develop algorithms using 3D laser, the technology used to detect pavement ravelling (Tsai & Wang 2015). The report concluded that proposed algorithms have demonstrated a promising capability for automatically detecting, classifying and measuring asphalt pavement ravelling.

Drenth *et al.* 2016 have used LCMS to identify good correlation between a static sand patch test method and traffic-speed measured mean-texture depth. The analysis of aggregate loss was used to rate the severity and extent of ravelling in Singapore.

A study conducted in New Zealand suggests that surface age and average equivalent standard axles per day may contribute to an open graded porous asphalt surface ravelling (Henning & Roux 2012).

The Australian Road Research Board (ARRB) currently undertakes road condition surveys for Australian road agencies using the LCMS. ARRB has stated that its Automatic Crack Detection (ACD) system, which is based on the LCMS, identifies ravelled surfaces by calculating the volume of aggregate loss over the surface area (Wix & Leschinski 2013).

3 DATA COLLECTION AND THE CALCULATION OF THE RAVELLING INDEX

The condition of the MRWA's road network is assessed by ARRB on a biennial basis using its intelligent pavement assessment vehicle (iPAVe), which measures both the structural and functional condition of the pavement concurrently and at traffic speed. It does this through a series of fully integrated data acquisition systems installed into the iPAVe, including an automatic crack detection (ACD) system, which uses the Laser Crack Measurement System (LCMS) manufactured by Pavemetrics.

The LCMS consists of two high-performance 3D sensors mounted to the rear of the iPAVe at a nominal height of 2.2m above the pavement in each wheel path. Each sensor comprises two main components: a line laser; and a high-speed 3D camera mounted at an angle to the laser light source. When combined, the two 3D sensors project a 4m-wide laser line transversely across the pavement, consisting of approximately 4,000 measurement points. Each camera captures the half of the transverse profile in its own wheel path and interprets any distortions to a straight line as variations in the vertical surface profile. Due to the high pixel resolution, height measurement accuracies of 0.5mm are possible.

A picture of the road surface is then produced by combining the contiguous transverse profiles that are sampled every 5mm of longitudinal travel into 5m-long by 4m-wide road sections. Both range and intensity images are generated, which are then merged to produce a 3D image of the road surface. During data processing, the software also automatically analyses the road section for a variety of pavement defects, including ravelling, which is reported as a Ravelling Index (RI).

The RI is calculated for each individual road section using the following six-step process (Pavemetrics 2021):

(1) a 3D curve-fitting algorithm is used to fit a smooth surface over the textured surface of the pavement
(2) the road section is divided into a maximum of 320 individual 250mm x 250mm squares
(3) the air void content (AVC) is determined for each square by calculating the air volume between the smooth surface and the pavement surface
(4) areas within each square where there is ravelling (loss of stone) are identified using the LCMS range image
(5) the air volume between the smooth surface and textured surface is re-calculated excluding the previously identified ravelling. This is known as the road porosity index (RPI)
(6) the ravelling index for each square is then calculated by subtracting the RPI from the AVC and dividing by the area of the square as shown in equation (1).

$$RI = (AVC - RPI)/0.0625 \ (cm^3/m^2) \eqno(1)$$

The results for each individual square are then combined to provide an average RI for the selected reporting interval e.g. 10m, 100m etc.

An example of how each road section is divided into individual 250mm x 250mm squares is shown in Figure 1, with the RI of each square overlayed in the image.

Note that the LCMS can determine if line marking is present and if this option is selected, it will limit the RI calculations to the pavement surface between the line marking (thus the number of squares can be fewer than 320, as is the case in Figure 1).

4 METHODOLOGY

4.1 *Trial study*

A trial study of LCMS-measured ravelling was undertaken by surveying 50km of road and processing raw laser data to obtain an RI value. The roads surveyed as part of the trial included a 30km section of H016 and a 20km section of H018.

The survey data was validated against the visual inspections conducted by experienced inspectors. It was found that RI had a good correlation with field observations at identifying varying levels of ravelling. An example result of this comparison is shown in Figure 2.

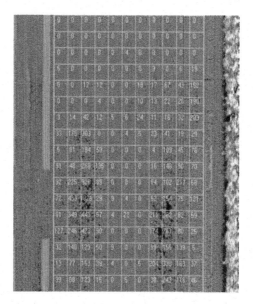

Figure 1. RI calculation methodology (Courtesy of Pavemetrics).

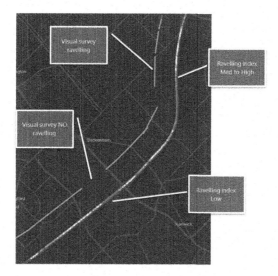

Figure 2. Ravelling data collected for H018 compared with visual inspection validation.

Ravelling data was also compared with the surface age of OGA to determine if there is a correlation between surface age and RI. It is expected that ravelling is correlated to the level of binder oxidisation in OGA. The level of binder oxidisation is proportional to the time OGA is exposed to the environment. Figure 3 shows the relationship between surface year (which indicates age) and RI, and the line of best fit with R^2 of 95%.

4.1.1 *Trial study conclusions and recommendations*
The trial study found that RI is an effective indicator in identifying ravelling on OGA surfaces and shows a good correlation with field observations and surface age. Given the positive results from the trial, it was recommended to collect ravelling data for the entire

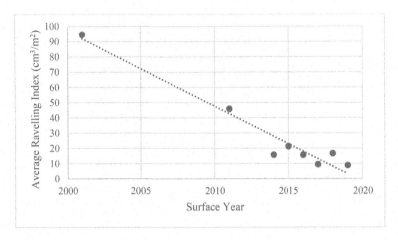

Figure 3.   Surface year against average ravelling index for H015 SLK 20 to 40, Lane 1.

OGA network to develop a ravelling progression model and use RI as an objective measure to identify resurfacing needs for OGA.

4.2   Development of surface age-based ravelling progression model

4.2.1   Collection of ravelling data
Traffic-speed ravelling data for the entire Perth Metropolitan Region network was collected in October–November 2020 as per the recommendation of the previous ravelling trial.

Ravelling data on all major OGA surfaced roads in the Perth Metropolitan Region network listed in Table 1 was analysed.

The analysis was performed on longer road lane segments with the same surface age. The segments were created by extracting surface age information from MRWA's corporate asset information system for inventory, known as the Integrated Road Information System (IRIS).

4.2.2   Determination of characteristic ravelling index value
This analysis aims to determine the characteristic RI value of a road segment using the data continuously collected by the survey.

The RI data reported in 10m sections was then aggregated onto the road segments with the same surface age using three different methods (length-weighted average, 75th percentile and median) to determine the best aggregation method. Segments with the surface year 2020 were excluded from this analysis, as it was unknown whether they were resurfaced before or after the 2020 survey.

With ravelling in OGA expected to occur uniformly over road segments with the same surface age, the length-weighted average was therefore anticipated to be better representative of the characteristic value for surface age-based segments. This proved true, with length-weighted average resulting in the best correlation for RI among three aggregation methods tested, giving the highest R^2 value.

4.2.3   Analysis of network-wide ravelling
Figure 4 shows the average RI for the OGA network in the Perth Metropolitan Region, with corresponding lengths of aggregated segments. The length of segments with a surface age of 12 years and above significantly decreases, suggesting that most OGA sections are resur-

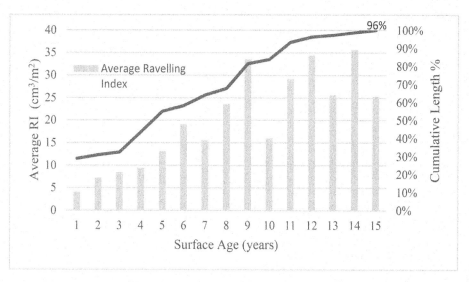

Figure 4.   Average RI and road length segments (cumulative) from surface age one to 15 years.

faced when they reach 'end of surface life' at a surface age of 10–12 years. 'End of surface life' occurs when the costs of maintaining the road surface become excessive and the road condition becomes a hazard for road users.

4.2.4   *Progression of ravelling index*

All OGA sections on the network were combined to undertake a regression analysis. Data from the tenth year was removed from the regression analysis. There was a significant reduction in the RI value, which was identified as completed resurfacing work but not yet updated in MRWA's IRIS inventory records. Figure 5 shows the data fitted with an exponential curve, which was found to be suited to the data given the expected acceleration of RI

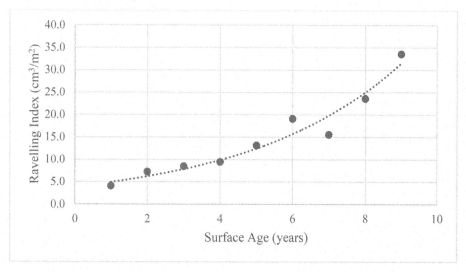

Figure 5.   Length-weighted average RI by surface age (one to nine years) with an exponential trendline.

increasing beyond a surface age of eight years. Experienced surface practitioners were also able to validate this information. Therefore, the relationship between RI and surface age can be expressed as shown equation (2), where y is the RI value in cm^3/m^2, and x is the OGA surface age in years.

$$y = 3.89e^{0.23x}, \quad R^2 = 0.95 \tag{2}$$

4.2.5 *Initial ravelling index value*

An analysis of the RI of a new OGA surface (surface age zero, which can include surfaces from zero to 12 months old) was conducted to determine a RI value of a newly surfaced section. The average ravelling index for new OGA surfaces was investigated by comparing the survey date with the schedule of completed works. Sites that were resurfaced after the ravelling data was collected were excluded from this analysis. The average RI for a new surface was 4.9, based on the analysis of two OGA roads. The average RI of all surfaces in year one is 4.1. Given the limited data available, a RI of 4.5 was used for modelling for new surfaces, which is the average of the currently limited year zero and year one data.

4.2.6 *Ravelling index value at failure*

The data was further refined to study if RI could assist with inspections, programming and prioritisation of resurfacing work. A 'failed' road section is defined as one that has deteriorated below the acceptable level of service and been programmed for the following year's Annual Works Program (AWP).

The sites planned for resurfacing in the AWP for the summer of 2021–22 were further analysed to determine the RI of 'failed' segments. Two selected road segments that were inspected in early 2021 and deemed to be in reasonable condition had rapidly failed and were displaying severe ravelling in August 2021. The RI of these sections at the time of the data collection survey (October to November 2020) was used as an indicative value to trigger further investigation and site inspection. It was found that the average RI of selected road sections prior to failure was 40. Hence, an RI of 38 was recommended as an investigatory level, including a 5% margin. The equation (2) also predicts an RI of 40 at a surface age of 10 years. It is recommended that an RI of 38 is used in conjunction with a surface age of nine years, depending on which value occurs first, to indicate that a road section should begin to be monitored for signs of ravelling.

4.3 *Model validation*

The model's predictive capability was tested by comparing the model predicted values against the actual measurements from a subsequent survey one year later. The model error for each road segment is the difference between the predicted RI calculated using equation (2) and the subsequently measured RI value. Road sections that were resurfaced in between surveys were removed from this comparison. As equation (2) predicts RI for a surface age for up to nine years, the segments used for validation were road segments with a surface age of nine years and under. Figure 6 shows the model error distribution.

Figure 6 shows that 49% of the road segments are within model error of RI ± 4. The previous analysis shows that the average annual rate of change of RI is about 3.5. Therefore, the model can predict the RI value for nearly half of the network within a year. Similarly, 77% of the segments can be predicted within ± three years. The ravelling progression for nine years old and above surfaces should be further investigated and validated to determine if a separate model can be developed for older surfaces. Further validation should be conducted for road-specific RI progression. Road segments that were observed to behave differently from the generic relationship developed were excluded from this validation.

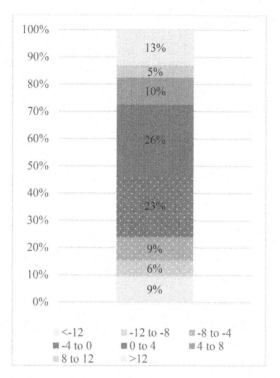

Figure 6.   Distribution of model error between predicted and actual RI .

5   CONCLUSIONS

The initial data collection trial showed that the LCMS could identify ravelling on OGA surfaces, showing a good correlation between the ravelling data and visual inspection data, which led to proceeding with the data collection survey of the entire OGA network in the Perth Metropolitan Region.

The data analysis from all OGA-surfaced roads showed good correlation between RI and surface age, indicating a correlation between ravelling and the oxidisation of binder within the surfacing. An exponential fit was used to model the progression of ravelling with surface age, with an R^2 value of 0.95. The exponential ravelling progression was found to be representative of OGA failures in the field, as confirmed by experienced practitioners.

Further, an analysis with a limited dataset of the average RI prior to failure, where a surface has deteriorated below the acceptable level of service, was found to be RI 40, with a value of 38 to be used to initiate monitoring.

A second set of survey data was also compared with predicted RI values and currently suggests that roughly 77% of the road sections could be predicted with an accuracy of within three years. The model's level of predictability proves that it is an excellent alternative to estimating long-term resurfacing needs compared to relying solely on visual inspections.

The above conclusions have led to the ravelling data being included in MRWA's existing dTIMS pavement performance model to provide a more comprehensive and accurate model for OGA surfaces. The new model is intended to be used to prioritise resurfacing sites and optimise treatment selections in the future.

6 RECOMMENDATIONS

It is recommended to adopt either RI 38 or a surface age of nine years for monitoring and investigation of OGA surfaces for signs of ravelling. There is still insufficient data to conclusively determine the RI value at the end of surface life. It is recommended this is further investigated with data analysis of annual surveys to continue refining the RI value at 'failure'.

It is also recommended that the OGA surface life should be modelled using equation (2) for a surface age of up to 10 years in the dTIMS model. The deterioration rates of surfaces aged 10–15 years should also be further investigated with more data to estimate the RI value at the end of OGA service life.

As the deterioration of road surface by ravelling appears to vary widely between different roads in the Perth Metropolitan Region, the effects of AADT, percentage of heavy vehicles, pavement strength and spread of traffic across lanes, should also be considered to further improve the ravelling progression model.

ACKNOWLEDGEMENTS

We would like to acknowledge the invaluable support from Shaan Ciantar, Kellie Keable, Qindong Li, Anthony Maroni, Paul Olsen, Nik Stace, Dao Tran and Chris Triplow.

Disclaimer: The views expressed in this paper are those of the authors and do not necessarily reflect the views of Main Roads Western Australia.

REFERENCES

Australian Asphalt Pavement Association 1997, *Open Graded Asphalt: Design Guide*, AAPA, Kew, Vic.
Austroads 2014, *Guide to Pavement Technology part 4B: Asphalt*, AGPT04B-14, 2nd edn, Austroads, Sydney, NSW.
Austroads 2019, *Guide to Pavement Technology part 5: Pavement Evaluation and Treatment Design*, AGPT05-19, edn 4.1, Austroads, Sydney, NSW.
Drenth K., Ju F.H. & Tan J.Y. 2016, 'Traffic-speed MTD Measurements of Asphalt Surface Courses', *International Conference on Maintenance and Rehabilitation of Pavements (MAIREPAV), 8th, 2016, Singapore*.
Henning T. & Roux D. 2012, 'A Probabilistic Approach for Modelling Deterioration of Asphalt Surfaces', *Journal of the South African Institute of Civil Engineering*, vol. 54, no.2.
Laurent J., Hebert J.F. & Talbot M. 2017, 'Using Full Lane 3D Road Texture Data for the Automated Detection of Sealed Cracks, Bleeding and Raveling', *World Conference on Pavement and Asset Management (WCPAM2017), Milan, Italy*, 11 pp.
Pavemetrics 2016, 'LCMS: 3D Road Scanning and Texture', *ERPUG 2016 Conference, Prague, Czech Republic*, European Road Profile Users' Group, Sweden, accessed 12 July 2022, <http://www.erpug.org/media/files/forelasningar_2016/23-Using%20high%20resolution%203D%20Richard%20Habel.pdf>.
Pavemetrics 2021, 'LCMS Data analyser manual', Pavemetrics, Québec, Canada.
Tsai Y & Wang Z 2015, *Development of an Asphalt Pavement Raveling Detection Algorithm using Emerging 3D Laser Technology and Macrotexture Analysis*, NCHRP IDEA Project 163, Transportation Research Board, Washington, DC, USA.
Wix R & Leschinski R 2013, '3D Technology for Managing Pavements', *International Public Works Conference, 2013, Darwin, Northern Territory, Australia*, Institute of Public Works Engineering Australia (IPWEA), Sydney, NSW, 9 pp.

Pavement management

Roads and Airports Pavement Surface Characteristics – Crispino & Toraldo (Eds)
© 2023 The Author(s), ISBN 978-1-032-55149-4

Estimation of pavement maintenance backlog and use in decision-making

Vesa Männistö*
Finnish Transport Infrastructure Agency, Helsinki, Finland

ABSTRACT: Due to scarce resources for periodic maintenance of road pavements, the current condition of paved road networks is not satisfactory in many countries. Many agencies are still using mainly technical parameters when describing the state of the road network. However, decision-makers are generally little aware of technical parameters, which makes justification of funding challenging. To cover this gap, pavement maintenance backlog is becoming rather widely used key performance indicator, which quantifies the monetary value of accumulated pavement maintenance works that should already have been taken care of. However, the definition of this indicator has been little researched, and in the most cases of maintenance backlog, no evidence-based definitions are available. The objective of this research work is to summarise the existing pavement maintenance backlog calculation methods, and to give a summary of pros and cons of the existing methods and recommendations for future development. The results show that there has been limited maintenance backlog related research in the road sector during the last decade, and the existing maintenance backlog calculation methods are still considered to be valid. Some other infrastructure sectors, such as energy, have been more active in this field by including routine maintenance costs and risk consideration into maintenance backlog discussions. The key issue in road maintenance backlog calculation is still how the maintenance or road condition standards are determined for backlog calculations. It is also important that these standards not only depend on technical parameters or road agency's costs, but consider also other points of view, such as road user and other costs incurred by clients and other stakeholders. More research is still needed to widen the scope of maintenance backlog calculation.

Keywords: Road, Maintenance Backlog, Deferred Maintenance

1 INTRODUCTION

1.1 *Background*

Pavement maintenance is a very important task. To ensure adequate funding, several key performance indicators have been used to report on the state of the paved road or street network. Most of the indicators are technical, but for wider use, especially for discussions with funders, politicians and other decision-makers, technical performance indicators do not suffice. Therefore, there has been a need for developing key performance indicators, which are more understandable for audiences with less technical knowledge.

---

*Corresponding Author: vesa.mannisto@vayla.fi

DOI: 10.1201/9781003429258-47

Maintenance backlog (MB) can be used as a common and comparable key performance indicator of road maintenance results for road assets. The term was already in use in 1980s in Northern America. In terms of the definition, there exist a few different ways to define pavement maintenance backlog. The most common used definition of maintenance backlog is the one recommended by multi-country European research project ERANET (Weninger-Vycudil et al. 2009) as follows:

"Maintenance backlog of the road infrastructure is the number of unfulfilled demands at a given point of time in explicit reference to the predefined standards to be achieved. Maintenance backlog can be expressed in functional (non-monetary) or monetary terms, and it refers to single components, sub-asset or to the whole road infrastructure asset of a given road network".

The calculation procedure recommended by Eranet 2009 is given in Figure 1. The key elements of this procedure are: (1) backlog is defined by referencing to predefined (usually condition-related) standards, (2) it is calculated for a given point of time and (3) it can be measured in both monetary and non-monetary terms. The calculation is, thus, based on the technical description of the transportation infrastructure. While this approach is appropriate for optimally allocating limited funds in the face of increasing investment backlog, it is not economically efficient because it does not account for other costs, for example, road user costs (Salem & al. 2010). The social costs, transport policy issues and financing possibilities are also partly or fully ignored in the analysis. These shortcomings were not discussed in detail in the Eranet study but listed in the study as future development needs (Weninger-Vycudil et al. 2009).

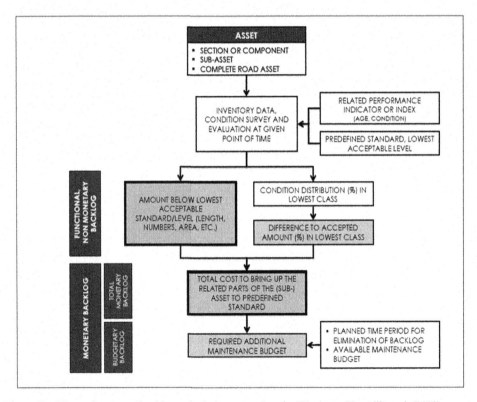

Figure 1.   The maintenance backlog calculation procedure by Weninger-Vycudil et al. (2009).

In addition to the most common definition described above, two other main MB calculation methods identified by Eranet were the following: (1) the amount of funding, which would improve all roads to excellent condition level in each time (MAPC 1987), or (2) the amount of funding needed to improve roads to a certain level of service (NTAMP 2008). A somewhat different definition is the one used by UK counties. In this method the backlog is defined as the gap between the actual accumulated asset consumption and the accumulated annual expenditure (Road Liasion Group 2017). Even a wider definition was used in Champaign, where both recurrent and periodic maintenance needs were summed to maintenance backlog (Vavrik 2008).

The use of maintenance backlog information by road authorities was found to be surprisingly limited. Only a few countries were identified as comprehensive users of backlog information. Moreover, it was found that the concept of maintenance backlog was just used by many organisations, but little information was available of the details how maintenance backlog is defined and calculated. Out of 90 publications identified discussing backlog, only 16 had any detailed information of how maintenance backlog had been calculated (Weninger-Vycudil *et al.* 2009).

The limited information and wide variety both in MB calculation methods and in the use of MB information indicates that the topic had not been thoroughly developed and implemented before the concluding work by Weninger-Vycudil *et al.* in 2009, and there is a need to update the state-of-the art of pavement maintenance backlog calculation and implementation issues.

1.2 *Road maintenance backlog in Finland*

Finnish Transport Infrastructure Agency (FTIA) is a public agency, which is responsible for asset management of public roads, railways and waterways in Finland. The FTIA was established in 2010 by merging road, railways and waterways administrations as a part of enhancing of public sector activities in Finland. Funding of FTIA is 100 percent based on public funding, and FTIA is controlled by the Ministry of Transport and Communication as a public agency.

Road backlog information has been utilised in Finland since early 2000s (Männistö 2006). The calculation method is the same what was recommended in the study described in the previous chapter (Weninger-Vycudil *et al.* 2009). Calculation of maintenance backlog was later extended to railways and waterways (Äijö & Virtala 2011), and today calculation and use of maintenance backlog information is a standard annual procedure. In 2021, the monetary backlog of the 78 000 km public road network was about 1,6 billion euros. Maintenance backlog of roads, railways and waterways is used as key performance indicator in the annual reporting of FTIA for the Ministry of Transport and Communication.

The use of current type maintenance backlog information has been widely accepted inside the FTIA, and among stakeholders. However, the topic is being actively discussed, and there is a growing demand to benchmark the methodology in use against current international practice. The following question have been raised to be investigated: (1) is the current method of backlog calculation still valid, (2) is all the backlog "real backlog", (3) what the optimal level of maintenance backlog is, and (4) what the wider effects of maintenance backlog are.

1.3 *Research questions for this study*

To find answers to some of the question given above, this paper reports a literature review in the topic of pavement maintenance backlog, trying to answer the main research questions as follows:

- What are the different methods for calculating pavement maintenance backlog?
- How is maintenance backlog information used in asset management?
- What topics need more investigations?

2 METHODOLOGY

With a focus on pavement maintenance backlog, a literature review is conducted in this paper. It is worth mentioning that during the literature search of this paper, no up-to-date systematic review in road maintenance backlog since the work of Eranet published in 2009 (Weninger-Vycudil et al. 2009) were identified.

Based on the suggestions from Sadelowski and Barroso (2007), the following research strategy was applied for the review, including:

(1) problem identification,
(2) formulation of the research questions and selection of the databases and websites,
(3) identification of the keywords with sufficient precision,
(4) determination of the inclusion and exclusion criteria for reviewing titles and abstracts,
(5) review of the full text of the selected papers, and
(6) data analysis and presentation.

2.1 Selection of the search strategy

To reach a wide range of academic journal articles, the citation databases TRIS and Scopus were used. Moreover, Google Scholar was also employed to ensure a broader delivery of unpublished studies, conference proceedings, thesis and dissertations. A database search was performed in March-April 2022 to locate studies published from 2009 onwards. A combination of the following keywords was used in the search

- (road* OR pavement*) AND
- (maintenance* OR investment*) AND
- (backlog* OR deferred*)

The search strategy for the review was primarily directed towards finding published papers (archival journals, conference proceedings, or technical reports) from the contents of three electronic databases, although each identified primary source has been checked for other relevant references. The search was restricted to papers published between 2009 and the present day. The year 2009 was chosen as the baseline as this was when the work of Eranet was published (Weninger-Vycudil et al. 2009). The inclusion criteria used in the search were, thus: (1) year of publication > 2009 and (2) peer reviewed papers only. Search results were stored in the Covidence software for deeper analysis.

3 RESULTS

3.1 Literature search

The selection of primary sources was initially based on a review of title, keywords, and abstracts. This was extended to include the conclusions section in the cases where the title, keywords and abstract provided insufficient information. All selected studies were then reviewed to identify whether a study can help to answer the specified research question. In particular, the main aim was to identify whether a study describes details how maintenance backlog was calculated and whether it reports the relationship to actual usage of backlog information.

Initially, 137 papers were identified as potentially relevant to the research questions; however, after removing duplicates and applying the inclusion/exclusion criteria, 22 papers remained in the set of relevant papers. The number of false positives in the initial set (papers

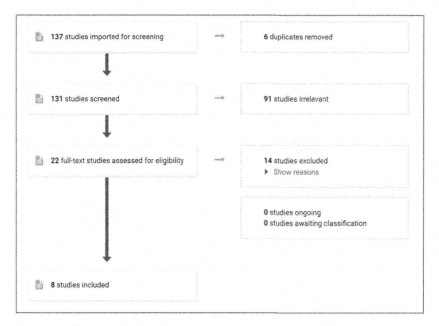

Figure 2/   The PRISMA chart of the research process.

that may have been relevant but on detailed investigation turned out not to be so) was disappointingly high. Therefore, the search was extended to other infrastructure and also grey literature in order to find more valid references. This extension revealed some interesting findings from the area of facility management systems, mainly in the energy sector. Finally, 8 studies were included for deeper investigation.

3.2   *Maintenance backlog issues in road maintenance*

3.2.1   *Definition of maintenance backlog*
Interestingly, the literature search did not reveal any comprehensive piece of road related research, which would have proposed a novel method for definition of road maintenance backlog. The papers investigated either had no precise information of how maintenance backlog is defined, or they primarily referred to the work by Eranet in 2009. Most papers investigated referred to the most common definition of maintenance backlog, which relates its calculation to existing condition classification and maintenance standards without giving any details of it.

One key factor in the MB calculation and interpretation of results is making an informed decision in setting an appropriate performance goal. It has been a challenge to highway engineers and asset managers because of the lack of a systematic procedure. The amount of resources required keeping the expected level of condition of the pavement network, or the performance goal, depends on the expected level. A lower performance goal helps reduce the needs for maintenance and rehabilitation, but it will adversely result in an increase in other costs, such as vehicle operating costs and deferred maintenance costs incurred from losing pavements that were eligible for less expensive treatment options, for example, preventive maintenance (Zhang *et al.* 2010). The amount of backlog is also directly dependent on the expected level of condition. If, as an illustrative example, the road users and road asset managers would be satisfied to the current level of service, the MB could be claimed to be zero.

Another key factor in the MB calculation is the definition of the maintenance strategy. Maintenance strategies and practices can differ significantly within and between highway agencies due to the asset portfolio size, overall condition, and the management policies affecting each asset group. Determining the consequences of delayed maintenance requires that the highway agency first articulate its preservation policy, including criteria to determine when to schedule maintenance activities. (NCHRP 2017) One key part in definition of the standard is the technical level at which the assets are after the maintenance treatment. Rantanen (2014) has introduced a concept of maintenance responsibility, which is the cost of maintenance treatment up to the level of new asset, in comparison to maintenance backlog, which is the cost up to adequate condition level.

The maintenance standard used in the MB calculation derives the amount of interventions, which are needed to clear the backlog. Definition of the funding needed for this clearance has also been vaguely reported. In most cases, unit prices of the measures are theoretical, average prices, indicating that all measures needed could be optimally executed at the given price. This, however, underestimates the MB in case of linear assets. For example, although pavement backlog is calculated based on 100-meter sections, it cannot be effectively be taken care of if paving works are executed in very short sections, and the funding needed for clearance of backlog is somewhat higher. Therefore, there is a need to explore, how unit prices should be defined to reach the realistic estimate for the MB.

Consideration of the MB is frequently seen to build a too narrow picture of the problem in hand. In many cases, a strategy for pavement rehabilitation primarily depends on initial construction and maintenance costs for identifying the most economically efficient pavement rehabilitation alternative. While this approach is appropriate for optimally allocating limited funds to decrease increasing of investment backlog, it is not economically efficient because it does not account for user costs (Salem *et al* 2010; Weninger-Vycudil *et al.* 2009).

3.2.2 *Further use of MB in road sector*

The method by ERANET has, however, been amended by including new assets to the estimation of the MB, such as railways (Statens vegvesen 2013) and railways and waterways (Äijö & Virtala 2011) in order to get more comprehensive view of the amount of backlog. Wehrle *et al.* (2022) discuss the MB from the point of view of inland water transportation and conclude that the MB in transport infrastructure mostly affects the neighbouring industries, since cargo has to be shifted to other modes of transport, and urgently needed goods experience delivery problems. Depending on the type of goods, different industries can be affected in different ways (Wehrle et al 2022).

Some fewer known synonyms for the MB were also identified, such as deferred preservation liability, which is defined as an estimate of the funding necessary to address the backlog of deferred pavement rehabilitation (FHWA 2014; SGA 2014). Therefore, calculating the liability, in monetary terms, of not performing these needed activities can be done by assuming cost of the M&R activity that will be assigned in the following year.

3.3 *Definition of maintenance backlog in other industries*

The searches on the MB or deferred maintenance related articles in other industries identified other aspects of maintenance backlog. One of those is the consideration of corrective and routine maintenance. The US congress defines it as maintenance that was not done when scheduled or planned (Howard 2022). Peters (2014) expanded the definition towards security issues by defining MB as number of unfulfilled maintenance demands about predefined security standards.

A summary of existing MB methods used in Finland was done by Kesälä and Koivula in 2012. They concluded that the concept of MB was still quite new and not well specified. Many Finnish cities and municipalities estimate the MB of their infrastructure using very simple calculations. In some cases, the MB starts to evolve after theoretical residual value of an asset fall below 70–80 percent of price of a new asset. An alternative method for the MB is, where it starts to build up when technical value of an asset is 75 percent of the price of a new asset (Rantanen 2017). Application of these two methods is, however, complicated, because there are no adequate models, which could indicate when the percent limits are reached, and the calculations are only estimates. (Kesälä & Koivula 2012).

Rødseth and Schjølberg (2017) and Rødseth (2018) took a further step by proposing that maintenance backlog can have a financial perspective that is both based on work orders and the technical condition of the facility based on the understanding of maintenance backlog from petroleum authorities (Petroleum-stilsynet 2016) and road authorities (Litzka *et al.* 2012) respectively. When providing a score of maintenance backlog, both the deviation of expected work completed from the work orders and the technical condition should be evaluated. This corresponds well to the different methodology to define backlog being used in facility management systems, where it describes all approved maintenance works, which are yet to be completed and helps management to make informed decisions on overtime, task allocation, recruitment, and subcontracting.

The work by Mike (2020) has come to similar conclusion. Backlog can be related to preventive or predictive maintenance activity since its primary aim is to make sure that critical assets do not break down and prevent work orders that are not ready to be scheduled. (Mike 2020).

In the model proposed by Rødseth and Schjølberg (2017) the risk influencing factor (RIF) structure in the Risk OMT (Risk modelling—Integration of Organisational, human and technical factors) adjusted the level of MB after evaluating the RIF of people such as the skills to the craft technicians and the RIF of tools they use in maintenance planning. The reason for this was that maintenance critical backlog is regarded as a potential for major accidents. The amount of MB was further included by Rødseth (2018) as one KPI in quantitative risk analysis (QRA) in the Integrating Planning Model (IPL). Rødseth concluded that improved maintenance quality can to some extent compensate for poor maintenance backlog. Still the organisation should strive to reduce MB to improve the overall risk picture of QRA.

4 CONCLUSION

The main conclusion of this systematic literature review is that there has been little progress in the field of developing road maintenance backlog calculation methods. Only few roads related research articles identified discussed the topic in any deeper detail but none of those had made any major progress in the further development backlog calculation methods. On the other hand, there has been some promising progress of maintenance backlog concept in other fields, such as energy and various facility management branches. Their idea of including costs of undone routine maintenance costs in the MB would be applicable in road sector as well. Moreover, the role of the MB as a part in risk analysis is also novel in road sector.

The key issue in utilisation of MB information is the selection of the maintenance standard, which is used as a basis for the MB calculations. Most of the methods used to define maintenance standards are based on heuristic selection of criteria or are based on optimisation of road agency's life cycle costs. It is, however, important to base optimisation of standards to other costs, such as vehicle operating costs, or other social costs (Salem *et al.* 2013; Weninger-Vycudil *et al.* 2009). Definition of the funding needed for clearing the backlog has also been less precisely defined. Often unit prices of the measures are average prices, indicating that all measures needed could be optimally executed at the given price, which lead to underestimation of the MB. Therefore, there is a need to explore, how unit prices should be defined to reach the realistic estimate for the MB.

One typical application of the MB has been to compare a road agency with other reasonably similar organisations. This type of comparison was seen problematic already by Eranet in 2009, because of obvious differences between organisations' road networks, and the methods of how the MB has been defined. No pieces of research were identified where this topic had been discussed at any detail. It is therefore recommended, that further research is still needed to build a solid ground for adequate possibilities for objective comparison of organisations' MB.

5 DISCUSSION

Although little research on the maintenance backlog has been done during the last decade, it is still seen as a valid key performance indicator in informing and justification of funding needs for road maintenance. The shortcomings of this very simplistic and summarising KPI must be understood, and it should not be used as an only indicator in decision-making.

REFERENCES

Äijö J. and Virtala P., 2011. *Maintenance Backlog of Transport Infrastructure*. Research Reports of the Federal Highway Administration (FHWA), 2014. EDC-4 Pavement Preservation When/Where Finnish Transport Agency 42/2011.

Howard S., 2022. Bridging the Gap: A Picture of State Transportation Funding in the United States. *Institute of Transportation Engineers. ITE Journal*; Washington Vol. 92(2), 35–38.

Kesälä A. and Koivula H., 2012. Maintenance Backlog. B.Sc. thesis. University of Lappeenranta.

Litzka J. and Weninger-Vycudil A., 2012. The Effect of Restricted Budgets for Road Maintenance. *Procedia - Social and Behavioural Sciences* 48, 484–494.

Männistö V., 2006. *Development of Road Asset Valuation in Finland*, Transport Research Arena Europe 2006, Göteborg, Sweden.

Metropolitan Area Planning Council (MAPC), 1987. *Pavement Management Forecasting Model - A Microcomputer Program for Lotus* 1-2-3. Boston (US), June 1987.

Mike D., 2020. *Maintenance Backlog: Take Control with a Priority Index*. https://www.prometheusgroup.com/posts/maintenance-backlog-take-control-with-a-priority-index.

National Cooperative Highway Research Program (NCHRP), 2017. Consequences of Delayed Maintenance of Highway Assets. NCHRP report 859.

Norfolk's Transport Asset Management Plan 2008–09 (NTAMP), 2008. www.norfolk.gov.uk.

Peer Exchange Report. https://www.fhwa.dot.gov/pavement/pubs/hif20057.pdf. Referred 14.5.2022

Peters R.W., 2014. Reliable Maintenance Planning, Estimating, and Scheduling; *Elsevier Science: Amsterdam*, The Netherlands, 2014.

Petroleumstilsynet, 2016. Hovedrapport – Utvikling-strekk 2016 – Norsk Sokkel.

Rantanen J., 2014. Korjausvelan Laskentaperiaatteiden Määrityshanke (Development Project for Calculation of the Maintenance Backlog). *Kuntaliiton Verkkojulkaisu (in Finnish)*.

Road Liasion Group, 2017. Well-maintained highways, code of practice for highway maintenance management. DOT London. https://www.ciht.org.uk/media/11915/well-managed_highway_infrastructure_combined_28_october_2016_amended_15_march_2017_.pdf. Referred 14.5.2022.

Rødseth H. and Schjølberg P., 2017. Maintenance Backlog for Improving Integrated Planning. *Journal of Quality in Maintenance Engineering* 23(2), 195–225.

Rødseth H., 2018. *Risk-based Maintenance Backlog. Safety and Reliability - Safe Societies in a Changing World - Haugen et al. (Eds), 645–651.*

Sadelowlski M. and Barroso J., 2007. *Handbook for Synthesising Qualitative Research*. Springer Publishing Company, Inc. New York, US.

Salem O.M., Deshpande A.S., Genaidy A. and Geara T.G., 2013. User Costs in Pavement Construction and Rehabilitation Alternative Evaluation *Structure and Infrastructure Engineering* 9(3), 285–294.

Smart Growth America (SGA), 2014. Repairing priorities 2014. Transportation Spending Strategies to Save Taxpayers' Dollars and Improve Roads. https://smartgrowthamerica.org/wp-content/uploads/2016/08/repair-priorities-2014.pdf. Referred 14.5.2022.

Statens vegvesen, 2013. *Hva vil det Koste å Fjerne Forfallet på Fylkesvegnettet?*. Resultat av Kartlegging. Statens Vegvesens Rapporter 183.

Vavrik W.R. *et al.*, 2008. *Rolling Wheel Deflectometer-based Pavement Management System Success*: Champaign County. TRB 2008 87th Annual Meeting, Washington DC, US.

Wehrle R., Wiens M., Neff F. and Schultmann, F. 2022. *Empirical Studies on the Impact of Inland Waterway Maintenance Backlogs on Riparian Industries.* Available at SSRN: https://ssrn.com/abstract=4042610 or http://dx.doi.org/10.2139/ssrn.4042610

Weninger-Vycudil A., Litzka J., Schiffmann F., Lindenmann H. P.,Haberl J., Scazziga I., Rodriguez M., Hueppi A. & Jamnik J., 2009. Maintenance Backlog Estimation and use. *Road Eranet.*

Zhang Z. & Jaipuria S., Murphy M., Sims T. and Garza T., 2010. *Pavement Preservation. Transportation Research Record.* 2150. 28–35. doi:10.3141/2150-04.

Roads and Airports Pavement Surface Characteristics – Crispino & Toraldo (Eds)
© 2023 The Author(s), ISBN 978-1-032-55149-4

Definition of a low-cost pavement management method based on a dual analytic hierarchy process

Nicholas Fiorentini*, Giacomo Cuciniello, Pietro Leandri & Massimo Losa
Department of Civil and Industrial Engineering (DICI), University of Pisa, Pisa, Italy

ABSTRACT: In countries such as Italy, funds available for road monitoring may be limited, especially for the secondary road network. In such circumstances, there is a need to define expeditious and low-cost procedures that allow the development of reliable and objective pavement management strategies. In this research, a methodology for the management of road pavements based on visual inspections, identification of road distresses, rating of road pavement conditions, and prioritization of maintenance needs, was defined. In order to appropriately recognize all types of road distresses, a manual that includes 19 road surface distresses organized by level of extension (3 levels) and level of severity (3 levels) was structured. The pavement conditions rating is performed according to a Dual Analytic Hierarchy Process (D-AHP), i.e., a double AHP process involving two aspects of road pavement management: structural performance for road serviceability and functional performance for users' safety. The D-AHP considers several expert opinions and allows defining the Road Condition Index (RCI). Priorities of maintenance needs were computed by defining a Priority Maintenance Index (PMI) that accounts for RCI and traffic flow. A case study was conducted on a 10-km two-lane road section. Findings seem to be encouraging, showing that low-cost procedures can effectively contribute to planning road maintenance in secondary road networks and limited available funds.

Keywords: Road Pavement Management, Maintenance and Rehabilitation Strategies, Monitoring and Inspection

1 INTRODUCTION

Road pavement management is of paramount importance worldwide. Inadequate management, unappropriated maintenance, and missing survey of roads markedly constrain mobility, significantly increase transportation operating costs and related human and property costs, increase accident occurrences, and aggravate segregation, poverty, and poor health condition of citizens (Burningham & Stankevich 2005). The concepts of road pavement management and Pavement Management Systems (PMS) have become very important in the past thirty years due to the increasing need for the maintenance and repair of road pavements and, a parallelly, a global trend in limiting the available funding for road authorities (Tavakoli *et al.* 1992). This aspect is significant in several European countries, such as Italy, where most of the existing roads achieve the end of their service life with the road authorities supported by minimal funds dedicated to their monitoring, inspection, and rehabilitation. In this scenario, detecting road damage and designing preventive maintenance interventions become essential (Fiorentini *et al.* 2021). Therefore, network-scale tools for supporting PMS are required.

*Corresponding Author: nicholas.fiorentini@phd.unipi.it

DOI: 10.1201/9781003429258-48

In the case of relevant infrastructures and funds available (e.g., motorways and freeways), Pavement Management Systems can be filled by the results of Non-Destructive survey methods, such as the Falling Weight Deflectometer, the Ground Penetrating Radar, the laser profilometric survey (Fiorentini *et al.* 2021; Gkyrtis *et al.* 2021; Nabipour *et al.* 2019), or, more recently, the satellite interferometric survey (Balz & Düring 2017; Fiorentini *et al.* 2020; Xing *et al.* 2019). These techniques, exploiting high accuracy, reliability, speed of execution, and high coverage, allow for obtaining reliable and accurate results.

Conversely, in the case of a minor road network, reliable low-cost survey methods need to be adopted. Such methods are typically based on the visual inspection of the existing pavement f conducted using distress manuals that define a correlation between the distresses detected and quality indexes. There are numerous examples in the literature where pavement management of local and county roads has been performed with visual inspections and quality indices (Hafez *et al.* 2019; Saha & Ksaibati 2016; Tavakoli *et al.* 1992). Likely, one of the most used indexes is the Pavement Condition Index (PCI). The method, developed by Shahin (2005), has been implemented as standard procedure by many agencies worldwide and published as ASTM standard (2018). The PCI provides a measure of pavement integrity and surface operational conditions based on a numerical scale, from 0 (very poor conditions) to 100 (excellent conditions). Such index is based on visual inspections, and the computation depends on the road distress type, severity, and extension (Ragnoli *et al.* 2018).

Although PCI is internationally recognized, the distresses considered are not necessarily applicable to contexts different from the one of development (e.g., Italy). Therefore, it is necessary to develop a local quality index for pavements, to support road authorities (Loprencipe & Pantuso 2017).

This paper aims to define a relatively low-cost pavement management methodology based on in-situ visual inspections, allowing for the definition of a quality index of flexible pavements of the secondary road network located in the Tuscany Region (Italy) that comprises approximately 1,000 km of rural two-lane roads.

Firstly, a manual that includes 19 road surface distresses, organized by extension (3 levels) and severity level (3 levels), was structured. Subsequently, through a Dual Analytic Hierarchy Process (D-AHP), the Road Condition Index (RCI) was defined. Finally, a Priority Maintenance Index that combines the RCI (indicative of pavement condition) and the traffic flow (indicative of road importance) was defined to identify road maintenance priorities. The study finalizes with a case study conducted on 10 km of a rural two-lane road.

2 METHODOLOGY

2.1 *Activity 1: Definition of a manual for road distress classification*

The first phase of this study concerns the definition of a manual of road distresses. Based on the knowledge of road authorities, technicians, and academics, 19 typologies of road distress were identified by several extensive pavement inspections (Table 1). This activity was

Table 1.   Typologies of road distresses included in the manual.

Road Distresses			
Crack sealing deficiencies	Potholes	Corrugation	Thermal cracking
Bleeding	Patching	Swelling due to tree roots	Longitudinal cracking
Surface disintegration	Rutting	Cracks due to underground utilities	Block cracking
Raveling	Roughness	Cracks due to manholes	Wheel track cracking
Debonding	Settlement	Edge cracking	

conducted using the PASER and FHWA distress manuals for flexible pavements as references (Miller & Bellinger 2014; Walker 2002). A detailed description corroborated by images is provided for each distress included in the manual. Besides this, each distress is classified according to different levels of severity and extension.

2.2 Activity 2: Computation of the Road Condition Index (RCI)

In order to provide a quantitative indicator of pavement condition, the RCI parameter was defined according to Equation 1:

$$RCI = K - \sum_{i=1}^{n} p_i \cdot s_i \cdot e_i \tag{1}$$

Where K = Numerical constant accounting for the optimal state of the pavement (in this research, $K = 100$); n = number of road distresses detected on the inspected road section; p_i = relative importance of the i-th road distress compared to the other road distresses; s_i = Coefficient accounting for the severity of the i-th road distress; and e_i = Coefficient accounting for the extension of the i-th road distress.

The p_i coefficient assumes a value between 0 and 100. The greater the p_i of road distress is, the higher is its relative importance. The sum of all p_i is equal to 100. The determination of the p_i was carried out through a D-AHP. Multicriteria analysis is a supportive tool in decision-making processes dealing with different, not directly monetizable, and heterogeneous factors. AHP allows comparing qualitative and quantitative, numerical (positive and negative), categorical, and ordinal variables. This methodology allows the decision-maker to analyze and evaluate different alternatives (e.g., different distresses of road pavement) concerning a specified road-related aspect.

In this research, the aspects (or criteria) to be considered to assess the impact of each road distress are two, thus creating a D-AHP:

- *Functional performance for users' safety*: the impact caused by each distress on road safety has been assessed. For instance, road surface raveling has a greater impact than longitudinal cracks on the users' safety of a road section;
- *Structural performance for road serviceability*: the impact caused by each distress on the road serviceability has been assessed. For instance, rutting has a greater impact on the structural performance of the pavement than road surface debonding.

Therefore, the D-AHP method allows assigning priorities to several alternatives, concerning more than one aspect, by relating qualitative and quantitative variables, otherwise not directly comparable, by combining multidimensional scales of measures into a single, homogeneous scale of values.

The D-AHP method relies on a series of pairwise comparisons between alternatives (i.e., road distresses), attributing them a score of relative importance, i.e., for each pair, the impact that an alternative has on the specified criterion compared to the other alternative. The relative importance score is assigned by the expert knowledge who performs the analysis. The score corresponds to a positive number between 0 and 9, according to the intensity of importance defined by Saaty (1987), shown in Table 2. The D-AHP ends with the computation of a weight to each distress, i.e., the p_i coefficient defined in Equation 1.

These pairwise comparisons are organized in a matrix called the Evaluation Matrix E. Matrix E is an $n \cdot n$ matrix. Each row and each column correspond to single road distress. In the present D-AHP, there are 19 typologies of road distress; therefore, E is a $19 \cdot 19$ matrix. The generic element of Matrix E, called a_{ij}, shows the comparison between the distress on the i-th row and the one on the j-th column. As stated before, the comparison is intended to evaluate the impact of the i-th distress compared to the j-th distress concerning one of the

492

Table 2. Coefficient defined by Saaty for pairwise comparisons.

Intensity of importance on an absolute scale	Relative importance of a road distress compared with another one	Reciprocals
1	Equal importance	If activity i has one of the above
3	Moderate importance of one over another	numbers assigned to it when compared with activity j, then j has the reciprocal
5	Essential or strong importance	value when compared with i
7	Very strong importance	
9	Extreme importance	
2 – 4 – 6 – 8	Intermediate values between two judgments	

specified criteria. For each criterion, a different AHP process has been performed. The features of Matrix E are:

- The elements a_{ij} with $i = j$, i.e., those positioned on the main diagonal, are equal to 1 since the comparison is made between identical road distresses, and therefore, they have equal importance;
- It is reciprocal, according to Equation 2:

$$a_{ij} = \frac{1}{a_{ji}} \tag{2}$$

- Moreover, Matrix E must be consistent, according to Equation 3:

$$a_{ij} \cdot a_{jk} = a_{ik} \tag{3}$$

Considering the high number of distresses, the respect of Equation 3 for all $a_{ij} \cdot a_{jk}$ pairs is not straightforward. Nonetheless, a certain scattering can be accepted in the evaluation of the consistency of Matrix E, according to the following procedure:

(1) Determination of Matrix E, as specified before;
(2) Computation of the Normalized Evaluation Matrix $\overline{E}$, in which each element $\overline{a}_{ij}$ is defined as follows (Equation 4):

$$\overline{a}_{ij} = \frac{a_{ij}}{\sum\limits_{i=1}^{n} aij} \tag{4}$$

Where a_{ij} is the element of E on the i-th row and j-th column.
(3) Computation of the Normalized Eigenvector $\overline{W}$ of the Matrix $\overline{E}$. The vector $\overline{W}$, called the Weight Vector w_i, is the result of the AHP process for a single criterion. Each element of $\overline{W}$ is defined as follows (Equation 5):

$$w_i = \frac{1}{m} \cdot \sum\limits_{j=1}^{m} \overline{a}_{ij} \tag{5}$$

Where w_i is the i-th element of $\overline{W}$;

(4) Computation of the Vector of the Weighed Sums, S_w, defined as follows (Equation 6):

$$S_w = E \times \overline{W} = \begin{pmatrix} d_1 \\ \dots \\ d_n \end{pmatrix} \tag{6}$$

(5) Computation of the Consistency Vector, CV, defined as follows (Equation 7):

$$CV = \begin{pmatrix} \dfrac{d_1}{w_1} \\ \vdots \\ \dfrac{d_n}{w_n} \end{pmatrix} = \begin{pmatrix} c_1 \\ \dots \\ c_n \end{pmatrix} \tag{7}$$

(6) Calculation of the Principal Eigenvalue, λ, defined as follows (Equation 8):

$$\lambda = \frac{1}{n} \sum_{i=1}^{n} c_i \tag{8}$$

Where c_i is the i-th element of the vector CV, and n is the number of elements of CV.
(7) Computation of the Consistency Index, CI, defined as follows (Equation 9):

$$CI = \frac{\lambda - n}{n - 1} \tag{9}$$

Where n is the order of the matrix E. In the present study, n is equal to 19.
(8) Computation of the Consistency Ratio, CR, defined as follows (Equation 10):

$$CR = \frac{CI}{RI} \tag{10}$$

Where RI is called Random Index, i.e., the average value of CI for a random generation of a large number of matrices of order n. The RI is tabulated according to the order n of the Matrix E (Table 3). In the present study, the RI value is equal to 1.53.

Table 3.   Values of RI.

n	2	3	4	5	6	7	8	9	10	11
RI	0	0.49	0.80	1.06	1.18	1.25	1.32	1.37	1.41	1.42
n	12	13	14	15	16	17	18	19	20	
RI	1.45	1.46	1.48	1.50	1.51	1.52	1.53	1.53	1.54	

Finally, the consistency of E relies on the CR value:

- If $CR < 0.10$, the inconsistency of Matrix E is negligible, and it is appropriate for the AHP analysis; the vector $\overline{W}$ is valid.
- If $CR > 0.10$, it is necessary to adjust the Matrix E since it shows excessive inconsistency; the vector $\overline{W}$ must be recalculated.

Therefore, the whole procedure is repeated with the other criteria to be analyzed; additional vectors $\overline{W}$ are determined. The p_i coefficients used in the computation of RCI are the outcome of weighting different $\overline{W}$. Indeed, considering that the present procedure is a Dual-AHP analysis, the coefficients p_i were computed as the weighted average of the coefficients

494

w_i obtained from multiple Matrices E, given the importance road authorities provide to each criterion. For the generic i-th road distress, p_i was computed as shown in Equation 11:

$$p_i = \alpha \cdot w_{FPi} + \beta \cdot w_{SPi} \tag{11}$$

Where α can range between 0 and 1, representing the importance of road functional performance, $\beta = 1 - \alpha$, represents the importance of road structural performance, w_{FPi} is the weight of the i-th distress concerning functional performance (road safety), and w_{SPi} is the weight of the i-th distress concerning structural performance (road serviceability).

2.3 Activity 3: Computation of the Maintenance Priority Index (MPI)

Once the road condition has been determined, it is necessary to identify which road sections have priority maintenance needs. Therefore, the Maintenance Priority Index (*MPI*) was defined. Such an index accounts for the potential expected benefits obtained by restoring the adequate conditions of a road section. The benefits are assessed in terms of the difference between the current *RCI* and an optimum reference value and traffic flow. A similar strategy can be found in (Saha & Ksaibati 2016; Tavakoli *et al.* 1992). Accordingly, *MPI* for the i-th road section can be carried out with the following Equation 12:

$$MPI_i = TVF_i \cdot (100 - RCI_i) \tag{12}$$

Where TVF_i is the Traffic Value Factor, which accounts for the traffic flow class of the road section. *TVF* is equal to 1 for AADT > 4000 vehicles/day/lane, *TVF* is 2/3 for 2000 < AADT < 4000 vehicles/day/lane, and *TVF* is 1/3 for AADT < 2000 vehicles/day/lane.

To appropriately allocate the available funds for maintenance interventions, road authorities could be guided by the following parameters: *RCI* Target (RCI_T), and *RCI* Limit (RCI_L). The RCI_T is the *RCI* value, which should have a restored pavement, and RCI_L is the lowest value of the *RCI* that can be accepted for a road section. The choice of RCI_T allows road authorities to pursue different strategies. Indeed, a road authority may assign greater importance to restoring the *RCI* of pavements to an optimal level (for instance, $RCI_T = 95 - 100$). Consequently, road interventions should be significant, albeit in a limited number depending on the available budget. Conversely, considering the same budget, a road authority may prefer to restore the largest number of road sections to a medium-high *RCI* (for instance, $RCI_T = 85 - 90$). Therefore, there should be more interventions than the previous strategy, albeit less important. The choice of RCI_L should rely on the available budget. A high budget should result in a low RCI_L (for instance, $RCI_L = 60 - 65$). Conversely, a low budget should result in a high RCI_L (for instance $RCI_L = 70 - 75$); consequently, a reduced number of road sections that requires maintenance activities will be classified as critical. Moreover, regardless of the *MPI*, a road section whose *RCI* is lower than RCI_L should be urgently restored.

3 RESULTS AND DISCUSSION

Table 4 below shows the coefficients relating to extension (e_i) and severity (s_i) of road distresses to be considered for computing *RCI* after in situ visual inspections.

The coefficients in Table 4 stem from the judgment of experts who participated in the D-AHP analysis. Figure 1 shows the relative importance of road distresses (p_i), i.e., the leading outcome of the D-AHP. In the present case study, it is worth noting that the relative importance of each road distress derives from the weighted average of two AHP. Therefore, the coefficients p_i are computed as the weighted average of the coefficients w_i obtained from

Table 4. Coefficient relating to the extension and severity of road distresses.

Road Distress	Extension			Severity		
	Low	Medium	High	Low	Medium	High
Crack sealing deficiencies, Debonding	0.5	0.8	1	0.4	0.7	1
Bleeding	0.6	0.9	1	0.8	0.8	1
Surface disintegration	0.5	0.7	1	0.5	0.7	1
Raveling	0.5	0.8	1	0.6	0.6	1
Potholes	0.5	0.8	1	0.4	0.8	1
Patching	0.6	0.8	1	0.3	0.6	1
Rutting	0.6	0.8	1	0.3	0.7	1
Roughness, Settlement, Corrugation, Roots, Utilities, Manholes	0.3	0.5	1	0.4	0.6	1
Edge cracking, Thermal cracking, Longitudinal cracking, Block cracking, Wheel track cracking	0.5	0.7	1	0.4	0.7	1

multiple Matrices E, given equal importance provided for functional performance and structural performance of a road pavement. According to Equation 11, $\alpha = \beta = 0.5$ were adopted.

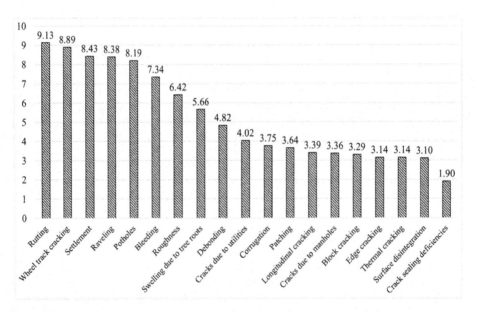

Figure 1. Relative importance of road distresses, considering both functional and structural performance.

Figure 1 shows differences in the relative importance of road distresses. Data demonstrate that the accumulation of permanent deformations (rutting) and the pavement failure due to fatigue (wheel track cracking) are more significant than other types of crack (for example, thermal cracking, edge cracking, or crack sealing deficiencies). This fact depends on the impact of rutting on structural performances. Indeed, the rehabilitation of extensive and severe rutting (e.g., subgrade rutting) requires the complete reconstruction of the pavement structure. Same considerations can be drawn for fatigue cracking when a pavement reaches the end of its service life. Furthermore, from a safety point of view, rutting (as well as settlements and potholes) has a significant impact since it affects the planarity of the pavement surface. Finally, raveling and bleeding are also of significant importance, although independent from the residual structural capacity.

As a case study, Figure 2 shows the outcomes of visual in situ inspections on the Regional Road SR206, located in the Tuscany Region, between km 30 + 000 and km 41 + 000. The SR206 is a rural two-lane road. The survey considered 10 road sections of 100 m each. Road sections were randomly selected with a frequency of 1 section per km.

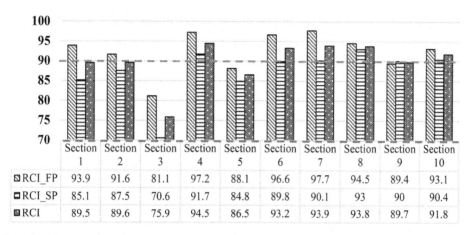

	Section 1	Section 2	Section 3	Section 4	Section 5	Section 6	Section 7	Section 8	Section 9	Section 10
RCI_FP	93.9	91.6	81.1	97.2	88.1	96.6	97.7	94.5	89.4	93.1
RCI_SP	85.1	87.5	70.6	91.7	84.8	89.8	90.1	93	90	90.4
RCI	89.5	89.6	75.9	94.5	86.5	93.2	93.9	93.8	89.7	91.8

Figure 2.   Computation of RCI (SR206).

For the present case study, $RCI_T = 90$ (green dashed line) and $RCI_L = 70$ (red dashed line) were considered. These values should refer to a road maintenance strategy with a limited budget. The D-AHP allows for a double assessment of road conditions. Indeed, it is possible to evaluate the conditions of the road sections considering both the functional performances (i.e., RCI_{FP} histogram) and the structural performances (i.e., RCI_{SP} histogram). Considering the equal importance of the two evaluation criteria (i.e., $\alpha = \beta = 0.5$), an overall RCI (RCI histogram) can also be calculated. Figure 2 shows that there are road sections where RCI_{FP} and RCI_{SP} display significant differences (e.g., Section 1 and Section 3, where RCI_{FP} is significantly greater than RCI_{SP}). It is advisable to expect maintenance interventions that mainly aim at structural rehabilitation in these road sections. The outcomes of Figure 2 show that the RCI of five road sections (i.e., Section 1, Section 2, Section 3, Section 5, and section 9) is below the target value and that Section 3 is in critical conditions ($RCI \cong RCI_{SP}$).

Traffic flow surveys carried out on the SR206 in 2015, 2016, and 2017 at km 41 + 600 and km 37 + 900 were considered to identify the priority of road maintenance needs. Such surveys report an average $AADT$ per lane equal to 6,600 vehicles. Therefore, the TVF parameter for all the road sections investigated is equal to 1. Once the RCI and TVF parameters have been determined, road maintenance priority can be determined according to Equation 12. Table 5 reports the list of critical road sections, sorted by MPI in descending order.

Once the acceptability ($RCI_L = 70$) and the target thresholds ($RCI_T = 90$) have been set, we can exclude from the MPI computation all the road sections that have an RCI lower than RCI_L. Indeed, such sections have the highest maintenance priority and do not require

Table 5. Identification of maintenance priority.

Road Section	RCI	TVF	MPI
Section 3	75.9	1	24.1
Section 5	86.5		13.6
Section 1	89.5		10.5
Section 2	89.6		10.5
Section 9	89.7		10.3

further network-level assessments. Specific field inspections need to be conducted and appropriate maintenance intervention designed and implemented. Besides, the sections that have an *RCI* higher than RCI_T were excluded from the computation of *MPI* as their condition are adequate and do not require maintenance. Accordingly, Table 5 shows the computation of *MPI* for road sections that have an *RCI* between RCI_L and RCI_T, i.e., which require maintenance priority with variable relevance, depending on the traffic flow and their *RCI*. Section 3 is the one that needs the highest maintenance priority, while the other road sections have a considerably lower *MPI*, demonstrating that the maintenance priority is less relevant. According to the availability of the allocated budget, road maintenance interventions can be carried out later.

4 CONCLUSIONS

This paper discusses the development and the implementation of a Road Condition Index to perform low-cost road pavement management. Such an index can quantify the road pavement condition by considering functional performance for users' safety and structural performance for road serviceability. These aspects have been included in the determination of the Road Condition Index by implementing a Dual – Analytic Hierarchy Process. This analytical process allowed the quantification of the impact, in terms of reduction of Road Condition Index, of 19 typical road distresses that may be detected on regional roads belonging to the Tuscany Region. Moreover, once low-cost field visual inspections are carried out and the Road Condition Index computed, practitioners can determine the expected benefits steaming from a road maintenance intervention; such analysis follows the computation of the Maintenance Priority Index, i.e., a parameter accounting for traffic flows and restoration of an appropriate Road Condition Index. A case study conducted on 10 km of a rural two-lane road has been presented, demonstrating that the proposed procedures are easy to implement, fast to perform, and reliable enough for ordinary road maintenance. The present research seeks to encourage regional and local road authorities to adopt low-cost tools for improving their pavement management duties.

ACKNOWLEDGEMENTS

This work was funded by the Academy Research Project (PRA) "Smart and Sustainable Use Phase of Existing Roads" (S-SUPER), promoted by the University of Pisa, Italy.

REFERENCES

ASTM D6433-18. Standard Practice for Roads and Parking Lots Pavement Condition Index Surveys. *ASTM Int.* West Conshohocken, PA, USA. 2018.
Balz T. and Düring R. Infrastructure stability surveillance with high resolution InSAR. In: *IOP Conf. Ser.: Earth Environ. Sci 012013.*Vol 57.; 2017. doi:10.1088/1755-1315/57/1/012013

Burningham S. and Stankevich N. Why Road Maintenance is Important and How to Get it done. *Transp. Notes Ser. No. TRN 4.* 2005.

Fiorentini N., Leandri P. and Losa M. Predicting International Roughness Index by Deep Neural Networks with Levenberg-Marquardt Backpropagation Learning Algorithm. In: Schulz K, ed. *Proc. SPIE 11863, Earth Resources and Environmental Remote Sensing/GIS Applications XII.* Madrid (Online Only): SPIE; 2021:118630P. https://doi.org/10.1117/12.2598005

Fiorentini N., Maboudi M., Leandri P. and Losa M. Can Machine Learning and PS-InSAR Reliably Stand in for Road Profilometric Surveys? *Sensors.* 2021;21(10):3377. https://doi.org/10.3390/s21103377

Fiorentini N., Maboudi M., Leandri P., Losa M. and Gerke M. Surface motion prediction and mapping for road infrastructures management by PS-InSAR measurements and machine learning algorithms. *Remote Sens.* 2020;12(23):3976. https://doi.org/10.3390/rs12233976

Gkyrtis K., Loizos A. and Plati C. Integrating Pavement Sensing Data for Pavement Condition Evaluation. *Sensors* 2021, Vol. 21, Page 3104. 2021;21(9):3104. https://doi.org/10.3390/s21093104

Hafez M., Ksaibati K. and Atadero R. Best Practices to Support and Improve Pavement Management Systems for Low-volume paved roads. *Int. J. Pavement Eng.* 2019;20(5). https://doi.org/10.1080/10298436.2017.1316648

Loprencipe G. and Pantuso A. A Specified Procedure for Distress Identification and Assessment for Urban Road Surfaces based on PCI. *Coatings.* 2017;7(5). https://doi.org/10.3390/coatings7050065

Miller J.S. and Bellinger W.Y. FHWA, Distress Identification Manual for the Long-Term Pavement Performance Program. Report FHWA-HRT-13-092. *Fed. Highw. Adm.* 2014;(May).

Nabipour N., Karballaeezadeh N., Dineva A., Mosavi A., Mohammadzadeh S.D and Shamshirband S. Comparative Analysis of Machine Learning Models for Prediction of Remaining Service Life of Flexible Pavement. *Mathematics.* 2019;7(12):1198. https://doi.org/10.3390/math7121198

Ragnoli A., De Blasiis M.R. and Di Benedetto A. Pavement Distress Detection Methods: A Review. *Infrastructures.* 2018;3(4). https://doi.org/10.3390/infrastructures3040058

Saaty R.W. The Analytic Hierarchy Process-what it is and How it is Used. *Math. Model.* 1987;9(3–5).

Saha P. and Ksaibati K. A Risk-based Optimisation Methodology for Pavement Management System of County Roads. *Int. J. Pavement Eng.* 2016;17(10). https://doi.org/10.1080/10298436.2015.1065992

Shahin M.Y. *Pavement management for airports, roads, and parking lots: Second edition.*; 2005.

Tavakoli A., Lapin M.S., Ludwig Figueroa J. PMSC: Pavement Management System for Small Communities. *J. Transp. Eng.* 1992;118(2).

Walker D. Pavement Surface Evaluating and Rating - Asphalt PASER Manual. *Wisconsin Transp. Inf. Cent.* 2002.

Xing X., Chang H-C., Chen L., Zhang J., Yuan Z., Shi Z., Xing X., Chang H-C., Chen L., Zhang J., Yuan Z. and Shi Z. Radar Interferometry Time Series to Investigate Deformation of Soft Clay Subgrade Settlement—A Case Study of Lungui Highway, *China. Remote Sens.* 2019;11(4):429.

Roads and Airports Pavement Surface Characteristics – Crispino & Toraldo (Eds)
© 2023 The Author(s), ISBN 978-1-032-55149-4

Innovation pathways within road performance management to upgrade investment analysis: A walkthrough of New Zealand and Philippine resilience landscapes

Jessica Vien S. Mandi*, Annisa Hasanah & Theunis F.P. Henning
Department of Civil and Environmental Engineering, The University of Auckland, Auckland, New Zealand

ABSTRACT: Road networks play an essential role in well-functioning communities. Governments worldwide strategically prioritise road transport needs to ensure that their systems – through road networks – are able to deliver social, economic, and environmental development and service benefits sufficiently. At the same time, performing resilience against disaster, climate, and other disruptive events has become an overarching goal of most countries. The quest for a best practice framework recognises the momentum to continually improve road performance while upholding system functionality and addressing various risks. However, vulnerabilities arising from the conflicting social, economic, and environmental considerations increase road asset deterioration. Such deterioration undermines the development of resilient and sustainable road networks, thus challenging the pursuit of the required level of service. This research aims to identify opportunities for innovation in performance management strategies to meet future road life-cycle analysis needs. The research embarks on a case study of New Zealand and the Philippines with a resilience lens to identify relevant practices in road performance management and understand the resilience landscapes in the countries. Both countries are prone to a range of natural hazards. Although their road networks are vastly different, two topics concerning road performance are pertinent: (1) mainstreaming resilience in asset management and (2) asset management practices. The analysis reveals, among amiable strengths, the long-term planning with asset management strategies institutionalised in New Zealand and the performance governance system measures in the Philippines. To some extent, applicable in both countries, multi-objective optimisation should be reinforced for greater resilience and sustainability outcomes. Findings of this research are like signposts showing pathways to upgrade investment analysis. Upgraded investment analysis, involving the adoption of more balanced performance measures, will prolong the service life, resilience, and sustainability of road networks.

Keywords: Durability and Long-Term Performance of Pavements, Performance Management, Performance Reporting, Performance Measures, Resilience, Climate Adaptation, Asset Management

1 INTRODUCTION: WHERE ROAD PERFORMANCE AND ASSET MANAGEMENT MEET RESILIENCE

Roads are critical infrastructure playing an essential role in well-functioning communities. Communities worldwide face many challenges arising from disaster and climate events, as

---

*Corresponding Author: jman281@aucklanduni.ac.nz

DOI: 10.1201/9781003429258-49

well as stresses such as ageing infrastructure, population growth, and increased traffic congestion. Increased severity of damages caused by the factors mentioned above has become the main driver for adapting resilience practices in the road asset management system. Infrastructure policy- and decision-makers are challenged by the need to shift traditional asset management to operationalise resilience. Concomitantly, existing knowledge gaps should be addressed to keep road performance good while facilitating community service when disruptions occur.

One gap in implementing the resilience-based asset management approach lies in having different measures on infrastructure asset performance. The differences within the road sector and whole-of-government agencies lead to multiple considerations in addressing the holistic needs of the society, the economy, and the environment. This premise implies another gap such that a new paradigm beyond single-level analysis is needed. Amid these gaps, overall government resources are limited, and road network demands are increasing. Additional work is required to identify the best application of practices to enable resilient road asset management.

In line with the Sendai Framework for Disaster Risk Reduction 2015-2030, nations have incorporated the performance of resilience to disaster, climate, and other disruptive events within their overarching goals. Coping, adaptive, and transformative mechanisms are incorporated into policies, strategies, and guidelines in order to reduce vulnerabilities and enhance well-being. As resilience requirements evolve along with the frequency and intensity of disruptions, resilience developments should be mainstreamed in vulnerable areas. The Intergovernmental Panel on Climate Change (IPCC 2022) refers to increasing adverse impacts due to climate change (i.e., flood/storm-induced damages in coastal areas of Asia, Australasia, North America, small islands, and cities by the sea, as well as damages to infrastructure in the global South, including Australasia, small islands, and cities by the sea). New Zealand, a developed country in Australasia, and the Philippines, a developing country in South Asia, are island countries prone to climatological, hydrometeorological, volcanic, and seismic events. Facilitating climate-responsive infrastructure asset management is a welcome undertaking for both countries as well as communities recognising the interdependence of our climate, society, economy, and environment.

2 CONTEXT: ROAD NETWORKS AND RESILIENCE IN NEW ZEALAND AND THE PHILIPPINES

Road infrastructure in New Zealand is about 94,000 km (Gardiner *et al.* 2009), owned by various entities including The New Zealand Transport Agency (NZTA) and local authorities. The extensive road network is exposed to climate change impacts. The Ministry for the Environment (2016) anticipated several climate change events, such as higher sea levels and higher storm surges in coastal areas, both low and high extreme temperatures, intense rain and wind events, and thawing glaciers or permafrost.

The business case approach has been widely used in New Zealand to justify infrastructure investment and strategic vision, especially for lengthy and highly complex projects or activities (The Treasury 2022). The National Disaster Resilience Strategy introduced by the Ministry of Civil Defence (2019) emphasises the opportunity for investment in resilience to support better business cases. In 2020, NZTA conducted a programme business case on land transport system resilience. According to the report, four options (status-quo, improved decision making, integrated investment model, and invest for resilience) may be implemented, and strategic investments can support system and community resilience (Tonkin +Taylor & Tregaskis Brown 2020). The current state of resilience application is on a strategic level. For example, the Government Policy Statement refers to 'improved network resilience' as a policy objective. Moreover, the Ministry of Transport (2018) structured resilience, along with security, as a key part of the New Zealand Transport Performance Framework (See Figure 1). This high-level environmental policy addresses resilience as an activity throughout the investment process, from strategic case development, investment and benefit logic mapping, to benefit-cost analysis (NZTA 2022).

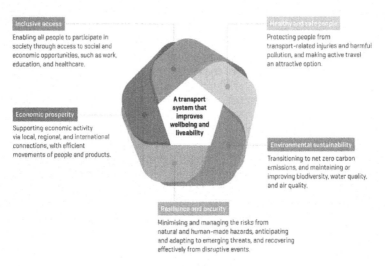

Figure 1.   Outcome Investment Priorities for the Transportation Sector (New Zealand Ministry of Transport 2018).

Philippine road networks are linked primarily to the responsibility of the Department of Public Works and Highways (DPWH) to develop the national roads and expressways, as well as local government units to connect provincial, municipal, city, and *barangay* roads. Some 33,213 km comprise the national roads (DPWH 2021a) out of an estimated total of 215,000 km (Lacambra *et al.* 2020). The National Disaster Risk Reduction and Management Plan 2020-2030, led by government agencies across sectors, acknowledged the centrality of risk and focused on all-hazards in the changing climate, including extreme and slow onset events and man-made activities (National Disaster Risk Reduction and Management Council [NDRRMC], 2020).

On a national scale, laying down the foundation for a high-trust and resilient society is a policy goal towards the vision of a predominantly middle-class country by 2040. The urgent directive for the National Economic and Development Authority (NEDA) is to focus on building resilience via system reforms and regional development (NEDA 2021a). Thereof, the DPWH uses the Performance Governance System (PGS) adopted from the Harvard Business School. The DPWH PGS 2017-2022 framework consists of a strategy map and an enterprise scorecard to guide the formulation of governance documents (DPWH 2018). It envisions improved Filipino life through quality infrastructure, for instance, via disaster-resilient structures (see Figure 2). The system allows organisations to track DPWH's progress in achieving the planned outcomes e.g., lives and properties protected from disasters (DPWH 2018).

Monitoring performance is a precursor to innovating current management practices. Learning from the corresponding experiences of New Zealand and the Philippines provides credence in determining resilience-informed improvement strategies in road performance management.

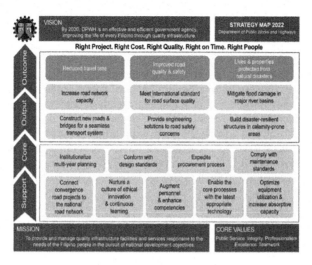

Figure 2.   PGS Framework (DPWH 2017), a Part of the Long-Term Vision of Better Filipino Family Life.

3   MOTIVATION / PROBLEM STATEMENT: CALL FOR UPGRADED INVESTMENT ANALYSIS

It is not possible to escape from risks. The quest for a best practice framework recognises the momentum to continually improve road performance while upholding system functionality and addressing various risks. However, vulnerabilities arising from conflicting social, economic, and environmental considerations increase road asset deterioration. Such deterioration undermines the development of resilient and sustainable road networks, thus challenging the pursuit of the required level of service. Hence, the importance of resilience-based asset management is growing exponentially in anticipating various natural and man-made hazards, and maximising the life-cycle of capital-intensive infrastructure.

Investment into resilience cannot be analysed in isolation. It has to be balanced against providing for the growing population and addressing the maintenance backlog of ageing infrastructure. The starting point for such analysis is a performance framework that quantifies the extent of infrastructure asset delivery needs. Infrastructure resilience performance monitoring, in particular, is an emerging area and only covering parts of the asset management perspectives.

4   AIM AND OBJECTIVES: THINKING ABOUT FUTURE LIFE-CYCLE ANALYSIS NEEDS

Owing to a holistic perspective to upgrade network resilience and sustainability, this research aims to identify resilience performance areas for meeting future road life-cycle analysis needs. To accomplish the aim, the research investigates road networks in New Zealand and the Philippines by examining the state-of-the-art policy in road performance management and understanding the application of resilience in the national management practices of both countries. Further, this research explores the gaps and opportunities in developing an infrastructure resilience performance management approach to upgrade investment analysis.

5 METHODOLOGY: REVIEW AND CONCEPTUALISATION FOR REAL-WORLD APPLICATION

Resilience-based asset management through upgraded investment analysis sparks the impetus for desirable resilience and sustainability interventions. Literature review and framework creation were done to build a harmonised real-world application of related analysis techniques in reporting infrastructure asset performance. Relevant literature – organisational policy documents from government websites, as well as science- and practice-informed publications from academic journals and Google searches – were considered. The resulting lessons from the review, applicable in New Zealand and Philippine policy scenes, apprised the development of the conceptual performance framework and potential improvement strategies for implementation by interested communities.

6 RESULTS AND ANALYSIS: HOW TO MAINSTREAM RESILIENCE PERFORMANCE ALONG THE PATH TO SUSTAINABILITY

Although both countries' road networks are vastly different, two topics concerning road performance are pertinent: (1) mainstreaming resilience in asset management and (2) asset management practices.

6.1 *Mainstreaming resilience in asset management*

Resilience can be defined as "the ability of a system to maintain minimum level of service (LOS) after magnitudes of disruptive events during its life-cycle within time and budget constraints" (Mohammed *et al.* 2017). Mohammed *et al.* (2019) identified two main features to reflect resilience in road asset investments: road condition during its life-cycle, as well as consequences of extreme and periodic events, and their required maintenance programme and recovery strategy to reduce adverse impact and life-cycle deterioration. Works of literature also discussed the role of infrastructure resilience in increasing capacity and durability during extreme events (Lu *et al.* 2017; Zhang & Wang, 2016), interdependent and interconnected networks (Gay & Sinha 2013; Baroud *et al.* 2015), and scenarios for multi-hazard assessments (Dehghani *et al.* 2013; Levenberg *et al.* 2016).

Figure 3 illustrates these principles graphically. It shows the scenario of two roads, A and B. Road B is an old deteriorated road that will experience significant damage from a climate or weather event. Because of the extent of damage to this road, it will be more costly to

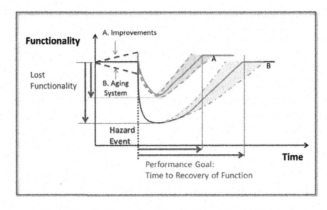

Figure 3.   The Loss of Functionality Curve for Disaster Impacts (Author/artist unknown 2020).

repair, and it will take longer to fix. Road A, on the other hand, may be better maintained and/or may also have some resilience improvement, e.g., better drainage. Both these strategies will result in a higher resilience, thus limiting the damage for an event of equal magnitude, and by having less damage, also being able to be repaired or accessed in a shorter time frame. However, resilience improvement or better maintenance costs more money, and the asset management system assists in determining situations where the return on investment justifies the additional costs.

An investment programme for resilience today brings protection against risk impacts and avoidance of high costs in the future. With existing budgetary pressures in the regular road asset management activity, adding investments for resilience may stimulate delays in maintaining and rehabilitating non-priority projects. In this situation, integrating resilience with reliability, availability, maintainability, and safety (RAMS) is a potential solution (Mohammed *et al.* 2019). By establishing an adequate investment in RAMS, the road network resilience within the existing budget limitation is argued to be achievable. According to multiple authors (Gay and Sinha 2013; Mostafavi 2017; Mohammed *et al.* 2019), a resilience investment programme for RAMS requires components such as funding to enable a central database, compiled asset data for assessment and modelling purposes, as well as the road asset and government resilience requirements.

6.2 *Asset management practices in New Zealand and the Philippines*

The United Nations (UN 2021) stated that the asset management principles for sustainable development are community-focused, risk-based, service-focused, forward-looking, value-based, and transparent. Both New Zealand and the Philippines embrace such principles at the infrastructure network level by practising business/public service continuity plans, emergency preparedness measures, traffic management/ engineering interventions, asset-based strategies/plans, and performance management initiatives.

Asset management has been institutionalised and is thus more mature in New Zealand. Prioritising maintenance and renewal programmes and requiring essential infrastructure works were implemented pursuant to the State Highway Asset Management Plan 2012-2015, which aimed at a resilient and secure transport network by decreasing the number and duration of closures (NZTA 2011). Prioritisation is made by reviewing different aspects (i.e., duration, cost, risk) vis-à-vis the provision of primary access to road users. A more comprehensive approach for critical infrastructure, asset risks, and resilience is covered by the State Highway Infrastructure Assets Management Plan, which provides maintenance and renewals assessment factors at a managerial level (NZTA 2015a). In addition, resilience programme business cases are conducted as a risk management strategy pursuant to the pavement life-cycle plan for overslips and floods, as well as other situations requiring emergency and restorative actions (NZTA 2015b).

Ensuring asset preservation for infrastructure development and environmental protection is a strategy embedded in the Philippine Development Plan 2017-2022 (NEDA 2021a). DPWH naturally undertakes asset management in agency-wide inventory processes, planning and management applications (reporting information on roads and bridges, including road slope and bridge repair), as well as condition monitoring and evaluation in accordance with its PGS. In 2020, the Department of Finance (DOF), Department of Budget and Management (DBM), and NEDA issued a joint circular to implement a national government asset management policy for all strategically non-financial government assets (DOF *et al.* 2020). Towards operationalising the asset management system, the DPWH is formulating an agency asset management plan to include the planned budgeted acquisition, disposal, rehabilitation/repairs, and needs assessment-based capability development activities (DPWH 2021b).

6.3 *Conceptual performance framework*

Practically, governments prioritise transport system targets in pursuit of the 2030 UN Sustainable Development Goals (SDGs) for eradicating road traffic accidents, developing rural roads, expanding public transport, and decreasing losses caused by disasters. The issue is whether or not the road asset investments are adequate, to the effect that important network links can withstand direct and knock-on impacts resulting from exposure and vulnerability to shocks and stresses, rapidly restore system performance, and mitigate asset damages. Since the spatial distribution and timing of these large-scale events cannot be precisely predicted, risk management practices should acclimatise to adaptation resilience besides maintenance optimisation and life-cycle demand management. In the meanwhile, Figure 4 provides the conceptual performance framework resulting from this research. The four attributes regarding asset delivery in performance management – stakeholder value, functionality, vulnerability, and criticality – are essential parts of robust decision-making for road network investments.

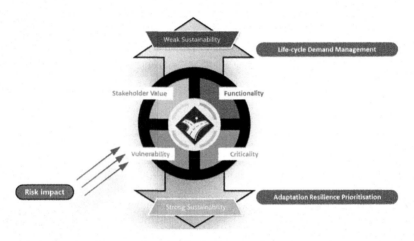

Figure 4.   Conceptual performance framework of this research.

The said attributes assist decision-making at the investment level while enlightening tactical and operational levels for resilience-informed performance management. The tactical infrastructure performance should compare to a systemic balancing of stakeholder value, functionality, vulnerability, and criticality. Sustainable investment levels addressing these attributes would level the playing field for immediate programme outputs and far-reaching network developments. The intent is to link life-cycle demand management with sustained investments to adaptation resilience prioritisation for longer-term outcomes. In turn, the operational programming builds on appropriate timing and type of investments backed up by evidence-based measures. The challenge then lies in how to utilise and manage data sets. Table 1 enumerates some resilience performance-related documents/measures of NZTA and DPWH for each attribute, followed by our remarks.

6.3.1 *Vulnerability and functionality*

Performance levels refer to "quality criteria that define the minimum condition of an asset" (Asian Development Bank [ADB] 2018, p. vii). Our framework treats vulnerability and functionality as the capacity or potential for diminishing and increasing infrastructure network performance, respectively. A systemic approach prioritises critical components that tend to create dependencies and redundancies, and addresses weak points that may produce

Table 1. Resilience performance-related documents/measures of NZTA and DPWH.

Attribute	NZTA Document/Measure	DPWH Document/Measure
Stakeholder value	Measures on reduction of risk, readiness, response, and recovery, including resilience-based mobility; Resilience and security; Value for money; *etc.*	Measures on protecting lives and properties, reducing travel time, and improving road quality and safety; Public service; Value engineering; *etc.*
Functionality	One Network Road Classification (ONRC) - functional classification	Department Order (road conversion)-functional classification
Vulnerability	Measures on social impact affecting people due to disruptions; Multi-hazard references; Economic losses due to unplanned disruptions	Measures on protecting lives and properties from disasters (mobility during and after disasters, flood mitigation, and disaster-resilient structures)
Criticality	Criticality ranking tool based on the current ONRC, access to lifeline utilities or evacuation routes, access to essential services, and overall criticality	No existing criticality framework; Monitored road parameters include functional classification, pavement surface type, condition rating, year of last resurfacing; *etc.*

critical vulnerabilities (Organisation for Economic Co-operation and Development [OECD] 2019).

On vulnerability, the UN Office for Disaster Risk Reduction pertains to the physical, social, economic, and environmental conditions that increase the asset's or system's susceptibility to hazard impacts. It is deemed prudent that the following be considered in modelling vulnerability: road network gaps, asset surface and structural conditions (physical); locational patterns including damages, losses, and other impacts due to hazard/climate events (environmental); as well as population concentration, travel disruptions, financial constraints, and disinvestments (socio-economic). Understanding infrastructure systems' dependencies and interdependencies in damage and recovery will extend existing models towards strategic pre-event risk mitigation and post-event recovery decision-making (He & Cha 2021).

On functionality, NZTA has a classification system involving criteria and thresholds, based on the functions that roads perform within the network. DPWH has a classification system criteria stating which communities the roads connect. New Zealand and Philippine road classes thereby reveal that movement volume and border access shape the notion of functionality. At the same time, disruptions have knock-on effects across borders with insufficient network resilience capacities, leading to negative impacts cascading down the society, economy, and environment, that could render road assets' functions nugatory.

6.3.2 *Stakeholder values and criticalities*

Performance levels are usually grouped into management and operational – the former reflecting ability to successfully manage outputs for the client, and the latter entailing the need to: satisfy road users on accessibility, comfort, travel speed, and safety; as well as minimise the total long-term cost to the road organisation and users, and minimise environmental impacts (ADB 2018). Resilience may be delivered throughout the infrastructure life-cycle (through robustness and redundancy designs and investments, operations continuity planning, and/or retrofitting for adaptability) (OECD 2019). Our framework sees that current management and future operational performance levels are akin to value- and criticality-based performance levels.

There are many types of values to contemplate in capital-intensive and long-lived infrastructure systems. Both NZTA and DPWH regard the concept of value for money in delivering infrastructure services. A simple definition of value is function over cost (NEDA 2009; DPWH 2015). Infrastructure resilience's value – in terms of cost and benefit – has not been institutionalised in New Zealand and the Philippines. NZTA (2019) noted measures informing the assessment of resilience performance benefit (i.e., availability of a viable alternative to high-risk and high-impact route, level of service and risk, redundancy, and temporal availability). The option value, which accounts for users' internalised probability perceptions and risk aversion, may figure vulnerability in the analysis of users' willingness to pay more for a backup transport service (Jenelius 2010). More so, varying the criteria and values affects the precision and calibration of criticality (Anderlini 2020). For instance, unprogrammed disaster recovery and reconstruction projects are mainly driven by pooled resources from the government, donor, community, or market. Boosting investment levels to recuperate from a disaster, therefore, suggests a short-term application of resilience or what stakeholders perceive as important, depending on users' choices, leadership or management visions, community resources, and matters beyond their control.

Relative to criticality, NZTA includes critical connectivity in the socio-economic criteria of its road functional classification. This inclusion at the system level demonstrates a good practice of identifying the relevant policies or principles used in prioritising investments and the inevitability of multi-criteria analysis. On the one hand, an equity-based criticality – maximising total benefits and minimising pollution exposure to vulnerable groups in road lanes extension – would still warrant composite indicators from multiple distributive principles (Jafino 2021). On the other hand, the functional criticality, attributed to connectivity and service, would most perceptibly affect social criticality in a case of multi-criteria prioritisation of public highway infrastructure investment projects (Anderlini 2020). Nonetheless, the scarcity of critical infrastructure location and performance data, impeding the development, testing, and validation of recovery metrics and decision frameworks (He and Cha 2021), makes it hard to monitor and evaluate performance levels properly. Such limited awareness yields inadequate prevention of and preparedness against unacceptable risks.

7 DISCUSSION: SAMPLE RESILIENCE PERFORMANCE AREAS AND POTENTIAL IMPLEMENTATION MECHANISMS

The conceptual performance framework invites the interpretation that the functionality and value for money attributes are relatively established in New Zealand and the Philippines. Consequently, more effort may be needed to guarantee that communities sufficiently report the vulnerability and criticality attributes towards forward-looking resilience-informed improvement strategies for better sustainability results. Applying resilience performance measures – threshed out in strong policy documents, asset management plans, road life-cycle options, and contractor performance guidelines – would fill the gaps in the foregoing context to tackle vulnerabilities and criticalities in the asset risk management processes.

Fundamental to the performance-based approach to systemic resilience are resilience metrics that monitor system performance and track programmatic contribution to the resilience outcome (National Academies of Sciences, Engineering, and Medicine 2021a). Monitoring and evaluating resilience over time entail the review of performance throughout and without disruptive events, as well as the planning and adaptation aspects. Existing data from asset management measures and traditional performance metrics may be used, and a resilience scorecard for indirect measurement scales may be adopted, in assessing and evaluating systems' state of functioning to achieve resilience results or outcomes (National Academies of Sciences, Engineering, and Medicine 2021b). Taking into account the conceptual performance framework of this research and the performance-based approach leading to resilient and sustainable road networks, Table 2 groups the observed road resilience performance elements using information and/or data reported by New Zealand and Philippine national governments.

Table 2. Observed road resilience performance elements based on New Zealand and Philippine governments.

Observed New Zealand National Road Resilience Performance Elements	Observed Philippine National Road Resilience Performance Elements
Earlier proposed measures for transport system resilience: Technical resistance (structural, procedural, interdependencies)	Presently documented/monitored road resilience-linked indicators (based on DPWH, NEDA, and NDRRMC): Service performance
- robustness, redundancy, safe-to-fail Organisational resilience (cross-cutting)	- accessibility, connectedness (per functional classification), reliability (based on condition and maintenance ratings), social and environmental safeguards, preparedness planning
- change readiness, networks, leadership and culture Presently monitored road resilience properties (based on NZTA): Functionality	Risk response
- availability state (degree of access) Vulnerability	- robustness (based on damages from the event and whether vulnerable roads are passable), equipped response, inter-organisational coordination
- outage state (duration of loss or reduced access); disruption state (combination of availability and outage states) Criticality	Resilience-linked investment
- State Highway ONRC (categorisation depending on whether busy, connection to destinations, route availability)	- rapidity (based on unhampered mobility), route redundancy, diversity (for improved connectivity), organisational flexibility, knowledge and change management

Both countries report resilience-related measures as part of performance monitoring. However, New Zealand has a more technical understanding of road resilience, allowing for resilience prioritisation scores and certain properties addressing the proposed performance attributes as in Figure 4. The following are some areas wherein resilience performance reporting might be improved.

Hughes and Healy (2014) suggested a qualitative approach for NZTA at general 'all-hazards' and 'specific-hazard' levels based on resilience dimensions (technical and organisational) and principles (robustness, redundancy, safe-to-fail, readiness, continuous management, leadership and culture, and networks). To excel investment programming for resilient transportation in the country, NZTA mainstreamed the proactive resilience business case with investment logic mapping to overview the causes and key problems of resilience investment (Tonkin+Taylor & Tregaskis Brown 2020). Notwithstanding, McWha and Tooth (2020) recognised the lack of technical guidelines on quantifying resilience when appraising infrastructure investment. In view of such background, communities should instil consistent objectives and specific relevance of their investment targets for unified tracking and reporting of holistic systemic resilience. What is more, hazard exposure and climate data should be constantly updated and integrated into asset management systems. These underlying mechanisms must apply similarly to the Philippines.

The DPWH PGS 2017-2022 protection measures include unhampered mobility in identified vulnerable areas during and after disasters, as well as the number of resilient bridges and linear metres of slope protection along primary roads (DPWH 2017). The said measures are a good start for resilience performance measurement. Despite plans to observe disaster resilience standards during rehabilitation and recovery (NDRRMC 2020), and increase resilience of critical infrastructure (Department of Environment and Natural Resources -

Climate Change Service 2018), there is space to codify infrastructure resilience standards and assign critical infrastructure levels on a national scale. On this note, infrastructure governance models may make resilience one of the decision-making criteria (OECD 2019). The DBM (2022) advised supporting the poorest, climate change and disaster risk vulnerable areas, social sector, and basic public services in increased infrastructure spending. Further to this policy, the data requirements of the Three-Year Rolling Infrastructure Programme, an investment programming tool basis for inclusion in the national budget (NEDA 2021b), could include resilience components in infrastructure proposals to strengthen the case for resilience. Furthermore, budgeting for updated design/operational manuals, appropriate financial instruments, and regular user surveys on infrastructure performance is recommended to enhance reporting of resilience outcomes (Lacambra *et al.* 2020).

8 CONCLUSION: UPGRADING INVESTMENT ANALYSIS BY MAINSTREAMING RESILIENCE PERFORMANCE IN ASSET MANAGEMENT

This research linked road performance, asset management, and resilience in the New Zealand and Philippine national governance context. It reveals, among amiable strengths, the long-term planning with asset management strategies institutionalised in New Zealand, and the performance governance system measures in the Philippines. To some extent, applicable in both countries, multi-objective optimisation should be reinforced for greater resilience and sustainability outcomes. Mainstreaming resilience performance in asset management – by balancing investments within the areas of stakeholder value, functionality, vulnerability, and criticality – would move organisations along the path to resilience and sustainability. In that vein, consistent identification of resilience objectives, stakeholders, investment case, model cost-benefit analysis, and priorities is essential to making resilience a meaningful, achievable, and measurable reality. The same will be reinforced by comprehensive asset- and risk-based solutions that are mindful of the evolving environment, and the resilience-guided implementation of investment analysis. These research findings are like signposts showing pathways to upgrade investment analysis. Upgraded investment analysis, involving the adoption of more balanced performance measures, will prolong the service life, resilience, and sustainability of road networks. At any rate, the successful journey does not end there. The climate risk horizon demands more than just asset-based and organisational adaptation means. We know that community climate-conscious risk reduction is supreme in the prudent now and transformational future.

ACKNOWLEDGEMENT

This research was prepared by The University of Auckland (postgraduate students Jessica and Annisa, together with their research supervisor Theunis) and primarily used online materials from the New Zealand and Philippine governments.

REFERENCES

Anderlini C.G. (2020). *Practical Definition of Criticality Regarding Road Infrastructure.* Global Initiative on Disaster Risk Management. Retrieved from https://www.gidrm.net/user/pages/get-started/resources/files/GIDRM_Criticality_RoadInfrastrucutre_MX.pdf

Asian Development Bank. (2018). *Guide to Performance-Based Road Maintenance Contracts.* Retrieved from https://www.adb.org/documents/guide-performance-based-road-maintenance-contracts

Author/artist unknown. (2020). *Example System Resiliency Curves – Coronavirus* [Online image]. Retrieved from https://imgur.com/gallery/3F82Ot1

Baroud H., Barker K., Ramirez-Marquez J.E. and Rocco C.M. (2015). Inherent Costs and Interdependent Impacts of Infrastructure Network Resilience. *Risk Analysis, 35*(4), 642–662.

DBM. (2022). *National Budget Call for FY 2023.* Retrieved from https://www.dbm.gov.ph/wp-content/uploads/Issuances/2022/National-Budget-Memorandum/NATIONAL-BUDGET-MEMORANDUM-NO-142-DATED-JANUARY-12-2022.pdf

Dehghani M.S., Flintsch G.W. and McNeil S. (2013). Roadway Network as a Degrading System: Vulnerability and System Level Performance. *Transportation Letters, 5*(3), 105–114.

Department of Environment and Natural Resources - Climate Change Service. (2018). *The Cabinet Cluster on Climate Change Adaptation and Mitigation and Disaster Risk Reduction Roadmap (2018-2022).* Retrieved from https://climatechange.denr.gov.ph/index.php/programs-and-activities/cabinet-cluster-on-ccam-drr/ccam-drr-performance-and-projects-roadmap

Department of Finance, Department of Budget and Management, & NEDA. (2020). *Implementation of a Philippine Government Asset Management Policy.* Retrieved from https://www.dbm.gov.ph/index.php/265-latest-issuances/joint-memorandum-circular/joint-memorandum-circular-2020/1727-joint-memorandum-circular-no-2020-001-dof-dbm-neda

DPWH. (2015). *DPWH Guidelines: Value Engineering.* Retrieved from https://1library.net/document/z383onmq-dpwh-guidelines-on-value-engineering-pdf.html

DPWH. (2017). *Adoption of DPWH PGS Strategy Map and Enterprise Scorecard 2017–2022.* Retrieved from https://www.dpwh.gov.ph/DPWH/sites/default/files/issuances/DO_082_s2017.pdf

DPWH. (2018). *DPWH PGS Frequently Asked Questions.* Retrieved from: https://www.dpwh.gov.ph/dpwh/pgs/pdf/DPWH_PGS_FAQs.pdf

DPWH. (2021a). *Philippine National Road Network.* Retrieved from: https://www.dpwh.gov.ph/dpwh/DPWH_ATLAS/06%20Road%20WriteUp%202021.pdf

DPWH. (2021b). *Creation of Philippine Government Asset Management Policy (PGAMP) Focal Persons.* Retrieved from: https://www.dpwh.gov.ph/DPWH/sites/default/files/issuances/SO_247_s2021.pdf

Gardiner L., Firestone D., Waibl G., Mistal N., Van Reenan K., Hynes D., Byfield J., Oldfield S., Allan S., Kouvelis B., Smart J., Trait A., and Clark A. (2009). *Climate Change Effects on the Land Transport Network volume one: Literature Review and Gap Analysis.* NZ Transport Agency Research Report 378.

Gay L.F. and Sinha S.K. (2013). Resilience of Civil Infrastructure Systems: Literature Review for Improved Asset management. *International Journal of Critical Infrastructures, 9*(4), 330–350. doi: 10.1504/IJCIS.2013.058172

He X. and Cha E.J. (2021). State of the Research on Disaster Risk Management of Interdependent Infrastructure Systems for Community Resilience Planning. *Sustainable and Resilient Infrastructure.* doi: 10.1080/23789689.2020.1871541

Hughes J.F. and Healy K. (2014). *Measuring the Resilience of Transport Infrastructure.* NZ Transport Agency Research Report 546.

IPCC. (2022). *Climate Change 2022: Impacts, Adaptation, and Vulnerability: Contribution of Working Group II to the Sixth Assessment Report of the IPCC.* [H.-O. Pörtner D.C. Roberts M. Tignor E.S. Poloczanska K. Mintenbeck A. Alegría M. Craig S. Langsdorf S. Löschke V. Möller A. Okem and B. Rama (eds.)]. Cambridge University Press. In Press.

Jafino B.A. (2021). An Equity-Based Transport Network Criticality Analysis. *Transportation Research Part A: Policy and Practice* (144), 204–221. doi: 10.1016/j.tra.2020.12.013

Jenelius E. (2010). *Large-Scale Road Network Vulnerability Analysis.* Department of Transport Science, KTH, Stockholm. ISBN 978–91–85539–63–5. Retrieved from https://www.diva-portal.org/smash/get/diva2:354583/FULLTEXT01.pdf

Lacambra C., Molloy D., Lacambra J., Leroux I., Klossner L., Talari M., Cabrera M.M., Persson S., Downing T., Downing E., Smith B., Abkowitz M., Burnhill L.A. and Johnson-Bell L. (2020). *Factsheet Resilience Solutions for the Road Sector in the Philippines.* Inter-American Development Bank. Retrieved from https://publications.iadb.org/publications/english/document/Factsheet-Resilience-Solutions-for-the-Road-Sector-in-the-Philippines.pdf

Levenberg E., Miller-Hooks E., Asadabadi A. and Faturechi R. (2016). Resilience of Networked Infrastructure with Evolving Component Conditions: Pavement Network Application. *Journal of Computing in Civil Engineering, 31*(3), 04016060.

Lu D., Tighe S. L. and Xie W. (2017). *Pavement Fragility Modelling Framework and Build-in Resilience Strategies for Flood Hazard.* Paper Presented at the Transportation Research Board 96th Annual Meeting, Washington DC, United States.

McWha V. and Tooth R. (2020). *Better Measurement of the Direct and Indirect Costs and Benefits of Resilience.* NZ Transport Agency Research Report 670.

Ministry for the Environment. (2016). *Climate Change Projections for New Zealand: Atmosphere Projections Based on Simulations from the IPCC fifth Assessment.* Wellington: Ministry for the Environment.

Ministry of Civil Defence. (2019). *National Disaster Resilience Strategy.* Retrieved from: https://www.civil-defence.govt.nz/assets/Uploads/publications/National-Disaster-Resilience-Strategy/National-Disaster-Resilience-Strategy-10-April-2019.pdf

Mohammed A., Abu-Samra S. and Zayed T. (2017). *Resilience Assessment Framework for Municipal Infrastructure.* Paper presented at the 2017 MAIREINFRA-The International Conference on Maintenance and Rehabilitation of Constructed Infrastructure Facilities. Seoul, South Korea.

Mohammed A., Abu-Samra S., Zayed T., Bagchi A. and Nasiri F. (2019). *Resilience-Based Asset Management Framework and its Application on Pavement Networks.* Paper presented at the Canadian Society for Civil Engineering Annual Conference, CSCE 2019, Laval, Canada.

Mostafavi A. (2017). A System-of-Systems Approach for Integrated Resilience Assessment in Highway Transportation Infrastructure Investment. *Infrastructures, 2*(4). doi: 10.3390/infrastructures2040022

National Academies of Sciences, Engineering, and Medicine. (2021a). *Mainstreaming System resilience Concepts into Transportation Agencies: A guide.* Washington, DC: The National Academies Press. doi: 10.17226/26125

National Academies of Sciences, Engineering, and Medicine. (2021b). *Resilience Primer for Transportation Executives.* Washington, DC: The National Academies Press. doi: 10.17226/26195

National Disaster Risk Reduction and Management Council. (2020). *National Disaster Risk Reduction and Management Plan 2020-2030.* Retrieved from https://ndrrmc.gov.ph/attachments/article/4147/NDRRMP-Pre-Publication-Copy-v2.pdf

NEDA. (2009). *Value Analysis Handbook.* Retrieved from https://neda.gov.ph/wp-content/uploads/2014/01/Value-Analysis-Handbook.pdf

NEDA. (2021a). *Updated Philippine Development Plan 2017-2022.* Retrieved from https://pdp.neda.gov.ph/updated-pdp-2017-2022/

NEDA. (2021b). *Updating of the 2017-2022 Public Investment Program.* Retrieved from https://neda.gov.ph/call-for-the-updating-of-the-2017-2022-pip/

New Zealand Ministry of Transport. (2018). *A Framework for Shaping our Transport System: Transport Outcomes and Mode Neutrality.* Retrieved from: https://www.transport.govt.nz//assets/Uploads/Paper/Transport-outcomes-framework.pdf

NZTA. (2011). *State Highway Asset Management Plan 2012–2015.* Wellington, N.Z: Waka KotahiNZ Transport Agency.

NZTA. (2015a). *State Highway Infrastructure Assets Management Plan.* Wellington, N.Z: Waka Kotahi NZ Transport Agency.

NZTA. (2015b). *Pavement Lifecycle Management Plan.* Wellington, N.Z: Waka Kotahi NZ Transport Agency.

NZTA. (2019). *Investment Performance Measures: Benefits and Measures.* Retrieved from: https://www.nzta.govt.nz/assets/P-and-I-Knowledge-Base/docs/Investment-performance-measures-for-download-update-2019-08.pdf

NZTA. (2022). *Business Case Approach Guidance.* Retrieved from: https://www.nzta.govt.nz/planning-and-investment/learning-and-resources/business-case-approach-guidance/

OECD. (2019). *Good Governance for Critical Infrastructure Resilience.* Retrieved from https://www.oecd.org/gov/risk/good-governance-for-critical-infrastructure-resilience-02f0e5a0-en.htm

The Treasury. (2022). *Better Business Cases^{TM} (BBC).* Retrieved from: https://www.treasury.govt.nz/information-and-services/state-sector-leadership/investment-management/better-business-cases-bbc

Tonkin+Taylor, and Tregaskis Brown. (2020). *National Resilience Programme Business Case.* Wellington. NZ Transport Agency Publication.

United Nations. (2021). *Managing Infrastructure Assets for Sustainable Development: A handbook for local and National Governments.* [N. Hanif C. Lombardo D. Platz C. Chan J. Machano D. Pozhidaev and S. Balakrishnan (eds.)]. Retrieved from https://www.un.org/development/desa/financing/document/un-handbook-infrastructure-asset-management

Zhang W. and Wang N. (2016). Resilience-Based Risk Mitigation for Road Networks. *Structural Safety.* 62, 57–65. doi: 10.1016/j.strusafe.2016.06.003

Predicting the coefficient of friction as a support tool for decision- making in pavement management system

José Breno Ferreira Quariguasi* & Francisco Heber Lacerda de Oliveira
Department of Transportation Engineering, Federal University of Ceará, Fortaleza, Brazil

Minh-Tan Do & Manuela Gennesseaux
Laboratory AME-EASE, Gustave Eiffel University, IFSTTAR Nantes, France

ABSTRACT: The pavement surface properties, represented by the coefficient of friction and the texture, are significant to ensure the safety of aircrafts during take-off and landing operations. These properties must be periodically measured to ensure that they are by the values established by civil aviation regulatory agencies. The results of the measurements of these properties may be used to support the decision-making process about maintenance services, such as rubber removal activation. However, there is not always a reasonable indication of when these maintenance measures should be applied, and, sometimes, there is a subjective approach in the decision-making process of these services. Therefore, this paper aims the development of two prediction models for the coefficient of friction measured at 6 meters from the runway axis (left side and right side). The models were developed with data from 29 coefficients of friction measurement, representing 679 observations and reports carried out between August 2014 and May 2020 on Fortaleza International Airport (SBFZ), the Northeast region of Brazil. Artificial Neural Networks - ANN techniques, written in Python programming language, were used, with the data distributed randomly and without repetition in 80% for training and 20% for testing. The prediction model for the runway left side reached, in the test phase, a Coefficient of Determination (R^2) of 69.11%, Mean Absolute Error (MAE) of 0.0379, and Mean Square Error (MSE) of 0.002. This model has only one hidden layer with 96 neurons and rectified linear as activation function (ReLU). Regarding the prediction model for the runway right side, the test phase obtained R^2 of 68.46%, MAE of 0.0373, and MSE of 0.0023. This model consists of only one hidden layer with 80 neurons and ReLU as activation function. The models may be helpful as a support tool for the decision-making process about maintenance services and allowing the monitoring of the friction condition on the runway.

Keywords: Safety, Runway, Maintenance

1 BACKGROUND

The adherence between the airplane tire and the pavement surface is important to assure operational safety during landing and take-offs. According to the Aeronautical Accidents Investigation and Prevention Centre (CENIPA 2021), most of the accidents between 2010 and 2019 in Brazil occurred during the phases of landing and take-off.

In this sense, it is fundamental to perform measurements on the pavement surface. Among the pavement surface characteristics to be analysed, it can be highlighted the macrotexture and the coefficient of friction.

---

*Corresponding Author: brenoquariguasi@det.ufc.br

The results of these measurements may also be used to support the decision-making process about maintenance procedures. However, some airport managers, historically, have made decisions about Maintenance and Rehabilitation based on immediate needs or experiences and this approach does not allow rational use of the available resources (FAA 2014).

Predicting the pavement performance and its evaluation is one of the critical phases of a Pavement Management Systems (PMS) (Hossain *et al.* 2019). Although, in the last decades, the development of pavement performance prediction models has been thrust. These models are a key component of a PMS. They are employed in several fields, such as estimation of maintenance and rehabilitation times, life cycle cost analysis, and user cost calculation, among others. For that reason, these models should predict values as close to the actual values as possible. In general, PMS uses data collected from laboratory or field observations to develop these models (Shekharan 1998).

Then, a tool that could predict the pavement condition may be useful to assist the decision-making process about the effects of maintenance procedures like rubber removal activation, for example. Considering it, this paper aims the development of two prediction models for the coefficient of friction measured on a runway as a support tool for a Pavement Management System.

2   LITERATURE REVIEW

Management of runway friction level is a major concern for airport authorities due to its importance to provide a safe surface for landing and take-off operations. Runway friction is responsible for important aspects, such as deceleration of the aircraft after landing, guarantee the directional control during the ground roll on take-off or landing, and wheel spin up at touchdown to achieve full rotational speed (Fwa *et al.* 1997).

Pavement surface characteristics, especially the friction, are fundamental to guarantee the safety on runway. Therefore, pavements are constructed to provide enough texture to assure friction, especially when the surface is wet (Chen *et al.* 2008).

An adequate friction characteristic can be achieved by providing enough microtexture and macrotexture. Macrotexture consists of wavelengths of 0.5 mm to 50 mm and is an important aspect that provides frictional properties at high speeds. Microtexture, for another hand, refers to wavelengths of 1 μm to 0.5 mm and assure the frictional properties for aircraft operating at low speeds (Henry 2000).

Several aspects may affect the friction and runway friction will change over time depending on frequency and type of aircraft operating on the runway and environmental factors. For example, the amount of rubber deposits on the surface will decrease the available friction, then it is important to carry out a rubber removal periodically. These rubber deposits occur mainly at the touchdown zone and can be very extensive (Chen *et al.* 2008).

Rubber removal procedures also affect the runway friction, as it ensures macrotexture depth and friction remain above the values required for guarantee the safety of landing and take-off operations. Although its performance has not caused an increase of more than 50% in the measurements studied (Sales *et al.* 2021).

Another important aspect that affects the pavement friction it is the temperature. The coefficient of friction is lowest during summer season and highest in spring, regardless the type of pavement mix. The rate of decrement in friction with ambient temperature is almost independent of the type of pavement surface (Anupam *et al.* 2013).

2.1   *Artificial Neural Networks*

Artificial Neural Networks (ANN) has become popular and useful model for tasks such as classification, clustering, pattern recognition and prediction in several areas, including transportation engineering. This technique has become relatively competitive to conventional regression and statistical models concerning usefulness. It happens due to its excellent properties of self-learning, adaptivity, fault tolerance, nonlinearity, and advancement in input to an output mapping (Abiodun *et al.* 2018).

In general, an ANN is composed of an input layer, hidden layers, and output layers (Shekharan 1998). Each layer is interconnected by neurons and each neuron has an output, which relates to its activation level, and it can spread out to other neurons (Najafi *et al.* 2016).

According to Lecun *et al.* (2015), a Multilayer Neural Network can be mathematically expressed by the equations 1 and 2:

$$z_j = \sum w_{ij} \times x_i \qquad (1)$$

$$y_j = f(z_j) \qquad (2)$$

Where w_{ij} are the weights of the neuron i, x_i are the input signal, z_j is the sum of the outputs, y_j is the output signal of the neuron and $f(.)$ is the activation function.

The signal is propagated from the input layer to the output layer and then the errors are computed. After, the signal is propagated in the opposite direction, from output to input. The weights are updated by the backpropagation algorithm. The backpropagation equation (3) can be applied repeatedly to propagate gradients through all modules (Lecun *et al.* 2012).

$$W_{(T)} = W_{(T-1)} - \eta \frac{\partial E}{\partial W} \qquad (3)$$

Flintsch *et al.* (1996) applied ANN for selecting pavement rehabilitation projects, the artificial neural network analysis reduces the level of effort necessary to find sections for the pavement preservation program, reduces subjectivity, and minimizes the probability of missing sections that should be programmed.

Fwa *et al.* (1997) used ANN to improve the consistency and continuity of the rubber removal decision-making process. Bosurgi and Trifiro (2005) developed models to predict sideway force coefficient and accidents using ANN. Najafi *et al.* (2016) applied ANN to predict the rate of wet and dry vehicle crashes based on surface friction, traffic level, and speed limit.

3 METHODOLOGY

The methodology adopted in this paper can be divided into four steps. The first step regards to literature review about features that may affect the coefficient of friction and prediction models previous developed.

Then, during the second step, coefficient of friction data were collected from measurements reports obtained from the National Civil Aviation Agency of Brazil (ANAC), climatological data were collected from Airspace Control Institute (ICEA), and data traffic were acquired from ANAC. The runway from the International Airport of Fortaleza was adopted by the authors because of the amount of available data. This airport is served by one asphalt concrete runway, which is 2,545 m long and 45 m wide.

The period of analysis ranges from 2014 to 2020, and all coefficient of friction measurements were carried out by a Griptester equipment at 65 km/h. A total of 29 reports were used, each report has coefficient of friction from 3 metres and 6 metres from the centreline, but only the values from 6 metres were adopted.

In the third step, the dataset was organised, cleaned, and then randomly divided into two distinct sets, 80% for training and 20% for testing. In this step, the data was also pre-processed using the z-score, as shown in equation 4.

$$z = \frac{x - \bar{x}}{\sigma} \tag{4}$$

Next, the models, developed using Python programming language, were trained using the Scikit-Learn library. It was tested several hyperparameters, such as one and two hidden layers, different activation functions, learning rates, and different inputs. Because the weights are initialized randomly, every architecture was performed five times and its mean and standard deviation were collected to find the best models. And, finally, the models errors were analysed.

4 RESULTS AND DISCUSSION

The first model, called Model 1 (M1) was developed for the left side of the runway. This model has 6 features as input, which are: distance of measurement, humidity, temperature, surface age, annual take-offs and landing, and rubber removal. Its architecture is composed of one hidden layer with 96 neurons, rectified linear as activation function, the learning rate of 0.1, and was performed 300 epochs. The results of Coefficient of Determination (R^2), Mean Absolut Error (MAE), and Mean Squared Error (MSE) are shown in Table 1 and in Figure 1.

Table 1. Results of the Model 1 (left side).

	Training	Test
R^2 (%)	74.66	69.11
MAE	0.034	0.0379
MSE	0.002	0.002

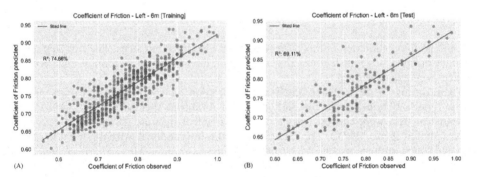

Figure 1. Scatter plots of the results between the coefficient of friction observed and predicted for the training (A) and test phases (B) for Model 1.

The results presented in Figure 1 show a trend between the values, which is expected. However, there is still a significant dispersion. Therefore, it is important to analyse the errors, which are shown in Figure 2.

516

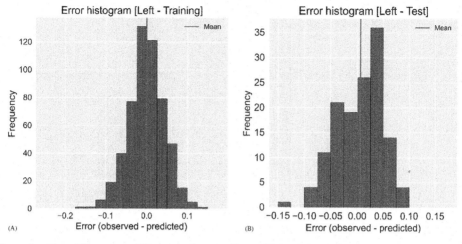

Figure 2.   Error Histogram of the results between the coefficient of friction observed and predicted for the training (A) and test phases (B) for Model 1.

Regarding the results in Figure 2(A), most of the errors range from -0.05 to +0.05. Figure 2(B), for another hand, most of the errors range from -0.025 to +0.05.

Table 2 shows the results of R^2, MAE, and MSE of the second model, called Model 2 (M2), which was developed to predict the coefficient of friction for the right side of the centreline. M2 has 7 features as input, which are: distance of measurement, humidity, temperature, local atmospheric pressure, surface age, annual take-offs and landing, and rubber removal. This model is composed of one hidden layer with 80 neurons, rectified linear as activation function, the learning rate of 0.1, and was performed 300 epochs. Figure 3 presents the scatter plots between the observed and predicted values.

Table 2.   Results of the Model 2 (right side).

	Training	Test
R2 (%)	77.59	68.46
MAE	0.034	0.037
MSE	0.002	0.002

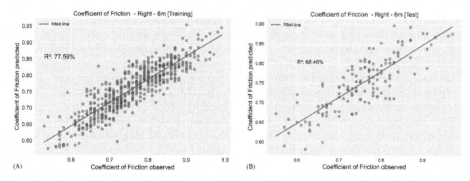

Figure 3.   Scatter plots of the results between the coefficient of friction observed and predicted for the training (A) and test phases (B) for Model 2.

Similar to the M1, Model 2 also presents a trend between the values with significant dispersion. Figure 4 shows error histograms from the M2.

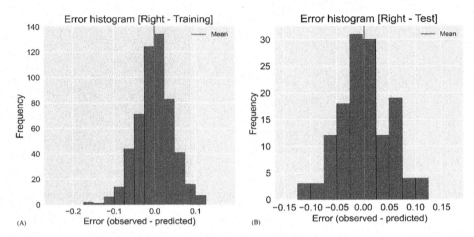

Figure 4.   Error Histogram of the results between the coefficient of friction observed and predicted for the training (A) and test phases (B) for Model 2.

According to Figure 4(A), most of the error range from -0.05 to +0.05. For the test phase, Figure 4(B), most of the error are concentred between -0.05 and +0.025. These results show that these models have the potential to be used as a support tool for the decision-making process in a Pavement Management System.

5   CONCLUSION

Two models were developed, using Python programming language and Scikit-Learn library, to predict the coefficient of friction on runway, called to Model 1 (M1) and Model 2 (M2). M1 was developed to predict the coefficient of friction on the left side 6 metres from the centreline and M2 to predict the coefficient of friction on the right side 6 metres from the centreline. Both models predict values of a Griptester at 65 km/h.

Model M1 has as input the distance of measurement, humidity, temperature, surface age, annual take-offs and landing, and rubber removal. Model M2, for another hand, has as input the distance of measurement, humidity, temperature, local atmospheric pressure, surface age, annual take-offs and landing, and rubber removal. Both models shown similar results, concerning the metrics analysed, Coefficient of Determination, Mean Absolut Error, and Mean Squared Error.

The models showed potential to be used as a support tool for airport managers and regulatory agencies as a way to plan maintenance procedures or monitor the runway friction level. However, the models still need to be improved by data from new reports.

Despite the capability of the models as predictive tools, it is still important to perform field measurements to monitor the friction level. Furthermore, these measurements may be used to improve such models and become them even better.

REFERENCES

Abiodun O.I., Jantan A., Omolara A.E., Dada K.M., Mohamed N.A., and Arshad H., 2018. State-of-the-Art in Artificial Neural Network Applications: A survey. *Heliyon* 4 (11).

Anupam K., Srirangam S.K., Scarpas A. and Kasbergen C., 2013. Influence of Temperature on Tire–Pavement Friction: Analyses. *Transportation Research Record* 2369 (1), 114–124.

Bosurgi G., and Trifirò F., 2005. A Model Based on Artificial Neural Networks and Genetic Algorithms for Pavement Maintenance Management. *International Journal of Pavement Engineering* 6 (3), 201–209.

CENIPA, 2021. Aeródromos - Sumário Estatístico 2010-2019. *Aeronautical Accidents Investigation and Prevention Centre* (CENIPA). Brasília.

Chen J.-S., Huang C.-C., Chen C.-H., and Su K.-Y., 2008. Effect of Rubber Deposits on Runway Pavement Friction Characteristics. *Transportation Research Record* 2068 (1), 119–125.

FAA, 2014. Airport Pavement Management Program (PMP). AC 150/5380-7B. *Federal Aviation Administration* (FAA).

Flintsch G.W., Zaniewski J.P., and Delton J. 1996. Artificial Neural Network for Selecting Pavement Rehabilitation Projects. *Transportation Research Record* 1524 (1), 185–193.

Fwa T.F., Chan W.T. and Lim C.T. 1997. Decision Framework for Pavement Friction Management of Airport Runways. *Journal of Transportation Engineering* 123 (6), 429–435.

Henry J.J. 2000. NCHRP Synthesis of Highway Practice 291: Evaluation of Pavement Friction Characteristics. *Transportation Research Record, National Research Council.*

Hossain M.I., Gopisetti L.S.P., and Miah M.S., 2019. International Roughness Index Prediction of Flexible Pavements Using Neural Networks. *Journal of Transportation Engineering, Part B: Pavements* 145 (1).

LeCun Y.A., Bottou L., Orr G.B., and Müller K.R., 2012. *Efficient BackProp. In: Montavon, Neural Networks: Tricks of the Trade.* Lecture Notes in Computer Science, vol 7700. Springer.

LeCun Y., Bengio Y. and Hinton G., 2015. Deep learning. *Nature* 521, 436–444.

Najafi S., Flintsch G.W. and Khaleghian S., 2016. Pavement Friction Management – Artificial Neural Network Approach. *International Journal of Pavement Engineering* 20 (2), 125–135.

Sales R.S., Oliveira F.H.L., and Prado L.A., 2021. Performance of Tire-asphalt Pavement Adherence According to Rubber Removal on Runways. *International Journal of Pavement Engineering.*

Shekharan A.R., 1998. Effect of Noisy Data on Pavement Performance Prediction by Artificial Neural Networks. *Transportation Research Record* 1643 (1), 7–13.

BIM method for municipal pavement

Alexander Buttgereit & Maria Koordt
Stadt Münster - Amt für Mobilität und Tiefbau

Markus Stöckner*
UAS Karlsruhe, Karlsruhe, Germany

Ute Stöckner
Steinbeis Transferzentrum Infrastrukturmanagement im Verkehrswesen

ABSTRACT: Municipal roads represent a considerable asset value for the road authorities, which can only be efficiently maintained in the future with data-supported lifecycle
management. To this end, a sustainable asset management system has been developed in
Münster for several years, one part is the Civil Engineering Infrastructure Management
Münster (TIMM). For this purpose, it was necessary to describe the entire process from an
organizational and engineering point of view and in addition to identify the related data
requirements as well as the respective data transfer points in the life cycle. Process descriptions and data transfer points were simultaneously aligned with the application of the BIM
method. It is shown how the requirements of value preservation from a business perspective
can be combined with the engineering requirements from a planning and maintenance perspective to form a cross-facility, joint control system. A major challenge is the digital mapping of the traffic infrastructure such as traffic areas, structures, sewers, traffic signals, etc. in
a uniform ontology so that authoring applications can use the data. For this purpose,
existing databases were examined and evaluated, and the need for additions was identified.
As a result, the scope and accuracy of the data are basically sufficient to process the main
tasks of the TIMM. Furthermore, it even provides decision-relevant results for the management tasks of an asset management from a strategic, tactical, and operational point of
view. However, the current situation also shows that the required data or specialized models
originate from different databases and systems and are neither comparable in terms of spatial referencing nor in terms of ontology and semantics. This task should be supported by the
application of the BIM method in the future. Since the requirements for data content and
data flows in the life cycle have already been worked out from the preliminary work, it is
now being outlined for a first pilot application. It is shown how the required specialist
models from different databases can be merged into an IFC-compatible model, used on a
project-specific basis and how they then must be fed back into the existing databases in the
as-built model. There are still a number of challenges to be solved in the process of the
implementation that is now in progress.

Keywords: Building Information Modeling, Asset Management, Pavement

---

*Corresponding Author: markus.stoeckner@h-ka.de

DOI: 10.1201/9781003429258-51

1 INTRODUCTION

The city of Münster is developing a civil engineering infrastructure management Münster (TIMM) for the existing infrastructure over the entire life cycle as part of an asset management system. The primary goal is to manage the necessary maintenance and rehabilitation (M&R) measures in the life cycle of the relevant asset parts efficiently and effectively. Technical aspects, such as asset condition, environmental aspects as well as risk management or asset value considerations are combined. Current work focuses on the treatment of pavement areas, and the BIM method is to be used as a supporting method. The implementation of the TIMM project for pavement areas requires a functional combination of engineering planning, economic evaluation and political objectives (Stöckner et al. 2020). This addresses different decision-making levels.

The political decision-making level decides on framework-setting plans that provide guidelines for the actions of road authorities. The technical levels decide on prioritization and action plans based on engineering and economic considerations. The operational level then implements these M&R-measures in the road network, considering execution issues and traffic considerations. Common to all levels is that such decisions and considerations can only be made with a reliable and transparent data base in combination with an information system, evaluation tools and a differentiated system of key performance indicators. To solve this task, the Office for Mobility and Civil Engineering of the City of Münster is increasingly applying asset management procedures. The background is provided by the DIN EN ISO 55000 ff. series of standards, which outlines the requirements for asset management of physical facilities on the basis of a general description (DIN EN ISO 55000 2017). The standards series requires an appropriate data basis for this purpose. The problem here is often different existing data sources, DIN EN ISO 55000:2017-05. p. 22 refers to this as follows: "Some asset data comes from planning and control systems, which are often not connected to other information systems. Integration of these information systems through the AM management system can provide new asset information that leads to improved organizational decision making."

Even if this challenge is obvious, it should be noted that many databases are currently available, but due to their different taxonomies, ontologies, and semantics, they require a great effort to provide the necessary data sets for specific tasks. In general, the following tendencies can be identified:

- The management of life cycle processes of transport infrastructure is increasingly improved and thus also more complex.
- The partners involved in the life cycle will use digital forms of collaboration to an increasing extent in the future.
- Changes in IT technology will make the processes of planning, execution, and maintenance/operation significantly more efficient and effective.
- In the future, digital twins will support vendor-free data exchange throughout the entire life cycle.

2 RELATED WORK AND BACKGROUND

To the moment, extensive efforts have been made in Münster to implement the TIMM project. The work relates to organizational, engineering and data processes. In preliminary work, the necessary process workflows including process stakeholders, needed data and data transfer points required in each process step were described in a structured way for the field of transportation infrastructure. As a result, the data requirements and the necessary data flow are known in principle. The existing data situation is comparatively positive for municipal conditions. For example, there is a network coding, an extensive inventory database as well as condition survey & assessment data for various years.

The BIM method (Building Information Modelling) is basically suitable for solving this task of collaboration within the TIMM structure. As a method, BIM takes into account project management processes on the one hand and the data-based modelling of the needed information on the other hand. The technical data modelling is based on a digital three-dimensional infrastructure model that has predefined objects and represents a geometric reference. Relevant information is assigned to this geometric model; for traffic areas, for example, built-in materials and material qualities. The information is acquired and transferred throughout the entire life cycle. Ideally, a standardized data basis is created across all phases of the life cycle. This leads to the relationships shown in Figure 1.

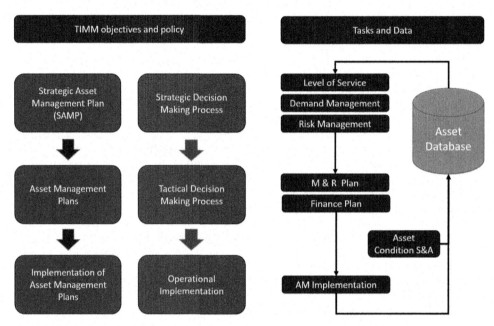

Figure 1.   Relationship of TIMM levels with tasks and database (own chart).

A comprehensive introduction of BIM for transportation areas is still pending. BIM is based on the IFC data standard, which is intended to ensure barrier-free, i.e., vendor-neutral data exchange. This is to be expected for municipal traffic areas with the current IFC 4.3 standard when it is approved as mandatory. In addition, it must be defined which data is required within the context of a life cycle and which data should be transferred to the successive stages of the life cycle. It is also relevant which tasks are to be assigned to the consultant or road authority in detail and how this can be reasonably integrated into the life cycle of a transportation facility. Basically, there is no alternative to the implementation of the BIM method because it could deliver a bis contribution to solve the problem of data availability of data within the lifespan of i.e., pavements (Stöckner/Niever 2018). This requires the implementation of adequate processes and data flow models. This should be clearly defined for all project stakeholders due to the new methods of digital collaboration. For example, there are some research projects at the national road administration level that deal with the implementation of the BIM method for maintenance planning or asset management of road networks and bridges. (Hajdin et al. 2020; König et al. 2019; Stöckner et al. 2019). The basic generic processes and the relevant data transfer points including the necessary information content were identified int these projects. For Germany, existing data standards, such as the Road Database directive (ASB) are regarded. The proper semantic is ensured by the object catalogue for road and transportation systems (OKSTRA®). Depending on the relevant evaluation systems, it is defined which data are relevant for a lifespan management. For the municipal sector, however, only individual pilot projects are currently visible They are carried out on a project-level basis and are not linked to the lifespan approach. The results from the extra-urban sector is valuable for the municipal sector, but is not sufficient for the complexity of municipal traffic areas: Examples are different user requirements as well as different infrastructure assets parts in the context of pavement design and construction. With the TIMM project, extensive preliminary work is available, so that the application and implementation of the BIM method can be developed with pilot projects and defined data handover from the as built model to a maintenance or operation model. For example, the complete process model including the associated data requirements for the lifespan is already available as an initial basis. It is beneficial that these process basics are already completely aligned with the tasks of the asset management of the city of Münster. In the context of the BIM pilot project described here, two primary objectives are being targeted:

1. The BIM method is to be established for road maintenance, i.e., for M&R in the existing road network. The pilot project addresses a single maintenance measure using a uniform and software-independent data basis. The starting point is the data handover of an actual as-built model, which is derived from the existing asset model. Furthermore, the communication processes and data drops must be clearly defined.
2. Handover of data for the next lifespan step: When the M&R measure is finished, the updated as-built model is generated, which represents the renewed construction. It must be ensured that all data necessary for the function of all processes and for the evaluations defined in the TIMM are included.

3 PREPARATION OF THE BIM PILOT PROJECT

3.1 *Planned approach*

An essential point is a correct understanding of the information model as defined by DIN EN ISO 19650-1 (cf. Figure 2). The information requirements of the road authority at the strategic level result from the requirements of the Asset Management Münster System (TIMM). The information requirements for the TIMM project for traffic areas are defined and lead to a desired asset information model. The exchange requirements for a particular

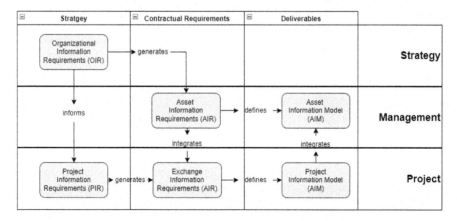

Figure 2.   Information model based on DIN EN ISO 19650-1 and König (2019).

project are also known, but naturally also depend on project-specific requirements. From this, a project information model will be derived out of the asset information model. After conducting the measure, the updated data will then later be transferred back the asset information model. The approach of the currently running pilot project is, to show in an initially simple project how the asset information model can be transferred from the existing legacy database into a BIM Model, then conducting a measure and bring back the updated data into the legacy database.

To solve this task, the existing process descriptions were enhanced into a reference process for BIM projects. Based on this, asset owner information requirements (AIA) and a BIM execution plan (BAP) were developed as a general template for corresponding BIM projects and the use in Münster. Based on the BIM use cases relevant for M&R measures, it was examined how the necessary digital models from the legacy databases can be transferred in a BIM model and the common data environment (CDE).

This step is currently being executed and, together with the modelling, represents the actual challenge. The next steps will deal with planning, tendering and contracting, project monitoring, acceptance of the measure including the new digital model to be handed over and the transfer of the data into the legacy databases.

3.2 *Process analysis*

The generic lifespan process is defined by asset management requirements. The main steps are level of service definitions, planning, tendering and contracting, implementation of measures, maintenance and operation phase, and demolition or renewal. It is assumed that the specialized models for planning are available as LOD 200 and that these are further developed to the required level of information needed, usually LOD 500, the "as-built model". From this, a so-called maintenance or operating model is derived, which should cover all aspects and information from the operation of traffic areas. Essentially, this includes information for M&R planning and also the inventory information that is necessary for the asset management tasks such as cleaning, winter service, green maintenance or structural maintenance.

In addition, an inventory model should contain all the information that is important for maintenance planning. Briefly summarized, these are all those properties that can, for example, provide information on the current asset condition and the expected remaining service life. The necessary groups of characteristics can be divided into material-related data, planning-relevant data such as traffic and climate, and surface condition data and structural conditions of the pavements. Thus, one starting point of the present project is to define the individual property sets and properties in detail and to map them in the BIM model that they can be used for the life cycle. To ensure this, two detailed process considerations were made. Firstly, the entire maintenance process was analysed and it was shown at which point it is necessary to transfer acquired data back into the legacy databases of the city of Münster. On the other hand, the process for the execution of planning, tendering and contracting as well as construction execution was described in detail, which basically represents a specified sub-process of the entire maintenance management. With this approach, it is shown how the BIM process of an individual measure can be integrated into the overall asset management process (cf. Figure 3).

The LOGO® product is used in Münster as the central element for storing road inventory data and road condition data. The LOGO® database is updated at three points: these are the transfer of current condition data after a new condition survey and assessment, the description of a current maintenance plan and the update of the inventory data after a maintenance measure has been carried out. The data required for a project is exported from the asset model and transferred to a project-specific common data environment. Detailed exchange information requirements are defined for this purpose.

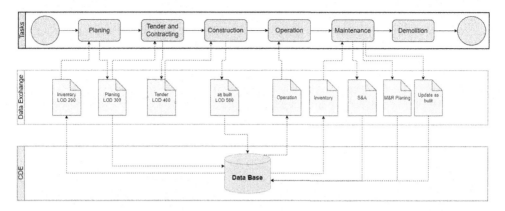

Figure 3.   Generic life cycle process as a starting point (own chart).

The detailed process for the pilot measure was described according to the BPMN method and contains project stakeholders, tasks and representation of the data handovers in the respective format. The essential elements of the information delivery manual were derived from this.

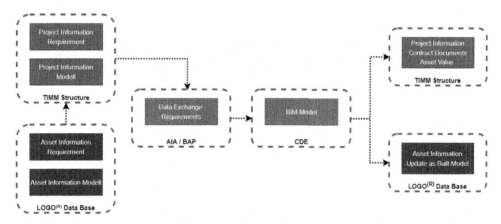

Figure 4.   Information Modell BIM-Project Münster (own chart).

4   USE CASES AND BIM

For the pilot project, the use cases were specified following the BIM4INFRA examples (BIM4INFRA 2020). The scope was kept simple with one maintenance measure: Requirements analysis, planning and execution specifications are fixed, so that the focus is on the use cases of the measure preparation, execution and the needed data transfers from and back to the legacy database. Due to the different usage requirements of municipal traffic areas, this results in a higher complexity than, for example, in the extra-urban area.

The overall idea is to create a digital twin for the entire municipal infrastructure (roads, water and wastewater, electricity and telecommunications, gas and district heating). This initially includes the geometry and structure of the traffic areas. For the single execution of a construction measure, all engineering information should be available there. For planning and execution, a coordination process for these different models is needed. These are naturally an infrastructure model, in which all elements of the supply and disposal in the road space are contained, if necessary, also a soil model, a terrain model and a model of the surrounding elements. It is obvious that this is not yet feasible at present, if only for reasons of data availability. For complex planning in the municipal area, however, it would be desirable to have these models available independently of the software. In the present project, the focus is on the inventory model of pavements, because the primary task is to map the information relevant for the life span and for the needed object types with property sets and properties.

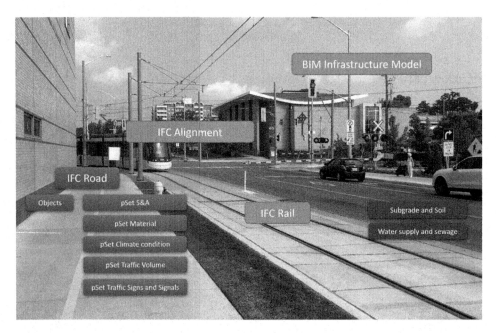

Figure 5.   BIM for municipal Assets (own chart).

For the handling of tasks in the life cycle management of pavements, the property sets that are required for operational and strategic maintenance as well as for road maintenance and operation are therefore of particular importance. These property sets can be described and defined comparatively well due to the known evaluation and planning methods. As already known, a geometric model is created for the modelling, to which the necessary properties are then assigned. To identify the required data, the following tasks were defined:

1. Pavement management model (PMS): The PMS model comprises all property sets that are necessary for a network-wide analysis of the maintenance requirements. The related tasks are the calculation of a medium-term financial requirement forecast and a prioritisation of the upcoming maintenance measures. The necessary property sets are the condition data of the surface condition and the structural condition, an age distribution, traffic load and function, structural data with material properties and further characteristic values still to be defined for a later prioritisation of M&R measures.
2. Operating model: Necessary information for an operating model are area-related requirements for cleaning, winter service and green maintenance, which can then be used to plan equipment and personnel deployment.
3. Maintenance planning model: The maintenance planning model provides the necessary object-related information for the preparation of tender documents. In contrast to the

EMS model, more detailed information on the existing structure, the material properties and, if necessary, the subgrade properties are required here.

These property sets can, in principle, be defined relatively quickly. However, especially for pavements, it must be taken into account that new developments are underway both in the area of condition assessment and in the methodology for determining the remaining service life, some of which are already being used in Münster. Examples include the determination of bearing capacity with the Falling Weight Deflectometer (FWD), the determination and control of existing structures with the Ground Penetrating Radar (GPR) and the use of so-called performance properties for the asphalt types used. In addition, a project on resource-efficient recycling management of municipal infrastructure, which also includes pavements, is currently being launched. The goal here is to find the most sustainable recycling process of the construction materials generated during maintenance measures. In this respect, another aim of the project is to describe the property sets as comprehensively as possible in order to map the requirements of the TIMM project, which go beyond known applications, as far as possible. Therefore, the focus is on using a comparatively simple measure as an example to demonstrate the strategic procedure for the implementation of the BIM method and then a transfer to more complex projects.

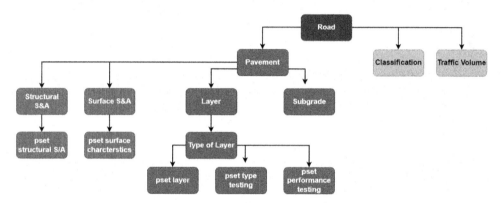

Figure 6.   Example of structuring the data requirements (own chart).

5   PROJECT DOCUMENTATION

As mentioned above, sample documents for the asset owner information requirements (AIA) and for the BIM execution plan (BAP) were developed based on the previous considerations. The current work focuses on the elaboration of the Information Delivery Manual (IDM).

Another work package point is the specification of the data environment (CDE) to be used as well as the possibilities of integrating the required authoring software. At this point, it must be noted that barrier-free data transfer does not yet work for some special software applications in BIM based asset management. This is due on the one hand to the IFC standard, which is still under development, and on the other hand to the individual software products themselves. It is assumed that this point will be resolved anyway with future developments. In the present case of dealing with a usually state-owned infrastructure, such documents can be standardised to a large extent and thus the necessary comparability in project handling for asset management of a complex transport route network can be achieved. In contrast to BIM applications in building construction, there is extensive comparability here.

6   SUMMARY AND CONCLUSION

In Münster, asset management for infrastructure TIMM has been developed in the past. A sub-section of this deals with transportation infrastructure as a component of civil engineering infrastructure management. Asset management requires a sufficient data basis, which must be available without barriers for various applications. In the case of pavements, this can best be achieved with the application of the BIM method. In the case of Münster, comprehensive preliminary work is available, particularly in the area of process description. These were used as an input for a first BIM pilot project, the aim of which is not the handling of a single possibly complex construction measure, but rather to demonstrate the use of the BIM method for the asset management of the municipal pavements in Münster as an example The project should demonstrate a scalable system for BIM Models with regard to the TIMM information requirements. In the case of a transport infrastructure, this can be formalised as far as possible, since a complex set of rules for planning, construction, operation and maintenance as well as deconstruction provides extensive specifications that are repeated in the individual projects with few adjustments. Based on this, the processes of maintenance management and the implementation of maintenance measures were further differentiated and, based on this, information requirements for asset management were derived. In addition, sample documents for the AIA and the BAP were established in accordance with the requirements of the TIMM project. In addition, it was demonstrated which property sets and property characteristics are necessary in the operational phase of asset management for traffic areas. The applicability of the project results will now be tested in a clearly defined maintenance project. In doing so, problem points in the formulation of the tender, in the setting up of a common data environment as well as in the connection of the specific authoring software for maintenance management still have to be taken into account, whereby it is assumed that not all points can be solved within the framework of this pilot project. The overarching goal is to advance the introduction of the BIM method in Münster and to demonstrate a feasible course of action.

REFERENCES

(ASB): *Federal Ministry of Transport and Digital Infrastructure (BMVI)*. ASB Road Database Directive. Version 2.03., 2014. https://www.bast.de/DE/Publikationen/Regelwerke/Verkehrstechnik/Unterseiten/V-ASB.html

(BIM4INFRA 2020): *Umsetzung des Stufenplans*, Digitales Planen und Bauen. www.bim4infra.de.

(DEGES) 2021: *Building Information Modeling (BIM)*. www.deges.de/building-information-modeling-bim/.

(DIN ISO 55000, 2017): Asset Management - Overview, Principles and Terminology (ISO 55000:2014); DIN ISO 55000: Version 2017-05. Beuth-Verlag, Berlin, 2018.

(DIN EN ISO 19650-1) *Organization and Digitization of Information About Buildings and Civil Engineering Works, Including Building Information Modelling (BIM) - Information Management Using Building Information Modelling*- Part 1: Concepts and Principles (ISO 19650-1:2018); German Version EN ISO 19650-1:2018. Beuth-Verlag, Berlin, 2018.

(Hajdin *et al.* 2020) *BIM-Erweiterung Durch Implementierung Der Nutzung Baustofftechnischer Daten Von Straßen und Brücken im AMS*; BIM4AMS; (EN: BIM Enhancement by Implementing the Use of Construction Material Data of Pavements and Bridges in AMS). D-A-CH Call 2019 (running project)

(König *et al.* 2019) BIM4ROAD - *Building Information Modeling (BIM) im Straßenbau Unter Besonderer Berücksichtigung der Erhaltungsplanung*; (EN: Building Information Modeling (BIM) in Road Construction with Consideration of Maintenance Planning) Research Report, unpublished, Bundesanstalt für Straßenwesen (BASt), Bergisch Gladbach, Germany.

OKSTRA®: Object Catalog for Road and Traffic. *Federal Highway Research Institute* (BASt) https://www.okstra.de/

(Stöckner/Niever, 2018) Stöckner M.; *Niever M: Building Information Modeling: BIM im Life Cycle Management (Teil 2). In: Forschungsgesellschaft für Straßen- und Verkehrswesen (FGSV) (Hrsg.): Deutscher Straßen- und Verkehrskongress. Deutscher Straßen- und Verkehrskongress* 2018 (Erfurt, 12.-14.09.2018), Köln: FGSV-Verlag 2018, S. 329-340. ISBN 978-3-86446-225-2

(Stöckner et al. 2019) *AMSFree - Exchange and exploitation of data from Asset Management Systems using vendor free format*, CEDRcall 2018. (Running Project)

(Stöckner et al. 2020) Stöcker M.; Stöckner U., Niever M.: *Tiefbau – Infrastruktur - Management Münster (TIMM); Konzeption und Aufbau; Im Auftrag der Stadt Münster,* Abschlussbericht 2020, unpublished.

Economic and political strategies

Achieving performance in major public infrastructure projects: The case of road works in Senegal

Mohamed Laye*
Head of Major Road Works Division in AGEROUTE Senegal, Business Science Institute
Associate Researcher

ABSTRACT: Road infrastructure provides direct support for the socio-economic development of nations, hence the importance of their role in the socio-economic development of nations. This importance means that overruns of time, costs and quality defects are undesirable results in the management of road projects. However, in Senegal, as worldwide, compliance with these performance criteria continues to be a challenge in the implementation of road projects as they are complex and surrounded by uncertainties. Thus, in the context of the road projects carried out by AGEROUTE in Senegal, we asked ourselves as a research question: "How to achieve the performance of road projects as part of a broader vision of the stakeholders?". The main objective of our research is to provide tools to achieve the performance of road projects through the roles of stakeholders (client, consultant, company, financial backer, users and residents). In order to achieve this objective, our research has been based on a quantitative approach methodology which has made it possible to diagnose the current state of play of the performance of road projects. On the other hand, our research has been based on a qualitative methodology of an inclusive and participatory approach through 34 semi-directional interviews with stakeholder representatives. The textual data thus collected were the subject of textual data analysis (TDA) using the Sphinx software. The TDA allowed for the emergence of managerial themes relating to the roles of each of the stakeholders. The management implications arising from these themes have been articulated at each stage of the road project lifecycle (identification, preparation, implementation and evaluation), according to the criteria (quality, time, cost, environmental protection and satisfaction of users and residents) and the axes (effectiveness, efficiency, relevance) of performance.

Keywords: Project Management, Performance, Road

1   INTRODUCTION

According to Hansen (1965: 3-14) public infrastructure is "infrastructure that provides direct support for productive activities and/or the movement of goods, or for the development of human capital". To this end, the United Nations (UN) states that 'productivity growth, income growth and improvements in health and education require investment in infrastructure"[1]. This is what the Senegalese authorities have understood by stating in the Emerging Senegal Plan (PSE) that "in order to meet the challenge of growth, Senegal needs to have structural infrastructure at the best standards" (PSE, 2014: 96).

---

*Corresponding Author: layemohamed@hotmail.com
[1]https://www.un.org/sustainabledevelopment/fr/infrastructure/

DOI: 10.1201/9781003429258-52

The increase in infrastructure capital requires the implementation of massive investments through large public infrastructure projects. However, the implementation of large infrastructure projects often experiences performance problems which result in overruns of time, costs and quality defects.

What are the reasons for so many failures in major infrastructure projects? According to the literature, the answer to this question is to be found in the intrinsic nature of the projects. Indeed, public infrastructure projects are characterized by their complexity and the uncertainty surrounding them.

Beyond complexity and uncertainty, inadequate consideration of the context in which public infrastructure projects are implemented and the culture of actors was also identified as a failure factor. This research started with the exploration of the different literary fields related to our research topic: project management, governance of public infrastructure and management of organizations. The aim of this exploration is to find the theories developed in project management with a view to contributing to the improvement of project performance through the respect of time, cost and quality criteria. However, three main gaps were identified:

- these theories are primarily directed towards the project manager;
- the theories have been developed in order to produce metarules[2] in the light of the performance issues faced by all projects around the world;
- in project management, in general, the three main performance criteria dealt with are quality, time and cost (Q-T-C). While distinguishing the success of project management from the success of the project, other success criteria will emerge. Thus, it seems important to develop the gold triangle (Q-T-C) towards a pentagon by seeking to achieve, in addition, the criteria relating to the satisfaction of users, residents and the saving the project environment.

After identifying gaps in literature, our research question is: "How to achieve project performance as part of a broader stakeholder vision?" We are looking at the particular case of the road projects carried out by AGEROUTE in Senegal.

The overall objective pursued through our research is to provide tools to achieve the performance of road infrastructure projects in Senegal through the roles of stakeholders. The specific objectives of our research are to:

- diagnose the state of implementation of the road projects carried out by AGEROUTE in Senegal, with a view to describing the state of play;
- determine and analyze the roles of the client, the service providers (consultant, company and prime contractor), the financial backer and other stakeholders (users, residents) in order to achieve the performance of the road project.

Our research is based on the epistemological positioning of pragmatic constructivism, abduction as a type of reasoning and quantitative and qualitative methodologies.

The discussion of the research results led to managerial recommendations on the roles of each of the stakeholders, at each stage of the project lifecycle, enabling road projects to be carried out more performantly.

2 CONCEPTUAL FRAMEWORK FOR STRATEGIC ANALYSIS OF THE PERFORMANCE OF A ROAD PROJECT

2.1 *What is a road project?*

Project Management Institute (PMI) describes the project as "a temporary effort to create a single product, service or result" (PMI, 2008: 4).

---

[2]A metarule is a rule that makes it possible to construct a rule (Poulingue, 2007: 97).

'The word 'road' derives from the Latin rupta (via), 'brown track, fried track'[3]. Then, the road means a lane of land communication designed to enable wheeled vehicles to be moved.

From the definition of project and road concepts, we will draw the definition of the road project:

It is a unique set of stakeholder activities coordinated by a client for the realization of a road. The project shall mobilize resources (means) that are defined, combined and deployed according to the final objective, with a view to producing a result (road) in accordance with the specified quality, as soon as possible, at the lowest cost and saving the environment.

The road project as defined, with activities that are carried out to produce a work, necessarily involves a process, called the project lifecycle, which can be summarized in four main steps, as shown in Figure 1 below.

Figure 1.   Road project implementation process.

2.2   *Who are the stakeholders of the road project?*

According to Freeman (2010: (1) "the stakeholder approach concerns the groups and individuals that may affect or be affected by the achievement of the organization's objectives, as well as the appropriate managerial behavior in response to these groups and individuals".

By adapting Freeman's approach (2010) to the context of the road project, we were able to identify the stakeholders that can influence or be influenced by the realization of the project.

To this end, on the one hand, the strategic stakeholders that can influence the performance of the project are: the client, the financial backer, the consultant, the company. On the other hand, the stakeholders affected by the implementation of the project are: users, residents and the physical environment.

Figure 2 below illustrates the different stakeholders involved in the road project and the relationship between them.

2.3   *What is the performance of the road project?*

Performance is a polysemic term. For our part, we will adopt the Gilbert performance model (1980), which seems to be more consistent with the performance of the road project. This model 'describes performance in a ternary relationship between the objectives pursued (targets, estimates, projections), the means to achieve them (human, material, financial or information resources) and the results achieved (goods, products, services, etc.)' (Gilbert T. 2007, cited by Moscarola 2018: 39).

---

[3]https://www.cnrtl.fr/etymologie/route (National Centre for Textile and Lexicales Resources)

Figure 2.   The different stakeholders in the road project.

Performance measurement is then carried out on three axes: **relevance**, **efficiency** and **effectiveness**:
Figure 3. below illustrates Gilbert's performance model.

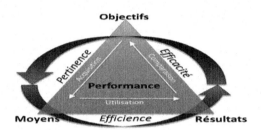

Figure 3.   Gilbert's performance model (1980), Source: adapted from Moscarola (2018: 39).

By adapting the Gilbert performance model (1980) to the context of the road project, we describe the terms of the ternary performance relationship as follows:

- **the objectives**: the contractual deadline, initial cost, specified quality and socio-environmental acceptability of the project;
- **the results**: the duration of the project, the final cost, the quality of the work and the usefulness of the project for the population through its environmental and social impact;
- **the means of action**: the financial, human and material resources mobilized to carry out the project.

Thus, in the context of this research, the performance of the road project is:
effectiveness: the project is carried out with results that correspond to the objectives.
efficiency: optimizing the financial resources used to achieve the results of the project. In other words, the best quality of the work is achieved at the lowest cost.
relevance: the use of adequate means to acquire the overall objective of the project to meet a transport need while saving the physical and social environment.

3   EPISTEMOLOGY AND METHODOLOGY OF RESEARCH

3.1   *Epistemology of research*

In the field of management sciences, the main epistemological traditions are: post-positivism, critical realism, pragmatic constructivism and Interpretativism. For our part, the answer to the epistemological question will move towards the pragmatic constructivism. Indeed, as Martinet (2017) recalled, "the fundamental purpose of pragmatic epistemology is to produce knowledge of scientific, desirable (ethical) and actionable (pragmatic) intent, but deliberately active and transformative to help people live better" (Martinet & Beaulieu *et al.* 2017: 27).

3.2   *Research methodology*

The selection of the methodology is influenced by the epistemology (the paradigm) chosen by the researcher. In Figure 4 below, we present the characteristics of the main contemporary epistemological traditions at each end of a continuum.

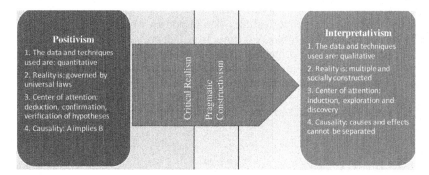

Figure 4.   Characteristics of epistemologies, Source: adapted from Walsh (2015: 36).

It follows from the choice of our epistemological orientation, the pragmatic constructivism in the middle of the continuum of the main epistemological traditions, that our center of attention will be induction, exploration, and the data, methods and techniques will be quantitative and qualitative according to the specific objective pursued as indicated in the introduction. Thus, the methodology of our research is divided into three steps shown in Figure 5 below:

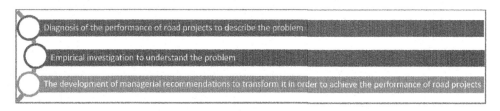

Figure 5.   The three steps of the research methodology.

3.3   *Diagnosis of the performance of road projects*

In order to establish the current state of performance of road projects in Senegal, a collection of input and exit data for major road projects carried out by AGEROUTE from 2007 to 2016 was carried out. These data, from a secondary source, relate to: the initial time, the final time, the initial cost, the final cost, the quality of the road, the length of the road, the origin of the company, the financial backer, the type of works (construction, rehabilitation or maintenance). For example, a list of forty-one (41) major road projects, with a market value of at least five (05) billion XOF, has been drawn up with a view to establishing the population of this part of our research.

The data thus collected were archived and formatted in an Excel file which was exported to the Dataviv' by Sphinx software[4] for data processing with a quantitative approach. Analysis of the results of the data processing has made it possible to establish, in an objective manner, the state of play of the performance of the road projects studied.

3.4 *Empirical investigation to understand the problem*

After diagnosing the "state of health" of the implementation of road projects in Senegal, it was necessary to listen to the actors involved in the implementation of these projects in order to understand why the results are what they are and how to improve them. This was done through interviews. Different types of interviews exist, the most suitable for our research is the personal interpersonal interview selected through face-to-face interviews.

The number of interviews was stopped at thirty-four (34), when we considered that there was saturation.

As shown in Figure 6 above, of the 419 responses obtained from the interviews, approximately one third (125 responses) came from interviewees in the Client category, 90 from the representatives of the companies and 77 from the category of the Consultant, which are responsible for the studies, the monitoring and supervising of the works. Financial backers (77 responses) are the bodies that finance projects. The users and residents' representatives (50 responses) who were interviewed came from the transporters' unions, local residents' associations and the Territorial Command (Prefect).

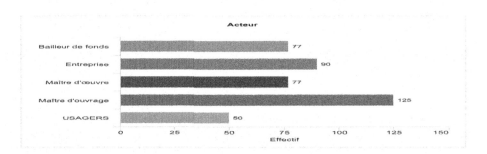

Figure 6.   Number of responses by interviewees.

Thus, the interviews were archived through a textual corpus. Subsequently, this textual corpus was formed for export to the Sphinx IQ2 textual data processing software and then to Dataviv' by Sphinx.

3.5 *Development of managerial recommendations*

There are many approaches to the processing and analysis of textual data from qualitative protocols: minutes, summary, content analysis, lexical analysis and semantic analysis. Computer processing was used with Sphinx IQ and Dataviv' by Sphinx. This is how we have been able to carry out the various approaches listed above in turn. The thematic and statistical processing of the textual body has led to an overall summary of the lexical analysis; the thematic distribution; coding by analysis of content, based on the construction of our

---

[4]Software for the survey and processing of quantitative and qualitative data.

own reading grid called 'manual book code'. In a final step, the semantic analyzer of the software was used to analyze the coded corpus following our ad hoc thesaurus.

The result of this stage is the generation of a distribution of the themes, which is a representation, in proportion to their frequency, of the various managerial topics referred to in the four hundred and nineteen (419) observations from the interviews. These topics are distributed according to the roles of each stakeholder. The synthesis of the distribution of themes and the review of the literature made it possible to determine and analyze the roles of each of the stakeholders, at each stage of the life cycle of the project, with a view to achieving the performance of the road project, thus allowing us to answer our research question.

4 PRESENTATION OF THE RESULTS OF THE RESEARCH

4.1 Results of the diagnosis of project implementation

The Dataviv' by Sphinx software was used to process the quantitative data collected by carrying out various flat analyses which led to the results below.

The left-hand side of Figure 7 below shows that there is a certain delay in the completion of the diagnosed projects.

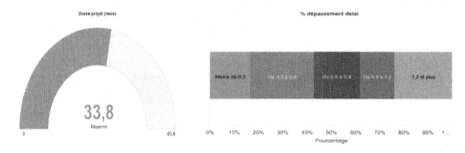

Figure 7.   Actual time of projects completion.

The average duration of projects is around thirty-four (34) months, whereas the initial average time was only twenty (20) months, i.e. an overrun of 0,7 — (34-20)/20 — times the initial deadline. This is illustrated by the right-hand side of Figure 7, which shows that only almost 20 % of projects have a deadline of less than 0,3 times the initial deadline. The majority of projects (around 60 %) exceeded the deadline by more than 0,6 times the initial deadline. With these results, nobody can refute the fact that the problem of exceeding deadlines in road projects is a real phenomenon.

The right-hand side of Figure 8 above suggests that the average final cost of the projects studied is FCFA 16,618 billion, bearing in mind that the average cost was initially 14,165

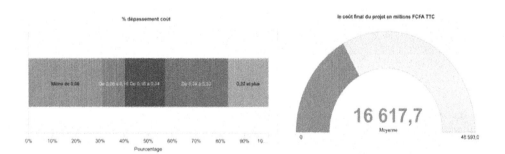

Figure 8.   Actual costs of project implementation.

billion. It thus appears that the phenomenon of cost overruns in road projects is verified. Indeed, the left-hand side of Figure 8 is evidence of this, as it shows that 60 % of projects end with a cost overrun ranging from 16 % to more than 32 %. Based on the analysis of the data, the ratios presented in Figure 9 below have been established.

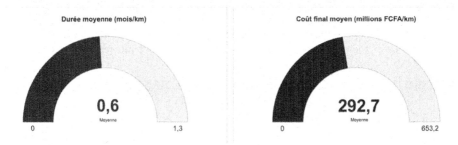

Figure 9. Project duration and average cost ratios.

The graph shows that, on the one hand, the average execution time of the major road projects analyzed over the decade (2007-2016) is 0,6 months/km. On the other hand, the cost per km ratio is around FCFA 293 million/km. These ratios could be useful in the studies of new projects in order to set realistic timeframes and budgets, depending on the linear, while taking into account the specificity of each project.

Figure 10 below shows that, overall, the quality of the works is considered to be good, for 76 % of the projects studied, according to our assessment criterion set out above.

Figure 10. Quality of the works.

On the basis of the results presented above, we can say that from 2007 to 2016, the model project with an average initial period of 20 months is carried out on average over a period of 34 months. The final amount is FCFA TTC 16,618 billion compared with an initial amount of FCFA 14,165 billion. The quality of the work is rather good.

4.2 *Results from the interview analysis*

The qualitative approach used to present and comment on the results of the interviews was to analyze the content of the body of interviews consisting of four hundred and nineteen (419) responses (comments) to the interview guide questions.

An example of a reply: " *The company must be selected and monitored. There must also be a good consultant, good experts (head of mission, etc.). For financial backer, it is the responsiveness of: diligent payments and speed in procurement procedures, in order to be effective. The rest is a follow-up with the company and the inspection mission*".

The analyses carried out were based on lexical[5] and semantic[6] exploration and then on the construction of a targeted or ad hoc thesaurus[7]. They served to understand the roles of stakeholders in the performance of road projects in Senegal.

4.2.1 *Lexical analysis*

The body collected consists of 48 720 words (see the word cloud in Figure 11 below). The processing was carried out using the software Sphinx-iQ 2.

Figure 11.   Overall lexical and semantic synthesis.

Referring to the abductive reasoning chosen in the epistemology of research, the interpretation of this word cloud devotes the first iteration from the world of facts to the world of knowledge.

4.2.2 *Semantic analysis*

Semantic analysis starts with the automatic reading of the responses by a semantic analyzer using a generic French thesaurus[8]. The result is the characterization of the four (04) classes of topics produced by the software as shown in Figure 12 below. The review of these classes shows that the four hundred and nineteen (419) responses focus on quality, project, enterprise and road.

4.2.3 *Construction of a targeted thesaurus*

In order to identify references to other explanatory variables (financial backer and other stakeholders) that do not appear in the topic class characterization (Figure 12), it is necessary to conduct a more precise and focused content analysis on what we seek to identify. In order

---

[5]Lexical: list of keywords or corpus concepts.
[6]Semantic network: all relationships between signifying elements leading to a clarification of the meaning of those elements according to the elements to which they are connected.
[7]Thesaurus: a set of meanings, ideas and concepts organized according to a tree classification from general to private. All these definitions are taken from the following reference: Sphinx Development, Training Support, 1986-2015, pp. 115-116.
[8]This thesaurus incorporated into the Sphinx software was published by Larousse in 1991. It is based on a 3 675-element ontology and implements semantic networks.

	Effectifs	Longueur moyenne	Les 5 mots spécifiques	Concept spécifique
Route	27	49 mots	route - problème - conseil - transport - ouvrage	Action
Projet	117	110 mots	projet - pays - infrastructure - routier - performance	Nature et technique
Entreprise	99	101 mots	entreprise - réclamation - travail - délai - chantier	Hiérarchie
Qualité	129	115 mots	qualité - étude - délai - travail - coût	Communication et information - ...

Figure 12.   Characterization of topic classes.

to achieve this goal, semantic modelling shall be put in place. It consists of building 'an ad hoc thesaurus grouping the words of the lexicon into families' (Moscarola & Boughzala 2016: 6). These families were organized and named with reference to knowledge of road projects. Therefore, the client, the financial backer and the other stakeholders find their place as shown in Figure 13 below.

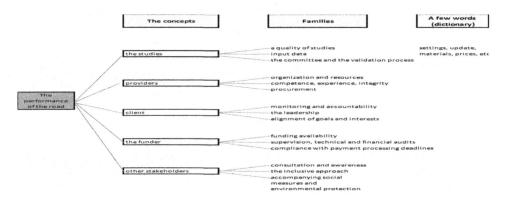

Figure 13.   The ad hoc thesaurus, source: adapted from Moscarola and Boughzala, (2016: 8).

This thesaurus corresponds to our own manual book code, or 'content analysis grid' (Bardin 1989; cited in Moscarola & Boughzala 2016: 7), which was used for manual coding. For this purpose, each response, depending on the words it contains, has been associated (coding) with one or more families. Each family has been related to a concept, and the concepts are the explanatory variables of the performance of the road project.

This stage marks the second iteration of abductive reasoning with the structured model (the ad hoc thesaurus) of the world of knowledge which has served as a framework for analyzing empirical observations from the empirical world of facts.

Manual coding resulted in the presence of the subjects relating to the various explanatory variables being identified with a frequency measurement indicating the weight of each topic. It is the distribution of themes.

The managerial themes relating to the roles of stakeholders to achieve the performance of the road project have been identified. This stage marks the third iteration of our abductive reasoning, with the distribution of themes that have emerged from empirical observations contributing to the increase of theoretical knowledge.

5 MANAGERIAL RECOMMENDATIONS

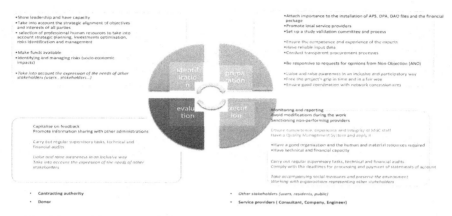

Figure 14. Managerial recommendations.

6 CONCLUSION

Through this research, we have looked at road projects carried out by AGEROUTE in Senegal, in view of the problem of delays, costs and shortcomings in quality, which is a concern in the implementation of major public infrastructure projects. Our results showed, on the one hand, that the majority (60 %) of the major road projects implemented by AGEROUTE from 2007 to 2016 exceeded the initial deadline by more than 49 %. Similarly, the majority (63 %) of these projects exceeded their initial cost by at least 16 %. It also appears that at least 16 % of these projects have received maintenance works before their fifth year of commissioning. On the other hand, through a participatory approach, our results have made it possible to determine the roles of each of the road project stakeholders in order to achieving the performance of the road project. These roles are broken down into different themes according to their frequency in the textual corpus of the interviews, which has been analyzed using the IT tools for textual data analysis. As managerial recommendations, the themes relating to the roles of each of the stakeholders have been articulated at each stage (identification, preparation, implementation, evaluation) of the project implementation process, indicating the performance criterion (time, cost, quality, user satisfaction, residents and environmental protection) and the focus of performance (relevance, effectiveness, efficiency) that are affected.

REFERENCES

AGEROUTE (2017). Manual of Procedures for Missions and Organisation of the Agency.
Ageroute (2019), *Annual Report*.
Bardin L. (1989). Content Analysis (5th edition, *Rev. and Augm.* Paris: PUF.
Beaulieu P. and Kalika M. (2017). *DBA's Draft Thesis*, Collection BSI, EMS, 276 p.
Boughzala Y., Moscarola J., and Hervé M. (2014). Sphinx -i: A New Tool for Textual and Semantic Analysis. Proceedings of the 12th International Days of Statistical Analysis of Text Data, 91–103.
Knight F., and Meyer V. (2018). CHAPTER 6. The Interviews. Françoise Chevalier ed., *DBA Research Methods*, 108–125.
Knight F., Cloutier M., and Mitev N. (2018). *The DBA Search Methods*. EMS Editions, 513 p.
Freeman R.E. (2010). *Strategic Management: A Stakeholder Approach*. Cambridge University Press.
Hansen N.M. (1965). "Unbalanced Growth and Regional Development". *Economic Inquiry*, 4 (1), pp. 3–14

Kalika M., (2015). *Knowledge Creation by Managers*, Colombelles, EMS, pp. 25–52.

Sphinx Développement (2015). *Training Support*, Le Sphinx France

Moscarola J. (2018). *Talking About the Data: Quantitative and Qualitative Methodologies*. EMS editions. 257 p.

Moscarola J. & Boughzala Y. (2016). "Analysing the Online Corpus of Opinions: Exploratory Lexical Analysis and/or Semantic Modelling?'. *13th International Days of Statistical Analysis of Textual Data*, Nice.

Ndiaye P. M. (2017). *Risk Factors in the Management of Development Projects in Senegal*, EMS BSI, 120 p.

PMBOK A. (2017). *A Guide to the Project Management Body of Knowledge*, 6th edition. Pennsylvania: Project Management Institute. Inc. USA.

Poulingue G. (2007). "Have the Members of the Montreal Club Influenced Project Management Research?". Future Management (2), 89–104.

Soumaré E.M. (2016). Performance of International Development Projects, *L'Harmattan*, 221 p.

Walsh I. (2015). *Discover New Theories: A Mixed Approach Rooted in the Data*. EMS editions, 126 p.

World Bank Group (2017), *Transport and Logistics White Paper in Senegal*

Pavements, energy efficient actors in public lighting

Sébastien Liandrat*
Cerema ITS Research Team, Clermont Ferrand, France

Valérie Muzet
Cerema ENDSUM Research Team, Strasbourg, France

Vincent Bour
AFE, Roissy, France

Jérôme Dehon
Schréder, Liège, Belgique

Jean-Pierre Christory
Consultant, Rambouillet, France

Brice Delaporte
Routes de France & Office des asphaltes, Paris, France

Joseph Abdo
Cimbéton, Paris, France

Florence Pero
Specbea, Paris, France

Thibaut Le Doeuff
CERIB, Épernon, France

ABSTRACT: The variety of road surfacing techniques meets most of the current challenges, particularly those related to physical and mechanical aspects such as skid resistance or durability. However, the light reflection capabilities of pavements are rarely considered in the design of lighting installations, as typical characteristics that are used are now being questioned.

The working group "Pavements and Lighting" is composed of many actors, both public and private, from very different background related to these topics. It is conducting a study that aim to develop tools and methods for managers, lighting designers and road builders to optimize lighting both in interurban and urban areas. The objective is to increase the knowledge on the photometric characteristics of pavements, the relationship between lighting and pavements and reduction of light pollution or urban heat phenomenon. One main deliverable of this study is a library with a complete characterization of the photometry of a large panel of current and innovative pavements available on the French market.

This library of urban and interurban pavements and the methodological approach will be available to all.

A series of 39 pairs of samples were tested with Cerema's gonioreflectometer in the initial state and after 30 months of aging in outdoor conditions, exposed to rain and sun. This panel

---

*Corresponding Author: sebastien.liandrat@cerema.fr

DOI: 10.1201/9781003429258-53

includes different surfacing materials: asphalt concrete (bituminous or synthetic binder), mastic asphalt, poured and precast cement concrete and natural stones.

The results show an enormous variability of photometric characteristics among the existing paving techniques as well as a significant evolution over time for some of them.

Lighting calculation were carried out to assess the impact of using actual photometric properties of the pavement in the lighting design. They demonstrated the importance of using the real pavement measurement rather than typical characteristics to avoid excessive lighting or poor uniformity, which can lead to potential road safety problems. Thus, a lighting installation optimized by considering a clear and diffusive pavement can lead to energy savings of up to 50% while maintaining optimal lighting quality, which highlights the importance of considering the pavement and light combination.

Keywords: Pavement Surface Properties, Road Photometry, Durability and Long-Term Performance of Pavements, Energy Efficiency

1 INTRODUCTION

1.1 *Pavements and lighting context*

A few decades ago, roads were only required to be robust, safe and comfortable for vehicle traffic. It was also always important that it remains economically acceptable, both during its construction and during maintenance phases. Over time, the number and variability of pavement families and techniques have also increased considerably to meet today's needs and uses. Thus, among the major families such as asphalt concrete, mastic asphalt, poured cement concrete, precast concrete and natural stone, there are still many variables. Aggregates, which are the essential component of many families and formulas, can vary in size, color and origin (natural, recycled, artificial) and then the particle size distribution can also differ. The binder can also vary, whether it is as often based on cement or bitumen, the additives can thus bring them different properties. New binders are also developed, sometimes synthetic or even of vegetable origin. Today, the road is still widely used and is asked to meet more and more societal needs and challenges. Indeed, the qualities described above such as comfort and safety have become prerequisites complemented by other objectives to be achieved.

Public lighting is one of these issues that refers both to traditional objectives in terms of safety and comfort of users but also to more recent societal concerns such as energy consumption and the issue of light pollution. The lighting installations are designed on the principle of "projected light flux" on a standard road surface. The important thing for the user is not only the projected luminous flux, but also the light reflected by the surface or the obstacle and thus the luminance. Perception is based on luminance contrast and this depends on the properties of the road surface. Road lighting installations are generally designed by calculating the performance in terms of luminance distribution as defined in (CIE 140 2000) and the EN 13201 European standard (EN 13201-3:2015). Since the photometric characteristics of pavements are generally not measured, a standard r-table as defined in (CIE 144 2001) is often used for lighting design. The "standard" r-tables are over 50 years old and several studies have shown that they are no longer representative of today pavements (Dumont *et al.* 2007, Gidlund *et al.* 2019, Jackett & Frith 2010, Muzet & Greffier *et al.* 2019).

1.2 *Pavements and light work group*

The *Pavement and Lighting* working group (P&L group or in French "Revêtements et Lumière") is composed of project managers and public authorities[1], professional

---

[1]AITF (Association des Ingénieurs Territoriaux de France)

associations and unions of lighting designers[2] and road builders[3], public and private research organizations[4], and expert consultants. After having organized and monitored demonstrators and operations on real sites to show the relevance of the issues and concepts of optimal lighting (Abdo *et al.* 2010; Christory *et al.* 2014), our objective is to develop a library of real and innovative urban pavements available on the market. This library is intended to facilitate the choice of decision makers and develop tools and methods for managers, lighting designers, and road builders to optimize lighting in interurban and urban areas.

The objective of this study was first to improve the knowledge of the photometric properties of a large number of urban pavements. Today, although measurements exist, it is essentially the reference r-tables that are easily accessible and therefore known by the actors concerned. Moreover, when they are carried out, photometric measurements are generally made very quickly after the pavement has been installed, whether or not it has received a surface finish. Finally, the quality of public lighting is rarely assessed by taking into account the actual photometric properties of the pavement (Chain 2007).

2 BASICS KNOWLEDGE

2.1 *Road photometry*

The surface of a pavement is classified according to its reflective properties (CIE 144 2001). The most characteristic parameter is the luminance coefficient q, given as:

$$q(\alpha, \beta, \gamma) = L(\alpha, \beta, \gamma)/(E_h) \tag{1}$$

It is the ratio of the observed luminance L in cd/m^2, which the observer sees, to the illuminance E_h in lux, which is incident on the surface. The standard observation height is 1.5 m and the observation angle α is constant at $1°$, which corresponds to an observation distance of 86 m. The lighting standards use the area of the road between 60 m and 160 m in front of the driver, as it is considered an important area obstacle detection. It has also been defined for interurban driving where the speed is about 80 to 90 km/h.

Since the 1980s, for practical reasons the luminance coefficient was replaced by the reduced luminance coefficient r in cd/m2/lux, which is derived from q:

$$r(\beta, \gamma) = 10^4 \times q(\beta, \gamma) \times \cos^3\gamma \tag{2}$$

A reduced coefficient table called r-table was defined, where the luminance coefficient r is given for a combination of fixed illumination angles β and γ. This table can be represented by a reflection indicatrix as shown in Figure 1. To simplify the description of photometric performance of road surfaces, the additional parameters Q_0 and S_1 are calculated. The average luminance coefficient Q_0, represents the degree of lightness of the measured surface. It is calculated as the average of the luminance coefficients over the specified solid angle, Ω_0. The specular factor S_1 represents the degree of specularity (shininess) of the observed surface.

---

[2]AFE (Association Française de l'Eclairage)

[3]Office des Asphaltes, ROUTES DE FRANCE, SPECBEA (Spécialistes de la Chaussée en Béton Et des Aménagements), CIMbéton (Centre d'information sur le ciment et ses applications), EUROBITUME

[4]Cerema (Centre d'Etudes et d'expertise sur les Risques, l'Environnement, la Mobilité et l'Aménagement), CERIB (Centre d'Etudes et de Recherches de l'Industrie du Béton), CTMNC (Centre Technique des Matériaux Naturels de Construction).

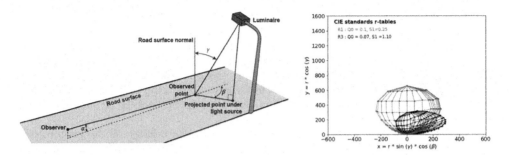

Figure 1. Left: Road photometry characteristics angles, observation (α), deviation (β) and incidence (γ) Right: Representation of a the pavement reflection indicatrix for the standards r-tables R1 and R3.

It is defined as the ratio of the reduced luminance coefficients of two specific illumination conditions.

$$Q_0 = \frac{1}{\Omega_0} \int q \ d\Omega \tag{3}$$

$$S_1 = \frac{r(\beta = 0, \tan\varepsilon = 2)}{r(\beta = 0, \tan\varepsilon = 0)} \tag{4}$$

The CIE (International Commission on Illumination) has defined different set of standard r-tables that are directly available in all lighting design software (CIE 066 1984). Since 2001, the CIE (CIE 144 2001) recommends a scaling of the chosen standard table according to the average luminance coefficient parameters Q_0. It is obvious that the design of a road lighting system should be based on the knowledge of the actual luminance coefficient for the actual road. Because the actual quantity of q is not known, nor is listed as reference values in the EN standard (it provides only the directions in which q should be known), designers use in the calculations as q reference values the ones given in CIE 144 scientific publication. In France the CIE r-table type R3 is mostly used. In some countries (USA, UK, etc.), R1 reference table is used for cement concrete.

2.2 *Road lighting*

Guidelines and road lighting standards in Europe give values for illuminance and luminance and their distribution on the road surface according to a grid of points whose number N depends on the pole spacing S and number of traffic lanes (Figure 2).

The standard road lighting quality criteria (CIE 140 2000; CIE ILV 2015) are derived from the luminance calculated on the grid, which are:

- The average luminance L_{ave}, calculated on every point of the grid
- Overall uniformity of luminance U_0, as the ratio between the minimum and the mean
- Lengthwise uniformity of luminance U_l, as the ratio between the minimum and the maximum along the axis of each driving lane;
- Threshold Increment TI, as the measure of the disability glare defined in (Anon 2011).
- Edge Illuminance Ratio EIR, as the ratio of the illuminance on a strip adjacent to the road, to the illuminance on the same strip lying on the adjacent driving lane.

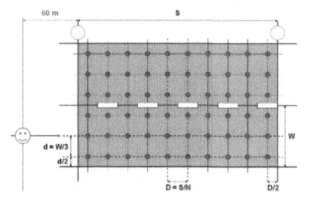

Figure 2.   Example of a calculation /measurement grid points (in red) in road lighting design/ evaluation, according to the standard (w is the traffic lane width, d the transversal distance between two lines of points and D the longitudinal distance between two columns of points).

The luminance is calculated from the axis of each lane of the road, embracing the whole width of the grid. For luminance, uniformities and edge illuminance ratio, the lowest value from all lanes is operative. For threshold increment, the highest value is operative.

This approach makes it possible to choose the lighting fixtures used and thus has an impact on urban energy consumption.

3   METHODOLOGY OF THE STUDY

3.1   *Pavements library constitution*

A representative panel of 39 urban and interurban paving materials including innovative French technologies was first established and is described in Table 1.

3.2   *Gonioreflectometer test campaign*

3.2.1   *Device presentation*
The Cerema gonioreflectometer in Clermont-Ferrand measures the reflection of road surfaces under the observation conditions of a motorist at different observation angles including 1°, according to CIE specifications, It gives the 580 reflected luminance coefficients $q(\beta,\gamma)$ of the r-table, and the parameters Q_0 and S_1. The mechanical positioning unit consists of a steel base on which is adapted a rotating measuring arm to change the sight angle β from 0 to 180° and a light source positioning system to vary the angle of incidence of light γ from 0 to 90° (see picture of Figure 3a). The mechanical movement system of the lamp and the arm carrying the photometer is computer controlled, so the data acquisition is fully automated.

The reference source is a type A halogen lamp (Philips PAR38 Spot bulb with a power of 120W, a nominal voltage of 24 V, a nominal flow rate is 1545 lm) with a color temperature of 2700K (warm light) and an angle of diffusion of 10°. The light source is a fixed on a metal arc with a radius of 2.05 m, whose movement corresponding to the incidence angle is ensured by a motor connected to an indexer-transformer.

The illuminated area is a 10×10 cm square, always at the center of the sample. The "sample holder device" consists of a turntable allowing adjusting the height of the sample, its lateral positioning and inclination to obtain the horizontality of the sample upper surface

Table 1. Description of the 39 formulas studied.

	Families	Aggregates size	Aggregates color	Binder	Surface finishes
Asphaltic concrete	4 medium coarse asphalt concrete ("BBSG")	9 formulas in 0/10	2 dark	8 bitumen	9 raw surfaces
	8 very thin asphalt concrete ("BBM")	3 formulas in 0/6	10 light	4 synthetic	3 sandblasted
Mastic asphalt	2 sidewalk asphalt, AT	2 formulas in 0/6	6 light	3 bitumen	2 raw surface, 2 shot blasted
	4 road asphalt, AC2Gr	4 formulas in 0/10		3 synthetic	2 sanded
Poured cement concrete	Not specified	6 formulas in 6/14	7 light	Cement	1 raw, 1 swiped, 1 smoothed,
		3 formulas in 4/16	3 dark		2 deactivated, 2 bush hammered, 1 sanded,
		1 formula in 6/10			2 sandblasted
Precast cement concrete	Not specified	5 formulas in 0/2	1 white	Cement	2 raw, 1 washed, 1 shot blasted,
		1 formula in 0/3	5 grey		1 aged, 2 flamed
		1 formula in 0/14	1 black		
Natural stones	3 granite 1 limestone	Not applicable	Not applicable	Not applicable	4 flamed

Figure 3. Left: the Cerema gonioreflectometer. Right: the samples outdoor during natural ageing.

(Figure 3b). This allows the measurement area to be centered on the sample without having to change the luminance meter and source settings.

3.2.2 *Test campaign presentation*
Sample pairs corresponding to the 39 selected surfacings described above were made. All the samples were then measured with the Cerema laboratory gonioreflectometer (Figure 2) in order to evaluate their photometric properties: table-r, Q_0, S_1 (Muzet & Colomb 2019).

All the pavements were first measured at their initial state (called T0). Half of the samples were then stored in fridges and half of them were installed outside in order to undergo a natural ageing during 30 months (called T30, Figure 3). This ageing implied sun, rain and any other weather condition exposition but not any mechanical influence such as traffic. The ageing applied corresponds to what could be observed on both the central lanes of roads and on urban pavements not used by cars, such as cycle paths, sideways or squares. It should be noted that this methodology does not take into account the possible evolution of the surface condition due to the use of the coating: wear, dirt, cleaning, etc.

3.3 *Lighting simulation parameters*

The next step of the study presented here consisted in the realization of simulations of public lighting using the various table-r measured previously. The quality of lighting was designed and/or analyzed according to the criteria defined in the standard (EN 13201 2015).

In all the cases presented in this study, the simulations correspond to a frequent situation where the dimensioning of the lighting is done for an already existing infrastructure, and there is a change of the luminaires called relamping. Thus the geometry of the lighting and the road are imposed.

The first scenario is the most frequent case where the lighting is dimensioned without knowing the photometric characteristics of the pavement in place. The lighting design was conducted using an R3 table and an R1 table, giving two possible LED luminaire. To study the impact of the real pavement photometry, lighting performance simulation were conducted by considering the photometric characteristics measured on the samples at their initial state (T0) and after ageing (T30) with the use of the LED luminaire corresponding to R3 and R1 pavement. The compliance with the lighting standard requirement in term of luminance level and homogeneity was studied for the two LED luminaire chosen with the R3 and R1 pavement.

The second scenario of the simulation consisted in designing the lighting directly according to the real photometric characteristics measured on the samples, T30 measurements were used.

The two scenarios were applied to a residential street composed of two driving lanes, illuminated by a single-sided lighting installation.

4 RESULTS

4.1 *Photometry results*

The graph of the Figure 4 presents the evolution of Q_0 and S_1 parameters during aging for each sample, the origin of the arrow is the T0 measurement and the arrowhead is the T30 measurement. Each sub-figure corresponds to a specific pavement type.

The results concerning natural stones are presented for information in the graph corresponding to the precast concrete samples but are not studied hereafter because only the results at T0 are available for the moment, the measurements of T30 will be made in summer 2022. Although 12 asphalt concrete samples were made, several were degraded and had to be remade and their results are therefore not presented here.

4.1.1 *General analysis*
The results of the photometric measurements of the pavements show that there is an enormous diversity of behaviors. This is especially true at T0 but remains true after 30 months of aging. These results confirm again that the 4 CIE standard tables do not reflect the diversity of all pavements (Dumont *et al.* 2007, Gidlund *et al.* 2019).

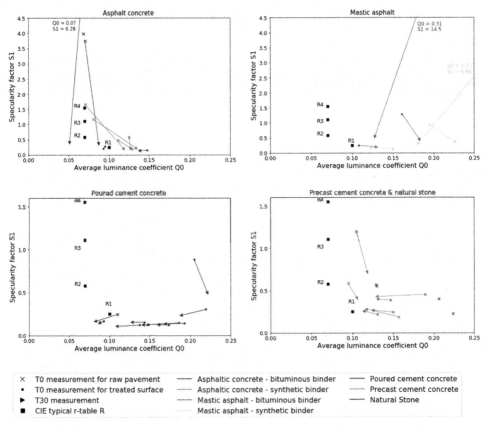

Figure 4.   Representation of the P&L photometric measurements (S_1 and Q_0) for all the pavements. T0 is the origin of the arrow and T30 measurement corresponds to the arrowheads. The reference CIE r-tables are also presented in black squares. The S_1 scale is different between the two top and bottom graphs.

It can be observed that in each sub-figure, all arrows are pointing downwards, which means that there is a decrease in specularity over time on the vast majority of pavements. This evolution is really important for asphalt concrete and mastic asphalt families which where initially very specular. It remains for cement concrete samples, both poured and precast, but this phenomenon is less pronounced. These results are in accordance with previous results (Dumont et al. 2007; Muzet & Abdo 2018; Muzet & Greffier et al. 2019).

After 30 months, all samples are classified as CIE R1 type (except 2 which are R2 type), mainly due to this specularity evolution. Such low values of specularity are not usually measured on conventional circulated road surfaces (Dumont et al. 2007, Gidlund et al. 2019) but it was shown in (Jackett & Frith 2010) that non circulated pavements were more diffuse after aging than the ones exposed to traffic.

The evolution with time of the average luminance coefficient Q_0 also differs depending on the pavement surface material. Except for initially very specular surfaces, for the asphaltic concrete and mastic asphalt samples, the luminance coefficient increased between T0 and T30. A general explanation could be that ageing with meteorological condition has a kind of tarnishing effect on these samples, slightly lightening their surface. On the contrary, the luminance coefficient of cement concrete almost systematically slightly decreased. Since the surface of these samples was quite light from the start, ageing seems to darken them slightly, which is rather confirmed by the visual evolution. As above, this behavior is not observed when the T0 samples were more specular.

At T30, all the concrete and precast concrete samples have characteristics close to the R1 reference table, this is particularly marked for precast concrete.

4.1.2 *Binder and finishes impacts*
Some of the above observations can be explained by the photometric characteristics of the binder used. Indeed, bitumen is initially a very shiny and specular material, which however evolves very quickly. This behavior remains true for synthetic binders although it is attenuated, in this case they are less specular in the initial state and evolve less, their characteristics at T30 is then comparable to that of bitumen.

Both for cast and precast concrete, samples made with cement show little change over time. This is due to the fact that the cement has a diffuse appearance almost from the beginning and does not change much.

The color of the binder can also explain the evolution of Q_0, except in cases where the specularity was very high at the beginning. Indeed the bitumen is very dark while the cement is clear.

Samples with a specific surface finishes are marked on the graphs with a circle instead of a cross. It can be seen that for both asphalt concrete and mastic asphalt samples, the specularity values are systematically lower for the ones with a specific surface finishes (sanded, sandblasted or shot blasted). In accordance with what was said earlier, the photometric properties of these samples evolve less over time than the others.

There is no observable difference between the finishing states for the poured cement concrete samples.

4.2 *Lighting simulation results*

4.2.3 *Tables presentation*
The tables in Figure 5 show the results of all the simulations (scenarios 1 and 2). Each table corresponds to a family of pavements. Natural stones are not presented because all the measurements have not yet been performed. For the scenario 1, the red boxes correspond to cases where the designed lighting would not meet the public lighting quality standards. For the scenario 2, since the lighting design considers the measured pavement photometry, the standard lighting quality requirement are always respected, so the difference of the installed luminous power with a R3 pavement is given in percentage (and therefore the potential economy in power). As previously, only 7 of the 12 asphalt concrete samples are presented here.

4.2.4 *Scenario 1 observations*
In the case of scenario 1, i.e. following the usual methodology of designing the lighting on the basis of the references table-r, the simulations show that the quality of the lighting actually obtained with the measured photometric properties would not comply with the standards for a large quantity of pavements. This is especially true when the lighting designing is carried out with the R3 r-table where many boxes appear in red.

The results for asphalt concrete show that the use of a synthetic binder (samples 9 to 12) leads to a more standard-compliant lighting. This is rather logical since their photometric characteristics are closer to the reference table-r (Figure 4). The results for mastic asphalt show a strong impact of surface finishing (shot blasting and sand blasting, samples 2, 3, 5 and 6) resulting in greater compliance with the standards. The results for cast concrete samples systematically show a non-conforming lighting compared to the standards. Although not presented here, the values show in fact that non-compliance with the standard is almost systematically due to poor lengthwise uniformity of luminance, a direct consequence of over-illumination (average luminance too high). This is consistent with the

Asphalt concrete

Sample Nb.	Agg. Color.	Surf. Fin.	Bind.	Age	Scenario 1 R3	Scenario 1 R1	Scenario 2 Power evolution
1	D	-	B	T0 / T30			69%
5	L	-	B	T0 / T30			-13%
6	L	SB	B	T0 / T30			-53%
9	L	-	S	T0 / T30			-46%
10	L	-	S	T0 / T30			-40%
11	L	-	S	T0 / T30			-51%
12	L	-	S	T0 / T30			-44%

Mastic asphalt

Sample Nb.	Agg. Color.	Surf. Fin.	Bind.	Age	Scenario 1 R3	Scenario 1 R1	Scenario 2 Power evolution
1	L	-	B	T0 / T30			-30%
2	L	ShB	B	T0 / T30			-61%
3	L	Sd	B	T0 / T30			-45%
4	L	-	S	T0 / T30			-58%
5	L	ShB	S	T0 / T30			-57%
6	L	Sd	S	T0 / T30			-56%

Poured cement concrete

Sample Nb.	Agg. Color.	Surf. Fin.	Bind.	Age	Scenario 1 R3	Scenario 1 R1	Scenario 2 Power evolution
1	L	Sw	C	T0 / T30			-51%
2	L	Sm	C	T0 / T30			-60%
3	D	Dea	C	T0 / T30			-19%
4	L	Dea	C	T0 / T30			-51%
5	D	BH	C	T0 / T30			-43%
6	L	BH	C	T0 / T30			-62%
7	L	Sd	C	T0 / T30			-54%
8	D	SB	C	T0 / T30			-36%
9	L	SB	C	T0 / T30			-47%
10	-	-	C	T0 / T30			-26%

Precast cement

Sample Nb.	Agg. Color.	Surf. Fin.	Bind.	Age	Scenario 1 R3	Scenario 1 R1	Scenario 2 Power evolution
1	-	-	C	T0 / T30			-46%
2	-	W	C	T0 / T30			-25%
3	-	ShB	C	T0 / T30			-34%
4	-	Ag	C	T0 / T30			-20%
5	-	-	C	T0 / T30			-33%
6	-	Fl	C	T0 / T30			-33%
7	-	Fl	C	T0 / T30			-33%

Abbreviations :

- D: dark
- L: light
- B: bitumen
- S: synthetic

- SB: sand blasted
- ShB: shot blasted
- Sd: sanded
- Sw: swiped

- Sm: smoothed
- Dea: deactivated
- BH: bush hammered
- W: washed

- Ag: aged
- Fl: flamed

Figure 5. Simulation results for each pavements families. For the scenario 1, a check of compliance with the standard requirement is conducted using R3 and R1 typical r-table (green color). For the scenario 2 that uses the T30 measured photometry, the difference of power consumption is computed in %.

photometric properties of the latter, namely a high Q_0 above R1 logically inducing high luminance values.

The results for the precast concrete samples show compliance with the standard in a large majority of cases. This can be explained by the proximity of the photometric properties of these pavements to the reference r-tables, especially R1 and after samples aging, as shown in Figure 4.

4.2.5 Scenario 2 observations

In the case of scenario 2, i.e. the design of the lighting based on the actual photometric properties measured on the samples, the standards are by definition systematically met. The results show very significant decreases in power, often between 30 and 60%. It is interesting to note that this gain remains valid even in cases where the standard was already met with scenario 1 (e.g. for precast concrete).

5 CONCLUSIONS

This study allowed the realization of photometric measurements on a large number of classical and innovative French urban surfaces. In particular, this allowed us to obtain results on pavement families not well represented in the literature, such as mastic asphalt, precast concrete or natural stone. Beyond the raw results, it was also an opportunity to highlight the impact of aging due to weather. The results show a very large variability and largely confirm the lack of relevance and representativeness of the CIE reference r-tables.

The simulations showed first of all that the classical method of lighting design based on the CIE references r-tables is limited. Indeed, it leads in many cases to a non-conformity with the standard. This is particularly pronounced for poured concrete, asphalt concrete and mastic asphalt samples. Nevertheless, the use of synthetic binder instead of bitumen had a positive impact.

The use of a surface treatment also has a beneficial effect on the asphalt concrete and mastic asphalt samples. No effect could be observed for poured cement concrete. This seems logical given the photometric properties of bitumen and its impact on the specularity of samples at T0.

Finally, the simulations performed by dimensioning the lighting according to the real photometric properties of the surfacings show a strong reduction of the luminous power to be installed in order to respect the standard. In almost all cases, this reduction is of the order of 30% to 60%, including when the standard was already met using the reference r-table. Similar simulations have been performed on 4-lane road profiles under interurban conditions, and the results also show the same trends (Muzet V. 2021).

This shows the importance of using the real pavement properties and thus making measurements. This ensures that the standard is met and thus avoids comfort and safety problems, but also results in significant energy savings.

The results also showed the importance of not performing these measurements of the properties of the samples in their initial state but rather to wait for a certain ageing. On this point, the bibliography shows a rather fast evolution in the first months after the implementation. In order to mitigate this phenomenon, the results also show the beneficial impact of surface treatments for bitumen-based formulas. These treatments at least prevented too great an evolution of the photometric properties over time but also allowed a better respect of the standard.

The library of pavements developed within the framework of the Pavements & Light working group shows its full interest here. It offers an overview of the photometric properties of a large number of different pavements and will be a source of information on them. It also demonstrates the interest in changing urban lighting design practices for multiple interests: user comfort and safety, but also optimization of energy consumption by managers.

Although not presented here, the work carried out on the different pavement samples also provides elements on other observation angles that could be relevant in an urban environment. Similarly, colorimetric and radiometric measurements have been carried out that could provide elements of answers for treating urban heat island phenomena (Lebouc L. 2022).

ACKNOWLEDGEMENTS AND FUNDINGS

We would like to thanks Frédérico Batista (CD 78), Salah Boussada (AITF/MEL), Jérôme Dherbecourt (Routes de France/EIFFAGE Routes), Florian Greffier (Cerema), Romain Lafon (Routes de France/EUROVIA), Christine Leroy (Routes de France), Emmanuel Loison (Routes de France/COLAS), Lionel Monfront (CERIB), and Didier Pallix (CTMNC).
This project received financial support of the "Pavements and Lighting group" members.

REFERENCES

Abdo J., Batista F., Carré D., Christory J.P., Depetrinji A., Gandon-leger P. and Peret M., 2010. Démarche Innovante 'Revêtements et Lumière' de l'idée à la pratique";. RGRA. 885, pp. 49–53.

ANON., 2015. EN 13201-2:2015. Road lighting - Part 2: Performance requirements.

Chain C., Lopez F., and Verny P., 2007. *Impact of Real Road Photometry on Public Lighting Design. In:* . Presented at the The CIE 26th Session, Beijing, China: CIE.

Christory J.P., Batista F., Gandon-Leger P. and Talbourdet P., 2014. *Pavements and Light for the Right Lighting: Contribution of Concrete Pavements.*

CIE 066, 1984. *Road Surfaces and Lighting (joint technical report CIE/PIARC)* [online]. CIE. Available from: http://cie.co.at/publications/road-surfaces-and-lighting-joint-technical-report-ciepiarc [Accessed 12 Mar 2020].

CIE 140, 2000. CIE 140:2000 *Road Lighting Calculations.* [online]. Available from: https://www.techstreet.com/cie/standards/cie-140-2000?product_id=1210058 [Accessed 18 Mar 2020].

CIE 144, 2001. CIE 144:2001 *Road Surface and Road Marking Reflection Characteristics. CIE.*

Dumont E., Fournela F., Muzet V., Paumier J.L. and Venin C., 2007. *Pavement Reflection Properties and luminance Distribution: Measurements on a Road Lighting Test-track and Comparison with Standard Calculations.*

CIE ILV, 2011. CIE S 017/E:2011 *International Lighting Vocabulary* [online]. Available from: http://cie.co.at/publications/international-lighting-vocabulary [Accessed 12 Mar 2020].

EN 13201-3:2015, 2015. EN 13201-3:2015. *Road lighting - Part 3: Calculation of performance.*

Gidlund H., Lindgren M., Muzet V., Rossi G., and Iacomussi P., 2019. Road Surface Photometric Characterisation and Its Impact on Energy Savings. Coatings. 9 (5), p. 286.

Jackett M. and Frith W., 2010. *Reflection Properties of New Zealand Road Surfaces for Road Lighting Design.* In: . p. 15.

Lebouc L. 2022. *Colorimetry of Pavements for Urban Planning.* SURF 2022

Muzet V. Liandrat S., Bour V., Dehon J. and Christory JP., 2021 *Is it Possible to Achieve Quality Lighting Without Considering the Photometry of the Pavements?.* Conference CIE 2021, NC Malaysia, CIE, Sep 2021, Kuala Lumpur, Malaysia. pp.11–25.

Muzet V., Colomb M., Toinette M., Gandon-Leger P. and Christory J.P., 2019. Towards an Optimization of Urban Lighting Through a Combined Approach of Lighting and Road Building Activities. *In: Proceedings of the 29th Quadrennial Session of the CIE* [online]. Washington DC, USA: International Commission on Illumination, CIE. pp. 789–800. Available from: http://files.cie.co.at/x046_2019/x046-PP23.pdf [Accessed 12 Mar 2020].

Muzet V., Greffier F., Nicolaï A., Taron A. and Verny P., 2019. Evaluation of the Performance of an Optimized Road Surface/lighting Combination. Lighting Research & Technology. 51 (4), pp. 576–591.

Muzet V. and Abdo J., 2018. On Site Photometric Characterisation of Cement Concrete Pavements with COLUROUTE Device. Light and Engineering. 26, pp. 88–94.

New procedure for quality assurance of municipal road survey and assessment campaigns

Christiane Krause & Ulrich Pfeifer
Senatsverwaltung für Umwelt, Mobilität, Verbraucher- und Klimaschutz, Berlin

Markus Stöckner*
UAS Karlsruhe, Karlsruhe, Germany

Ute Stöckner
Steinbeis Transferzentrum Infrastrukturmanagement im Verkehrswesen, Karlsruhe

ABSTRACT: Municipal requirements of roads' survey and condition assessment differ from supra-local standards. One of the reasons are the needs of often just small road cross sections, where well-known and certified Survey and Assessment systems machines (S&A) are not able to work efficiently. Therefore, smaller vehicles with S&A-Systems were developed, but there is no method for their quality assurance comparing to certified systems. For this reason, an advanced method in ensuring quality control for those systems was developed.

The need arose, because the Senate of Berlin is establishing an asset management system (AMS) for the whole municipal road network. One main goal of the overall project is to provide reliable and comparable information about pavement surface condition to all districts of Berlin. The AMS is going to support the decision-making process for maintenance measures for short term planning as well as condition prediction for estimating future financial demand.

On the main road network already surface condition characteristics were collected by established certified S&A-Systems. The next task is completing the data basis with comparable data for the secondary road network, not necessary using certified S&A-Systems.

Requirements benchmarking the quality level of a certified reference survey were developed with regard a.o.to permissible deviations of survey characteristics. The approach requires a small road condition survey and assessment by system operators in a selected part of the Berlin road network, but opening the call to all system operators for municipal road S&A, independent from their procedures in surveying and data processing. The additional effort for this quality assurance in a tendering process seems to be moderate for road authorities and system operators.

So, for Berlin's secondary road network there is the current tendering for a road condition campaign set up without the common quality confirmation, given as time-limited operating license by German Federal Highway Research Institute. Further consideration to adapt the process for including bicycle and pedestrian infrastructure S&A is following up.

Keywords: Road Survey, Quality Assurance, Municipal Road Network

---

*Corresponding Author: markus.stoeckner@h-ka.de

DOI: 10.1201/9781003429258-54

1 INTRODUCTION

The task of road authorities is to plan, build, operate and maintain road infrastructure with the aim of usability, value retention and thus sustainability. This is solved with asset management (AM) approaches. The fundamentals of AM are defined in the DIN EN ISO 55000 standard and are based, among other things, on high-quality data, ideally a digital image of the road infrastructure. AM thus focuses on managerial and technical approaches to resource allocation and utilization with the goal of better decision making based on high-quality information and clearly defined objectives (AASHTO 2013). Over the past few decades, road administrations have been driving the adoption of strategic AM to derive long-term infrastructure goals and support investment planning. Alongside this is the goal of more reliable and safer infrastructure.

Content principles and guidance can be found in (FHWA 2012) and (PIARC 2017). Reliable data are needed for an AM system (IAMS) to obtain information on the current state of the transportation infrastructure, among other things, and to forecast its future condition state based on this information. These data then serve as the basis for the decision-making process on the timing, scope and costs of maintenance measures (Hajdin et al. 2020). The pavement management system (PMS) procedures used in many cases for this purpose build on data from the condition survey and, in combination with other inventory data within the AM, provide the necessary statements for coordinated maintenance planning, but also require a complete and accurate database for this purpose (Stöckner et al. 2022).

2 PROBLEM DEFINITION

The Berlin Senate is currently establishing an pavement management system (PMS) for the entire municipal road network. One of the main objectives of the overall project is to provide all districts of Berlin with reliable and comparable information on the condition of road surfaces. The PMS is intended to support decision-making for maintenance measures in short-term planning as well as condition prediction to estimate future financial requirements. Core components are the development of an IT-supported evaluation tool for operational and strategic maintenance planning as well as therefore ensuring a sufficient database. To this end, the first step was to record and evaluate the condition of the superordinate road network at the level of condition variables. For this purpose, a S&A system, which is certified by the German Federal Highway Research Institute (in German: Bundesanstalt für Straßenwesen, BASt) was used. The certification by the BASt takes the form of a time-limited operating license (German abbreviation: ZbBz) for one year by referencing the kind of measurement, e.g. longitudinal and transverse evenness and substance characteristics of surface (BASt 2019, 2020a, 2020b). The measurement systems certified in this way are considered to be quality-assured and thus to be suppliers of high-quality and reproducible recording results.

However, the measurement systems were developed for out-of-town areas, so that it can be assumed that they reach their limits of application in small-scale municipal networks. The municipal secondary road network, that is now to be recorded, is expected to limit the practical trafficability due to small-scale road structures of different road hierarchies as well as regular obstructions by bottlenecks, e.g. by delivering, disposing or parking vehicles. In addition, turning facilities in dead end roads are not available everywhere. In practice, therefore, smaller vehicles with S&A systems have been developed, but these are operated without the certification mentioned above; thus, there is no comparative quality assurance compared with certified systems.

For an efficient road condition monitoring this raises the question of S&A systems' applicability on the one hand side and on the other hand side the question of possibilities for different systems and methods, which are not subject to regulated quality assurance. There is no procedure known for quality assurance of road condition monitoring systems in the municipal sector that is specifically designed for close-meshed networks. For this reason, a procedure for quality assurance of S&A systems in the municipal sector was developed in the current tendering procedure, which can also be used for other municipal applications.

3 SYSTEMS AND METHODS FOR ROAD CONDITION ASSESSMENT

Basically, there are various options available for recording the condition of roads' surface properties, which -depending on their methodology- serve different quality standards and also go with different demand in time and financial effort. The methods can be divided into visual-sensitive survey, visual-image-based survey, metrological survey and mixed forms. Typically, longitudinal and transverse evenness as well as surface damages are municipally recorded.

3.1 *Visual-sensitive survey*

The inspection by a team for visual-sensitive condition survey is the classic method. However, a visual-sensitive recording of condition variables by trained staff is time-consuming and cost-intensive. The integration of the assessment into the recording process involves the risk of biased results due to human judgement. Further it does not allow to adjust the quality level of built components or single assessment aspects retroactively.

3.2 *Visual-image-based survey*

The pure visual or visual-image-based condition survey by evaluating an inspection video represents a significant acceleration of the procedure due to the faster recording process itself. The visual image-based condition survey has the disadvantage of often only roughly recognizable cracks and unevenness, depending on weather conditions and picture quality characteristics the statements are naturally limited to the damage area's extent, neglecting its specificity (level of severity). And without a reviewable damage documentation as result of the recorded pictures the disadvantages of visual-sensitive survey (evaluation's human bias and lack of readjustment) also apply.

3.3 *Metrological survey*

As a third method of surveying the road condition there is the certified metrological recording and evaluating, which is a largely automated, repeatable and thus quality-assured procedure as basis for an objective road condition assessment. Going with S&A systems, this is also internationally the standard method for highways and rural roads (Hajdin et al. 2020). Requirements for repeatability are formally regulated, e.g. in the German technical regulation in abbreviation ZTV ZEB-StB (FGSV 2006/2018), which require the certification by the Federal Highway Research Institute (BASt): The so-called time-limited operating license (in German abbreviation ZbBz) represents a recognized quality standard for survey sections with lengths of 100m for highways and rural roads as well as 20m for areas in town. Within the scope of condition surveys in accordance with ZTV ZEB-StB (FGSV 2006/2018) the S&A systems ensure to achieve comparable and repeatable measurement results.

The certification with the German ZbBz as a well introduced and quality-assured process represents a quality-assuring process due to its requirements on the measuring devices for the recording of the condition variables as well as due to the associated standardized data processing within the scope of the network projection and condition assessment itself. With the certification of the measuring devices, the results are therefore considered to be objective, reliable and valid and are successfully used for highways and rural roads as well as for main roads in town.

3.4 *Mixed forms*

For to be able to carry out road condition surveys even in municipal road spaces, smaller vehicles have been developed; but for various reasons no comparable certification is available for them. In addition, in the field of laser measurement technology laser point clouds in conjunction with picture recordings are available on the market. So far, these are mainly offered for the digitization of the road inventory (real surfaces and road equipment) and together with measurement functions in conjunction with 360° panoramas used as a substitute for on-site appointments. A visual image-based survey of surface damages (see above) can also be carried out in these systems and is also offered as a service in some cases. An evaluation as section-related condition variables is possible. Some providers are already able to deliver additional evaluations of evenness based on the laser measurement data. Since this procedure is based on a metrological approach on the one hand side, but on the other hand the process-related quality assurance -compared to the certified ZbBz procedure as described above- is lacking, this type of recording is considered as a mixed form.

4 KEY QUESTION AND APPROACH

Basically, a metrological certificated S&A may be seen as the preferred variant, since the results are traceable and repeatable. Thus, a high quality of the data is achieved, which is ensured by -e.g. the German- approval system. This means that the referencing and accuracy of the data can be ensured independently of the observer and the measuring device used. To this end, certification by the BASt involves both a technical system test and a dynamic test of the measuring devices, which also includes safety aspects relating to the driver and vehicles.

Due to smaller dimensions of non-certified vehicles in many cases, the Berlin Senate Administration has therefore decided to waive the time-limited operating license when tendering for the survey services for secondary road network. At the same time, however, due to the above-mentioned requirements of the pavement management tool (EMS Berlin), the data quality should be trustworthy and, if possible, comparable to the main roads data, recorded with ZbBz certification. This gives rise to two questions:

4.1 *How can the quality level be demonstrated and ensured independently of the provider even without the ZbBz certification?*

Without to elaborate system tests, such as those carried out by the BASt during certification, from the current point of view, this can only be done by comparing results. Therefore, the results of a certified reference S&A system are necessary for to compare the non-certified measurement results with a previously known condition. This gives operators with not certified devices the possibility to demonstrate the quality of their systems. This leads to the second question of the associated requirements.

4.2 *Which requirements for the quality level of road condition survey by certified S&A systems are fulfilled in everyday practical application in a municipal, subordinate road network for condition recording and evaluation?*

The possible external influence on the quality of the certified S&A results (confer obstructions due to narrow cross-sections, parked vehicles, etc. and the associated deviation in the driving lines) is to be examined in order to be able to define the quality level to be required for the upcoming condition survey. For this purpose, comparative results are required for different driving tours by certified S&A systems.

5 PROCEDURE AND STUDIES

First, a reference route was selected and driven on with two different certified S&A systems by two operators. In order to enable a test as broad as possible, the reference route contains different construction methods as well as different road area situations. The measurement results of the two systems were compared with each other for to check out, if the accuracy of the certified systems is also given in subordinated municipal areas and thus also to clarify what variation exists. In addition, three measurement trips, recorded with the same measurement system, were compared. Based on this, the measurement repeatability with the same measurement system as well as the comparative accuracy with different systems were expected to see.

The reference route with a length of around 20km comprises sections in the construction methods asphalt, concrete and cobblestone (see Table 1).

Table 1. Sections by construction method.

Construction Method	Sections by Numer	Sections by Percentage
Asphalt	939	87%
Concrete	89	8%
Cobblestone	51	5%
Other	2	0,2%
Total	1.080	100%

For each comparison track within the reference route twenty evaluation sections with a municipal standard section length of 20m were used in accordance to the BASt procedure for the ZbBz-certification. Longitudinal and transverse evenness were recorded for all construction methods as well as construction method-specific damage characteristics on the surface. For the comparing survey results, the condition variables and characteristics were recorded by two different operators. Operator 1X drove the route once, while Operator 3X took the route three times (trip 1-3). The time difference between the two recording times is about half a year. The following procedure was used for to compare the results:

5.1 *Comparison of sections' localization by the location of single damages*

Considering the location of single damages shows the influence of the driving path on the section localization. The following two local situations shown in Figure 1 are furthermore

| Local Situation concerning Figure 2 | Local Situation concerning Figure 3 |

Figure 1. Two examples for local situations of the reference route.

considered as examples. Figure 2 shows surface pictures, taken by the tree trips of Operator 3X: Driving along, the recording of Trip 1 and Trip 2 were each started new, Trip 3 is continuing the ended Trip 2. So Figure 2 shows exemplarily the resulting offset of the single damages (within the permissible tolerance range): Compared to Trip 1 und Trip 2 there is in Trip 3 a 20m-section offset against another of 1m up to 7m. This can be seen as well from the surface pictures as from the stationing (STATION) of the picture recording.

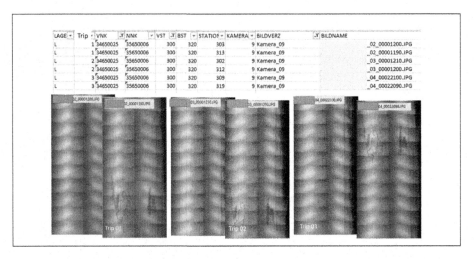

Figure 2.   Comparison of sections' localization based on the location of single damages (Operator 3X).

Comparing the surface pictures on base of the two detectors' results, there is shown the reproducibility of localization and picture representations on the basis of the position of characteristic surface features in Figure 3: There is an offset of about 2m between Operator 1X and Trip 3 by Operator 3X, even under consideration of an everyday practical use of device.

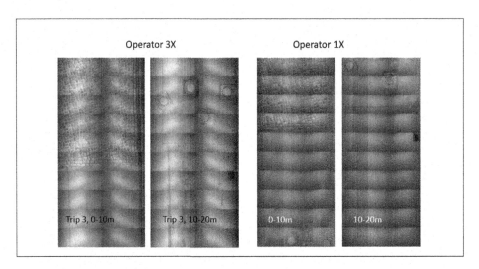

Figure 3.   Comparison of sections' localization based on the location of single damages (Operator 3X to Operator 1X).

5.2 *Statistical comparison by selected condition variables of evenness*

The condition variables (ZG) Overall Roughness AUN [cm³], Average of Rut Depth MSPT [mm], Average of Theoretical Water Depth MSPH [mm] and Transverse Gradient QN [%] are considered, all corresponding to German technical regulations ZTV ZEB-StB 2006 (Stand 2018). The regulations refer to two levels of requirement: One for ZbBz-certification and self-monitoring, one -on lower level- for controlling tests. Referring the level of requirement, they define maximum deviations for the difference between the variables of the compared recordings (average Δ and standard deviation σ). Both levels are tested. The limit values for AUN are also taken for Average and Maximum Distance Under the 4m-Straightedge PGR_AVG [mm] and PGR_Max [mm], because there is no statement in the mentioned technical regulations.

For this purpose, there are recordings of twenty sections of the reference route used, but -instead of the 100m-sections for highways- only with the regular municipal length of 20m per section. The furthermore considered comparison tracks within the reference route do have a length of 400m. Exemplarily the results of comparison track V22 are shown. In the following figures the adjunct "2020" marks values of Operator 1X.

In addition to the graphical representation of the twenty condition variables of the 20m-sections in Figure 4 the feature-dependent limit values of normalization are shown, according to ZTV ZEB-StB 2006 (2018) for the evaluation of the condition variables. Shifts in the section evaluation of single characteristics are visualized, depending on the driving operation mentioned above.

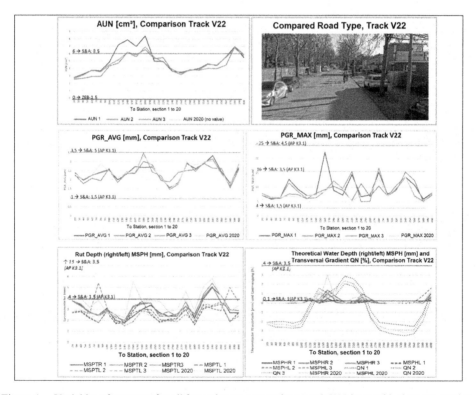

Figure 4.   Variables of evenness for all four trips on comparison track V22 in graphical representation.

All test values -in Table 2 shown in green table elements - are complying with the limit values, the non-colored fields are outside the limit values. Shown in blue table elements,

Table 2.   Evaluation by limit values for ZbBz-certification and self-monitoring according to ZTV ZEB-StB 2006(2018), example comparison track V22.

	Limit Values for				Comparison Track V22					
	Certification and Self-Monitoring		Controlling Test		Trip 1 (Operator 3X) to Trip 2020 (Operator 1X)		Trip 2 (Operator 3X) to Trip 2020 (Operator 1X)		Trip 3 (Operator 3X) to Trip 2020 (Operator 1X)	
ZG (condition value)	Δ ZG	σ ZG	Δ ZG	σ ZG	Δ ZG	σ ZG	Δ ZG	σ ZG	Δ ZG	σ ZG
Longitudinal Evenness: PGR_AVG when AUN ≥ 3,0 cm³					-0,06	0,22	-0,12	0,33	-0,08	0,17
Longitudinal Evenness: PGR_Max, when AUN ≥ 3,0 cm³	0,4	0,8	0,6	1,0	1,66	3,97	0,34	1,88	0,86	2,84
Transversal Evenness: MSPTR [mm]					-0,71	0,7	-0,61	0,6	-0,52	0,61
Transversal Evenness: MSPTL [mm]					-0,1	0,53	0,32	1,03	-0,18	0,4
Transversal Evenness: MSPHR [mm]	0,7	1,5	1,0	2,5	-0,26	0,56	-0,14	0,4	-0,18	0,48
Transversal Evenness: MSPHL [mm]					-0,26	0,65	-0,11	0,47	-0,23	0,63
Transversal Gradient QN [%]	0,3	0,5	0,3	0,8	0,53	0,15	-0,2	0,14	0,16	0,15

there are strictest test conditions used. The PGR_MAX value, what is not itself part of ZbBz-certification (see above), is complying with any limit value only in rare cases, which is characteristic for the analysis of the selected comparison tracks as a whole.

5.3   *Statistical comparison by selected substance characteristics of surface*

The comparison of the surface characteristics is carried out analogously to the BASt certification on the basis of the condition values. Figure 5 shows the graphical representation of the twenty characteristic-dependent condition values of the 20m-sections considered. The comparative track V88 (asphalt) shows values for the characteristics patches (FLI) and applied patches (AFLI) that are rated with condition value ZW = 2,0 and better because of lack of occurrence and therefore do not meet the requirements for a test section for certification for these two characteristics. The detected cracks (ZW_RISS), on the other hand, are represented by all four compared trips.

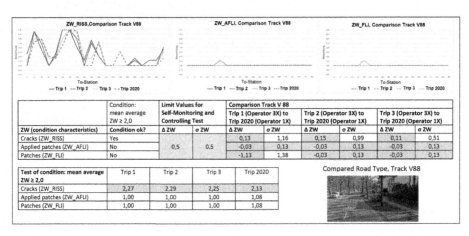

	Condition: mean average ZW ≥ 2,0	Limit Values for Self-Monitoring and Controlling Test		Comparison Track V 88: Trip 1 (Operator 3X) to Trip 2020 (Operator 1X)		Trip 2 (Operator 3X) to Trip 2020 (Operator 1X)		Trip 3 (Operator 3X) to Trip 2020 (Operator 1X)	
ZW (condition characteristics)	Condition ok?	Δ ZW	σ ZW	Δ ZW	σ ZW	Δ ZW	σ ZW	Δ ZW	σ ZW
Cracks (ZW_RISS)	Yes			0,13	1,16	0,15	0,99	0,11	0,51
Applied patches (ZW_AFLI)	No	0,5	0,5	-0,03	0,13	-0,03	0,13	-0,03	0,13
Patches (ZW_FLI)	No			-1,13	1,38	-0,03	0,13	-0,03	0,13

Test of condition: mean average ZW ≥ 2,0	Trip 1	Trip 2	Trip 3	Trip 2020
Cracks (ZW_RISS)	2,27	2,29	2,25	2,13
Applied patches (ZW_AFLI)	1,00	1,00	1,00	1,08
Patches (ZW_FLI)	1,00	1,00	1,00	1,08

Compared Road Type, Track V88

Figure 5.   Surface characteristics for all four trips on comparison track V88 (asphalt), graphical representation and evaluation by limit values for self-monitoring according to ZTV ZEB-StB 2006 (2018).

6   CONCLUSION AND IMPLEMENTATION IN TENDERING

Overall, the results confirm the application of metrological road condition recording by means of ZbBz-certified vehicles also in the municipal area of subordinate roads in asphalt and concrete construction method. As expected, deviations in the results of single recording and evaluation processes show the influence of driving lines, which are caused in the municipal area as a result of local punctual road installations such as manholes and inlets in connection with the 20m sections and local varying traffic situations. The presented procedure is based on the BASt certification, but applies stricter standards than the BASt procedure with the same number of sections: Due to the shortening of the section lengths from 100m in the certification to the 20m section lengths common in municipalities, the relative proportion of individual damages within an evaluation section increases and therefore damages' spatial shifts between two sections can have an influence on the section evaluation. As a consequence, it can be shown as an answer to question 2 with the explanations given above that the system of certified metrological condition recording in everyday practical use does not completely fulfill the requirements on limit values for controlling tests or even self-monitoring in all respects, but it does fulfill them on the whole. Thus, the requirements for the quality level to be achieved in road condition monitoring without ZbBz-certification can be formulated as follows.

For the Berlin Senate Administration, the following requirements were defined for quality assurance of road condition survey in the secondary road network by bidders with vehicles without ZbBz certification and for use in the Berlin EMS:

- condition recording at the level of condition variables
- repeatability of condition recording independent from persons, for that also
- traceability of the resulting data of condition assessments by independent third parties, e.g. documentation of relevant processing steps, in its type depending on the recording method used
- section generation for condition survey with reference to the Berlin node edge stationing system (called Detailnetz Berlin)

To check whether these requirements are fulfilled, a reference route with recording and evaluation by bidders without ZbBz is used: Based on the recordings, front and surface pictures are to be submitted together with the associated section-by-section condition values as results of the network-reference. In analogy to the BASt certification procedure, these are compared with the results of three ZbBz-certified road recordings. However, comparing to the BASt certification larger deviations are allowed:

For the characteristics of **evenness**, a comparison is made in each case on the basis of the evaluation of the mean value and standard deviation of the differences between the condition variables of the reference and the best-fitting comparison measurement (Δ ZG and σ ZG) of twenty consecutive 20m sections. The following parameters are used as relevant for evaluation:

- Longitudinal evenness as PGR_AVG with minimum requirement to the variable differences: Δ ZG $\leq$ 0,6 and σ ZG $\leq$ 1,0
- transversal evenness as MSPTR, MSPTL, MSPHR, MSPHL without minimum requirement to the variable differences, but with 0 points in tendering procedure's offer evaluation, when Δ ZG > 1,0 and σ ZG > 2,5
- transverse gradient QN with minimum requirement to the variable differences: Δ ZG $\leq$ 0,3 and σ ZG $\leq$ 0,8

For the **substance characteristics of surface**, a comparison is made in each case on the basis of the evaluation of the mean value and standard deviation of the differences between the characteristics of the reference and the best-fitting comparison measurement (Δ ZW and σ ZW) of twenty consecutive 20m sections. The following parameters are used as relevant for evaluation:

- depending on the construction method as substance characteristics of surface are used for asphalt construction as RISS and FLI (see above) and for concrete construction as longitudinal and transverse cracks (LQRL), corner demolition (EAB) and edge demolition

565

(KAS), each with minimum requirement for the differences of the condition characteristics Δ ZW $\leq$ 0.7 and σ ZW $\leq$ 0.7

The highest achievable score for each feature in the bid evaluation is reached when the quality level of the ZbBz certificated vehicles is achieved. As a result, a feature-related estimation of the expected data quality can be made, which can be taken into account in data's further in the EMS System for asset management of Berlin's urban roads.

7 FUTURE PROSPECTS

As explained above, a prototype for the quality assessment of the results of condition surveys without certificated measurement systems has been developed, the practical use is currently still pending. At the time it is possible to transfer the test procedure e.g. to bicycle and pedestrian paths, because the system approval can be carried out on the basis of results of ZbBz-certified systems, irrespective of the type of the traffic unit. This means that the database of the EMS Berlin can be equipped with trustworthy condition variables. Even by extending the MS Berlin e.g. including bike paths or further traffic facilities, the method allows a comparability of the data among each other due to known correlations. Questions of possible different requirements for built facilities can be reflected in the quality level of the assessment classes. Thus, they are independent from the condition recording and can also different be answered over the time.

REFERENCES

AASHTO (2013): Transportation Asset Management Guide: *Designed to Replace Pages of "Standard Specifications for Highway Bridges" and Contains Revisions to "Standard Specifications for Structural Supports for Highway Signs, Luminaires and Traffic Signals". A Focus on Implementation.* Washington, DC, 2013.

BASt (2019): *Prozessbeschreibung Zeitbefristete Betriebszulassung (ZbBz) und Systemprüfung (Sp) Von Schnellfahrenden Messsystemen zur Erfassung Der Substanzmerkmale (Oberfläche)* (Download by Bundesanstalt für Straßenwesen (BASt) »https://www.bast.de/DE/Strassenbau/Qualitaetsbewertung/Pruefungen/pdf/Prozessbeschreibung-Substanz.pdf?__blob=publicationFile&v=2«, last check on 2022-04-21; availability German only)

BASt (2020a): *Prozessbeschreibung Zeitbefristete Betriebszulassung (ZbBz) und Systemprüfung (Sp) von schnellfahrenden Messsystemen zur Erfassung der* Längsebenheit (Download by Bundesanstalt für Straßenwesen (BASt) »https://www.bast.de/DE/Strassenbau/Qualitaetsbewertung/Pruefungen/pdf/Prozessbeschreibung-l%C3%A4ngs.pdf?__blob=publicationFile&v=3 «, last check on 2022-04-21; availability German only)

BASt (2020b): *Prozessbeschreibung Zeitbefristete Betriebszulassung (ZbBz) und Systemprüfung (Sp) von Schnellfahrenden Messsystemen zur Erfassung Der Querebenheit* (Download by Bundesanstalt für Straßenwesen (BASt) »https://www.bast.de/DE/Strassenbau/Qualitaetsbewertung/Pruefungen/pdf/Prozessbeschreibung-quer.pdf?__blob=publicationFile&v=2«, last check on 2022-04-21; availability German only)

FGSV (2006/2018): *Zusätzliche Technische Vertragsbedingungen und Richtlinien Zur Zustandserfassung und – bewertung Von Straßen (ZTV ZEB-StB 06). Forschungsgesellschaft Für Straßen- und Verkehrswesen,* FGSV-Verlag, Köln, 2006; Stand 2018.

FHWA (2012): *Executive Brief: Advancing a Transportation Asset management Approach*, 2012.

Hajdin, Rade *et al.* (2020): *BIM-Erweiterung Durch Implementierung Der Nutzung Baustofftechnischer Daten Von Straßen und Brücken im AMS*; BIM4AMS; D-A-CH Call 2019 (running)

PIARC (2017): Asset Management Manual: A Guide for Practitioners. URL https://road-asset.piarc.org/en – last check on 2022-04-19

Stöckner, Markus; Rade, Hajdin; Markus König (2022): *BIM im Asset Management für Verkehrsanlagen Sachstand zur Forschung.* FGSV-OKSTRA Symposium, Hamburg, 2022.

Roads and Airports Pavement Surface Characteristics – Crispino & Toraldo (Eds)
© 2023 The Author(s), ISBN 978-1-032-55149-4

Comparing environmental impacts of alternative roads: Mechanical and sustainable considerations

G. Bosurgi
Department of Engineering, University of Messina, Messina, Italy

C. Celauro & G. Lo Brano
Department of Engineering, University of Palermo, Palermo, Italy

O. Pellegrino & G. Sollazzo*
Department of Engineering, University of Messina, Messina, Italy

ABSTRACT: In recent years, sustainability and "green" issues have gained increasing attention. When considering the environmental impacts of transport infrastructures, road pavements prove to have a major environmental impact, due to huge resource consumption (both materials and fuels), emissions, consequences on the ecosystems, etc. Consequently, researchers, governments and industries have focused on developing methods for evaluating the sustainability of the different technological solutions for ensuring sustainable development.

In this paper, different pavement solutions (including warm mixing technologies, WMA) have been considered and their performance evaluated in terms of both mechanical efficiency and sustainability. First, based on selected traffic and environmental conditions, mechanically equivalent pavements were designed, and then a Life Cycle Assessment (LCA), including production and construction phases, was performed. The numerical results proved that WMA, although it may ensure a reduction in mixing temperature, and thus in fuel consumptions during production, determines related environmental impacts and emissions that are still very marked as a consequence of the need for specific additives (i.e. zeolite) involved in this technology.

Keywords: LCA, Road Pavements, Environmental Impacts, WMA

1 INTRODUCTION

In recent years, the public opinion has been widely made aware by researchers of the emergency related to topics of sustainability and impacts on and damages to our planet and the available resources. As it is well known, sustainability means the possibility of economic and social development to be performed in compliance with the future needs of the next generations, preserving resources and quality of life on the earth (Jabareen 2008).

In terms of impacts on the environment, the transport sector has for sure a major role, owing to the huge volumes of involved resources (both raw materials and fuels among other things), the impacts on ecosystems, and emissions in all processes, such as production, construction, use, maintenance, and disposal. Among the various contributions, road pavements, which represent the core part of the strategical road networks all around the

---

*Corresponding Author: gsollazzo@unime.it

DOI: 10.1201/9781003429258-55

countries, contribute a great deal to the overall impacts. For example, road construction determines a contribution of up to 10% of the total greenhouse gases (GHG) emissions (i.e. a reference indicator for sustainability assessment).

However, owing to this importance, by changing the point of view, road pavements may become a strategical player in inverting the current trend. Consequently, increasing attention is focused on investigation of alternative solutions to traditional materials, methods and approaches, with the aim of increasing the sustainability of the sector and reducing consumption of energy and resources, together with waste production (Lee et al. 2010; Plati 2019). Such solutions implicitly also aim to assure the highest quality of the adopted materials and components, to maximize the mechanical performance of the pavement. This is obviously essential for preserving durability and, even more, safety and comfort of driving.

To evaluate the sustainability of different products and processes, a comprehensive analysis of emissions and impacts may be performed through a life cycle assessment (LCA). This methodology, regulated by the International Organization for Standardization (ISO) through the 14040 series publications (ISO 2006a,b), regards the whole life cycle of the products. In the road sector, an LCA can assess the sustainability of all the phases involved from production to disposal of the different products and components, including pavements. An LCA may take into consideration production of raw materials and mixtures, construction, effective exercise and final disposal, listing all the resources involved and accurately evaluating the impacts on the environment in "cradle-to-grave" approaches. This is not the only available perspective, but the analysis may be focused on specific steps only (such as production, construction, exercise) for specific assessments and evaluations.

Thus, several studies have been carried out to investigate, through LCA, impacts of traditional and alternative materials, additives, production technologies, construction, maintenance and rehabilitation processes (for example, Balaguera et al. 2018; Celauro et al. 2015, 2017; Farina et al. 2017; Franzitta et al. 2020; Gulotta et al. 2019; Santero et al. 2011a,b; Sollazzo et al. 2020). Similar research may represent references for practical applications of the methodology and to support decision-makers towards more sustainable choices and processes. Among the various solutions aiming to reduce emissions and impacts due to production and construction phases, the reduction of production and laying temperatures appears as a good opportunity. In this perspective, Warm Mix Asphalt (WMA) was introduced to represent a more sustainable option than traditional Hot Mix Asphalt (HMA). However, there is the need to perform specific and market/context-relevant comparative analyses, in order to specifically quantify emissions and impacts of the two products for sounder considerations and operative decisions in current practice. In any case, this goal is of particular interest, considering the size of the asphalt mixture market. For example, in 2018 in Italy, more than 26 million tons of asphalt mixtures were produced for maintenance and construction needs (SITEB 2019); in Europe, in 2017, over 250 million tons were produced (EAPA 2017).

In this paper, two pavement solutions (one with HMA and one with WMA) are compared and their performance evaluated in terms of both mechanical efficiency and sustainability. First, based on selected traffic and environmental conditions, mechanically equivalent pavements were designed, and then an LCA, including production and construction phases, was performed for sustainability evaluations. The numerical outcomes proved that WMA emissions and impacts are still comparable with those of HMA, due to the need for specific additives (i.e. zeolite) involved in this technology, despite the reduction in mixing temperature, and thus in fuel consumption during production.

After a brief discussion of the LCA framework, the following sections present first goal and scope definition and inventory analysis, and then a life cycle impact evaluation and result analysis for the two alternative and mechanically equivalent solutions are provided and discussed.

2 METHODOLOGICAL APPROACH

LCA is a strategical analytical approach useful in evaluating impacts on the environment of specific products, processes, and technologies. In general, LCAs are cradle-to-grave analyses of the products or services and include four basic steps:

1. Goal definition and scoping: identification of the goal of the study, the system boundaries, the functional unit (FU), the target audience, etc.;
2. Inventory analysis: quantification of input and output of the processes involved in the life cycle considered, in terms of resource consumption, waste flows, and emissions;
3. Impact assessment: quantification of the effects on the environment and the ecosystem of the FU considered;
4. Interpretation: processing of conclusions and strategic recommendations to improve the assessment and highlight the advantages and drawbacks of each specific process and product.

2.1 *Goal and scope definition*

The main goal of this study is to estimate the emissions and evaluate the environmental impacts of the production and construction phases of the pavements for a typical rural road in the Italian context. This study may be useful as a benchmark for the Italian, mainly, and Southern European scenario and for providing accurate results to be included in from "cradle-to-grave" LCAs of asphalt pavements. To achieve this goal, two alternative solutions were investigated and compared, also to highlight the effective advantages (in terms of emissions and impacts) of reducing mixing and laying temperatures of the asphalt concrete.

The study was performed in compliance with the LCA regulations provided by the international standards of series ISO 14040 (ISO 2006a,b) and by the Environmental Product Declaration (EPD) product category rules (EPD 2018). According to the reference methodology framework, input and output flows of each productive step are considered for the different processes in which they occur, considering an attributional approach (EU 2010).

2.2 *Functional unit and boundary conditions*

The investigated product is 1 km of asphalt concrete pavement for a C2 class road, according to Italian Standards. The road section is 9.5 m wide, including two lanes of 3.5 m and two shoulders of 1.25 m. The pavement is made up of four layers: a sub-base in unbound aggregates and three asphalt concrete layers (surface, binder, and base). According to the aims of the study, the asphalt layers are considered in the FU in two alternative compositions: HMA, as a traditional asphalt mixture, and WMA, as a mixture including zeolite to reduce mixing and laying temperatures from 165°C to 135°C. The raw materials included were limestone aggregates, bitumen, and zeolite. To compare the two solutions, they had to be mechanically equivalent in terms of their on-site performances.

In this regard, first the volumetric and mechanical features of the two solutions were analytically evaluated. The stiffness modulus was calculated through the well-known Witzack equation, for different weather conditions (from spring to winter). A target life of 20 years was considered. Based on reliable traffic scenario forecasting (average daily traffic 3200 veic/d, 10% trucks and 1% rate of growth), the traffic actions were evaluated and used to perform a mechanistic calculation using KenPave software (Huang 1993). The calculation hypotheses included:

- Subgrade in granular soil with CBR = 25 and elasticity module of 2500 kg/cm2;
- Sub-base in unbound limestone aggregates, with characteristics in compliance with Chapter 1 of the Italian Special Tender Specifications for road works (MIT 2001);

- Aggregates for HMA and WMA consisting in a set of large and fine limestone and basalt aggregates and fillers. Their size distribution is differentiated for each layer, according to the design fuses of the Special Tender Specifications;
- Traditional Bitumen, type 50/70;
- Attack or anchorage hand consisting of a cationic bituminous emulsion (65% - 70% binder) with a minimum residual bitumen content of at least 1 kg/m^2;
- Zeolite for WMA production: i.e. a mineral with a crystalline and microporous structure.

The design thickness for the 4 layers is shown in Table 1, while the volumetric features of the various mixtures for the two FU are plotted in Table 2. In Table 3 the specific gravity and mass values of the mixtures and of the various components are provided. Then, Table 4 provides the seasonal variation of |E*| for the asphalt mixtures of the two pavements, according to their thermal susceptivity evaluated according to Huang (1993).

Table 1. Layer composition of the two pavements.

Layer	Thickness [cm]	HMA	WMA
Surface	5	HMA	HMA + 5% zeolite
Binder	6	HMA	HMA + 5% zeolite
Base	10	HMA	HMA + 5% zeolite
Subbase	20	Unbound aggregates	Unbound aggregates

Table 2. Bitumen and air voids for the asphalt mixtures of the two pavements.

	HMA		WMA	
Layer	b (%)	v (%)	b (%)	v (%)
Surface	5.0	5.0	5.0	4.5
Binder	4.5	6.0	4.5	5.0
Base	4.2	7.0	4.0	5.5

Table 3. Specific gravity and mass values for the various mixtures.

Pav	Layer	γ_{agg} [kg/m^3]	γ_{bit} [kg/m^3]	γ_{tot} [kg/m^3]	γ [kg/m^3]	Tot mass [kg]	Agg. mass [kg]	Bit. gross mass [kg]	Net bit. mass [kg]	Zeo. mass [kg]
HMA	Surf.	2870	1035	2646.56	2514.23	1194260.79	1137391.23	56869.56	56869.56	-
	Binder	2870	1035	2666.43	2506.44	1428671.25	1367149.52	61521.73	61521.73	-
	Base	2870	1035	2678.58	2491.08	2366527.39	2271139.53	95387.86	95387.86	-
WMA	Surf.	2870	1035	2646.56	2527.47	1200546.37	1143377.50	57168.87	54310.43	2858.44
	Binder	2870	1035	2666.43	2533.11	1443869.88	1381693.66	62176.21	59067.40	3108.81
	Base	2870	1035	2686.79	2539.01	2412063.27	2319291.61	92771.66	88133.08	4638.58

Table 4. Seasonal variation of |E*| in MPa for the asphalt mixtures of the two pavements.

	Surface				Binder				Base			
Pav.	Spr.	Sum.	Aut.	Win.	Spr.	Sum.	Aut.	Win.	Spr.	Sum.	Aut.	Win.
HMA	7416	2343	4621	11176	5391	1718	3370	8102	5157	1642	3223	7758
WMA	4343	2140	3103	6381	4206	2072	3005	6175	4643	2263	3302	6842

For defining the system boundaries, generally the whole life cycle of asphalt pavements includes 5 stages (Figure 1): production, construction, use, maintenance, and disposal. In this research, the analysis focuses on the production and construction stages. Therefore, the considered processes are the following: bitumen supply (including oil extraction and its transformation); aggregate supply (including extraction and transformation); zeolite supply; aggregate, bitumen, and zeolite transportation to the production plant; plant production processes; transportation to the construction site; construction activities for adequate laying.

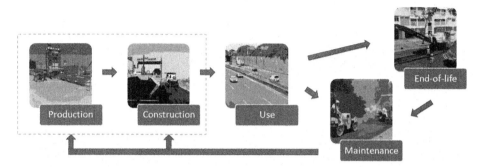

Figure 1.   Scheme of life cycle of asphalt pavements.

2.3   *Emissions and impact assessment methodologies*

Impact calculations were based on the impact categories and characterization factors of the EPD 2018 method. The following five emission substances were calculated according to the above method in order to describe the performance of the considered FU: 1) CO: carbon monoxide (g); 2) Pb: lead (mg); 3) Hg: mercury (µg); 4) NO_x: nitrogen oxides (g); 5) PM_{10}: particulates (g).

Furthermore, for deepening the analysis and also evaluating some reference values directly related to the effects on the environment of the FU, the following two impact categories were evaluated: 1) GW: Global warming (100 years) (kg CO2eq); 2) AD: Abiotic depletion (kg SBeq).

3   DATA ANALYSIS AND LIFE CYCLE INVENTORY ANALYSIS

3.1   *Production phase*

First, the production stage of the asphalt concrete was investigated. The mix designs of the two alternatives mixtures were defined according to the Italian standards and literature references. In this regard, HMA included only limestone aggregates and bitumen, while in the WMA 5% zeolite was considered for reducing mixing and laying temperatures (Memon *et al.* 2020).

To consider emissions and impacts of the production phase, data regarding the quantities of the materials involved, transport distances and energy required at the asphalt plant are needed. Table 5, for example, lists the reference distances for a hypothetical ideal "average" construction site. Considering the energy required for asphalt plant operations, owing to the lack of related primary data, it was evaluated based on average data obtained from literature (Bueche & Dumont 2012; Chehovits & Galehouse 2010; Franzitta *et al.* 2020; Jiang *et al.* 2021). In detail, 320 MJ/t is assumed as a reference value for HMA, while 278.5 MJ/t is considered for WMA, considering around 13% reduction, according to literature (Calabi-Floody *et al.* 2020).

Table 5.   Transportation distances in the production and construction phases.

Transport	Phase	Distance [km]
Quarry – asphalt plant	Production	35
Refinery – asphalt plant	Production	172.5
Fuel production site – asphalt plant	Production	10
Zeolite plant – asphalt plant	Production	10
Demolition site (RAP) – asphalt plant	Production	10
Quarry – construction site	Construction	10
Asphalt plant – construct<ion site	Construction	10

3.2   Construction phase

Considering the construction phase, first the typical machinery pool available for potential companies involved in these working activities was defined. The list of the selected machineries is provided in Table 6, in which for each piece of equipment the engine power and the estimated fuel consumption are provided. Then, the basic activities required to perform the pavement construction are identified and listed in Table 7. For each activity, based on the estimated productivity rates and the performance features of the equipment involved, the total required times and the litres of fuel consumed are provided.

Table 6.   Machinery details and fuel consumption.

Machinery	Code	Model	Engine Power [HP]	Fuel consumption [l/h]
Excavator	E	CAT 352FT	417.05	84.91
Truck	T	Mercedes Arocs 3240 (8x4)	421.00	20.00
Dozer	D	CAT 1150M	138.12	28.12
Roller compactor (dynamic)	Rd	CAT CS79B	178.61	36.36
Paver	P	CAT AP655F	171.00	34.81
Roller compactor (static)	Rs	CAT CS74B	178.61	36.36
Emulsion sprayer	S	IVECO 150E24	132.00	26.87

Table 7.   Construction activities.

Construction phase	Machineries	HMA Time [h]	HMA Fuel cons. [l]	WMA Time [h]	WMA Fuel cons. [l]
Site cleaning	1xE, 3xT	2.3	147.5	2.3	147.5
Topsoil removal	1xD, 1xE, 20xT	49.2	10708.6	49.2	10708.6
Sub-grade compaction	1xRd	11.3	412.6	11.3	412.6
Sub-base construction	20xT, 1xD, 1xRd, 1xS	29.6	5096.7	29.6	5096.7
Base construction	7xT, 1xP, 1xRs, 1xS	144.0	8228.9	146.7	8385.2
Binder construction	7xT, 1xP, 1xRs, 1xS	97.0	5319.4	98.0	5374.9
Surface construction	7xT, 1xP, 1xRs	80.6	4461.9	81.0	4485.4

3.3 *Unitary emissions and impacts*

Numerical calculations rely on secondary data derived from the Ecoinvent database (Wernet *et al.* 2016). Table 8 provides the unitary emissions values per unit of mass (zeolite, bitumen, and aggregates), of energy (plant and machineries), and of distance (in tkm). Similarly, Table 9 lists the unitary values for the same units.

Table 8. Unitary emissions for the considered components.

| Substance | Unit | Per unit of mass [kg] | | | Per unit of energy [MJ] | | For unit of distance [tkm] |
		Zeolite	Bitumen	Aggregates	Plant	Machineries	Transport
CO	g	4.96	0.64	0.02	0.70	0.32	0.34
Pb	mg	9.24	0.12	1.59E-03	0.01	0.01	0.29
Hg	μg	288.79	19.64	0.06	1.04	1.79	3.06
NOx	g	12.02	1.54	0.05	1.42	1.08	0.73
PM10	g	1.44	0.04	0.05	1.42E-03	0.01	0.06

Table 9. Unitary impacts for the considered components.

| Environ. Cat. | Unit | Per unit mass [kg] | | | Per unit of energy [MJ] | | Per t-km [tkm] |
		Zeolite	Bitumen	Aggr.	Plant	Machineries	Transp.
GW	kg CO2 eq	4.71	0.43	2.41E-03	0.09	0.09	0.16
AD	kg Sb eq	1.35E-04	8.88E-08	2.02E-08	8.00E-09	1.45E-08	4.51E-06

4 NUMERICAL RESULTS AND DISCUSSIONS

Based on the methodology described in the previous paragraphs, the various quantities and unitary emissions and impacts, the overall emissions and impacts of the two compared pavement solutions were calculated. To simplify interpretation of the results and to utilize the comparative analysis, the results are normalized with respect to the values obtained for HMA. The results in terms of emissions are provided in Figure 2, while the impacts (Global Warming and Abiotic Depletion) are shown in Figure 3.

It is clear that, despite the common belief, the selected WMA technology (representing the most common solution) does not appear a more sustainable solution than HMA. On the contrary, in fact, considering the various emissions, the CO and NOx values are almost comparable, while for WMA Pb, Hg and PM10 may be considered more critical. Pb emissions, for instance, are 41% higher for WMA than for HMA. Similarly, even considering impacts, the same considerations emerge. Indeed, both GW and AD, instead of decreasing as expected, significantly rise in the WMA scenario. AD, in particular, reaches almost 50% more for WMA than for HMA. This result is due to the nature of zeolite and the consequences of its use and production on ecosystems and in terms of resource availability. In this connection, considering production values, adoption of zeolite (despite its very reduced quantity) determines an extraordinary increase in AD for raw materials, becoming more than 13 times higher for WMA than for HMA.

These results may be helpful for evidencing the need to perform careful investigations and analyses on different materials and technologies before any practical application, in order to be effectively aware of their sustainability or, at least, to see whether the emissions and impacts caused are acceptable in the overall perspective. Furthermore, it is also strategical to

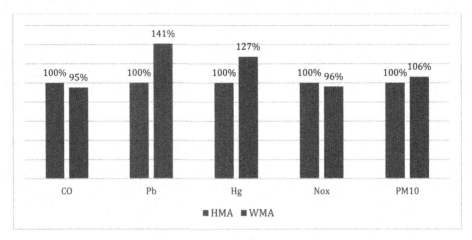

Figure 2.   Emissions for HMA and WMA normalized with respect to HMA values.

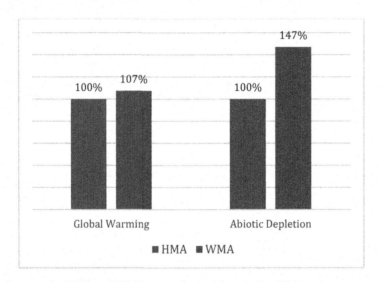

Figure 3.   Impacts for HMA and WMA normalized with respect to HMA values.

identify the most critical emissions and impact categories to consider in the LCA for the various products and, mainly, to rank and compare the various numerical results for an effective interpretation of the most dangerous outcomes. Finally, as evidenced also in previous studies, the quality and reliability of the inventory data is fundamental for ensuring accuracy and effectiveness of the assessment results. In this regard, relying on accurate primary data derived from specific survey campaigns related to the assessment boundaries and scope is absolutely preferable respect to secondary data from literature and generic databases.

5   CONCLUSIONS

In this paper, a comparative LCA of two alternative and mechanically-equivalent pavements for rural roads (C2 class according to Italian standards) was performed to verify whether the expected more sustainable solutions (such as WMA) are effectively preferable to traditional ones (such as HMA). The numerical results proved that, despite the common expectations and feelings (for professional involved experts too), the selected WMA technology (representing the most common solution) does not represent a more sustainable material than HMA, when the analysis is extended to multiple perspectives.

In this connection, in current practice WMA is considered preferable to HMA owing, above all, to the reductions in mixing and laying temperatures and the consequent reduction in emissions. However, as evidenced in this research, when the analysis is extended and deepened, other initially hidden consequences and impacts may become evident. In particular, zeolite (despite the very reduced quantities involved) may determine major effects on the sustainability rating of WMA. This may evidence the need for performing more accurate and complete evaluations before selecting this technology too, in order to reduce the potential unexpected drawbacks.

The analysis also showed that it is important to perform (and perhaps officially regulate and impose, by specific policies) appropriate LCA analyses before accepting products and materials, to effectively verify that those considered sustainable are effectively so. Furthermore, it is clearer that it is important to deeply analyse the different usable additives and thus enrich and extend available databases for LCAs with similar detailed and reliable data, as the additives may play a major role in the overall analysis.

REFERENCES

Balaguera A., Carvajal G.I., Jaramillo Y.P.A., Albertí J. and Fullana-I-Palmer P. 2018. Technical Feasibility and Life Cycle Assessment of an Industrial Waste as Stabilizing Product for Unpaved Roads, and Influence of Packaging. *Science of the Total Environment*, 651, 1272–1282.

Calabi-Floody A.T., Valdés-Vidal G.A., Sanchez-Alonso E. and Mardones-Parra L.A. 2020. Evaluation of Gas Emissions, Energy Consumption and Production Costs of Warm Mix Asphalt (WMA) Involving Natural Zeolite and Reclaimed Asphalt Pavement (RAP). *Sustainability*, 12(16), 6410.

Celauro C., Corriere F., Guerrieri M. and Casto B.L. 2015. Environmentally Appraising Different Pavement and Construction Scenarios: A Comparative Analysis for a Typical Local Road. *Transportation Reseach Part D: Transportation and Environment*, 34, 41–51.

Celauro C., Corriere F., Guerrieri M., Lo Casto B. and Rizzo A. 2017. Environmental Analysis of Different Construction Techniques and Maintenance Activities for a Typical Local Road. *Journal of Cleaner Production*, 142, 3482–3489.

EAPA. 2017. *EAPA'S Position Statement on the Use of Secondary Materials, by-Products and Waste in Asphalt Mixtures*; European Asphalt Pavement Association: Brussel, Belgium.

EDP. 2018. *Asphalt mixtures (Europe, Australia). Product Category Rules (PCR) of the Environmental Product Declaration (EPD)*. The International EPD®System.

EU, 2010. *ILCD Handbook. International Reference Life Cycle Data System. General Guide for Life Cycle Assessment Provisions and Action Steps*. Publication of the European Union: Luxembourg, 2010.

Farina A., Zanetti M.C., Santagata E. and Blengini G.A. 2017. Life Cycle Assessment Applied to Bituminous Mixtures Containing Recycled Materials: Crumb Rubber and Reclaimed Asphalt Pavement. *Resources, Conservation and Recycling*, 117, 204–212.

Franzitta V., Longo S., Sollazzo G., Cellura M. and Celauro C. 2020. Primary Data Collection and Environmental/Energy Audit of Hot Mix Asphalt Production. *Energies*, 13 (8), 2045.

Gulotta T., Mistretta M. and Praticò F. 2019. A Life Cycle Scenario Analysis of Different Pavement Technologies for Urban Roads. *Science of the Total Environment*, 673, 585–593.

Huang Y.H. 1993. *Pavement Analysis and Design*. Prentice-Hall, Incorporated. Englewood Cliffs, NJ United States.

ISO, 2006a. *Environmental Management-Life Cycle Assessment-Principles and Framework; ISO 14040_2006 (E)*; International Organization for Standardization: Geneva, Switzerland.

ISO, 2006b. *Environmental Management-Life Cycle Assessment-Requirements and Guidelines; ISO 14044_2006 (E)*; International Organization for Standardization: Geneva, Switzerland.

Jabareen Y. 2008. A New Conceptual Framework for Sustainable Development. *Environment, Development and Sustainability*, 10(2), 179–192.

Jiang Q., Wang F., Liu Q., Xie J. and Wu S. 2021. Energy Consumption and Environment Performance Analysis of Induction-heated Asphalt Pavement by Life Cycle Assessment (LCA). *Materials*, 14, 1244.

Lee J.C., Edil T.B., Tinjum J.M. and Benson C.H. 2010. Quantitative Assessment of Environmental and Economic Benefits of Recycled Materials in Highway Construction. *Transportation Research Record, Journal of the Transportation Research Board*, 2158, 138–142.

MIT. 2001. Norme Tecniche di Tipo Prestazionale per Capitolati Speciali d'Appalto. Ricerca realizzata nell'ambito della convenzione tra l'Ispettorato per la Circolazione e la Sicurezza Stradale ed il CIRS, Roma.

Plati C. 2019. Sustainability Factors in Pavement Materials, Design, and Preservation Strategies: A Literature Review. *Construction and Building Materials*, 211, 539–555.

Santero N.J., Masanet E. and Horvart A. 2011a. Life-cycle Assessment of Pavements. Part I: Critical review. *Resources, Conservation and Recycling*, 55, 801–809.

Santero N.J., Masanet, E. and Horvath A. 2011b. Life-cycle Assessment of Pavements Part II: Filling the Research Gaps. *Resources, Conservation and Recycling*, 55, 810–818.

SITEB. 2019. *L'assemblea annuale di SITEB*. Rass. Bitume, 92, 19–25. (In Italian)

Sollazzo G., Longo S., Cellura M. and Celauro C. 2020. Impact Analysis Using Life Cycle Assessment of Asphalt Production from Primary Data. *Sustainability*, 12 (24).

Wernet G.; Bauer C.; Steubing B.; Reinhard J.; Moreno-Ruiz E.; Weidema B.P. 2016. The Ecoinvent Database Version 3 (Part I): Overview and Methodology. *International Journal of Life Cycle Assessment*, 21, 1218–1230

Safety and risk issues

Detection of irregularities within the asphalt surface behind the paver screed using image analysis

Leandro Harries*, Stefan Helfmann & Stefan Böhm
Department of Transportation Infrastructure Engineering, Technische Universität Darmstadt, Darmstadt, Germany

ABSTRACT: Asphalt pavers are used in road construction projects all over the world. The goal of the BASt research project Robot-Strassenbau 4.0 (FE: 88.0159/2017) was to optimize the paving process and increase occupational health and safety by partially automating the activities of particularly vulnerable employees at the paving screed.

One of these diverse activities consists of observing the surface of the asphalt mix being paved directly behind the screed and, if necessary, intervening in the screed control system.

In the course of these investigations, characteristic values for potentially visually detectable irregularities and segregations were first defined and classified into three super-ordinate categories (segregations, scrapings or imprints and grain fragmentation). Sample plates were prepared in the laboratory that exhibited equivalent failure patterns.

The surfaces of the asphalt sample plates were recorded using a flash/camera system. Here, the flash light was aligned with a defined angle. The viewing angle of the camera was quasi orthogonal. Due to the introduced lateral flash light, irregularities in the surface cast a shadow, which is used in the image-analytical evaluation to detect the irregularities. The lower illuminance and, consequently, the less pronounced shadows cast at considered points located farther from the source of the flash light can be taken into account by means of the known angle of incidence and the squared distance.

For the image analysis of the asphalt surface, a tool was developed that is capable of cropping the image captures and, in the case of non-orthogonal captures, transforming the image. Using the gray value view, an average threshold value is calculated and illumination values are converted using a curve fitting.

To validate the measurement program, the flash/camera system was attached to the asphalt paver during a field trail. The flash system used is capable of "flashing over" shadow casts as a result of solar radiation. The irregularities in the surface could be successfully detected by the image analysis.

Keywords: Paving, Image-Detection, Irregularities

1 OBJECTIVE OF THE STUDIES

Paving hot mix asphalt (HMA) with a paver is a highly complex, manpower-intensive process. Minor changes in the machine parameters or in the asphalt mix can already lead to paving errors. Some of these paving errors become apparent as irregularities in the surface. These, in turn, can have a wide variety of causes. Depending on the type and level of the

---
*Corresponding Author: sekretariat@vwb.tu-darmstadt.de

DOI: 10.1201/9781003429258-56

irregularity, the performance properties of the asphalt pavement can be deteriorated (Rosauer 2010). Segregation of the asphalt mix, for example, causes a locally different composition of the asphalt mix as well as a locally different compaction of the asphalt layer, so that early damage can occur and as a result the service life of the asphalt layer can be shortened. Economic considerations further revealed that, depending on the degree of segregation, 10 % to almost 50% additional costs (based on the production costs of the asphalt layer) may then be incurred over the service life as a result of additional measures that become necessary (Stroup-Gardiner & Brown 2000). Any irregularities that occur must be recognized by the paving personnel and appropriate actions must be taken. The classification of the irregularities and the quality of the action to be taken depend essentially on the experience of the paving personnel. The aim of this project is to use image analysis of the surface to automatically detect irregularities. In addition, a corresponding catalogue of measures is to be developed to obtain possible recommendations for immediate action by the screed setting personnel to eliminate the irregularities. The recording of the surface is to be realized by a system consisting of a camera and a flash unit. Previous approaches like in (Xun et al. 2021) and (Lin et al. 2019) use a camera system without flash. This has the crucial disadvantage that the images are strongly dependent on the prevailing sun and shadow conditions and thus no accurate environmental independent evaluation of the surface can be conducted.

2 SURFACE IRREGULARITIES

When considering irregularities in the asphalt surface as a result of paving errors, a fundamental distinction must be made between segregation, scrapings or imprints and grain fragmentation.

In the case of segregation, a distinction must also be made between mechanical and thermal segregation. In general, asphalt mixes with a wide grain size distribution and lower binder content tend to segregate more easily (Brown et al. 1989). First segregation phenomena can already occur when filling the asphalt mix into the silo as well as during transport of the asphalt mix. When paving with feeders, remixing of the asphalt mix occurs, so that the risk of segregation is significantly reduced (Ulrich 2009). If the paver or screed is set and operated incorrectly, avoidable segregation occurs at different positions in the surface of the paved layer. Segregation in the central area is mainly caused by a too low HMA level in front of the paver's screed. A porous, rough stripe appears in the pavement (cf. Figure 1 left). With increasing working width, segregation phenomena may also occur in the

Figure 1.   Left - segregation (longitudinal); Right - segregation (bitumen accumulation) (Fliegl 2017).

outer area (VÖGELE 2005). Missing material guide plates and inadequately adjusted material feed of the level controllers favor these. Segregation can also occur at isolated points in the surface. These are either spots with bitumen accumulation (cf. Figure 1 right), which appear when the vibration unit "pulls" bitumen upwards due to a frequency that is set too high, or due to thermal segregation that has already occurred during transport or in the paver's bunker. Also, an insufficiently heated screed can cause fine aggregate accumulations on the screed front wall or the tamper which then detach in irregular intervals. (Kappel 2016; Utterodt 2013)

Imprints can occur when the paver stops during paving (cf. Figure 2 left). The screed sinks into the HMA, the weight of the screed presses into the material and creates an imprint with the rear edge of the screed across the pave width. However, the imprint can already be caused by improper docking of the tractor units when paving without a feeder. When operating with vario-screeds at a very high angle of attack, imprints in the form of longitudinal stripes may occur due to the uneven height levels of the extending units and the base screed. (Kappel 2016; VÖGELE 2005)

The risk of grain fragmentation exists in particular when paving low layer thicknesses. Grain fragmentations (cf. Figure 2 right) are easily recognizable when the aggregate color or a whitish flour appears on the surface, although all components were coated with black bitumen. The reason for this is that the compaction energy of the paving screed is too high for the paving thickness, which causes the grains to be crushed. Mostly it is due to a too large tamper stroke or a too high tamper frequency. (Cai et al. 2021; Stroup-Gardiner & Brown 2000)

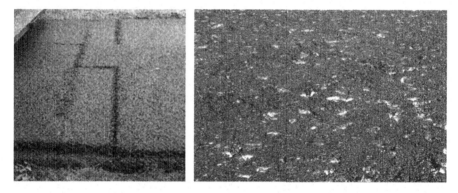

Figure 2. Left - imprint of the screed due to paver stoppage; Right - grain fragmentation due to over compacting (Kappel 2016).

Table 1, all irregularity patterns from the literature mentioned are collected. Possible causes for each irregularity are also listed. Some of these causes cannot be eliminated when paving with the paver because the problems have already occurred in the mixing plant or during transport of the HMA (no immediate action possible). For correctable causes due to paver maladjustment, appropriate immediate actions are suggested in the last column.

Table 1. Classification of possible irregularities during paving with the paver and suggested immediate actions to eliminate the irregularity.

Type of irregularity	Positioning of the irregularity	Potential causes of the irregularity	Proposed immediate action
Segregation (thermal/mechanical)	Isolated spots	Mixing plant delivers highly inhomogeneous material	No immediate action possible
		Cooling of the material during transport	No immediate action possible
		Insufficient heating capacity of the screed heating system	Adjust heating temperature of screed
		Vibratory frequency set too high (bitumen accumulation)	Adjust vibratory frequency
	Transverse to the direction of paving	Material remains in the bunker for too long	No immediate action possible
	Longitudinal to the direction of paving (centered)	Roof profile too steep	Adjust roof profile
		HMA level in auger channel too low	Increase material supply or reduce paving speed
		Choice of auger blades unfavorable	No immediate action possible
	Longitudinal to the direction of paving (in the area of the screw bearing)	Auger level too low	Adjust auger level
Scrapings /Imprints	Longitudinal to the direction of travel (between the extending part and the base screed)	Angle of attack too steep	Adjust screed tow points
	Longitudinal to the direction of paving	Inadequate cleaning of the screed before/after operation	No immediate action possible
	Transverse to the direction of paving	Stoppage of the paver	No immediate action possible
Grain Fragmentation	Isolated spots	Compaction energy too high for layer thickness	Adjust tamper resp. vibratory unit
		Maximum grain size too large for paving thickness	No immediate action possible

3 PHOTOMETRIC PARAMETERS

In order to assess an area A with regard to its "exposure", the illuminance E_v measured in lux can be used. This is a photometric quantity which is adapted to the sensitivity of the human eye and is defined as the luminous flux Φ_v measured in lumens per illuminated area A and is calculated according to equation (1).

The luminous intensity I_v on the other hand is calculated according to equation (2) as quotient of the luminous flux Φ_v and the solid angle Ω. A transformation of the equations into each other yields the illuminance E_v depending on the angle of incidence ε and the distance r to the light source (cf. equation (3)).

$$E_v = \frac{\Phi_v}{A} \tag{1}$$

$$I_v = \frac{\Phi_v}{\Omega} \tag{2}$$

$$E_v = \frac{I_v}{r^2} \cos(\varepsilon) \tag{3}$$

$$\Omega = \frac{A}{r^2} \qquad (4)$$

The solid angle depends on the geometry of the flash reflector (cf. Figure 3 right) and is a measure of the "focusing" of the luminous flux. To simplify the calculation, a spherical light source may be assumed. Accordingly, it is calculated according to equation (4).

To experimentally determine the luminous flux E_v, the flash unit is set up in a darkened room (cf. Figure 3 left) and a discrete area is exposed (cf. Figure 3 middle). The illuminance E_v in lux is determined with a light meter. The luminous flux in lumens can then be determined by transforming equation (1).

According to the experimental setup shown in Figure 3, an area of approximately 1.23 m2 is exposed. The measurement of the light meter results in an illuminance of 38.531 lux (for comparison: the illumination intensity with a clear sky and a sun inclination of $16°$ is 20.000 lux (DIN e.V (Publ.) 1985)). The luminous flux is thus calculated as:

$$E_v A = \Phi_v = 38.531 \cdot 1.23 = 47.39 \; Lumen$$

To determine the luminous intensity, I_v, first calculate the solid angle according to equation (4). The exit area according to Figure 3 (right) corresponds to $A = 0.013 \; m^2$ where the orthogonal distance to the (assumed) point light source is $r = 0.095 \; m$. Therefore, the solid angle results to:

$$\Omega = \frac{A}{r^2} = \frac{0.013}{0.095^2} = 1.44$$

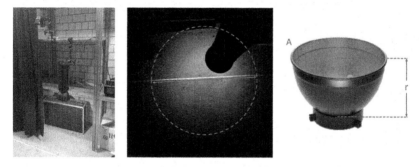

Figure 3.   Left - experimental setup; middle - illuminated area; Right - exit area and light source distance of the flash reflector.

Figure 4 shows the described phenomenon pictorially. On the left of the picture, one can see a block with lateral incidence of light, due to the surrounding sunlight. This can be clearly seen in the resulting shadow cast. In the right picture, this shadow cast can be "over-flashed" with the help of the flashlight. Irregularities in the area of the shadow could thus be analysed. However, the use of the flash results in the characteristically radially decreasing exposure surface, which can be eliminated by the method described below.

A uniform exposure of the surface is fundamentally important for the capturing. This is the only way to guarantee that the evaluation of images only must be calibrated once and

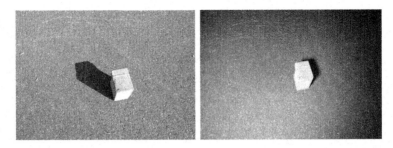

Figure 4.   Left - brick without flash; Right - brick with flash.

that external influences such as shadows do not affect the result. It is therefore advisable to work with strong flash units that guarantee uniform surface illumination.

4   IMAGE ANALYSIS

A target-oriented effective approach is the two-dimensional location of larger occurrences of black pixels. Black pixels are counted row by row and column by column. A visual evaluation in Figure 5 shows clusters of black pixels in the form of spikes.

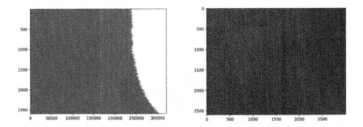

Figure 5.   Adding up gray values row by row.

For a localization, a metric must now be determined that indicates what constitutes an installation irregularity in an image. The mean and median values are suitable for initial tests. Since paving errors should generally be an exception and binarized images of the asphalt surface should ideally be very homogeneous, both metrics should allow deviations to be detected. Here, the mean value is more sensitive to strong, individual outliers. Values that are above these metrics can be used to identify areas with an excess of black pixels. If we look at the sums of gray values for each pixel row, we see that they increase continuously from the top to the bottom. This is due to the decrease in illumination from the flash. Therefore, this effect has to be eliminated.

To achieve this, first a captured image with a homogenous surface and thus no vertical anomalies was picked out and converted into a gray scale image. In this way, only one value per pixel has to be considered, while the relevant brightness information is preserved. However, binarization is not used, since this would result in the loss of information that is needed in the subsequent process. Adding up the resulting gray values row by row (cf. Figure 6 left) produces a gradient similar to equation (3). With the underlying row sums, the parameter a can now be determined. Figure 6 (left) shows the resulting curve in orange. Once the parameter of the function is determined, it can be used for all future captures as long as the strength of the flash system used remains constant. For the correction of the values in a row, the quotient of a base value and the output of the determined function of this row can now be used. Each pixel value of a row is divided by the quotient corresponding to this row, which brings all values close to the base value. The lowest row sum was chosen as the base value. This ensures the lowest influence of the flash apparatus. Figure 6 (right) shows an image before and after correction.

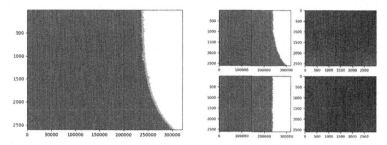

Figure 6. Left - Row sums with approximated function orange; Right-top - comparison before; right-bottom - after the correction.

Considering the distribution of the column sums, the paving irregularities within the asphalt surface can be visually recognized. Using a peak detection algorithm provided by the Python library SciPy (SciPy community 2022) the irregularities can be detected autonomously as shown in Figure 7. The red lines indicate the area of the irregularity. For calibrating the image detection process several variables have to be taken into account. Here, the parameters *prominence* and *width* were used for calibration. *Prominence* determines how far data points have to protrude from the rest to be considered as a peak while width sets a minimum width a peak has to have.

5 LABORATORY ANALYSIS OF IRREGULARITY PATTERNS

In order to systematically analyze irregularity patterns within the pre-compacted asphalt surface, various asphalt slabs (cf. Table 2) were produced in the laboratory using the rolling sector compactor and the asphalt mix AC 11 DS (according to TL Asphalt-StB (FGSV 2007a)) which was also used for field testing. For this purpose, the compaction program "pre-compaction" (FGSV 2007b) was carried out with which a degree of compaction k (according to TP Asphalt-StB Part 8 (FGSV 2012)) of approx. 90 % is achieved. This corresponds approximately to the pre-compaction of a tamper-vibratory (TV) screed. Thus, the

Table 2. Irregularity patterns for laboratory asphalt slabs.

Irregularity	Slab Nr.	Position/Style
Scrapings/Imprints	1.1	Longitudinal scraping
	1.2	Longitudinal imprint
Segregation	2.1	Longitudinal segregation (grain distribution)
	2.2	Spot segregation (grain distribution)
	2.3	Spot segregation (bitumen accumulation)
Fraction	3.1	Spot fraction
	3.2	Longitudinal fraction

properties of the surface appearance should be equivalent to those of the surface paved in situ. The asphalt slabs produced were then captured using the camera-flash system described above and subsequently evaluated.

Figure 7 shows all seven results of the examination of the asphalt slabs produced in the laboratory. As already described, the red markings indicate the detected area of irregularities. With the exception of Slab 2.1, irregularities could be detected for all slabs. The coarse grain accumulation in the longitudinal direction of Slab 2.1 is also difficult to detect with the human eye. This is aggravated by the fact that, due to the comparatively small areas observed in the laboratory, the scatter in pixel counting is relatively large. This phenomenon should only occur to a lesser extent in in-situ measurements. In order to distinguish punctual irregularities from irregularities in longitudinal direction, these oughts to be detected in vertical as well as in horizontal direction. The results of Slab 2.3 show that the punctiform bitumen accumulation was detected.

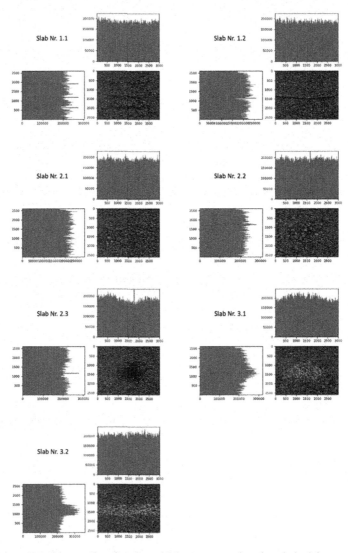

Figure 7.   Image detections results of evaluated laboratory produced asphalt slabs.

6 FIELD TEST

The test paving was carried out in Limburg, Germany, on MOBA Mobile Automation AG's test site using a BF 700 C-2 paver from FA BOMAG and a AC 11 DS (according to the guideline TL Asphalt-StB (FGSV 2007a)) which was also used for laboratory testing.

The emergence angle of the flash setup (cf. Figure 8) result from the actual experimental setup to $\varepsilon = 51°$. The distances from the light source to the point of emergence can be calculated from the hypotenuse theorem. Since the light source is not orthogonal to the emergence surface, it is not a circular but an elliptical surface that is illuminated. In the center of these ellipses, the illuminance is calculated according to equations (2) and (3) as follows:

$$E_{v1} \approx \frac{\Phi_v}{\Omega(h_1^2 + b_1^2)} cos(\varepsilon_1) \approx \frac{47.39}{1.44(1,10^2 + 2,05^2)} cos(51) \approx 3.826 \, Lux$$

Due to the increased distance (compared to laboratory conditions) between the flash source and the illuminated surface, the illuminance is below the illumination intensity of the sun under clear skies and a steep sun inclination. In order to perform a reliable image analysis under all weather conditions, a more powerful flash unit would be desirable.

Figure 8.   In-situ camera system setup.

Before the image analysis can be performed, the acquired capture must be transformed (cf. Figure 9). A corresponding tool is provided by the library OpenCV (OpenCV community 2022). For the correct localization of an irregularity, the pseudo-orthogonal image is required (cf. Figure 9, right). Alternatively, a much longer boom arm would have been necessary that could directly capture an orthogonal image. However, this would have been more cost-intensive and would have caused difficulties with smaller paving radii due to tilting behind the paver.

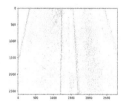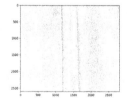

Figure 9.   Left: gray-scale image capture (untransformed); Right: gray-scale image capture (orthogonally transformed).

As can be seen in Figure 10, both the irregularity in the longitudinal direction and the spot irregularity were successfully detected. The scattering due to the individual grains is much less significant due to the much larger area. The detection of irregularities in Figure 10 (right) in the vertical and horizontal directions indicates that the irregularity occurs at a single spot. In contrast, the imprint in the longitudinal direction (cf. Figure 10 left) can be seen over the entire image and therefore a peak was only detected in the horizontal direction.

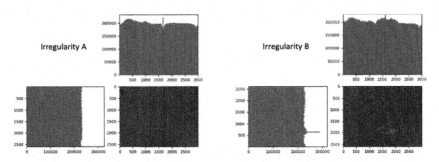

Figure 10.   Image detections results of evaluated in-situ pre-compacted asphalt surface; left – longitudinal irregularity; right – spot irregularity.

7   CONCLUSIONS

In this paper, surface irregularities that can occur while operating with a paver were first defined and categorized. Using the rolling sector compactor and AC 11 DS as HMA, asphalt slabs were produced in the laboratory and evaluated by image analysis, which display irregularities of the defined categories. For validation purposes, irregularities were provoked in an in-situ test with an actual paver and the matching asphalt mix AC 11 DS which were subsequently evaluated. In summary, the following conclusions can be drawn.

1. Irregularities in the surface when paving with the paver can be classified into the following categories: segregations, scrapings or imprints and grain fragmentation.
2. The developed method for detecting irregularities within the pre-compacted asphalt surface can reliably detect them under laboratory conditions as well as under in-situ conditions.
3. Sufficient "over-flashing" of shadows caused by sunlight is possible. However, for larger distances from the light source to the exposed area, very powerful flash units are required for this purpose.

In order to react fully autonomously to irregularities, errors must be automatically classified into the defined categories beyond mere detection in order to act according to the catalog of measures. Further investigations are necessary for this. Amongst this, the surface properties of other asphalt mixes need to be documented.

REFERENCES

Brown E.R., Collins R. and Brownfield J.R., 1989. Investigation of Segregation of Asphalt Mixtures in the State of Georgia. Transportation Research Record 1217.
Cai X., Wu K. and Huang W., 2021. Study on the Optimal Compaction Effort of Asphalt Mixture Based on the Distribution of Contact Points of Coarse Aggregates. *Road Materials and Pavement Design* 22, 1594–1615. https://doi.org/10.1080/14680629.2019.1710238

DIN e.V (Publ.), 1985. *DIN 5034-2:1985-02 Tageslicht in Innenräumen*; Grundlagen. Beuth Verlag GmbH, Berlin.

FGSV, 2012. *TP Asphalt-StB - Teil 8: Volumetrische Kennwerte von Asphalt-Probekörpern und Verdichtungsgrad*, Ausgabe 2012. ed. FGSV Verlag, Köln.

FGSV, 2007a. *TL Asphalt-StB*, Ausgabe 2007. ed. FGSV Verlag, Köln.

FGSV, 2007b. *TP Asphalt-StB - Teil 33: Herstellung von Asphalt-Probeplatten im Laboratorium mit dem Walzsektor-Verdichtungsgerät (WSV)*, Ausgabe 2007. ed. FGSV Verlag, Köln.

Fliegl M., 2017. Untersuchungen im Asphaltbau (Fliegl Bau- und Kommunaltechnik GmbH).

Kappel M., 2016. *Angewandter Straßenbau: Straßenfertiger im Einsatz*, 2. Auflage. ed. Springer Vieweg, Wiesbaden.

Lin C., Jiachin S., Tongjin W., Fan Y. and Tiantong Z., 2019. A Method to Evaluate the Segregation of Compacted Asphalt Pavement by Processing the Images of Paved Asphalt Mixture. *Construction and Building Materials* 224.

OpenCV community, 2022. *OpenCV documentation (Vers. 4.5.5)* [WWW Document]. https://docs.opencv.org/4.5.5/. URL (accessed 1.8.22).

Rosauer V., 2010. *Abschätzung der herstellungsbedingten Qualität und Lebensdauer von Asphaltdeckschichten mit Hilfe der Risikoanalyse*. Technische Universität Darmstadt, Darmstadt.

SciPy community, 2022. *SciPy Documentation (Vers. 1.8.0)* [WWW Document]. https://docs.scipy.org/doc/scipy/. URL (accessed 1.16.22).

Stroup-Gardiner M. and Brown E.R., 2000. *Segregation in Hot-Mix Asphalt Pavements*. National Cooperative Highway Research Program.

Ulrich A., 2009. *Prozesssicherer Automatisierter Strassenbau*. Straße und Autobahn 60.

Utterodt R., 2013. *Der Einfluß schwankender Mischguttemperaturen und Materialfüllstände vor der Einbaubohle auf ausgewählte Funktionseigenschaften einer Deckschicht aus Splittmastixasphalt*. Technische Universität Berlin, Berlin.

VÖGELE, 2005. *VÖGELE Booklet on Paving*. Joseph Vögele AG, Mannheim.

Xun Z., Lige X. and Feiyun X., 2021. Asphalt pavement paving segregation detection method using more efficiency and quality texture features extract algorithm. *Construction and Building Materials* 227.

Roads and Airports Pavement Surface Characteristics – Crispino & Toraldo (Eds)
© 2023 The Author(s), ISBN 978-1-032-55149-4

An exploration of friction supply and demand: The management of road surface friction as a factor in vehicle safety

Peter D. Sanders*
Dresden, Germany

Helen E. Viner
Enodamus Ltd, Bracknell, UK

ABSTRACT: The relationship between road surface friction and vehicle safety is self-evident and this relationship means that the effective management of friction is critical in maintaining a safe environment for all road users. Many road authorities measure low speed friction with devices utilising the sideway-force measurement principle and assess texture depth as a proxy for high-speed friction (SF-Tex friction management policies). In the UK, historical accident studies have demonstrated a link between friction as measured using Sideway-Force Devices (SFDs) and incident occurrence, but recent studies have demonstrated that this correlation has become less robust for most road types. Changes to the vehicle fleet resulting from the widespread adoption of peak friction exploiting driver aids such as ABS/ESC have the potential to alter the prevailing frictional demand (amount of friction required for any given vehicle to complete any given manoeuvre in a controlled manner) of the vehicle fleet compared to that of previous decades and thus the relationship between friction measurement and incident occurrence. This paper explores the key aspects that link road surface friction measurements and vehicle safety, namely, friction, supply, characterization, demand, and management. The link between friction characterisation methodologies, specifically those based on SFDs, and the frictional demands placed on the road surface are discussed. As a case study, the current UK friction management policy is considered in relation to the friction measurements made under that policy, and the frictional demands of the UK vehicle fleet. This paper demonstrates that there is sufficient justification to challenge whether SF-Tex friction management policies characterise the correct aspects of friction in order to effectively manage vehicle safety. A discussion about the implications of this finding on future research programmes is presented.

Keywords: Friction, Skid Resistance, Road Management

1 INTRODUCTION

As road surface characteristics contribute to the friction available to road users, managing friction is critical to maintaining a safe environment for road users.

Many road authorities measure low speed friction with devices utilising the sideway-force measurement principle (Sideway Force Devices (SFDs)). (The description "low speed" reflects the low slip speed between the SFD test tyre and the road surface, compared with typical vehicle speeds.) Accident studies (e.g. Hosking 1986; Parry 2005) have demonstrated a link between such measurements and collision rates. In "high speed" tests, the slip speed

---

*Corresponding Author: peter.sanders@live.co.uk

DOI: 10.1201/9781003429258-57

between road and tyre is closer to typical vehicle speeds. In this regime, the surface macro-texture plays an important role in maintaining friction, and has also been linked to reduced collision risk (Roe *et al.* 1991; Sabey & Storie 1968).

For this reason, the UK operates a friction management policy that controls friction in both of these areas (Highways England *et al.* 2020) and (Department for Transport 2019), an approach that will be referred to as a "Sideway-Force-Texture depth (SF-Tex)" friction management policy. However, recent studies (Viner *et al.* 2021; Wallbank 2016) have demonstrated weaker correlations between skid resistance and accident occurrence than observed previously. Additionally, recently published literature (Sanders 2021) has revealed novel information regarding the measurements made by SFDs.

This paper explores the link between road surface friction characterisation methodologies and vehicle safety and, in the light of the above, aims to determine the efficacy of SF-Tex friction management policies in affecting vehicle safety. It considers:

1. **Friction supply**; the maximum level of friction that it is possible to generate between a vehicle tyre and a road surface, the vocabulary that should be used, and the amount of information needed to meaningfully describe road/tyre friction.
2. **Friction characterisation**; the direct measurement of road surface friction, or its inference from direct measurements.
3. **Friction demand**; the amount of friction required for any given vehicle to complete any given manoeuvre in a controlled manner.
4. **Friction management**; how the above points can be manipulated to affect vehicle safety.

2 FRICTION SUPPLY

The topic of friction supply often refers to the maximum levels of friction which can be generated between a tyre and road surface. In this paper, friction supply also entails the development of a common language to describe friction, and an understanding of the information needed to meaningfully discuss it, namely:

- **Tyre/road friction**, the reaction force generated to oppose motion between a vehicle tyre and road surface, generated in response to, and required for, vehicle manoeuvres.
- **The operational velocity (op. vel.)**, the effective speed of the vehicle when resolved in the direction of the frictional forces acting on the tyre (this is not necessarily the same as vehicle speed).
- **The percentage wheel slip (% wheel slip)**, the amount of slipping between the tyre and road surface expressed as a percentage of the op. vel., i.e. 100% wheel slip = full sliding, and 0% wheel slip = free rolling.
- **The presence of contaminants**, e.g. water, ice, or snow between the tyre and road.

The complexity of friction supply is demonstrated in Figure 1, showing the relationship between % wheel slip, op. vel. and friction, under wet conditions; a friction profile. The friction profiles in this paper were generated from data collected using the National Highways Pavement Friction Tester (PFT) (a longitudinal, variable slip test device) under work reported by Sanders (2021).

Figure 1 demonstrates the difference in measured friction at different op. vels. and % wheel slips. For example, the friction measured at 15% wheel slip is markedly greater than that measured at 100% wheel slip, as are measurements made at lower op. vels. Friction profiles are heterogeneous, this is demonstrated in Figure 2 which presents four friction profiles.

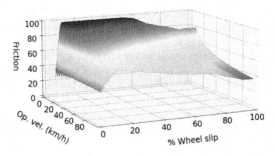

Figure 1. A friction profile demonstrating the information required to describe wet road surface skid resistance under different conditions of operational velocity and, % wheel slip.

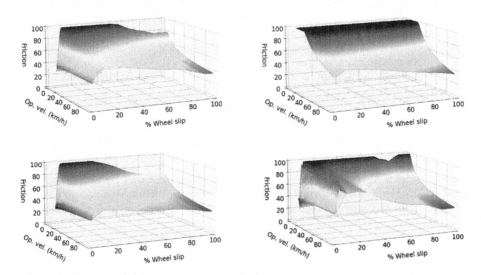

Figure 2. Friction profiles with similar friction values at 100% wheel slip and 100km/h op. vel. but markedly different behaviours at areas on the friction profile diverging from this location.

The friction profiles in Figure 2 represent data obtained at four locations: Top Left, a Hot Rolled Asphalt (HRA) with Sensor Measured Texture Depth (SMTD) 1.25mm; Top Right, a Thin Surface Course System with a 14mm nominal aggregate size (14mm TSCS) and SMTD 0.92mm; Bottom Left, HRA with SMTD 1.21mm; Bottom Right 14mm TSCS with SMTD: 1.5mm. These friction profiles have similar friction values at 100% wheel slip and 100km/h op. vel. (~30 units), but markedly different values at areas divergent from this point. For example:

- Measurements at 10% wheel slip and 100km/h op. vel. are (in reading order) approximately 62, 50, 40, and 75 units respectively, and
- Measurements at 100% wheel slip and 20km/h op. vel. are (in reading order) approximately 60, 100, 50, and 70 units respectively.

Clearly, the friction value measured at one location on the friction profile is not necessarily representative of the friction at a different location on the friction profile, highlighting the necessity for % wheel slip and op. vel. to be provided to meaningfully discuss tyre/road friction.

3 FRICTION CHARACTERISATION

Friction characterisation methods allow the friction supplied to be quantified, and can be grouped into two categories; direct measurement, and inference from other pavement properties.

3.1 *The direct measurement of friction*

In Australia, Belgium, Czech Republic, England, France, Germany, Hungary, Ireland, New Zealand, Northern Ireland, Scotland, Slovenia, Spain, and Wales, the frictional characteristics of the road network are characterised by SFDs. In many countries, SFDs utilising a wheel angle of 20 degrees, and a standard vehicle speed of 50 km/h are used (SFDs$_{(20,50)}$), for example:

- In many EU countries, and Great Britain, the sideway-force coefficient routine investigation machine is used.
- In Germany the SeitenKraftMessverfahren (SKM) is used.
- In Australia the intelligent Safety Assessment Vehicle (iSAVe) is used.
- In Belgium the Odoliograph is used.

As demonstrated above, knowledge of the location on the friction profile at which devices operate (the measurement properties) is critical in understanding their measurements.

Until recently, the working hypothesis for the measurement properties of SFDs$_{(20,50)}$ was that they make measurements at 34% wheel slip and 50km/h op. vel.[1] Whilst theoretically sound, that hypothesis was (to the authors knowledge) not supported by empirical observation. Sanders & Browne (2020) demonstrated that the measurement properties of SFDs$_{(20,50)}$ could also be described at 100%, and 6% wheel slip with the same validity as the 34% wheel slip hypothesis.

Sanders (2021) tested these hypotheses through an experiment comparing the measurements made by an SFD$_{(20,50)}$ with those made using the PFT (which reports % wheel slip and op. vel. directly). This study concluded that SFDs$_{(20,50)}$ characterise friction at 100% wheel slip and 17 km/h op. vel.

3.2 *The inference of friction from other pavement properties*

A second method of friction characterisation is its inference from other pavement properties such as texture depth, using models developed through experimentation.

The ROSANNE model (Equation 1, Cerezo *et al.* (2015)), reports the friction measured by any device to a common Skid Resistance Index (SRI). The ROSANNE model is the result of four studies carried out to develop procedures to normalise skid resistance measurements made by different devices to a common scale; Leu & Henry (1978), Descornet *et al.* (2006), Vos et al (2009), and Cerezo *et al.* (2015). A key feature of the ROSANNE model is the segregation of measurement devices into the following device families:

- Sideway-force devices, reported to SRI$_{SF}$,
- Longitudinal low % wheel slip devices, reported to SRI$_{LFLS}$,
- Longitudinal high % wheel slip devices, reported to SRI$_{LFHS}$,

$$SRI_x = \mathrm{B} \cdot \mathrm{F} \cdot exp\left[\frac{S - S_{ref}}{a \cdot MPD^b}\right] \tag{1}$$

---

[1]See Sanders & Browne (2020) for the derivation of this hypothesis.

Where; SRI_x is the Skid Resistance Index for device family x, a b and B are experimentally derived device specific parameters, F is the measured friction, S is the vehicle speed, S_{ref} is the reference speed at which SRI_x is reported, MPD is the texture depth of the pavement.

A second model of note is the "PPR815 model" (Equation 2, Sanders *et al.* (2017)) which estimates the friction at any location on the friction profile based on pavement texture depth, and measurements made using a sideway-force coefficient routine investigation machine.

$$\mu_x = [a_x \cdot ln(ln(Op.\ Vel.))] + [b_x \cdot ln(SMTD)] + [c_x \cdot SC(50)] - d_x \qquad (2)$$

Where; μ_x is the friction at % wheel slip x and Op. Vel. (km/h), a_x b_x c_x and d_x are model coefficients, SMTD is the road texture depth, SC(50) is the friction value characterised by the Sideway-force coefficient routine investigation machine.

Whilst these models have furthered the understanding of road surface friction characterization, neither can accurately transpose skid resistance measurements made at one % wheel slip to another. For the ROSANNE model this is evident in that the model segregates devices based on the % wheel slip under which they operate. For the TRL815 model the accuracy of the model was explored by Sanders *et al.* (2017), finding an accuracy of 20 units to a 95% degree of confidence.

4 FRICTION DEMAND

Friction demand describes the friction required by vehicles to complete manoeuvres, during which they access different areas on the friction profile. Understanding friction demand and friction supply is critical in managing vehicle safety; in cases where demand exceeds supply, vehicles do not have access to the reaction forces necessary to complete the intended manoeuvre.

An example of vehicle demand is provided in Figure 3 which has been annotated with a dotted trail showing the areas on the friction profile accessed during a theoretical braking manoeuvre in which full sliding occurs. Points spaced further apart show locations on the profile that are traversed quickly, and vice-versa.

Literature pertaining to the friction demands of vehicles is scarce and gathering such information through experimentation would require a substantial research effort. However, some inferences regarding friction demand can be made through information on vehicle driver aids.

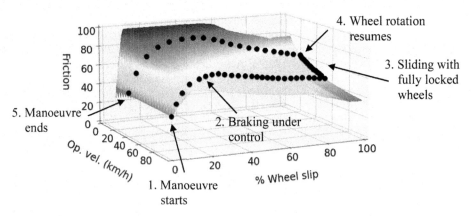

Figure 3. A friction profile annotated with a theoretical braking manoeuvre where full skidding occurs. The distance between each point on the dotted trail indicates a uniform elapsed period of time.

Anti-lock Braking Systems (ABS) have been mandated on all new production cars sold within the EU[2] since 2004 and Electronic Stability Control systems (ESC) have been mandated since 2014. ABS and ESC both exploit peak friction by ensuring that vehicle tyres remain in the 10% to 20% wheel slip range (Bullas *et al.* 2020). For vehicles using such aids, the manoeuvre presented in Figure 3 should be extremely rare (occurring only in the presence of ice or snow) as the % wheel slip would be limited to 20%.

To estimate the prevalence ABS enabled vehicles in the UK, vehicle registration data has been queried (Department for Transport 2022). These data show that in 2020 almost 100% of cars were registered after 2004 and so would have had ABS fitted. For ESC, Hynd *et al.* (2019) estimated that in 2020, approximately 70% of the car fleet would have been fitted with ESC and that by 2027 this would increase to 90%. It can therefore be concluded that the friction demanded of the UK road network is limited to 20% wheel slip and that a similar trend prevails in other countries which have mandated use of vehicle driver aids over a similar period.

5 FRICTION MANAGEMENT

Based on the work presented in this paper thus far, the UK SF-Tex policy can be summarised as making direct measurements of friction at 17km/h op. vel. and 100% wheel slip, and using texture depth (SMTD) measurements as a proxy for friction at high op. vels. and 100% wheel slip. In other words, it controls friction performance, at all op. vels., at 100% wheel slip.

This view has a subtle but important difference to the view held before the publication of Sanders (2021). In that previously accepted view, friction was considered to be directly measured at 50km/h op. vel. and 34% wheel slip with SMTD measurement being used as a proxy for high speed friction. Figure 4 presents these two viewpoints on a plan view of a friction profile, comparing the locations within the profile that are controlled by the UK friction management policy with the areas accessed by vehicles.

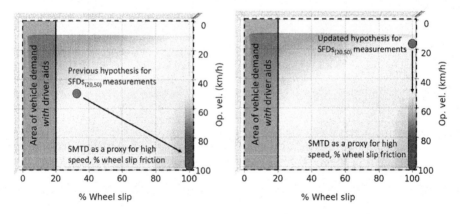

Figure 4. A representation of the location within the friction profile controlled by the SF-Tex policy under the previously accepted view (Left) and the updated view (Right) of SFD$_{(20,50)}$ measurements. The dark shaded area represents the area of vehicle demand *with* driver aids, The broken line encompasses the area of vehicle demand *without* driver aids, and the fading red bar represents the level of influence of SMTD.

---

[2]At the time of writing this is still the case in the UK.

It is taken as axiomatic that, to effectively manage vehicle safety, road surface friction should be characterised at locations on the friction profile that correspond with the frictional demands of vehicles. Figure 4, highlights a potential discrepancy between these areas. Furthermore, while the PPR818 model can predict the friction provided by road surfaces at all % wheel slips and op. vels., the uncertainty was estimated as 20 units. The implication of this uncertainty has been investigated by using this model to predict the friction supply at 18% wheel slip (a % wheel slip representative of friction demand in emergency cases) and an op. vel. of 50 km/h for two hypothetical materials, namely:

1. A motorway section comprising a non-porous asphalt material with a texture depth of 0.95mm SMTD and a SC(50) 0.35; and
2. an approach to a pedestrian crossing comprising a non-porous asphalt material with a texture depth of 0.95mm SMTD and SC(50) of 0.55.

These materials represent the extremes of friction requirements under the current UK friction management policy (Highways England *et al.* 2020) and (Department for Transport 2019). Figure 5 presents the predicted friction values where the series markers present the average value and the error bars represent the uncertainty of the model. Here it can be seen that the error bars for each material overlap the average value for the other by approximately 5 units.

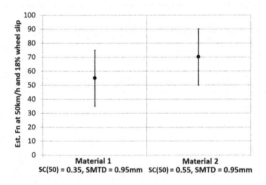

Figure 5. A comparison of friction predictions at 50km/h and 18% wheel slip made using the PPR815 model for two materials. The error bars represent a 20 unit uncertainty.

6 DISCUSSION AND CONCLUSION

Under the previously accepted view (that SFDs$_{(20,50)}$ characterise friction at 34% wheel slip) it was reasonable to assume that, with standards in place to control friction performance at the locations shown in Figure 4, good performance could reasonably be expected over the rest of the profile. But, under the updated view (that SFDs$_{(20,50)}$ characterise friction at 100% wheel slip) this assumption appears less robust, as characterisations are grouped towards one end of the profile.

This analysis is supported by a recent collision study (Wallbank 2016) which demonstrated weaker relationships between skid resistance data and collision rates for much of the English SRN, compared with a similar study 18 years prior. This change may be linked to the changes in vehicles, discussed above, that has led to them accessing different parts of the friction profile to those being measured.

The observations presented in this paper challenge whether SF-Tex friction management policies characterise the aspects of friction required to effectively manage vehicle safety.

7 IMPLICATIONS FOR THE FUTURE OF SKID RESISTANCE/FRICTION RESEARCH

The conclusion above prompts the following questions.

What areas on the friction profile are used by vehicles conducting various manoeuvres and where do these manoeuvres occur on the road network?

The first part of this question pertains to the theoretical braking manoeuvre presented in Figure 3 which considers the areas of the friction profile accessed by a vehicle without ABS. Experimental works are recommended to generate a library of similar figures which represent the areas accessed by vehicles conducting a wide range of manoeuvres. This would act as the basis for identifying the most influential areas on the friction profile for various manoeuvres. For example, for severe braking at higher speeds by ABS-equipped vehicles, a range of 80–100 km/h op.vel. and 10–20 % wheel slip may be appropriate. Appropriate ranges for cornering vehicles could be established, with consideration of how this is influenced when combined with braking or acceleration.

The second part of this question pertains to the manoeuvres carried out at different locations. For example, it would be expected for turning manoeuvres to be prevalent on bends and roundabouts, whereas straight line braking would be expected on approaches to junctions. Combining these data would enable friction demand to be better understood.

What is the relationship between our current friction characterisation techniques, or alternative techniques which could be developed, and the areas on the friction profile used by vehicles?

To align better the friction demands of vehicles with friction supply, it may be necessary to investigate alternatives to the current SF-Tex approach. These fall into three categories:

1. The development of targeted, multi-variate statistical models.

   The PPR815 model attempts to predict friction across a large range of op. Vel. and % wheel slip. It is apparent, from Figure 4 and Equation 2, that this is a complex shape, which is being modelled by few parameters. A more targeted model, to predict friction at a restricted range of op.vel and % wheel slip may be capable of greater accuracy.

   If successful, this would enable a more accurate assessment of road condition to be made from existing surveys, i.e. SC(50) and SMTD data. The significance of texture depth is of particular interest, given recent observations that SMTD is a significant predictor of collision rate on dry roads on the English SRN, but not for wet roads (Fairall *et al.* 2021). A probabilistic approach could be taken, in which the probability of a given length of road providing given levels of performance could be estimated. Such an approach could be extended with the addition of other variables. For example, Andriejauskas *et al.* (2022) showed that fluctuations in skid resistance over time at specific locations can be predicted with the addition of weather data.

2. The use of alternative friction measurement equipment.

   There are numerous devices available, which characterise surface friction in various areas of the friction profile. A change to the established national survey fleet would require a strong business case, linked to road safety objectives, e.g., National Highways' Strategic Business Plan has an objective to reduce the number of people killed or seriously injured on the SRN, to 50% of the 2005–09 level, by 2025 (National Highways 2020). Establishing how direct friction measurements, in a different part of the friction profile, could lead to improved road safety will be crucial to justifying a future course of action.

3. The use of crowd-sourced vehicle data.

An alternative approach, based on the use of in-vehicle sensor data, is detailed by Hammond *et al.* (2021). The methodology identified locations where the friction demands of vehicles were not being met by the friction supplied by the road surface, i.e. providing a direct link between the friction supplied by a road surface and the friction demands of vehicles using it. Effectively, the work demonstrated the viability of using crowd-sourced vehicle data for managing pavement friction and could be scaled, using larger datasets to manage whole networks.

How can the effect of friction policy on the risk to road users be verified?

To some degree, this question can be answered through accident studies similar to those discussed previously. But accidents are rare, and subject to a large number of contributing factors. Identifying the link between friction and accident occurrence is therefore inherently challenging and would require a large amount of observation in order to generate robust results. Alternatively, large-scale use of in-vehicle sensor data, as described above and detailed by Hammond *et al.* (2021), could act as a baseline dataset against which the effect of friction management policies on vehicle safety/the risk to road users could be assessed.

REFERENCES

Bullas J,. Andriejauskas T,. Sanders P, D. & Greene M, J. (2020), *PPR962 The Relationship Between Connected and Autonomous Vehicles, and Skidding Resistance – A Literature Review*, TRL, Wokingham.

Cerezo V., Viner H., Greene M., Schmidt B. & Scharnigg K. (2015), *Analysis of Data from the First Round of Tests and Initial Development of the Common Scale, Rolling Resistance, Skid Resistance, and Noise Emission Measurement Standards for Road Surfaces (ROSANNE project)*, Available at: https://rosanne-project.eu/.

Department for Transport (2019), *Road Pavements - Bituminous Bound Materials, Manual of Contract documents for Highway Works*, Volume 1, Series 900, Department for Transport, Department for Transport, London.

Department for Transport (2022), *'Licensed Cars 2020'*, https://www.gov.uk/government/statistical-data-sets/veh02-licensed-cars. Accessed: 12/01/2022.

Descornet G., Schmidt B., Boulet M., Gothie M., Do M.T., Fafie J., Alonso M., Roe P., Forest R. & Viner H. (2006), Harmonisation of European Routine and Research Measuring Equipment for Skid Resistance, Forum of European National Highway Research Laboratories.

Fairall K., Wallbank C. & Greene M. (2021), *PPR988 The Relationship Between Vehicle Data, Collision Risk and Skid Resistance*, TRL, Wokingham.

Hammond J., Bell M., Wallbank C. & Sanders P.D. (2021), *PPR988 The Relationship Between Vehicle Data, Collision Risk and Skid Resistance*, TRL, Wokingham.

Highways England, Transport Scotland, Welsh Government, Department for Infrastructure (2020), *Skidding Resistance*, CS 228, Design Manual for Roads and Bridges.

Hosking J.R. (1986), *RR76 Relationship Between Skidding Resistance and Accident Frequency: Estimates Based on Seasonal Variation*, Transport and Road Research Laboratory, Wokingham.

Hynd D,. Wallbank C,. Kent J,. Ellis C,. Kalaiyarasan A,. Hunt R. & Seidl M. (2019), *PPR868 Costs and Benefits of Electronic Stability Control in Selected G20 Countries*, TRL, Wokingham.

Leu M.C. & Henry J.J. (1978), *Unified Theory of Rubber and Tire Friction*, Transportation Research Board, Washington DC.

National Highways (2020), *Strategic Business Plan 2020-2025*, https://nationalhighways.co.uk/strategic-business-plan.

Parry A.R. & Viner H.E. (2005), *TRL622 Accidents and the Skidding Resistance Standard for Strategic Roads in England*, TRL, Wokingham.

Roe P.G., Webster D.C. & West G. (1991), *RR296 The Relation Between the Surface Texture of Roads and Accidents*, Transport and Road Research Laboratory, Wokingham.

Sabey B.E. & Storie V.J. (1968), *LR173 Skidding in Personal-injury Accidents in Great Britain 1965 - 1966*, Road Research Laboratory, Wokingham.

Sanders P.D. (2021), *PPR980 Characterising the Measurements Made by Sideways-force Skid Resistance Devices: An Experimental Study*, TRL, Wokingham.

Sanders P.D. & Browne C. (2020), *PPR957 Characterising the Measurements Made by Sideways-force Skid Resistance Devices: A Desk Study and Proposal for an Experimental Study*, TRL, Wokingham.

Sanders P.D., Militzer M. & Viner H.E. (2017), *PPR815 Better Understanding of the Surface Tyre Interface*, TRL, Wokingham.

Viner H., Smith S. & Boden K. (2021) *The LASR Approach: A new Methodology for Prioritising Local Authority Skid Resistance*, https://www.lasr-approach.org/

Vos E., Groenendijk J., Do M.T. & Roe P. (2009), Report on Analysis and Findings of Previous Skid Resistance Harmonisation Research Projects, *Tyre and Road surface Optimisation for Skid resistance And Further Effects (TYROSAFE Project)*, Available at: http://tyrosafe.fehrl.org/.

Wallbank C., Viner H., Smith L. & Smith R. (2016), *PPR806 The Relationship Between Collisions and Skid Resistance on the Strategic Road Network*, TRL, Wokingham.

Relationship between driving conditions, pavement surface characteristics and rolling resistance

Veronique Cerezo* & Joao Santos
Université Gustave Eiffel, France

Mohamed Bouteldja
CEREMA

Xavier Potier
DDTM

ABSTRACT: Rolling resistance is a physical phenomenon related to the dissipation of energy that occurs during the passage of a tire on a road pavement. This loss of energy generates forces opposed to the vehicle movement, which in turn increase fuel consumption and CO_2 emissions. Rolling resistance can represent until 30% of the resistive forces depending on the vehicles' characteristics and driving conditions (rural or urban roads, motorways). Previous studies identified the main influencing factors related to the vehicle (i.e. tire pressure, load, speed, temperature) and pavement (texture, roughness). Nevertheless, the influence of the driving conditions is less known.

This paper presents a study performed at the University Gustave Eiffel's test track which explores the impact of vehicle speed, tire pressure and gearbox ratio on rolling resistance. The test track combines a banked corner 1300 m long with a straight section of 1000 m, paved with twelve different pavement surface mixtures exhibiting various microtexture and macrotexture levels. A passenger vehicle was instrumented with various sensors to measure its dynamical behavior (speeds, accelerations), the forces and torques applied to the wheel and the driving commands (gearbox ratio, engine speed, steering wheel angle). A wide experimental test plan was performed to collect data at a frequency of 100 Hz, which was posteriorly post-processed to filter noise. Finally, a statistical analysis was carried out to assess the impacts of tire pressure and engine speed on rolling resistance forces. Variations of the rolling resistance forces between 2 and 10% were observed depending on the vehicle speed and pavement surface texture.

Keywords: Rolling Resistance, Pavement Surface Properties, Driving Conditions

1 INTRODUCTION

Rolling resistance is a physical phenomenon related to the dissipation of energy that occurs during the passage of a tire on a road pavement. This loss of energy generates forces opposed to the vehicle movement, which in turn increase fuel consumption and CO_2 emissions (Bryce *et al.* 2014). Rolling resistance can represent until 30% of the resistive forces depending on the vehicles' characteristics and the driving conditions (rural or urban roads, motorways). Three physical phenomena can be identified that explain rolling resistance: 1) deformation of

---

*Corresponding Author: veronique.cerezo@univ-eiffel.fr

DOI: 10.1201/9781003429258-58

the tire in the tire/road contact area, 2) aerodynamic drag of the rotating tire, and 3) slip between the tire tread and the pavement surface (Das & Redrouthu 2014). According to (Michelin 2003) (Clark & Dodge 1979), rolling resistance is responsible for 5 to 20% of the fuel consumption of a passenger car and 15 to 40% of trucks' fuel consumption.

The state of the art about the impact of these factors on rolling resistance was presented by (Sandberg et al. 2011) (Michelin 2003) (Ydrefors et al. 2021). Overall, all studies are unanimous in asserting that rolling resistance depends on several factors related to the vehicle (i.e. tire pressure, load, speed, rubber temperature, torque) and the pavement (roughness, texture, stiffness). Tire inflation pressure is one of the most influencing factors due to the fact that an increase of the tire pressure entails a stiffening of the tire and a decrease of the hysteresis effect (i.e. energy losses due to the tire deformation). That, in turn, originates a decrease of the rolling resistance force (F_{rr}). On the contrary, the increase of the tire load generates an increase of F_{rr} due to the growth of the contact area and the energy losses in the contact.

Speed is another factor that affects negatively F_{rr}. On one side, the aerodynamic drag of the tire increases with speed, and on the other side the frequency of the deformations increases. The combination of these two phenomena explains the higher energy losses due to hysteresis and the increase of F_{rr}. Further, it has also been observed that the relation between F_{rr} and speed is exponential (Pacejka 2006).

Not least relevant to rolling resistance is the impact of temperature. Specifically, the temperature of three elements affects rolling resistance: air (T_{air}), pavement ($T_{pavement}$) and tire rubber (T_{rubber}). Generally, an increase of tire temperature leads to a decrease of F_{rr} due to a lower hysteresis effect of the rubber and an increase of the inflation pressure. Moreover, ground and air temperature affect the rolling resistance measurement and correction laws are proposed to estimate it at the reference temperature of around 25°C (Ejsmont &Taryma 2013). The ISO standard uses a linear correction law with a coefficient K_t ranging between 0.004 and 0.018 depending on the type of vehicle, tire, road surface and speed for temperatures ranging from 5 to 35°C.

Lastly, road pavement characteristics also affect the generation of F_{rr} (Sandberg et al. 2011) (Sohaney et al. 2013). These authors showed that F_{rr} increase with the increase of macrotexture[1] according to a linear dependency. They also demonstrated the impact of microtexture[2] with a combined effect of friction and hysteresis.

From the studies mentioned above, it is clear that all the factors described previously can have opposite effects on rolling resistance. This makes it harder to understand and decouple their individual contribution. Moreover, other factors related to vehicle dynamics can also affect the generation of F_{rr} (e.g., tire camber, torque, etc.) but are less analysed in literature.

This paper presents a study conducted in controlled conditions on a test track to explore the impact of vehicle speed, tire pressure and gearbox ratio on rolling resistance. A factorial test plan was designed and posteriorly realized with an instrumented light vehicle. Finally, the contribution of the various factor to F_{rr} is explored by means of a few statistical analyses.

2 EXPERIMENTAL CAMPAIGN

2.1 *Test track*

The data collection campaign was performed on a test track located at Université Gustave Eiffel (Nantes' campus).

The test track comprises a curve and a straight line featuring fifteen different pavement surfaces covering a wide range of microtexture and macrotexture values. Microtexture is assessed through British Pendulum friction measurements according to the standard ISO EN

---

[1]Surface Irregularities with wavelengths ranging from 0.5 to 50mm
[2]Surface irregularities with wavelength smaller than 0.5mm

Figure 1.   Test track of Université Gustave Eiffel (Nantes' campus).

13036-4 (ISO 13036-4, 2012), which provides a friction coefficient (BPN) ranging between 0 and 100. In turn, macrotexture is assessed through laser profilometer measurements according to the standard ISO EN 13473-1 (ISO 13473-1, 2019), which provide a Mean Profile Depth (MPD) value expressed in mm. Among the existing pavement surfaces, it was decided to select four surfaces with the characteristics described in Table 1. Three surfaces (E1, E3 and M2) are commonly found on European roads network and the fourth surface (F) exhibits very high values of micro and macrotexture.

Table 1.   Pavement surfaces characteristics.

Test track	Pavement mixture	MPD (mm)	BPN
E1	Semi-coarse asphalt concrete 0/10	0.87	51
E3	Stone Mastic Asphalt 0/10	1.13	67
M2	Very thin asphalt concrete 0/6	1.30	55
F	Surface dressing with calcined bauxite	1.42	85

2.2   Instrumented vehicle

A light vehicle Clio 2 mounted with Michelin energy saver 185/60 R15 tyres was used for the tests. It is equipped with different types of sensors, namely a dynamometric wheel, accelerometers, inertial unit, steering angle sensor, speed sensor, temperature sensors (rubber and pavement) and a tire deformation sensor (Figure 2). They are controlled periodically to check their accuracy and robustness and the data is acquired at a frequency of 100 Hz. The dynamometric wheel is a 6-component wheel force transducer for passenger car containing a precision load cell based on strain gauge technology.

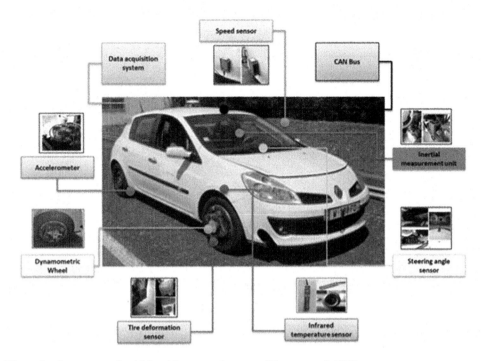

Figure 2. Instrumented vehicle with mounted sensors (Sharma *et al*. 2021).

F_{rr} is calculated by considering the ratio between the rolling resistance moment M_y (in N. m) and the effective radius R_{eff} (i.e. loaded radius) of the tire (in m), as shown by equation (1). The rolling resistance moment is obtained with the dynamometric wheel while the effective radius R_{eff} is calculated by considering equation (2).

$$F_{rr} = \frac{M_y}{R_{eff}} \tag{1}$$

$$R_{eff} = R_{static} \pm def_{dyn} \tag{2}$$

With R_{static}: static loaded radius (in mm) and def_{dyn}: change in tire deformation during rolling (in mm). The static loaded radius is equal to 290 mm. The values of def_{dyn} are measured continuously with laser sensor mounted on the wheel. (Sharma, 2020) analysed that the mean variation of def_{dyn} for a standard inflation of 2.2 bars was around 0.8% in relation to the value of reference (R_{static}) whereas the mean variation of def_{dyn} for an infla-tion pressure of 1.3 bars was around -2.4%. Finally, a low pass Butterworth filter of order 2 with a cut-off frequency of 2 Hz is applied on raw data during the data processing to remove noise.

2.3 *Parameters considered*

This study focussed on the following parameters: speed, tire inflation, pavement surface and engine speed by combining various longitudinal speeds and gearbox ratios. Their values are summarised in Table 2. A factorial plan was performed with three tests for each combina-tion, which resulted in a total of 168 tests.

Table 2. Parameters of the experimental test plan.

Name	Values
Speed (km/h)	50 and 80
Tire pressure (bars)	1.5 and 2.2
Gearbox ratio	2nd, 3rd and 5th at 50 km/h
	3rd and 5th at 80 km/h
Pavement surfaces	E1, M2, E3 and F

3 RESULTS

3.1 *Effect of tire pressure*

Figure 3 presents the effect of tire inflation pressure on F_{rr}. For all the tested conditions, the same trend can be observed, in the sense that F_{rr} increases as the inflation pressure decreases. This result is consistent with the literature and can be explained by the fact that when the pressure is higher, the tire is more rigid and the energy losses due to rubber deformation are less important. When considering the reference tire pressure as 2.2 bars, variations of F_{rr}

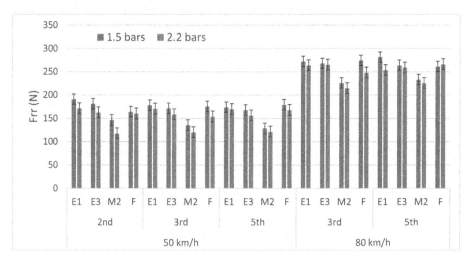

Figure 3. Effect of tire inflation pressure on rolling resistance force (F_{rr}) for various speeds and gearbox ratio.

between -2 and 25% with an average increase of 7% are observed (Figure 4). However, there is no clear trend between these changes and parameters like pavement roughness, speed and gearbox ratio.

3.2 *Effect of speed*

Figure 5 presents the effect of speed on F_{rr}. It can be clearly seen that F_{rr} increases as the vehicle speed also increases. Specifically, an increase of 80 to 100 N can be observed between 50 and 80 km/h for all pavement surfaces and driving conditions (i.e., gearbox ratio).

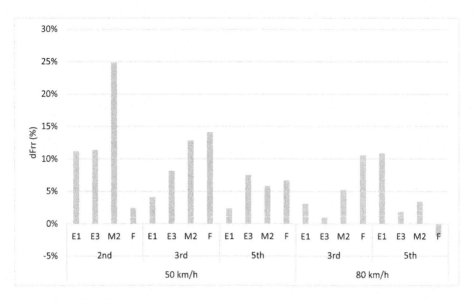

Figure 4. Variation of F_{rr} induced by a decrease of the tire inflation pressure from 2.2 to 1.5 bars.

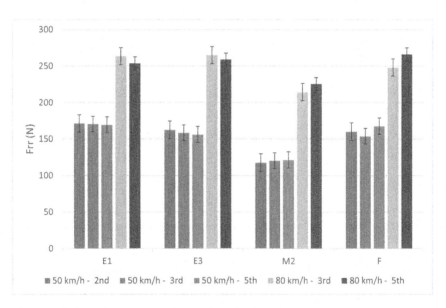

50 km/h - 2nd   50 km/h - 3rd   50 km/h - 5th   80 km/h - 3rd   80 km/h - 5th

Figure 5. Effect of speed on rolling resistance force (F_{rr}).

3.3 *Effect of gearbox ratio and engine speed*

Various combinations of gearbox ratio and longitudinal speed are tested. They entail different values of engine speeds (see Table 3). The results of the effects of engine speed on F_{rr} for various values of tire inflation pressure and vehicle speed are displayed in Figure 6. It appears that F_{rr} depends slightly on engine speed (i.e. small gearbox ratio for a given speed) but this dependency is also connected to inflation pressure and speed. In this context, it is difficult to identify the real contribution of engine speed to F_{rr} generation.

605

Table 3. Engine speed corresponding to each speed/gearbox ratio combination.

Longitudinal speed (km/h)	Gearbox ratio	Engine speed (rpm)
50	2	3500
50	3	2500
50	5	1500
80	3	3500
80	5	2500

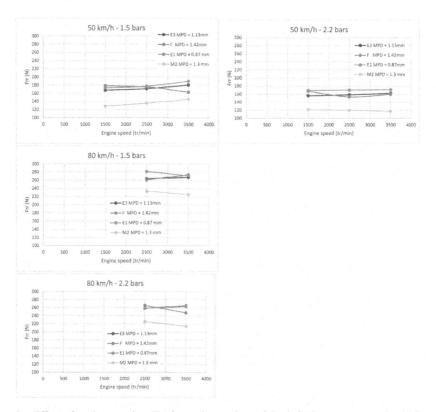

Figure 6. Effect of engine speed on F_{rr} for various values of tire inflation pressure and vehicle speed.

3.4 *Effect of pavement surface*

As mentioned previously, the experiments were performed on four different pavement surfaces with high macrotexture values (MPD > 0.8 mm). Figure 7 displays the effect of pavement surface macrotexture on F_{rr}. It shows that the relationship between F_{rr} and MPD is not linear (coefficient of correlation between Frr and MPD is equal to -0.12), and furthermore F_{rr} tends to decrease as macrotexture increases (excluding the last and highest macrotexture value). This result contradicts the common pattern documented in the literature and is likely to be explained by the reduced set of pavement surfaces considered in the study. Thus, additional tests are needed on pavement surfaces featuring a wider range of macrotexture values to increase the assertiveness of the conclusions that can be drawn. Moreover, several studies underlined the fact that MPD may not be the most suitable parameter to explain variations of rolling resistance. Thus, future tests should also consider other descriptors of texture (Ech *et al.* 2009) (Do and Cerezo 2015).

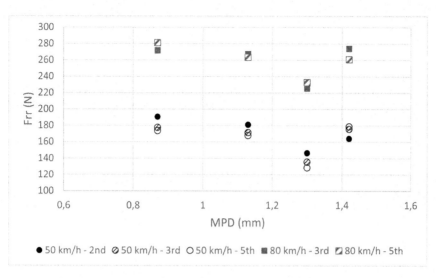

Figure 7.   Effect of pavement surface texture on F_{rr}.

3.5   *Correlation analysis*

To complete the analyses, the correlation between all the parameters is determined and the results are presented in Figure 8. The heatmap shows that F_{rr} is highly correlated with speed (r = 0.92), while the correlation with the remaining parameters is almost neglectable. This result might be explained by the limited size of the dataset.

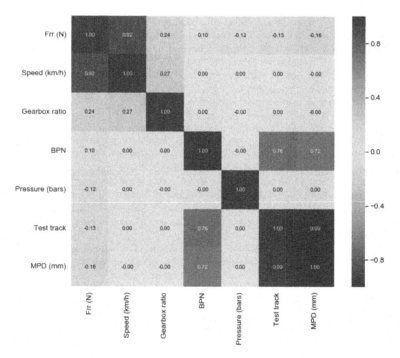

Figure 8.   Correlation heatmap.

4 CONCLUSIONS

This paper presents the results of an experimental campaign performed at Université Gustave Eiffel's test track with an instrumented vehicle. It aimed at assessing the effect of some well-known parameters on rolling resistance force, such as speed and pavement surface macrotexture, as well as others less well-known, such as engine speed. The results of the study confirm that speed is one of the most important parameters to explain rolling resistance. They also confirm the relevance of tire inflation pressure. However, the results of the impacts of the various pavement surfaces on rolling resistance are less conclusive. Such results are likely to be explained by the lack of diversity of the macrotexture values considered. To overcome this shortcoming, additional experimental campaigns are scheduled to complete the initial dataset and to consider other texture parameters that may be more suitable than MPD.

REFERENCES

Bryce J.; Santos J.; Flintsch G.; Katicha S.; McGhee K. and Ferreira A. 2014. Analysis of Rolling Resistance Models to Analyse Vehicle Fuel Consumption as a Function of Pavement Properties. In *Asphalt Pavements*; CRC Press: Boca Raton, FL, USA, pp. 263–273.

CEN 2011. CEN/TS 13036-4: *Road and Airfield Surface Characteristics – Test Methods. Part 4: Method for Measurement of Slip/skid Resistance of a Surface: The Pendulum Test.*

CEN 2019. CEN ISO 13473-1: *Characterization of Pavement Texture by Use of Surface Profiles - Part 1: Determination of Mean Profile Depth (Corrected version 2021-06)*

Clark S.K. and Dodge R.N., 1979. *A Handbook for Rolling Resistance of Pneumatic Tires.* The University of Michigan, Technical report, 89 pages.

Das S. and Redrouthu B., 2014. *Tyre Modelling for Rolling Resistance. Master's Thesis, Chalmers University of Technology*, 61 pages.

Do M.-T. and Cerezo V., 2015. *Road Surface Texture and Skid Resistance.* Surface Topography: Metrology and Properties (STMP), vol.3, Issue 4, October.

Ech M. Morel S. Yotte S. Breysse D. and Pouteau B., 2009. An Original Evaluation of the Wearing Course Macrotexture Evolution using the Abbot Curve, *Road Materials and Pavement Design*, 10:3, 471–494.

Ejsmont J. and Taryma S., 2013. *Deliverable 3.3: Parameters Influencing Rolling Resistance and Possible Correction Procedures*, Collaborative Project FP7-SST-2013-RTD-1, ROlling resistance, Skid resistance, ANd Noise Emission Measurement Standards for Road Surfaces (ROSANNE), 47 pages.

ISO EN 13036-4, 2012. *Road and airfield surface characteristics — Test methods — Part 4: Method for measurement of slip/skid resistance of a surface — The pendulum test.*

Michelin 2003. The Tyre Rolling Resistance and Fuel Savings.

Pacejka H.B., 2006. *Tyre and Vehicle Dynamics.* Second Edition. Butterworth-Heinemann - Elsevier Science, Isbn: 980-0-7506-6918-4.

Sandberg U., Bergiers A., Ejsmont J., Goubert L., Karlsson R. and Zöllzer M., 2011. Deliverable 4: Road Surface Influence on Tyre/road Rolling Resistance, *MIRIAM SP1*, 68 pages.

Sharma A., 2020. Contribution to Variable-gain Observer Synthesis for Nonlinear Systems: Application to Rolling Resistance Estimation, PhD Thesis, Ecole Centrale de Nantes, 281 pages.

Sharma A., Bouteldja M. and Cerezo V., 2021. Vehicle Dynamic State Observation and Rolling Resistance Estimation via Unknown Input Adaptive High Gain Observer, *Mechatronics*, 79. https://doi.org/10.1016/j.mechatronics.2021.102658

Sohaney R., Rasmussen C. and Robert O. 2013. Pavement Texture Evaluation and Relationship to Rolling Resistance at MnROAD. *Report number MN/RC 2013-16. MN Department of Transportation, St.* Paul, Minnesota, US, 139 pages.

Ydrefors L., Hjort M., Kharrazi S., Jerrelind J. and Stensson Trigell A., 2021. Rolling Resistance and its Relation to Operating Conditions: A Literature Review, *Proc IMechE Part D:* J Automotive Engineering, 235(12), 2931–2948.

ISO EN 13473-1, 2019. *Characterization of Pavement Texture by Use of Surface Profiles — Part 1: Determination of Mean Profile Depth*

Roads and Airports Pavement Surface Characteristics – Crispino & Toraldo (Eds)
© 2023 The Author(s), ISBN 978-1-032-55149-4

A risk analysis in the event of a fire in a twin-tube road tunnel with considerations for resilience

Ciro Caliendo*, Isidoro Russo & Gianluca Genovese
Department of Civil Engineering, University of Salerno, Salerno, Italy

ABSTRACT: The paper is an extension of our previous study in which it was investigated the loss of resilience of a twin-tube motorway tunnel in the event of a traffic accident in the north tube with the resulting blockage of either the right lane or both lanes. The functionality of the tunnel system was assumed to be recovered by using the remaining undisrupted lane of the tube interested by the disruptive event (i.e., only one lane is closed) or reorganizing the traffic flow by using the parallel tube for bi-directional traffic (i.e., both lanes of the north tube are closed).

In this study, the average travel speed of traffic flow was used as a resilience metric and coupled this with the risk level of tunnel users in the event of a simultaneous occurrence of a fire in the left lane of the north tube or in the adjacent tube, respectively. A Quantitative Risk Analysis, which is based on the results of Computational Fluid Dynamics (CFD) modeling, was developed to assess the risk of users under different combinations of fire scenarios in the aforementioned undisrupted lane or in the parallel tube when the north tube characterized by a traffic accident is partially or completely closed. The results of the QRA were found to be coherent with those of the tunnel resilience based on the average travel speed. A lower risk level was found with the partial closure of the tube in contrast to the complete one, and by activating the Variable Message Signs (VMSs) to suggest an alternative route to Heavy Good Vehicles (HGVs) only. This study can increase our knowledge on the operative conditions of road tunnels, and can help to make a more suitable choice in the recovery process of functionality accounting for the exposure to risk and resilience at the same time.

Keywords: Resilience of Road Tunnels, Average Travel Speed, Quantitative Risk Analysis

1 INTRODUCTION

Road tunnels, as elements of a transportation network, are crucial for the mobility of people and goods. To fulfill their social and economic functions, tunnels should be maintained accessible to traffic for as long as possible. Nevertheless, the occurrence of a traffic accident in a tunnel might make the structure unavailable for vehicular transit even for several hours or days, thus compromising the operating conditions not only of the tunnel itself but also of the road infrastructure containing it. As a result, the recovery process of the functionality of a road tunnel affected by a traffic accident, also in relation to the possible availability of alternative routes identified on the nearby transportation network, is an increasingly relevant issue nowadays.

A transportation system, or more specifically the road tunnels included in it, that have an adaptive behavior to a disruption (e.g., a traffic accident) is usually defined as resilient. There is no single shared definition of resilience. In the field of transportation networks,

---

*Corresponding Author: ccaliendo@unisa.it

DOI: 10.1201/9781003429258-59

resilience might be described as *"the ability of a system to rapidly recover its functionality after a disturbing event"* (Freckleton *et al.* 2012).

The resilience of an infrastructure network is mainly linked to its functionality, so the resilience analysis of such a system should be performed using traffic-related metrics. Since the disruption of the ordinary operating conditions of a transportation facility due to, for example, the occurrence of a traffic accident might lead to high traffic congestion along it, the average travel speed appears to be a suitable traffic-related resilience metric.

Quantitative resilience analyses are often conducted by means of traffic simulation methods that, by comparing several scenarios and identifying the corresponding criticalities, are useful support tools for road and tunnel management agencies in their decision-making stages.

The resilience of an infrastructure network has been examined from several points of view. Knoop *et al.* (2008), through traffic simulation models, studied both the robustness of a road network and the efficiency of certain recovery measures using the delay time as an indicator. Antoniou *et al.* (2011) and Omer *et al.* (2011) developed frameworks to assess the resilience of a road system by using a travel time-related resilience metric. Liao *et al.* (2012), by using simulation techniques, focused on the efficiency of several recovery measures by adopting average density, vehicle queue length, and average speed as indicators. Kaviani *et al.* (2017), by using the travel time as an indicator, identified the optimal position of road guidance devices to redirect traffic after the occurrence of a natural disaster. Amini *et al.* (2018) set up a method to prove the efficacy of diverting traffic on the resilience of a road system. Zhao (2021) proposed a genetic algorithm to identify the best post-disaster recovery strategy of a road network, while Abudayyeh *et al.* (2021), by using the travel time as an indicator, developed an optimization framework for designing traffic signal timing to improve the resilience of a transportation system. Sohouenou and Neves (2021) dealt with the robustness of a transportation system, also addressing the issue of the rapidity of the recovery process of its functionality. Borghetti *et al.* (2021) focused on the resilience of road tunnels, while also considering emergency management. Khetwal *et al.* (2021) proposed a stochastic simulation model to predict tunnel resilience, which was computed using a functionality metric related to the traffic capacity loss and its duration. PIARC (2021) summarized the studies carried out so far on the resilience of road tunnels. Lastly, Caliendo *et al.* (2022a) proposed a traffic macro-simulation model to assess the resilience of a twin-tube road tunnel and the efficiency of certain recovery strategies using the delay time, resilience loss, resilience index, and recovery speed as indicators.

The aforementioned chronological literature review shows that very few studies have focused on the resilience of road tunnels in relation to the reduction in the average travel speed of traffic flow due to the occurrence of a disruptive event such as a traffic accident. This highlights the lack of knowledge on tunnel resilience that this study will try to address.

However, it is worth noting that the average travel speed should also be coupled with the risk level of tunnel users that might be exposed to a fire, which might occur on the undisrupted lane of the partially closed tube or in the adjacent tube in the event of the complete closure of the tube involved in a traffic accident.

With reference to our previous studies in the field of fire safety engineering (Caliendo *et al.* 2022b, 2021a, 2021b, 2021c, 2020, 2018, 2013, 2012; Caliendo & De Guglielmo 2019, 2017a, 2017b, 2016; Caliendo & Genovese 2021), the risk level of tunnel users was not coupled with the resilience metric based on the average travel speed of traffic flow, which justifies this paper.

In the light of the above considerations, the paper is organized as follows: the next section describes the tunnel investigated. Subsequently, the traffic macro-simulation model is presented and implemented to evaluate the average travel speed. Then, a Quantitative Risk Analysis (QRA), which is based on the findings of Computational Fluid Dynamics (CFD) modeling, is performed. The results are reported and commented on, and comparisons are made. Finally, several conclusions, recommendations, and future studies are discussed.

2   MATERIALS AND METHODS

2.1   *Characteristics of the twin-tube road tunnel and the nearby road network*

The twin-tube tunnel investigated is located along an Italian motorway. Each of its tubes is 850 m long and has two lanes used for traffic in one driving direction under the ordinary operating conditions of the tunnel. The tubes present a horseshoe-shaped cross-section of 55.2 m^2, and are 6.8 m high and 9.5 m wide (with two sidewalks of 1 m, and two lanes of 3.5 m). They are straight and flat, without emergency lanes, and with an emergency exit – connecting the tubes – situated at the midpoint of the tunnel length.

The occurrence of a traffic accident in the north tube was assumed to cause several potential scenarios including the partial or complete closure of the disrupted tube, with adverse effects (i.e., reduction in the average travel speed) for traffic flows traveling on both the two motorway carriageways and the nearby road network.

The study area, which was already extensively presented and discussed in Caliendo *et al.* (2022a), is 8 km wide and 25 km long. Figure 1 schematically shows a portion of it containing: (i) the motorway section between the Nodes B and C including the twin-tube tunnel studied; (ii) the north tube used by vehicles traveling towards the north direction (i.e., from the Nodes B to C) that might be affected by a traffic accident leading to its partial or complete closure; (iii) the south tube used by vehicles traveling towards the south direction (i.e., from the Nodes C to B) that might be affected by two-way traffic in the case of the complete closure of the north tube; (iv) the traffic by-passes situated at the tunnel tube portals that allow for the use of the south tube for traffic in two driving directions when the north tube is completely closed; (v) the alternative route that, through the motorway junctions (i.e., the B-E and L-C sections), might be used by the vehicular flow traveling towards the north direction when the north tube is partially or completely closed; other than links B-E and L-C, it also includes a major rural road (i.e., the E-I section) and a minor rural road (i.e., the I-L section) that are both characterized by two-way traffic. Figure 1 also reports the value of capacity per lane, assumed in accordance to HCM (2010) by taking into account similar roads.

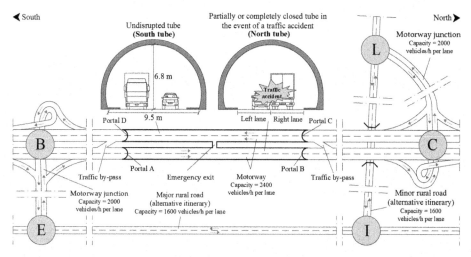

Figure 1.   Schematic representation of a portion of the road network containing the twin-tube tunnel investigated with the cross-sections of the two tubes (the image is not to scale).

2.2 Speed limits

The speed limits for both passenger cars and Heavy Good Vehicles (HGVs), transiting on the different types of road sections included in the infrastructure network studied, were assumed in compliance with the Italian Highway Code. The speed limits along the motorway are therefore 130 and 100 km/h for passenger cars and HGVs, respectively; while they are 70 km/h for passenger cars and 50 km/h for HGVs in the case of both major and minor rural roads. Whereas, a speed limit of 40 km/h is imposed on motorway junctions for both passenger cars and HGVs.

2.3 Scenarios investigated

The traffic accident leading to the partial or complete closure of the north tube was assumed to take place during a morning peak hour (i.e., at 7:00 a.m.) to simulate the worst consequences on vehicular flows (i.e., a significant reduction in their average travel speeds). Moreover, the north tube was assumed to remain partially or completely closed for different durations: 1, 2, or 3 hours (i.e., from 7:00 to 8:00 a.m., from 7:00 to 9:00 a.m., or from 7:00 to 10:00 a.m.).

The scenarios investigated are: (i) Scenario 0 representing the twin-tube tunnel studied under its ordinary operating conditions (i.e., in the absence of any disruption and with the two lanes of each tube used for one-way traffic); (ii) Scenario 1 corresponding to the case where the north tube is partially closed (i.e., the traffic accident blocked the right lane and only the left lane remained open for one-way traffic flow traveling towards the north direction), the speed limit fixed at the entrance portal of the north tube was 60 km/h, and the capacity per the left lane was 1700 vehicles/h; (iii) Scenario 2 representing the case where the north tube was completely closed (i.e., the traffic accident blocked both the two lanes of the north tube) and the vehicular flow traveling through it was reorganized by using the south tube for traffic in two driving directions, the traffic by-passes were supposed to be activated by the rescue teams within 10 minutes from the occurrence of the traffic accident in the north tube, the speed limits were assumed to be equal to 60 km/h at the two portals of the south tube, the vehicular capacity per lane was assumed in the south tube to be equal to 1600 vehicles/h; (iv) Scenario 3 is similar to Scenario 1 except for the deviation of only Heavy Goods Vehicles (HGVs) towards an alternative itinerary, which was suggested by the Variable Message Signs (VMSs) placed before the Node B; (v) Scenario 4 is similar to Scenario 2 except for the rerouting of only HGVs towards an alternative route.

2.4 Traffic volume

The hourly traffic volumes on the road sections examined during both peak hours (i.e., from 7:00 to 9:00 a.m.) and off-peak hour (i.e., from 9:00 to 10:00 a.m.) in the morning were taken from the traffic database of the road management agencies and are summarized in Table 1.

Table 1. Hourly traffic volumes on the road sections examined during both peak hours (7:00–9:00 a. m.) and off-peak hour (9:00–10:00 a.m.). In brackets, the percentage of heavy vehicles (i.e., HGVs and buses). The arrows indicate the travel directions.

	Hourly traffic volumes [vehicles/h per lane]						
	Motorway		Major rural road	Motorway junctions			
Traffic direction	B⇒C	C⇒B	E⇔I	B⇒E	E⇒B	L⇒C	C⇒L
Peak hours: 7:00-9:00 a.m.	1100	1038	776 (2.5%)	499	504	571	550
(Percentage of heavy vehicles)	(25%)	(25%)		(20%)	(20%)	(20%)	(20%)
Off-peak hour: 9:00-10:00 a.m.	275	260	194 (2.5%)	125	126	143	259
(Percentage of heavy vehicles)	(25%)	(25%)		(20%)	(20%)	(20%)	(20%)

2.5 Resilience metrics

The average travel speed over time computed with reference to each duration in which the north tube was partially or completely closed (i.e., 1, 2, or 3 hours, respectively) was used as a resilience metric. It is therefore worth highlighting how higher reductions in the average travel speed, after the occurrence of a traffic accident in the north tube, corresponded to greater resilience losses. However, in this paper, the resilience index (R_{th}) was used to describe the resilience loss. This parameter – which is defined as follows: $R(t_h) = \int_{t_0}^{t_h} F^*(t)dt/(t_h - t_0)$, where F^* is the functionality of the tunnel system, t_0 is the time instant at which the disruptive event takes place, and t_h is the time instant at which the tunnel system recovers its full functionality – can assume values contained in the range 0-1. A resilience index equal to one indicated a zero-resilience loss (for more details see Caliendo *et al.* 2022a).

3 TRAFFIC MACRO-SIMULATION MODEL

PTV Visum 17 (2017), which has been widely applied to investigate more or less extensive road systems, was used as a traffic macro-simulation code.

The main steps to run the code were: (i) importing the transportation facility geometry; (ii) assigning the speed limit and capacity to each road section imported; (iii) defining the analysis period of the simulations that was assumed to be 7 hours (i.e., from 6:00 a.m. to 1:00 p. m.), namely 1 hour before the traffic accident occurs (i.e., from 6:00 to 7:00 a.m.) so that the system is already charged when the disruption takes place, and 3 hours (i.e., from 10:00 a.m. to 1:00 p.m.) after the end of the disruption with the aim of considering the time that the system takes to recover its ordinary functionality; (iv) setting the hourly traffic volume going from an origin point to a destination point in order to model traffic demand; (v) applying a dynamic traffic assignment procedure with the aim of modeling the probable time dependence of the traffic demand (e.g., peak and off-peak hours) and transport supply (e.g., the potential closure of one or more lanes and/or road links); (vi) specifying the simulation time interval that was assumed to be 5 minutes so that the code provides results, in our case the average travel speed, for every 5 minutes of analysis; (vii) setting the convergence criterion of the simulation results that consisted in stopping the runs (10 runs were needed) when the difference in the predictions of two successive runs was lower than 5%. The proposed traffic macro-simulation model was calibrated by comparing the predicted and measured traffic volumes and validated by comparing the queue length measured following a real traffic accident in the north tube with that predicted through the PTV Visum 17 code.

4 AVERAGE TRAVEL SPEED PROFILES

The average travel speeds over time, which are output of the PTV Visum 17 code, were also used to build the corresponding spatial profiles. Figure 2 shows the average speed spatial profiles along the motorway section (i.e., from the Nodes B to C) containing the north tube, after the occurrence of the traffic accident in the north tube with reference respectively to its partial or complete closure of 1, 2, or 3 hours, and for all the aforementioned scenarios investigated.

From Figure 2, it is possible to note that for Scenario 0 (corresponding to the ordinary operating conditions without the occurrence of a traffic accident in the north tube) the average speed profile, which includes both passenger cars and HGVs, is almost constant with a speed value equal to about 120 km/h. For Scenarios 1 and 3, referred to the partial closure of the north tube due to a traffic accident that blocked the right lane leaving only the left one undisrupted, from Figure 2a, it is possible to note that: (i) as a natural consequence of the disruption, the average speed decreases towards the entrance portal; (ii) at the tunnel

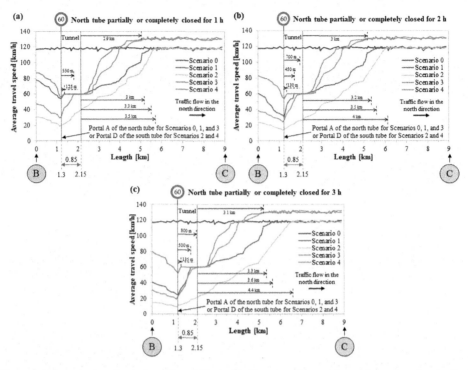

Figure 2. Average travel speed profiles of the traffic flow traveling on the motorway section in the north direction (i.e., from the Nodes B to C) over the time in which the north tube is partially or completely closed for: (a) 1 h, (b) 2 h, and (c) 3 h. Scenario 0 corresponds to the ordinary operating conditions of the twin-tube tunnel; Scenario 1 refers to the partial closure of the north tube; Scenario 2 corresponds to the complete closure of the north tube; Scenarios 3, and 4 are the same as Scenarios 1 and 2 respectively, but with the HGVs only diverted to the alternative route.

entrance (i.e., 1.3 km downstream of the Node B), the average speed is approximately equal to the speed limit of 60 km/h when the duration of the disruption is equal to 1 hour for both traffic flow constituted by all types of vehicle (i.e., Scenario 1 including both cars and HGVs) and cars only (i.e., Scenario 3); (iii) the average speed of 60 km/h remains constant along the entire tunnel length; (iv) downstream of the tunnel, the passenger cars reach the speed of about 130 km/h, while the traffic flow formed by all the vehicles reaches a slower speed (i.e., 120 km/h); (v) passenger cars reach the mentioned speed (i.e., 130 km/h) within less distance than that needed by all the vehicles to reach their speed (i.e., 120 km/h); (vi) then along the motorway the aforementioned speeds remain constant.

From Figure 2a, it can also be noted, for Scenarios 2 and 4 corresponding to the complete closure of the north tube due to a traffic accident respectively with and without HVGs diverted to an alternative route, that: (i) upstream of the entrance portal, there are greater speed reductions; (ii) at the tunnel entrance, the average speed is less than 60 km/h; (iii) the traffic flow constituted only by cars (i.e., Scenario 4) reaches the speed of 60 km/h in the tunnel in contrast with that of the traffic flow formed by all the vehicles (i.e., Scenario 2); (iii) downstream of the tunnel, the passenger cars reach the speed of 130 km/h and within less distance compared to the flow of all the vehicles for which a speed of 120 km/h is reached; (iv) then the mentioned speed remains constant.

Evidently, the reductions of average speeds are much more significant when the duration of the disruptive event in the north tube is assumed to be equal to 2 and 3 hours, respectively (see Figures 2b and 2c).

In the light of the above considerations, the partial closure of the north tube leads to a lower reduction in the average speeds than the complete closure. Moreover, the activation of the alternative route for only HGVs allows to further contain the reduction of speeds. This means that for the aforementioned condition (i.e., the partial closure of the tube and with HGVs redirected towards an alternative itinerary) a shorter delay time is also expected. This indicates a greater resilience index (i.e., a minor resilience loss). The results of the resilience index are reported in Figure 3.

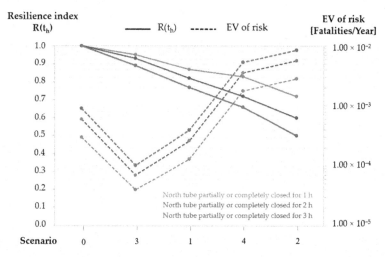

Figure 3. EV of risk and resilience index related to the ordinary operating conditions of the tunnel system (i.e., Scenario 0), as well as to the partial (i.e., Scenarios 1 and 3) or complete (i.e., Scenarios 2 and 4) closure of the north tube for 1, 2, and 3 hours.

5 QUANTITATIVE RISK ANALYSIS

A Quantitative Risk Analysis (QRA) was carried out to assess the risk level of tunnel users in the event of a fire that might occur: (i) during the ordinary operating conditions in the north tube (i.e., Scenario 0, without the occurrence of a traffic accident); (ii) on the left lane in the case of the partial closure of the north tube (i.e., Scenarios 1 and 3, with the simultaneous occurrence of a traffic accident that blocked the right lane); (iii) in the adjacent tube used for bi-directional traffic in the case of the complete closure of the north tube (i.e., Scenarios 2 and 4, with the simultaneous occurrence of a traffic accident that blocked both the two lanes of the north tube). Specifically, the fires were assumed to take place at five different locations (i.e., 145, 280, 420, 570, and 710 m from portal A under Scenarios 0, 1, and 3, or from portal D in Scenarios 2 and 4). The fire sources, in Scenarios 0, 1, and 2, consist of two cars, a van, a bus, and two different types of HGVs able to develop a Heat Release Rate (HRR) of 8, 15, 30, 50, and 100 MW, respectively; while the fire of only light vehicles (i.e., HRRs of 8 and 15 MW) was simulated in Scenarios 3 and 4 since the HGVs are assumed to be diverted to the alternative route.

The Fire Dynamics Simulator (FDS) (McGrattan *et al.* 2019) was used as a CFD code to simulate the aforementioned fires. The main input data to implement the FDS code include: tube geometry, sizes and location of the fire source, dimension and position of the queued vehicles, HRR growth curve, combustion product yields, and pressure difference between the tube portals. The people egress process after the occurrence of a certain fire, instead, was reproduced using the Evac code (Korhonen 2018), which is the evacuation module of the FDS code. The main input data to run the Evac code regard the number of users and their initial position along the tube, as well as the pre-movement time (i.e., the sum of the detection and reaction times), escape path, and walking speed of each evacuee.

The results provided by the FDS + Evac codes, which are not reported here to save space, were used to perform a QRA, based on a probabilistic approach, aimed at assessing user safety in the case that the aforementioned fires occur. The main steps to run the QRA were: (i) defining the event tree to compute the probability of occurrence of each hazard (i.e., fires with an HRR of 8, 15, 30, 50, or 100 MW); (ii) estimating the annual frequency of traffic accidents (i.e., the average number of crashes per year) that is the initial event in the event tree; (iii) computing the annual frequency of occurrence of a given fire in each of the mentioned positions; (iv) identifying the individual annual cumulative frequencies of a certain fire by associating the annual frequency of occurrence of that fire in each of its five locations with the corresponding number of potential deaths provided by the FDS + Evac codes; (v) calculating the final annual cumulative frequency of a given scenario by combining the individual annual cumulative frequency of all the fires included in that scenario. The main result of the QRA was the so-called societal risk, which was represented graphically through the F-N curves (F is the final annual cumulative frequency of having a certain number of deaths N), the results of which are not reported here to save space, but the corresponding Expected Value (EV) of risk (i.e., $EV = \int_1^{+\infty} F(N)dN$) was reported.

5.1 *Expected value of risk and resilience index*

Figure 3 shows the EV of risk and the resilience index related to the ordinary operating conditions of the tunnel system (i.e., Scenario 0), as well as to the partial (i.e., Scenarios 1 and 3) or complete (i.e., Scenarios 2 and 4) closure of the north tube for 1, 2, and 3 hours.

From Figure 3, it can be noted how the EV of risk increases as the duration of the partial or complete closure of the north tube increases (i.e., passing from 1 h to 2 h, and from 2 h to 3 h); on the contrary, the resilience index decreases with an increasing of the duration of the disruption, which denotes a greater resilience loss (i.e., a higher reduction in the average travel speed). Moreover, it is possible to observe how the smallest EV of risk is measured when the north tube is partially closed for a duration of 1 hour and the HGVs only are diverted to the alternative route (i.e., Scenario 3), and that the corresponding resilience index assumes the highest values (i.e., close to 1), which means the lowest resilience loss (i.e., the smallest reduction in the average travel speed).

Therefore, the results of the QRA, expressed in terms of EV of risk, are coherent with those of the resilience analysis based on the average travel speed.

6 SUMMARY AND CONCLUSIONS

In this study, the average speed over time of traffic flow was used as a metric to assess the resilience of a twin-tube motorway tunnel affected by a traffic accident causing the partial or complete closure of the north tube. The functionality of the tunnel system after the occurrence of the traffic accident was assumed to be recovered by using the remaining undisrupted lane of the tube interested by the disruptive event (i.e., only 1 lane is closed) or reorganizing the traffic flow by using the parallel tube for traffic in two driving directions (i.e., both lanes of the north tube are closed in this case). Moreover, the average travel speed was coupled with the risk level of tunnel users in the event of a fire. A QRA, based on the output of CFD modeling, was implemented to evaluate the risk level of tunnel users subject to different combinations of fire scenarios occurring on the undisrupted lane or in the adjacent tube when the north tube affected by a traffic accident was partially or completely closed, respectively.

The results of the QRA, expressed in terms of EV of risk, were found to be coherent with those of the tunnel resilience based on the average travel speed. Specifically, the partial closure of the north tube led to both a lower risk level as well as a lower reduction in the average travel speed. Moreover, further benefits in terms of both user safety and resilience were found to be obtained by activating the VMSs that indicate an alternative itinerary to HGVs only.

This paper can improve our knowledge on the operative conditions of road tunnels, and can help to make a more suitable choice in the recovery process of functionality accounting for the exposure to risk and resilience at the same time. However, further studies might be needed to also consider the uncertainty analysis in numerical simulations.

REFERENCES

Abudayyeh D., Nicholson A. and Ngoduy D., 2021. Traffic Signal Optimisation in Disrupted Networks, to Improve Resilience and Sustainability. *Travel Behaviour and Society* 22, 117–128.

Amini S., Tilg G. and Busch F., 2018. Evaluating the Impact of Real-time Traffic Control Measures on the Resilience of Urban Road Networks. *21st International Conference on Intelligent Transportation Systems (ITSC)*, Maui, Hawaii, USA, 519–524, doi:10.1109/ITSC.2018.8569678.

Antoniou C., Koutsopoulos H.N., Ben-Akiva M. and Chauhan A.S., 2011. Evaluation of Diversion Strategies using Dynamic Traffic Assignment. *Transportation Planning and Technology* 34 (3), 199–216.

Borghetti F., Frassoldati A., Derudi M., Lai I. and Trinchini C., 2021. Resilience and Emergency Management of Road Tunnels: The Case Study of the San Rocco and Stonio Tunnels in Italy. *WIT Transactions on the Built Environment* 206, 81–92.

Caliendo C., Ciambelli P., Del Regno R., Meo M.G. and Russo P., 2020. Modeling and Numerical Simulation of Pedestrian Flow Evacuation from a Multi-storey Historical Building in the Event of Fire Applying Safety Engineering Tools. *Journal of Cultural Heritage* 41, 188–199.

Caliendo C., Ciambelli P., De Guglielmo M.L., Meo M.G. and Russo P., 2018. Computational Analysis of Fire and People Evacuation for Different Positions of Burning Vehicles in a Road Tunnel with Emergency Exits. *Cogent Engineering* 5 (1), 1530834.

Caliendo C., Ciambelli P., De Guglielmo M.L., Meo M.G. and Russo P., 2013. Simulation of Fire Scenarios Due to Different Vehicle Types with and Without Traffic in a Bi-directional Road Tunnel. *Tunnelling and Underground Space Technology* 37, 22–36.

Caliendo C., Ciambelli P., De Guglielmo M.L., Meo M.G. and Russo P., 2012. Numerical Simulation of Different HGV Fire Scenarios in Curved Bi-directional Road Tunnels and Safety Evaluation. *Tunnelling and Underground Space Technology* 31, 33–50.

Caliendo C. and De Guglielmo M.L., 2019. *Risk Level Evaluation of Dangerous Goods Through Road Tunnels.* Proceedings of the World Conference on Pavement and Asset Management, Baveno, Italy.

Caliendo C. and Guglielmo M.L., 2017a. Simplified Method for Risk Evaluation in Unidirectional Road Tunnels Related to Dangerous Goods Vehicles. *International Journal of Civil Engineering and Technology* 8 (6), 960–968.

Caliendo C. and De Guglielmo M.L., 2017b. Quantitative Risk Analysis on the Transport of Dangerous Goods Through a Bi-directional Road Tunnel. *Risk Analysis* 37 (1), 116–129.

Caliendo C. and Guglielmo M.L., 2016. Quantitative Risk Analysis Based on the Impact of Traffic Flow in a Road Tunnel. *International Journal of Mathematics and Computers in Simulation* 10, 39–45.

Caliendo C. and Genovese G., 2021. Quantitative Risk Assessment on the Transport of Dangerous Goods Vehicles Through Unidirectional Road Tunnels: An Evaluation of the Risk of Transporting Hydrogen. *Risk Analysis* 41, 1522–1539.

Caliendo C., Genovese G. and Russo I., 2022b. A Simultaneous Analysis of the User Safety and Resilience of a Twin-Tube Road Tunnel. *Applied Sciences* 12 (7), 3357.

Caliendo C., Genovese G., and Russo I., 2021a. Risk Analysis of Road Tunnels: A Computational Fluid Dynamic Model for Assessing the Effects of Natural Ventilation. *Applied Sciences* 11 (1), 32.

Caliendo C., Genovese G., and Russo I., 2021b. A Numerical Study for Assessing the Risk Reduction Using an Emergency Vehicle Equipped with a Micronized Water System for Contrasting the Fire Growth Phase in Road Tunnels. *Applied Sciences* 11 (11), 5248.

Caliendo C., Russo I., and Genovese G., 2022a. Resilience Assessment of a Twin-Tube Motorway Tunnel in the Event of a Traffic Accident or Fire in a Tube. *Applied Sciences* 12 (1), 513.

Caliendo C., Russo I., and Genovese G., 2021c. Risk analysis of One-way Road Tunnel Tube used for Bi-directional Traffic Under Fire Scenarios. *Applied Sciences* 11 (7), 3198.

Freckleton D., Heaslip K., Louisell W., and Collura J., 2012. Evaluation of Resiliency of Transportation Networks after Disasters. *Transportation Research Record* 2284 (1), 109–116.

HCM, 2010. *Highway Capacity Manual*. Transportation Research Board, Washington D.C., USA.

Kaviani A., Thompson R.G., and Rajabifard A., 2017. Improving Regional Road Network Resilience by Optimised Traffic Guidance. *Transportmetrica A: Transport Science* 13 (9), 794–828.

Khetwal S.S., Pei S., and Gutierrez M., 2021. Stochastic Event Simulation Model for Quantitative Prediction of Road Tunnel Downtime. *Tunnelling and Underground Space Technology* 116, 104092.

Knoop V., van Zuylen H., and Hoogendoorn S., 2008. The Influence of Spillback Modelling When Assessing Consequences of Blockings in a Road Network. *European Journal of Transport and Infrastructure Research* 8 (4), 287–300.

Korhonen T., 2018. *Fire Dynamic Simulator with Evacuation: FDS+Evac Technical Reference and User's Guide*. VTT Technical Research Centre of Finland, Espoo, Finland.

Liao S.Y., Hu T.Y., and Ho W.M., 2012. Simulation Studies of Traffic Management Strategies for a Long Tunnel. *Tunnelling and Underground Space Technology* 27 (1), 123–132.

McGrattan K., Hostikka S., Floyd J., McDermott R., and Vanella M., 2019. *Fire Dynamics Simulator: User's Guide*. National Institute of Standards and Technology, Fire Research Division, Maryland, USA.

Omer M., Mostashari A., and Nilchiani R., 2011. Measuring the Resiliency of the Manhattan Points of Entry in the Face of Severe Disruption. *American Journal of Engineering and Applied Sciences* 4, 153–161.

PIARC, 2021. *PIARC Literature Review - Improving Road Tunnel Resilience, Considering Safety and Availability*. Technical Committee on Road Tunnels, the World Road Association, Paris, France.

PTV Visum 17, 2017. *User Manual*. PTV AG, Karlsruhe, Germany.

Sohouenou P.Y.R., and Neves L.A.C., 2021. Assessing the Effects of Link-repair Sequences on Road Network Resilience. *International Journal of Critical Infrastructure Protection* 34, 100448.

Zhao F., 2021. Research on Resilience Recovery Strategy Optimization of Highway After Disaster based on Genetic Algorithm. Journal of Physics: Conference Series, 2083, doi:10.1088/1742-6596/2083/3/032014.

Roads and Airports Pavement Surface Characteristics – Crispino & Toraldo (Eds)
© 2023 The Author(s), ISBN 978-1-032-55149-4

Improving the criteria for plans for bridge inspection

Bodnar Larysa*

Transport Facilities Center, M.P. Shulgin State Road Research Institute State Enterprise (DerzhdorNDI SE), Kyiv, Ukraine

ABSTRACT: In order to make management decisions on the bridge operation effective, it is needed to have reliable information about bridge technical state. Obtaining qualitative and quantitative indicators of bridge operational properties is achieved by bridge inspection and further processing of bridge inspection results by calculations and expertizing. A bridge inspection is an important part of bridge life cycle, its results make it possible to plan the effective use of funds for bridge maintenance and therefore considerable attention should be given to this type of work. A timely bridge inspection makes it possible to ensure reliable and trouble-free bridge maintenance. The best practices show that the cost of the bridge inspection is repaid by optimizing the cost of bridge maintenance. Only in the case of a comprehensive timely assessment of the bridge state and timely implementation of repair works, effective bridge maintenance is possible.

The paper considers the regulatory framework on the requirements for the bridge inspection, including classification and frequency, etc. A detailed analysis of implementation of bridge inspections on public roads of Ukraine is conducted. Information on bridge inspections is collected and accumulated in a common database, in the software Analytical Expert Bridge Management System (AESUM). AESUM database has been used in Ukraine since 2006 and contains 16°141 bridges on public roads, including 5°820 bridges on state roads and 10 321 bridges on local roads. About 76% of bridges on state roads and only about 15% on local roads have been inspected so far. The analysis showed that the normative bridge inspection interval is mostly not maintained. Therefore, the paper proposes a number of criteria for prioritizing while making plans for bridge inspection. The detailed justification of the offered criteria is carried out. The analysis was performed mainly by tools developed within the individual modules of the AESUM software.

Keywords: Bridge, Bridge Inspection, Technical State, Software, AESUM

1 INTRODUCTION

In order to make management decisions on the bridge operation effective, it is needed to have reliable information about bridge technical state. Obtaining qualitative and quantitative indicators of bridge operational properties is achieved by bridge inspection and further processing of bridge inspection results by calculations and expertizing.

The best practices show that the cost of the bridge inspection is repaid by optimizing the cost of bridge maintenance. Only in the case of a comprehensive timely assessment of the bridge state and timely implementation of repair works, effective bridge maintenance is possible.

---

*Corresponding Author: LaraGor@ukr.net

DOI: 10.1201/9781003429258-60

Based on research of Bezbabicheva *et al.* (2016), ensuring reliable and safe bridge maintenance is a necessary condition for the functioning of transport logistics of the state. To maintain bridges, it is important to correctly determine their technical condition due to DSTU23 (2012), and it is possible by carrying out their inspections. Forecasting the lifetime of bridge elements by Yatsko (2016) allows planning the strategy of bridge maintenance and timely completion of repair works allows achieving the scheduled operating time of transport structures, Dekhtyar (2011).

To ensure the proper bridge maintenance the contractor of maintenance should schedule bridge inspections in accordance with DBN6 (2009) within intervals cited in Table 1, if only intervals haven't been changed before based on previous bridge inspection results. Bridge inspection intervals can be influenced by consequences of a road accident on the bridge, sudden destruction of the sustaining structural element or the impact of any other unforeseen events that significantly deteriorate the bridge condition. Thus, for example, in accordance with DBN6 (2009) bridges in nonoperable state (5) should be inspected yearly, if only another interval hasn't been set based on previous bridge inspection results.

Table 1. Bridge inspection intervals.

Bridge type	Bridge age, years				
	1–20	21–40	41–60	61–80	80 and more
	Bridge inspection intervals, years				
Steel, composite	5	4	3	2	1
Reinforced concrete	7	6	5	3	1

The analysis by Bodnar, Koval, Stepanov (2016) has shown that the scheduled bridge inspection intervals aren't maintained mainly. During the last 20 years, higher-quality and more durable structures and materials have been used for bridge construction, and the quality of construction works has increased due to new technologies and construction control. Therefore, it is necessary to extend bridge inspection intervals for bridges aged 1–20 years. For example, for steel and composite bridges bridge inspection interval should be extended up to 7 years, and up to 10 years for reinforced concrete bridges. Bridge inspection interval updating should ensure its maintaining and provide cost savings.

System use of bridge inspection results by State Road Agency of Ukraine (Ukravtodor) is provided by AESUM software, Bodnar (2010).

Therefore, Ukravtodor decided to develop proposals on bridge inspection intervals based on analysis of bridge inspection results delivered by AESUM software also.

2 MAIN PART

AESUM software contains a bridge inspection module of following modes:

- using actual data for analysis of completed bridge inspections, inspection frequency, etc.;
- long-term (10 year) planning for bridge inspection;
- annual planning for bridge inspection.

Comparison bridge inspection results over the past 10 years (Table 2, Figure 1) with tentative plan of bridge inspections for next 10 years, shows that quantity of bridge inspections completed is 23 times less that planned performances.

Table 2.   Quantity of bridges inspected on public roads in 2012– 2021.

№	Region of Ukraine	Quantity of bridges inspected										
		2012	2013	2014	2015	2016	2017	2018	2019	2020	2021	Mean
1	Crimea	32	6								4	4
2	Vinnytsia	50	56			21	7	2		4	16	16
3	Volyn	18	16		24	14	7	4	7	6	11	11
4	Dnipro	35	32	3	2	14	15	13	10	77	25	25
5	Donetsk	11		4	31	16	33	12	6	26	16	16
6	Zhytomyr	11	5	26	7	43	1		4	6	11	11
7	Zakarpattia	1	18	7		31	12	12	11	29	17	17
8	Zaporizhzhia		4	3		1			11	13	5	5
9	Ivano-Frankivsk		3	6	6	11	29	7	8	17	12	12
10	Kyiv	26		4	18	5	66	32	39	28	27	27
11	Kropyvnytskyi			4		7	5	2	2	45	7	7
12	Luhansk		7	2	5		3		13	9	5	5
13	Lviv	6	10	8	7	10	27		9	34	11	11
14	Mykolaiiv	1	12	2		12		3	3	3	6	6
15	Odesa			8	2			4	7	4	4	4
16	Poltava	61	49	70	69	76	6		5	17	36	36
17	Rivne	4	65	7		6	3	5	4	5	10	10
18	Sumy				2	9	6	5	3	1	5	5
19	Ternopil		15	15	13	14			1	7	11	11
20	Kharkiv	31	56	5	3	3	3	4	3	9	13	13
21	Kherson	18	14	1	2	4	9	1	4	7	7	7
22	Khmelnytskyi	34	21	7	1	10	8	6	5	3	12	12
23	Cherkasy	1			5	6	6	8	10	9	5	5
24	Chernivtsi	8			3	9	10	10	10		5	5
25	Chernihiv	20	26		5	10	3	1	5	6	8	8
26	The city of Sevastopol										0	0
	Total	368	415	182	205	332	259	131	180	365	420	286

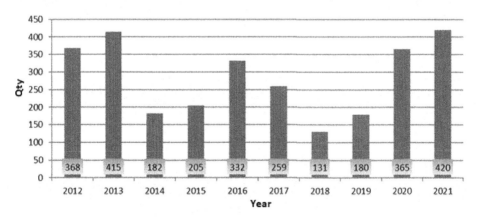

Figure 1.   Quantity of bridges inspected on public roads in 2012–2021.

It should be noted that for the first year (2022) tentative plan (Table 3, Figure 2) contains postponed or cancelled bridge inspections what makes a difference so impressive; but even if excluding 2022, the planned performances are 19 times higher than actual.

Table 3. Quantity of bridges to be inspected on public roads in 2022–2031.

№	Region of Ukraine	Quantity of bridges to be inspected										
		2022	2023	2024	2025	2026	2027	2028	2029	2030	2031	Mean
1	Crimea	349	48	53	174	70	181	199	76	120	188	146
2	Vinnytsia	833	131	145	426	164	478	476	173	261	456	354
3	Volyn	366	19	26	127	48	233	142	51	80	144	124
4	Dnipro	598	127	170	325	191	297	407	171	230	339	286
5	Donetsk	506	77	88	206	131	259	288	124	183	218	208
6	Zhytomyr	528	28	31	153	45	315	211	80	85	165	164
7	Zakarpattia	1255	352	409	641	446	855	746	425	535	679	634
8	Zaporizhzhia	369	58	68	215	87	187	232	89	132	221	166
9	Ivano-Frankivsk	1101	82	109	393	152	635	510	138	270	447	384
10	Kyiv	407	55	53	136	97	272	171	85	95	172	154
11	Kropyvnytskyi	501	32	41	169	65	336	183	61	125	242	176
12	Luhansk	548	51	68	184	88	335	246	87	116	201	192
13	Lviv	1596	961	961	1258	990	1295	1269	991	1056	1355	1173
14	Mykolaiiv	243	51	61	120	74	157	136	66	101	133	114
15	Odesa	861	106	115	378	152	483	454	149	213	404	332
16	Poltava	558	72	79	211	95	320	296	92	130	226	208
17	Rivne	550	24	25	180	51	320	219	61	74	202	171
18	Sumy	462	42	52	92	59	325	177	55	101	123	149
19	Ternopil	789	64	75	218	92	586	276	79	155	269	260
20	Kharkiv	790	48	100	219	131	465	342	107	219	258	268
21	Kherson	85	13	16	40	21	50	50	20	27	45	37
22	Khmelnytskyi	536	76	88	227	101	348	263	102	159	251	215
23	Cherkasy	482	39	51	219	61	264	255	59	139	237	181
24	Chernivtsi	587	65	64	187	81	401	223	101	162	232	210
25	Chernihiv	486	35	38	140	57	296	210	53	113	160	159
26	The city of Sevastopol	36	3	5	15	8	11	24	4	11	15	13
	Total	15422	2659	2991	6653	3557	9704	8005	3499	4892	7382	6476

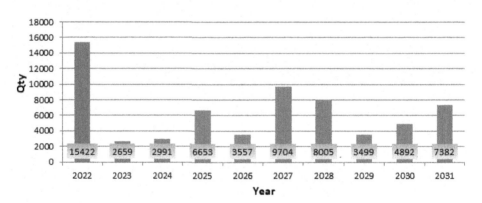

Figure 2. Quantity of bridges to be inspected on public roads in 2022–2031.

There is a need for significant funding for execution of plan, so overcoming of difficulty cannot be completed in one year. This is a very important and responsible task. A long-term program of bridge inspections is proposed. The bridge inspections should be gradually distributed over time period, taking into account both the current bridge state and the financial capacity of the bridges for a period of 10 years; in the AESUM it is a mode on planned surveys for 10 years. However, it is proposed that the plan be adjusted yearly based on bridge inspections schedule, which takes into account the road works prioritization due to limited funding.

2.1 Criteria for bridge inspection planning

Bridge inspection sequencing according to the proposed criteria should be set while yearly planning. Bridge inspection sequencing becomes urgent with lack of funding and if there are lots of bridges to be inspected. Criteria for bridge inspection planning are top-down sorted by importance.

2.2 A bridge or any of its determinant elements are in nonoperable state (5)

Bridge determinant elements are spans, supports and foundations. Bridge or any of its determinant elements being in nonoperable state should be inspected yearly in accordance with DBN6 (2009), if only the interval hasn't been changed before based on previous bridge inspection results.

For such bridges a special inspection must be carried out and results should be used for design documentation for reconstruction and repair works in the future.

Of course, as soon as a bridge is found to be in nonoperable state, emergency measures should be taken to repair this bridge or build a new one. However, until repair works or reconstruction started and there no bridge closure, special inspection should be conducted within year or another interval based on bridge inspection results. Now AESUM database contains (Table 4) data of 267 bridges in nonoperable state (5) and 399 bridges with determinant elements in nonoperable state (5). Normative inspections intervals of these bridges are exceeded and there is a high probability of accident.

Table 4. Quantity of bridges in limited operable state (4) and in nonoperable state (5).

Bridges total	Quantity of bridges in limited operable state (4)		Quantity of bridges in nonoperable state (5)	
	total	including non-inspected bridges after reconstruction	total	including non-inspected bridges after reconstruction
5820	1208	193	233	38

2.3 Special inspection intervals

There are cases, when there is a necessity, for example, for experimental bridges (should be mentioned separately at the planning), to deviate from the normative inspection interval. For example, an inspection interval for brand new piers must be half a year.

2.4 No data on bridge inspection in AESUM database

For the bridges on state roads the filling level of AESUM database on bridge inspection results varies from 34% to 100% (Table 5). This may not be caused by the lack of funding only, but the staff professionalism level of bridge maintenance contractor.

If there is no data on bridge state, it means that bridge operates like a "black box". Such bridges may become nonoperable and there is a high probability of accident.

Table 5. Quantity of bridges inspected on state roads and data recorded to AESUM database as of 01.05.2022.

№	Region of Ukraine	Quantity of bridges to be recorded to AESUM database			
		total in region	inspected and data recorded	no data recorded	Rate, %
1	Crimea	198	94	104	47.47
2	Vinnytsia	215	215	0	100.00
3	Volyn	179	179	0	100.00
4	Dnipro	353	350	3	99.15
5	Donetsk	173	167	6	96.53
6	Zhytomyr	180	126	54	70.00
7	Zakarpattia	396	206	190	52.02
8	Zaporizhzhia	149	148	1	99.33
9	Ivano-Frankivsk	365	358	7	98.08
10	Kyiv	222	212	10	95.50
11	Kropyvnytskyi	172	92	80	53.49
12	Luhansk	209	201	8	96.17
13	Lviv	459	156	303	33.99
14	Mykolaiiv	97	88	9	90.72
15	Odesa	389	143	246	36.76
16	Poltava	112	111	1	99.11
17	Rivne	305	140	165	45.90
18	Sumy	210	210	0	100.00
19	Ternopil	243	135	108	55.56
20	Kharkiv	243	243	0	100.00
21	Kherson	54	54	0	100.00
22	Khmelnytskyi	232	207	25	89.22
23	Cherkasy	183	172	11	93.99
24	Chernivtsa	204	135	69	66.18
25	Chernihiv	253	250	3	98.81
26	The city of Sevastopol	25	23	2	92.00
	Total	5820	4415	1405	75.86

2.5 *A bridge is in limited operable state (4)*

Usually, bridges in limited operable state (4) already have signs of threatened failure of element(s).

No data on bridge inspection after completion or repair or reconstruction works in AESUM database.

A bridge inspection must be conducted after completion or repair or reconstruction works (Table 6). in accordance with DBN6 (2009). But contractors often violate the normative requirement.

Despite the bridge renewal, bridge inspection results may detect defects and damages that occurred during the design and construction stages.

2.6 *No data on bridge state change after completion or repair or reconstruction works in AESUM database*

Repairs significantly improve the technical state of bridges. Bridge state changes shall be recorded into AESUM database to avoid wrong decision on bridge maintenance and bridge inspection planning.

Table 6. AESUM database updating on bridge inspection results.

| | | | Bridges in AESUM database | | Bridges which don't meet the requirements | | | | | | | |
| Roads | Bridges total | | | | total | | by passage clearance | | by load | | by both passage clearance and load | |
	Qty	length, m	Qty	length, m	Qty	length, m	Qty	length, m	Qty	length, m	Qty	length, m
international roads	746	39067.84	579	32712.56	520	26060.91	258	15513.3	61	1968.40	201	8579.19
national roads	13	151.35	13	151.35	13	151.35			4	51.25	9	100.10
state roads total	759	39219.19	592	32863.91	533	26212.26	258	15513.3	65	2019.65	210	8679.29

Table 4 shows the need for bridge inspection of 38 bridges in nonoperable state (5) and of 193 bridges in limited operable state (4).

2.7 *Bridge is located on the international road*

Bridges on international roads should bear the highest loads of local and transit traffic, to meet all the safety and reliability requirements both Ukrainian and international.

There is the need for bridge inspection of 168 bridges on international roads.

2.8 *Problematic superstructure desk design*

The analysis of bridge maintenance experience by Strakhova and Kholoden (2005) shows that shortest lifetime has girder bridge with prestressed stringcrete beams.

2.9 *Road class*

Road class (Tables 7 and 8) should be taken into account while making plans for bridge inspection.

2.10 *Road category*

Road category shows traffic volume and maximum vehicle axle load and should be taken into account while making plans for bridge inspection (Table 9).

Table 7. Bridge distribution in respect to road class.

| Bridge length | Indicator | State roads | | | | | Local roads | | | Total |
		international	national	regionnal	territorial	total	provincial	district	total	
up to 25 meters	qty	562	653	797	1766	3778	3852	4550	8402	12180
	length, m	6347	6614	8322	18460	39743	40234	48259	88493	128236
from 25 to	qty	588	341	245	529	1703	810	934	1744	3447
100 meters	length, m	32091	17655	12196	25468	87410	38272	41219	79491	166901
over 100 meters	qty	135	54	74	76	339	106	69	175	514
	length, m	25964	10630	15362	13740	65696	15397	9998	25395	91091
Total	**qty**	**1285**	**1048**	**1116**	**2371**	**5820**	**4768**	**5553**	**10321**	**16141**
	length, m	**64402**	**34899**	**35880**	**57668**	**192849**	**93903**	**99476**	**193379**	**386228**

Table 8.   Overbridge distribution in respect to road class.

| Overbridge length | Indicator | State roads | | | | | Local roads | | | Total |
		international	national	regional	territorial	total	provincial	district	total	
up to 25 meters	qty	39	7	1	5	52	5	1	6	58
	length, m	590	82	12	91	775	79	7	86	861
from 25 to 100 meters	qty	291	121	53	65	530	26	28	54	584
	length, m	17165	6875	3025	3457	30522	1354	1490	2844	33366
over 100 meters	qty	34	8	4	7	53	0	3	3	56
	length, m	5607	948	558	807	7920	0	401	401	8321
Total	**qty**	**364**	**136**	**58**	**77**	**635**	**31**	**32**	**63**	**698**
	length, m	**23362**	**7905**	**3595**	**4355**	**39217**	**1433**	**1898**	**3331**	**42548**

Table 9.   Bridge distribution in respect to road category.

| Road category | Bridges | |
	qty	length, m
I	98	8 274.04
Ib	266	15 968.34
II	1 710	70 439.67
III	2 854	82 215.51
IV	9 504	185 003.14
V	1 709	24 325.23
Total	**16 141**	**386 225.93**

2.11   *Overbridge*

The overbridges must be maintained twice as good carefully because the failure influences both road and railway traffic.

2.12   *Bridge length*

According to DBN22 (2009) bridges are classified as followed (Table 10, Figure 3):

- small bridges with total length of up to 25 m, with spans of 6 - 25 m;
- medium bridges with total length of 25 - 100 m;
- large bridges with total length of more than 100 m and bridges with spans of over 60 m;
- unclassified bridges with spans over 100 m; turning bridges, drawbridges, etc.

Table 10.   Bridge distribution in respect to bridge length.

Bridge length classes	Bridge length	Qty	%
Small	up to 25 meters	12 274	76.04
Medium	from 25 to 100 meters	3 353	20.77
Large	from 100 to 500 meters	505	3.13
Unclassified	over 500 meters	9	0.06

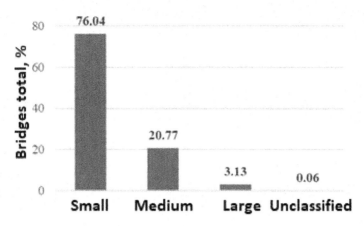

Figure 3.   Bridge distribution in respect to bridge length.

The bridge length affects the possibility of its renewal. The longer the bridge, the costly its reconstruction or repair is.

Analysis shows there is the need for bridge inspection of 169 large bridges.

2.13   *Wooden bridge*

Timber is a relatively short-lived material, so wooden bridges need special attention. There are 53 timber bridges on local roads.

2.14   *Bridge carriageway pavement*

The bridge inspection collects information about the main defects of the main groups of elements, namely - the bridge deck, girder structure, piers, foundations, regulatory structures, bridge approaches, channel. Bridge carriageway pavement is definitely a component of the group of the bridge deck and its condition, is accordingly a component of the assessment of the bridge state. Therefore, timely bridge inspections in general contribute to maintaining the bridge reliability, bridge durability, as well as maintaining bridge carriageway pavement in good condition.

3   CONCLUSIONS

Significant resources are required to perform bridge inspections, so rational bridge inspection planning is a responsible task.

On the other hand, it is necessary to consider the significant impact of bridge inspections on effective bridge maintenance strategy planning that will promote rational use of financial resources aimed on safe and efficient use of the road network.

The sequence of inspection of bridges according to the proposed criteria should be established during the annual planning. The sequence of bridge surveys becomes relevant in the absence of funding and the presence of a large number of bridges that need to be surveyed. In this work, a number of criteria for planning bridge inspection are proposed, which were taken into account in the automated module of the software Analytical Expert Bridge Management System (AESUM).

REFERENCES

Bezbabicheva A.I., Kirienko M.M., Cherepniov I.A. and Topchyi V.L., 2016. *Safe Operation and Reliability of Bridge Structures on Ukrainian Roads as Necessary Elements of Transport Logistics / Environmental Engineering.* - 2016. - № 1(5). - P. 29–39.

Bodnar L.P., 2010. AESUM Software. *Current State and the Concept of Further Development: Collection of Scientific Works "Roads and Bridges".* - Kyiv, 2010. - Vol. 12. - P. 31–39.

Bodnar L.P., Koval P.M. and Stepanov S.M. 2016. *Bridge Inspections and Improvement of Criteria for Inspection Plans /.: Collection of Scientific Works "Roads and Bridges".* - Kyiv: DerzhdorNDI SE, 2016. – Issue 16. - P. 21–27.

DBN6, 2009. Transport Facilities. Bridges and Culverts. Inspection and Testing: ДБН I.2.3-6:2009. - [In force since 2009-11-11]. - *Kyiv: Ministry of Regional Development and Construction of Ukraine, 2009.* - 48 p. (in Ukrainian).

DBN22, 2009. Transport Facilities. Bridges and Culverts. The Basic Requirements for the Design: ДБН I.2.3-22:2009. - *Kyiv: Ministry of Regional Development and Construction of Ukraine, 2009.* - 57 p. (in Ukrainian).

Dekhtyar A.S., 2011. *Optimal Terms and Scopes of Repairs of Reinforced Concrete Bridges / Dekhtyar A.S.// Diagnostics, Durability and Reconstruction of Bridges and Building Structures: Collection of Scientific Papers.* - Lviv.: Kameniar, 2001. - Vol. 3. - P. 83–86.

DSTU23, 2012. Guidelines for Assessment and Prediction of the Technical Condition of Road Bridges: ДСТУ-Н Б.I.2.3-23:2012. - [Valid from 2013-12-01]. - Kyiv: State Standart of Ukraine, 2012. (In Ukrainian).

Strakhova N.Y. and Kholoden T.M., 2005, *Reliability of Road Reinforced Concrete Bridge Spans Constructed According to Type Designs: Collection of Scientific Works "Roads and Bridges".* - *Kyiv*: DerzhdorNDI SE, 2005. - Vol. 3 - P.195–202.

Yatsko F.V., 2016. *Methodology of Predicting the Service Life of Reinforced Concrete Bendable Elements of Bridges in Computer-aided Design / Yatsko F.V. // Bridges and Tunnels: Theory, Research, Practice: Collection of Scientific Papers.* DNUZT named after Academician Lazaryan. - Dnipropetrovsk, 2016. P.107–114.

Minimizing road impacts
(noise, vibration, pollution, etc.)

Roads and Airports Pavement Surface Characteristics – Crispino & Toraldo (Eds)
© 2023 The Author(s), ISBN 978-1-032-55149-4

Optimizing the pattern of grooves of the Next Generation Cement Concrete Surface (NGCS) for less tyre/road noise and less fuel consumption

Luc Goubert*
Belgian Road Research Centre, Brussels, Belgium

ABSTRACT: The Next Generation Cement Concrete Surface (NGCS) is a relatively new method for improving existing cement concrete road surfaces to reduce drastically the tyre/road noise and the rolling resistance. It consists of applying a regular pattern of narrow and wider longitudinal grooves in the existing concrete by means of a drum equipped with diamond disks. Originating from the USA and further optimized in Germany, the idea has been picked up in Flanders which has built its own test tracks in 2015. One of the two test tracks yielded an initial noise reduction of 4,5 dB with respect to the reference surface SMA 0/10, which is unequalled for a cement concrete surface. A Flemish project, GHRANTE, aimed among others to investigate if a new material (geopolymers) might be used instead of concrete and if extra noise reduction would be possible by an optimization of the groove pattern. This paper summarizes the results of this empirical optimization process. First the pattern of the most successful NGCS test track was approximated with a pattern consisting of a juxtaposition of two sinus functions. This pattern was applied on the outside of eight segments, casted in epoxy resin in order to perfectly fit on the 2.0 m diameter drum of the TU Gdansk. By making this drum turn and pushing a test tyre on the segments with the NGCS pattern, the tyre road noise and the rolling resistance can be measured under controlled circumstances. Measurements were carried out with six different test tyres at three speeds. The results obtained on the drum and the test track were very similar. Then several new NGCS patterns were designed by changing some parameters. They were tested on the drum in the same way and the results were compared with the first NGCS profile. The methodology and the measurement results are presented in this paper as well as the conclusions one can draw from them.

Keywords: Next Generation Cement Concrete Surface, Tyre/road noise, Rolling resistance

1 INTRODUCTION

After the discovery of the mechanisms behind the generation of tyre/road noise (Sandberg & Descornet 1980), efforts have been done to develop pavements with a reduced noise emission. Tyre road noise is dominating already at low speeds the vehicle noise and should therefore be targeted. Reducing the megatexture while keeping some macrotexture is the principle in a nutshell. It turned out to be easier for asphalt concrete roads than for cement concrete roads. Moreover, asphalt roads were made porous adding an extra noise reduction by the noise absorption effect and single and double layer porous asphalts are still used in some countries, e.g. in The Netherlands, where they are the standard pavement on highways. One can obtain a stunning 7 dB noise reduction with two layer porous asphalt. Porous asphalt is not used in most

---

*Corresponding Author; l.goubert@brrc.be

DOI: 10.1201/9781003429258-61

countries and neither in Flanders, as the pores tend to clog (reducing the noise reduction with time) and the winter maintenance is more difficult. Porous asphalt has a shorter lifetime (9-11 y) than ordinary asphalt (typical 12-15 y). An alternative approach since the 1990s was to use dense asphalt or SMA but with a small aggregate and applied as a thin layer. Good initial noise reductions were measured (up to 5-6 dB) but the durability (both technical and acoustical) is less good with a lifetime of about 9 y (Kragh *et al.* 2011; Sandberg *et al* 2010). Concrete pavements have a long technical lifetime (typically 25 y or even more) and the acoustic properties tend to stay more stable, but the initial acoustical properties used to be poor, due to the excess of megatexture and/or the transversal grooves for increased wet grip but resulting in a terrible tonal noise. The last decades techniques have been developed to make existing concrete pavement less noisy, e.g. by grinding, typically reducing the noise with 6 dB (Descornet *et al.* 2000). The ground surface has the same acoustic properties as a typical dense asphalt concrete, so it is not really noise reducing. Another technique which is now standard on Flemish highways is the exposed aggregate cement concrete. There is a single layer and a double layer version (with application of smaller sized aggregates at the top layer with D_{max} of 6,3 mm) and the best performing one, the double layer is not really noise reducing either as it also has typically the same noise emission as a dense asphalt concrete. However, some interesting developments could be seen the last years in Germany, Australia and the USA, where a pattern of narrow and wider longitudinal grooves in concrete (NGCS, "Next Generation Cement Concrete Surface") allowed to get a noise reduction of 3 dB, similar to a single layer porous asphalt (Scotfield 2011; Vorobieff *et al.* 2013).

The basic idea leading to this study was that it might be possible that the pattern could still be improved, as the optimization was so far based on a limited number of lab tests and on a not very reliable noise model, which was moreover not designed for this task (Goubert *et al.* 2015). A second issue is that the diamond grooving does not yield the optimum realization of the pattern: one gets thin, sharp "fins", breaking off irregularly leading to unwanted megatexture in the driving direction.

For this study, an experimental "trial and error" approach has been followed, comprising measurements on variations of the best existing NGCS pattern on the 2.0 m drum of the TU Gdansk. Both tyre/road and rolling resistance measurements are carried out with six tyre types at three speeds. In total eight different NGCS variations will be tested, but only five will be reported in this paper as the results for the last three are not available yet at the moment of the drafting of this paper.

2 MEASURING TYRE/ROAD NOISE AND ROLLING RESISTANCE ON A DRUM

The Technical University of Gdansk is equipped with an indoor facility for the measurement of tyre/road noise and rolling resistance of road surfaces, consisting of a powered steel drum with 2.0 m diameter (Figure 1, left hand side). On the outer surface of the smooth drum, the

Figure 1. TU Gdansk 2.0 m drum (left hand side), segments with the road surface under investigation applied to it (middle) and microphones mounted for measuring the tyre/road noise (right hand side).

pavement is applied which will be tested (Figure 1, middle). One spins the drum by means of an electrical engine and one keeps it turning at a fixed rotational speed, corresponding to a desired speed of the surface of the drum. Then the tyre mounted on a wheel fixed on a hydraulic arm is pushed against the turning surface (Figure 1, right hand side) and the emitted rolling noise is measured with two microphones.

The measurement conditions are fully complying with the CPX method as outlined in ISO 11819-2:2017. For this project, the measurements are carried out at surface speeds 50, 80 and 110 km/h and a set of six tyres (Figure 2) is used:

- Pirelli Cinturato P7 235/45R17
- Bridgestone Ecopia EP500 175/60R19
- GoodYear Efficient Grip 225/60R16
- AVON SUPERVAN AV4 195R14
- UNIROYAL TigerPaw SRTT 225/60R16
- Continental Conti.eContact BLUECO 195/50R18

The AVON and UNIROYAL tyres comply with ISO/TS 11819-3:2017. The rolling resistance of the tyre/road surface combination is evaluated by measuring the torque. The method is described in ISO 28580:2018.

Figure 2.   The six tyres used for the test; from left to right: the continental tyre, the uniroyal tyre, the avon tyre, the goodyear tyre, the bridgestone tyre and the pirelli tyre.

3   NEXT GENERATION CEMENT CONCRETE SURFACE (NGCS)

3.1   *What is NGCS?*

The surfacing of the type "Next Generation Cement Concrete Surface" (NGCS) consists basically of a pattern of fine "grindings" and combined with wider and deeper "groovings" in the longitudinal direction (Figure 3).

Based on experiences in the USA and Germany in October 2015, this new surfacing was applied by grinding with diamond disks on the N44 in Maldegem (Belgium), which is an old concrete road, dating from 1958 (Bergiers & Vanhooreweder 2022; Vanhooreweder & De

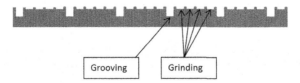

Figure 3.   Schematic transversal cross section of an NGCS.

Winne 2018; Vanhooreweder *et al.* 2020). The road consisted of concrete slabs with a thickness between 23 and 25 cm. Two test tracks of 100 m each were applied with slightly different patterns (profile 1 and profile 2). The diamond blades had the same diameter and thickness, the only difference is the thickness of the spacers between the blades for the application of the grindings (Figure 4).

The figures related to the two NGCS profiles can be found in Table 1. Pictures of the two surfaces is shown in Figure 5.

Measurements of the type Close-Proximity (CPX) were carried out with the CPX trailer of the Agency for Roads and Traffic 1, 5, 11 and 18 months after construction at 80 km/h and results for the SRTT tyres showed for profile 1 an initial noise reduction of 4,5 dB with respect to reference SMA 0/10, which boiled down to 2,5 dB nine months later. Then the

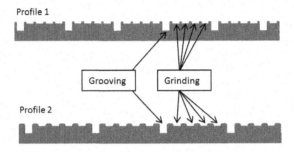

Figure 4.   Two NGCS patters of the test tracks on the N44 in Maldegem.

Table 1.   Figures related to the two NGCS profiles tested on the N44 in 2015.

Profile	N44 kp	Grooving: segment blade	Grinding: segment blade	Spacer
1	16.2–16.3	Thickness: 2.8 mm Diameter: 363 mm Depth groove: ± 4 mm	Thickness: 2.8 mm Diameter: 356 mm Depth groove: ± 1 mm	1,5 mm
2	16.4–16.5	Thickness: 2.8 mm Diameter: 363 mm Depth groove: ± 4 mm	Thickness 2.8 mm Diameter: 356 mm Depth groove: ± 1 mm	3,0 mm

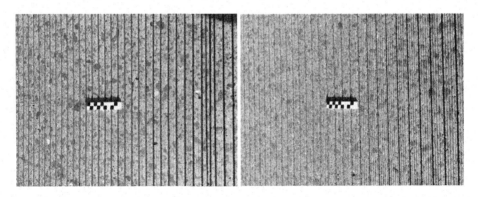

Figure 5.   The two test tracks shortly after the application of the surfacing: profile 1 (left hand side) and profile 2 (right hand side).

noise reduction stabilized. For profile 2 the initial noise reduction was almost zero and it remained that way during its lifetime. With the Avon tyre no noise reduction was measured on either profile, which is not unusual/unexpected.

3.2 *Bisinusoidal approximation of the NGCS profile*

Grinding with diamond disks leads in theory to a "rectangular" pattern as shown in Figure 4, but it looks in practice as shown in Figure 6.

It appears that the grindings are not as rectangular as one would expect but are rather "rounded". Note the numerous fins which have locally disappeared due to the breaking off, already at the time of construction, as this measurement was done in October 2015.

The measured profile can be approximated quite well with pieces of two sinus functions (Figure 7).

Drum measurements revealed that the results on the bisinusoidal approximation of profile 1 showed a very similar acoustic behaviour as measured on the ground/grooved profile on the road: the overall CPX level at 80 km/h was for 95.1 dB on the drum and 95.2 dB on the road. The noise spectra are very similar as well. The rolling resistance coefficient (Crr) for the Avon tyre at 50 km/h is 1.08 and on the drum it was 1.11. The RRC on the test track could only be measured with the Avon tyre. At 80 km/h the Crr on the drum is 1.1, while on the test track it was 1.20, which is due to increased air drag at higher speeds.

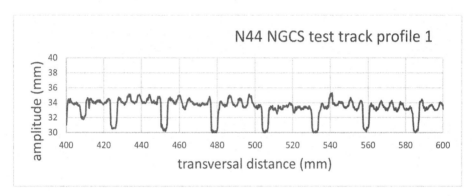

Figure 6.   Transversal profile measured on the N44 test track with NGCS profile 1.

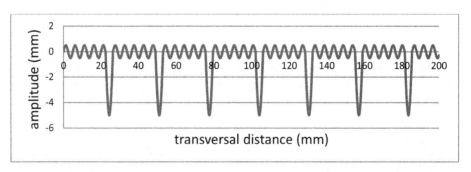

Figure 7.   Transversal NGCS profile, approximation of profile 1 by means of the composition of two sinus functions.

4   NGCS OPTIMIZATION ATTEMPTS

NGCS profile 1 was used as a starting point and – as shown above – is already performing quite well with respect to tyre/noise emission and rolling resistance. In this project seven new variations of the NGCS (each time differing slightly from profile 1) will be tested, four of which are already available at the moment of the drafting of this paper.

4.1   *Profiles 2 and 3*

Profile 2 which is tested on the drum is identical to profile 1, except that the amplitude and the texture wavelength of the grindings has been altered: both are increased with 40% (Figure 8). The idea is that a larger amplitude might further reduce air pumping and a larger texture wavelength would make the grindings more resilient to wear.

Profile 3 is also basically the same as profile 1, but with modified grindings. In this case the amplitude is as well increased with 40%, but the texture wavelength of the grindings is decreased with 40%. The aim is to have an even better initial noise reduction, combined with a lower rolling resistance (Figure 9).

4.2   *Profiles 4 and 5*

Profile 4 is shown in Figure 10 and is identical to profile 3, except for the increased amplitude, which is again set to the value as in profile 1.

Profile 5 is identical to profile 2, apart from the increased amplitude, which is again set to the value as in profile 1. In the course of the testing procedure, it has been decided not to test

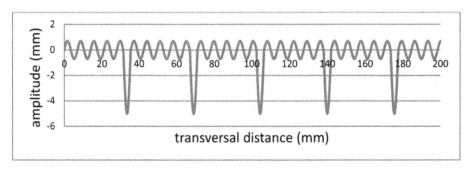

Figure 8.   Profile 2. The grindings in profile 2 have an increased amplitude (+40%) and an increased texture wavelength (+40%).

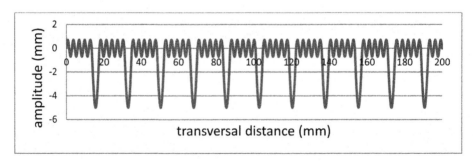

Figure 9.   Profile 3. The grindings in profile 3 have an increased amplitude (+40%) and a decreased texture wavelength (-40%).

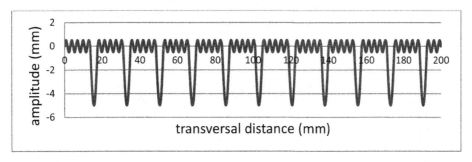

Figure 10.   Profile 4 is identical to profile 3, but without increased amplitude.

profile 5 and to abandon it, as it was assumed that it would not perform better than the other profiles.

4.3   *Profile 6*

Profile 6 has grindings with wider "fins", i.e. 40% wider than in the reference profile 1, while the canyons have the same with as in profile 1. From the experiment with the test tracks on the N44 we know that too wide fins decrease the initial acoustic performance, but wider fins wear off at a lower pace and hence the acoustic performance deteriorates less quickly. This profile is to seek for a better compromise between these two mechanisms (Figure 11).

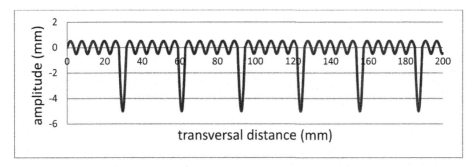

Figure 11.   Profile 6 with wider fins but the same canyons for the grindings (compared to profile 1).

5   TYRE/ROAD NOISE MEASUREMENT RESULTS

Figure 12 shows the noise reductions obtained on profile 2 with respect to profile 1. Negative noise reductions mean a noise *increase*, hence a less good performance of profile 2. It appears that only for the Avon tyre at 50 km/h, profile 2 performs slightly better than profile 1. For the other tyre/speed combinations the performance is worse or even much worse. An increase of the wavelength of the fine grindings combined with an increase of the amplitude of the fine grindings does hence not yield the desired effect of a noise reduction.

The trend visible in Figure 12 appeared unfortunately to be the same for all alternative NGCS profiles tested so far. None of them appeared to be successful and outperforming profile 1.

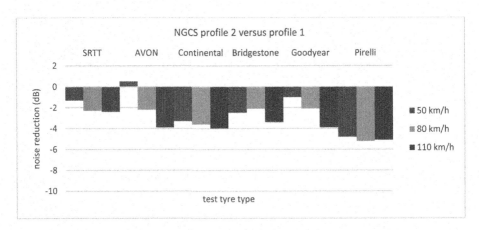

Figure 12. Noise reductions measured on NGCS profile 2 on the drum with respect to profile 1, measured on the drum for the six considered tyres and the three measurement speeds.

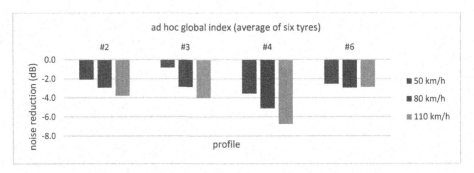

Figure 13. Noise reductions per tested NGCS variant and per speed, but averaged over the six test tyres, creating an ad hoc global index for the assessment of the acoustic quality.

To summarize and better visualize the noise measurement results on the drum so far, for each tested NGCS variant the noise reduction with respect to profile 1 is averaged over the six tyres. The result is shown in Figure 13. Best performing new variant at low speeds is profile 3, but it is still 1 dB noisier than the reference profile 1. At high speeds profile 6 performs best of the four tested alternatives, but still 3 dB worse than the reference profile.

6 ROLLING RESISTANCE MEASUREMENT RESULTS

The rolling resistance coefficient (Crr) was also tested on the drum for all combinations tyre/speed/profile. Generally, the Crr is only very slightly dependent on the speed (Goubert 2015) and this was confirmed here as well. The results for the six test tyres measured at 80 km/h on profile 1 and the four variants are shown in Figure 14. For all the tyres profile 1 yields the lowest Crr.

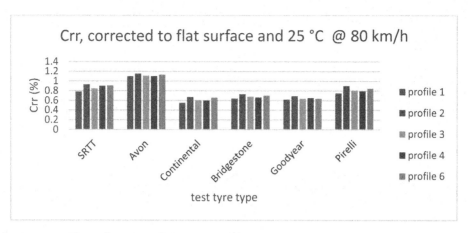

Figure 14. Rolling resistance coefficients (Crr) for the different tyre/profile combinations at 80 km/h.

7 CONCLUSIONS

We succeeded to find a procedure to make segments with a precise NGCS profile allowing to do tyre/road noise and rolling resistance measurements with a set of six tyres at three speeds on the TU Gdansk drum.

First a bi-sinusoidal approximation of the best performing NGCS profile tested in 2015 on the N44 in Maldegem was made and tested on the drum, which was called profile 1 and serves as a "starting point"/reference for the variations on the NGCS patterns. There was an excellent correspondence between the tyre/road noise emission for the SRTT and Avon tyre measured on the test track and on the drum and a fair correlation of the rolling resistance coefficient for the Avon tyre at 50 km/h.

Four alternatives for the reference profile 1 were tested so far, but profile 1 outperforms them all in nearly all circumstances, both for tyre/road noise as for rolling resistance. An increase or decrease of the wavelength of the small grindings combined with an increase of the amplitude worsened the tyre/road noise emission and the rolling resistance. Even worse results were obtained with only a decrease of the wavelength. Increasing only the width of the "fins" of the small grindings led to a moderate increase of the tyre/road noise. Wider fins are more resilient to wear, but the "price" paid as an increase of the initial tyre/road noise emission appears rather high.

ACKNOWLEDGEMENTS

This research was funded by the cluster "Strategic Initiative Materials (SIM)" of the Flemish Region through the GHRANTE project (HBC.2017.0608).

REFERENCES

Bergiers A. and Vanhooreweder B. 2022. Test Section With Next Generation Concrete Surface (NGCS) in Belgium: Rolling Resistance and Noise, *The 9th Symposium on Pavement Surface Characteristics (Surf)*, Milano, Italy, 12–14 September 2022

Descornet G., Faure B., Hamet J.F., Kestemont X., Luminari M., Quaresma L. and Sandulli D. 2000. Traffic Noise and Road Surfaces: State of the Art. Siruus Project: *Silent Road for Urban and Extraurban Use, Belgian Road Research Centre (BRRC)*

Goubert L., Bergiers A. Sandberg U. Min T.D.; Karlsson R. and Maeck, J. 2015. State-of-the-art Concerning Texture Influence on Skid Resistance, *Noise Emission and Rolling Resistance, Rosanne Project Deliverable D4.1*, Downloadable from www.rosanneproject.eu

International Standardization Organization (ISO) 2017. *11819–2:2017 Acoustics – Measurement of the Influence of Road Surfaces on Traffic Noise – Part 2: The close-proximity method*

International Standardization Organization (ISO) 2017. *ISO/TS 11819-3:2017 Acoustics – Measurement of the influence of road surfaces on traffic noise – Part 3: Reference tyres*

Kragh J., Nielsen E., Olesen E., Goubert L., Vansteenkiste S., De Visscher J., Sandberg U. and Karlsson R. 2011. *Final Report Optimization of Thin Asphalt Layers*, Opthinal Project Deliverable

Sandberg U. and Descornet G., 1980. *Road Surface Influence on Tyre/Road Noise Part 1*, Internoise – The International Conference on Noise *Control Engineering, Miami, Florida, USA*

Sandberg U., Kragh J.; Goubert L.; Bendtsen H., Bergiers A.; Biligiri K.P., Karlsson R., Nielsen E. Olesen E. and Vansteenkiste S. 2010. *Optimization of Thin Asphalt Layers – State-of-the-Art Review, ERA-NET Road Report*

Scotfield L. 2011. *Development and Implementation of the Next Generation Concrete Surface, Final Report, ACPA*

Vanhooreweder B. and De Winne P. 2018. Next Generation Concrete Surface (NGCS). Finally A Quiet and Sustainable Road Pavement?, *13th International Symposium on Concrete Roads, Berlin*

Vanhooreweder B., Beeldens A., Bergiers A., Goubert L. and Rens L. 2020. Next Generation Cement Concrete (NGCS), Evaluatie Proefvak N44 Maldegem uit 2015, *Final Report, Agency for Roads and Traffic of the Flemish Government, Belgium (Unpublished)*

Vorobieff G. et al. 2013. Development of Low Noise Cement Concrete Pavement in NSW, ASPA Conference

640

Roads and Airports Pavement Surface Characteristics – Crispino & Toraldo (Eds)
© 2023 The Author(s), ISBN 978-1-032-55149-4

Understanding user experience of in-vehicle noise based on road surface texture data

Tadas Andriejauskas*, Alex Wright, Rob Lee & Craig Thomas
TRL Ltd., Wokingham, UK

Stuart McRobbie & Neng Mbah
National Highways, Guildford, UK

ABSTRACT: Although the smoothness of road surfaces has a significant impact on the experience of road users in relation to ride comfort, in-vehicle noise also plays an important role in user experience. However, understanding the influence of the noise generated inside the vehicle as a result of the tyre-road interaction requires the use of complex noise measurement systems, which are difficult to deploy at the network level. In the UK work is being undertaken to develop a better understanding of how road surface quality affects user experience, including ride quality and noise, with the aim to better quantify these in network level surveys. A research study has been carried out for National Highways to understand how road surface texture, which is straightforward to measure over the network, influences in vehicle noise, and hence user experience. In this research measurements of in-vehicle noise and road surface texture were carried out on test routes, along with user surveys in which participants used an App to rate the levels of noise experienced on the same routes. By comparing the user data, noise and texture profile measurements an approach was developed to predict user experience of in-vehicle noise using the texture measurements. This paper describes the approach taken to carrying out this study and the development of a novel approach, based on the analysis of specific noise and texture wavebands, which enables road surface texture data to be correlated to the components of in-vehicle noise that affect users, and hence that adversely influence user experience. The encouraging results enabled the study to understand the potential for the development of a metric/index to broadly estimate user's experience of in-vehicle noise, based on texture profile measurements.

Keywords: User Experience, In-Vehicle Noise, Surface Texture

1   INTRODUCTION

Studies of road user satisfaction on the UK Strategic Road Network, carried out by Transport Focus (TF) and the Office of Rail and Road (ORR), have suggested that the current approach taken to assess and quantify road surface condition do not fully reflect the experience of condition reported by road users (Transport Focus & Highways England 2017; Transport Focus & Office of Rail and Road 2017). In the light of these National Highways has commissioned research to determine how the ability to relate the experience of road users with objective measurements of road pavement surface condition could be improved. This study has aimed to obtain quantitative data on the experience reported by users on specific lengths of the network and compare that with measurements of profile collected on the same

---

*Corresponding Author: tandriejauskas@trl.co.uk

DOI: 10.1201/9781003429258-62

lengths. When considering road profile, the surface smoothness and bumpiness are typically considered as the major contributors to the experience of road users, in relation to ride comfort. However, the influence of smoothness and bumpiness on the experience reported by road users can be influenced by other factors such as the in-vehicle noise experienced during the journey. When attempting to establish thresholds to quantify lengths of the network having poorer levels of ride quality, it can be difficult to isolate or quantify these factors and the extent to which they influence or confound the overall ride comfort and journey experience reported by users. Therefore, as part of the research relating the experience of road users with objective measurements of road surface condition, a focussed investigation was undertaken into the influence of noise on user experience.

The vehicle noise experienced by the user is dominated by the noise generated from tyre/road interaction at medium-high speeds (Sandberg & Ejsmont 2002). The road surface texture (at macrotexture and megatexture scale) is one of the main influencing factors affecting the noise transmitted into the vehicle (BS EN ISO 13473-1:2019). It arises from features that have smaller wavelengths of surface shape that those influencing road smoothness or bumpiness. In addition, research studies (Kragh *et al.* 2013; Sandberg & Ejsmont 2002) have shown that different wavelengths of texture can have either a positive or negative correlation with the sound pressure level, depending on its frequency band. It has been found that there is a positive correlation between noise levels below ~1 kHz and surface texture wavelengths larger than 16-20 mm, and a negative correlation between high frequency noise levels and texture with smaller wavelengths. Hence, the research study aimed to understand the feasibility of developing a metric to quantify the likely level of in-vehicle noise levels reported by users, based on road surface texture data. The following research questions were raised and are discussed further in this paper:

- How is user experience influenced by in-vehicle noise?
- Can in-vehicle noise be quantified using network survey data (texture)?

2 METHODOLOGY

The experimental methodology included the measurement of in-vehicle noise and road surface texture over the broad range of surface types present on the English Strategic Road Network (SRN), and on-road trials with participants to collect their experience (ratings) of in-vehicle noise. This was followed up with the data processing, alignment, and analysis to address the research questions on road surface texture, in-vehicle noise and its relationship with user experience.

2.1 *Reference data collection*

A route of approximately 150 km in length (Figure 1) was selected to collect reference in-vehicle noise and road surface texture data. The route included sections of motorways (M) and A roads, and included a range of surfaces (Thin Surface Course System Asphalt, Hot Rolled Asphalt, 14 mm and 10 mm materials and Retextured concrete). The route was defined in terms of a set of Sections using the referencing system currently in place on the SRN, which are referred to as "HAPMS Sections". These are defined in the network database using coordinates to specify the Section start and end points. Surveys that collect coordinate information alongside the measurements (such as in this study) can then be fitted to the network and compared.

Road surface texture profile was surveyed at 0.5 mm spacings using the National Highways HARRIS3 laser profilometer in the nearside wheelpath. The texture profile was used to calculation SMTD (Sensor Measured Texture Depth), PSD (Power Spectral Density) in 1/3 octave bands and skewness parameters over 10 m lengths.

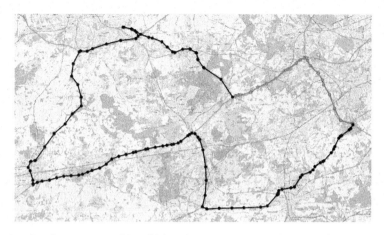

Figure 1.   Road surface texture and in-vehicle noise survey route (orange part of the route was used for user trials) (source: Ordnance Survey Maps 2021).

Reference noise measurements were collected using a sound level meter Norsonic 140EXP installed in a Skoda Octavia, following the recommendations of BS 6086:1981, ISO 5128-1980. A microphone was mounted to minimise the effect of vehicle vibration by attaching it to a metal frame covered with sound absorbing material. The measurements were collected to represent the noise levels experienced by the left (nearside) ear of the front passenger (Figure 2). The in-vehicle noise survey was carried out maintaining vehicle speed in a range of 60-70 mph whenever possible. This speed was selected to represent the majority of driving conditions on the Strategic Road Network. The route was surveyed 7 times.

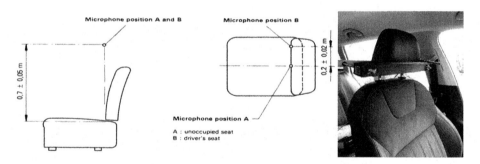

Figure 2.   Microphone installation (in accordance with BS 6086:1981, ISO 5128-1980).

Sound level measurements collected global sound levels and sound spectrum. The sound level meter was set to measure the fast response sound level with A frequency weighting. 1/3 octave band filtering was applied to the collected noise data spectral analysis. The sound was also itself recorded during the survey, to support in post-processing analysis. In addition to the sound measurement equipment set-up, an external GPS was used to collect location data to align with the sound readings. A dashcam was also fitted to the front windscreen to collect audio and video data throughout the survey, to help alignment and filtering of the collected sound level meter data, e.g., removing lengths where the vehicle had to move out of lane or additional noise sources or sudden events (such as loudly passing motorcycles, goods vehicles, etc.) could have adversely affected the results.

2.2 User trials

User trials were carried out using an App that was specifically developed to obtain spatial, quantifiable information on road user experience of in-vehicle noise. During the trial participants (front seat passenger) were asked to record their experience of in-vehicle noise every 20 seconds of their journey using buttons on the App. The question appeared on the app with 5 buttons to select from (Figure 3). The Noise App also featured an extra button which allowed the users to report when the vehicle had changed lane.

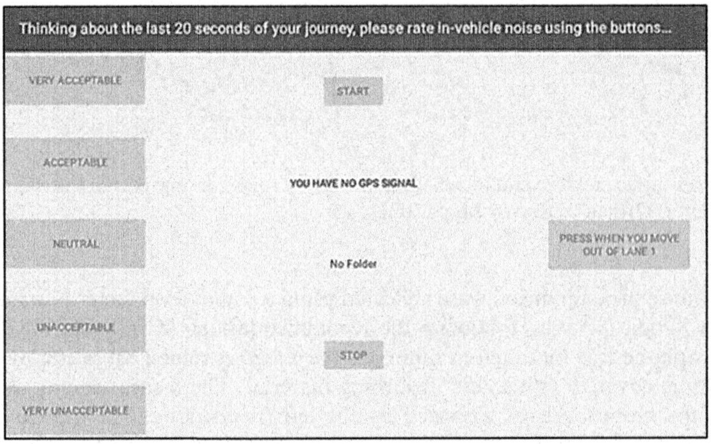

Figure 3. App used to collect user ratings on experienced in-vehicle noise.

Trials were carried out using a Skoda Octavia in Lane 1 at target speeds of 65-70 mph (whenever it was possible). The target speed was selected to be the most representative of typical driving speeds on the SRN. User trials were carried out on approximately 23 km of the test route (see Figure 1). As the trials were carried out during the COVID-19 pandemic, 3 "bubbles" formed of 2 participants each were recruited. In the first run one participant acted as the driver while the other used the App to record their experience, the participants then exchanged roles. Each participant carried out 2 repeat trial journeys, giving a total of 12 sets of user experience data.

2.3 Data processing and alignment

All data collected (noise survey data, HARRIS3 survey data, user trial data) and additional data was processed and aligned to create a several datasets:

- HARRIS3 texture data (SMTD, PSD in 1/3 octave bands, skewness) was fitted to HAPMS sections over 10 m lengths.
- The in-vehicle noise data (A-weighted global noise levels LAFspl, sound spectrum levels in 1/3 octave band) was collected on a time basis (sampling rate 125 ms). The Noise data was aligned with the distance and coordinates provided by the on-board GPS tracker and again fitted to the HAPMS sections and reported over 10 m lengths. Lengths where the data was collected in lanes other than Lane 1, or where additional noise sources (e.g., motorcycles) could have affected the results, were removed from the dataset. Survey speed data was also reported (calculated from GPS).
- User ratings were filtered to remove locations where a vehicle was out of lane. User ratings were then fitted to the HAPMS sections and reported over 10 m lengths. Again, speed was also reported for each user trial run.

644

- Pavement construction data (surface material, surface layer thickness, construction date) were obtained from the National Highways database, and fitted to HAPMS Sections. This was supported by a visual survey (carried out by manually analysing the HARRIS3 downward and forward-facing images) to identify (where possible) surface material type, max aggregate size and grading.

3   RESULTS

3.1   *User experience vs noise*

Before undertaking a comparison between the user and noise data, the in-vehicle noise values were filtered to compensate for differences in speed between the noise survey and the user survey. As there were several noise survey runs, this filtering was achieved by selecting the measurements from noise survey runs on each length that were collected at a speed close to the average user survey run over the same length. The noise measurement was then further corrected for speed using a linear correction, that was derived from the survey data.

The user ratings were also normalised to adjust for individual sensitivity. Analysis of the user ratings found that some users used only 2 ratings over the whole route, while others used 4 ratings. To accommodate for user sensitivity the data from each user trial run was normalised, using its mean and standard deviation values. The normalised user ratings were then averaged for each 10 m length to obtain an overall normalised user rating for each length. Figure 4 plots the normalised user ratings against normalised in-vehicle noise levels. There is a reasonable visual trend, but a large degree of variation. The difference in noise levels between the concrete (average normalised in-vehicle noise level 72.08 dBA) sections and other surface types (average normalised in-vehicle noise level 68.08 dBA) can be seen in the two "groups" of data - a significant proportion of the "higher" noise group is associated with the concrete lengths. Combining all surface types together assists in achieving a wide range of user experience ratings (as there were few "very noisy" flexible lengths on the route). The correlation can be considered as good with R^2 of 0.71 This comparison therefore shows, with reasonable confidence, that user experience of noise can be understood using this App based approach to obtaining user experience data. It also shows that the experience reported by the users (herein assessed over a range of acceptability) can be related to the noise physically measured in the vehicle.

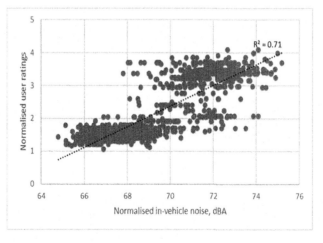

Figure 4.   Relationship between normalised user ratings and in-vehicle noise levels.

645

3.2 *Estimating noise using road surface texture*

The relationship between the texture and noise can be difficult to characterise using simple parameters derived from the texture. For example, low correlation ($R^2 = 0.06$) was found between the SMTD and in-vehicle noise (LAFspl) at 105-115 km/h. Slightly better agreement is obtained when comparing the texture skewness parameter with in-vehicle noise ($R^2 = 0.31$). This broadly confirms that road surfaces having negative textures will generate lower noise levels.

Building on previous road surface texture and vehicle noise studies (Kragh *et al.* 2013; Sandberg & Ejsmont 2002), a spectral analysis of the in-vehicle noise and texture wavelengths was undertaken. Correlations between couples of texture wavelengths (1/3 octave centre bands ranging from 1 to 500 mm) and sound spectra (1/3 octave band centre frequencies covering range from 31.5 to 10,000 Hz) were calculated. Relationships were obtained from linear regression analyses for each surface type at driving speeds of 105-115 km/h (these speeds are closest to normal driving speeds at SRN and correspond to the speeds used in the user trials). Strong correlation (greater than 0.5 either positive or negative) was found only for 10 mm aggregate sizes and HRA pavements. For other materials the correlation was weak. That suggested there is no common set of single frequencies that can be extracted to characterise the noise of all surface types. However, we propose that a combination of different correlation couples could be used to improve the performance. 2D contour plots were used to select the most relevant texture wavelength and acoustic frequency bands and multiple linear regression analysis were carried out. The selection of the frequency bands of interest drew on further analysis of the data collected in the user trials.

The sound pressure levels in each frequency band were extracted from the noise measurements and compared with the user ratings. Regression analysis was carried out between user ratings and each frequency band so that an understanding could be obtained of which frequency bands had most effect on the users in the trials. Figure 5 shows that user ratings were most affected by the noise levels in 1/3 octave bands 100-500 Hz, with Pearson correlation coefficients R of 0.70-0.83. Therefore, comparisons were carried out between couples in this range. It was found that acoustic frequency bands of 160 Hz and 500 Hz gave the highest multiple R for all surface types. A similar approach was used to select the most relevant texture wavelengths, which found that 4 mm, 63 mm and 315 mm texture wavelengths have the strongest relationships with 160 and 500 Hz acoustic frequency bands, and potentially could be used as a combined approach for estimating noise using texture data.

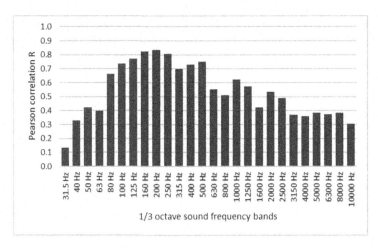

Figure 5. Pearson correlation between normalised user ratings and sound pressure levels in 1/3 octave frequency band.

Table 1 summarises the multilinear regression analysis in the selected bands, broken down by all surfaces (excluding concrete) and by each surface type. The global noise level, LAFspl, has a very strong relationship with the acoustic frequency in the 160 Hz and 500 Hz bands for all surfaces (excl. concrete), TSCS, HRA and 10 mm materials, but is weaker for 14 mm materials and concrete. When multilinear relationships are considered between the texture (wavelengths 4, 63 and 315 mm) and the acoustic measurements (160 Hz and 500 Hz bands) it can be seen that the individual relationships are weaker. However, this improves when the texture is compared with the LAFspl. Finally, Table 1 shows the relationship with the LAFspl when SMTD and skewness are included. Although the individual relationship is weak, SMTD and skewness are influencing factors on tyre/road noise generation and their inclusion results in a slight improvement in correlation for most of the surfaces – the right column in Table 1.

Table 1. Analysis of the multilinear regression between the 4, 63, 315 mm texture wavelengths, global noise level LAFspl and sound frequencies in 160 Hz and 500 Hz.

		Multilinear regression R			
Surface type	LAFspl against 160 and 500 Hz	160 Hz against 4, 63 and 315 mm wavelengths	500 Hz against 4, 63 and 315 mm wavelengths	LAFspl against 4, 63 and 315 mm wavelengths	LAFspl against 4, 63, 315 mm wavelengths, SMTD and skewness
All surfaces (excl. concrete)	0.94	0.53	0.54	0.63	**0.72**
TSCS	0.97	0.61	0.53	0.67	0.74
HRA	0.95	0.67	0.74	0.80	0.80
10 mm materials	0.96	0.79	0.75	0.83	0.85
14 mm materials	0.87	0.21	0.30	0.30	0.33
Concrete	0.76	0.12	0.23	0.28	**0.34**

3.3 Application to the prediction of user experience levels

The multilinear relationships from Table 1 have been applied to the surface texture data to estimate/predict the global (LAFspl) in-vehicle noise level that would be experienced on each length of the test route. To calculate these values we have applied the combined relationship obtained using the texture wavelengths, SMTD and skewness to predict LAFspl on all surfaces (excluding concrete) and the combined relationship obtained using the texture wavelengths, SMTD and skewness to predict LAFspl on concrete. The combination of these two sets of predicted values are shown in Figure 6 (concrete lengths highlighted grey). The differences between the measured and predicted values are also shown. There is very good agreement in the shape of the data. It can be seen that the measured in-vehicle noise values have more variability than the predicted values. This is partially a result of the collation of noise data from 7 noise survey runs (selecting the survey run in each 10 m length that had a speed closest to the noise user trial speed, as discussed above), which has introduced some steps and variation into the measured noise data

For the concrete surfaces, the predicted values do not capture the small variations shown in the measured data within the concrete surfaces – but generally higher noise levels are predicted. On the majority of the route, the predicted values are higher than the measured values, with average residuals between measured and predicted sound pressure levels of -0.28.

Bearing in mind the original objective of the study – to understand user experience of interior vehicle noise - the ability to predict the noise from the texture data also provides the

647

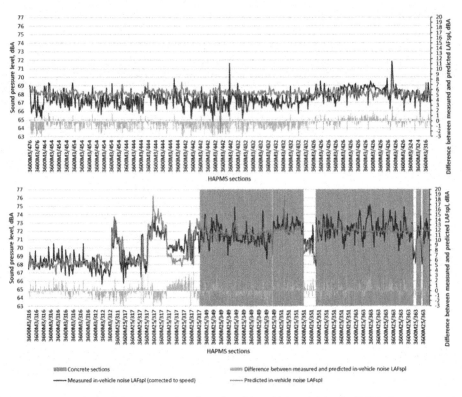

Figure 6. Comparison of the measured and predicted in-vehicle noise levels (LAFspl) over the noise user trial route (use of multilinear relationships for "all other surface types" and "concrete").

ability to predict how users would rate the noise levels. This can be achieved using the relationship obtained in Figure 4 and applying this to the predicted noise levels. Figure 7 compare the predicted and true user ratings reported on the test route. The correlation between the normalised true user ratings and predicted user ratings is 0.71.

When the predicted and true user ratings are plotted along the survey route (Figure 8) it can be seen that our approach shows agreement between the predicted and true noise ratings on the lower noise sections. It also clearly predicts poorer user noise ratings on the concrete lengths. However, it does under-report (i.e., predicts better user ratings) the specific level of user ratings on the noisier surfacings (especially concrete). There are also other localised mismatches (e.g., green bubble in Figure 8) where our predicted user ratings are much higher than the users reported. This disagreement can be partly explained by the fact that the user trial speed which was a little slower than the 105-115 km/h speed used to develop the in-vehicle noise->texture relationships. However, examination of the blue data on Figure 8 shows that the user ratings did not agree with the measured in-vehicle noise on these localised lengths. A similar, but opposite, situation is also noted in the middle of the concrete length where the measured noise levels decreased but users still reported high ratings. This may be due to "user fatigue" where users experiencing persistent lower or higher noise levels have not reported the sudden increase or decrease in noise or it may reflect specific frequency components that have a different effect on user experience. Nevertheless, overall results suggest that the method can predict in-vehicle noise and user rating to a reasonable accuracy even where no previous knowledge on surface type (except if it is asphalt or concrete) is known.

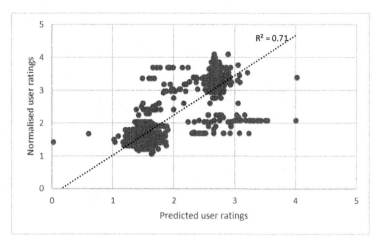

Figure 7. Relationship between recorded and predicted user ratings (use of multilinear relationships for "all other surface types" and "concrete").

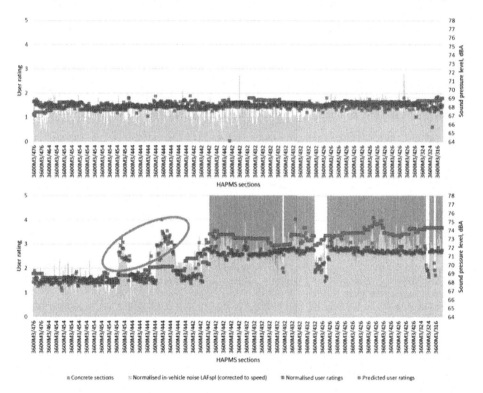

Figure 8. Comparison of the recorded and predicted user ratings over the noise user trial route (use of multilinear relationships for "all other surface types" and "concrete").

4 CONCLUSIONS

This research study has aimed to better understand the experience of in-vehicle noise by road users, and how this relates to the surface texture. It has shown that user ratings of in-vehicle noise can be collected using an App approach. In general, the user ratings obtained using the App have good agreement with measured in-vehicle noise levels.

The relationship between road surface texture and in-vehicle noise is complex and is dependent on surface type and vehicle speed. However, this work suggests that vehicle noise levels can be estimated using a multilinear relationship of combined texture power spectral density in the 4 mm, 63 mm and 315 mm 1/3 octave bands, and SMTD and texture skewness. This suggests that there is a potential to develop a general relationship between in-vehicle noise and road surface texture that is broadly independent of surface type. In addition, it is likely that the robustness of this relationship could be improved if information on the surface type, including aggregate size, grading (open graded, dense graded, semi-dense graded, etc.) and surface condition (e.g. ageing, fretting) is available. The study has developed this relationship for speeds of 105-115 km/h, which correspond to the most common traffic speeds on the SRN.

As the study has shown the potential to establish relationships between in-vehicle noise and user experience, and between road surface texture and in-vehicle noise, it is feasible to add a further step that estimates the rating that users would give on in-vehicle noise, based on measurements of road surface texture. Clearly, as this is a broad estimate, there will be uncertainties in the prediction of the user rating. However, it is reasonable to assume that it would be feasible to use road surface texture as a proxy to estimate user experience of noise, at least for particular groups of road surfacing materials. The development of stronger individual relationships between in-vehicle noise and road surface texture parameters would improve confidence in this ability.

ACKNOWLEDGEMENT

This study was funded and supported by National Highways (SPaTS contract 1-1095).

REFERENCES

BS 6086:1981, ISO 5128-1980 *Method of Measurement of Noise Inside Motor Vehicles*. British Standards Institution, 1999.

BS EN ISO 13473-1:2019 Characterization of Pavement Texture by Use of Surface Profiles. Part 1: Determination of Mean Profile Depth. British Standards Institution, 2021.

Kragh J., Iversen L.M. and Sandberg U., 2013. *Road Surface Texture for Low Noise and Low Rolling Resistance. Nordtex Final Report*. Veijdirektoratet, Copenhagen, Denmark.

Sandberg U. and Ejsmont J.A., 2002. Tyre/road Noise Reference Book., Informex, Kisa, Sweden.

Thomas C., Dhillon N., Nesnas K. and Wright A., 2020. *A Practical Investigation to Understand How Road Users Experience Ride Quality on the SRN, TRL Client Project Report CPR2767*.

Transport Focus, & Highways England. (2017). *Road Surface Quality: what Road Users Want from Highways England*.

Transport Focus, & Office of Rail and Road. (2017). *Measuring Performance of England's strategic Roads: What Users Want Transport Focus*.

Roads and Airports Pavement Surface Characteristics – Crispino & Toraldo (Eds)
© *2023 The Author(s), ISBN 978-1-032-55149-4*

Effect on noise reduction of repaving only the top layer in a double-layer porous asphalt pavement

Ulf Sandberg*

Swedish National Road and Transport Research Institute (VTI), Linköping, Sweden

ABSTRACT: Since 2010 a double-layer porous asphalt pavement has been used on motorway E4 through the Swedish city Huskvarna. The pavement has been a great success despite the challenge to use this kind of pavement in a country where studded tyres are the main tyres used in wintertime, resulting in excessive surface wear and subsequent clogging of pores. After 7 years the pavement had lost half of its initial 7.5 dB noise reduction and was therefore repaved. Then the top layer of the major part of the 2.7 km long section was milled-off and then repaved only with a new top layer, while the bottom layer was reused. Normally, both layers would be repaved, but in this case only a few hundred meters were repaved fully, which makes it possible to compare the two principles (repaving both layers or only the top layer). Reusing the bottom layer in this way of course saves money, but the question is how effective this pavement will be when the bottom layer already was partly clogged in its first 7 years of service. A question is also whether the connection between the old and new layer could be durable enough without the creation of a partly dense binder between the new top and the old bottom. Now this new pavement section has been in service more than halfway towards its expected end-of-life. Consequently, it is possible to start evaluating how well the new paving principle has worked out. The results so far indicate that the noise reduction which is lost by reusing the old bottom layer is less than 1 dB (of the initial 7-8 dB) over at least the first four years of operation, which may be seen as a reasonable sacrifice for saving a major part of the repaving cost. There has been no problem with separation between the two layers. However, it is unlikely that the bottom layer will be useful for another cycle of operation. Nevertheless, the principle of reusing the bottom layer (each second cycle) may make the use of double-layer porous asphalt more attractive from a cost-benefit point of view.

Keywords: Porous Asphalt, Double-Layer, Noise Reduction, Recycling, Repaving, Top Layer

1 INTRODUCTION

One of the most common ways to mitigate road traffic noise is to pave the road with a wearing course of porous asphalt (PA). The porosity reduces the so-called air pumping noise in the tyre/road interface and also provides a certain sound absorption. A general problem is that the pores get clogged by tyre rubber and road wear particles, which eventually eliminates the advantage of the porosity. To be effective, especially for sound absorption, but also for providing room for temporary or permanent storage of wear particles, it is important that the porous pavement is relatively thick—generally the thicker, the better. Therefore, double-layer porous asphalt has become common and nowadays, in Sweden only double-layer porous asphalt is used for noise reduction purposes. The technology related to paving double-layer porous asphalt was presented

---

*Corresponding Author: ulf.sandberg@vti.se

DOI: 10.1201/9781003429258-63

around 30 years ago by Dutch researchers and has been known and implemented for about 20 years now and is currently more or less standard internationally. Much of the knowledge was developed within the huge Dutch Innovation Programme (for Noise) in the years 2002-2008 and the main principles are still the same. The very comprehensive reports from that programme are no longer easy to find on-line, at least not in English, but summaries are found in (Goubert *et al.* 2005) and in a workshop report (Bendtsen *et al.* 2005).

Compared to a reference pavement of SMA 11 or DAC 11 (the virtual "average" of those two pavements is the reference in the European traffic noise prediction model Cnossos-EU), usually a noise reduction of 6 dB ($\pm$ 1 dB) is achieved initially, which by time is reduced by approximately 0.5 dB per year, mainly due to clogging and ravelling. In the Netherlands, the leading nation in the use of this type of pavement, usually, an acoustic lifetime of 6-9 years is achieved with pavements having 8 mm max. aggregate size in the top layer. Pavements having 6 mm max. aggregate size are also used, which give somewhat higher initial noise reduction but at the expense of lifetime.

2 REPAVING POLICY

After a certain (acoustic) lifetime, which in Sweden has been 7 years so far, the pavement needs to be replaced. Note that in Sweden studs in tyres in wintertime wear road surfaces much more than non-studded tyres, which means that pavements in Sweden must be repaved much earlier. The normal policy and practice would then be to mill off both the two porous asphalt layers (which may not be so porous anymore) and replace them with two new porous layers. However, in the latest repaving of double-layer porous asphalt in Sweden, only the 30 mm thick top layer was replaced with a new layer and the 50 mm thick old bottom layer was re-used for most of the road section. Since a part of the road section was repaved with new layers for both the top and bottom, it is possible to compare the two paving policies: repaving only the top layer on top of the old bottom layer, versus repaving both the top and bottom layers. To partly recycle the pavement in this way will of course mean a substantial cost saving, as more than 60% of the material is re-used, while simultaneously the operation is simplified as only one paving operation is needed, but is the potential loss in noise reduction worth it?

An earlier study by the author has demonstrated how important the bottom layer is for noise reduction (Sandberg & Mioduszewski 2012). Therefore, one may suspect that if the bottom layer is not of the highest quality and essentially free of clogging, the practice of re-using the bottom layer may be detrimental to noise reduction.

The author is unaware of any other attempts to re-use the lower layer of double-layer porous asphalt, as explained in this paper. Nevertheless, it is possible that Dutch researchers have tried the technology, but not reported it publicly in English.

3 PURPOSE AND LIMITATIONS OF THE STUDY

The main purpose of this study is to compare the acoustical performance of a double-layer porous pavement in Sweden, where the top layer was exchanged with a new layer, but the bottom layer was re-used, with another section of the same road where both layers were replaced with (similar) new ones. Both sections carry the same traffic and are similar in all essential ways, except for the bottom layer. When this is written it is almost five years since the paving took place (2017) but in this paper, measurement results from only the four first years (2017-2021) are available.

This paper reports the results of the first four years of operation (which may cover about 60% of the lifetime) but when the oral presentation of this paper is made, it is hoped that noise data are available also for the fifth year (70-80% of the expected lifetime). Nevertheless, the plan is to cover the full lifecycle in the following years and report the final results at a later congress.

The performance of this kind of pavement is highly dependent on how it is laid, and it is possible, or even likely, that with another paving contractor, results may be different.

Finally, it shall be noted that the results reflect the situation in Sweden, with the extreme wear by studded tyres in wintertime. In warmer climates, the lifecycle of such pavements will be significantly longer due to less wear resulting in less clogging and rutting. But the principle of re-using the bottom layer is universal.

4 DATA FOR THE TEST SECTIONS

In 2010 a double-layer porous asphalt (DPA) pavement was constructed on the E4 motorway through the Swedish city Huskvarna (near Jönköping) which is the test object in this study. The motorway has two lanes per direction, where the right lane is designated K1, and the left lane designated K2. This pavement (which included a few special sections) was repaved in 2017, when the first generation of the DPA had been in service for 7 years and at that time had lost about half of its initial noise reduction.

The main section of the newly (in 2017) paved DPA (here referred to as DPA 2) is approx. 2.3 km long, of which the northern 1.7 km is shown in Figure 1 with the newly laid top layer. It was repaved by milling off the top layer and adding a totally new top layer of the same construction as in the initial DPA pavement. That means that the bottom layer of the 7 years old DPA pavement was re-used as the bottom layer of the renewed pavement. Just south of this location is a 300 m long section of the motorway (here referred to as DPA 1) where both the top and bottom layers were repaved in 2017. It means that the only difference between the two sections is the bottom layer, which is either old or new (in 2017).

Some facts of the test objects are listed below:

- Length: 2.3 km DPA 2, and 0.4 km DPA 1
- Posted speed limit: 90 km/h
- AADT: 24000 vehicles (15% heavy vehicles). Distribution between lanes: 70% in K1 and 30% in K2, the main part of the heavy traffic in K1. These are five years old data, probably traffic has increased since then.
- Layer thickness (nominal): top layer 30 mm, bottom layer 50 mm
- Aggregates in the top layer: max. size 11 mm
- Aggregates in the bottom layer: max. size 16 mm
- The air voids content was initially approx. 25%
- The binder is a highly modified polymer bitumen from Nynas, named Endura D1
- Cross slope is 3% (note that this is higher than the more common 2 or 2.5%)
- DPA 1 is located immediately south of DPA 2, in both southern and northern direction. Therefore, they would carry the same traffic
- The contractor for this test road is Svevia AB.

Figure 1. Part of the DPA 2 section on E4 in Huskvarna just after repaving.

In most other European countries, a maximum aggregate size of 6 or 8 mm in the top layer would have been preferred, because it would have resulted in better noise reduction, but in Sweden and other countries which allow studs in tyres in wintertime, it is considered that aggregates smaller than 11 mm would result in too much ravelling and other wear. Thus, a part of the potential noise reduction had to be sacrificed for durability reasons.

5 PAVING OPERATIONS

The regular paving of a DPA pavement is to use a paver with two screeds. The first one lays the bottom layer, and the second screed lays the top layer. This technique is called "hot-on-hot" since the two asphalt layers are hot when they are combined into one pavement (Sandberg & Masuyama 2005). A variant of this is to use two normal pavers, operating in conjunction – one-after-the-other. It has traditionally been considered that in order to have a durable connection between the bottom and the top layer it is necessary to lay them hot-on-hot.

However, the contractor on this motorway, Svevia AB, has used a completely different technique, namely, to lay the bottom layer with a normal paver one day and the top layer the next day. So far, it has been done always in summertime at temperatures above 20 °C. This means that the road lane must be closed a longer time than when laying hot-on-hot, which can work if there is an alternative route for the traffic or if traffic is not too high. It saves the cost of a paver with two screeds. This alternative paving procedure has appeared to have no disadvantages in terms of poor connection between the bottom and the top layers. No such problems have occurred during the presently 15 years after the first paving in this way was made.

For the test object DPA 2 (when only the top layer was replaced), the operations during the repaving were as follows:

- The top layer was milled off (nominally 30 mm)
- The bottom layer (the old one) was cleaned by a standard road-cleaning machine
- The new top layer was paved, without any added median binder

A close-up picture of the top layer surface appears in Figure 2, at an age of four years.

Figure 2.   Picture of the surface of the top layer of the DPA in Huskvarna in May 2021 (age four years). The coin has a diameter of 25 mm. The surface looks essentially the same for both DPA 1 and DPA 2 with signs of some pores open to the surface (maybe 20 or so within the area of the picture).

Unless otherwise mentioned, all noise measurements reported in this paper were made using a CPX trailer from the Gdansk University of Technology (GUT); in recent years the version marked "Tiresonic Mk4"; see Figure 3. Tests have been made in all essential details according to the CPX (Close Proximity) method specified in ISO 11819-2. The CPX measurements were made at least once per year and have in most cases covered the entire length of the tested sections, except for run-in and run-out parts, which means lengths of 100-2700 m.

During the noise measurements, two reference tyres have been used: SRTT and Avon AV4, denoted P1 and H1, respectively, in ISO/TS 11819-3. See Figure 4. The P1 tyre is assumed to represent car tyres and the H1 tyre assumed to be a "proxy" for truck tyres.

The tyre load during measurements was fixed at 3200 N and the inflation pressure was adjusted to 200 kPa in cold conditions, as required in ISO 11819-2. Measurements have been performed at 50 and 80 km/h and have been made in the right wheel track in all lanes. In the slow (K1) lane, measurements were made also in the left wheel track and between the wheel tracks.

To compensate for variations in air temperature during the measurements, the correction procedure in ISO/TS 13471-1 has been applied. In addition, texture measurements (Mean Profile Depth – MPD) have been made occasionally in accordance with ISO 13473-1, using the VTI RST vehicle. This has also included rut depth, megatexture and IRI.

Figure 3.   (left). CPX noise tests with the GUT Tiresonic Mk4 trailer on the DPAC pavement on E4 in Huskvarna. The test tyre is mounted in the middle of the chamber.

Figure 4.   (right). Tread patterns of the two reference tyres used during the CPX noise tests. From left to right: SRTT 16" (P1) and Avon AV4 (H1).

Texture measurements were made in some years, using the RST vehicle of VTI, but did not distinguish between the two sections studied here, so these data are not useful in this paper. Instead, the texture measurements made in 2021 in the annual road surface testing programme of the Swedish Transport Administration can be used. They are reported in the database PMSv3 [PMSv3 2022] and, consequently, the author has picked out the texture and rut depth data from that database.

At one occasion (2021) when parts of the tested sections were closed for visual inspection in one direction at the time, permeability ("drainage") measurements were made, in accordance with EN 12697-40. Additionally, measurement of rolling resistance was made at one occasion, using the "R2 trailer" of GUT. However, this measurement did not distinguish between the two sections so it will not give any additional information to how they may differ. The same applies to wet friction made with the Saab Friction Tester.

7 RESULTS

7.1 *Noise*

Noise measurements have been made each year, starting one month after the laying of the new pavements. The results in terms of overall A-weighted (CPX) noise levels are shown in Figure 5 as noise reduction compared to the reference pavement (which is at 0 dB). However, it must be noted that the measurements in 2017 have been found to be influenced by water remaining under the surface, probably in the bottom layer, from rain 24 hours earlier. The CPX standard allows measurements 24+ hours after rain but in this case the weather was rather chilly, cloudy, and windless. Under such unfortunate conditions, 24 hours is not enough to dry up a thick pavement like this 80 mm thick DPA. It is estimated that both noise reductions in 2017 (at 0 age) are 1.0 – 1.5 dB too low due to this. The results shown are averages at 80 km/h for both tyres as well as for both the southern and northern direction of lanes and for both the slow and fast lanes.

The reference pavement is a mix of SMA 16 pavements in "normal" condition, of age 2-8 years, measured each year. This is selected as reference since it is the totally dominating pavement on the Swedish national road system. Separate measurements by VTI have shown that this reference pavement is 1.1 dB "noisier" than the European virtual standard Cnossos-EU, which is a mix of SMA 11 and DAC 11.

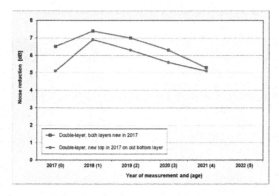

Figure 5.   Noise reduction at 80 km/h versus measurement year and age, compared to similar measurements on a mix of Swedish SMA 16 pavements. Note that the measurements in 2017 were influenced by poorly dried-up pavements and it is estimated that the reduction levels then are 1.0 – 1.5 dB too low due to this.

Comparisons of frequency spectra are shown in Figures 6 and 7. The first one shows the situation after one year of service (age one year), comparing DPA 1 and DPA 2, and with the reference pavement also shown. Due to the mentioned problem with insufficient dry-up of the pavements when they were new (in 2017), it is not meaningful to show the spectra of 2017 as the comparison may be influenced by different degrees of dry-up.

Figure 7 shows the same spectral comparison when the pavements had been in service for four years (in 2021). See further the discussion.

7.2 *Texture and rut depth*

Texture and rut depth measurements, allowing to distinguish between the two pavement sections, were made only in 2021 [PMSv3 2022]. These results are shown in Tables 1 and 2. However, a few measurements were made in 2018 when the pavements were one year old.

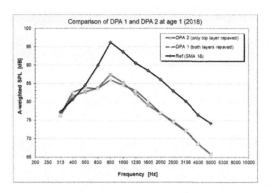

Figure 6.　Frequency spectra measured on DPA 1 and DPA 2 at an age of one year, with the mix of SMA 16 pavements as reference. Average for the two tyres, the two lanes and the two directions.

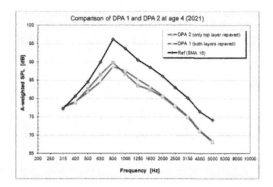

Figure 7.　Frequency spectra measured on DPA 1 and DPA 2 at an age of four years, with the mix of SMA 16 pavements as reference. Average for the two tyres, the two lanes and the two directions.

Table 1.　Results of texture measurements (MPD values) in 2021, at an age of four years.

Lane	Wheel track	DPA 1 (both layers repaved)	DPA 2 (only top layer repaved)
	Right	1.60	1.60
Slow (K1)	Middle	1.90	1.80
	Left	1.55	1.55
	Average	1.68	1.65
	Right	1.45	1.20
Fast (K2)	Middle	1.77	1.65
	Left	1.45	1.25
	Average	1.56	1.37
Slow & fast	Overall average	1.62	1.51

657

Table 2. Results of rut depth measurements in 2021, at an age of four years, in mm.

Pavement	Slow lane (K1)	Fast lane (K2)	Average
DPA 1 (both layers repaved)	6	3.5	4.8
DPA 2 (only top layer repaved)	6	4.8	5.4

Table 3. Results of permeability measurements (time of water outflow in seconds).

Lane	Wheel track	DPA 1 (both layers repaved)	DPA 2 (only top layer repaved)
Slow (K1)	Right	60	67
	Middle	34	61
	Left	60	53
	Average	51	60
Fast (K2)	Right	58	80
	Middle	52	44
	Left	32	52
	Average	47	59
Slow & fast	Overall average	49	59

Then the MPD values were 1.80 mm in the wheel tracks and 1.85 mm between the wheel tracks. This would likely be the case for both pavement sections as they were still rather new and were nominally the same pavement layer when they were laid. Note that the MPD values of the reference mix of SMA 16 pavements is around 1.2 mm.

7.3 *Permeability*

Permeability measurements were made when the pavements were four years old, and the results are shown in Table 2. Measurement method was according to EN 12697-40. More frequent measurements could not be done since they require closure of one-half of the road. Note that higher values mean a slower water outflow, i.e., a lower permeability.

It appears that the DPA 1 has a faster outflow (greater permeability) than DPA 2 by about 20%. It is logical, as the latter has a bottom layer which was 11 years old and potentially more clogged, compared to DPA 1 which has a bottom layer four years old (same age as the top layers).

8 DISCUSSION

8.1 *Noise*

The results in the diagrams show only average for the two tyres, the two lanes and the two directions. This is both because there are no dramatic differences between those 2x2x2 combinations and that looking at averages reduces misjudgements due to uncertainties.

It is immediately obvious how efficient in noise reduction the DPA:s are and this applies to both of them. With an initial noise reduction of 7-8 dB, decreasing to 5 dB after four years of service, it means that the decrease of noise reduction by time is approx. 0.5 dB per year. This is very good for conditions when wear of studded winter tyres is substantial. The frequency spectra show that the most efficient reduction is at the peak and its two adjacent third-octave bands. There is no reduction at the frequencies below 600 Hz since sound absorption is effective only at 630 Hz bands and above, and the air pumping mechanism is important from and above about 1000 Hz (Sandberg & Ejsmont 2002).

It appears, not unexpectedly, that DPA 2 gives slightly less noise reduction than DPA 1. The difference is on the average 0.6 dB during the years 1-4. The frequency spectra, especially in Figure 7, show that DPA 2 is more similar in shape to the reference pavements than DPA 1, which suggests that it is somewhat less porous than DPA 1. Figure 7 shows that both DPA:s are approaching the shape of the reference SMA 16 which means that they are starting to become clogged. The somewhat lower spectral levels at frequencies below 600 Hz at the age of four could be due to the progressive wear of the texture.

It is also noteworthy that the differences between DPA 1 and DPA 2 at the frequency bands 800 to 1600 Hz, show signs of sound absorption shifting its peak from 800 Hz to 1250 and 1600 Hz. This is consistent with the observations that thick pavements like double-layers have a sound absorption peak at 630-800 Hz, while thinner pavements have a shift in that peak to higher frequencies; see e.g. (Sandberg & Ejsmont 2002). This would suggest that the DPA 2 has a thinner *effective* layer for sound absorption, namely the 30 mm top layer, while the DPA 1 still has an effective bottom layer with its thickness of 50 mm in addition to the top layer of 30 mm.

8.2 *Texture, permeability and rut depth*

Note that the MPD values of the reference SMA 16 pavements are around 1.2 mm, which means that the texture of the DPA:s is generally rougher. However, one must note that much of the profile of the DPA:s' texture is directed downwards, into deep valleys, where the tyres cannot envelope the profile. This is often referred to as a negative texture and is acoustically positive. Changes in texture, rut depth and permeability may occur in different ways:

- Heavy vehicles may cause the two layers to get compacted which may reduce the voids and cause rutting. The pores may get narrower. This would occur essentially only in the slow lane, which is dominated by the slower moving heavy vehicles.
- Light vehicles (mainly cars, including SUV:s), of which more than 50% are equipped with studded tyres, will wear the top of the surfaces and part of the dirt worn off may be pressed down into the pores. The wear of the top of the surface will be seen as an increasing rut depth. This would occur both in the fast lane, which is dominated by the faster moving light vehicles, and to some extent also in the slow lane.
- Self-cleaning of pores occurs due to the passing vehicles and is more efficient the faster they move. This will reduce the clogging of the pores. Due to the faster speeds, it may in general be more effective in the fast lane, but average speeds are not very different in the two lanes due to the posted speed being 90 km/h (as an extra measure to reduce noise emission at the site).

Consequently, the observation that the slow lane has developed the larger rut depth is logical as it is exposed to both the heavy and light vehicles. Also logical, the reduction of texture (MPD) is larger in the fast lane, which is most probably due to the wearing action of the studded tyres. Since the initial MPD values were 1.80 mm or somewhat higher, which is still reflected in the area between the wheel tracks, and it is now substantially reduced in the fast lane (1.2 – 1.45 mm), it is another indication of the studded tyre effect.

Permeability is mainly influenced by the interconnected air voids and the openness of the pores. It is the experience of the author that on a new PA or DPA it will be around 10 s. Now that it is around 50-60 s, it indicates that clogging is serious, but not yet total. That it is generally a lower permeability (higher outflow times) in the wheel tracks than between the wheel tracks is an indication of clogging being the worst there. The fact that permeability is about the same in both the slow and fast lanes suggests that clogging is about equal in both lanes.

With regard to the difference between DPA 1 and DPA 2, one can see two trends: (1) DPA 2 has lower macrotexture (MPD), and (2) DPA 2 also has lower permeability (higher water outflow times). Both observations may have the same cause, namely that the clogging is higher in DPA 2 than in DPA 1, and this of course reduces permeability, but it will also reduce the MPD values as the space under the peaks in the profile will be more filled with dirt.

A conclusion is, therefore, that the DPA 2 is more clogged than DPA 1 now after four years of service. This is likely to be because the bottom layer is 11 years old as compared to four years for the bottom layer of the DPA 1.

9   CONCLUSIONS AND RECOMMENDATIONS

This paper has analysed the differences between two double-layer pavements which are similar in all respects except one, namely that one of them (DPA 2) makes use of a bottom layer from an earlier lifecycle of DPA at the site, while the other one (DPA 1) had both layers repaved when repaving took place in 2017.

Already when the pavements were 0-1 years old, the pavement with a recycled bottom layer showed somewhat lower noise reduction than the other one and this has persisted during the first four years of operation. As an average, the loss in noise reduction has been approx. 0.6 dB due to recycling the bottom layer. This is significant but not alarming.

The study suggests, not unexpectedly, that the re-used bottom layer is slightly clogged already at the start of this cycle which shows up as less sound absorption.

No problem has been noticed in the adhesion between the two layers of any of the DPA:s. Neither has ravelling been excessive so far. The technical quality and condition of the two DPA:s are good and do not suggest that serious failure may occur anytime soon.

The economic benefit of recycling the bottom layer one time is significant. However, it comes at the cost of noise reduction loss. It would be interesting to calculate the net cost-benefit of this measure. This author would not be surprised if the economic benefit would more than balance out the cost of loss of noise reduction.

Two notes of caution are necessary: First, this study has only covered the first four years of the pavements' lifecycle, which is expected to be either six or seven years before they must be repaved to provide the noise reduction required by the city. Secondly, it is not likely that the bottom layer can be recycled in one more cycle; after having been used for 13-14 years it is not expected to provide significant sound absorption anymore. Recycling the bottom layer each second lifecycle of a double-layer porous asphalt is well enough.

It is recommended that further research on this recycling technology is conducted as follows:

- Extend the study to cover the full lifecycle (planned)
- Calculate the economic benefit of re-using the lower asphalt layer, versus the economic cost of less noise reduction
- Make similar trials with other paving contractors
- Try to clean the lower (re-used) paving layer more effectively before the new layer is applied
- Make computer holography analyses of bore cores from the pavement layers and analyse the clogging as a function of depth under the surface (possible in this study but not yet done)

ISO AND EUROPEAN STANDARDS MENTIONED IN THE TEXT

EN 12697-40:2018. *Bituminous mixtures - Test methods - Part 40: In situ drainability.* European standards, Brussels, Belgium.

ISO 11819-2:2017. *Acoustics — Measurement of the influence of road surfaces on traffic noise — Part 2: The close-proximity method.* International Organization for Standardization (ISO), Geneva, Switzerland.

ISO/TS 11819-3:2017. *Acoustics — Measurement of the influence of road surfaces on traffic noise — Part 3: Reference tyres.* International Organization for Standardization (ISO), Geneva, Switzerland.

ISO/TS 13471-1:2017. *Acoustics — Temperature influence on tyre/road noise measurement — Part 1: Correction for temperature when testing with the CPX method.* International Organization for Standardization (ISO), Geneva, Switzerland.

ISO 13473-1:2019. *Characterization of Pavement Texture Utilizing Surface Profiles - Part 1: Determination of Mean Profile Depth.* International Organization for Standardization (ISO), Geneva, Switzerland.

ACKNOWLEDGEMENTS

This research is made within a project following-up the performance of the E4 Huskvarna low noise pavement, sponsored by the Swedish Transport Administration as part of the research programme of Swedish agency BVFF. The author is grateful for the financial contributions for the measurements and analyses, but also want to thank road contractor Svevia AB for fruitful cooperation.

It is acknowledged that Dr Piotr Mioduszewski at GUT has been responsible for all CPX measurements, made with the CPX trailer of GUT, and also has been a co-author of some earlier papers reporting results from the motorway section in Huskvarna.

REFERENCES

Bendtsen H., Hasz-Singh H. and Bredahl-Nielsen C., 2005. *Workshop on Optimization of Noise Reducing Pavements*, Road Directorate, Danish Road Institute, Denmark (https://www.vejdirektoratet.dk/api/dru-pal/sites/default/files/publications/workshop_on_optimization_of_noise_reducing_pavements.pdf)

Goubert L., Hooghwerff J., Thec P.; Hofman R., 2005. *Two-layer Porous Asphalt: An International Survey in the Frame of the Noise Innovation Programme (IPG)*. Paper #1793, Proc. of Inter-Noise 2005, Rio de Janeiro, Brazil.

PMSv3, 2022: *Tool for Handling of Road Data of the Swedish National Road System.* Swedish Transport Administration, Borlänge, Sweden (in Swedish). Website: https://bransch.trafikverket.se/for-dig-i-bran-schen/vag/Information-om-belagda-vagar-verktyget-PMSv3/.

Sandberg U. and Ejsmont J.A., 2002. *Tyre/Road Noise Reference Book.* Informex HB, Kisa. Sweden (www.informex.info)

Sandberg U. and Masuyama Y., 2005. *Japanese Machines for Laying and Cleaning Double-layer Porous Asphalt - Observations From a Study Tour.* Report produced by direction of DWW/IPG order number 64520946, Swedish National Road and Transport Research Institute (VTI). Full text available from https://www.diva-portal.org/.

Sandberg U. and Mioduszewski P., 2012. *The Importance for Noise Reduction of the Bottom Layer in Double-layer Porous Asphalt.* Proc. of Acoustics 2012, Hong Kong Institute of Acoustics, Hong Kong (abstract at https://asa.scitation.org/doi/10.1121/1.4708027, full text from https://www.diva-portal.org/).

Roads and Airports Pavement Surface Characteristics – Crispino & Toraldo (Eds)
© 2023 The Author(s), ISBN 978-1-032-55149-4

Pavement maintenance influence on road traffic noise emission

Manfred Haider*, Reinhard Wehr & Roland Spielhofer
AIT Austrian Institute of Technology GmbH, Vienna, Austria

ABSTRACT: Pavement surfaces have a substantial influence on the noise emission of road vehicles via their texture, aggregate size, and void content among other parameters. While a lot of research focuses on these parameters of pavements in the newly built or slightly worn state, degradation over time, maintenance actions and repair can substantially alter the character of the pavement surface. This paper presents an investigation of the connection between the maintenance state of the pavement and the changes in tyre/road noise emission with respect to the initial or reference condition. The results are based on monitoring and maintenance classifications as well as noise measurements according to ISO 11819-2 (CPX trailer) in the Austrian motorway network.

Keywords: Noise, Pavement, Degradation

1 INTRODUCTION

Road traffic noise is one of the major impacts of the road network on the population in general and the environment. The European Environmental Noise Directive (END 2002) therefore requires comprehensive mapping of road traffic noise sources originating from the high-level road network of each EU country and the development of action plans to reduce noise emission and immission to keep the numbers of affected citizens as low as possible. The noise levels affecting residents strongly depend not only on the noise generated by the vehicles using the road network, but also on the geometry and relative location of acoustic emitters and receivers, which determine the noise propagation and attenuation. However, any noise that is not generated or can be reduced already at the source or close to the source in general leads to at least some reduction in immission levels irrespective of the propagation path. Noise reductions at or close to the source can be realized by influence and restrictions imposed on the vehicles and vehicle speeds, which can lead to restrictions in the primary mobility function of the road network. The most important infrastructure-based noise abatement tools available to road authorities and road operators, are noise barriers and low-noise pavements. While noise barriers act on the sound propagation, low-noise pavements have the advantage of directly reducing the tyre/road noise portion of the emitted sound, which is typically dominant at motorway speeds. Therefore, there is an increasing trend to use pavements with noise reduction potential as stand-alone measure or in combination with noise barriers to keep noise immissions within the allowed limits.

However, to ensure the effectiveness of low-noise pavements as noise abatement measure, the noise emission parameters of these pavements have to be measured and subsequently established in the noise calculation method. The noise calculation method used for road traffic noise in Austria containing these parameters is described in RVS 04.02.11 (2021). Compliance of the real-life pavements with these parameters can be verified with approval

---

*Corresponding Author: manfred.haider@ait.ac.at

                                                    DOI: 10.1201/9781003429258-64

testing, which is typically carried out after installation. However, two groups of effects can change the noise performance of pavements over time: on the one hand, changes in the texture or accessible void content of pavements influence the otherwise undisturbed surface, and on the other hand the development of cracks and surface defects create discontinuities in the pavement. The available knowledge on the first group of effects is growing, as more long-term studies are undertaken (e.g. Wehr *et al.* 2021). However, in the case of cracks and defects, the surface is often no longer deemed representative and excluded from such studies. Nevertheless, depending on the maintenance strategy, at least a certain part of the road network may not be in the reference state assumed for that pavement type from the point of view of acoustics.

2 ROAD TRAFFIC NOISE EMISSION

2.1 *Generation mechanisms*

As described in (Sandberg & Ejsmont 2002), there are multiple noise generation mechanisms for tyre/road noise, which can be classified into two main groups:

- Noise generation through tyre vibrations: The interaction of the profile of the rolling tyre with the topmost elements of the road surface texture induces vibrations in the tyre, which lead to sound radiation. For passenger car tyres this generation mechanism dominates the low-frequency part of the emitted noise below 1 kHz. Low-noise pavements focusing on this group of generation mechanisms are designed to exhibit a smooth surface to the tyre tread elements to avoid the excitation of vibrations. This does not require a completely smooth surface on the microtexture level, only the contact points with the tyre should be at approximately the same level and close enough together to avoid penetration of tyre tread elements into the space between them.
- Noise generation through air pumping: When the tyre rolls over the pavement surface small volumes of air are at least partially enclosed between the two, which are compressed and decompressed in the process. For passenger car tyres this generates the components of tyre/road noise above 1 kHz. Low-noise pavements addressing this mechanism are designed with enough macrotexture and void content to avoid the compression effect by providing room for the air to escape.

Ideally low-noise pavements will try to achieve both types of effects at the same time, leading to pertinent design requirements concerning the pavement texture. Sound absorbing low-noise pavements like porous asphalt combine this with a high void content in a system of connected voids which leads to absorption of the generated sound and a further reduction of air pumping.

2.2 *Measurement*

The international standardization organization ISO has developed two complementary methods for the determination of the pavement influence on road traffic noise emission. The so-called Statistical Pass-By or SPB method (ISO 11819-1 1997) characterizes the typical noise emission of a pavement test section 100 m in length by means of sound measurements at a defined receiver location at the roadside during a required number of vehicle pass-bys within a certain speed range. The vehicle category and speed are recorded, and a statistical regression analysis is performed to determine the sound pressure level at the chosen reference speed for each category. This method is often used to establish noise emission parameters for use in noise propagation calculations. It has the advantage that it captures the whole emitted road vehicle noise but is limited to a specific relatively short test section. Moreover, it requires the absence of reflecting objects close to the microphone and a traffic flow that allows the separation of individual vehicle pass-bys. Therefore, it can be time-consuming and the possible locations are limited.

For this reason, a complementary method has been developed in the form of the Close-Proximity-Method (CPX, ISO 11819-2 2017), which is used in this investigation. CPX is based on measuring the noise emission very close to the tyre/pavement contact of a dedicated reference tyre, which is most commonly realized as a noise measurement trailer with an enclosure lined with sound-absorbing material on the inside. This trailer is equipped with the reference tyre and the microphones are placed within approximately 20 cm from the tyre contact patch, so that the tyre/road noise at typical motorway speeds typically surpasses any background noises, which makes the method very robust. With this setup and auxiliary measurement equipment for vehicle speed and air and pavement temperature, measurements can be performed at or close to traffic speed. This makes CPX especially suitable for long-distance measurements and network surveys, but also for approval testing where the necessary conditions of SPB cannot be met.

3 ROAD SURFACE DEFECTS

The Austrian guidelines for pavement management include two key documents concerning surface defects. RVS 13.01.11 (2009) contains a pavement distress catalogue with qualitative descriptions of different pavement damage types together with possible causes and consequences. RVS 13.01.16 (2013) deals with the assessment of surface defects and cracks with a view to deriving key performance indicators for pavement management. These guidelines are used to classify the occurring road surface defects and to plan suitable maintenance actions as a result of this analysis. However, it should be noted that not all types of surface distress which are noteworthy from the point of view of road maintenance necessarily have an influence on the noise emission of road vehicles passing over it.

Road surface defects which influence the tyre/road noise emitted by road vehicles need to impact one of the generation mechanisms described above. To change the acoustic performance of the pavement, the defect or repaired defect should have one of the following effects:

- Increase of surface roughness in the macro- and megatexture range
- Adverse change in road surface texture (e.g. reduction of void content)
- Introduction of step changes in the level of the road surface
- Removal of low-noise pavement surface and exposure of less optimized surfaces below
- Replacement of low-noise pavement surface with other surfaces as a repair measure

These changes also have to occur in the typical wheel paths of the road vehicles, ruling out any effects only found at lane edges.

The following broad classification can be attempted:

- Road surface defects which do not lead to substantial level differences or road material loss: This group comprises all types of longitudinal and lateral cracking, from single cracks to alligator cracking, where crack widths do not attain the size of tyre tread elements and the tyre can still roll unimpeded. This also includes loss of binder or bleeding, as well as limited loss of aggregates. Rutting which only introduces lateral level differences also falls into this category. These defects can still lead to a deterioration of acoustic performance due to changes in texture and void content.
- Road surface defects which introduce step changes in level: This comprises potholes, all types of large-scale material loss like missing patches, any movement or damage of concrete slabs leading to the creation of steps or joint damage and rutting or shoving when it creates longitudinal corrugation. In these cases, there is a potential for added impact noise when the road vehicles travel over the steps.

- Road surface defects which lead to a replacement of the exposed type of surface: This is often the case when defects in low-noise pavements are not repaired with the same type of surface or when one of the defects mentioned above exposes another pavement layer.

Moreover, it is important to note that a deteriorated state of the pavement with respect to acoustic performance will only be relevant if the deterioration does not lead to a pavement replacement for other reasons, which depends on the general pavement management strategy.

4 IMPACT ON NOISE EMISSION

In order to investigate a link between road surface defects and noise emission, long-distance CPX measurement results have been combined with data on the maintenance status of individual 100 m sections of road based on data gathered within the national Austrian research project ROSALIA (2020-2022). The results are shown in Figures 1 and 2.

On the vertical axis of both diagrams the overall L_{CPX} level determined according to ISO 11819-2 using only the reference tyre for passenger cars P1 at the measurement speed of 80 km/h for 100 m sections can be seen. The horizontal axis shows the percentage of road surface affected by either road surface defects except cracks (Figure 1) or cracking (Figure 2) for three very common pavement categories concrete (predominantly exposed aggregate concrete), standard stone-mastic asphalt (SMA S1) and SMA with higher void content (SMA S3). Both graphs show that the majority of test sections can be found at low percentages of deteriorated pavement area due to the motorway network being a road network requiring a high level of service due to the high traffic loads and therefore maintenance is carried out frequently. The test sections at low percentages also show a relatively large spread in CPX levels, probably indicating variations stemming from the original construction of the surfaces more than those due to deterioration. Concrete and SMA S1 pavements in Figure 1 with higher percentages of damaged area also tend to show higher CPX levels in

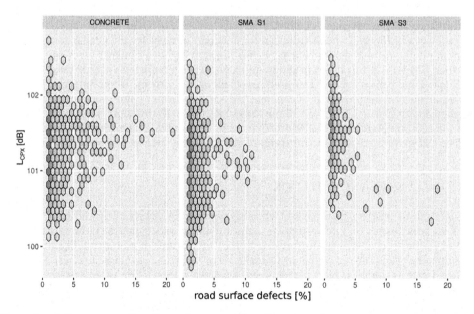

Figure 1.   Influence of road surface defects on road traffic noise emission measured as CPX level according to ISO 11819-2 (2017).

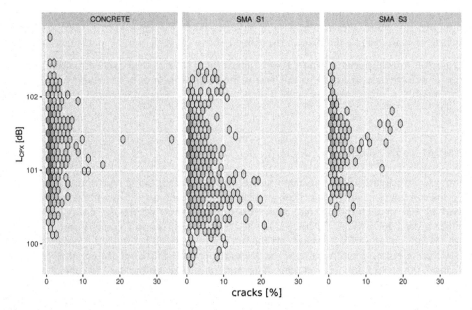

Figure 2.   Influence of cracks on road traffic noise emission measured as CPX level according to ISO 11819-2 (2017).

the lower part of the distribution. At higher CPX levels no test sections with correlated high percentages can be found, which may be a result of maintenance actions removing them from the available population. As a consequence, the highest percentages are found at medium-range CPX-levels. The SMA S3 data in Figure 1 show a different picture, based on a lower number of test sections. Here the large spread of CPX levels at low percentages is the same, but the only sections with high percentage of damaged pavement area all exhibit very low CPX values. This might indicate that this pavement type can maintain a good acoustic performance despite pavement deterioration.

For cracks in Figure 2 the distribution for concrete surfaces is very similar to the corresponding diagram in Figure 1, maybe indicating that the two damage types often occur together. SMA S1 in this diagram shows an inconclusive picture with a wide range of both CPX levels and percentages of distressed pavement area occurring. Interestingly SMA S3 in this figure mirrors the behaviour of concrete and SMA S1 from Figure 1, with more cracking associated with higher CPX levels in the lower half of the distribution. This could indicate that the acoustic performance of SMA S3 is more susceptible to changes in the small-scale surface texture (macro- and microtexture) of the pavement than to other types of defects.

5   CONCLUSION AND OUTLOOK

The analysis of the data gathered so far has shown that the effects of the different types of road surfaces distress on tyre/road noise are not uniform and should be differentiated with respect to the specific type of distress and to the key features of pavements like texture or void content which have a potential to make them work as low-noise surfaces. The connection is not as simple as higher level of surface defects does not necessarily lead to higher noise emission. This may be partially true in the case of already close to optimal low-noise surfaces in the low-noise part of the population, but with higher percentages of damaged areas pavements are also more likely to be replaced for non-acoustic reasons, especially in a high-level road network. Further investigations should therefore also include lower-level road networks with longer maintenance intervals, but also differentiate between the specific damage types with the aim of including only the acoustically relevant ones.

ACKNOWLEDGEMENTS

Part of the work presented in this paper was performed in an Austrian national research project called ROSALIA (FFG No. 879369), funded by the Austrian Ministry for Climate Protection, Environment, Energy, Mobility, Innovation and Technology represented by FFG (Austrian Research Promotion Agency), the Austrian federal states and ASFINAG.

REFERENCES

Environmental Noise Directive/END 2002: *Directive 2002/49/EC of the European Parliament and of the Council Relating to the Assessment and Management of Environmental Noise*, 2002

ISO 11819-1, *"Acoustics - Measurement of the Influence of Road Surfaces on Traffic Noise - Part 1: Statistical Pass-By method"*, International Standard published by ISO (www.iso.org), 1997

ISO 11819-2, *"Acoustics - Measurement of the Influence of Road Surfaces on Traffic Noise - Part 2: The Close-Proximity Method"*, International Standard published by ISO (www.iso.org), 2017

ROSALIA project, 2020 – 2022, *Tyre/road Noise Measurements on Road Surfaces – Evaluation and Updating*, at https://projekte.ffg.at/projekt/3791280

RVS 04.02.11, *Environmental Protection - Noise and Air Pollution – Calculation of Sound Emissions and Noise Protection*, FSV, Vienna, 2021

RVS 13.01.11, *Pavement Distress Catalogue for Flexible and Rigid Pavements*, FSV, Vienna, 2009

RVS 13.01.16, *Assessment of Surface Defects and Cracks on Asphalt and Concrete Roads*, FSV, Vienna, 2013

Sandberg U, Ejsmont J.A., 2002. *Tyre/road Noise Reference Book*, Informex.

Wehr R., Fuchs A., Breuss S., 2021. Statistical Tyre/road Noise Modelling based on Continuous 3D Texture Data, Acta Acustica 2021, Volume 5, p. 52ff

Impacts of the shift from distressed pavements to low noise pavements in motorways – A case study in Portugal

Maria João Rato & Maria Inês Ramos
BRISA, Quinta da Torre da Aguilha, Portugal

Elisabete Freitas*
University of Minho, ESISE, Portugal

Rosa Daniela Domingues, Isabel Gonzalez, Maria Margarida Braga,
Luís Fernandes & Ana Falcão
BRISA, Quinta da Torre da Aguilha, Portugal

Eduardo Fernandes
Geovia, Lisboa, Portugal

ABSTRACT: Road traffic noise is a relevant environmental problem, resulting essentially from the contact mechanisms between tyre and pavement surface. According to the current legislation, noise management actions must primarily intervene at the source. BRISA is employing efforts to determine pavement influence as a parameter of source noise reduction in order to address the well-being of the population surrounding highways and simultaneously comply with European directives regarding Environmental Noise evaluation and management. This Project evaluates the environmental noise effects of replacing a wearing course of Porous and Bituminous Asphalt at end-of-life for a course of SMA12, using two different methodologies for tyre-road noise measurement: the Statistical Pass-By method and the Close Proximity method.

Keywords: Tyre-Road Noise, CPX/SPB, SMA

1 INTRODUCTION

Tyre-road noise is influenced by several factors, namely driver behaviour (speed control and tyre pressure), tyre characteristics (structure, dimension, rubber stiffness, tread, wear, and age), pavement surface characteristics (macro and mega texture, irregularity, porosity, stiffness, age, wear, and water presence) and weather conditions (temperature and wind) (Sandberg & Jerzy 2002). There are specific test methods for measuring tyre-road noise, which must be complemented with other surface characterisation tests such as texture, sound absorption, and surface layer stiffness determined by mechanical impedance.

In Portugal, a few studies were carried out based on those methodologies, namely the Statistical Pass-By method (SPB) and the Close ProXimity method (CPX), only with an exploratory nature or to support research activities (Antunes *et al.* 2008; Freitas *et al* 2008, 2009, 2019). There is yet no technical documentation that defines reference values or surfaces, regarding the conformity of production or performance over time. Recently, the EU Green Public Procurement Criteria for Road Design, Construction and Maintenance was

---

*Corresponding Author: efreitas@civil.uminho.pt

DOI: 10.1201/9781003429258-65

adapted to Portuguese conditions (APA 2020), in the framework of the National Strategy for Green Public Procurement (ENCPE 2020). Nevertheless, it is a guiding document where minimum applicable requirements for the design of low noise pavements are indicated. Despite the lack of references for tyre-road noise assessments, the framework dictated by the European directives on environmental noise assessment and management (2002/49/EC and (EU) 2015/96, of June 25th and May 19th, respectively, in which noise predictions are based on road-noise, must be respected. Therefore, a gap must be fulfilled concerning tyre-road noise characterization for the Portuguese conditions.

In this context, *Brisa – Concessão Rodoviária*, S.A., the biggest Portuguese highway concessionaire, has developed efforts to determine pavement influence as a source noise reduction method and gather the information necessary to apply the CNOSSOS noise prediction method to its highway network, following the work developed by Anfosso-Ledee and Goubert (2019).

This study analyses the effect of replacing wearing courses of Porous Asphalt (PA 12.5) and Asphalt Concrete (AC 14), at end-of-life, with a high-performing wearing course of Stone Mastic Asphalt (SMA 12), through SPB and CPX methods.

2 STUDY SECTIONS AND TEST METHODS

2.1 *Study methodology*

For this exploratory study, three highway sections were selected where pavement interventions were foreseen, i.e., replacing the existing wearing course with one of the SMA 12 type. Before and after the interventions, tyre-road noise was evaluated through two different methodologies: Statistical Pass-By Method (SPB) and Close ProXimity Method (CPX).

2.2 *Description of the study sections*

The main characteristics of the pavement wearing course of the three highway sections are summarised in Tables 1–3, before and after replacing the wearing course. These characteristics include grading curve, bitumen content, air void content, and macrotexture.

The Mean Profile Depth (MPD) values obtained for the pavements of the three highway sections before replacing the wearing course are presented in Table 4.

Table 1. Characterisation of the mixtures of wearing courses before and after intervention (Highway A).

Highway A Old Pavement - AC 14 Surf 35/50					Highway A Rehabilited Pavement - SMA 12 Surf PMB 45/80-65						
Grading Curve AC 14 Surf 35/50			Bitumen Content	Air void content	Macrotexture MTD	Grading curve (Quality Control) SMA 12 Surf PMB 45/80-65				Bitumen content	Air Void content	Macrotexture
Sieves # mm	Grading Specification (% pass)		Specification limits			Sieves # mm	Grading Specification (% pass)	Passing values %	STD - passing %	% avg	% avg	MTD$_{avg}$
16	100	100	≥ 5,0%	6,0% ± 2,0	≥ 0,7 mm	16	100	100	100	± 0	6,3 ± 0,2	6,3 ± 1,2	1,3 ± 0,1
14	100	90				14	100	100	97	± 1			
12,5	88	80				12,5	100	95	92	± 2			
10	77	67				10	100	80	79	± 3			
8						8	80	60	60	± 3			
6,3						6,3	60	43	43	± 3			
4	52	40				4	28	22	23	± 2			
2	40	25				2	22	18	18	± 2			
1						1			16	± 2			
0,500	19	11				0,500	19	15	15	± 1			
0,250						0,250			13	± 1			
0,125	11	6				0,125			11	± 1			
0,063	8	5				0,063	11	8	8,8	± 1,1			

Table 2. Characterisation of the mixtures of wearing courses before and after intervention (Highway B).

Highway B Old Pavement - AC 14 Surf 35/50					Highway B Rehabilited Pavement - SMA 12 Surf PMB 45/80-65						
Grading Curve AC 14 Surf 35/50		Bitumen Content	Air void content	Macrotexture MTD	Grading curve (Quality Control) SMA 12 Surf PMB 45/80-65				Bitumen content	Air Void content	Macrotexture
Sieves # mm	Grading Specification (% pass)		Specification limits		Sieves # mm	Grading Specification (% pass)		Passing values %	STD - passing %	% avg	% avg	MTDavg
16	100	100			16	100	100	100	± 0			
14	100	90	≥ 5,0%	6,0% ± 2,0	≥ 1,1 mm	14	100	100	99	± 1	6,5 ± 0,1	6,2 ± 0,01	1,4 ± 0,1
12,5	90	70			12,5	100	95	97	± 3			
10	78	62			10	100	80	80	± 3			
8					8	80	60	60	± 4			
6,3					6,3	60	43	41	± 4			
4	39	28			4	28	22	25	± 3			
2	30	22			2	22	18	19	± 1			
1	25	17			1			17	± 1			
0,500	20	12			0,500	19	15	15	± 1			
0,250					0,250			13	± 1			
0,125					0,125			11	± 1			
0,063	10	6			0,063	11	8	8,6	± 0,7			

Table 3. Characterisation of the mixtures of wearing courses before and after intervention (Highway C).

Highway C Old Pavement - PA 12,5 PMB 45/80-65					Highway C Rehabilited Pavement- SMA 12 Surf PMB 45/80-65						
Grading Curve PA 12,5 PMB 45/80-65		Bitumen Content	Air void content	Macrotexture MTD	Grading curve (Quality Control) SMA 12 Surf PMB 45/80-65				Bitumen content	Air Void content	Macrotexture
Sieves # mm	Grading Specification (% pass)		Specification limits		Sieves # mm	Grading Specification (% pass)		Passing values %	STD - passing %	% avg	% avg	MTDavg
16	100	100			16	100	100	100	± 1			
14			≥ 4,0%	22 - 30%	≥ 1,2 mm	14	100	100	98	± 2	6,4 ± 0,1	5,6 ± 1,4	1,3 ± 0,1
12,5	100	80			12,5	100	95	95	± 2			
10	80	55			10	100	80	85	± 3			
8					8	80	60	65	± 3			
6,3	48	28			6,3	60	43	45	± 3			
4	28	14			4	28	22	25	± 2			
2	21	10			2	22	18	19	± 1			
1	14	6			1			16	± 1			
0,500					0,500	19	15	14	± 1			
0,250					0,250			13	± 1			
0,125					0,125			11	± 1			
0,063	5	2			0,063	11	8	8,7	± 1,1			

Table 4. Macrotexture Information (MPD).

Highway	Min. (mm)	Max. (mm)	Mean (mm)
A	0,5	1,4	0,8
B	0,6	2,3	1,3
C	1,7	3,6	2,5

2.3 Statistical Pass-By method (SPB)

The Statistical Pass-By method (SPB) is a standardised method published by ISO 11819-1:1997, aiming to determine an indicator that considers the noise emitted by pass-by road traffic.

In this way, it is possible to obtain a quantitative classification of road pavement surfaces related to road traffic noise to satisfy the necessities expressed by road infrastructure managers, designers, contractors, pavement manufacturers, and other parties interested in predicting and controlling road traffic noise.

To determine the sound pressure levels that characterise a given pavement surface (wearing course), a reference speed for light and heavy vehicles is adopted. The method is applicable at constant traffic speed, i.e., free flow conditions (without interference from other vehicles) circulating at speeds equal to or greater than 50 km/h, meaning, for highways, a speed of 90 km/h for heavy vehicles, and 120 km/h for light automobiles. The SPB method requires several in situ measurements, under normal driving conditions, of the maximum sound pressure level (Lmax) and circulating speed of a passing vehicle, using a sound meter (class 1 as specified in IEC 61672-1) positioned at 7,5 m from the centre line, and a kinemometer (radar).

Maximum sound pressure levels differ according to the class of the vehicle. Thus, at each vehicle pass-by, the maximum A-weighted sound pressure level is recorded, the speed and the vehicle type (light, heavy dual-axle, and heavy multi-axle vehicles). After the passage of at the least 100 light vehicles and 80 heavy vehicles, a linear regression is established between the logarithm of the speed and the maximum sound pressure level. Subsequently, the corresponding sound level for a certain reference speed is determined according to the road type. The resulting SPB Indicator (SPBI) from this method is an index value, in dB(A) based on the noise levels of different vehicle classes.

In this work, since the method requires measuring each vehicle per si, without the interference of others, only the events which fulfilled such criteria were selected. Therefore, passages that were influenced by the noise from other sources were excluded. Only two classes of vehicles (light and heavy) were considered.

2.4 *Close ProXimity method (CPX)*

With the advantage of measuring the tyre-road noise continuously, the Close ProXimity method (CPX) was used as defined in the EN/ISO 11819-2:2017 standard. In the present case, the noise measurement was performed close to one of the wheels of the testing vehicle, where two microphones were placed according to the mounting scheme defined by the standard. An analysis software processed the signals recorded during testing, and the noise emission (A-weighted) was evaluated in 20-metre sections as the arithmetic mean of the sound levels recorded by each microphone, and by the corresponding sound spectrum in 1/3 octave bands (L_{CPX}). In this study, only the tyre representative of light vehicles (P) was considered.

The measurements were taken along the three sections for reference speeds of 50 km/h, 80 km/h, and 100 km/h, although, in the latter case, they were not carried out in all sections tested.

3 PRESENTATION AND ANALYSIS OF THE RESULTS FOR THE SPB METHOD

3.1 *Measured noise level on each highway*

Table 5 shows the results obtained by the SPB method on the three highway sections, before and after replacing the wearing courses. The comparison of the SPBI shows a significant reduction of 3 and 4 dB(A), respectively on the highway sections A and B, which was similar for light and heavy vehicles.

Highway A provided consistently lower tyre-road noise values. This performance must be further investigated. One possible cause might be the applicability of the SPB method concerning the geometric requirements. The north of Portugal is characterized by high road

Table 5. Results obtained by SPB testing.

Highway	Intervention	Vehicle class	% Light/ Heavy vehicle	Average L$_{max}$ dB (A)	Standard sound signal deviation dB(A)	Standard speed deviation (km/h)	Average speed (km/h)	L$_{veh}$ (dB (A)(120 km/h and 90 km/h)	SPBI dB (A)
A	**Before**	Light	95%	77	3,7	17,7	108	78	78
		Heavy	5%	80	3,7	13,5	85	80	
A	**After**	Light	95%	77	6,5	14,4	109	75	75
		Heavy	5%	78	1,8	12,3	88	77	
B	**Before**	Light	66%	82	1,9	18,3	114	82	85
		Heavy	34%	86	1,2	4,5	87	87	
B	**After**	Light	67%	82	1,3	17,9	122	82	86
		Heavy	33%	89	1,8	6,6	92	88	
C	**Before**	Light	93%	87	4,5	13,7	124	86	86
		Heavy	7%	87	2,4	5,6	90	87	
C	**After**	Light	93%	82	3,4	13,9	120	81	82
		Heavy	7%	84	3,6	8,3	94	82	

slops, consequently, high embankments, short shoulders and recovery areas. Before the intervention, Highway C, on porous asphalt provided the same noise level as Highway B, on asphalt concrete, which shows that it had lost most of the absorption capacity that characterizes this type of mixture. The analysis of the noise spectra will help explain these remarks.

3.2 *Spectrum analysis*

Several mechanisms and factors determine the sound spectra resulting from tyre-road contact, the main ones being vibrations promoted by the pavement texture in the tyres and the pavement maintenance state (for frequencies lower than about 1000 Hz), and the air movements resulting from the interaction of the tyre tread with the irregularities of the pavement (for frequencies higher than about 1000 Hz) and by the sound absorption (Bühlmann & Ziegler 2012).

Figure 1 shows the sound spectra for each highway.

Before and after replacing the wearing course, the analysis of the frequency spectrum per octave band shows an identical behaviour for low frequencies, except for highway C where a decrease in sound pressure levels per frequency was observed for light vehicles. For high frequencies, on highway A the behaviour at low frequencies is also identical, while on highways B and C, there was an increase in the sound pressure levels per frequency, which was more accentuated for heavy vehicles.

4 PRESENTATION AND ANALYSIS OF THE RESULTS FOR THE CPX METHOD

4.1 *Measured noise level on each highway*

The measured noise level is significantly affected by the test vehicle speed. For results comparison for the three defined speeds from the noise levels determined in each 20-metre segment and the corresponding speed, L$_{CPX}$ - log10(speed) regression lines were defined, whose slope (m) is used to correct the measured L$_{CPX}$ for a given reference speed. For the situations before and after the intervention, Figure 2 shows the obtained data and the fit lines. The figure also shows, for the three highways, the obtained regression line parameters, slope (m) and ordinate at the origin (b), and the coefficient of determination (R^2).

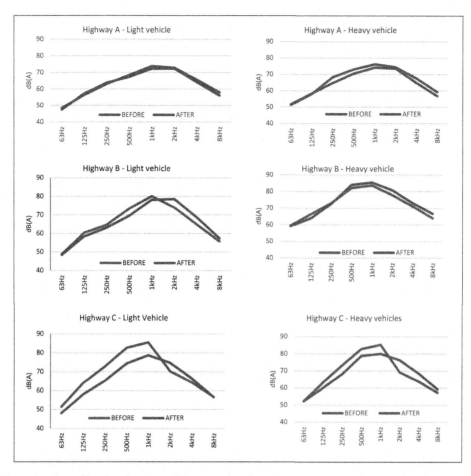

Figure 1.   Sound level vs frequency (1/1 octave band).

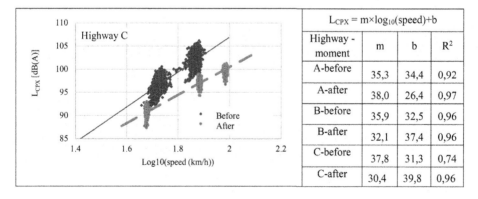

$L_{CPX} = m \times \log_{10}(speed) + b$			
Highway - moment	m	b	R^2
A-before	35,3	34,4	0,92
A-after	38,0	26,4	0,97
B-before	35,9	32,5	0,96
B-after	32,1	37,4	0,96
C-before	37,8	31,3	0,74
C-after	30,4	39,8	0,96

Figure 2. L_{CPX} at 50, 80 and 100 km/h in Highway C / Regression line coefficients for all highways.

673

It can be observed that the parameter representing noise increase with speed changes significantly for highway C, indicating that the impact of the intervention in terms of noise reduction is higher at higher speeds. Highway B shows a similar trend to that of C. However, for highway A this trend is reversed.

All L_{CPX} values were adjusted for the reference speeds (50, 80, and 100 km/h). Figure 3 shows the values obtained for the 80 km/h reference speed before and after intervention on highway C. In addition to facilitating the comparison of noise levels obtained along a section at different pavement life moments, this type of visualisation helps identify zones of homogeneous and heterogeneous behaviour, which can be related to performance explicative factors such as texture. In this section before the intervention, noise variability along it is notorious, reaching 7,6 dB(A).

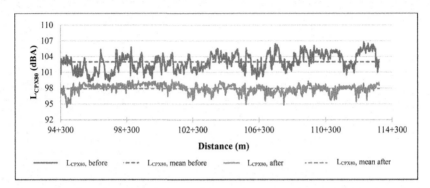

Figure 3. L_{CPX} at 80 km/h in highway C before and after intervention (example).

After the intervention, besides the observed reduction of the average LCPX by 5 dB(A), the noise variability was also reduced to 5 dB(A). The coefficient of variation after the intervention reduced for each section, which indicates that tyre-road noise became more homogeneous and that the effect of the intervention in some locations is much higher than the average effect determined by the difference of the mean L_{CPX}. If L_{CPX} per segment is considered, there are differences before-after intervention reaching 12 dB(A).

For an overall evaluation of the effect of changing the wearing course in the three highway sections, the mean L_{CPX} was determined at each reference speed in both traffic directions (see Figure 4). Highway C benefitted the most from the course change, while highway B presented only a small reduction.

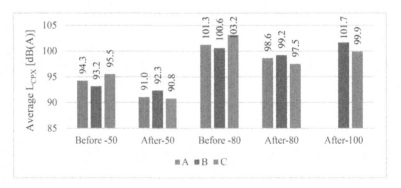

Figure 4. L_{CPX} Noise Levels at 50, 80 and 100 km/h in Highways A, B and C, before and after intervention.

4.2 *Spectrum analysis.*

Figure 5 presents the sound spectra for a speed of 80 km/h, per direction (C-crescent, D-decrescent), before and after the intervention, for the three highways, and for a speed of 100 km/h after intervention for highways B and C.

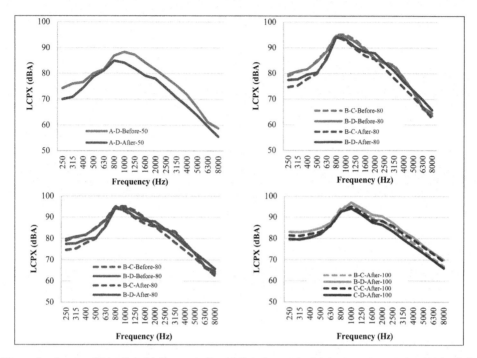

Figure 5. L_{CPX} at 80 km/h in highways A, B and C, before and after intervention, and at 100 km/h in highways B and C, after intervention.

On highway A, the effect of changing the wearing course was predominant at high frequencies, meaning that noise reduction was essentially due to favourable air movements provided by the texture. This effect is opposite to that observed for highway C, where after the intervention there was a significant noise reduction at low frequencies and a small increase at high frequencies. The new wearing course provided a reduction in tyre vibrations and negatively affected the air movement mechanisms resulting from tyre-road interaction. The effect in the case of highway B was closer to that of highway A.

Considering the reference speed of 100 km/h, as it is closer to the operational speed in this type of road, it should be noted that the noise variation at each frequency, resulting from the intrinsic variability of construction conditions, was in average 4 dB(A).

5 SPB AND CPX METHOD COMPARISON

The two methods were able to provide data to determine the resulting noise change after the replacement of the existing wearing course by one of the SMA 12 type and also investigating the factors affecting it.

The results of both methods allow to establish the same hierarchy in terms of ordering the highways according to the level of noise reduction after replacement of the wearing course, specifically: highway C, highway A, and highway B.

Also, there is some similarity regarding the level of noise reduction after replacement of the wearing course of the three highways, obtained by both methods. Specifically, for light vehicles, there is a reduction of about 3 dB(A) on highway A, a reduction of 5 dB(A) on highway C, and a variation between - 1.4 dB(A) and + 1 dB(A) on highway B.

Overall, the results obtained by both methods on highways B and C confirm the noise difference of approximately 20 dB(A) for light vehicles, corroborating the relationship obtained in international studies [Rosanne 2016]. Thus, in an expedite way the noise levels obtained by one methodology can be reasonably estimated in function of the values measured by the other.

Only in a more detailed analysis of noise level reduction by frequency ranges can be seen a greater dissonance between the two methods. In fact, the results of the CPX method indicate that replacement of the wearing course generates a greater reduction in noise levels at high frequencies in the case of highways A and B, and at low frequencies in the case of highway C. The results of the SPB method, in turn, indicate that wearing course replacement does not allow for a reduction in noise levels at high frequencies in any of the highways, and that it allows for a reduction at low frequencies in highway C. This observation is due to the effect of heavy vehicles which, in this analysis, are considered only in the SPB method, and to the effect of sound propagation.

6 CONCLUSIONS

From the data obtained via the SPB and CPX methods, the effect in terms of noise reduction of the replacement of the existing wearing course by SMA 12 was assessed in three highways. While in two of them there was a noise reduction between 3 and 4 dB(A), in the other one the effect was negligible.

Both SPB and CPX methods point to similar global noise reduction levels caused by replacement of the wearing course. However, the more detailed analysis of noise reduction levels by frequency indicates some dissonance between the results obtained by the two methods, caused by the sound propagation effect in the results from the SPB method. Therefore, it seems that both methods can be used complementarily, since the SPB method allows for the observation of noise reduction levels in a wide range of vehicle types and considers a greater diversity of factors, and the CPX method allows for the characterisation of a long stretch of road in a short period of time.

For future studies, and for the SPB method, the need for a larger sample size was identified, given the assumptions associated with the test method. Also, to relate the observed noise reduction with variations of the wearing course characteristics, these must be evaluated after intervention through measurement in continuum, for comparison with the MPD values obtained prior to the intervention.

The analysis of the data resulting from the SPB and CPX methods suggests that these approaches can contribute to obtaining baseline data on pavement characteristics for predictive noise models. These data can be very useful for the obtaining of models better adjusted to existing conditions once being collected in situ and consequently more adapted to effective noise propagation (Anfosso-Ledee & Goubert 2019). Therefore, the development of methodologies for obtaining pavement characterisation parameters for use in noise simulation models is envisaged as a future challenge, designing and outlining tests based on the SPB and CPX methods with this objective in mind, and considering the pavement typology used in the network operated by Brisa.

REFERENCES

APA, *Critérios de Contratação Pública Ecológica, no Âmbito da ENCPE 2020*, Para Conceção, Construção, Reabilitação e Conservação de Estradas, Agência Portuguesa do Ambiente, 2020. In Portuguese.

Antunes M., Coutinho S., Patrício J., Freitas E., Paulo J. and Coelho J., 2008. Avaliação do Ruído de Tráfego: Metodologia para a Caracterização de Camadas de Desgaste Aplicadas em Portugal. Evaluation of Pavement Surfaces Characteristics, Proceedings of the Seminar, pp 137–145, Guimarães, Portugal.

Anfosso-Ledee F. and Goubert L., 2019. The Determination of Road Surface Corrections for CNOSSOS-EU Model for the Emission of Road Traffic Noise, Universitätsbibliothek der RWTH, Aachen.

Bühlmann E. and Ziegler T., 2012. Interpreting Measured Acoustic Performance on Swiss Low-noise Road Surfaces Using a Tyre/Road Interaction Model. *Proceedings of the Acoustics 2012*, Hong Kong.

Freitas E.F. and Pereira P., 2008. Contribution of Portuguese Pavement Surfaces to Traffic Noise. Transport Research Arena Europe *2008*, Ljubljana, Slovenia.

Freitas E.F., Pereira P., Picado-Santos L. and Santos A., 2009. Traffic Noise Changes Due to Water on Porous and Dense Asphalt Surfaces, *Road Materials and Pavement Design*, 10(3), pp 587–608.

Freitas E.F., Silva L. and Vuye C., 2019. *The Influence of Pavement Degradation on Population Exposure to Road Traffic Noise, Coatings*, 9(5), 298.

ISO 11819-1, 1997. *Acoustics — Measurement of the Influence of Road Surfaces on Traffic Noise — Part 1: Statistical Pass-By Method*, International Organization for Standardization. Switzerland.

ISO 11819-2, 2017. *Acoustics — Measurement of the Influence of Road Surfaces on Traffic Noise — Part 2: The Close-proximity Method*, International Organization for Standardization. Switzerland.

ROSANNE, 2016. Collaborative Project FP7-SST-2013-RTD-1 Seventh Framework Programme Theme SST.2013.5-3: Innovative, Cost-effective Construction and Maintenance for Safer, Greener and Climate Resilient Roads.

Sandberg U. and Jerzy E., 2002. Tyre/road noise. Reference book. Informex.

© 2023 The Author(s), ISBN 978-1-032-55149-4

Innovative pavement contracts to reduce CO2-emissions

Even K. Sund* & Thor A. Lunaas
Norwegian Public Roads Administration, Trondheim, Norway

ABSTRACT: The Norwegian Public Roads Administration (NPRA) has an ambition of reducing CO2-emissions from maintenance works by 50% by 2030. Therefore, a contract strategy involving several new incentives for reaching this goal has been developed. The most important of these is weighting of CO2-emissions in monetary terms when awarding paving contracts. This paper describes how this strategy has been implemented for pavement maintenance contracts. An essential component of the strategy is the use of Environmental Product Declarations (EPD's). The first part of the paper describes the framework for EPD's and how an EPD-calculator has been developed specifically for the paving industry. We then go on to describe how the strategy has been implemented in NPRA pavement contracts, and the results achieved so far. In 2021 about 25% of all NPRA pavement maintenance contracts included CO2-weighting, while in 2022 almost 90% of all contracts uses this mechanism for allocation. All bituminous materials (asphalt) in the contracts must be documented by EPD's in accordance with EN 15804 + A2:2019. The calculation of CO2-emissions is based on the project specific EPD's. The bidding prices are adjusted according to differences in calculated total CO2-emissions, using a fixed price for CO2. This mechanism has been developed further by combining CO2-weighting with estimated life-cycle costs based on performance related parameters regarding permanent deformations and wear from studded tires. Improved properties for resisting deformations and wear guaranteed by the contractor is used to calculate an increase in expected pavement life, thus reducing the calculated annual cost. This combination of mechanisms implies that there are good incentives for the contractor to optimize the materials with the aim of achieving longer pavement service life as well as lower CO2-emissions. This approach has been used in two pilot-contracts in 2021 and is being used in four pilot contracts in 2022.

An important factor for successful implementation of the new contract strategy is trust, transparency, and accountability regarding calculated CO2-emissions. In Norway this has been achieved by the development of a tool for calculating project specific EPD's. There is also a system in place for third party verification of compliance with the EPD-values for CO2-emissions throughout the contract period.

Experiences with the new contract strategy have been positive, with a reported reduction of 13% in CO2-emissions per ton asphalt from 2020 to 2021.

Keywords: Innovative Pavement Contracts, CO2-Emissions, Life-Cycle Costs

1 INTRODUCTION

The Norwegian Public Roads Administration (NPRA) is an administrative body and a provider of national public services, subordinate to the Ministry of Transport. The NPRA is responsible for most of the Norwegian national road network (*riksveger*), which consists of

---

*Corresponding Author: even.sund@vegvesen.no

DOI: 10.1201/9781003429258-66

10,500 km of main roads. The NPRA is responsible for planning, construction, operating, and maintaining this part of the road network. The NPRA is to strive to reduce the environmental impacts of construction, operation, and maintenance of the road network, and to provide useful services through cost-effective use of public funds. This contributes to Norway's fulfilment of its climate and environmental goals. For construction, maintenance, and operations the goal is to cut greenhouse gas emissions by 50% by 2030 (Norwegian Ministry of Transport 2021). The NPRA has developed a contracting strategy for operations and maintenance that implies that all contracts must contain requirements related to the environment, climate, and sustainability in accordance with overall goals and the nature of the tasks involved in the contract. This paper describes how this strategy has been implemented for pavement maintenance contracts. An essential component of the strategy is the use of Environmental Product Declarations (EPD's). The first part of the paper describes the framework for EPD's and how an EPD-calculator has been developed specifically for the paving industry. We then go on to describe how the strategy has been implemented in NPRA pavement contracts, and the results that have been achieved so far.

2 FRAMEWORK FOR EPD IN NORWAY

An EPD is a concise document that summarizes the environmental profile of a component, a finished product, or a service in a standardized and objective way. The requirements for how an EPD is made are specified in the standard ISO 14025 *Environmental labels and declarations – Type III*. An EPD is created based on of a life-cycle analysis (LCA) according to ISO 14040 *Environmental management – Life cycle assessment – Principles and framework* and ISO 14044 *Environmental management – Life cycle assessment – Requirements and guidelines*. The content of an EPD must also comply with requirements and guidelines in ISO 14020 *Environmental labels and declarations – General principles* and are recommended to meet the requirements of ISO 14021 *Environmental labels and declarations – Self-declared environmental claims*. The standardized methods ensure that environmental information within the same product category can be compared across material types and products, making it possible for the customer to assess and make choices based on the environmental declarations.

The Norwegian EPD foundation (*EPD-Norge*) is a program operator for EPD type III according to ISO 14025. The program has established a system for verification, registration, and publication of EPD's as well as maintenance of registers for EPD and Product Category Rules (PCR). The PCRs for construction products for the European market are in compliance with EN 15804 *Sustainability of construction works – Environmental product declarations – Core rules for the product category of construction products* + A2:2019. The framework of the EPD-Norge programme is shown in Figure 1.

The light grey boxes show activities relating to administration and the EPD-forum, the green boxes show activities relating to the development of EPD's and verification and the dark grey boxes show activities relating to PCR development. The blue box represents PCR hearings.

3 EPD-CALCULATOR FOR THE NORWEGIAN ASPHALT INDUSTRY

To make implementation of the use of EPD's in the Norwegian asphalt industry more practical the Norwegian Contractors Association (EBA), together with EPD-Norge and other partners including the NPRA, in 2017 developed a cloud-based computer program for generating EPD's for asphalt materials. The program can be used by each individual supplier (contractor) to generate project specific EPD's including transportation and in-situ construction works. This makes the process of producing project specific EPD's effective and

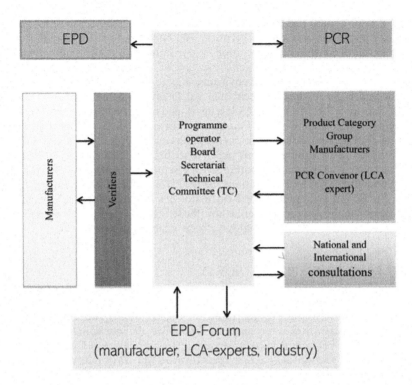

Figure 1.   Framework of the EPD-Norge programme (EPD-Norge 2019).

simple, while at the same time ensuring transparency and trustworthiness through a common database containing basic environmental data (e.g., CO_2 per kg for each type of material). This database is administered by the system operator (*LCA.no*) and cannot be accessed or changed by the users of the system. The development of the EPD-generator has been an important prerequisite for the rapid implementation of e.g., CO_2-weighting when allocating pavement contracts by the NPRA and other public clients. A system overview of the EPD-generator is shown in Figure 2.

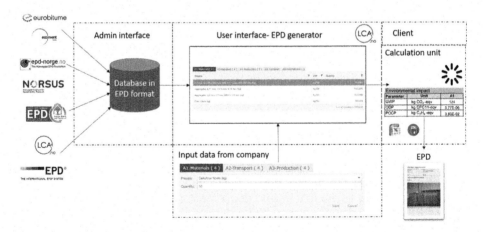

Figure 2.   System overview of the EPD-generator computer program for asphalt (Iversen 2021).

4 USE OF EPD'S IN PAVEMENT CONTRACTS

In recent years the annual production of asphalt in Norway has been about 7 million metric tonnes. Of this about 20 – 25% has been warm mix asphalt, i.e. asphalt produced at lower temperatures than normal hot mix asphalt (EBA 2020). In 2020 the average CO_2-emissons in NPRA pavement maintenance contracts was about 62 kg CO_2/tonne asphalt, when including the production and construction installation stages (A1 – A5):

- A1: Production of raw materials
- A2: Transportation of raw materials
- A3: Production of asphalt mix
- A4: Transport of asphalt mix
- A5: In situ paving works

The average contribution of the different stages to CO_2-emissions for asphalt in the NPRA pavement maintenance contracts in 2020 was as shown in Figure 3. The figures are based on EPD's from all contracts.

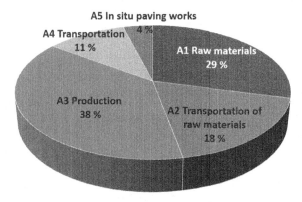

Figure 3.   Average proportion of CO_2 emissions for different stages (A1-A5) of asphalt in NPRA contracts 2020.

The NPRA have actively supported and cooperated with the Norwegian asphalt industry in their efforts to reduce CO_2-emissions from production of asphalt pavements. The best tool is to use incentives and mechanisms in the contracts. Together with EBA the NPRA have gradually phased in the use of EPD's in pavement maintenance contracts, beginning in 2018. The steps have been:

1. Requirement for contractors to deliver any available EPD's
2. Requirement for contractors to deliver EPD's in all contracts
3. Use of CO_2-weighting in allocation of some contracts, based on EPD's
4. Use of CO_2-weighting in allocation of more contracts, and increasing the price of CO_2
5. Use of CO_2-weighting combined with life-cycle costs in pilot contracts

The mechanisms used in CO_2-weighting in allocation of contracts is as follows:

- The EPD's are used to calculate a CO_2-budget for each bidder on the contract, including the production and construction installation stages (A1 – A5)
- The bidder with the lowest total CO_2-budget has no addition to the bidding sum
- The other bidders have their bidding sums increased by adding 5 NOK per kg CO_2 they are over the lowest bidder's CO_2-budget ($\approx$ 0,5 €/kg CO_2).

- The contract is allocated to the bidder with the lowest adjusted bidding sum
- The contract sum is equal to the original bidding sum of the winning contractor

At the end of the contract an account of actual CO_2-emissions must be provided by the contractors. If the final account deviates more than 5% from the original CO_2-budget, then a bonus (5 NOK/kg CO_2 for lower total emissions) or a penalty (10 NOK/kg CO_2 for higher total emissions) is triggered.

There has been some discussion about the price of CO_2 used for calculating the adjusted bidding prices. In the first year this mechanism was used a price of 2 NOK/kg CO_2 was used, but this was found to be too low to have the desired effects. After consultation with the contractor association (EBA) the price was increased to its current level of 5 NOK/kg CO_2. It is important that the price of CO_2 is high enough to allow contractors to make the necessary investments, e.g. in production plants, and still be competitive.

In 2021 about 25% (7 of 27) of NPRA pavement maintenance contracts used CO_2-weighting. Two of these were pilot contracts which combined CO_2-weighting and life-cycle costs. In 2022 $C0_2$-weighting is being used in almost 90% (24 of 27) of the contracts, of which four use combination of CO_2-weighting and life-cycle costs (pilot contracts).

Using EPD's directly in allocation of contracts requires that both the client and the bidding contractors can trust in their validity and that they can be verified throughout the contract period. Therefore, the NPRA requires that the contractors submit extra background information about the EPD's, which is used to verify them throughout the contract period. This includes, among other things, information about:

- Type and amount of binder, including any biogenic binder
- Amount of recycled asphalt, including bitumen content
- Type an amount of additives
- Type and amount of aggregates within each fraction size
- Information about transportation of raw materials
- Sources of energy used in production plant and planned production temperature
- Information about transport of asphalt from production plant to site
- Information about equipment used for laying and compacting asphalt on site

In addition, the NPRA uses a third-party verification process, which is described in a later section of this paper.

5 COMBINING CO2-WEIGHTING WITH LIFE-CYCLE COSTS IN ALLOCATION OF CONTRACTS

Although the focus on CO_2-emissions from asphalt production is important, it is even more important that the CO_2-emissions over the total life cycle of pavements is as low as possible. It is also important to achieve as low life-cycle costs as possible. A long pavement life is usually beneficial for both goals, and it is important to avoid sub-optimal solutions. Therefore, the NPRA have tried out a mechanism where CO_2-weighting is combined with estimated life-cycle costs in contract allocation. There were two such pilot contracts in 2021, and four more are being carried out in 2022.

In these pilot contracts the bidders are free to choose their own mix designs, irrespective of requirements given in the pavement design manual (NPRA 2021) regarding materials and composition. They must fulfil functional requirements related to stability against permanent deformations, resistance to rutting from studded tires, good durability, and acceptable friction. For the pilot contracts rutting is assumed to be the critical parameter for length of pavement life. Rutting is caused by permanent deformation and wear from studded tires. A normal pavement life is expected if the values given in Table 1 are achieved for stability (wheel track) and resistance to wear from studded tires (Prall-value).

Table 1. Requirements related to rutting for achieving normal pavement life.

Parameter	Standard	AADT 5 000-10 000	AADT > 10 000
Stability (Wheel Track PRD_{AIR})	NS-EN 12697-22	7%	5%
Stability (Wheel Track WTS_{AIR})	NS-EN 12697-22	0,06	0,04
Wear (Prall-value)	NS-EN 12697-16	25	22

Table 2. Rules for calculating changes in pavement life.

Damage mechanism/Test method	Percentage of vehicles using studded tyres		
	Low < 40%	Medium 40 – 60%	High >60%
Prall (change in pavement life in % for each unit change in Prall-value for AADT > 10 000)	0,5%	1%	2%
Prall (change in pavement life in % for each unit change in Prall-value for AADT 5 000 – 10 000)	0	0,5%	1%
Wheel Track (increased pavement life in % for each unit decrease in WT-value (PRD_{AIR}))	12%	10%	8%

In the pilot contracts the normal pavement life is assumed to be 10 years. The bidders may choose to offer different asphalt qualities regarding rutting. If so, the calculation rules used to transform these to changes in pavement life are shown in Table 2.

As an example, one may consider a road section with medium level of vehicles using studded tyres and an AADT > 10 000. If a contractor e.g., offers a Prall value = 19 this is an improvement of 3 units compared to the requirement for achieving a normal pavement life of 10 years. Using the rules in Table 2 this is transformed to an increase of 3% in pavement life, i.e. 0.3 years. If the same contractor also offers a WT-value (PRD_{AIR}) = 3.5%, this represents a 1.5 unit improvement compared to the value given in Table 1. This means a calculated improvement in pavement life of 1.5 * 10% = 15%, i.e. 1.5 years. The total calculated increase in pavement life is thus 1.8 years compared to the assumed normal pavement life of 10 years.

When allocating the contract, the bidding sum is first corrected for CO2-emissions as described in the previous section of this paper. Then the annual cost is calculated as follows:

$$Annual\ cost = \frac{Bidding\ sum\ adjusted\ for\ CO2\ [NOK]}{Adjusted\ pavement\ life\ [years]} \tag{1}$$

The contract is allocated to the bidder with the lowest calculated annual cost. The contract sum equals the original bidding sum. In the case of deviation between offered quality in the bid and actual quality in-situ, the penalty is calculated as the value of lost pavement life.

6 THIRD PARTY VERIFICATION OF EPD'S

The increased use of EPD's in pavement contracts has triggered a need for independent, third- party verification. Therefore, the NPRA has involved the notified body according to

Figure 4.   Asphalt with added biogenic binder on E14 near Meråker (Photo: Ellinor Hansen, NPRA).

the Construction Products Regulation, *Kontrollrådet*, to act as an independent verifier of the EPD's that are used in the contracts. *Kontrollrådet* is established as a private foundation. It is accredited for certification of products, quality systems and environmental systems, including Class P – Aggregates and Class S – Asphalt and Bituminous Mixtures. One of their tasks is to certify asphalt production plants according to NS-EN ISO 9001 and 14001. During their annual visits to the production plants, they now also do a verification of the EPD's used in the contracts, e.g. energy sources, aggregate types, bitumen grade, additives etc. They report back to the NPRA about their findings. Together with the contractor organisation (EBA) the NPRA have taken the initiative to develop a simple certification process for asphalt EPD's. The aim is to have this ready for use before the 2023 paving season.

7   CONCLUSION - RESULTS OF THE NEW CONTRACT STRATEGY

The main result of the new contract strategy is that the contractors have a real incentive to reduce CO_2 emissions when doing pavement maintenance for the NPRA. In the last two years (2020 – 2021) the NPRA has seen a reduction in the average CO_2 emissions per tonne asphalt of about 13% in pavement contracts, from 62 kg CO_2 per tonne asphalt in 2020 to 54 kg CO_2 per tonne asphalt in 2021.

The NPRA also observes that innovative approaches are being implemented by the contractors to minimise the CO_2 emissions for all stages of the paving process. One example is the use of biogenic binders to replace some of the bitumen. Many producers have also changed to alternative energy sources for heating of materials in their production plants, e.g. LNG, biogas/bio-oil or wooden pellets. They are also focusing on optimising the supply chain of materials used in asphalt production, e.g. reducing the transport distances for aggregates. Reduction in production temperature, ensuring low water content in aggregates and increased use of recycling are other examples of typical measures being taken.

The pilot contracts combining CO_2-weighting with life-cycle costs inspires contractors to utilise their best knowledge and craftmanship to produce pavements with long pavement lives with a minimum life cycle cost and low CO_2 emissions.

Allocating contracts based on other criteria than lowest bid obviously means that the NPRA may be paying more for pavement maintenance in a given year, but contributing to reducing CO_2 emissions in the long run with the lowest possible life-cycle costs. In 2022 the NPRA are paying about 8.5 million NOK ($\approx$ € 850,000) more than if the lowest bids had been allocated for every contract. This represents about 0.8% of the total value of the pavement maintenance contracts this year. In addition come the costs associated with any bonuses for further reduction of CO_2 during the contract period. Weighting of CO_2 alone or in combination with life-cycle costs was decisive for allocation of 7 of 27 contracts in 2022, i.e. the contract was not awarded to the bidder with the lowest price.

It has been a quite steep learning curve for all parties involved. The NPRA have taken measures to increase the knowledge needed in the organisation to be able to implement the new contract strategy and acknowledge that this effort must be continued in the years to come. The NPRA still needs to gather more experience with the new contract strategy, and to further refine and develop it together with the pavement industry.

REFERENCES

EBA (Norwegian Contractors Association), 2020. *Annual Statistics for Asphalt Production Published on* www. eba.no/vei-og-jernbane/asfalt.

EPD-Norge, 2019. *General Programme Instructions for the Norwegian EPD Foundation*, version 5:2019

Iversen, Ole M.K., 2021. *EPD-verktøy: Hvilken Informasjon får vi. Presentation Made at Conference «Miljødagen»*, Oslo 9. Nov 2021 Arranged by Norsk Asfaltforening (in Norwegian).

Norwegian Ministry of Transport, 2021. *National Transport Plan 2022 – 2033*, Meld. St. 20 (2020-2021) Report to the Storting (white paper) – English summary.

NPRA, 2021. *Vegnormal N200 Vegbygging* (in Norwegian).

Roads and Airports Pavement Surface Characteristics – Crispino & Toraldo (Eds)
© 2023 The Author(s), ISBN 978-1-032-55149-4

Large scale noise survey of the Walloon road network using CPX method: Second measurement campaign

Anneleen Bergiers*

Department of Surface Characteristics and Noise, Belgian Road Research Centre, Brussels, Belgium

Sébastien Marcocci

Noise Division, Direction Environmental and Landscape Studies, Department Expertise Hydraulics and Environment, SPW Mobility and Infrastructures, Namur, Belgium

ABSTRACT: A public tender was issued by the Wallonian Road Administration (SPW Mobility and Infrastructures) to realize a measurement campaign using CPX method (ISO 11819-2) to determine the acoustical quality of road pavements on the Walloon major road network. The task was assigned to the Belgian Road Research Centre (BRRC) after the procedure of public procurement. It was performed twice: measurement campaign CPX1 in 2018-2019 and measurement campaign CPX2 in 2020-2021.

The survey allows the Walloon road administration to assess the state of maintenance, e.g. wear and damage to the road surfaces and to complete the actual database (composed of skid resistance characteristics, ...) to help in the global strategy for road resurfacing of the major road network. In further studies results will be used as input for noise mapping.

CPX2 consists of the realization of CPX measurements on about 3745 km of roads. In contrary to CPX1, only the P1 tyre (SRTT) which is representative for passenger cars, was used during CPX2. While in CPX1 mainly motorways were measured, also a lot of national roads were included in CPX2.

A speed of 80 km/h was used and only the slow lane was measured in two directions. To be able to analyze and compare visual aspects with results, localized photos were taken during the measurement campaigns and linked to the noise measurements per 20 m road segment.

An acoustical classification system was established and used for the end analyses. A visualization of results on map was realized by BRRC. In addition to being didactic, this map makes it possible to visualize the distribution of results labeled "very noise reducing" to "very noisy".

Finally a comparison between CPX1 and CPX2 was made to study the evolution of acoustical quality in time.

Keywords: Noise, CPX, Road Network

1 INTRODUCTION

In 2020, SPW Mobility and Infrastructures reconducted the tender "Réalisation d'une campagne de mesures de caractérisation acoustique des revêtements sur le réseau wallon par la méthode CPX", which was assigned to the Belgian Road Research Centre (BRRC) in 2018.

The project (CPX2) includes the realisation of CPX measurements on a total length of approximately 3745 km of roads and motorways. The measurements are carried out in

---

*Corresponding Author: a.bergiers@brrc.be

DOI: 10.1201/9781003429258-67

accordance with ISO 11819-2 (ISO, 2017a). In contrast to the 2018-2019 assignment (CPX1), only measurements using the P1 tyre (SRTT), corresponding to a light vehicle tyre, were carried out. These were done at a constant speed of 80 km/h when the maximum traffic speed is 90 or 120 km/h and only on the slow lane (V1) in both driving directions. A limited part of the measurements was carried out at 50 km/h, i.e. in dangerous measurement situations or in the presence of speed limits (70 to 50 km/h).

Due to the discontinuation of measurements with tyres representative of heavy vehicle traffic, the number of road kilometres measured with the P1 tyre was increased. In addition, part of the measurements was devoted to a second run on a section of the motorway network. From a timing point of view, the project took place in two phases:

- the first one ran from 1 July 2020 to November 2020, after which the weather conditions were no longer suitable (rain, air temperature) due to autumn and winter;
- the second took place from the end of March 2021 to June 2021, with the main objective to carry out a new run on the motorway network.

2 METHODOLOGY

2.1 *The Close ProXimity (CPX) method*

The CPX measurements (ISO, 2017a) were carried out, for the first CPX1 and second CPX2 campaign, with the BRRC device illustrated in Figure 1, left, with the reference tyre P1 shown in Figure 1, middle, and described in ISO/TS 11819-3 (ISO, 2017b).

2.2 *Geo-tagged photos*

Coupled with the device described above, the IMAJBOX® of BRRC was used to take geo-tagged photos. One photo is linked to the noise measurements per 20 m of road segment. The IMAJBOX® is attached to the windscreen of the measurement vehicle (Figure 1).

Figure 1.   Measurements with the BRRC CPX trailer (left); interior of BRRC CPX trailer with P1 tyre (middle); mounting the IMAJBOX® on the measurement vehicle (right).

3 MEASUREMENT PARAMETERS

3.1 *CPX measurement uncertainty*

In general a measurement carried out according to the CPX methodology has some uncertainty due to variations in the procedure (e.g. lateral position of the tyres, speed variation), the sound and speed measurement equipment, variable weather conditions, different background noises from external sources, possible undesired contributions from the pulling vehicle and the choice of reference tyres. All these factors are described in ISO 11819-2 (ISO, 2017a) and give a 95% expanded uncertainty of about 1.0 dB.

There is also an uncertainty due to the use of a different set of tyres. ISO/TS 11819-3 (ISO, 2017b) lists the following influence parameters, giving the 95% expanded uncertainty of approximately 0.6 dB for the P1 tyre:

- variations between different tyres of the same type;
- variations caused by changes in tyre properties due to wear and ageing of the rubber (not related to hardness);
- uncertainty about the correction of the rubber hardness;
- uncertainty about the air temperature correction.

It is important to take this into account when making comparisons (e.g. between CPX1 and CPX2 and between different runs).

3.2 *Pneumatic characteristics*

In view of the second measurement campaign, BRRC acquired new P1 tyres. They have been run in over more than 400 km as required by the standard. It should be noted that the entire 2020-2021 measurement campaign was carried out with the same set of tyres.

ISO 11819-2 (ISO, 2017a) refers to ISO/TS 11819-3 (ISO, 2017b), which defines a range in which a correction based on rubber hardness (H_A) is made, namely: 62-73 Shore A for P1. The correction formula is included in the "Corrections" section of this document. A correction of 0.2 dB(A) is made for Shore A hardness which differs from reference 66. The results of measurements carried out with tyres with a hardness higher than 66 Shore A, as is the case for CPX1 and CPX2, are corrected downwards.

As prescribed in the standard, BRRC regularly performs a rubber hardness measurement during the measuring season. The 2021 CPX2 results have, therefore, been corrected by - 0.74 to - 0.76 dB(A) more than the original 2020 CPX2 results. This is important for the comparison of the measurements, which will be detailed in the section "Analysis of the effect of winter conditions".

Since a comparison with the CPX1 results is also made in the section "Comparison of CPX2 and CPX1", the hardness of the CPX1 measurement tyres is also briefly discussed here. In the CPX1 measurement campaign a different set of tyres was used and these tyres were harder than the tyres in the CPX2 campaign. The CPX1 results were corrected by - 0.4 to - 1.68 dB(A) more than the CPX2 results.

It should be noted that the hardness correction for P1 tyres in ISO/TS 11819-3 (ISO, 2017b) has recently been questioned by the ISO/TC 43/SC 1/WG 33 working group, which is revising the relevant standards for CPX measurements. Based on new data and research, it was finally decided in the revision of ISO/TS 11819-3 (ISO, 2021) in 2021 to reduce the hardness correction for the P1 tyre from 0.2 to 0.12 dB/Shore A. In this project, the same original hardness correction was applied to all measurement results in order to maintain the same methodology and thus not to compromise the comparability of CPX1-CPX2 and 2020-2021.

The current hardness measurement method with the durometer is very sensitive to the operator, which is not desirable for a parameter that has such a large influence on the final result. BRRC is currently investigating whether digital hardness testers can improve hardness measurements.

4 ANALYSIS AND REPORTING OF RESULTS

4.1 *Corrections*

As detailed in the previous section, all measurement results were corrected for tyre rubber hardness and air temperature according to the procedure described in ISO 11819-2 (ISO 2017a), ISO/TS 11819-3 (ISO, 2017b) and ISO/TS 13471-1 (ISO 2017c):

$$L_{CPX, corr} = L_{CPX, measured} - \gamma \, (T - T_{ref}) - 0, 2 \, (H_A - H_{A, ref}) \qquad (1)$$

$$\gamma = -0.14 + 0.0006v \qquad (2)$$

$$v = 50 \text{ or } 80\text{km/h}; \; T_{ref} = 20\,^{\circ}C; \; H_{A,ref} = 66$$

According to the standard, the use of a different temperature correction was only necessary for concrete:

$$\gamma \,(\text{concrete}) = -0.10 + 0.0004v \qquad (3)$$

4.2 *BRRC GIS tool*

Given the large amount of data generated and in order to facilitate its representation, the IT division and the GIS team were asked to automate part of the process.

The CPX measurement results are available per 20 m road segment. Every 5 m, a picture was taken with the IMAJBOX®. Both the photos and the CPX results contain GPS coordinates. For each 20 m road segment of the CPX, the nearest image was linked. This was then linked to the SPW GIS road network by also searching for the nearest point, in order to incorporate various information such as the type of road surface or the district concerned. The link between CPX, photos and GIS road network was established using a GIS tool of BRRC.

For each road measured, an Excel table per axis and driving direction is provided to the SPW. In addition, BRRC has transposed all the results onto cartographic media in order to facilitate the reading of the results.

5 RESULTS

5.1 *CPX2 campaign*

The same classification and colour code as in the 2018-2019 measurement campaign were used. The 80 km/h classification is based on that used in Flanders by the Flemish Road Agency "Agentschap Wegen en Verkeer" (AWV). The 50 km/h classification has been deduced by BRRC. Figure 2 shows the result in detail, divided by direction des routes.

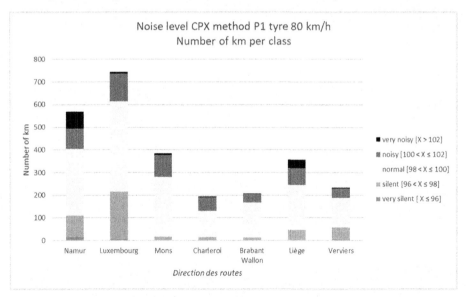

Figure 2. Class and directional distribution of the roads for the measurements with P1 tyre at 80 km/h; display in number of km per class.

Of all the measurement results in Wallonia, approximately 1% of the roads are "very quiet", 16.7% are "quiet", 58.1% are "normal", 19.2% are "noisy" and 5% are "very noisy".

As explained above, some measurements were carried out at 50 km/h due to the context. As the number of kilometres is limited (about 55 km), they are not presented here.

Figure 3 shows the overall map of CPX2 results for P1 tyre at 80 km/h in both positive and negative driving directions.

5.2 *Comparison of CPX2 and CPX1*

All motorways were measured with the P1 tyre at 80 km/h, both in the first measurement campaign in 2018-2019 (CPX1) and in the second measurement campaign (CPX2) in 2020-2021. This observation makes it possible to carry out an initial reflection on the evolution of the acoustic quality over the two campaigns. For more detailed CPX1 results, please consult (Bergiers & Marcocci 2020a, 2020b).

In order to visualise the results obtained during the two campaigns, they were transposed to the same map.[1] Figure 4 shows part of the measurement results of CPX1 and CPX2 for the "positive" driving direction.

Overall, the noise levels appear to have increased. It seems that the class "silent" has sometimes switched to the class "normal". Even a small increase in noise level can cause road segments to suddenly move up to the next class, even though they were initially still in the most silent class. However, there are also areas where the noise level has decreased, for example for the part of the A004 motorway under the responsibility of the Direction des Routes du Brabant wallon. Discussions with the managers of the Walloon Brabant motorway network have shown that roadworks were carried out there between the construction of CPX1 and CPX2.

It should also be noted that the set of tyres for CPX1 was not the same as for CPX2. The hardness corrections introduce an additional uncertainty, as mentioned in the section "Measurement parameters". The large difference in hardness between the CPX1 and CPX2 measurement tyres and the recent study showing that too much hardness correction was prescribed in the standard (see "Measurement parameters") may have led to the CPX1 results being overcorrected by about 0.7 dB(A) downwards compared to the CPX2 measurement results. 0.7 dB(A) is of the same order of magnitude as the differences that could be caused by changes in the road surface. A recalculation of the CPX1 and CPX2 measurement results with the proposed new hardness correction could help to improve the comparison.

The uncertainties in the measurement result, which are described in the standard, state that it is normal to note differences between measurements and that these are not necessarily directly related to actual changes in the acoustic quality of the road surface. Before drawing conclusions on this basis, the results should be studied in more detail calculating the absolute differences (not only by class, see further "Conclusions and perspectives") with due caution. Obtaining more measurement data over time can also help to draw more informed conclusions.

5.3 *Analysis of the effect of winter conditions*

As part of the CPX2 2020-2021 campaign, a selection of motorways was measured twice (in 2020 and 2021) to study the evolution of the acoustic quality of road surfaces, particularly after the winter period, namely the motorway routes A003, A004, A007, A015 and A027.

---

[1]Note about the map: Please note that the map only shows one measurement direction! The second CPX2 measurement is represented by a line in the middle. The first CPX1 measurement is indicated by a thicker line below the line of the second measurement. On both sides of the second measurement, the result of the first CPX1 measurement is thus visible (see legend on the map) for only one measurement direction: positive.

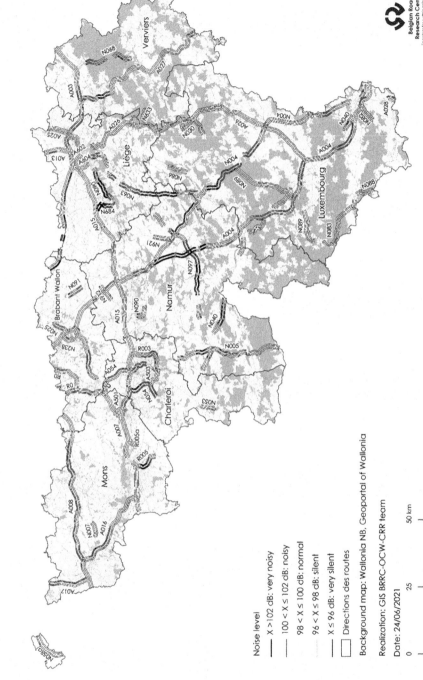

A. Overall map of CPX2 results for P1 tyre at 80 km/h in positive and negative driving directions

Noise level

— x >102 dB: very noisy

— 100 < x ≤ 102 dB: noisy

— 98 < x ≤ 100 dB: normal

···· 96 < x ≤ 98 dB: silent

···· x ≤ 96 dB: very silent

☐ Directions des routes

Background map: Wallonia NB, Geoportal of Wallonia

Realization: GIS BRRC-OCW-CRR team

Date: 24/06/2021

0   25   50 km

Figure 3.   Overall map of CPX2 results for P1 tyre at 80 km/h in positive and negative driving directions.

E. Comparison of the first measurement campaign 2018-2019 CPX1 and the second measurement
campaign 2020-2021 CPX2 - Results CPX P1 tyre at 80 km/h positive driving direction

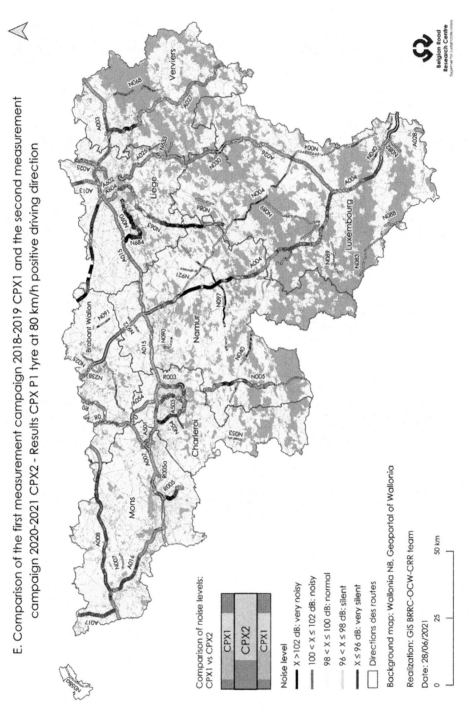

Figure 4.   Comparison between the first (2018-2019) and second (2020-2021) measurement campaigns.

A period of about six to ten months elapsed between the first and second measurement. The first measurement was carried out from July to October 2020; the second from March to May 2021.

Figure 5 shows the measurement results for these five motorways in 2020 and 2021 respectively.

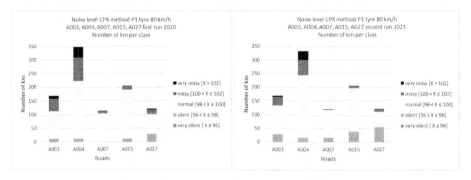

Figure 5.   Distribution by class and motorway of measurements with P1 tyre at 80 km/h on motorways A003, A004, A007, A015, A027 first run 2020 (left) and second run 2021 (right); displayed in number of km per class.

The results of the second measurement appear to be largely unchanged, if not slightly quieter. A decrease in noise levels is contrary to expectations if the road surface has not been renewed. Instead, an increase or stabilisation is expected. The following influences can explain the decrease in noise:

• As the tyre hardness increases, the noise levels are corrected downwards (see "Pneumatic characteristics" and "Corrections"). The measurement tyres were about 4 Shore A harder in the 2021 measurements than in the 2020 measurements. Due to the new lower hardness correction mentioned in the revision of the standard (see "Measurement parameters"), the measurement result for 2021 has probably been corrected by about 0.3 dB(A) too much compared to 2020.

• The measurements in 2020 were mainly carried out at higher temperatures (approx. 16 – 24 °C) than the measurements in 2021 (approx. 6 – 16 °C). The noise levels are corrected downwards (cf. "Corrections") for temperatures below 20 °C (e.g. - 1.29 dB(A) for 6 °C and 80 km/h) and upwards for temperatures above 20 °C (e.g. + 0.37 dB(A) for 24 °C and 80 km/h). The uncertainty of the temperature correction may also have influenced the results.

Given the uncertainties of the measurement method (see "Measurement parameters"), the change in tyre hardness between July 2020 and March 2021 and the limited number of data sets over time, it cannot be stated that there is a significant difference. It should be emphasised that it is difficult to draw firm conclusions on the basis of only two data sets. The more data available, the more reliable the resulting analysis.

6   CONCLUSIONS AND PERSPECTIVES

Beyond the current project, further investigation of the results of CPX1 versus CPX2 could be made by subtracting the noise levels of CPX2 and CPX1. This would result in a more accurate picture of the changes that have been made. Currently, only the transition from one class to another is visible, not the increases that have taken place within the same class. The magnitude of the difference is not clear. According to the current classification, the difference could be around 0.1 to 3.9 dB(A).

A full list of roadworks involving road surface changes/repairs that have taken place between CPX1 and CPX2 and between CPX2 run 1 (2020) and CPX2 run 2 (2021) may help to interpret the results and better understand some of the changes in noise levels.

Additional measurement campaigns over time can increase the reliability of conclusions regarding the development of noise quality over time. To determine the desired interval between measurement campaigns, the experience of the Flemish Road Agency (AWV) in recent years could be used. If the interval is too long, there is a risk that certain changes or developments will be missed. For some new road surfaces, for example, there is a rapid increase in noise levels in the first year, after which there is a stabilization.

On the basis of the available data from CPX1 and CPX2, a more detailed statistical analysis of the influence of the age of the road surface on the noise level can be made (and thus not only according to the class to which it belongs). Average sound levels could be determined per road surface type, as well as the dispersion of the result using the standard deviation and the minimum and maximum sound levels measured for a given road surface type.

The measurement results of CPX1 and CPX2 could be recalculated on the basis of the new hardness correction proposed for the P1 measurement tyres (0.12 dB(A)/shore A instead of 0.2 dB(A)/shore A) in order to reduce the uncertainty of the final result. The application of the same hardness correction is recommended for comparability. Either 0.2 dB(A)/shore A is retained for all measurements (as is currently the case), or CPX1 and CPX2 are recalculated with the new hardness correction.

Specific road sections could be selected to monitor the evolution of a certain type of road surface in more detail and to study it further. This may be useful for example for the types of road surface that will be used a lot by the SPW in the future. On these selected road sections, additional measurements could be organised at regular intervals to monitor the development.

As only the acoustic quality of the road surface was measured, it is not possible to establish the link with other characteristics of the road surface that may have changed. If the SPW has results from texture measurements on the same roads, these can be used to better understand the results and changes in acoustic quality.

ACKNOWLEDGEMENTS

A special word of thanks to all the BRRC collaborators who made this exciting project possible, in particular to Sebastien Defrance and Lilia Pleskach! We would also like to thank Bruno Schepers from the Direction des études environnementales et paysagères for his active participation in the project.

REFERENCES

Bergiers A. & Marcocci S. (2020a). Cartographie de la Qualité Acoustique du Réseau Routier Wallon. *Revue Générale des Routes et de L'aménagement (RGRA)*, (970), 56–60.

Bergiers A. & Marcocci S., (2020b). Large Scale Survey of the Walloon Road Network using CPX Method, *e-Forum Acusticum*. 2020, 7-11 December 2020.

International Organization for Standardization. (2017a). *Acoustics: Measurement of the Influence of Road Surfaces on Traffic Noise*. Part 2: The Close-proximity Method (ISO 11819-2). https://www.iso.org/standard/39675.html

International Organization for Standardization. (2017b). *Acoustics: Measurement of the Influence of Road Surfaces on Traffic Noise*. Part 3: Reference Tyres (ISO/TS 11819-3). https://www.iso.org/standard/70808.html

International Organization for Standardization. (2017c). *Acoustics: Temperature Influence on Tyre/road Noise Measurement*. Part 1: Correction for Temperature When Testing the CPX Method (ISO/TS 13471–1). https://www.iso.org/standard/25630.html

International Organization for Standardization. (2021). *Acoustics: Measurement of the Influence of Road Surfaces on Traffic Noise*. Part 3: Reference Tyres (ISO/TS 11819–3). https://www.iso.org/standard/82067.html

Test section with Next Generation Concrete Surface (NGCS) in Belgium: Rolling resistance and noise

Anneleen Bergiers*
Department of Surface characteristics and Noise, Belgian Road Research Centre, Brussels, Belgium

Barbara Vanhooreweder
Agency for Roads and Traffic, Flemish Government, Brussels, Belgium

ABSTRACT: In October 2015, a Belgian test section of Next Generation Concrete Surface (NGCS) was realized on the national road N44 in Maldegem (Belgium). The NGCS technique can be used for existing concrete surfaces and new concrete pavements. It consists of a combination of diamond grinding and longitudinal grooving. In this project the former road pavement consisted of old concrete plates with exposed aggregate surface. Two different profiles with a length of 100 m each were realized. A negative texture in the longitudinal direction is created using diamond blades. Every deep groove is alternated with four shallow grindings.

CPX- and SPB-noise measurements were performed by the Flemish Agency for Roads and Traffic. CPX measurements were repeated several times over a period of six years to follow up the evolution in time. SPB measurements were performed five times over a period of 20 months. While CPX noise reductions for light vehicles are promising (2 to 6 dB with respect to the former situation), they only reveal a limited influence for heavy traffic. However, SPB measurements demonstrate noise reductions up to 4 dB for heavy traffic.

Rolling resistance measurements were performed by the Belgian Road Research Centre in February 2019.

A reduction in rolling resistance with respect to the former situation is observed.

As a result of the study the Flemish Agency for Roads and Traffic integrated NGCS Profile 1 in the Flemish road surface tender specifications.

Keywords: Noise, Rolling Resistance, Concrete

1 INTRODUCTION

Various finishing techniques already exist for concrete roads, e.g. exposed aggregate concrete (EAC), brushing and grinding (Goubert 2022; Rens 2014). Recently a new technique, adopted from USA (Scofield 2016, 2010), has been further developed to reduce rolling noise. The technique can be used for existing concrete surfaces and new concrete pavements. It consists of a combination of diamond grinding and longitudinal grooving.

In Belgium a test section was realized in October 2015 to investigate this new technique. The test location is a national road (N44) in Maldegem (See Figure 8). The test section was followed up in time by performing noise, rolling resistance, skid resistance and texture measurements. Information about the skid resistance and texture measurements can be found in (Goubert 2022; Vanhooreweder *et al.* 2020, 2018).

---

*Corresponding Author: a.bergiers@brrc.be

DOI: 10.1201/9781003429258-68

2 TEST SECTION

The existing road consisted of old concrete plates with thickness 23 – 25 cm from 1958 with exposed aggregate surface. Two different profiles with a length of 100 m each were constructed. A negative texture in the longitudinal direction is created using diamond blades. Every deep groove is alternated with four shallow grindings. Both profiles were realized with diamond blades of the same thickness and diameter. The spacers are different (see Table 1).

Figure 1 shows a principal sketch of the transversal section (AWV 2021b) which is the result of the settings described in Table 1. By using the spacer of 1.5 mm parameter x in the Figure is equal to 1.1 mm for Profile 1 and by using spacer 3.0 mm parameter x is equal to 2.6 mm for Profile 2.

Table 1. Thickness and diameter of segment blades and width of spacer for both profiles.

Profile	N44 Reference point	Longitudinal grooving: Segment blade	Diamond grinding: Segment blade	Spacer
1	16.2 – 16.3	Thickness: 2.8 mm Diameter: 363 mm Depth groove: ± 4 mm	Thickness: 2.8 mm Diameter: 356 mm Depth groove: ± 1 mm	1.5 mm
2	16.4 – 16.5	Thickness: 2.8 mm Diameter: 363 mm Depth groove: ± 4 mm	Thickness: 2.8 mm Diameter: 356 mm Depth groove: ± 1 mm	3.0 mm

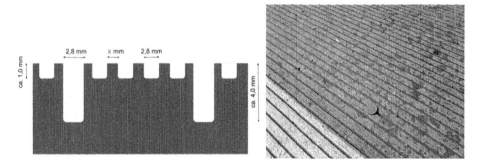

Figure 1. Principal sketch of the transversal section (left). Photo of the surface (with 1 euro coin as reference) (right).

3 NOISE MEASUREMENTS

To determine the effect of this new surface finishing technique on the noise level, before and after the construction noise measurements were carried out with the Close-Proximity (CPX) and Statistical Pass-By (SPB) method.

3.1 *Close Proximity (CPX) method*

3.1.1 *Measuring method*
The *Close-Proximity* method (CPX) is an acoustic measuring method whereby the contact noise between tyre and road surface is measured by driving on the road with a measuring trailer. The purpose of the CPX method is to evaluate both the noise production and the homogeneity of a road pavement over a certain route.

The rolling noise of the reference tyres is measured with 2×2 microphones assembled near the tyre/road pavement contact surface in two acoustically insulated housings fitted on a trailer chassis. This measuring trailer is driven over the road surface at a reference speed of 50 or 80 km/h. The measurements are carried out in dry weather.

As result the noise level per 20 metres and the spectrum can be represented. The standardisation of this measuring method is set out in standard ISO 11819-2 (ISO 2017a). At the beginning of the project the draft standard ISO/DIS 11819-2 (ISO, 2012) was used until the final version was published in 2017. This final standard refers to ISO/TS 11819-3 (ISO, 2017b), which defines a correction based on rubber hardness (H_A) of the tyres.

At the beginning of the project no rubber hardness correction was applied. The hardness correction for P1 tyres has recently been questioned by the ISO/TC 43/SC 1/WG 33 working group, which is revising the relevant standards for CPX measurements. Based on new data and research, it was finally decided in the revision of ISO/TS 11819-3 (ISO, 2021) in 2021 to reduce the hardness correction for the P1 tyre from 0.2 to 0.12 dB/Shore A.

Temperature correction was applied according to ISO/DIS 11819-2 (0.03 dB/°C) and was adapted since the publication of the standard ISO/TS 13471-1 (ISO, 2017c) in 2017 (0.092 dB/°C).

3.1.2 *Measurements results*

On 5 August 2015 pre-measurements were carried out on both trial sections with the CPX trailer with both reference tyres, SRTT (P1) and AVON AV4 (H1), at 80 km/h. The SRTT tyres are representative for passenger cars. The AVON AV4 tyres are representative for heavy vehicles. The measurements show that with P1 tyre the CPX level of the original exposed aggregate surface is between 101.1 and 101.4 dB. This is 2.1 to 2.4 louder than the Belgian acoustic reference pavement SMA-C2 (stone mastic asphalt with maximum aggregate size of 10 mm). The acoustical quality of this Belgian acoustic reference road pavement is the result of a statistical analysis on a large number of SMA-C2 pavements in Flanders. With H1 tyre the CPX level is between 99.6 and 100.0 dB.

The CPX measurements were repeated 1, 5, 11, 18, 45, 56 and 72 months after applying the NGCS method. The difference in $L_{CPX:P,80}$ and $L_{CPX:H,80}$ compared to the initial road pavement and the Belgian acoustic reference road pavement SMA-C2 is shown in Figures 2 and 3. The initial road pavement has only been measured before applying the technique and the CPX results are compared to this original value which does not change in time. The results after 45 and 56 months reveal a decrease of the noise level. This is caused by the fact that rubber hardness correction has been introduced in 2017. It only has been applied to measurements after this date. The hardness of P1 tyres was high at 45 months (71 - 72 shore A). Results were overcorrected with 0.2 dB/Shore A.

For the first profile with P1, immediately after treatment $L_{CPX:P,80}$ was determined at 95.2 dB. This is 6 dB quieter than before and 4.1 dB quieter than the Belgian acoustic reference pavement SMA-C2. The measured level corresponds to an AGT (thin noise reducing asphalt top layer) pavement type I (AWV, 2021a). One year later the noise level has increased by approximately 1 to 2 dB and then remains stable[1] up to 6 years. For the H1 measurements $L_{CPX:H,80}$ remains practically the same.[1]

For the second profile, after the application of the NGCS method $L_{CPX:P,80}$ of 99.3 dB was measured and the noise level remains stable. This noise level is comparable with the SMA-C2 acoustic reference pavement. In comparison with the former situation this is a reduction of 2.1 dB. With H1 tyre the CPX level remains around that of the existing pavement.

Figures 4 and 5 show the spectrum of both profiles before, immediately after and 6 years after the NGCS treatment. Here it can clearly be seen that for the first profile, with the narrowest spacers, there is a large reduction of the noise level in all frequency bands but a clear dip is also visible at

---

[1]Experience of AWV: In general the difference in CPX-level between various pavements with the AVON AV4 (H1) tyre is smaller than with the SRTT (P1) tyre.

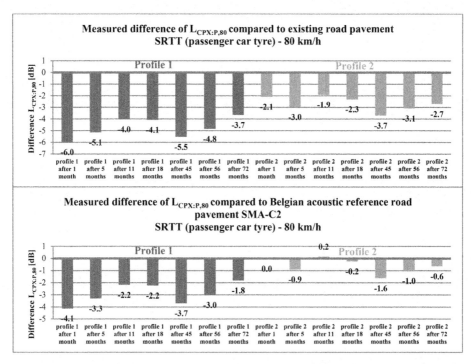

Figure 2.   Measured difference $L_{CPX:P,80}$ at 80 km/h for cars compared to existing road pavement above) and Belgian acoustic reference road pavement (below).

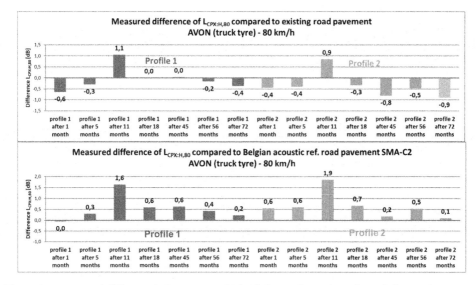

Figure 3.   Measured difference in $L_{CPX:H,80}$ at 80 km/h for trucks compared to existing road pavement (above) and Belgian acoustic reference road pavement (below).

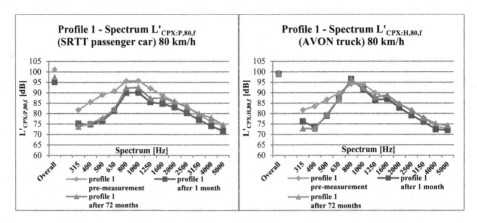

Figure 4. CPX spectra at 80 km/h before, immediately after and 6 years after NGCS treatment for Profile 1 and reference tyre P1 (left), for reference tyre H1 (right).

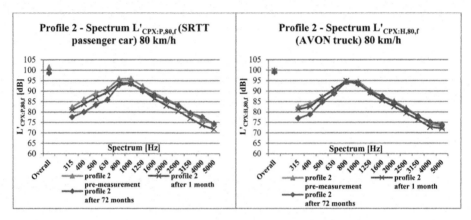

Figure 5. CPX spectra at 80 km/h before, immediately after and 6 years after NGCS treatment for Profile 2 and reference tyre P (left), for reference tyre H1 (right).

both the mid and low frequencies (1250 Hz and 315 – 630 Hz). The much finer and negative texture ensures that tyres vibrate much less and emit less noise at the low frequencies. With the presence of the grooves the air escapes more easily than with a dense pavement. This gives a reduction at the higher frequencies. The same phenomenon is also seen in the H1 figures. For Profile 2 a rather constant reduction of the noise level is visible for all frequencies. After 6 years there is mainly an increase at the higher frequencies. This is probably due to the wear of the fine texture. The reason for the decrease at low frequency (315 – 630 Hz) remains unclear to the authors.

3.2 Statistical Pass-By (SPB) method

3.2.1 Measuring method

The Statistical Pass-By (SPB) method is a measurement of the noise of many individual passing light and heavy vehicles at a certain speed. The registration of the noise is carried out at one set position along the road. The purpose of the SPB method is to accurately establish the acoustic properties of certain road surfaces for light, medium-heavy and heavy vehicles, and this for different standard speeds. The SPB method is set out in the ISO standard 11819-1 (ISO, 1997). A temperature correction, which did not change over time, has been applied to the results.

3.2.2 Measurements results

The measured SPB noise levels are shown in Figure 6 and Figure 7.[2] For the finest, first profile the noise reduction amounts to approximately 6 dB for light vehicles and 4 dB for heavy vehicles. This remains quite stable in the course of time.

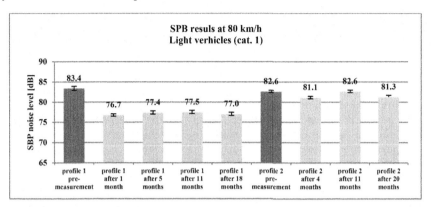

Figure 6. SPB results at 80 km/h for light vehicles before, 1, 5, 11 and 18/20 months after construction.

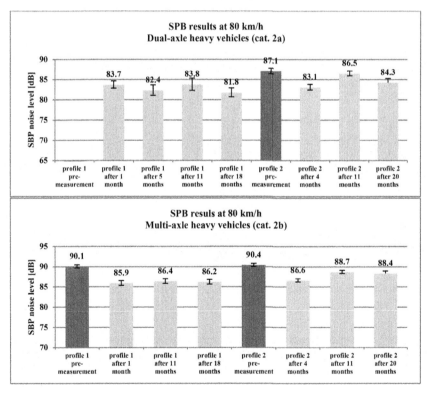

Figure 7. SPB results at 80 km/h for dual-axle (above) and multi-axle heavy) vehicles (below) before, 1, 5, 11 and 18/20 months after construction.

---

[2]No SPB-results are shown for dual-axle heavy vehicles before construction (pre-measurement) as not enough vehicles of this category were measured.

For the second profile the noise level remains approximately the same for cars. For heavy traffic, a reduction of almost 4 dB is measured. One year after the treatment the reduction still amounts to about 2 dB. Afterwards no new SPB measurements were performed.

4 ROLLING RESISTANCE MEASUREMENTS

4.1 *Measuring method*

Rolling resistance (RR) is an important characteristic which is worldwide acknowledged for tyres. Road surfaces however also play a significant role in the RR story, which is shown in a lot of research which is still ongoing (Bergiers & Maeck 2018; Goubert 2022). The use stage within the pavement's lifetime contributes largely to the carbon footprint of the road. Decreasing the RR is a straightforward way to reduce emissions from road traffic stage within the pavement's lifetime.

Rolling resistance measurements are performed with the BRRC trailer according to the ROSANNE draft standard (Anfosso *et al.* 2016), see Figure 8 (left). The trailer is designed as a quarter-car with a common car-suspension. It is connected to the measurement vehicle with bolts and has a fixed and a movable frame. At the end of the movable frame the extra load is attached. The total vertical force exerted on the tyre by the trailer is 2 kN.

The principle of these measurements is shown in Figure 8 (right). The tyre can lean backwards and forwards thanks to the hinged connection with the fixed frame. During the measurements the tyre is pulled backwards as a result of the rolling resistance force (R). This results in the angle θ which is measured. The rolling resistance coefficient C_r is the proportion of the horizontal force (rolling resistance force, R) to the vertical force (vertical load, F_z). This corresponds to the tangent of the measured angle θ which may be approximated by θ, itself expressed in radians as it is a small angle.

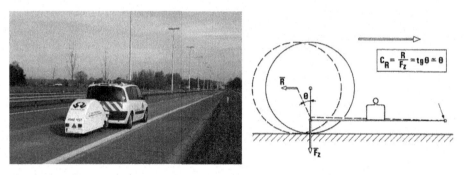

Figure 8.   Rolling resistance trailer BRRC on test section (left). Measurement principle trailer (right).

The trailer has an enclosure that can prevent the tyre air drag from affecting the results. It is equipped with several sensors to register different parameters during the measurement, e.g. inclination θ of the wheel carrier with respect to the frame of the trailer, inclination of the frame of the trailer with respect to the horizontal plane, inclination between the trailer and the towing vehicle, ... Corrections are performed afterwards.

4.2 *Measurement results*

Three reference surfaces were selected: two adjacent surfaces (road N44) with old concrete plates with thickness 23 – 25 cm from 1958 with exposed aggregate cement concrete (EACC)

702

surface (like the situation before applying the technique) and an SMA-C2 surface (ref. 3) nearby with maximum aggregate size of 10 mm (road A11). A mean value of the results of the two adjacent surfaces (ref. 1 and ref. 2) with old concrete plates was calculated and considered as the concrete reference.

Measurements were performed on 20 February 2019. Air temperature was 4 to 8 °C. No temperature corrections were performed. Both profiles and the three reference surfaces were measured with Avon H1 tyre at 50 and 80 km/h. Minimum five measurement runs were performed on each surface at each speed. The results at 50 km/h are shown in Figure 9 and the RR reductions with respect to the references for all speeds are shown in Table 2.

Profile 1 shows the highest RR reductions with respect to all references. Profile 1 and 2 show a lower RR than all references.

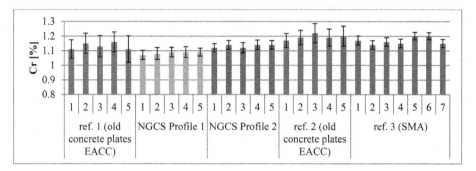

Figure 9.   Rolling resistance coefficient Cr for H1 at 50 km/h for the various reference surfaces. The horizontal scale shows the various measurement runs (1, 2, 3, … , 7).

Table 2.   Relative RR difference with respect to the references (expressed in %) at 50 and 80 km/h.

Reference Speed	Concrete plates EACC		SMA–C2	
	50 km/h	80 km/h	50 km/h	80 km/h
Profile 1	- 6.8%	- 5.5%	- 7.1%	- 6.3%
Profile 2	- 2.7%	- 0.8%	- 3.0%	- 1.7%

Drum measurements were performed on Profile 1 in the frame of the GHRANTE project (Goubert 2022). A fair correlation with the results at 50 km/h was found between on site and drum measurements.

5   CONCLUSION

CPX measurements performed with P1 tyre at 80 km/h after applying the Profile 1 NGCS technique on the concrete plates reveal an initial noise reduction of 6 dB with respect to the former situation (concrete plates with EACC surface) and 4.1 dB compared to the acoustic reference pavement SMA-C2. $L_{CPX:P,80}$ was determined at 95.2 dB. The noise reduction is comparable to the effect of a thin noise reducing asphalt pavement type I (AWV 2021a). One year later the noise level has increased by approximately 1 to 2 dB and then remains stable up to 6 years. Measurements with H1 tyre reveal only a small noise reduction of about 0.6 dB with respect to the former situation and this remains stable in time over 6 years. The acoustic quality for trucks is comparable to that of the acoustic reference pavement SMA-C2.

For Profile 2 $L_{CPX:P,80}$ was determined at 99.3 dB, which remains stable in time. This noise level is comparable with the SMA-C2 acoustic reference pavement. In comparison with the former situation this is a reduction of 2.1 dB. With H1 tyre only a small reduction of about 0.4 dB is measured with respect to the former situation which remains stable in time over 6 years.

SPB measurements on Profile 1 show initial noise reductions of 6 dB for light vehicles and 4 dB for heavy vehicles. These remain stable over 20 months. Almost no noise reduction for light vehicles is noted for Profile 2, while an initial noise reduction of 4 dB is measured for heavy vehicles. After one year 2 dB remains.

Measurements with rolling resistance trailer demonstrate RR reductions with respect to the former situation and the SMA-C2 pavement. Profile 1 reveals the lowest rolling resistance.

The test section demonstrated that the new technique is favourable for traffic noise, in particular light vehicles, and for rolling resistance of concrete roads. As a result of the study the Flemish Agency for Roads and Traffic integrated NGCS Profile 1 in the Flemish road surface tender specifications (AWV 2021b). $L_{CPX:P,80}$ of maximum 97.5 dB is required, which is 1.8 dB less than the acoustic reference pavement SMA-C2.

ACKNOWLEDGEMENTS

This pilot project and paper were only possible thanks to the collaboration of AB roads, AWV, BRRC, Febelcem and Robuco.

REFERENCES

Anfosso Lédée F., Cerezo V., Karlsson R. et al. 2016. *Experimental Validation of the Rolling Resistance Measurement Method Including Updated Draft Standard, Rosanne, WP3, D3.6*, 10 November 2016, available on http://rosanne-project.eu/ (last consulted on 22 April 2022).

AWV (Agentschap Wegen en Verkeer), 2021a. SB250 version 4.1a, Section 6, paragraph 2.6.2.6 D Rolling Noise (Rolgeluid), 24 June 2021, https://wegenenverkeer.be/zakelijk/documenten

AWV (Agentschap Wegen en Verkeer), 2021b. SB250 version 4.1a, Section 12, paragraph 1.6.3 NGCS, 24 June 2021, https://wegenenverkeer.be/zakelijk/documenten

Bergiers A. and Maeck J., 2018. Status-quo and outlook for rolling resistance in Europe, 8th Symposium on Pavement Surface Characteristics, SURF 2018, PIARC, Brisbane (Australia), 2–4 May 2018.

Goubert L. 2022. Optimizing the Pattern of Grooves of the Next Generation Cement Concrete Surface (NGCS) for Less Tyre/road Noise and Less Fuel 2 Consumption, *9th Symposium on Pavement Surface Characteristics*, SURF 2022, Milano (Italy), 12-14 September 2022.

ISO 11819–1, 1997. *Acoustics – Measurement of the Influence of Road Surfaces on Traffic Noise – Part 1: Statistical Pass-By method.*

ISO/DIS 11819–2, 2012. *Acoustics - Method for Measuring the Influence of Road Surfaces on Traffic Noise - Part 2: The Close-Proximity Method.*

ISO 11819–2, 2017a. *Acoustics - Measurement of the Influence of Road Surfaces on Traffic Noise - Part 2: Close-proximity Method.*

ISO/TS 11819–3, 2017b. *Acoustics - Measurement of the Influence of Road Surfaces on Traffic Noise - Part 3: Reference Tyres.*

ISO/TS 13471–1, 2017c, *Acoustics - Temperature Influence on Tyre/road Noise Measurement - Part 1: Correction for Temperature When Testing with the CPX Method.*

ISO/TS 11819–3, 2021. *Acoustics - Measurement of the Influence of Road Surfaces on Traffic Noise. Part 3: Reference Tyres.*

Scofield L., 2010. *Safe, Smooth and Quiet Concrete Pavement, Paper 78 First* International Conference on Pavement Preservation US. https://www.pavementpreservation.org/icpp/paper/78_2010.pdf last consulted on 21 April 2022.

Scofield L., 2016. *Development and Implementation of the Next Generation Concrete Surface, Report – Living Document.*

Rens L., 2014. *Duurzaam Stiller. Over Verkeerslawaai en Geluidsarme Betonwegen.* Oktober 2014. https://www.febelcem.be/fileadmin/user_upload/dossiers-ciment-2008/nl/I7-NL-DuurzaamStiller.pdf last consulted on 21 April 2022.

Vanhooreweder B. De Winne P.,Scheers A. (AWV), Rens. L (Febelcem) and Beeldens A. (AB-Roads), 2018. *"Next Generation Concrete Surface (NGCS). Finally A Quiet and Sustainable Road Pavement?"*, 13th International Symposium on Concrete Roads, 2018.

Vanhooreweder B. (AWV), Beeldens A (AB-Roads), Bergiers A. and Goubert L. (BRCC) and Rens L (Febelcem), 2020. *Next Generation Concrete Surface (NGCS) Evaluatie proefvak N44 Maldegem uit 2015: Eindrapport.*

Roads and Airports Pavement Surface Characteristics – Crispino & Toraldo (Eds)
© 2023 The Author(s), ISBN 978-1-032-55149-4

Acoustic characterisation of motorway pavements: New perspectives to optimise road asset management

Gaetano Licitra*
ARPAT, Pisa, PI, Italy

Lara Ginevra Del Pizzo, Antonino Moro & Francesco Bianco
IPOOL srl, Pisa, PI, Italy

Davide Chiola & Benedetto Carambia
Movyon spa, Campi Bisenzio, Firenze, Italy

ABSTRACT: The acoustical properties of road surfaces are an essential research topic for the optimisation and reduction of the environmental impact of the road infrastructure. Indeed, the main source of road traffic noise is represented by tyre/road interaction from vehicle speeds typical of urban contexts, up to motorway speed limits. Under this light, techniques like the Close ProXimity method (CPX) were developed to study the performance of pavements from an acoustical point of view. In this study, the CPX method was used to evaluate the performance of pavements designed by AutoStrade Per l'Italia (ASPI) and laid on several sites along the Italian motorway infrastructure. In particular, the dataset included different porous and semi-porous asphalts of different age, with some containing crumb rubber added as a modifier. On-site CPX evaluation included both the broadband and one-third octave band levels and the results were then used to correlate the acoustic performance with other traditional mechanical and superficial properties, such as drainability, friction and road texture. Results show that the pavements analysed show a great variability from an acoustical point of view: this fact could lead to new perspectives for the optimization of the road asset management, adding new performance indicators to the traditional ones.

Keywords: Pavement Monitoring, Acoustic Performances, CPX Method.

1 INTRODUCTION

The acoustical properties of road surfaces are an essential research topic for the optimisation and reduction of the environmental impact of the road infrastructure. The impact of Road Traffic Noise (RTN) on the quality of life of citizens living nearby the road infrastructure is far from negligible. Indeed, an exposure to high levels of TRN leads to a series of issues, such as sleep disorders (Muzet 2007; Skrzypek *et al.* 2017), learning impairments (Hygge *et al.* 2002; Lercher *et al.* 2003), cardiovascular, hypertension and ischemic heart disease (Babisch *et al.* 2012) and annoyance (Guski *et al.* 2017).

RTN is caused by different sources, which can be summarised in three categories (Sandberg & Ejsmont 2002):

- Mechanical noise caused by the engine,
- Air turbulence caused by the vehicle motion,
- Tyre/road noise

---

*Corresponding Author; g.licitra@arpat.toscana.it

DOI: 10.1201/9781003429258-69

For speeds higher than 35 km/h, Tyre/Road Noise (TRN) becomes the dominant source; therefore, a reasonable approach to RTN reduction involves the design of low-noise pavements aimed at reducing this noise source in urban and extra-urban contexts.

TRN results from the combination of aerodynamic and vibrodynamic phenomena. Aerodynamic noise is generally relevant at frequencies higher than 1 kHz, while vibrodynamic noise covers frequencies lower than 1 kHz (Li 2018). Techniques like the Close ProXimity method (CPX) were developed to study the performance of pavements from an acoustical point of view (ISO 11819-2:2017).

The European Green Public Procurement (GPP) defines noise criteria in road construction, based on the experiences of several European countries and suggests the use of the Close ProXimity method (CPX) to evaluate the acoustic performance of low-noise pavements, setting limits for the noise emission level as a function of the speed limits on the road section. These limits are provided both for design and for durability (Garbarino *et al.* 2016).

As well known, pavement characteristics such as road texture, porosity and layer thickness are strongly involved in tyre-road noise (Del Pizzo *et al.* 2020).

Under this light, the goal of this study was to characterise porous pavements on motorways from an acoustical point of view and to highlight correlations between noise emission levels and road surface properties, in order to find new inputs to optimise road asset management.

In particular, the broadband and one-third octave band CPX levels were analysed to correlate the acoustic performance with other road surface properties, such as drainability and road texture.

2 EXPERIMENTAL PLAN AND MATERIALS

In order to achieve the objective of the work, a specific experimental plan was designed. In particular, a measurement campaign involving CPX and surface parameters monitoring of 4 different motorway stretches in Italy was carried out during winter 2021. Along the four sites, a total of 11 different road surfaces were used to perform a correlation analysis between the superficial parameters and noise emission levels measured with the CPX method.

The different road surfaces analysed, both dense and porous mixes, are listed in Table 1. The set of porous pavements is divided in porous and semi-porous mixes, including two containing Crumb Rubber (CR), with different age at monitoring.

Table 1. List of pavements used in this study. legend: CR: Crumb Rubber, DAC: Dense Asphalt Concrete.

Site	ID	Pavement Type	Age at monitoring [yrs]
I	1	Type 1 Porous asphalt	1.84
I	2	Type 1 Porous asphalt	1.80
II	3	Type 1 Porous asphalt	8.44
III	4	Type 2 Porous asphalt	0.57
III	5	Type 1 Porous asphalt	2.07
III	6	Type 1 Porous asphalt	5.76
III	7	Optimized Porous asphalt w/t CR	5.76
III	8	Type 1 Porous asphalt	6.65
IV	9	Semi-porous asphalt	1.30
IV	10	Semi- porous asphalt w/t CR	1.30
IV	11	DAC	6.16

For each pavement, the following parameters were measured:

- International Roughness Index (IRI) measured with the ARAN system - ASTM E1926-08:2021 [mm/m],
- Mean Profile Depth (MPD) - ISO 13473-1:2019 [mm],
- Drainability - ASPI internal test method [dm3/min],
- Transversal Friction Coefficient (TFC) measured with the SUMMS2 system.
- CPX levels in broadband (dB(A)) and one-third octave bands at the reference speed of 80 km/h and normalized to the reference air temperature of 20°C and tyre hardness of 66 Shore(A) – ISO 11819-2:2017.

3 RESULTS AND DISCUSSIONS

Figure 1 summarizes the results related to the acoustic characterization of the analysed pavements in terms of CPX levels in dB(A) at the reference speed of 80 km/h. Taking into account only Type 1 Porous mixes which were available at different ages, it appears that noise levels increase with pavement age, in agreement with models available in literature (Licitra *et al.* 2018). Indeed, a linear regression using the experimental data from Type 1 Porous mixes, provides a goodness-of-fit measured in terms of adjusted r^2_{adj} of 0.86. An important limitation of this analysis regards the different geographical location of the surfaces, which were placed in different sites, characterised therefore by different traffic volumes and climatic conditions.

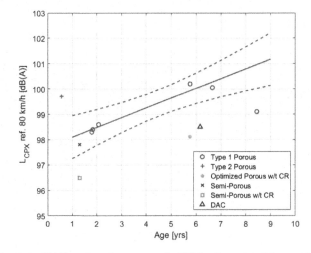

Figure 1.   CPX levels in dB(A) at the reference speed of 80 km/h vs age of the monitored pavements in years. A linear regression was performed considering only Type-1 Porous asphalts.

In terms of noise ranking, all pavements show a good performance referring to GPP values; in particular the lowest broadband noise levels were achieved by the Semi-Porous w/t CR (ID 10 in Table 1).
In order to study the correlation between CPX levels and road surface parameters, the measurement results were cross-correlated. The correlation matrix obtained from the 11 different data points is shown in Table 2.
Analysing the correlation between surface parameters, drainability values are well correlated with MPD, while no other correlation is evident. Moreover, TFC is the only parameter

Table 2. Correlation matrix between surface parameters and noise emission measured with the CPX method.

	Var	Surface parameters					one-third octave band L_{CPX} [Hz]												
		Drain	TFC	MPD	IRI	L_{CPX}	315	400	500	630	800	1000	1250	1600	2000	2500	3150	4000	5000
Surf. param.	Drain	–	-0.26	0.90	0.58	0.39	0.72	0.73	0.76	0.68	0.41	-0.55	-0.83	-0.70	-0.60	-0.71	-0.81	-0.82	-0.80
	TFC	-0.26	–	-0.41	0.18	-0.75	-0.55	-0.56	-0.65	-0.77	-0.66	-0.15	0.46	0.34	0.22	0.18	0.23	0.27	0.28
	MPD	0.90	-0.41	–	0.35	0.39	0.85	0.85	0.87	0.76	0.38	-0.43	-0.79	-0.75	-0.67	-0.74	-0.82	-0.84	-0.83
	IRI	0.58	0.18	0.35	–	-0.15	-0.03	-0.03	0.06	0.01	0.02	-0.37	-0.33	-0.42	-0.42	-0.47	-0.55	-0.55	-0.57
	L_{CPX}	0.39	-0.75	0.39	-0.15	–	0.56	0.57	0.64	0.86	0.95	0.22	-0.43	-0.15	0.08	0.03	-0.10	-0.14	-0.12
one-third octave band L_{CPX} [Hz]	315	0.72	-0.55	0.85	-0.03	0.56	–	0.99	0.97	0.86	0.44	-0.47	-0.81	-0.66	-0.53	-0.61	-0.65	-0.65	-0.62
	400	0.73	-0.56	0.85	-0.03	0.57	0.99	–	0.98	0.87	0.43	-0.50	-0.77	-0.62	-0.48	-0.56	-0.61	-0.62	-0.59
	500	0.76	-0.65	0.87	0.06	0.64	0.97	0.98	–	0.93	0.53	-0.44	-0.82	-0.66	-0.51	-0.58	-0.65	-0.66	-0.64
	630	0.68	-0.77	0.76	0.01	0.86	0.86	0.87	0.93	–	0.78	-0.15	-0.71	-0.52	-0.32	-0.38	-0.49	-0.52	-0.50
	800	0.41	-0.66	0.38	0.02	0.95	0.44	0.43	0.53	0.78	–	0.35	-0.44	-0.20	0.02	-0.02	-0.16	-0.22	-0.20
	1000	-0.55	-0.15	-0.43	-0.37	0.22	-0.47	-0.50	-0.44	-0.15	0.35	–	0.50	0.47	0.45	0.55	0.52	0.46	0.45
	1250	-0.83	0.46	-0.79	-0.33	-0.43	-0.81	-0.77	-0.82	-0.71	-0.44	0.50	–	0.89	0.76	0.83	0.87	0.87	0.85
	1600	-0.70	0.34	-0.75	-0.42	-0.15	-0.66	-0.62	-0.66	-0.52	-0.20	0.47	0.89	–	0.95	0.97	0.97	0.96	0.96
	2000	-0.60	0.22	-0.67	-0.42	0.08	-0.53	-0.48	-0.51	-0.32	0.02	0.45	0.76	0.95	–	0.98	0.95	0.94	0.94
	2500	-0.71	0.18	-0.74	-0.47	0.03	-0.61	-0.56	-0.58	-0.38	-0.02	0.55	0.83	0.97	0.98	–	0.98	0.96	0.97
	3150	-0.81	0.23	-0.82	-0.55	-0.10	-0.65	-0.61	-0.65	-0.49	-0.16	0.52	0.87	0.97	0.95	0.98	–	0.99	0.99
	4000	-0.82	0.27	-0.84	-0.55	-0.14	-0.65	-0.62	-0.66	-0.52	-0.22	0.46	0.87	0.96	0.94	0.96	0.99	–	1.00
	5000	-0.80	0.28	-0.83	-0.57	-0.12	-0.62	-0.59	-0.64	-0.50	-0.20	0.45	0.85	0.96	0.94	0.97	0.99	1.00	–

that shows a good negative correlation with broadband CPX levels, as shown in Figure 2. On the other hand, other parameters show poor correlation with broadband levels.

However, drainability and MPD show high correlation coefficients with noise emission in one-third octave bands, with a sign change in correlation at 800 Hz. The highest positive correlation values are found at 500 Hz (drain.: 0.76, MPD: 0.87), while the best negative correlation was obtained at 4000 Hz (drain.: -0.82, MPD: -0.84). Figures 3 and 4 summarise the behaviour of CPX levels VS drainability values, while Figures 5 and 6 show CPX levels VS MPD values. Since IRI values are influenced by irregularities that take into accounts long wavelengths, this parameter does not correlate with noise emission. Indeed, noise emission is more influenced by macrotexture and the shortest megatexture wavelengths ($\leq$ 100 mm).

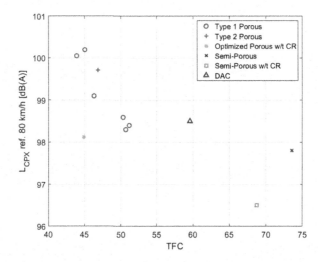

Figure 2. CPX levels in dB(A) at the reference speed of 80 km/h vs Transverse Friction Coefficient (TFC).

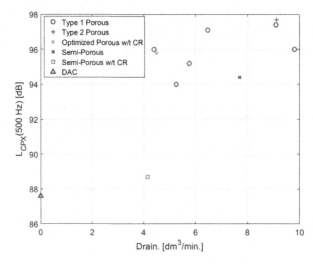

Figure 3. CPX levels at 500 Hz in dB at the reference speed of 80 km/h and drainability in dm3/min.

710

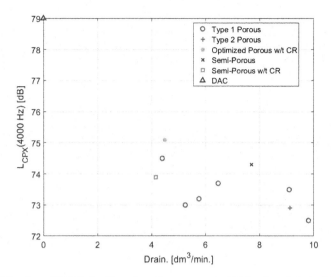

Figure 4.   CPX levels at 4000 Hz in dB at the reference speed of 80 km/h and drainability in dm³/min.

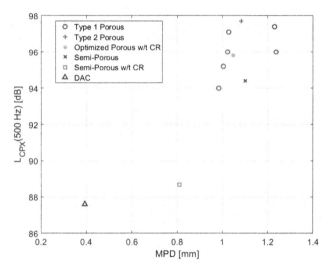

Figure 5.   CPX levels at 500 Hz in dB at the reference speed of 80 km/h and MPD in mm.

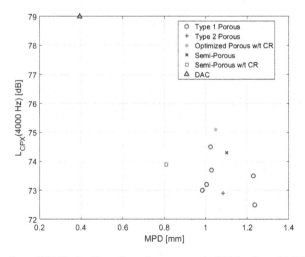

Figure 6   CPX levels at 4000 Hz in dB at the reference speed of 80 km/h and MPD in mm.

4 CONCLUSIONS

In this work, the acoustic performance of several motorway pavements was measured with the CPX method. The results were used to correlate noise emission with road surface parameters measured on-site, such as drainability, road texture in terms of MPD and IRI, and road friction.

All the pavements show a good acoustic performance taking into account GPP reference values, as shown in Figure 1. Good correlations were found between TFC values and CPX broadband levels, while a one-third octave analysis showed that drainability and MPD are strongly linked to different noise generation mechanisms present at low and high frequencies. Therefore, noise emission monitoring could aid the study of road surface condition and could be added to road asset monitoring as new performance indicators, using a non-invasive and dynamic technique such as the CPX method.

ACKNOWLEDGEMENTS

The activities presented in this paper were sponsored by Autostrade per l'Italia S.p.A. (Italy), which gave both financial and technical support within the framework of the Highway Pavement Evolutive Research (HiPER) project. The results and opinions presented are those of the authors.

REFERENCES

Babisch W., Swart W., Houthuijs D., Selander J., Bluhm G., Pershagen G. and Sourtzi P., 2012. Exposure Modifiers of the Relationships of Transportation Noise With High Blood Pressure and Noise Annoyance. *The Journal of the Acoustical Society of America*, 132(6), 3788–3808.

Del Pizzo L.G., Teti L., Moro A., Bianco F., Fredianelli L. and Licitra G., 2020. Influence of Texture on Tyre Road Noise Spectra in Rubberized Pavements. *Applied Acoustics*, 159, 107080

Garbarino E., Quintero R.R., Donatello S. and Wolf O. (2016). *Revision of Green Public Procurement Criteria for Road Design, Construction And Maintenance*. Procurement Practice Guidance Document.

Guski R., Schreckenberg D. and Schuemer R., 2017. WHO Environmental Noise Guidelines for the European Region: A Systematic Review on Environmental Noise and Annoyance. *International Journal of Environmental Research and Public Health*, 14(12), 1539.

Hygge S., Evans G. W. and Bullinger M., 2002. A Prospective Study of Some Effects of Aircraft Noise on Cognitive Performance in Schoolchildren. *Psychological Science*, 13(5), 469–474.

Lercher P., Evans G.W. and Meis M., 2003. Ambient Noise and Cognitive Processes Among Primary Schoolchildren. *Environment and Behavior*, 35(6), 725–735.

Li T., 2018. Literature Review of Tire-pavement Interaction Noise and Reduction Approaches. *Journal of Vibroengineering*, 20(6), 2424–2452.

Licitra G., Moro A., Teti L., Del Pizzo A. and Bianco F. (2019). Modelling of Acoustic Ageing of Rubberized Pavements. *Applied Acoustics*, *146*, 237–245.

Muzet A., 2007. Environmental Noise, Sleep and Health. *Sleep Medicine Reviews*, 11(2), 135–142.

Sandberg U. and Ejsmont J., 2002. *Tyre/road Noise Reference Book*. INFORMEX.

Skrzypek M., Kowalska M., Czech E.M., Niewiadomska E. and Zejda J.E., 2017. Impact of Road Traffic Noise on Sleep Disturbances and Attention Disorders Amongst School Children Living in Upper Silesian Industrial zone, Poland. *International Journal of Occupational Medicine and Environmental Health*, 30(3), 511.

Thermal and microclimatic behavior of innovative pavements for the life Cool & Low Noise Asphalt project

Maïlys Chanial*
Université Paris Cité, LIED, Paris, France
Paris City Hall, Water and Sanitation & Road and Transportation Division, Paris, France

Sophie Parison & Martin Hendel
Université Paris Cité, LIED, Paris, France
Université Gustave-Eiffel, ESIEE Paris, département SEN, Noisy-le-Grand, France

Laurent Royon
Université Paris Cité, LIED, Paris, France

ABSTRACT: In the wake of the heatwave of 2003 along with the context of climate change, strong awareness has risen within the City of Paris regarding the need for urban cooling. Several experiments have been carried out including pavement-watering. In this regard, the LIFE "Cool & Low Noise Asphalt" project aims to study the performance of pavement-watering on innovative pavements to reduce pedestrians heat stress on three test sites in the summer since 2019. Statistically significant air temperature reductions of up to -0.4°C were observed in addition to reductions of up to -0.6°C of UTCI-equivalent temperature. The impact of resurfacing the pavement and the combination of resurfacing and watering are also studied. In addition to the microclimatic analysis, a thermal analysis is also performed. A thermo-fluxmeter placed at 5 cm deep in the pavement allows the study of the watering impact on the temperature and heat flux.

Keywords: Urban Heat Island, Evaporative Cooling, Pavement Thermal Behavior.

1 INTRODUCTION

The climate change and the progressive intensification of the heat waves by 2050 are forcing the cities to adapt (Lemonsu *et al.* 2013). In the last few years, interest in heat waves mitigations tools has grown. Many ways exist to mitigate heat in the cities, such as vegetation, urban watering or the use of cool materials (Santamouris 2013).

The City of Paris is highly interested in these issues, and has been experimenting several public spaces cooling methods (reflective and evaporative materials, etc.). Among them, the urban watering has been studied, since 2012, using non-potable water in public spaces in order to limit pedestrian heat stress. The results of these study has shown reductions up to 3°C in pedestrian stress as measured by UTCI (Hendel *et al.* 2016; Parison, *et al.* 2020a).

More recently, the Cool & Low Noise Asphalt project, launched in 2017 and co-funded by the European Union's Life programme, aims to test the effectiveness of three innovative pavements compared to traditional asphalt concrete. These pavements have been developed

---

*Corresponding Author: mailys.chanial@gmail.com

to improve the effects of urban watering thanks to a surface texture that allows greater water retention.

In the present study, microclimatic results obtained from 2019 to 2021 from rue de Courcelles site will be presented. The microclimatic impact of watering, resurfacing the pavement and their combined effects will be presented as well as the thermal impact of watering on the temperature and heat flux deep in the pavement during summer 2021.

2 METHODOLOGY

Instrumentation and watering protocols

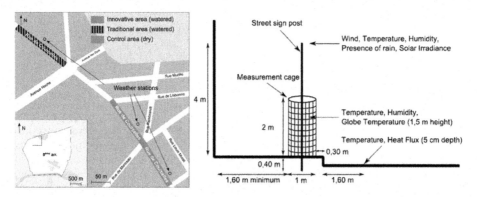

Figure 1. (left) experimental site on rue de Courcelles, (right) weather station diagram.

The study takes place in three streets of Paris (France): rue Frémicourt, Lecourbe and Courcelles. Here, we will focus on a single street, rue de Courcelles site, located in the 8th arrondissement. The street is oriented in a N NW-S SE direction and is divided into three 200-m long portions referred to as innovative, traditional and control (see Figure 1(left)). The portion referred as "Innovative" was repaved with the alternative pavement, between the summers 2018-2019. As for the innovative portion, the "Traditional" portion was also repaved but with a standard asphalt concrete. Both of those portions are watered during the watering campaigns. The "control" portion is the point of comparison; therefore it has not been repaved and remains dry during the watering campaign.

To evaluate the microclimatic and thermal impact of the experiment, each portion is equipped with a weather station on the North sidewalk. Figure 1(right) and Table 1, describe the instrumentation of the weather station which are, among others, used to assess pedestrian heat stress using the Universal Thermal Climate Index (UTCI) (Błazejczyk *et al.* 2013). Its calculation results from a heat balance of the human body and physical assumptions

Table 1. Weather station instruments and measured parameters.

Parameter	Instrument	Height	Uncertainty
Air Temperature, T_a	Sheltered Pt 100	1.5/4 m	0.1°C
Relative humidity, RH	Sheltered capacity hygrometer	1.5/4 m	1.5% RH
Black globe temperature, T_g	Black globe Pt 100 – ISO 7726	1.5 m	0.15°C
Wind speed, v	2D ultrasonic anemometer	4 m	2%
Net radiation, R_n	Net radiometer with thermopile	4 m	5% daily
Temperature/Heat flux	Thermofluxmeter (Type T)	−5 cm	0.1 °C / 1%

related to the environment (air temperature, wind speed, relative humidity and mean radiant temperature calculated following the ASHRAE method) (Ashrae Standard 2001).

The watering campaigns are performed during summer, between early June and mid-September, and are triggered only under certain meteorological conditions corresponding to relaxed criteria compared to heatwave alert in Paris. Those criteria are described in Table 2. When triggered, watering is carried out using cleaning trucks operated by Paris' Sanitation Division from 7 am to 11:30 am (UTC + 2) every 1.5 h and from 2 pm to 6:30 pm every 30 min (Hendel *et al.* 2014).

Table 2.   Trigger criteria for the watering campaigns.

Minimum temperature 3-day average	Maximum temperature 3-day average	Wind	Cloud cover
16 °C	25°C	<10 km/h	Sunny (<3 oktas)

2.1   *Microclimatic analysis*

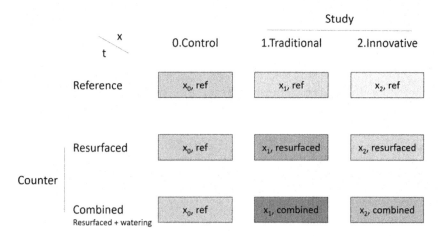

Figure 2.   Control, Traditional and Innovative portions in their three states: reference, resurfaced and watering.

The analysis of the microclimatic impacts of watering, resurfacing and their combination is carried out using the statistical method described by Parison *et al.*, 2020a based on a Before-After-Control-Impact (BACI) design (Smith 2012). This method requires to consider the microclimatic differences between the study portion (traditional x_1, or innovative x_2 pavement portions) and the control portion x_0, on radiative days. Radiative days refer to the days which meets the weather criteria described in Table 2. The three portions are considered during three states or time periods t:

- The reference state: before resurfacing or watering
- The resurfaced state: after resurfacing of x_1 or x_2
- The watering state: during pavement-watering

Portions and states are illustrated in Figure 2. Of note, the state of the control portion x_0 remains unchanged throughout the project and stays in the reference state while the other portions vary over time.

Three impacts can be studied independently and are obtained from Equation 1. Comparing the Resurfaced and Reference states gives the impact of resurfacing the traditional and innovative portions. Comparing the Watering and Resurfaced states gives the impact of watering the resurfaced portions alone. Finally, comparing the Watering and Reference states gives the impacts of the combination of watering and resurfacing.

$$\begin{cases} I_{x,resurfaced} = \Delta M_{x,resurfaced} - \Delta M_{x,ref} \\ I_{x,watering} = \Delta M_{x,watering} - \Delta M_{x,ref} \\ I_{x,combined} = \Delta M_{x,combined} - \Delta M_{x,ref} \end{cases} \quad (1)$$

With $M_{x,t}$ the considered meteorological parameter at portion x and time period t, and $I_{x,counter}$, the studied impact.

Finally, a statistical test is used to evaluate the statistical significance of the detected impact with a significance level of 0.05. The test used is a linear fixed-effects model (FEM).

2.2 Thermal analysis

A thermo-flux meter installed at a depth of 5 cm in the pavements of the innovative and traditional portions allows the measurement of temperature and heat flux. The purpose of these indicators is to characterize the thermal behavior of these pavements.

The temperature amplitude in the pavement provides information on the impact of watering. In addition, the watering can be visualized directly on the heat flux, which allows to evaluate the approximate drying time of the pavements.

3 RESULTS AND DISCUSSION

3.1 Microclimatic impact

In this part, the watering impacts only will be discussed, followed by the pavement resurfacing and the combination of resurfacing and watering. Of note, the results that will be presented in this study are obtained from the combined analysis of the three summers of experimentation: summer 2019, 2020 and 2021.

3.1.1 Watering effects

The summer 2019, 2020 and 2021 gather a total of 16 watering days and 19 reference radiative days. The reference radiative days refer to radiative days without watering campaigns.

Figure 3 illustrates the microclimatic impacts of watering on the air temperature and UTCI at 1.5 m height (pedestrian height). The watering effects are represented in blue from 6am to 6am the next day. The 95% confidence interval is represented in green. Statistically significant effects (stat. sign.) are those outside the confidence interval. The same color code will be used for the following results.

Table 3 summarizes the maximum, minimum, mean and duration of statistically significant effects on the air temperature and UTCI. We observe mean reduction of the air temperature of 0.4 °C and 0.3°C with stat. sign. effects lasting about 9 h and 10 h per days, respectively for the innovative and traditional portions. Respectively, we obtain mean reduction of the UTCI-equivalent temperature of 0.6°C and 0.7 °C during approximatively 6 h and 12 h. Also, we observe maximum reductions of 0.8°C on the air temperature and of 1.9°C on the UTCI for the innovative portion. For the traditional portion, maximum reductions of 0.6°C on the air temperature and of 2.1°C on the UTCI are observed.

The results show a positive impact of watering and are consistent with previous studies (Chanial *et al.* 2020; Hendel *et al.* 2016; Parison *et al.* 2020a, 2021).

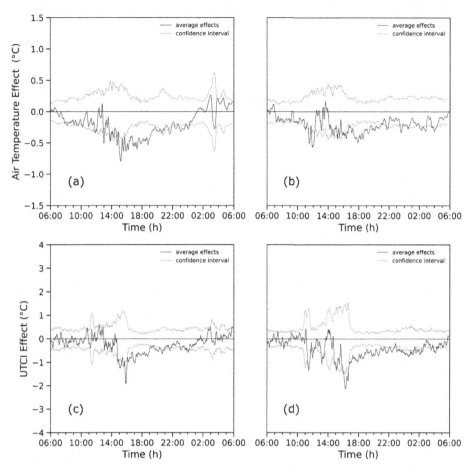

Figure 3. Average effect of watering at 1.5 m height from 2019 to 2021 for rue de Courcelles: air temperature (top) and UTCI (bottom), for innovative (left) and traditional (right) areas.

Table 3. Minimum, maximum and mean values and duration of stat. sign. effects of watering for air temperature and UTCI ($n_{watering} = 16$, $n_{ref} = 19$).

	Innovative portion		Traditional portion	
	T_{air}	UTCI	T_{air}	UTCI
Minimum effect	−0.8 °C	−1.9 °C	−0.6 °C	−2.1 °C
Maximum effect	−0.2 °C	+0.5 °C	−0.1 °C	+0.3 °C
Mean effect	−0.4 °C	−0.6 °C	−0.3 °C	−0.7 °C
Total duration	8 h 48 min	6h36min	9 h 42 min	12 h 25 min

3.1.2 *Pavement resurfacing impact and combined effects*

Figure 4, show the microclimatic impacts of resurfacing the pavement and the combined (resurfacing + watering) effects for the innovative portion on the air temperature and UTCI at 1.5 m height. On the left, the resurfacing impact and on the right, the combined impact.

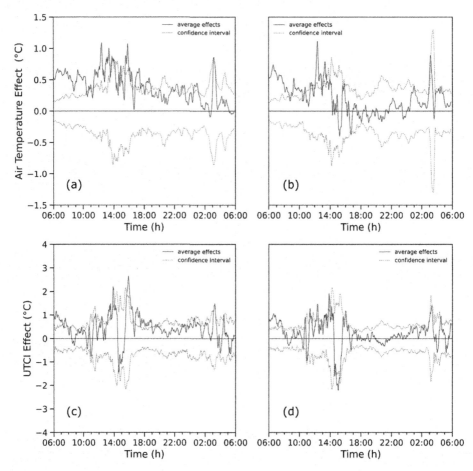

Figure 4. Average effects of resurfacing (left) and the combination (right) on air temperature (top) and UTCI (bottom).

The analyse of the pavement resurfacing impact show average effect of +0.6°C on the air temperature and of +0.9°C on the UTCI with maximum effects up to +1.1°C on the air temperature and up to +2.7°C on the UTCI. Regarding the combined effects (watering and resurfacing), we obtained average effects on air temperature of +0.5°C with maximum effect up to +1.1°C. We observe for the UTCI, average effect of +0.6°C and maximum effect of +1.9°C. As can be seen by comparing Figures 4c and 4d, a similar – though dampened – signal is observed in both analyses.

These results indicate a degradation of microclimatic conditions. Two hypotheses could explain these result.

The first hypothesis would be that the degradation may be caused by a lower pavement albedo following resurfacing. While, in situ albedo measurements are not available before and after resurfacing, a laboratory albedo measurement has been carried out with a UV-Vis-NIR Cary 5000 spectrophotometer. The results indicate an albedo of 0.04 for the innovative pavement before sandblasting. After, this value increases to 0.06. Given the low albedo obtained, it seems plausible in our opinion that the pre-existing pavement, which was several years old, had a higher albedo value, e.g. in the order of 0.15. Following resurfacing, pavement albedo has likely dropped, resulting in the observed degradation of pedestrian heat stress. Further analyses, e.g. comparing results year by year to test for a possible ageing effect, and in situ measurements of albedo will help determine how likely this interpretation is.

The second hypothesis is related to the BACI experiment conducted. The average profiles are built using the BACI design (Smith 2012), which supposes that the control site is unchanged throughout the experiment. This may be a stronger-than-anticipated limitation of the model as this condition may not be verified in an urban setting such as Paris over the course of several years (2018 to 2021)(Hurlbert 1984).

3.2 *Thermal impact*

At the end of summer 2021, three watering campaigns were performed. Only the second campaign is represented here in the interests of clarity. It should be noted that, Table 4 et Table 5 gathers the results of the entire watering campaigns of summer 2021.

Table 4. Mean Amplitude of temperature at -5cm in the pavement, rue de Courcelles

	Innovative portion		Traditional portion	
	Mean amplitude	**Standard deviation**	**Mean amplitude**	**Standard deviation**
Watered day	13.3°C	0.9°C	11.8°C	0.6°C
Reference day	19.3°C	0.0°C	15.5°C	0.5°C

Table 5. Drying time between 1:30 pm to 5:00 pm, rue de Courcelles.

	Innovative portion	**Traditional portion**
Drying time	31 min	32 min

Figures 5a and 5b show the evolution of the pavement temperature at 5 cm depth for different radiative days, watered (in shade of blue) or not (in red). The temperatures are normalized to 0°C in order to highlight the differences of the amplitude for each day represented. Amplitudes are defined as the differences between minimum and maximum temperatures for a given day.

Below, Table 4 gathers the mean amplitude of temperature and the associated standard deviation for watered and reference days. The results show that watering allows a temperature reduction of 3.7°C to 6 °C.

Heat fluxes are illustrated in Figures 5c and 5d. Spikes can be observed during watered days. Those spikes coincide with the watering circle. Indeed, when the pavement is watered we observe a high reduction of the flux. Once the pavement is dry, the heat flux increases again. Thus, it is possible to estimate the drying time of the pavement.

Table 5 show the average drying time of the innovative and traditional pavement calculated from 1:30 pm to 5:00 pm. The traditional pavement seems to have a slightly longer drying time compared to the innovative one. Nevertheless, these two drying time being very close, it is necessary to evaluate the statistical significance of those results.

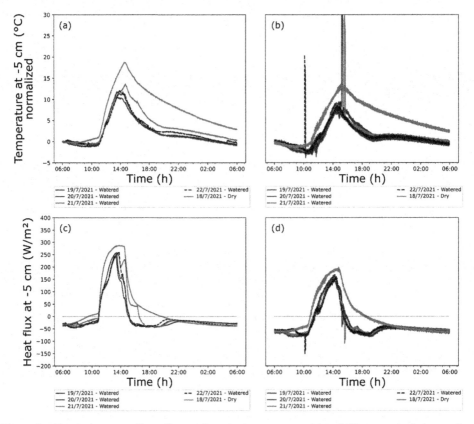

Figure 5. Temperatures and heat flux at -5 cm in the pavement. (a) and (b), respectively innovative and reference pavement, illustrate the evolution of the temperature in the pavement over a day from 6:00 to 6:00 for the second campaign 2021 (18/07/2021 to 22/07/2021). (c) and (d) illustrate the evolution of the heat flux over the same period and for the same pavements.

4   CONCLUSION

The microclimatic and thermal impacts of watering an innovative and a traditional pavement were studied for the rue de Courcelles test site in Paris, France within the framework of the LIFE Cool and Low Noise Asphalt project. The microclimatic effects of pavement resurfacing and the combination of watering and resurfacing, were also studied.

The results of the microclimatic impact of watering over the three summers of experimentations (2019, 2020 and 2021) shows statistically significant microclimatic effects. Indeed, the watering campaigns leads to a reduction of air temperature (max. effects: -0.6°C to -0.8°C) and heat stress (max. effects: -1.9°C to -2.1°C) at 1.5 m height.

The first results of the microclimatic impact of resurfacing and of the combination of resurfacing and watering are at this stage not fully understood. It seems to indicate a degradation of the microclimatic conditions due to lower albedo following pavement resurfacing. Complementary analyses of the ageing effects and in situ albedo measurements of the pavement should help in the interpretation of these results.

Regarding the heat flux at 5 cm depth, the watering impact is clearly observed by an abrupt reduction of the signal. This observation allows us to estimate the approximate drying time of the pavements. The results seem to indicate that the traditional pavement dries slightly slower than the innovative one (32 min against 31 min). These two values being very close, it is necessary to evaluate the statistical significance before concluding. An estimation of the water evaporation rate of both of the innovative and traditional portion based on a heat balance would also allow a more rigorous comparison of these values. Finally, the comparison of the temperature amplitudes at -5cm in the pavement shows that watering tends to reduce the amplitudes.

To go further in this study, the innovative pavements are studied in laboratory in order to observe their thermal behavior (Parison *et al.* 2020).

REFERENCES

Ashrae Standard. (2001). ASHRAE Handbook 2001 Fundamentals. In *Ashrae Standard.* https://doi.org/10.1017/CBO9781107415324.004

Błazejczyk K., Jendritzky G., Bröde P., Fiala D., Havenith G., Epstein Y., Psikuta A., & Kampmann B. (2013). An introduction to the Universal thermal climate index (UTCI). *Geographia Polonica, 86*(1), 5–10. https://doi.org/10.7163/GPol.2013.1

Chanial M., Parison S., Hendel M., & Royon L. (2020). Etude du Comportement Thermique et Microclimatique d'un Revêtement Innovant. *XXXIIIème Colloque de l'Association Internationale de Climatologie,* 151–156.

Hendel M., Colombert M., Diab Y., & Royon L. (2014). Improving a Pavement-watering Method on the Basis of Pavement Surface Temperature Measurements. *Urban Climate.* https://doi.org/10.1016/j.uclim.2014.11.002

Hendel M., Gutierrez P., Colombert M., Diab Y., & Royon L. (2016). Measuring the Effects of Urban Heat Island Mitigation Techniques in the field: Application to the Case of Pavement-watering in Paris. *Urban Climate.* https://doi.org/10.1016/j.uclim.2016.02.003

Hurlbert S.H. (1984). Pseudoreplication and the Design of Ecological Field Experiments. *Ecological Monographs, 54*(2), 187–211. https://doi.org/10.2307/1942661

Lemonsu A., Kounkou-Arnaud R., Desplat J., Salagnac J.L., & Masson V. (2013). Evolution of the Parisian Urban Climate Under a Global Changing Climate. *Climatic Change.* https://doi.org/10.1007/s10584-012-0521-6

Parison S., Chanial M., Hendel M., & Royon L. (2021). Comportement Thermique et Microclimatique de l'Arrosage d'un Revêtement Innovant. *XXXIVème Colloque de l'Association Internationale de Climatologie,* 1, 1–6.

Parison S., Hendel M., Grados A., & Royon L. (2020b). Analysis of the Heat Budget of Standard, Cool and Watered Pavements Under Lab Heat-wave Conditions. *Energy and Buildings,* 228, 110455. https://doi.org/10.1016/j.enbuild.2020.110455

Parison S., Hendel M., & Royon L. (2020a). A Statistical Method for Quantifying the Field Effects of Urban Heat Island Mitigation Techniques. *Urban Climate,* 33(April 2019). https://doi.org/10.1016/j.uclim.2020.100651

Santamouris M. (2013). Using Cool Pavements as a Mitigation Strategy to Fight Urban Heat Island - A Review of the Actual Developments. In *Renewable and Sustainable Energy Reviews.* https://doi.org/10.1016/j.rser.2013.05.047

Smith E.P. (2012). BACI Design. *Encyclopedia of Environmetrics, 1,* 141–148. https://doi.org/10.1002/9780470057339.vab001.pub2

Sustainability and performances issues
about materials and design

Roads and Airports Pavement Surface Characteristics – Crispino & Toraldo (Eds)
© 2023 The Author(s), ISBN 978-1-032-55149-4

Performance review of surfacing materials incorporating enhanced levels of reclaimed asphalt on the UK strategic road network

Michael Wright*
Practice Manager, Atkins, Cambridge, UK

Matthew Wayman
Senior Technical Advisor, National Highways, Guildford, UK

Wenxin Zuo
Senior Pavement Engineer, Atkins, Birmingham, UK

ABSTRACT: Asphalt can be reclaimed from expired bituminous-bound pavement layers. The use of reclaimed asphalt (RA) in surface course provides significant economic and environmental benefits by effectively using recycled materials in high value applications. In the UK, specifications previously permitted up to 10% of RA to be incorporated into the surface courses.

A range of trials incorporating RA content (>10%) into thin surface course systems have been installed on National Highways Strategic Road Network (SRN) since 2004 as documented by Carswell *et al.* (2010). These trials demonstrated the feasibility of incorporating higher RA content and the required quality controls. A number of trial sites have been in-service for a significant period of time, which now provide an opportunity to understand the performance of surface course materials containing increased levels of RA.

This paper summarises current practice in the UK and trials undertaken incorporating RA content (>10%). The paper conducts detailed analysis of the performance of the M25 J6-7 & 7-8 Clockwise (installed in 2007 incorporating 23% RA) and M25 J6-7 Anti-clockwise (installed in 2009 incorporating 40% RA). This paper presents analysis of network condition surveys in the form of Visual Condition Surveys, TRAffic-speed Condition Surveys (TRACS) and skid resistance measurements undertaken by Sideway-Force Coefficient Routine Investigation Machine (SCRIM®). In addition, the paper introduces recent trials undertaken in the UK containing 50% RA. The paper present quantifiable evidence regarding the performance of surface course materials containing RA > 10%, which has informed updates to national specifications.

Keywords: Surface Course Design, Recycling & Reuse of Materials, Sustainability

1 INTRODUCTION

Asphalt can be reclaimed from expired bituminous-bound pavements and is widely considered to be 100% recyclable. The use of RA in surface courses provides significant economic and environmental benefits by effectively using recycled materials in the highest-value application. Recycling into surface courses makes best use of high polished stone value (PSV) aggregate, reduces resource demand and minimises transportation.

---

*Corresponding Author: Michael.Wright@atkinsglobal.com

DOI: 10.1201/9781003429258-71

In the UK, the Specification for Highway Works (HE 2018) and PD 6691:2015 (BSI 2015) allow up to 10% of reclaimed asphalt (RA) to be incorporated into surface courses without a departure from standard. The introduction of a 10% RA limit was understandably cautious based on knowledge at the time of key factors such as skid resistance and durability in order to ensure a safe and serviceable network. The incorporation of 10% RA has subsequently become standard practice in the UK.

Trials of various surface courses incorporating RA contents >10% have also been installed on the Strategic Road Network (SRN) (Carswell et al. 2005, 2012; Nicholls et al. 2007; Schiavi et al. 2007; Wayman & Carswell 2010). These trials have demonstrated the feasibility of incorporating RA into surface courses and informed the publication of best practice guidelines (Carswell et al. 2010). These trial sites have been in service for a significant period of time and now provide an opportunity to understand the performance of surface course materials containing increased levels of RA.

1.1 *Objectives*

The objectives of this paper:

- Summarises surface course trials undertaken on the SRN incorporating RA > 10%.
- Presents detailed analysis of the performance of junctions 6–7 and 7–8 of the M25 clockwise (installed in 2007, incorporating 23% RA) and junctions 7–6 of the M25 anti-clockwise (installed in 2009, incorporating 40% RA).
- Introduces recent surface course trials in the UK containing 50% RA and presents early life results obtained M25 J26-25.

The paper summarises experience and provide quantifiable evidence regarding the performance of surface course materials containing increased levels of RA in the UK. This research has informed and will continue to inform future revisions to national specifications including PD 6691 (BSI 2022).

2 ANALYSIS OF EXISTING SITE TRIALS IN THE UK

2.1 *History of sites with RA ≥ 10%*

Trials of various surface courses incorporating RA ≥ 10%, have been installed on the SRN in the UK. In 2004, a trial was undertaken on the A405 at Bricket Wood, as documented by Carswell et al. (2005). The trial contained both 10% and 30% RA. In 2006, a further scheme on the M4 near Cardiff utilised approximately 25% RA (Nicholls et al. 2007). Trials on the M25 around Reigate Hill were undertaken in 2007 (Schiavi et al. 2007) and 2009 (Wayman & Carswell 2010). The 2007 trial was between J6–J7 and J7–J8 (clockwise carriageway) and incorporated 23% RA. This was followed by a trial in 2009 between J7–J6 (anti-clockwise carriageway) using 40% RA.

In recent years, a number of trial 50% RA content have been adopted on the A1 at Mill Hill (Flint & Bailey 2018) and on the A40 by FM Conway in partnership with Transport for London (TfL), in 2016 and 2017, respectively. This led to trials of 50% RA on SRN on the M25 J26-25 in 2019 and M3 J6 in 2021 A summary of sites installed containing levels of RA > 10% is provided in Table 1. This table summarises the key properties or the RA utilised and key performance results of the mix design incorporating the RA.

2.2 *Site analysis & basis for selection*

To understand the performance of surface course materials containing RA, detailed analysis of J6–J7 and J7–J8 M25 clockwise (23% RA) and J7–J6 M25 anti-clockwise (40% RA) was

Table 1. Summary of surface courses trials containing RA (> 10%).

Site	Installed Date	Commercial Traffic per lane per day (cv/l/day)	Supplier	Client	RA Content	Surface Course	RA Source	RA Properties			Mix Design			Reference
								RA Pen (dmm)	RA Softening Point (oC)	RA PSV	ITSM (MPa)	Water Sensitivity (%)	Wheel-tracking @60oC	
A405 Bricket Wood[1]	2004	1423(Lane 1)	Lafarge	Highways Agency	10 & 30%	Axoflex & Stone Mastic Asphalt (SMA)	Porous Asphalt#	Not Declared	Not Declared	Not Declared	Not Declared	Not Declared	Not Declared	Carswell et al. (2005), Carswell et al. (2010), Ojum (2017)
M4 Cardiff[2]	2006	3123(Lane 1)	Cemex	National Assembly for Wales	25%	Thin Surface Course	Porous Asphalt#	16	69.6	Not Declared	Not Declared	Not Declared	Not Declared	Cemex (2006), Carswell et al. (2010).
M25 J6-7 & 7-8 Clockwise	2007	5658(Lane 1)	Tarmac	Highways Agency	23%	14 mm Masterpave (Ultipave)	SMA	19	67.8	59	4561	99%	0.9 mm/hr BS 598-10	Schiavi et al. (2007)
M25 J6-7 Anti-clockwise	2009	1701(Lane 2)	Tarmac	Highways Agency	40%	20 mm Masterpave (Ultipave)	Porous Asphalt	Not Declared	Not Declared	Not Declared	5699	96%	PRDAir 4.7%	Wayman and Carswell (2010)
M20 J2-3 Southbound	2016	306(Lane 2)	Tarmac	Highways England	30%	14 mm Ultipave	Existing TSCS	18	67.4	61	5699	72%	PRDAir 4.7%	Markham (2017)
A1 (Mill Hill)	2016	2599(Lane 1)	FM Conway	Transport for London	50%	14 mm SurePave	SMA	18	67.0	63	4519	99%	PRDAir 4.7%	Flint and Bailey (2018)
A40	2017	2706(Lane 1)	FM Conway	Transport for London	50%	14 mm AC Surface Course	NotDefined	24	65.0	63	2528 - 3812	100%	PRDAir 11%	-
M3	2019	3005(Lane 2)	Tarmac	Highways England	20%	14 mm Ultipave	Existing TSCS	16	68.8	62	5754	82%	PRDAIR 4.5%	
M25 J26-25	2019	7639(Lane 1)	FM Conway	Highways England	50%	14 mm Surephalt	Existing TSCS	34	63.0	65	5699	92%	PRDAIR 4.7%	
M3 NB J6	2021	4471(Lane 1)	FM Conway	National Highways	50%	14 mm Surephalt	10/14 SMA	11/23	68.2/71.2	65	2075	111%	PRDAIR 4.0%	

1) Sections of the A405 trial replaced in 2011, 2012 and 2016.
2) M4 Cardiff site was replaced in 2016 following 10 years in service.
3) . # denotes Polymer Modified Binder in RA.

727

undertaken as presented in Section 3.0 and 4.0 of this paper, respectively. These sites were selected for detailed analysis due to the high RA content and extensive data available in terms of the RA properties, mix design and performance. In particular, M25 clockwise J6–J7 and J7–J8 (23% RA) experienced the highest traffic levels of all sites, at 5658 commercial vehicles per lane per day (cv/lane per day), as shown in Table 1. This site also incorporated a control section in order to compare performance with and without RA directly. The J7–J6 M25 anti-clockwise (40% RA) site was also selected as it represented the highest level of RA incorporated into a surface course on the SRN at the time of construction. The early life performance monitoring of 50% RA on M25 J26-25 is presented in Section 5.0. This site is reported as it represents the first surface course on the SRN to incorporate 50% RA. Detailed analysis of all the sites reported in Table 1 is presented elsewhere (Wright et al. 2020).

2.3   *Pavement monitoring & data analysis*

This paper presents a review of monitoring surveys using traffic-speed condition survey (TRACS), Sideway-force Coefficient Routine Investigation Machine (SCRIM®) and Visual Condition Surveys (VCS). The survey methods and parameters assessed are summarised in Table 2. The results of the laboratory testing of extracted cores including recovered binder properties are presented by Wright et al. (2020).

Table 2.   Monitoring undertaken on surface course incorporating RA.

Survey	Description
Traffic-speed condition survey (TRACS) 	On the SRN in England, pavement surface condition is assessed using TRACS in accordance with CS 229 (HE 2020a). TRACS uses survey vehicles equipped with lasers, video image collection and inertia measurement apparatus to enable surveys of the road surface condition to be carried out at variable speeds of up to 100 km/h. TRACS measure a number of parameters including texture, transverse profile, longitudinal profile and cracking. As part of this paper the surface texture and rutting have been considered in detail. The TRACS data can be assessed by means of four condition categories, as defined in section E/1 of CS 230 (HE 2020b).
Skid Resistance (SCRIM®) 	SCRIM® measures surface friction properties under wet- road skidding conditions. Measurements are made at a standard test speed of 50 km/h in the nearside wheel path. Processed SCRIM® measurements are used to derive a characteristic skid coefficient (CSC), which is an estimate of the pavement surface's underlying skid resistance and is adjusted for seasonal variations. The data are compared against defined investigatory levels (ILs).
Visual Condition Survey 	VCS were undertaken to assess surface conditions. A combination of video drive-over and walked VCSs were utilised depending on the site characteristics. The VCS was analysed in accordance with the seven-point scale inspection panel method defined by Nicholls et al. (2010).

3 M25 CLOCKWISE AROUND REIGATE HILL (23% RA)

3.1 *Introduction & mix design*

In 2007, a trial of a surface course containing 23% RA was undertaken on J6–J7 and J7–J8 of the M25 around Reigate Hill in the clockwise direction and is documented by Schiavi et al. (2007). The trial comprised Tarmac's 14 mm MasterPave (HE 2018: clause 942) TSCS material with the inclusion of up to 23% RA generated from the existing porous asphalt surface. A control section of 14 mm MasterPave (without RA) was also installed. Carswell et al. (2012) reported the 'scheme was undertaken by adding part of the RA through the hot elevator and part cold with the virgin aggregate superheated to compensate'. The RA was a blend of three aggregate types with a PSV of 59. Test results of the 14 mm screened RA reported a penetration of 19 dmm at 25°C and a softening point of 67.8°C. Appropriate binder penetration calculations were undertaken as part of the mix design. The key material properties are presented in Table 3.

Table 3.   Mix design properties for 14 mm MasterPave control and containing 23% RA.

Mix Properties at Target Binder content	Control	23% RA
Added Binder Content (%)	6.0	5.8
Found Binder Content (%)	6.0	6.1
Binder Drainage (%)	0.0	0.0
Air Voids (%)	3.1	3.0
Stiffness – ITSM (BSI 2003) MPa	3,348	4,561
Water Sensitivity – ITSM (BSI, 2004) (Ratio)	0.90	0.99
Wheel Tracking Rate @60oC (BSI, 1990)	2.6	0.9

3.2 *Analysis of TRACS rut depth*

Analysis of the mean rut depths is presented Figure 1. The results indicate that the level of rutting for the 14 mm MasterPave with 23% RA and the control was consistent, with no significant difference between the two sections. The rutting depths measured throughout were generally less than 6 mm, indicating a 'lower level of concern' in accordance with CS 230 (HE 2020b). Some variability is present in the data, with the reported values in 2019-2021 marginally lower than the values reported in 2016–2018. This is associated with the nature of traffic speed surveys and within the survey tolerances. No data are available for

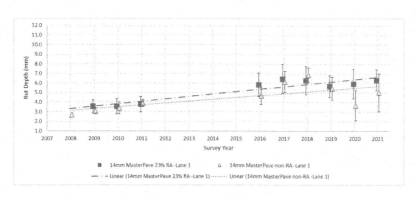

Figure 1.   Rut depth in lane 1.

2012–2015. The data show a 0.22 mm increase in rut depth per year for the 14 mm MasterPave control, as compared with a 0.27 mm increase in rut depth per year reported for the 14 mm MasterPave containing 23% RA. These values are low considering that both sections have been subject to traffic loading in excess of 24 million commercial vehicles.

3.3 Analysis of TRACS Sensor-Measured Texture Depth (SMTD)

Analysis of the SMTD is presented in Figure 2. Texture depth records are available from 2008 to 2011 and 2016 to 2021. The figure shows that the SMTD values were generally consistent for both sections (without and with RA) throughout the study period. The average SMTD values were generally greater than 1.1 mm prior to 2016, categorised as 'sound' in terms of texture as per Table 2. The SMTDs in both the control and trial sections showed a steady increase over the first few years and then reduced to a relatively consistent level of 1.0–1.2 mm after approximately 8 years of service. This is broadly consistent with the SMTD values reported by Nicholls et al. (2010) based on 9 years of monitoring for a 14 mm stone mastic asphalt (SMA). The mean SMTD was greater than 1.0 mm after 12 years of service, indicating a lower level of concern. The values have since remained above 0.8 mm, the level at which an increase in collision risk is noted (Roe et al., 1991). A slight increase in SMTD has been observed in the trial section in recent years. Nicholls et al. (2010: p. 5) concluded 'The general trend is for an increase in texture depth with age, particularly towards the end of the life in service'.

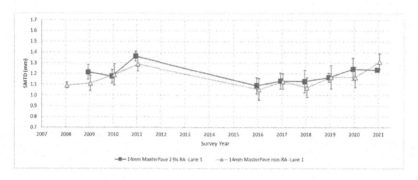

Figure 2. Texture depth in lane 1.

3.4 Analysis of CSC

Analysis of the CSC by surface type from 2009–2021 (per 100 m section) is presented in Figure 3. The results indicate that all values have remained above the 0.35 IL throughout the 14 years in service. The trial section (14 mm MasterPave containing 23% RA) showed results

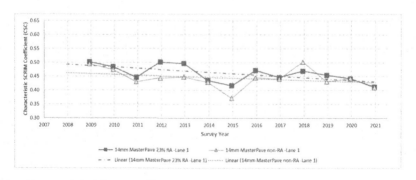

Figure 3. Characteristic SCRIM coefficient (CSC) in lane 1.

consistent with the control. The results are in line with the mean summer SCRIM coefficients reported by Nicholls et al. (2010) for SMA, where values were found to decrease with age.

3.5 VCS and video analysis

The visual condition of the site was historically assessed by Carswell et al. (2012). Video surveys were conducted in 2014, 2016 and 2021. The results presented in Table 4. No significant difference was noted between the two surfaces over time. The VCS reported both sections were in moderate/acceptable condition after 14 years in service, with chipping loss and fretting at the joint present as would be expected for a surface course of this age.

Table 4. Summary of visual condition survey.

| Time (Months) | VCS Condition Category (Nicholls et al. 2010) | | | | |
	38	58	84	108	165
14 mm MasterPave Control	Good	Moderate	Moderate	Moderate/ Acceptable	Moderate/ Acceptable
14 mm MasterPave 23% RA	Good	Moderate	Moderate	Moderate/ Acceptable	Moderate/ Acceptable

4 M25 ANTI-CLOCKWISE AROUND REIGATE HILL (40% RA)

4.1 Introduction & mix design

In 2009, a surface course trial containing 40% RA was undertaken on J7–J6 of the M25 anti-clockwise (Wayman & Carswell 2010). The trial comprised Tarmac's 20 mm MasterPave-R with the inclusion of 40% RA produced from the existing porous asphalt surface. The key material properties for the 20 mm MasterPave-R 40% RA are presented in Table 5. The mix design demonstrated full compliance with applicable certification requirements.

Table 5. 20 mm Masterpave-R 40% RA properties.

Mix Properties		Result
Target Binder Content	(%)	5.5
Binder Drainage	(%)	0.0
Air Voids (SSD) (BSI 2003)	(%)	5.5
Stiffness (ITSM) (BSI 2003)	MPa	5699
Water Sensitivity - BBA method	(Ratio C3)	0.91
Wheel Tracking Rate @ 60°C (BSI 2007)	WTSAIR (mm/103)	0.06
BS EN 12697-22 (Procedure B)	PRDAIR (%)	4.7

4.2 Analysis of TRACS rut depth

Analysis of the rut depth measurements is presented in Figure 5. The maximum rut depth records demonstrate that the measured mean maximum rut depths were less than 6 mm, indicating 'no visible deterioration' in accordance with CS 230 (HE 2020b). Throughout this period, lane 2 was subject to a significant traffic load (greater than 6 million commercial

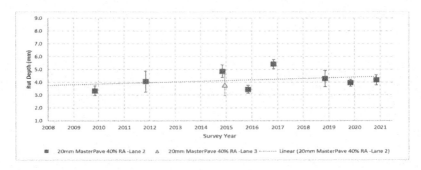

Figure 5.   Rut depth in lanes 2 and 3.

vehicles trafficking this section). A 0.07 mm increase in rut depth was reported per year between 2010 and 2021. This is significantly lower than experienced else-where on the SRN.

4.3   Analysis of TRACS SMTD

Analysis of the SMTDs is presented in Figure 6. The average SMTD was generally greater than 1.1 mm prior to 2013, categorised as sound in terms of texture (HE 2020b). The measured average SMTDs were all greater than 0.8 mm after 12 years of service, indicating a lower level of concern. Surface macro-texture measurements (BSI 2010) undertaken in March 2020 reported values of 1.3 mm and 1.6 mm, indicating a sound level of retained texture.

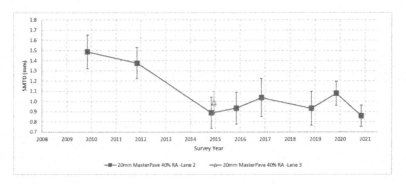

Figure 6.   Texture depth in lanes 2 and 3.

4.4   Analysis of CSC

The available CSCs records are plotted in Figure 7. The data indicates all the measured CSC values were significantly above the 0.35 IL after 12 years in service.

4.5   VCS and video analysis

The visual condition of the site was historically assessed by Carswell et al. (2012), which concluded that the trial section was scored as good after 13 months and moderate after 33 months. The assessment of video data in 2014 confirmed the moderate condition. Video survey data from 2020 and 2021 conclude the 40% RA section is in an acceptable condition

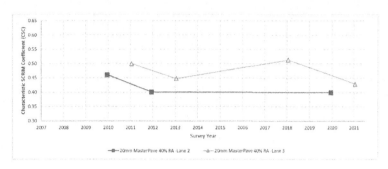

Figure 7.   Characteristic SCRIM coefficient (CSC) in lanes 2 and 3.

with some fatting up present in the wheel paths. A summary of the VCS is presented in Table 6.

Table 6.   Summary of visual condition survey.

	VCS Condition Category (Nicholls et al. 2010)				
Time (Months)	0	13	33	120	142
20 mm MasterPave-R 40% RA	Excellent	Good	Good	Acceptable	Acceptable

5   M25 J26-25 (50%) RA

5.1   *Introduction & mix design*

In 2019, a surface course trial containing 50% RA was installed on the M25 J26-25. The trial represents the first use of a surface course containing 50% RA on the SRN. The trial comprised FM Conways' SurePhalt 14 mm surf SP RC with the inclusion of 50% RA obtained from donor sites on the M25 network and comprised of existing Thin Surface Course Systems. The donor site was selected following a detailed coring survey at 20 m spacings. The trial site is subject to 7,639 commercial vehicle per lane per day. The key material properties for the SurePhalt 14 mm surf SP RC 50% RA are presented in Table 7.

Table 7.   Trial mixture approval test results 14 mm SurePhalt SP RC 50% RA.

Property Assessed	Value
Air voids content (BSI, 2003)	Vmean 3.2%
Binder Drainage BS EN 12697-18:2004 - Schellenberg Method (%)	BDMax 0.3%
Water Sensitivity BS EN 12697-12:2008 (BSI, 2008)	ITSR 106%
Stiffness Modulus - BS EN 12697-26:2004 (E) - Annex C (BSI, 2004)	3397 MPa
Fracture Toughness (N/mm3/2)	29
Resistance to Permanent Deformation (BSI, 2007)	
Small-size device, Procedure B, 60°C	
Mean Wheel-tracking slope (mm/ 103 Cycles)	WTSAIR 0.1
Mean proportional rut depth, PRDAIR, at 10 000 cycles (%)	PRDAIR 9.3

5.2 Analysis of TRACS rut depth

Analysis of the TRACs data (Table 8) demonstrate that the measured mean maximum rut depths were less than 6 mm in the early life, indicating 'no visible deterioration' in accordance with CS 230 (HE, 2020b). With the 50% RA site performing equivalent to the control surface.

5.3 Analysis of TRACS SMTD

Analysis of the SMTDs (Table 8) demonstrates consistent values for the trial and the control sections. The average SMTD values are greater than 1.1 mm since 2020, which are categorised as "Sound" in accordance with CS 230 (HE, 2020b).

5.4 Analysis of CSC

The CSCs (Table 8) are well above the 0.35 Investigatory Level throughout the first two years. The RA section has demonstrated consistent results with the control.

5.5 VCS and video analysis

The VCS of the 50% RA section was assessed to be in "Moderate" condition in 2020. The surface had visible good texture. A localised pothole was observed at chainage 2020 m with minor fretting between chainage 1980 and 2030 m. The condition of the control was deemed to be "Suspect/Poor". A number of defects identified were associated with the underlying substrate.

Table 8. TRACS, SCRIM® & VCS monitoring M25 J26-25 (50%) RA.

	SurePhalt 14 mm 50% RA		SurePhalt 14 mm Control	
	Year 1	Year 2	Year 1	Year 2
TRACS Rut Depth (mm)	4.2	3.8	4.1	3.6
TRACS SMTD (mm)	1.1	1.2	1.2	1.3
SCRIM® CSC	0.46	0.46	0.45	0.45
VCS	Acceptable	Acceptable	Suspect	Suspect/Poor

6 CONCLUSIONS

A number of surface course trials incorporating RA were installed on the UK SRN between 2004 and 2016. The RA contents ranged from 10% to 40%. These trials and further research led to the development of best practice guidance (Carswell et al. 2010). These trial sites have been in service for a significant period of time and now provide an opportunity to understand the performance of surface course materials containing greater than 10% RA.

Analysis of traffic speed condition data collected on the clockwise M25 (J6–J7, J7–J8) indicates that the installed 14 mm MasterPave with 23% RA has performed well. The site has been subject to 5658 cv/lane per day and is currently in a moderate/acceptable condition after more than 14 years in service. The performance of the 23% RA section was found to be directly comparable to the control section in terms of rut depth, texture depth and skid resistance. The CSC results measured in 2021 were in excess of 0.45, which is significantly above the IL.

Analysis of traffic speed condition data collected on the anti- clockwise M25 (J7–J6) indicated that the Tarmac 20 mm MasterPave-R with the inclusion of up to 40% RA installed in 2009 has performed well in both lanes 2 and 3. After 12 years in service, this section was found to be in an 'acceptable' condition. This performance is consistent with the typical performance reported by Nicholls et al. (2010) and exceeds the average service life of 10 years on this section of the SRN. In this period, the site was subject to 1701 cv/lane per day. Despite this trafficking, only a 0.07 mm increase in rut depth per year between 2010 and 2021 has been reported. The CSC results measured in 2021 were in excess of 0.45, which is significantly above the IL of 0.35.

The trials indicate that the surface courses containing higher levels of RA have performed equivalently to control sections containing no RA. The trials indicate that the inclusion of RA when subject to the appropriate characterisation, mix design and quality control processes has had no detrimental impact on skid resistance or durability. This is consistent with laboratory findings presented by Dunford et al. (2013) regarding known RA sources.

The results of this study support an incremental increase in the maximum permitted RA content in the UK from 10% to 20% for the majority of surface courses. This is reflected in the updated revision of PD 6691:2021 (BSI, 2022). This revision follows a significant period where additions of 10% RA have been successfully incorporated. In order to ensure durability and long-term skid resistance performance with an incremental increase in RA, it is important that the RA feedstock is produced from existing surface courses. In particular, the following properties are specified in PD6691 (BSI, 2022) for incorporations of more than 10% RA:

- A homogenous feedstock with consistent properties should be ensured.
- Minimum penetration (P15) should be achieved.
- PSV & AAV shall be tested on a representative sample from each feedstock unless the properties of the original aggregate can be satisfactorily determined from records
- Combined binder penetration should be calculated in accordance with BS EN 13108-1:2016 (BSI, 2016).

The recent trials of surface courses containing 50% RA present an opportunity to assess the performance of high RA surface courses. The trial installed on the M25 J26-25 indicates compliance with the required skid resistance and SMTD levels in the early life. The sites containing high levels of RA will continue to be monitored to inform future specifications.

ACKNOWLEDGEMENTS

The authors would like to acknowledge National Highways for funding this research project. The authors are also grateful to Connect Plus, Connect Plus Services, Tarmac and FM Conway.

REFERENCES

BSI (2003) BS EN 12697- 8:2003: *Bituminous Mixtures*. Test Methods for Hot Mix Asphalt - Determination of Void Characteristics of Bituminous Specimens. BSI, London, UK.

BSI (2004) BS EN 12697- 26:2004: *Bituminous Mixtures*. Test Methods – Stiffness. BSI, London, UK.

BSI (2007) BS EN 12697-22:2003+A1: 2007: *Bituminous Mixtures*. Test Methods for Hot Mix Asphalt – Wheel Tracking. BSI, London, UK.

BSI (2008) BS EN 12697-12:2008: *Bituminous Mixtures*. Test Methods for Hot Mix Asphalt – Determination of the Water Sensitivity of Bituminous Specimens. BSI, London, UK.

BSI (2010) BS EN 13036-1:2010: *Road and Airfield Surface Characteristics*. Test Methods. Measurement of Pavement Surface Macrotexture Depth using a Volumetric Patch Technique. BSI, London, UK.

BSI (2015) PD 6691:2015: Guidance on the use of BS EN 13108, *Bituminous Mixtures – Material Specifications* (+A1:2016). BSI, London, UK.

BSI (2016) BS EN 13108-1:2016: *Bituminous Mixtures*. Material Specifications. Asphalt Concrete. BSI, London, UK.

BSI (2022) PD 6691:2022: Guidance on the use of BS EN 13108, *Bituminous Mixtures – Material Specifications*. BSI, London, UK.

Carswell I., Nicholls J.C., Elliott R.C., Harris J. and Strickland D. (2005) *Feasibility of Recycling Thin Surfacing back into Thin Surfacing Systems*. TRL, Crowthorne, UK, TRL Report 645.

Carswell I., Nicholls J.C., Widyatmoko I., Harris J. and Taylor R. (2010) *Best Practice Guide for Recycling into Surface Course*. TRL, Crowthorne, UK, TRL Road Note 43.

Carswell I., Karlsson R., Raaborg J. and Kuttha D. (2012) *Re-road – End of Life Strategies of Asphalt Pavements. D2.6 Main Report on Results of Comparative Site Monitoring*. VTI (Swedish National Road and Transport Research Institute), Linköping, Sweden.

Cemex (2006) *M4 Junction 32–33 Resurfacing Scheme*. Cemex, Rugby, UK.

Dunford A., Carswell I. and Gershkoff G. (2013) *The Potential Effect of Reclaimed Asphalt on the Friction Characteristics of Surface Course Materials*. TRL, Crowthorne, UK, TRL PPR 670.

Flint M. and Bailey H. (2018) *The Application of High Recycled Content Mixtures on Strategic Roads*. Proceedings of the 16th Annual International Conference, Liverpool, UK.

HE (2018) *Manual of Contract Documents for Highway Works*. Volume 1, Specification for Highway Works. Series 900, Road Pavements – Bituminous Bound Materials. HE, UK.

HE (2020a) DMRB, CS 229, *Data for Pavement Assessment*. HE, UK.

HE (2020b) DMRB, CS 230, *Pavement Maintenance Assessment Procedure*. HE, UK.

Markham D. (2017) *Installation of 14 mm Ultipave Containing 30% Reclaimed Asphalt on the M20*.

Nicholls. J.C., Hannah A. and Carroll A. (2007) *A Practical Example of Recycling Reclaimed Asphalt Back Into the Surface Course*. Proceedings of the 6th International Conference on Sustainable Aggregates, Pavement Engineering and Asphalt Technology, Liverpool, UK, Paper PA/INF/5687/06..

Nicholls J.C., Carswell I., Thomas C. and Sexton B. (2010) Durability of Thin Asphalt Surfacing Systems. *Part 4: Final Report After Nine Years' Monitoring*. TRL, Crowthorne, UK, TRL Report 674.

Ojum C. (2017) *Sub-Task 5: Best Practice for Recycling Asphalt Pavements. Task 1-111 Collaborative Research Project*. Aecom, Nottingham, UK..

Roe P.G., Webster D.C. and West G. (1991) *The Relation Between the Surface Texture of Roads and Accidents*. TRL, Crowthorne, UK, TRL RR296.

Schiavi I., Carswell I. and Wayman M. (2007) *Recycled Asphalt in Surfacing Materials: A Case Study on Carbon Dioxide Emission Savings*. TRL, Crowthorne, UK, TRL PPR 304.

Wayman M. and Carswell I. (2010) *Enhanced Levels of Reclaimed Asphalt in Surfacing Materials: A Case Study Evaluating Carbon Dioxide Emissions*. TRL, Crowthorne, UK, TRL PPR 468.

Wright M., Zuo W., Bateman D. and Taylor L. (2020) *Task 1-583 Comprehensive Review of Asphalt Recycling with a View to Increasing Recycled Content in the Medium-Term. Subtask 3 – Technical Demonstration of Solutions. 3.2 – Thin Surface Course Recycling*. Highways England, Birmingham, UK.

Challenges for the development and implementation of asphalt mixtures containing recycled waste plastic for pavement surfacing

Greg White*
School of Science, Technology and Engineering, University of the Sunshine Coast, Queensland, Australia

Gordon Reid
Macrebur, Lockerbie, Scotland, UK

ABSTRACT: With the increased focus on recycling of waste materials in infrastructure construction and maintenance, there is an ever-increasing interest in the recycling of waste plastic in the production of asphalt mixtures for road and other pavement surfacing. However, there are many types of plastic and only some are compatible with asphalt production. Some compatible plastics are capable of extending the mineral aggregate in asphalt mixtures, while others can also improve the mixture properties, by increasing the resistance to rutting and cracking. However, the most valuable plastics can extend and improve the bituminous binder in the asphalt mixture, effectively replacing the synthesized polymers that are commonly used to improve moisture resistance, temperature susceptibility, crack resistance and deformation resistance. Despite these potential benefits, there are many challenges associated with the categorization of different plastics and their associated effects, as well as the sourcing of a consistent and uncontaminated plastic supply. Other challenges include the digestion and stability of plastic in the bituminous binder phase when the wet mixing process is used. It is also essential to confirm and demonstrate that asphalt mixtures containing recycled plastic do not increase fume generation during construction, or chemical leachate of road surfaces during service. These challenges must be resolved if the potential for recycling plastic in road and other pavement asphalt layers is to be fully maximised in the future. This paper summarises the potential benefits of waste plastic as an asphalt binder modifier and explores the challenges associated with the implementation of waste plastic as a mature technology in asphalt binder and mixture production. The potential to overcome those challenges is also considered.

Keywords: Recycled, Plastic, Surface, Asphalt, Challenges

1 INTRODUCTION

The desire for communities, nations and companies to be more sustainable has significantly influenced the desire to incorporate industrial by-products and other waste materials into civil engineering infrastructure (Jamshidi & White 2020). In pavement construction, the greatest opportunity exists when materials that are otherwise considered waste products can be used to replace, extend or enhance materials that are otherwise virgin ingredients (Van Den Heuvel & White 2021). For example, using fly ash and geopolymers to replace Portland

---

*Corresponding Author: gwhite2@usc.edu.au

DOI: 10.1201/9781003429258-72

cement in concrete production, crushed glass in lieu of natural sand in concrete or asphalt production, and crumbed tyre rubber to modify and extend bitumen in asphalt production (Jamshidi & White 2020). Where a relatively economical waste or by-product can fully or partially replace an expensive raw ingredient, the economical advantages can be as significant as the environmental ones (Van Den Heuvel & White 2021).

Incorporating waste plastic into asphalt production is a potential technology that has gained significant interest in recent years. The interest in recycled plastic asphalt (RPA) mixtures has likely been driven by the growing consciousness regarding waste plastic, particularly that associated with single use plastic bags and plastic drinking bottles (Grady 2021). Where the plastic can be used to extend and/or modify the bituminous binder, this also has significant economic benefits because processing the plastic is much less costly that the binder and/or conventional polymers that it can replace (Milad et al. 2020).

Laboratory research has shown that processed waste plastic can be incorporated into asphalt production for pavement construction. The plastic must be collected, cleaned, shredded or otherwise processed (White & Reid 2018) before being either wet-mixed into the bituminous binder prior to asphalt production, or dry-mixed by metered feeding into the asphalt production plant (White & Hall 2021). Some research has also demonstrated the technical viability of de-polymerising the waste plastic and adding the various polymers to the bituminous binder (Raouf et al. 2018), as well as pre-coating the aggregate particles with melted plastic (Chowdhury et al. 2020). However, these approaches are unlikely to be economically viable on a practical scale (White & Magee 2019).

As explained below, when an appropriate plastic is used to extend the bitumen used in asphalt production, the plastic can also significantly enhance the mechanical properties of the bituminous binder (Willis et al. 2020). Commonly reported enhancements associated with RPA include greater resistance to deformation and a higher contribution to the structural strength of the pavement (Lanotte & Desidery 2022). Less commonly, the fracture and crack propagation resistance are improved, as well as the resistance to moisture damage (Eskandarsefat et al. 2022). In some cases, the enhancement in the properties of the bituminous binder and asphalt mixture are so significant, the use of plastic can replace the need for conventional polymer modification in some heavy duty pavement applications (White 2020).

The parallel benefits of bituminous binder and asphalt mixture performance-related properties, as well as more sustainable pavements, makes RPA is attractive. However, there are a range of significant challenges that must be overcome before RPA becomes commonplace (White 2020). These challenges generally relate to post-construction risks and properties that can not be easily measured in the laboratory, such as environmental age-related durability and fuming/leaching of plastic chemicals into the environment.

This paper explores the challenges associated with the broad implementation of incorporating waste plastic into asphalt production for pavement construction. The identified challenges address differences in plastic types, product and production consistency, fuming and leaching potential, as well as the internationalisation of research findings and relative age-related durability of asphalt surfaces. Before those challenges are considered, the potential benefits of RPA are first summarised.

2 BENEFITS OF RECYCLED PLASTIC ASPHALT

Since around 2010, there has been a significant increase in the number of research publications. A simple Scopus (Elsevier 2022) search for "plastic asphalt" articles indicates an average of 19 publication per annum during the five-year period 2000-2004, compared to an annual average of 139 articles for the five year period 2017-2021.

Most of the research publications have focused on the properties of either the bituminous binder, or on the mechanical properties of the asphalt mixture (Willis et al. 2020). Many of the research efforts have compared the properties of the bituminous binder or asphalt mixture modified with plastic, to those of unmodified control samples (White 2021). Some have compared the effects of plastic to those associated with other common modifiers, such as styrene-butadiene-styrene, ethyl vinyl acetate, polyphosphoric acid and crumb tyre rubber (Milad et al. 2020).

In many cases, the RPA is generic in nature, with a particular source of waste plastic procured by the researcher, which is then shredded or otherwise processed in the laboratory. The amount of plastic added and the process for production is determined by the researcher. In contrast, some research has focussed on particular proprietary plastic products. The plastic product manufacturer provides the finished plastic, ready to be incorporated into the bituminous binder or the asphalt mixture, usually at a dosage recommended by the manufacturer. However, the manufacturer rarely discloses the type, source, or processing of the plastic. The products known as MR 6, MR 8 and MR 10, supplied by MacRebur Limited, are an example (Macrebur 2022), with different research efforts having investigated different aspects of the products over time. The three products are produced from different types of plastic and while the plastic type of each has remained unchanged over time, the form of each product has evolved (Figure 1).

(a)   (b)

(c)   (d)

Figure 1.   Different plastics known as (a) MR 6 (original), (b) MR 8 (original), (c) MR 6 (current) and MR 8 (current).

739

3 CHALLENGES FOR RECYCLED PLASTIC ASPHALT

As explained above, despite the significant interest in RPA as a strategy for more sustainable pavement construction, the technology is not yet broadly accepted by the pavement construction and maintenance industry. This reflects a number of challenges associated with RPA, that need to be addresses prior to RPA becoming considered a 'normal' practice. These challenges include different plastic types and polymers, product and production consistency, fuming and leaching concerns, the difficulty associated with internationalisation of research findings and environmental age-related durability of pavement surfaces. Other challenges include the resistance by commercial competitors, such as some bitumen and polymer suppliers, as well as the general difficulty associated with educating the various industry stakeholders located in different jurisdictions around the world. These commercial and educational challenges are not considered here, with the focus on the more technological challenges.

3.1 *Not all plastics are equal*

There are thousands of types of plastic produced for a large range of industrial, commercial and household applications (Wypych 2012). Many of these are incompatible with bitumen, many do not have an appropriate density or melting point to be used in asphalt production, and many produce bituminous binders and asphalt mixtures with detrimental properties for pavement construction. Consequently, the number of plastics that are actually viable for the production of RPA are limited, with low and high density polyethylene (PE) the most viable (Masad et al. 2020). However, even within this limited range of viable plastics, the various sources of plastic have significantly different effects.

The difference in plastic products available for RPA production limits the applicability of any particular research findings to only that particular type of plastic. For example, a conclusion based on high density PE is unlikely to be useful when considering PET, or even low density PE. This is similar to different asphalt mixture designs being required for different source of aggregate, or for the different effects of conventional polymers, such as SBS and EVA. This creates a challenge when attempting to translate the effects associated with a specific plastic type and source, to the broader concept of RPA. It also creates a challenge when different RPA research draws different conclusions, which may simply reflect differences in the plastics used. This challenge is exacerbated for proprietary plastic suppliers that do not disclose the type and source of their plastics.

As an example, the effects of the proprietary products MR 6, MR 8 and MR 10 were compared. Each contains different plastic types, and each has different effects on bituminous binders and asphalt mixtures (White 2020). The product MR 6 is intended to have primarily plastomeric and significantly increases the modulus and deformation resistance of asphalt mixtures. By comparison, MR 10 is more elastomeric, but still significantly increases the mixture stiffness. In contrast, MR 8 is intended to be an economical bitumen extender and has the least effect on the mechanical properties of binders and mixtures (White & Reid 2018). This is shown in Figure 2, where the different effects of the three products on binder softening point and on mixture wheel tracking resistance to deformation is clear. Similarly, Figure 3 shows the significant improvement in binder recovery and mixture ductility associated with all plastic types, but the effect was significantly greater for MR 10 than it was for MR 6 or MR 8.

This challenge can not be simply addressed. In some cases, proprietary plastic suppliers may elect to disclose their plastic sources and types. Alternatively, some range of plastic properties could be published, similar to that specified for different grades of bitumen and polymer modified binder. For researchers, it is important that the type and source of plastic, or the proprietary product, is disclosed so that conflicting conclusions can be better explained. However, it is unlikely that the challenge associated with different types of plastic

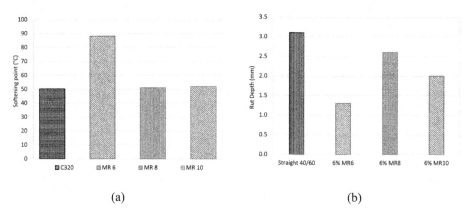

(a)   (b)

Figure 2. Effect of different plastic products on (a) binder softening point and (b) mixture deformation resistance. C320 and 40/60 refer to similar unmodified base bitumens supplied to a viscosity and a penetration specification, in Australia and the UK, respectively.

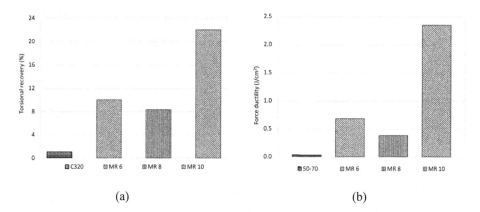

(a)   (b)

Figure 3. Effect of different plastic products on (a) binder torsional recovery and (b) mixture force ductility. C320 and 50-70 refer to similar unmodified base bitumens supplied to a viscosity and a penetration specification, in Australia and the UK, respectively.

being associated with different effects, whether positive or negative, is unlikely to be resolved, but must be mitigated by transparent disclose of plastic types and sources, and by explaining the limitations of the associated research conclusions.

3.2 *Product and production consistency*

Even with a single plastic source of a consistent plastic type, or a particular proprietary plastic product identified, there remains a challenge associated with product and production consistency. Product consistency relates to production of the plastic product, while production consistency relates to the incorporation of the plastic into the RPA.

Variables in plastic product include contamination of the plastic, the source of plastic and the processing of the plastic. Some plastics are sourced from industrial waste streams, and these are likely to be uncontaminated and consistent (Austroads 2021). However, some proprietary products are sourced from domestic and other streams of mixed plastic, and these have a higher risk of changes in the source and contamination (Hall & White 2021). Furthermore, when a local circular economy is desired by the end-user, the sourcing of plastic from reliable sources outside of that community is often resisted, forcing the supplier to find a local alternate. For example, Hall & White (2021) presented a case study on the introduction of a circular economy based on plastics extracted from the household waste stream being re-used to enhance asphalt surfacing for the roads in the same county in the UK. In this case, household plastic waste that had similar effects on asphalt mixtures as MR 6 were identified, verified and then used in side-by-side field trials. The economic and environmental savings associated with this mini-circular economy were significant (Hall & White 2021).

To provide users with increased confidence in RPA, some suppliers have developed quality assurance systems. For example, MacRebur operate under an accredited quality system allowing each package of finished plastic modifier to be traced to a specific production batch and the associated sources of recycled waste plastic (White & Reid 2019). This system also includes a range of melting temperature and particle size tests prior to the acceptance of incoming waste plastic feedstock (Hall & White 2021).

The consistent incorporation of plastic the RPA also presents a challenge. There are generally only two approaches to RPA production, the first is wet mixing into the bitumen, and the second is dry mixing through the asphalt production plant (White & Hall 2021). Wet mixing provides thorough distribution of the plastic through the bituminous binder, and then relies on normal asphalt mixture production to coat all the aggregate particles. However, wet mixing limits the plastics that can be used to those with melting points below typical bitumen production temperatures, and lower density plastics have been reported to float out of the bitumen, resulting in phase separation during hot storage (Willis et al. 2020). This creates a challenge for storage and hot transportation of plastic modified bituminous binder, although some research has demonstrated that chemical additives, such as reactive elastomeric terpolymers, have reduced this risk (Joohari et al. 2022). Dry mixing avoids the risk of binder phase separation at high temperatures, but there remains a challenge associated with complete melting and distribution of the plastic through the asphalt mixture. For example, a semi-digested MR 10 pellet was identified in a sawn fatigue beam produced from drum-plant mixed asphalt in Australia, as shown in Figure 4. Although it is likely this was an anomaly associated with insufficient asphalt production time, it can also be viewed as an indication that MR 10 does not melt and digest into bitumen. However, this has been separately proven to not be the case. Furthermore, and similar to wet mixing, where the plastic is intended to melt and become part of the bituminous binder, the melting point of the plastic is limited to the production temperature of asphalt, which is generally 190°C, or 160°C with warm mix technologies. It has also been observed that laboratory asphalt mixers

Figure 4.   Example of inadequately mixed MR 10 pellet in a sawn fatigue beam specimen.

do not distribute the plastic as efficiently as full-scale asphalt production plants. Furthermore, continuous drum mixing plants are less efficient at distributing dry mixed plastic, compared to batch mixing plants (White et al. 2019). Despite these observations, some studies have shown that wet mixed RPA has mechanical properties that were not significantly different to otherwise identical dry mixed RPA (White & Hall 2020). The properties of binder extracted from both mixtures were also not significantly different (White & Hall 2021).

Confidence in the consistency of the plastic products, and the reliable and repeatable production of RPA is a significant challenge. This must be addressed by the plastic and RPA suppliers. The level of risk and the transparency available to the end user varies significantly. For example, some products are offered as proprietary processed plastics, with the customer being an asphalt producer. In contrast, other products are supplied by asphalt producers as proprietary RPA, with the pavement owner being the customer. These two circumstances are distinctly different, and they require different approaches to provide the respective customers with confidence in the plastic product and asphalt production consistency.

3.3 *Fuming and leaching potential*

With the significant focus on the health of construction staff, as well as the minimisation of adverse impacts on the natural environment, there is significant interest in the fume and leachate generation from RPA (Willis et al. 2020). The potential for fume generation occurs during construction, when the RPA is supplied, paved and compacted at 120-180°C. In contrast, the potential for microplastic and chemical leachate occurs during the service life of RPA.

It is well established that asphalt production and construction generate fumes. It is also well established that road surfaces generate leachate and other runoff contamination (Townsend 1998). With the aim of causing 'zero harm' to personnel and the environment, there is a perception that RPA should be required to generate no fumes and no leachate. Although well intentioned, that is not a reasonable expectation. Rather, RPA should be required to be associated with no more or worse fume and leachate generation than otherwise normal asphalt mixtures that have been routinely used and accepted for decades.

Various studies have found that asphalt mixtures generate fumes during production, paving and compaction (Bywood & McMillan 2019), as well as chemical leachate in-service (Brantley & Townsend 1999). However, other research has shown that these fumes and leachates are no worse for RPA than for otherwise similar conventional asphalt mixtures (White 2019). In fact, some research has shown that RPA generated less harmful fumes than conventional unmodified asphalt mixtures (Boom et al. 2022). Similarly, recent research has demonstrated that microplastic leachate is a significantly bigger issue than RPA, with significant microplastics detected in road dust being attributed to tyre rubber, brake pad dust and external pollution such as roadside rubbish (Monira et al. 2022).

It is clear that asphalt mixtures for road construction generate fumes and leachates. Although further work is required, there is no evidence that RPA generates more or worse fumes or leachate than otherwise comparable and normal asphalt for pavement surfacing. However, convincing stakeholders that 'no additional harm' is a more appropriate measure than 'zero harm', remains a challenge to the broad acceptance of RPA.

3.4 *Internationalisation of findings*

The tendency of pavement authorities in various jurisdictions to want research findings to be based on their own test methods is a significant challenge for the broad implementation of new technologies, such as the effects of different plastics on bituminous binders and asphalt mixtures (Austroads 2019). Although the desire for the use of local and familiar test methods is understandable, it hinders the internationalisation of research finds, even where the test methods are similar in nature.

A good example is asphalt mixture Marshall properties. The Marshall mixture design method and associated properties, known as 'stability' and 'flow' are common across the world (White 1985). Almost all jurisdictions test samples at 55°C or 60°C and all use the same Marshall apparatus. Even in jurisdictions where performance-related tests are preferred, such as Superpave (Speight 2016) in the USA, Marshall test methods are still published and available (ASTM 2019). Despite this, authorities in one jurisdiction often require testing to be performed to their local test method, despite virtually identical Marshall test methods from another jurisdiction having already been used. Other examples include tests for binder softening point, viscosity at 60°C and penetration at 25°C (Austroads 2008). These are almost identically specified around the world and use the same equipment in most countries, but different jurisdictions have their own local standards and often resist international findings that used other versions of the same test method, just because it was published by another jurisdiction.

Overcoming this challenge can only increase the efficiency associated with the internationalisation of results across jurisdictions. The use of an Australian test method should not hinder the reliability or acceptance of research findings in the USA, the UK or Europe, and vice versa. Although it is unlikely that test methods will be standardised across the globe, it would be beneficial to document the similarity of key test method across jurisdictions, with the aim being an increased acceptance of international research findings relating the RPA.

3.5 *Environmental ageing and durability*

There are many sound and reliable test methods for the physical or mechanical properties of bituminous binders and asphalt mixtures. Examples include the softening point and penetration of binders, as well as the Marshall properties of asphalt mixtures. Some physical properties are also indicative of relative performance of binders and mixtures, such as the binder MSCR protocol. For asphalt mixtures, there are a range of tests for stiffness, deformation resistance, fatigue resistance, moisture damage resistance and workability. However, the protocols and test methods for durability of mixtures due to environmental ageing is a challenge for all sustainable pavement materials, including RPA (White & Abouelsaad 2022).

Although structural pavement design is generally limited to considering asphalt fatigue cracking and subgrade rutting, age related fretting and ravelling often determine the time at which resurfacing is required (Abouelsaad & White 2021). This is particularly important for local roads, which are only lightly trafficked, and for airport pavements, which support heavy wheel loads, but have significant areas of the pavement that are rarely trafficked (Abouelsaad & White 2020).

Most practical protocols for accelerated laboratory aging of asphalt mixtures include placing compacted samples in a conventional oven for a predetermined timeframe. For example, test method R30 requires five days at 85°C (AASHTO 2002). The combination of temperature and duration are intended to reflect field aging over an asphalt surface life cycle. The degree of ageing is usually based on an index, which is the ratio of a particular property after aging, to the same property before aging. The property used as the basis of the aging index varies, but modulus is commonly reported (Abouelsaad & White 2021). However, some researchers have used the chemical and rheological properties of extracted binder as the basis of the index (White & Abouelsaad 2022). For a particular asphalt mixture and a particular ageing protocol, the different properties used to calculate the ageing index will result in different relative effects of environmental ageing, which is a significant challenge for assessing the expected life of RPA compared otherwise identical asphalt mixtures.

The relative age-related life expectancy of RPA, compared to otherwise identical conventional asphalt mixtures remains a significant challenge. That is because surface life is a significant factor in whole or life costs, including financial and environmental costs. As an example, if using recycled plastic saved 10% of the financial cost and reduces the associated carbon emissions by 5%, but resulted in a one year surface life reduction, then assuming a 10 year surface life expectancy, which is common for airport pavements (Jamieson & White 2020), the whole of life financial cost is no better, and the environment cost is actually worse than achieving the full 10 year surface life with a conventional asphalt mixture (Van Den Heuvel & White 2021).

4 CONCLUSION

There is significant interest in providing more sustainable asphalt surfaces for road and other pavements, with recycled waste plastic a viable and attractive option. This has resulted in a plethora of research activities, publications and field trials since about 2015, and there are many examples to demonstrate the potential for waste plastics to increase the stiffness, deformation resistance and less commonly the moisture resistance and fracture resistance of asphalt mixtures. Some researchers have compared these effects to those commonly associated with conventional polymers for bituminous binder modification, particularly plastomeric polymers such as EVA. However, there remains many significant challenges before recycled plastic becomes a mature technology in road and airport pavement surfacing practice. These challenges include the diversity of different plastics available, despite only relatively few being compatible with bitumen and asphalt production, installing confidence in the consistency of plastic products and the uniformity of both wet and dry mixed plastic into asphalt mixtures. The perceived fuming and leaching risks are also a challenge, along with the resistance to accepting test results based on international test methods, even for test methods that are fundamentally standardised across the globe. Finally, because sustainability benefits of asphalt mixtures are sensitive to any change in the rate of environmental ageing and weathering, which affects the expected time between pavement resurfacing, a better understanding of the relative durability of asphalt mixtures with and without recycled plastic is critical. These challenges will not be simple to overcome but researchers, pavement practitioners, environmental scientists and regulators must work together to understand and educate the industry and its stakeholders if the potential benefits of more sustainable asphalt for pavement surfacing are to be fully realised in the future.

REFERENCES

AASHTO 2002, *Standard Practice for Mixture Conditioning of Hot Mix Asphalt (HMA)*, Method R30, American Association of State Highway and Transportation Officials, Washington, District of Columbia, USA, 1 January.

Abouelsaad A. & White G. 2020, 'Fretting and Ravelling of Asphalt Surfaces for Airport Pavements: A Load or Environmental Distress?', *Nineteenth Annual International Conference on Highways and Airport Pavement Engineering, Asphalt Technology, and Infrastructure*, Liverpool, England, United Kingdom, 11-12 March.

Abouelsaad A. & White G. 2021, 'Review of Asphalt Mixture Ravelling Mechanisms, Causes and Testing', *International Journal of Pavement Research and Technology*, doi. 10.1007/s42947-021-00100-7, article-in-press.

ASTM 2015, *Standard Test Method for the Marshall Stability and Flow of Asphalt Mixtures*, ASTM-D6927, 1 April, American Society for Testing and Materials, Washington, District of Columbia, USA.

Austroads 2008, *Guide to Pavement Technology: Part4F: Bituminous Binders*, Report AGPT04F/08, Austroads, Sydney, New South Wales, Australia, September.

Austroads 2019, *Viability of Using Recycled Plastics in Asphalt and Sprayed Sealing Applications*, Report AP-T351-19, Austroads, Sydney, New South Wales, Australia, October.

Austroads 2021, *Use of Road-grade Recycled Plastics for Sustainable Asphalt Pavements: Overview of the Recycled Plastic industry and Recycled Plastic Types*, Report AP-R648-21, Austroads, Sydney, New South Wales, Australia, March.

Boom Y.J., Enfrin M., Grist S. & Giustozzi F. 2022, 'Fuming and emissions of waste plastics in bitumen at high temperature', *Plastic Waste for Sustainable Asphalt Roads*, Giustozzi, F. & Nizamuddin, S. (Eds.), Woodhead Publishing, doi.org/10.1016/C2020-0-02914-2.

Brantley A.S. & Townsend T.G. 1999, 'Leaching of pollutants from reclaimed asphalt pavement', *Environmental Engineering Science*, vol. 16, no. 2, pp.105–116.

Bywood P. & McMillan J. 2019, 'Bitumen Contents and Fumes: A Review of Health Risks Associated with Exposure to Bitumen Contents and Fumes', *Institute for Safety, Compensation and Recovery Research*, March.

Chowdhury P.S, Kumar S & Sarkar D. 2020, 'Performance Characteristic Evaluation of Asphalt Mixes with Plastic Coated Aggregates', *Transportation Research*, vol. 45, pp. 793–803.

Elsevier 2022, Scopus, <https://www.scopus.com/standard/marketing.uri>, accessed 30 March 2022.

Eskandarsefat S., Meroni F. Sangiorgi C. & Tataranni P. 2022, 'Fatigue Resistance of Waste Plastic-Modified Asphalt', *Plastic Waste for Sustainable Asphalt Roads*, Giustozzi, F. & Nizamuddin, S. (Eds.), Woodhead Publishing, doi.org/10.1016/C2020-0-02914-2.

Grady B.P. 2021, 'Waste Plastic in Asphalt Concrete: A Review', *SPE Polymers*, vol. 2, no. 1, pp. 4–18.

Hall F. and White G. 2021 'Using Local Waste Plastics in Asphalt Modification to Improve Engineering Properties of Roads', *36th International Conference on Solid Waste Technology and Management*, Annapolis, Maryland, USA, 14-16 March.

Jamieson S. & White G. 2020, 'Review of Stone Mastic Asphalt as a High Performance Ungrooved Runway Surfacing', *Road Materials and Pavement Design*, vol. 21, no. 4, pp. 886–905.

Jamshidi A. & White G. 2020, 'Evaluation of Performance of Challenges of use of Waste Materials in Pavement Construction: A Critical Review', *Applied Sciences*, vol. 10, no. 226, pp. 1–13.

Joohari I.B., Maniam S. & Giustozzi F. 2022, 'Enhancing the Storage Stability of SPS-plastic Waste Modified Bitumen using Reactive Elastomeric Terpolymer', *International Journal of Pavement Research and Technology*, Article-in-press, doi.org/10.1007/s42947-021-00132-z.

Lanotte M. & Desidery L. 2022, 'Rutting of Waster Plastic-modified bitUmen', *Plastic Waste for Sustainable Asphalt Roads*, Giustozzi, F. & Nizamuddin, S. (Eds.), Woodhead Publishing, doi.org/10.1016/C2020-0-02914-2.

Macrebur 2022, Macrebur: *The Plastic Road Company*, <https://macrebur.com/>, accessed 30 March 2022.

Masad E., Roja K.L., Rehman A. & Abdala A. 2020, *A review of Asphalt Modification Using Plastics: A Focus on Polyethylene*, Technical Report, Texas A&M, Doha, Qatar, doi: 10.13140/RG.2.2.36633.77920.

Milad A., Ali A.S.B., & Yusoff N.I.M. 2020, 'A Review of the Utilisation of Recycled Waste Plastic Materials as an Alternate Modifiers in asphalt Mixtures', *Civil Engineering Journal*, vol. 6, pp. 42–60.

Monira S., Ali Bhuiyan M., Haque N. & Pramanik B.K. 2022, 'Road Dust-associated Microplastics from Vehicle Traffics and Weathering', *Plastic Waste for Sustainable Asphalt Roads*, Giustozzi, F. & Nizamuddin, S. (Eds.), Woodhead Publishing, doi.org/10.1016/C2020-0-02914-2.

Raouf M.R., Eweed K.M., and Rahma N.M. 2018. 'Recycled Polypropylene to Improve Asphalt Physical Properties', *International Journal of Civil Engineering and Technology*, vol. 9, no. 12, pp. 1260–1267.

Speight J.G. 2016, 'Test Methods for Aggregate and Asphalt concrete', *Asphalt Materials Science and Technology*, pp. 205–251.

Townsend T.G. 1998, 'Leaching Characteristics of Asphalt Road Waste', *Hot Mix Asphalt Technology*, vol. 3, no. 4, pp. 21–27.

Van Den Heuvel D. & White G. 2021, 'Objective Comparison of Sustainable Asphalt Concrete Solutions for Airport Pavement Surfacing', *International Conference on Sustainable Infrastructure*, a virtual event, 6-10 December.

White G. 2020 'A Synthesis of the Effects of two Commercial Recycled Plastics on the Properties of Bitumen and Asphalt', *Sustainability*, vol. 12, no. 8594, pp. 1–20.

White G. & Abouelsaad, A. 2022, 'Improved Accelerated Ageing of Asphalt Samples in the Laboratory', *9th Symposium on Pavement Surface Characteristics*, Milan, Italy, 12-14 September.

White G & Hall F 2020, 'Comparing Wet Mixed and Dry Mixed Binder Modification with Recycled Waste Plastic', *RILEM International Symposium on Bituminous Materials*, Lyon France, 14-16 December.

White G & Hall F 2021, 'Laboratory Comparison of Wet-mixing and Dry-mixing of Recycled Waste Plastic for Binder and Asphalt Modification', *100th Transportation Research Board Annual Meeting: a virtual event*, Washington, District of Columbia, USA, 5-9 January.

White G. & Magee C. 2019, 'Laboratory Evaluation of Asphalt Containing Recycled Plastic as a Bitumen Extender and Modifier', *Journal of Traffic and Transportation Engineering*, vol. 7, no. 5, pp. 218–235.

White G., Menzies C. & Kidd A. 2019, *'Comparing Asphalt Properties for Samples Produced in the Laboratory and Different Production Plants'*, International Airfield and Highway Pavements Conference, Chicago, Illinois, USA, 21-24 July.

White G. & Reid G. 2018 'Recycled Waste Plastic for Extending and Modifying Asphalt Binders', *8th Symposium on Pavement Surface Characteristics*, Brisbane, Queensland, Australia, 2-4 April.

White G. & Reid G. 2019, 'Recycled Waste Plastic Modification of Bituminous Binder', *7th International Conference on Bituminous Mixtures and Pavements*, Thessaloniki, Greece, 12-14 June, pp. 3–12.

White T.D. 1985, 'Marshall Procedures for Design and Quality Control of Asphalt Mixture', *Proceedings Asphalt Pavement Technolog*y, no. 54, pp. 265–285.

Willis R., Yin F., & Moraes R. 2020, *Recycled Plastics in Asphalt Part A: State of the Knowledge*, Report NAPA-IS-142, National Asphalt Pavement Association, Greenbelt, Maryland, USA, October.

Wypych G. 2012 (Eds), *Handbook of Polymers*, ChemTec Publishing, Elsevier, doi: 10.1016/B978-1-895198-47-8.50001-1.

Waste plastic as potential high-value solution for asphalt road applications

Bocci Edoardo
Faculty of Engineering, eCampus University, Milan, Italy

Caraffa Tullio & Stimilli Arianna*
Direzione Operation e Coordinamento Territoriale, Anas S.p.A., Rome, Italy

Bocci Maurizio
Department of Civil, Building and Environmental Engineering, Polytechnic University of Marche, Ancona, Italy

ABSTRACT: ANAS, the main Italian road agency, manages over 30,000 km of both local and national roads. In all cases, maintenance actions involve the reconstruction of the bituminous layers that require new aggregates and binder. This aspect, together with the need of acting in compliance with the new environmental regulations that increasingly push towards the recovery of waste and/or recycled materials for the construction of new bituminous mixtures, entails the study of innovative solutions that can be considered sustainable both from an environmental, technical and economic point of view.

Among the waste materials currently available on the market (open-loop materials), the so called "light plastics" recovered from municipal solid waste (MSW) represent a potential solution still under development. Plastics constitute a large percentage of the waste material that is difficult to recycle in other fields. However, waste plastics have a high potential for recovery in bituminous mixtures since their polymeric nature could act as a modifying agent in replacement and/or integration of the synthetic polymers usually added to the bitumen. This action can significantly improve the overall mixture performance and decrease production costs.

In order to validate this potential, ANAS has developed a study on the reuse of "light plastics" in the production of hot bituminous mixtures. The study includes a preliminary laboratory analysis to verify the effects on mechanical, volumetric and durability properties. Three different pre-treated wasted plastic materials were tested in different dosages to determine the optimum mix design and the most suitable mixing process. The results described in this paper, baseline to set up full-scale trial sections, represent the kick-off step to develop efficient, innovative and sustainable technological options for the production of new bituminous mixtures both from an environmental and economic point of view.

Keywords: Waste Plastic, Hot-Mix Asphalt, Recycling.

1 INTRODUCTION

The management of municipal solid wastes (MSW) is one of the major concerns for European Union. In particular, the Waste Framework Directive (2018) provides that the EU members will be prepared to reuse and recycle 55%, 60% and 65% by weight of the MSW by

---

*Corresponding Author: a.stimilli@stradeanas.it

DOI: 10.1201/9781003429258-73

2025, 2030 and 2035, respectively. According to the annual report by Italian Superior Institute for Environmental Protection and Research, waste plastics represent 8.6% of the MSW. In Italy, about 1.5 million tons of waste plastics generated in 2020 and the yearly production has been increasing (ISPRA 2021). For this reason, the identification of solutions for waste plastic recycling is currently a key target. Due to their chemical and physical properties, recent international studies highlighted the possibility to include waste plastic in hot-mix asphalt (HMA) for road applications, either as bitumen modifier or as replacement for aggregate and binder (Willis et al. 2020; Wu & Montalvo 2021). Waste plastics may include several types of polymeric materials, such as high-density polyethylene (HDPE), low-density polyethylene (LDPE), polyethylene terephthalate (PET), linear low-density polyethylene (LLDPE), polypropylene (PP), polystyrene (PS), polyvinyl chloride (PVC), ethyl vinyl acetate (EVA) and others. These plastics are characterised by different melting point, ranging from 80 °C to 250 °C (Ma et al. 2021), i.e. straddling around the production temperature of HMA (140-180 °C). The waste plastics with low melting points (PE, PP, EVA, PVC) can be recycled through a wet process, which consists in adding them into the hot bitumen as a modifying agent. On the other side, the dry process, which consists in the addition directly into the mixture during HMA production at the plant, is suitable for all the waste plastics. The bitumen modified with waste plastics (wet process) showed improved rutting resistance, moisture resistance, and fatigue resistance of the binder blends. However, plastics with high melting point tend to increase more viscosity but reduce more ductility of the binder blends. This reflects into two potential concerns: the phase separation between binder and plastic and the low-temperature performance of the HMA (Ma et al. 2021; Nizamuddin et al. 2021). In the dry process, waste plastics may act as aggregates substitute or bitumen modifier, according to the melting point. The great advantage of the dry processing technique lies in the easiness, as no new equipment is required at the asphalt plant. Moreover, it does not increase the amount of fumes that are released during HMA production and laying, while in the wet process the bitumen releases fumes during modification (Vasudevan et al. 2012). In terms of HMA properties, increased rutting and moisture resistance of the mix was observed when waste plastics were added through dry process (Angelone et al. 2016; Giustozzi et al. 2022; Radeef et al. 2022). Several studies have focused on the dry processing of specific plastic wastes, obtained by separation of the different materials and, in some cases, further treatment (Badejo et al. 2017; Lastra-Gonzalez et al. 2016; Moghaddam et al. 2014; Noor et al. 2022; Willis et al. 2020) but few researchers tried to recycle the non-separated waste plastic in HMA. Movilla-Quesada et al. 2019 named "plastic scrap" the non-separated plastic waste and observed that it can partially replace bitumen without penalising the mix performance. In particular, they obtained higher strength and stiffness and lower permanent deformations compared to the reference mix without plastic scrap. Dalhat et al., 2019 contemporary used selected waste plastics through wet process (binder modification) and non-separated waste plastics through dry process. They found that the combined recycling of waste plastics allowed achieving a better performance compared to a reference mix with crumb-rubber modified bitumen in terms of viscoelastic properties, resistance to rutting and fatigue. However, they did not investigate how the different materials in the non-separated plastic waste interact with the hot aggregate and bitumen during mix production. In front of the few experiences, further studies on the recyclability of non-separated plastic wastes in HMA are required.

2 OBJECTIVE AND EXPERIMENTAL PROGRAMME

The overall objective of this study was to evaluate the potential use of plastic waste into asphalt mixes for binder layer. The experimental plan was structured in two subsequent steps (hereafter indicated as Step 1 and Step 2) and involved different plastic treatments and dosages, as well as different bitumen contents. Step 1 aimed at selecting the most suitable plastic treatment for the incorporation into dense graded HMA mixtures (binder layer) via dry process. Step 2 dealt with the mechanical characterisation of the HMA produced with the plastic treatment selected in the Step 1 and including two different plastic dosages. Two additional mixtures prepared without any plastic and with neat and Styrene-Butadiene-Styrene (SBS, 3.8% by bitumen weight) modified bitumen respectively, were used as reference for comparison. In Step 2, the bitumen content of the mixtures with plastic was properly reduced in order to obtain an air voids content similar to that of the reference HMA. Volumetric and mechanical properties were assessed in terms of air voids, Indirect Tensile Strength (ITS), Cracking Tolerance (CT) index, Indirect Tensile Stiffness Modulus (ITSM) and fatigue resistance. Plastic homogeneity and melting capacity were also investigated through visual analysis of aggregate mixtures before bitumen addition and after bitumen extraction. In order to minimise the variables potentially involved in the performance development, all mixtures were prepared with the same standardised mixing process commonly applied in the laboratory to produce HMA. Aggregates and bitumen were heated in the oven at 170° C for 3 and 1 hours, respectively. Applying the same approach used in asphalt plant with mixture modifiers, plastic wastes were added directly into hot aggregate mix before bitumen inclusion.

2.1 *Materials*

The so-called "light plastics" recovered from municipal solid waste (MSW) were selected after three types of treatment, i.e. shredding, densification and pelletising, as coded in Figure 1. They differ in size and treatment, but are consistent in terms of chemical composition and melting point as the original waste plastic was exactly the same. The shredded plastic was put in water to eliminate the heavy elements potentially incorporated in the material, which also visually showed high heterogeneity. Only the floating fraction (around 90%) after drying was used in the HMA mixtures. The densified plastic underwent a treatment only aimed at unifying the larger pieces of plastics, whereas the pelletised plastic has a granulate texture and derived from the densified ones through extrusion. The waste plastic content ranged between 0.5% and 2% by aggregate weight.

Figure 2 depicts the aggregate distribution of the HMA dense graded mixtures for binder layers produced in this study obtained by combining limestone virgin aggregates and filler, in compliance with the technical standards of the main Italian road agencies. The neat bitumen used in the study was a 70/100 penetration bitumen with a penetration of 73 dmm and a softening point of 47 °C. The same bitumen was used to prepare the first reference mixture (coded as "Neat"). The second reference mixture (coded as "PmB") included a commercial polymer modified bitumen 45/80 characterised by a lower penetration value (53 dmm) and a higher softening point (i.e. 74 °C). The optimum bitumen content was found to be equal to 4.9% by aggregate weight for the reference mixtures without plastic, both in the case of neat and SBS modified bitumen. In the case of mixtures with plastic wastes, in Step 1 the investigated bitumen contents were 4.9% (for all the plastic dosages) and 4.2% (only for the

A (shredded plastic)   **B (densified plastic)**   **C (pelletised plastic)**

Figure 1.   Waste plastics recovered from municipal solid waste (MSW).

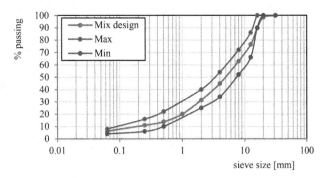

Figure 2.   Aggregate gradation distribution.

highest plastic dosage, i.e. 2.0%) by aggregate weight. In Step 2, the bitumen content was optimised with the aim to guarantee the same volumetric properties for all the tested mixtures and varied from 4.0% to 4.9% depending on the plastic amount. The experimental plan involved the preparation of gyratory compacted specimens with standard compaction parameters (vertical pressure 600 kPa; rotation speed: 30 revolutions/min; vertical mould inclination: 1.25°). According to the Italian specifications for binder layers, 120 gyrations were applied to prepare specimens with 100 mm diameter and about 60 mm height. All the mixtures with plastic wastes (coded as "*plastic treatment_ % plastic _%bitumen*") were produced with a neat bitumen in order to better identify the potential modifying power of the plastic.

2.2 *Test protocols*

The volumetric properties were investigated in terms of air voids content by using the maximum and bulk densities of the mixtures measured according to EN 12697-05 (procedure C). Along with the volumetric assessment, visual inspections of the aggregates with plastic before bitumen addition and after bitumen extraction were carried out to verify the capability of each treated plastic to effectively melt and homogeneously dissolve into the HMA, so demonstrating the plastic coating power of the aggregate and the potential extent of mix modification. A servo-pneumatic testing machine was used to characterise the mixtures in terms of Indirect Tensile Stiffness Modulus (ITSM) at 20 °C, according to EN 12697-26 Annex C. During the test, repeated load pulses with a rise time of 124 ms and a pulse repetition period of 3.0 s were applied. For each specimen, the load was adjusted using a closed-loop control system in order to achieve a target horizontal (diametral) deformation of 5 μm.

The Indirect Tensile Strength (ITS) was measured at the temperature of 25 °C by means of an electromechanical press, imposing a constant rate of deformation of 50 ± 2 mm/min until specimen failure occurred (EN 12697-23). According to ASTM D8225-19, from ITS test data the Cracking Tolerance (CT) Index, representing the ductile features of the HMA mixtures, was calculated as follows:

$$\text{CT-Index} = \frac{t}{62} \cdot \frac{G_f}{\frac{P}{l}} \cdot \left(\frac{l}{D}\right) \tag{1}$$

where t and D are specimen thickness and diameter respectively; the fracture energy G_f is determined as the area under the force-displacement curve divided by the fracture surface ($t \cdot D$); l and P/l are the displacement and the slope of the load-displacement curve when the

Table 1. Summary of the experimental programme.

	Step 1				
Property	Air void	ITS	CT Index	ITSM	Melting capacity
Norm	EN 12697-8	EN 12697-23	ASTM D8225	EN 12697-26	–
Tested mixes	Neat_4.9	PmB_4.9	A_1.0_4.9	B_0.5_4.9	C_0.5_4.9
			A_2.0_4.2	B_1.0_4.9	C_1.0_4.9
				B_2.0_4.9	C_2.0_4.9
				B_2.0_4.2	C_2.0_4.2

	Step 2				
Property	Air void	ITS	CT Index	ITSM	Fatigue
Norm	EN 12697-8	EN 12697-23	ASTM D8225	EN 12697-26	EN 12697-24
Tested mixes	Neat_4.9	PmB_4.9	C_0.5_4.7		C_2.0_4.0

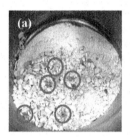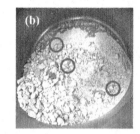

Figure 3. Melting capacity - aggregates after a first mixing @ 170 °C.

load is reduced to 75% of the peak. Generally, the lower the CT Index, the higher the material brittleness.

Cyclic indirect tension fatigue tests (ITFT) were performed for evaluating the fracture resistance due to cycling loading according to EN 12697-24 Annex E. The tests were carried out at a temperature of 20 °C in controlled stress mode, by applying a pulse load with 0.1 s loading time and 0.4 s rest time. Three horizontal stress amplitudes were fixed in order to obtain a fatigue life between one hundred and one million cycles. The fatigue failure of the specimen $N_{f=50}$ was assumed in correspondence to the number of load applications when the stiffness modulus decreased to half its initial value. All the tests provided 4 repetitions under the same conditions. Table 1 summarises the experimental programme.

3 RESULTS

3.1 *Step 1*

3.1.1 *Melting capacity*
One key aspect to judge the feasibility of adding plastic into a HMA is surely the plastic capability to melt and disperse into the mix to properly coat the aggregates and react with the bituminous phase. In this sense, a preliminary mixing attempt was performed using an intermediate plastic dosage (1% by aggregate weight) for all the plastic treatments. At the end of the first mixing (hot aggregates + cold plastic) it was noticed that, when shredded plastic A was used (Figure 3a), some plastic fragments did not properly melt, creating evident lumps. This negative effect significantly reduced with the densified plastic B (Figure 3b)

and almost disappeared with the pelletised plastic C (Figure 3c). Anyway, it should be observed that in no case the plastic completely dissolved: some plastic residues, even if small, could be found among aggregates also when using pelletised plastic. This behaviour was confirmed by the visual inspection carried out on HMA residue after bitumen extraction (Figure 4). The aggregates recovered from HMAs prepared with plastic A showed pieces of plastic of significant dimensions (some cm). With the plastic B, the recovered aggregates were characterised by the presence of clusters composed of aggregates and plastic not completely melted. With plastic C, only residues retained at 1 mm sieve (characterised by darker colour than the limestone sand) were detected and no visible cluster or lumps were identified. Based on this preliminary analysis that showed the difficulty to properly melt the plastic particles with large size, the investigation of mixtures with shredded plastic A was limited to only one plastic dosage (1%), whereas HMAs with plastic B and C were prepared with all the three dosages originally planned.

3.1.2 Volumetric properties

The air voids content (V_m) values are summarised in Figure 5. The results clearly show that the higher the plastic amount, the lower the air voids, due to a sort of lubricating effect of plastic that helps mixture workability. Air voids were significantly lower compared to both traditional and modified HMA. At the same time, no significant differences were detected as function of plastic treatment at equal plastic dosage. Lowering the bitumen percentage of the mixture at equal plastic dosage (i.e. from 4.9% to 4.2% by aggregate weight) allowed raising the air void content. In particular, the plastic C, which was able to melt almost completely in the mixture, contrarily to plastic A and B, showed lower air voids. Therefore, the plastic treatment which melts more (i.e. C) is able to affect more significantly mixture compactability when the amount of bitumen decreases, probably because the melted plastic acts as a replacement of the bitumen.

3.1.3 Indirect Tensile Strength Modulus (ITSM)

Figure 6 shows mixture stiffness as function of plastic treatment and dosage and bitumen dosage. It can be observed that the waste plastic determined a significant stiffening effect compared to the reference mixtures with neat and polymer-modified bitumen. Generally, the higher the plastic amount, the higher the stiffness. Between the different waste plastic treatments, higher ITSM values were obtained for plastic A, whereas plastic C showed the lower values, under the same dosage of waste plastic and bitumen. Considering that the higher the plastic amount, the lower the air voids, the stiffening effect may partially depend on the higher density grade reached with higher plastic amounts. Moreover, the higher ITSM values obtained for the HMA with waste plastic compared to the reference mix PmB can also be due to the different polymer dosage and type (in the PmB there is 3.8% SBS by bitumen weight). It is interesting to notice that even with a lower bitumen amount, HMA

Figure 4.   Melting capacity – aggregates and plastic after bitumen extraction.

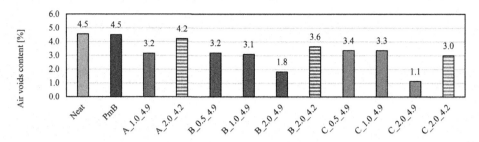

Figure 5.   Phase 1: air voids content.

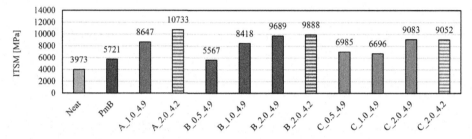

Figure 6.   Phase 1: ITSM values.

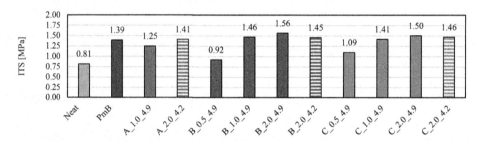

Figure 7.   Phase 1: ITS values.

mixtures were able to guarantee same stiffness performance, demonstrating that the plastic has the potential to compensate for the absence of a certain amount of bitumen as previously observed with the volumetric analysis.

3.1.4   *Indirect Tensile Strength (ITS) and cracking tolerance index*

In accordance with the local technical standards, the performance analysis needed for evaluating the acceptability of a mixture is primarily based on the Indirect Tensile Strength (ITS). The results are summarised in Figure 7.

Compared to the reference mixture with neat bitumen, all materials including plastic showed higher ITS values, which rose with the increase in plastic dosage. In particular, ITS similar to that of the mix with PmB were obtained with plastic dosages of 1% or 2%. This demonstrates the potential of plastic to improve the mechanical resistance of the mixture. It is worth nothing that the higher values detected were still within the limits prescribed by the Italian technical standards for binder layers regardless of the plastic treatment, which did not significantly influence the ITS.

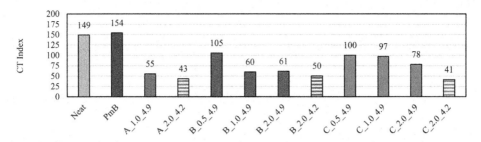

Figure 8. Phase 1: CT Index values.

Analogously to what already observed in terms of ITSM, the reduction of the bitumen content in the mixture did not appear detrimental in terms of mechanical performance, as comparable ITS values were recorded for both 4.2% (striped columns) and 4.9% (full columns) bitumen contents regardless of the plastic treatment. ITS test data were also analysed in terms of CT index, which represent the ductile properties of the HMA specimens after failure (Figure 8). The analysis of the CT indexes showed higher brittleness (lower CT) for all the mixtures incorporating plastic, compared to the reference mixtures with neat or modified bitumen. Particularly, HMAs prepared with shredded plastic A showed the highest brittleness tendency. Consistently with the previous results, the higher the plastic dosage, the higher the material brittleness, although the brittleness rate tended to decrease with the increase in plastic amount and no significant difference was detected between 1 and 2% of plastic. Moreover, it can be noted that the brittleness increased with the decrease in bitumen content. This behaviour is probably a consequence of the higher impact of plastic in a lower amount of bitumen (plastic/bitumen ratio).

3.2 Step 2

3.2.1 Air voids and mechanical properties

In the second part of the experimental programme, as the air voids content in the HMA showed to decrease when increasing the waste plastic content (Figure 5), new mixtures including the plastic C (0.5% and 2%) were produced with reduced bitumen contents (respectively 4.7% and 4.0% by mix weight) in order to obtain air voids comparable with the reference mixtures.

Figure 9 shows the voids content and the mechanical properties (ITSM, ITS and CT Index) of the mixtures investigated in Step 2. From the graphs, similar air voids percentages, ranging between 4.5 and 5%, were observed. The same trend of stiffness already detected in Step 1, with an increase in the ITSM values when the plastic dosage increased, was noted. Moreover, the results show that the increase in ITS values already detected in Step 1 cannot be consequence only of the lower air void contents. In fact, the increase in the plastic content determined an increase in the mixture strength even with the same amount of air void percentage. At the same time, the CT Index decreased significantly when the plastic amount increased, confirming a brittle tendency due to the presence of plastic.

3.2.2 Cyclic Indirect Tension Fatigue Tests (ITFT)

The resistance to repeated loading was evaluated through the cyclic indirect tension tests. The results are plotted in Figure 10, where the number of cycles to failure is reported as a function of the initial maximum horizontal deformation ($\varepsilon_{init-horiz-max}$). It is worth noting that the regression lines drawn in Figure 10 are only indicators of the fatigue life, which is also correlated to other variables not completely taken into account in this kind of test. Particularly, the ITFT defines the crack initiation characteristics of the studied mixtures

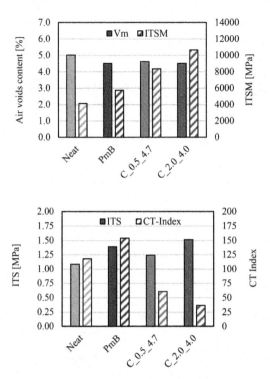

Figure 9.   Phase 2: Air voids content, ITSM, ITS and CT Index values.

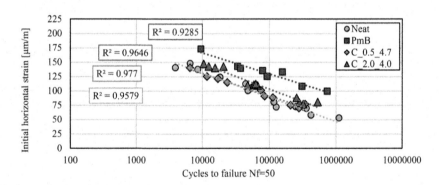

Figure 10.   Phase 2: Fatigue behaviour.

(resistance to early cracking). In this sense, the graph of Figure 10 should be analysed in correlation with the CT Index results.

   Although the CT Index indicated that the presence of plastic into the mixture determines a more brittle behaviour, ITFT results suggest that the presence of plastic in a certain amount (i.e. 2%) is able to delay the crack initiation compared to a traditional HMA leading to a fatigue resistance more similar to a polymer modified HMA, whereas the low plastic dosage (i.e. 0,5%) cannot provide any significant contribution. Anyway, even with the highest dosage of plastic (i.e. 2%) the performance ensured by a SBS-modified bitumen is better and associated with higher ductility and elasticity. Therefore, HMAs with a certain amount of plastic demonstrated the potential to postpone the crack formation, also due to the higher stiffness provided by the plastic, but once the crack was formed, it was observed that this fracture was associated with a sample collapse, clear symptom of the brittle behaviour of the material.

4 CONCLUSIONS

The present research aimed at exploring the possibility of recycling plastic from MSW in HMA without separating the several materials included in this waste. Quick and rather economic processing, i.e. shredding, densification and pelletising, were applied on the heterogeneous plastic wastes, which were included in laboratory-produced HMA in different dosages up to 2% by aggregate weight. The experimental campaign allowed observing that all the plastic types partially melt during HMA mixing, but pelletised plastic (type C) allows avoiding the formation of large aggregate-plastic clusters that can affect the mix mechanical behaviour. The presence of waste plastic determines an increase in stiffness and strength of the HMA (the higher the amount, the higher the effect), which positively influenced the resistance to crack formation in cyclic fatigue tests. However, the waste plastics also entailed a reduction of the ductile features in the post-failure behaviour. Finally, test results showed that waste plastics can act as a bitumen extender, since they allow reducing the bitumen dosage without penalising the volumetric and mechanical properties.

At the light of the promising findings, future investigations will focus on low temperature behaviour, moisture sensitivity, ageing, rheology, and surface characteristics of the HMA including waste plastics, with particular reference to the potential of plastics to improve adhesion and grip performance. Moreover, trial sections will be built in order to validate laboratory results and assess plant production, field constructability and workability.

REFERENCES

Angelone S., Cauhapé Casauz M., BorghiM., and Martinez F.O., 2016. Green Pavements: Reuse of Plastic Waste in Asphalt Mixtures. *Materials and Structures* 49, 1655–1665.

Badejo A.A., Adekunle A.A., Adekoya O.O., Ndambuki J.M., Kupolati K.W., Bada B.S., and Omole D.O., 2017. Plastic Waste as Strength Modifiers in Asphalt for a Sustainable Environment, *African Journal of Science, Technology, Innovation and Development* 9 (2), 173–177.

Dalhat M.A., Wahhab A.H.I., and Al-Adham K., 2019. Recycled Plastic Waste Asphalt Concrete via Mineral Aggregate Substitution and Binder Modification. *Journal of Materials in Civil Engineering* 31 (8), 04019134.

European Parliament and Council of the European Union, 2018. *Waste Framework Directive*. Directive 2018/851, May 30th 2018.

Giustozzi F., Enfrin M., Xuan D.L., Boom Y.J., Masood H., Audy R., and Swaney M., 2022. Use of Road-grade Recycled Plastics for Sustainable Asphalt Pavements: Final Performance and Environmental Assessment Part A. *Austroads*, Report AP-R669-22

Istituto Superiore per la Protezione e la Ricerca Ambientale (ISPRA), 2021. Rapporto Rifiuti Urbani (in Italian). *Sistema Nazionale per la Protezione dell'Ambiente*. Report 355/2021.

Lastra-González P., Calzada-Pérez M.A., Castro-Fresno A., Vega-Zamanillo A., and Indacoechea-Vega I., 2016. Comparative Analysis of the Performance of Asphalt Concretes Modified by Dry Way with Polymeric Waste. *Construction and Building Materials* 112, 1133–1140.

Ma Y., Zhou H., Jiang X., Polaczyk P., Xiao R., Zhang M., and Huang B., 2021. The Utilization of Waste Plastics in Asphalt Pavements: A Review. *Cleaner Materials* 2, 100031.

Moghaddam T.B., Soltani M., and Karim M.R., 2014. Experimental Characterization of Rutting Performance of Polyethylene Terephthalate Modified Asphalt Mixtures Under Static and Dynamic Loads. *Construction and Building Materials* 65, 487–494.

Movilla-Quesada D., Raposeiras A.C., Silva-Klein L.T., Lastra-González P., and Castro-Fresno D., 2019. Use of Plastic Scrap in Asphalt Mixtures Added by Dry Method as a Partial Substitute for Bitumen. *Waste Management* 87, 751–760.

Nizamuddin S., Boom Y. J., and Giustozzi F., 2021. Sustainable Polymers From Recycled Waste Plastics and Their Virgin Counterparts as Bitumen Modifiers: A Comprehensive Review. *Polymers*, 13(19), 3242.

Noor A., and Rehman M.A.U., 2022. A Mini-review on the Use of Plastic Waste as a Modifier of the Bituminous Mix for Flexible Pavement. Cleaner Materials 4, 100059.

Radeef H.R., Hassan N.A., Katma H.Y., Mahmud M.Z.H., Abidin A.R.Z., and Ismail C.R, 2022. The Mechanical Response of Dry-process Polymer Wastes Modified Asphalt Under Ageing and Moisture Damage. *Case Studies in Construction Materials* 16, e00913.

Vasudevan, R., Sekar, A.R.C., Sundarakannan, B., and Velkennedy, R., 2012. A Technique to Dispose Waste Plastics in an Ecofriendly Way – Application in Construction of Flexible Pavements. *Construction and Building Materials* 28 (1), 311–320.

Willis R., Yin F., and Moraes R., 2020. Recycled Plastics in Asphalt Part A: State of the Knowledge. *National Asphalt Pavement Association*, Report IS-142.

Wu S., and Montalvo L., 2021. Repurposing Waste Plastics into Cleaner Asphalt Pavement Materials: A Critical Literature Review. *Journal of Cleaner Production* 280, 124355.

Roads and Airports Pavement Surface Characteristics – Crispino & Toraldo (Eds)
© 2023 The Author(s), ISBN 978-1-032-55149-4

The use of rejuvenating bio-emulsion from soil bean oil in recycled aggregates for chip seals

Ataslina Silva* & Mateus Brito
Department of Transportation Engineering, Federal University of Ceara, Fortaleza, Brazil

Caio Falcão, Suelly Barroso & Ronald Williams
Institute of Transportation; Iowa State University, Ames, USA

Antônia Uchôa & Sandra Soares
Department of Transportation Engineering, Federal University of Ceara, Fortaleza, Brazil

ABSTRACT: The asphalt mixture milling process is one of the main factors of environmental issues in the industry of engineering, which generates significant amounts of Recycle Pavement Asphalt (RAP). Several engineering solutions were explored in the scope of the use of RAP aggregates in new mixtures. On the other hand, soybean oil partially epoxidized is a material of renewable origin that has shown potential for application in paving. Thus, the objective of this research was to investigate the use of these materials in the construction of Chip Seal (CS) using a bio-emulsion from soybean oil in the rejuvenation of aged binder. For this, granite is replaced with aggregate recycled from RAP, attempting to the arbitrary rates of 0%, 33%, 66%, and 100%. The rejuvenating bio-emulsion was applied to the alternative material at a rate of 0.1 L/m^2. The performance of the CS was evaluated by the Wet Track Abrasion Test (WTAT), which allows an adaptation for ravelling assessment. Results show that the rates of 33% and 66% underperformed during the WTAT procedure. However, there was a large improvement in performance for both rates after the addition of bio-emulsion, including the samples with the total substitution for recycled aggregate from RAP. It is concluded that the use of soybean oil bio-emulsion has viability in CS construction using RAP, where the improvements are understood for some activation of the aged binder.

Keywords: Sub-Epoxidized Soybean Oil, Bio-additive, Surface Treatment, RAP, Sustainability, Chip Seal

1 INTRODUCTION

In the past years, there has been a growth in sustainable concerns about the need for roadways restoration. This is on the application of the milling process at the end of the service lives of asphalt pavements, which generates a great amount of residue (Arabzadeh et al., 2021). The Recycled Asphalt Pavement (RAP) is an important step toward sustainability in roadway rehabilitation and pavement construction (Podolsky et al. 2020), which consists of aggregates covered by aged-binder.

The use of RAP in industry of engineering involves the establishment of innovative solutions. This is due to the complex mechanism observed in the mobilization of aged-binder

---

*Corresponding Author; ataslina@det.ufc.br

DOI: 10.1201/9781003429258-74

in the construction of the new pavement. The use of high temperatures in the production of recycled asphalt mixtures generates a very stiffer material and can induce failures such as stripping and fatigue cracking (Al-Qadi et al. 2009). In addition, the aged material is more brittle, and its permanence on aggregates surfaces from RAP can yield issues like dis-aggregation (Loise *et al.* 2021; Mesquita Júnior et al. 2018) indicated that the application of softening agents and rejuvenators are the most effective ways to recover the original properties of the aged binder.

The production of rejuvenators from biomaterials is an environmentally friendly alternative to industry and has been investigated in research with epoxied oils from cashew nut (Cavalli et al. 2018), cottonseed (Uchoa et al. 2021), palm (Nogueira et al. 2019) and soybean (Chen et al. 2018; Li et al. 2021; Podolsky et al. 2021). The Soy Bean Oil (SEO) has demonstrated good properties for bitumen modification. Danov et al. (2017) showed several benefits of the partial epoxidation of vegetable oils, where the Sub-Epoxidized Soybean Oil (SESO) was successfully applied by Podolsky et al. (2020) to a rejuvenation of RAP.

The environmental issues of roadway engineering have been a concern in developing countries, as in the case of Brazil. The National Department of Transportation (DNIT) just released a resolution for using the RAP in all constructions or rehabilitation, in which the milling process is included (DNIT 2021). The application of RAP in cold techniques of pavement requires the use of recycling agents (Zhang et al. 2020). Mesquita Júnior et al. (2018) verified the performance viability of partial replacements in Brazilian Chip Seals (CS), where some of these proportions resulted in divergences in desegregation indexes.

These types of thin layers are largely used in Brazilian low-traffic roadways, and the use of recycling additives in aged-binder from RAP has not been investigated yet for this scenario. The objective of this research is to investigate the use of SESO on replacements on CS aggregates for RAP observing the performance parameter of desegregation.

2 MATERIALS AND METHODS

It was made reference samples of Chip Seal (CS) using a granitic aggregate and an Asphalt Emulsion (AE). Both materials were supplied by local industry, based in Maracanau, county of Ceara, Brazil. The type of AE selected is the most common asphalt used in CS pavements (Silva 2018), consisting of cationic emulsion with a rapid setting (known as RR-2C). In the same way, Granitic Aggregate (GA) has a large use in this type of coating on local roads.

The aggregates from RAP were selected from the fractions of the sieves no. 9.5 mm and 12.5 mm. The bio additive named Sub-Epoxidized Soybean Oil (SESO) was supplied by Iowa State University labs (Podolsky et al. 2021), located in Ames, city of Iowa, United States of America. The product was obtained by a partial epoxidation of the soybean oil and was emulsified in water in the proportion of 1/3. The basic features observed were a viscosity of 408 cP at 25°C, a density of 0.95 g/cm^3, no toxicity, and white color (Podolsky et al. 2020).

2.1 *Performance-Based Uniformity Coefficient (PUC)*

The properties of aggregate material, like porosity, skid resistance, mineralogy, roughness, granulometry, shape, and angularity, can induce significant influence on the life cycle of CS (Aktas et al. 2013; Mesquita Júnior et al. 2018; Wei et al. 2020). The size and granulometric distribution of the particles had special attention by researchers with the performance of CS causing the incidence of the failures like raveling and stripping. Lee e Kim (2009) introduced the Performance-Based Uniformity Coefficient (PUC), which is a method from failure criteria established by McLeod (1969), and apply a concept from soil mechanic that consists of dividing the passing percentages of specific sieves for achieving uniformity by granulometric adjustments (Silva et al 2017). Equation 1 describes the PUC calculation:

$$PUC = \frac{PEM}{P2EM} \qquad (1)$$

Where:

PUC: Dimensionless parameter, which has to be smaller as possible.

M (%): It is a median, corresponding to 50% of passing percentages. It is an intrinsic parameter in the formula but is used for calculating the other terms.

PEM (%): The PEM is known as the percentage of the particles where the stripping failure can occur. This is relative to smaller sizes of aggregate, specifically at $0,7 \times M$ and which bitumen can emerge.

P2EM (%): The P2EM is referred to the bigger sizes of particles and is situated at $1,4 \times M$. The determination of its sieve mesh depends on the localization of the median mesh, and the probability of detachment of particles is referred to in 100- P2EM.

The parameter of PUC is not included in local specifications, but it was considered in this study for granitic and RAP aggregates. Between the different types of specifications in Brazil that vary from the department of transportation and regional practices, DNIT has standards for CS design. Silva et al. (2017) found so many high values from DNIT granulometric curves in the PUC parameter, so, this is why this study will not consider this standard specification on CS design.

2.2 Chip seal design procedure

The design procedure chosen for this study is by the local specification for chip seal (SOP 2020), which is applied to tropical road conditions. It was chosen the traffic Lane B ($2,5 \times 105$ to $7,5 \times 105$ repetitions of the 8.2-ton axle). This specification lane comprehends particles sized from 10 to 16 mm in diameter, then, the PUC parameter was applied over these limits to meet the uniformity of the lane about interval applied for materials, including the RAP.

Control samples were dosed with 100% of granitic aggregates and with replacements for RAP at the proportions of 33%, 66%, and 100%. In according with SOP (2020), the tray method was performed respectively for every proportion of replacement. Figures 1 (a) and

Figure 1.   Rate of aggregates for CS design: (a) Tray method. (b) Unitary specific bulk.

(b) illustrate this method, which consists of the determination of the rate of aggregates by the size of particles and unitary specific bulk.

The aggregates are positioned manually on the tray to fill all the empty spaces without overlapping. The apparatus is shown in Figure 1 (b) and is applied for a volumetric rate conversion. After determining the aggregate rates, the asphalt emulsion rates were performed for the different CS compositions.

The SESO bio-emulsion application was made only over the RAP aggregates to isolate variables. It remained separated for an approximate period of 72 hrs. Thus, samples were prepared with and without the use of bio-emulsion, considering that SESO did not interact with the asphalt emulsion, and this material was normally applied in the dosage for all samples.

2.3 *Aggregate loss procedure*

The performance tests are specific procedures for simulating the occurrence of failure on thin layers of coatings. It aims to test the effectiveness of design on these pavements' life cycles. In chip seals, some performance tests were applied as a tool to evaluate the use of alternative materials (Loiola 2009; Pereira 2013). The main types of distresses known for CS are bleeding and aggregate loss.

Several studies had spent special consideration on Aggregate Loss (AL) due to its influence on the different periods of the CS service life (Moraes e Bahia 2012). More specifically, the features of construction, traffic opening conditions, properties of materials, and excesses in the rate of aggregates are some of the reasons for AL occurrence (Adams et al. 2013; Aktas et al. 2013; Kim *et al.* 2015; Loiola 2009; Silva 2018).

In Brazil, the ABNT 14746 (ABNT NBR 14746, 2014) describes performance procedures for slurry seal and micro surfacing designs. This standard includes the Wet Track Abrasion Test (WTAT) equipment (Figure 2 (a)), and is similar to the standard ASTM D7000-11

Figure 2.   Wet abrasion adapted procedure: (a) WTAT equipment. (b) CS samples.

(ASTM 2011), which employs a sweeper for an evaluation of adhesion phenomena among CS materials. Loiola (2009) made comparisons between field cores and samples from the laboratory for disposing of an adaptation of WTAT methodology to evaluate the incidence of AL in CS pavements. Thus, this adaption of AL performance evaluation was used in this research to evaluate bio-emulsion performance. Figure 2 shows the WTAT device and the samples of CS, which consist of round plates of 25 cm, cured at controlled temperature and made in triplicate.

The aggregates were properly washed and dried before carrying out the tests. Pereira (2013) developed a range of classifications of performance in double CS based on the experience of the Brazilian state of Ceará. This classification will be applied to this research and is represented in Table 1.

Table 1. Classification of Desegregation for SCS. Pereira (2013).

Desegregation (D)	Classification
D < 10%	Very good
10% ≤ D < 20%	Good
20% ≤ D < 30%	Fair
D ≥ 30%	Poor

3 RESULTS

The Performance Uniformity Coefficient (PUC) was calculated for the different mixtures of conventional aggregate and RAP. Table 2 gathers these results from the granulometric curves of the mixtures.

Table 2. PUC calculation.

Granulometric parameters	SCS Mixtures			
	100% AG	67% AG + 33% RAP	33% AG + 67% RAP	100% RAP
M (mm)	14.0	12.0	13.0	12.5
0,7×M (mm)	9.8	8.4	9.1	8.75
1,4×M (mm)	19.6	16.8	18.2	17.5
PEM (%)	3.0	0.0	0.0	0.0
P2EM (%)	99.0	99.0	99.0	100.0
100-P2EM (%)	1.0	1.0	1.0	0.0
PUC	0.03	0,00	0.00	0.00

It can be seen that the Lane B of local specification (SOP 2020) predicts aggregate sizes with good uniformity. The PUC of granitic aggregate resulted in a value of 0.03, indicated by Zaman (2013) with low susceptibility to failure. The mixtures including aggregate from RAP are quite more uniform and showed PUC values near zero. From the point of view of granulometric properties, it is expected a small incidence of Aggregate Loss (AG), in special for those mixtures containing RAP.

The different rates of materials are summarized in Table 3.

Table 3. Materials rates of chip seal design.

| CS Mixtures | Rate of aggregate | | Rate of emulsion |
	(L/m^2)	(Kg/m^2)	RR-2C (L/m^2)
100% GA	9,16	12.78	1,07
33% GA + 67% RAP	9,05	11.40	1,06
67% GA + 33% RAP	8,87	12.22	1,04
100% RAP	8,36	10.40	0,98

As can be seen, there was a decrease in the consumption of materials due to the replacements for RAP. In general, the properties of recycled aggregates, such as angularity, shape, and density may have imposed a greater influence. The costs involved in 1 Km of road construction can be simulated for a platform of 7 m, which the 100% RAP CS can save approximately 1,500$/Km (19.3%) of the costs when compared to 100% GA mixture.

All samples of the single-chip seal were submitted to the WTAT procedure. Figure 3 represents the average of AG for each replacement of RAP and the use of bio-emulsion.

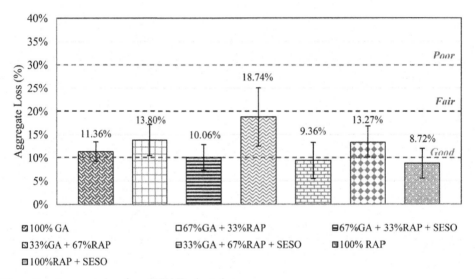

Figure 3. Aggregate loss from WTAT adapted.

The results of PUC showed that particle sizes were classified as a good performance, thus, the overall desegregation is attributed to the materials' properties. Figure 3 shows variable percentages of AG according to replacements for RAP. The use of SESO was capable to reduce these percentages. As can be seen, in 67%GA + 33%RAP, the reduction was nearly 3.74%, for 33%GA + 67%RAP, the reduction was 9.38%, and for 100% RAP was 4.55%. Figure 4 illustrates these differences, which can be seen in a better performance for 100% RAP in both types of mixtures.

The bigger reductions of AL occurred in these conditions, specifically at the greater replacements for RAP. The main hypothesis for this behavior is about possible chemical modifications of residual aged emulsion, which occurred by a bond with the RR-2C applied to the design. In addition, the use of SESO gives a classification of "good" for all samples.

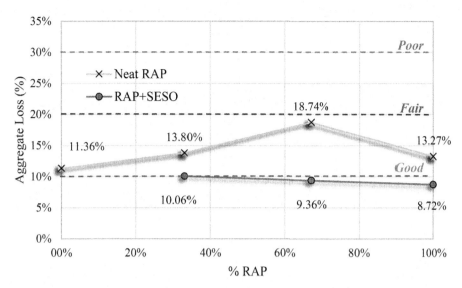

Figure 4.   Percentual of aggregate loss.

4   CONCLUSIONS

Environmental concerns have emerged in recent years with the scarcity of materials and intensification of degrading human activities. Thus, the development of renewable solutions is important to improve engineering capabilities and optimize the sources. As can be seen, the SESO was tested as a bio rejuvenator for aged-binder from RAP.

The use of this material on green chip seal construction was investigated in this research. The main results showed that SESO can reduce the incidence of aggregate loss, which is one of the main distresses identified in the field. The performance evaluation carried out in this study approached the desegregation from the point of view of uniformity, which was expressed by low values of PUC.

The reduction of desegregation can indicate that SESO induced an interaction with aged-binder remained on aggregates from RAP surfaces. This interaction has to be investigated in further future work. In this way, is expected that biomaterials technologies can be progressively introduced in paving services involving RAP and chip seals.

ACKNOWLEDGMENTS

This study was financially by Coordination for the Improvement of Higher Education Personnel (*CAPES*), the National Council of Technological and Scientific Development (CNPq), and the Ceara Foundation for Support to Scientific and Technological. The authors also acknowledge the Iowa State University (ISU) for cooperation.

REFERENCES

ABNT NBR 14746. *Micro Surfacing and Slurry Seal – Determination of Aggregate Loss by Wet Abrasion (WTAT)*. Sao Paulo. [in Portuguese]. 2014.
ASTM D7000-11, Standard Test Method for Sweep Test of Bituminous Emulsion Surface Treatment Samples, *American Society for Testing and Materials International*, West Conshohocken, PA, 2011.

Adams J. and Kim Y.R. Mean Profile Depth Analysis of Field and Laboratory Traffic-loaded Chip Seal Surface Treatments. *International Journal of Pavement Engineering*, 15, p. 645–656. 2013.

Aktas B., Karasahin M., Saltan M., Gürer and Uz, C.V.E. (2013) Effect of Aggregate Surface Properties on Chip Seal Retention Performance. *Constr. Build. Mater.*

Al-Qadi I.L., Carpenter S.H. Roberts G., Ozer H., Aurangzeb Q., Elseifi M. and Trepanier J.. (2009). *Determination of Usable Residual Asphalt Binder in RAP*. Rantoul, IL: *Illinois Center for Transportation.*

Arabzadeh A.A., Staver M.D., Podolsky J.H., Williams R.C., Hohmann A.D. and Cochran E.W. 2021. At the Frontline for Mitigating the Undesired Effects of Recycled Asphalt: An Alternative Bio Oil-Based Modification Approach. *Constr. Build. Mater*. 310.

Cavalli M.C., Zaumanis M., Mazza E., Partl M.N. and Poulikakos L.D. "Effect of Ageing on the Mechanical and Chemical Properties of Binder from RAP Treated with Bio-based Rejuvenators," *Compos. Part B Eng.*, vol. 141, no. May 2017, pp. 174–181, 2018, 10.1016/j.compositesb.2017.12.060.

Chen C, Podolsky J.H., Williams R.C. and Cochran E.W. Laboratory Investigation of Using Acrylated Epoxidized Soybean Oil (AESO) for Asphalt Modification, *Constr. Build. Mate*r. 187 (2018) 267–279.

Danov S.M., Kazantsev O.A., Esipovich A.L., Belousov A.S., Rogozhin A.E. and Kanakov E.A. (2017). Recent Advances in the Field of Selective Epoxidation of Vegetable Oils and Their Derivatives: A Review and Perspective. *Catalysis Science & Technology*, 7(17), 3659–3675.

DNIT 2021. Resolução n° 14/2021. Reaproveitamento de RAP em Obras de Restauração, Adequação de Capacidade e Ampliação.

Kim Y.R. and e Im J.H. *Extending the Use of Chip Seals to High Volume Roads by Using Polymer-Modified Emulsions and Optimized Construction Procedures.* (2015). NCDOT Project. 2011-03.

Lee J.S. and Kim Y.R. Performance-Based Uniformity Coefficient of Chip Seals. (2009). *Transportation Research Record*, v. 2180, p. 53–60.

Li Q., Zhang H. and Chen Z. Improvement of Short-term Aging Resistance of Styrenebutadiene Rubber Modified Asphalt by Sasobit and Epoxidized Soybean Oil, *Constr. Build. Mater.* 271 (2021), 121870, https://doi.org/10.1016/j. conbuildmat.2020.121870.

Loiola P.R.R. *Estudo de Agregados e Ligantes Alternativos para o Emprego em Tratamentos Superficiais em Rodovias (Masther's thesis).* Federal University of Ceara. Fortaleza. [in Portuguese]. 2009.

Louise V., Calandra P., Abe A.A., Porto M. and Rossi C.O. Additives on Aged Bitumens: What Probe to Distinguish Between Rejuvenating and Fluxing Effects? *Journal of Molecular Liquids.* 2021.

McLeod N.W.A. General Method of Design for Seal Coats and Surface Treatments. *Proc., Association of Asphalt Paving Technologists*, Vol. 38, pp. 537–630. 1969.

Mesquita Júnior G., Silva R.C., Barroso S.H.A. and Kim Y.R. Evaluation of the Integration of Alternative Materials in Laboratorial Tests of Chip Seals. *International Society for Asphalt Pavements.* Fortaleza. 2018.

Moraes R. and Bahia H. Effects of Curing and Oxidative Aging on Raveling in Emulsion Chip Seals. Transportation Research Record: *Journal of the Transportation Research Board*, v. 2361, n. 2361, p. 69–79, 2013.

Nogueira R.L.; Soares J.B. and Soares S.A. (2019) Rheological Evaluation of Cotton Seed Oil Fatty Amides as a Rejuvenating Agent for RAP Oxidized Asphalts. *Constr. Build. Mater.* 223:1145–1153. https://doi.org/ 10.1016/j.conbuildmat.2019.06.128.

Silva R.C. *Avaliação da Dosagem dos Tratamentos Superficiais por Penetração de Rodovias Baseada na Exsudação e na Perda de Agregados (Masther's thesis).* Federal University of Ceara. Fortaleza. [in Portuguese]. 2018.

Silva R.C., Barroso S.H.A.A. and Kim Y.K. Introdução do Coeficiente de Uniformidade Para Avaliação de Revestimentos Asfálticos do Tipo Tratamentos Superficiais. *Revista Transportes.* 2017.

Superintendência de Obras Públicas. Governo do Estado do Ceará. *Superintendência de Obras Públicas. Especificações Gerais para Serviços e Obras Rodoviárias.* Volume 1. Ceara State Government. Fortaleza. [in Portuguese]. 2020.

Pereira S.L.O. (2013) *Avaliação dos Tratamentos Superficiais Simples, Duplo e Triplo de Rodovias Através do Emprego de Diferentes Agregados da Região Metropolitana de Fortaleza (Masther's thesis).* Federal University of Ceara. Fortaleza. [in Portuguese]. 2013.

Podolsky J.H., Sotoodeh-Nia Z., Manke N., Hohmann A., Huisman T., Williams R.C. and Cochran E.W., 2020. Development of High RAP–High Performance Thin-Lift Overlay Mix Design Using a Soybean Oil-Derived Rejuvenator. *Journal of Materials in Civil Engineering*, ASCE, ISSN 0899-1561.

Podolsky J.H., Sotoodeh-Nia Z., Manke N., Hohmann A., Huisman T., Williams R.C. and Cochran E.W. Development of High RAP–High Performance Thin-Lift Overlay Mix Design Using a Soybean Oil-Derived Rejuvenator. (2021). *Journal of Materials in Civil Engineering.*

Uchoa A.F.J.; Da Silva W.R.; Feitosa J.P.M; Lopes R.N.; Brito D.H.A.; Soares J.B. and Soares S.A. (2021) Bio-based Palm Oil as an Additive for Asphalt Binder: Chemical Characterization and Rheological Properties. *Constr. Build. Mater.* 285:122883. https://doi.org/10.1016/j.conbuildmat.2021.122883.

Wei M.; Wu S.; Cui P.; Yang T. and Lv Y. (2020) Thermal Exchange and Skid Resistance of Chip Seal with Various Aggregate Types and Morphologies. *Applied Science.* Doi:10.3390/app10228192.

Zaman M., Gransberg D., Bulut R., Commuri S. and Pittenger D. (2013). Develop Draft Chip Seal Cover Aggregate Specification Based on Aims Angularity, Shape and Texture Test Results. Transportation Research Record: *Journal of the Transportation Research Board.*

Zhang J., Xiaomeng Z.X., Liang M., Jiang H., Wei J. and Yao Z. Influence of Different Rejuvenating Agents on Rheological Behavior and Dynamic Response of Recycled Asphalt Mixtures Incorporating 60% RAP dosage. (2020). *Construction and Building Materials.*

Roads and Airports Pavement Surface Characteristics – Crispino & Toraldo (Eds)
© 2023 The Author(s), ISBN 978-1-032-55149-4

Cellulose fibres for better performing road pavements

Patrizia Bellucci*
ANAS S.p.A – Direzione Ingegneria e Verifiche, Roma, Italy

Pierluigi Bernardinetti
ANAS S.p.A – Centro Sperimentale Stradale, Cesano di Roma, Italy

Giuseppe Chidichimo
Consorzio TEBAID, Rende (CS), Italy

Giuseppe Meli
ANAS S.p.A – Centro Sperimentale Stradale, Cesano di Roma, Italy

Filippo Praticò
*Department of Information Engineering, Infrastructure and Sustainable Energy,
Università Mediterranea di Reggio Calabria, Reggio Calabria, Italy*

Cristiano Sartori
ANAS S.p.A – Centro Sperimentale Stradale, Cesano di Roma, Italy

ABSTRACT: Sustainable road pavements are vital for the present and future transportation industry, especially when it comes to low noise pavements, whose mechanical and acoustic properties tend to degrade rapidly with age. Therefore, there is an increasing need to improve their performance and sustainability, especially following the introduction of the Minimum Environmental Criteria (CAM) for roads, which should be met when implementing noise mitigation measures. In order to improve the durability of low noise pavements, while keeping their acoustic properties unchanged, cellulose fibres from brooms were used to set up processes and products able to extend the lifetime of asphalt concretes.

The study includes experiments on the preparation and integration of functionalized cellulose fibres into the asphalt (engineering approach), as well as tests on their ability to positively affect the performance of bitumen, mastic and bituminous mixtures.

The results achieved so far demonstrate that cellulose fibres must be opportunely treated to achieve the expected improvement of the mixture and that only an accurate design of the mix can bring products with satisfactory mechanical and functional properties. The latter include the acoustic features (low rolling noise) and gas emissions (polycyclic aromatic compounds and volatile organic compounds) when laying down the road surface.

Such results are expected to benefit both researchers and practitioners.

Keywords: Low Noise Pavements, Durability, Acoustic Properties

1   INTRODUCTION

Sustainable road pavements are vital for the present and future transportation industry, especially when it comes to low noise pavements, whose mechanical and acoustic properties

---

*Corresponding Author: p.bellucci@stradeanas.it

DOI: 10.1201/9781003429258-75

tend to degrade rapidly with age. As a matter of fact, there is an increasing need to reduce noise levels in outdoor urban areas since according to the World Health Organization, 20% of the European population is exposed to noise levels exceeding 65 dB(A) during the day, whereas the maximum recommended level is 55 dB(A). Mitigating noise in such environments generally excludes the use of solutions that might interfere with the urban context, such as noise barriers, for many reasons. First of all, the proximity of receivers to the noise source. Secondly, the visual impact: noise barriers reduce the visibility of the surroundings and air circulation, causing local temperature rise (especially in summer) and social denial. Therefore, noise mitigation measures acting directly on the source, such as low noise pavements, are recommended and specifically mentioned in the Environmental Minimum Criteria (CAM) for roads. However, their durability is relatively poor, making them more expensive than ordinary road pavements.

The attempt to increase the durability of low noise pavements, in particular porous surfaces, has been the subject of numerous projects with more or less promising results.

Currently, low noise pavements can reduce noise by 3-5 dB (approximately corresponding to about LCPX = 90 dB(A) at 50 km/h), depending on the type of material used and texture, even if higher reductions were obtained in some European projects at prototypal level (cf. PERSUADE and LIFE NEREIDE projects). More importantly, in the attempt to improve noise-related performance, disappointing results were often obtained in terms of durability. In this context, the balance between durability and quietness has become crucial. This is why in the last years several projects have been proposed to identify a trade-off between technical, environmental and economic issues (e.g., LIFE NEREIDE, LIFE C-LOW-N, LIFE E-VIA, LIFE SNEAK-in progress).

Some of these projects have demonstrated the feasibility of peculiar solutions with good results in terms of noise mitigation, but with questionable sustainability and durability outcomes.

Here the target is not only to improve the lifetime of the solution but also its sustainability, by lowering costs and the impact on human health during the construction phase, keeping acoustic and safety performance unchanged. This should lead to noise attenuation comparable or better with respect to the best current commercial products, extended lifetime of the road pavement and lower emissions of CO_2 and costs.

2 THE IASNAF PROJECT

IASNAF, the acronym of Innovative Asphalts with Natural Fibres, is a project developed as part of a partnership between ANAS and the inter-university consortia TEBAID and NITEL.

The objective of this project is to design and test new formulations of bituminous mixtures and chemical technologies with natural cellulose fibres, to produce composite materials with high technological performance to improve road pavement durability.

The use of fibres of different types and materials in road pavements is well known, but only carbon fibres were considered capable of conferring high-quality mechanical properties to pavement compounds so far. However, the use of carbon fibres, given their high cost, is obviously prohibitive in sectors such as road pavements, where very high quantities of material are required. On the contrary, the use of low-cost cellulose and lignin-cellulose fibres, directly extractable from fast-growing plants (broom, hemp), or from waste materials (recycled paper and cardboard, agro-industrial waste: straw, rice husks), when properly treated on the surface, provides high-quality composite materials, as well as carbon fibres.

The addition of cellulose fibres in the mixtures allows prevention of fractures, increases the distribution of loads and provides shear resistance against rutting.

In the IASNAF project the challenge is twofold:

- To achieve a new type of "asphalt", introducing functionalized cellulose fibres into the mixture, more performing in terms of noise emissions and durability;
- To optimize and specify the relative production/construction processes, in particular those related to the "modification" of the binder/conglomerate and laying of the bituminous mixture.

The project involves experimenting and testing different combinations of materials and wear layers, as shown in Table 1.

Table 1. Main materials and components.

Bituminous binder	Fiber type	Optimized bituminous binder	Bituminous mixture
B50/70	–	**B**	1. Dense-graded friction course, DGFC (UCB)
B50/70	F[1]	**BF**	2. Porous asphalt, PA (DBBF)
B50/70	F	**BF**	3. DGFC (UCBF)
B50/70	FF[2]	**BFF**	4. PA (DBFF)
B50/70	FF	**BFF**	5. DGFC (UCBFF)
HD[3]	–	**HD**	6. PA (DHD)
HD	–	**HD**	7 DGFC (UCHD)
HD	F	**HDF**	8. PA (DHDF)
HD	F	**HDF**	9. DGFC (UCHDF)
HD	FF	**HDF**	10. PA (DHDFF)
HD	FF	**HDF**	11 DGFC (UCHDFF)
EB[4]	–	**EB**	12. PA (DEB)
EB	–	**EB**	13. DGFC (UCEB)
EB	FF	**EBFF**	14. PA (DEBFF)
EB	FF	**EBFF**	15. DGFC (UCEBFF)

Notes
[1]F = Cellulose Fibres
[2]FF = Functionalized Cellulose Fibres
[3]HD = Hard Modified Bitumen
[4]EB = Epoxy modified bitumen

Operationally, the project entails a first step in which the different formulations of the bituminous binder are designed and tested, before their application in the mixtures. Then, further tests are carried out to check the functional and mechanical performance of the diverse asphalt concrete samples, as described in the following paragraphs.

3   THE USE OF CELLULOSE FIBRES IN BITUMEN

Cellulose fibres extracted from fast-growing plants such as hemp, broom and jute are natural polymers with excellent chemical-physical and mechanical properties. In recent years, the interest in cellulose fibres to produce composite materials, with improved mechanical properties and lower weight than traditional materials, has widely increased. Several projects on the matter have been funded in Italy (Chidichimo et al. 2016) with the aim to develop composite materials with improved mechanical properties suitable to the industrial sector, such as automotive, furniture, construction, etc. However, cellulose fibres have a marked hydrophobic character that makes them unsuitable for mixing with other components. In fact, they are strongly hydrophilic, due to the presence of numerous hydroxyl groups, and therefore they have a low affinity with hydrophobic surfaces (Corrente et al. 2016). For this reason, functionalization techniques have been developed to convert the hydrophilic character of the cellulosic surface into a hydrophobic one. A schematic representation of functionalized fibres (indicated as FF in the following) is shown in Figure. 1, where the cellulose core is the elliptic portion limited by the continuous black line.

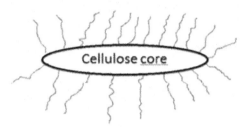

Figure 1.   Functionalized cellulose fibre.

These small cellulose particles can be obtained by mechanical moulding fresh cellulose extracted from broom or also from waste materials such as paper, cardboard, cotton clouts, residues from the textile industry, etc. The diameter of the core is usually of the order of a few tens of microns, while the length can vary from 100 microns to millimetres. The pendants, emerging from the surface, are chemical groups inserted in the cellulose core by green chemical processes where hydroxyl groups of the cellulose surface react with adequate organic molecules carrying an appropriate reactive group such as, for example, a siloxane or isocyanate functional group. The function of the superficial pendants can be twofold. On the one hand, they can serve to make the surface of the cellulosic particles compatible with that of other materials to which the fibres are to be added as reinforcing microstructures. On the other hand, functionalizing pendants can be structured so that their free ends are equipped with chemical groups capable of reacting with chemical groups of the other components of the composite material. Functionalizing cellulose fibres allows for the creation of strong bonds between the fibres and the other particles of different materials. In this way, it is possible to achieve composites with improved mechanical resistance compared to single isolated components. Whatever the approach used to functionalize cellulosic fibres, their application in the field of road pavements to improve their mechanical strength and durability seems promising. For this reason, ANAS, in collaboration with the research Consortiums NITEL (Italy) and TEBAID (Italy), has recently started a research project to evaluate the effects of integrating functionalized cellulosic fibres into various types of bitumen and related asphalts, with the objective to develop sustainable low-noise pavements, both in terms of cost and durability.

From the technological perspective, the research program has been broken down into two steps. The first step consists in evaluating the effects on bitumen and asphalt concrete of functionalized cellulose fibres with hydrophobic pendants. The second step consists in evaluating the effects induced on asphalt concretes by embedding functionalized cellulose fibres in a modified bitumen with a small quantity of epoxy resin. The latter step, which is still in a very preliminary stage, is based on the attempt to create within the bitumen a chemical network between surface pendants of the fibres and the epoxy resin, using functional groups with amino end groups, which can react with epoxy resin terminals. This idea is also supported by the fact that the chemical groupings of the amino type are also promoters of adhesion between the bitumen and aggregates. It is,

771

therefore, possible to hypothesize the possibility that limited quantities of epoxy resins added to the bitumen together with fibres, functionalized with amino groups, and small quantities of free amines can create a chemical network inside the asphalts, which, starting from the concrete, develops throughout the asphalt, making it much more wear-resistant.

3.1   Cellulose fibres functionalization

In the IASNAF project, cellulose fibres were extracted from the Spanish broom (Spartium Junceum) by subjecting the freshly cut vegetable to basic leaching. The branches of the plant were treated for half an hour with a 5% NaOH solution at a temperature of 80 ° C. This allowed the separation of the cellulose fibres from the rest of the plant. Next, fibres were first grinded by an SG Granulator 16 N/20 N (Shini Plastic Technologies) to a coarse gradation (3 mm). Then, finer gradations were obtained by using an Ultra Centrifugal Mill ZM 200 (Retsh Srl, Torre Boldone, BG, Italy) (125 and 500 microns). The functionalization was done by reacting the fibres with a 4,4'-diphenylmethane diisocyanate (4,4'-MDI, where MDI stands for Methylene diphenyl diisocyanate), according to the reactive scheme shown in Figure 2.

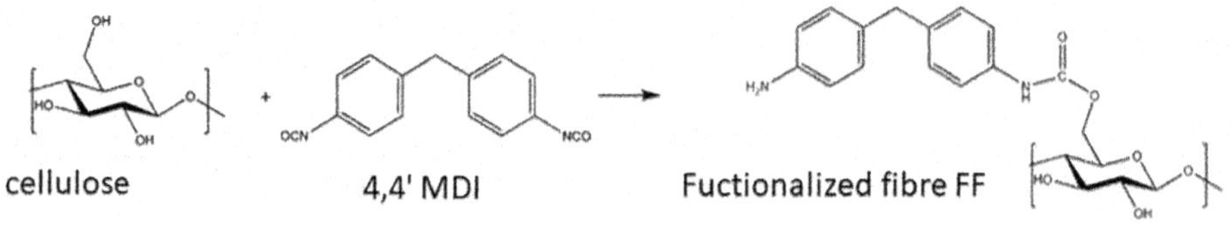

cellulose   4,4' MDI   Fuctionalized fibre FF

Figure 2.   Scheme of the functionalization reaction of the fiber by 4,4'-MDI.

Figure 2 shows the reaction between the isocyanic group (-NCO) and the -OH groups of the cellulose fibres and the consequent formation of the urethane bond. The second iso-cyanate group in the 4.4 MDI molecule, following washing processes with a weakly basic solution, used to remove excess and unreacted isocyanate was converted into an amino group. The reaction was carried out in a home-made reactor in which the functionalization process was carried out for about 20 minutes, sprinkling the powder fibres with 5% (w / w) of micro drops of a solution composed of acetone and -isocyanate 4,4'-MDI (v / v ratio equal to 10/1). Next, the fibres was placed in an oven at 80 ° C to complete the functionalization reaction and evaporate the acetone. After 12 hours, the functionalized fibres was washed with slightly acidic water to remove any excess isocyanic functionalizing agent and placed back in the oven to remove the washing water. Water Contact Angle (WCA) measurements indicated the modification of the hydrophilic nature of the functionalized cellulosic fibres, with an increase in the contact angle from 75°, relative to the raw fibre, to 150°, for the functionalized fibres, confirming the character purely hydrophobic of the latter.

3.2   Dispersion of FF into 50/70 bitumen and HD bitumen

The dispersion of the FF fibres into the bitumen was carried out after the fibres were dried at 105 °C for at least 2 hours. Next, FF Fibres were embedded into HD bitumen at a tem-perature of about 185 °C and at a temperature of 160 °C into the 50/70 bitumen. The following Figure. 3 illustrates the fibres (having a length of 120 microns and diameters of about 20 microns) just added (A) in an HD bitumen and its perfect dispersion (B) after only 30 seconds of stirring. It is evident that functionalized cellulose fibres "dissolve" very well

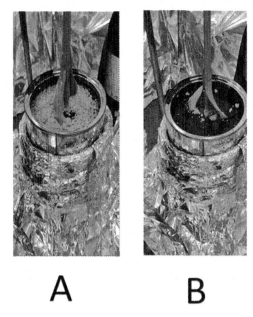

A   B

Figure 3. Mixing of functionalized cellulose fibres in an HD bitumen. Immediately after the addition of fibres to bitumen (A) and after 15 seconds from the addition under the stirring velocity of 2 turns/sec.

even at considerable concentrations (6% in the case of Figure 3) in the bitumen, although the mixing system used was extremely rudimentary.

4 EMBEDDING FUNCTIONALIZED CELLULOSE FIBRES INTO THE ASPHALT

4.1 *Materials and mix design*

To check the potential of embedding cellulose fibres into bitumen, several tests have been performed, involving two types of bitumen (B50/70 and HD), two types of mixtures (PA - Porous Asphalt and DGFC-Dense-Graded Friction Course) and three different types of fibres (SF – Short Cellulose Fibres, SFF-Short Functionalised Cellulose Fibres, SSFF – Very Short Functionalised Cellulose Fibres), as shown in Table 2.

Table 2. Asphalt binders, fibres, and mixture type.

Type of bitumen	Type of fibres	Type of mixture
B50/70	SF, SFF, SSFF	DGFC
HD	SF, SFF, SSFF	PA

For the two mixtures the following mix designs were used:
 Dense-Graded Friction Course (UC, ANAS Course Type B):

- nominal maximum aggregate size of about 6 mm;
- asphalt binder percentage of 5.25% (by the weight of mixture);
- asphalt binder type B50/70.

Porous asphalt (PA, D, ANAS Fuse, Cf. Praticò & Vaiana 2012);

- nominal maximum aggregate size of about 14 mm;
- asphalt binder percentage of 5.7% (by the weight of mixture);
- asphalt binder type HD.

Based on five different classes of aggregate gradations (aggregate source), the aimed gradation was matched through the optimization of their percentages. This applied to both DGFCs and PAs.

4.2 *Test methodology*

After the production and compaction at their final number of gyrations (N_{max}, 210 for DGFCs, 130 for porous asphalts), the mixtures were tested for:

- volumetrics (G_{mbCOR}, Porosity$_{COR}$, AV_{COR}, $G_{mb\ PAR}$, G_{sb}, AV_{PAR}, AV_{COR}, G_{mbDIM}, AV_{DIM}, G_{mm}, cf. Praticò et al. 2009);
- geometry (diameters, thicknesses in different points of the specimens);
- surface properties (mean texture depth, MTD, pendulum test value, PTV);
- mechanics (tensile strength Rt at 25°C and at 10°C, corresponding indirect tensile coefficient at 25°C, Marshall stability, Marshall flow, and Marshall quotient).

4.3 *Results achieved*

Only preliminary results are available so far. The experimental activity concerned two aspects: the investigations on asphalt binder modification with fibres and the related outcomes on bituminous mixtures (asphalt concretes). The latter, in turn, includes two types of mixtures, i.e., dense-graded friction course and porous asphalt. Figure 4 illustrates how

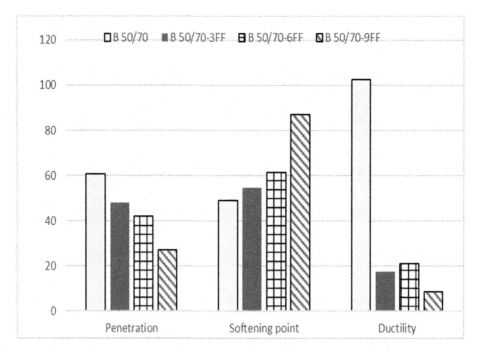

Figure 4.   Impact of functionalized fibres on binder characteristics.

functionalized fibres impact the main characteristic of the asphalt binder. Overall, higher percentages yield harder asphalt binders. The impact on ductility appears quite important and here further studies and efforts are needed.

Figure 5 focuses on the impact of functionalized fibres on acoustic absorption. Better results are achieved at low frequencies, while the acoustic behaviour on the mid-band frequencies seems approximately unchanged.

Figure 6 focuses on the surface versus volumetric properties of PAs and DGFCs, where higher air void contents bring to a higher value of macrotexture, while friction properties appear quite similar.

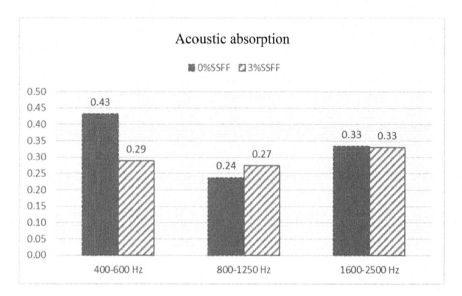

Figure 5. Example of the acoustic absorption spectrum for two cases (PA- preliminary results).

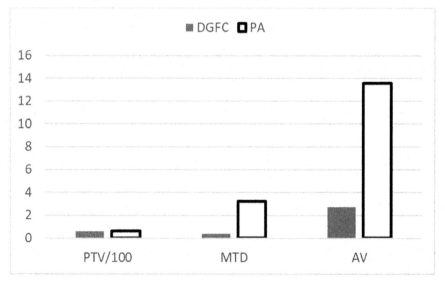

Figure 6. PTV/100, MTD, and AV at Nmax.

5 DISCUSSION AND CONCLUSIONS

Based on the preliminary results achieved from the ISNAF project, the following outcomes can be observed.

For bitumen, it is possible to state that:

- When mixing B50/70 and FF, higher percentages of functionalized fibres yield lower penetrations at 25°C and lower ductility.
- The storage stability of bitumen with fibres appears to improve.
- The lower the fibres length, the lower the decrease of penetration and ductility, as well as the increase of the softening point and dynamic viscosity. In summary, the hardening of the bitumen is lower with short functionalized fibres (SFF) than with functionalized fibres (FF), the remaining factors being constant.
- As for the elastic recovery, the addition of SFF to B50/70 improves its performance (i.e., increase).
- When adding fibres to B50/70, the effects on ductility, softening point, and penetration do not depend appreciably on the functionalization.
- When mixing hard bitumen (HD) with fibres, plain fibres seem to have higher hardening power than functionalized fibres. This includes the reduction of ductility.
- When using HD, the addition of SFF and SF (3%) has negligible consequences on the elastic recovery.

For DGFCs, it is possible to observe that (cf. Figure 4):

- The addition of SFF to B50/70 brings higher tensile strength and indirect tensile coefficients (T = 25°C) at least for SFF percentages up to 6% (by weight of bitumen).
- At the same time, for T = 10°C, the addition of SSFF does not cause an increase in tensile resistance, instead quite negligible reductions are observed.
- Minor consequences in terms of PTV and macrotexture seem to derive from the addition of fibres. More precisely, when adding fibres, a slight increase of air voids is caused, and this implies minor increases in macrotexture (+0.1 mm).

For porous asphalts, it is noted that:

- Further studies are needed to assess the dependence of the indirect tensile strength on fibres percentage.
- Minor consequences in terms of PTV and macrotexture seem to derive from the addition of fibres. In more detail, the addition of fibres seems to cause minor increases in terms of mean texture depth for percentages lower than 3%, while for higher percentages the mean texture depth seems to decrease.

REFERENCES

Chidichimo G., Aloise A., Beneduci A., De Rango A., Pingitore G., Furgiuele F., Valentino P., Polyurethanes Reinforced with Spartium Junceum Fibres, (2016) *Polymer Composites*, 37 (10), pp. 3042–3049, DOI.doi.org/10.1002/pc.23501

Corrente G.A., Scarpelli F., Caputo P., Rossi C.O., Crispini A., Chidichimo G., Beneduci A., Chemical–Physical and Dynamical–mechanical Characterization on Spartium Junceum L. Cellulosic Fibres Treated with Softener Agents: a Preliminary Investigation, *Scientific Reports, 11 (1) (2021), art. no.* 35. DOI: 10.1038/s41598-020-79568-5

LIFE SNEAK project – https://www.lifesneak.eu

LIFE NEREIDE project – https://www.nereideproject.eu

LIFE C-LOW-N – https://www.lifesneak.eu

LIFE E-VIA – https://life-evia.eu

PERSUADE project– PoroElastic Road SUrface: An Innovation to Avoid Damages to the Environment -www.persuadeproject.eu

Praticò F.G., Moro A. and Ammendola R., Modeling HMA bulk specific gravities: A Theoretical and Experimental Investigation (2009) *International Journal of Pavement Research and Technology*, 2 (3), pp. 115 – 122.

Praticò F.G. and Vaiana R., Improving Infrastructure Sustainability in Suburban and Urban Areas: Is Porous Asphalt the Right Answer? And how?, (2012) *WIT Transactions on the Built Environment*, 128, pp. 673 – 684, DOI: 10.2495/UT120571.

Roads and Airports Pavement Surface Characteristics – Crispino & Toraldo (Eds)
© 2023 The Author(s), ISBN 978-1-032-55149-4

Environmental efficiency assessment of RCA inclusion as partial replacement of virgin aggregates in HMA and WMA

Rodrigo Polo-Mendoza*
Faculty of Science, Charles University, Prague, Czech Republic
Department of Civil and Environmental Engineering, Universidad del Norte, Barranquilla, Colombia

Emilio Turbay & Gilberto Martinez-Arguelles
Department of Civil and Environmental Engineering, Universidad del Norte, Barranquilla, Colombia

Rita Peñabaena-Niebles
Department of Industrial Engineering, Universidad del Norte, Barranquilla, Colombia

ABSTRACT: Road infrastructure is one of the economic sectors with the most significant environmental impact on the planet. Nonetheless, the construction and preservation of pavements are essential for the sustained economic development of the communities. Therefore, it is necessary to find and implement methods, techniques, or designs that reduce the environmental burden caused by this industry. Hence, in this research effort, two sustainability alternatives are explored to optimize the design of traditional hot mix asphalt (HMA) from an eco-efficiency perspective, namely, recycle concrete aggregate (RCA) and warm mix asphalt (WMA). Specifically, the life cycle assessment (LCA) procedure was used to determine the optimal conditions for WMA production through chemical additive technology and the partial substitution of natural aggregates (NAs) by RCA. Accordingly, several asphalt mix designs were proposed to identify the influence of the design variables on the total environmental damage generated during the manufacturing processes. Coarse RCA contents of 0, 10, 20, 30 and 40% were evaluated for HMA and WMA. Regarding the WMAs, dosage combinations of RCA and chemical additive (0.1, 0.2, 0.3, and 0.4% by weight of asphalt binder) were assessed. The main findings of this case study were: (i) amounts of coarse RCA equal to or higher than 20% induce a more harmful environmental effect than that of NAs, this due to the increase in the asphalt binder demand that this waste material provokes; (ii) the additional environmental impacts associated with the chemical additives production are insignificant compared to the savings obtained by reducing mixing and compaction temperatures.

Keywords: Asphalt Mixtures, Life Cycle Assessment, Recycled Concrete Aggregate, Warm Mix Technology

1 INTRODUCTION

The transport infrastructure is one of the crucial factors in ensuring the growth of societies' economies (Abudinen et al. 2017). Therefore, the demand for pavement materials has increased over time in the paving industry. Nonetheless, this situation also augments the carbon footprint on the planet (Alaloul et al. 2021; Plati 2019). Specifically, asphalt

---

*Corresponding Author: rpoloe@uninorte.edu.co

DOI: 10.1201/9781003429258-76

materials have a significant part of the environmental impact caused by the construction and maintenance of road infrastructure (Cross et al. 2011; Lee et al. 2020). Consequently, the state-of-the-art has explored paths to develop novel materials and implement new technologies to reduce these environmental burdens.

Two of the most notable techniques to reduce the environmental burden of the asphalt pavement are the warm-mix asphalt (WMA) and the incorporation of recycled concrete aggregate (RCA) (Plati 2019; Polo-Mendoza et al. 2022). The WMA is an asphalt mixture relatively similar to the traditional hot-mix asphalt (HMA), except for one meaningful difference: the production temperatures are approximately 20-30°C lower (Behnood 2020; Hamdar et al. 2018; Hasan et al. 2017). On the other hand, the RCA has been widely employed as a partial replacement for natural aggregates (NAs) in the mix design of asphalt materials (Del Ponte et al. 2017; Loureiro et al. 2022; Salehi et al. 2021).

However, not in all cases do these techniques develop an environmental benefit. For example, in the case of RCA, due to the higher absorption (caused by the attached mortar), the optimal asphalt content (OAC) usually increases (Martinez-Arguelleset al. 2019; Sanchez-Cotte et al. 2020). Regarding the WMA, it is possible that the chemical additive production can generate an environmental impact higher than the savings induced by the reduction in the mixing and compaction temperatures (Polo-Mendoza et al. 2022).

In that order of ideas, in this investigation, several designs of HMA and WMA are assessed to determine the optimal incorporation rate of RCA and chemical additives. Figure 1 shows the sustainable technologies employed for this purpose. A total of 25 mixture designs were evaluated. Initially, laboratory tests were carried out to establish the influence of the added materials on the behaviour of asphalt mixtures. Then, the life-cycle assessment (LCA) methodology was implemented to establish the environmental burdens of each evaluated scenario. In this way, it was possible to demarcate the conditions that allow obtaining the asphalt mixtures with a lower environmental impact (LEI) and higher environmental impact (HEV).

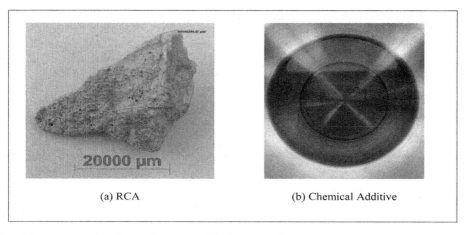

(a) RCA   (b) Chemical Additive

Figure 1. Sustainable technologies employed in this research.

2 MATERIALS AND METHODOLOGY

2.1 *Materials*

The NAs used in this study come from exploiting quarries in Arroyo de Piedra (State of Atlántico, Colombia). These quarries offer sedimentary rocks of marine origin. The virgin aggregates extracted from this site are widely utilized to construct highways in the northern region of Colombia. On the other hand, the RCA employed was obtained from the demolition and recycling process of a rigid pavement located on Hamburgo Avenue in the City of Barranquilla (likewise in the State of Atlántico). All aggregates (natural and recycled) used for this research were crushed and sieved to satisfy the Colombian technical specification for producing asphalt mixtures with a nominal maximum aggregate size of 25 mm (INVIAS 2013).

This study used a 60/70 penetration grade asphalt binder as the base binder. The physical properties of the asphalt binder are listed in Table 1. Further, a chemical additive based on fatty acids was employed as the WMA technology.

Table 1.   Properties of the asphalt binder.

Test	Standard	Unit	Value
Penetration (25 °C, 100 g, 5 s)	ASTM D-5	0.1 mm	67
Penetration Index	NLT 181/88	–	−0.11
Softening point	ASTM D-36-95	°C	51.5
Specific gravity	AASHTO T 228-04	–	1.005
Viscosity to 135° C	AASHTO T-316	Pa-s	0.38
Ductility (25 °C, 5 cm/min)	ASTM D-113	cm	>105
Mass loss after the RTFOT	ASTM D-2872	%	0.05

Following the Marshall method, 25 mixture designs were developed. These designs are the result of combining two variables: coarse RCA content (0, 10, 20, 30 and 40%) and chemical additive content (0, 0.1, 0.2, 0.3 and 0.4%). The percentage of RCA refers to the proportion of virgin coarse aggregate that the coarse RCA replaces. Meanwhile, the percentage of chemical additive corresponds to the addition by weight of the asphalt binder. These dosages were selected according to literature recommendations (Diab et al. 2016; Plati 2019) and experiences from previous research efforts (Polo-Mendoza et al. 2022; Vega et al. 2019).

The control asphalt mixture corresponds to the design of 0% of coarse RCA content and 0% of chemical additive. Besides, the designs that contain 0% of chemical additives are HMAs, whilst all others are WMAs. For all asphalt mixtures, the OAC was established according to the guidelines of the Marshall method. For the HMAs, the mixture and compaction temperatures were defined through the rotational viscometer test (ASTM 2015). On the other hand, the production temperatures for the WMAs were obtained from the test of determining the degree of particle coating of asphalt mixtures (AASHTO 2018). The results demonstrated that the inclusion of RCA does not affect the manufacturing temperatures. Figure 2 presents the outcomes provided by these tests.

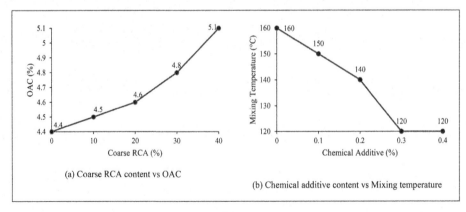

(a) Coarse RCA content vs OAC

(b) Chemical additive content vs Mixing temperature

Figure 2.   Effect of RCA and the chemical additive in the asphalt mix design. **Note**: The compaction temperature was 20°C lower than the mixing temperature in all cases.

2.2 Methodology

The LCA methodology was implemented to evaluate the environmental impact associated with the fabrication of the asphalt mixtures considered in this study (i.e., the 25 different mix designs). For these purposes, the guidelines provided by the International Organization for Standardization (ISO) were followed through the ISO-14040 and ISO-14044 standards (ISO 2006a, 2006b). According to these normative, the execution of LCA must be divided into four phases, namely (i) goal and scope definition phase, (ii) life cycle inventory (LCI) analysis phase, (iii) life cycle impact assessment (LCIA) phase, and (iv) interpretation phase.

2.2.1 Goal and scope definition phase

The central objective of this investigation is to determine the optimal conditions (from an environmental approach) to incorporate the WMA chemical additive technology and RCA inclusion into the design of asphalt mixtures. In this way, the main goal of the LCAs is to find the optimal incorporation rate of chemical additives and RCA that generates the maximin environmental benefits, i.e., the minimum environmental burdens.

Because this research aims to estimate the environmental impacts of producing a material (asphalt mixtures), it was decided to use the cradle-to-gate approach. According to the Federal Highway Administration (FHWA), this approach involves the environmental burdens generated by the extraction of raw materials and the industrial activities related to the manufacturing process (including transportation of materials and supplies) (Harvey et al. 2016; Vega et al. 2019). Consequently, this research considers three stages within the system boundaries: materials production, materials transport to the asphalt mixing plant, and asphalt mix production. Besides, the functional unit was defined as 1 ton of asphalt mixture.

The data source is composed of primary data and secondary data. The primary data were obtained from the investigations that preceded this research effort (Martinez-Arguelles et al. 2019; Polo-Mendoza et al. 2022; Sanchez-Cotte al. 2020; Vega et al. 2019; Vega et al. 2020; Vegaet al. 2020a, 2020b). Meanwhile, the secondary data only correspond to asphalt binder and chemical additive production, whose details were adopted from the USLCI and Ecoinvent v.3 databases, respectively (NREL 2009; Weidema et al. 2013).

2.2.2 LCI analysis phase

Table 2 presents the LCI employed for this study. On the other hand, SimaPro 9.1.1 software was used to analyze the LCI and carry out the LCAs. Due to its potential and versatility, SimaPro is the leading computational tool in this industry field (Herrmann & Moltesen 2015; Vidergar et al. 2021). Table 2 also shows the SimaPro Unit Processes (SPUP) utilized to model the system boundaries adopted for this research. These SPUP were modified to obtain a representative case study of the northern region of Colombia. The details of these considerations can be found in (Polo-Mendoza et al. 2022).

2.2.3 LCIA phase

In this research, the Tool for the Reduction and Assessment of Chemical and Other Environmental Impacts (TRACI v.2.1) was selected as the impact assessment method. TRACI classifies the environmental burden in ten impact categories (ICs), namely ozone depletion (OD), global warming (GW), smog (SM), acidification (AC), eutrophication (EU), carcinogenic (CA), non-carcinogenic (NCA), respiratory effects (RE), ecotoxicity (EC), and fossil fuel depletion (FFD) (Bare 2011; Bare et al. 2012; Dong et al. 2021; Ryberg et al. 2013).

2.2.4 Interpretation phase

Following the ISO-14040 and ISO-14044 standards (ISO 2006a, 2006b), the LCAs finishes with the characterization results. The characterization represents the estimation of the environmental impacts in all ICs according to the LCI and LCIA presented (Polo-Mendoza et al. 2022; PRé Sustainability 2020). These outcomes are shown in Table 3.

Table 2. LCI employed for this study.

LCA stage	LCA sub-stage	SPUP
Materials production	NAs extraction	Gravel, crushed {RoW}	production	Cut-off, U
	Truck loading of NAs	Loader operation, large, INW/RNA
	RCA processing	Diesel, burned in building machine {GLO}	processing	Cut-off, U
	Asphalt binder production	Bitumen, at refinery/kg/US
	Additive production	Fatty acid {GLO}	market for	Cut-off, U
Materials transport to the asphalt mixing plant	Aggregates transportation Asphalt binder transportation Additive transportation	Transport, freight, lorry 16-32 metric ton, EURO4 {RoW}	transport, freight, lorry 16-32 metric ton, EURO4	Cut-off, U
Asphalt mix production	NAs processing	Diesel, burned in building machine {GLO}	processing	Cut-off, U
	Mixing process	Heat, district or industrial, other than natural gas {RoW}	heat production, heavy fuel oil, at industrial furnace 1 MW	Cut-off, U

Table 3. Characterization results. **Color coding**: green - LEI; red - HEI.

OD (mg CFC-11eq)	Chemical Additive (%)					GW (kg CO2eq)	Chemical Additive (%)				
Coarse RCA (%)	0	0.1	0.2	0.3	0.4	Coarse RCA (%)	0	0.1	0.2	0.3	0.4
0	4719	4501	4283	3848	3850	0	22415	21388	20361	18308	18311
10	4711	4498	4280	3844	3846	10	22406	21371	20354	18300	18305
20	4704	4492	4271	3837	3844	20	22400	21364	20355	18289	18304
30	4738	4520	4303	3868	3868	30	22536	21503	20469	18402	18402
40	4748	4531	4313	3878	3878	40	22603	21564	20526	18448	18448

(a) Ozone Depletion. (b) Global Warming.

SM (kg O3eq)	Chemical Additive (%)					AC (kg SO2eq)	Chemical Additive (%)				
Coarse RCA (%)	0	0.1	0.2	0.3	0.4	Coarse RCA (%)	0	0.1	0.2	0.3	0.4
0	915	873	831	747	751	0	141	134	128	115	117
10	913	870	830	745	747	10	139	130	123	113	115
20	913	868	829	739	744	20	135	129	120	110	111
30	919	877	835	751	751	30	141	135	128	115	115
40	921	879	837	753	753	40	142	135	129	116	116

(c) Smog. (d) Acidification.

Table 3.   Continued

EU (g Neq)	Chemical Additive (%)					CA (E-6 CTUh)	Chemical Additive (%)				
Coarse RCA (%)	0	0.1	0.2	0.3	0.4	Coarse RCA (%)	0	0.1	0.2	0.3	0.4
0	12158	11602	11045	9931	9932	0	163	156	149	134	135
10	12152	11598	11042	9927	9931	10	162	155	142	130	133
20	12148	11593	11040	9923	9930	20	162	153	141	127	129
30	12209	11653	11097	9983	9984	30	164	157	150	135	135
40	12237	11681	11124	10010	10011	40	165	158	150	135	135

(e) Eutrophication.   (f) Carcinogenic.

NCA (E-6 CTUh)	Chemical Additive (%)					RE (g PM2.5eq)	Chemical Additive (%)				
Coarse RCA (%)	0	0.1	0.2	0.3	0.4	Coarse RCA (%)	0	0.1	0.2	0.3	0.4
0	1158	1106	1054	950	953	0	10797	10300	9803	8810	8812
10	1156	1102	1053	945	950	10	10769	10298	9800	8804	8809
20	1152	1100	1050	940	947	20	10766	10295	9796	8801	9904
30	1168	1116	1064	959	959	30	10841	10344	9848	8854	8854
40	1173	1121	1069	965	965	40	10865	10368	9871	8878	8878

(g) Non-Carcinogenic.   (h) Respiratory Effects.

EC (CTUe)	Chemical Additive (%)					FFD (MJ surplus)	Chemical Additive (%)				
Coarse RCA (%)	0	0.1	0.2	0.3	0.4	Coarse RCA (%)	0	0.1	0.2	0.3	0.4
0	35037	33442	31847	28656	28657	0	42359	40422	38485	34611	34614
10	35035	33440	31844	28650	28654	10	42355	40420	38483	34603	34610
20	35030	33438	31840	28648	28653	20	42349	40418	38481	34601	34606
30	35272	33677	32082	28890	28892	30	42622	40685	38748	34874	34874
40	35396	33801	32206	29015	29016	40	42757	40820	38883	35009	35009

(i) Ecotoxicity.   (j) Fossil Fuel Depletion.

3   DISCUSSION

In Table 3, it is evident that the design with the LEI is the asphalt mixture fabricated with 20% of coarse RCA and 0.3% of chemical additive. Meanwhile, the mix design with HEI is generated with an incorporation rate of 40% and 0% for the coarse RCA and chemical additive, respectively. These results are maintained for all ICs. Accordingly, the following findings can be drawn:

- The optimal content of coarse RCA was 20%. Therefore, dosages higher than 20% generates a greater environmental burden than the control case. This is because the additional consumption of asphalt binder opaques the environmental savings caused by the reduction in the depletion of virgin raw materials.

- The savings generated by reducing mixing and compaction temperatures are substantially more significant than the environmental burden associated with the chemical additive production. Nevertheless, this study achieved the minimum manufacturing temperatures at 0.3% (of the chemical additive by weight of asphalt binder). Hence, extra quantities of chemical additives do not induce an advantage from a sustainable perspective.
- Because a chemical additive content of 0.4% generates more contamination than that of 0.3%, it is found that the production of WMA chemical additive induces a higher environmental impact than refining asphalt binder from the crude oil. In fact, a slight extra amount of chemical additive made these variations in the associated environmental burden noticeable. These findings are consistent with the reported in previous research (Polo-Mendoza et al. 2022).

4 CONCLUSIONS

Laboratory tests and environmental assessments were carried out throughout this research to determine the optimal conditions for designing WMA with RCA contents. As a result, it was possible to demarcate some appropriate patterns to implement eco-friendly strategies in the road infrastructure industry. Therefore, this investigation achieves the design of alternative asphalt materials to the traditional HMA. Based on this study, the following conclusions were drawn:

- The higher the RCA content, the more elevated the OAC.
- Although RCA presents a greater porosity and absorption than NAs, these recycled material does not produce a demand for higher mixing and compaction temperatures according to the particle coating method.
- Initially, the more the chemical additive content increased, the lower the manufacturing temperatures of the asphalt mixtures. Nonetheless, at a specific dosage, this reduction is stabilized.
- The industry activities needed for the chemical additive production generate a more elevated environmental impact than the asphalt binder refining process.
- From an environmental approach, the optimum content of the chemical additive is obtained when reductions in asphalt mixture production temperatures are made constant. In this particular case, this was 0.3%.
- From an environmental approach, it was determined that the optimal inclusion of coarse RCA in HMAs and WMAs was 20%.
- According to the conditions established in this study, the optimum asphalt mixture from an environmental perspective was a WMA with 20% and 0.3% of RCA and chemical additive, respectively.

5 RECOMMENDATIONS FOR FUTURE RESEARCH WORKS

The following recommendations are proposed to enhance this research effort in future investigations: (I) create mathematical and computational models that can adequately estimate the environmental impacts caused by the production of asphalt mixtures; (II) propose methodologies to optimize the design of WMA with RCA content from a sustainability perspective; (III) improve the interpretation phase by developing other assessment processes, such as normalization, grouping, weighting, or data quality analysis;

(IV) consider the economic profitability of asphalt mixtures as a decision criterion within a comprehensive sustainability optimization; (V) evaluate the incorporation of other recycled aggregates (for example, crumb rubber and reclaimed asphalt pavement) for the partial substitution of NAs.

6 DECLARATION OF COMPETING INTEREST

The authors declare that they are unaware of any possible conflicts of interest that may have influenced the development of this research.

ACKNOWLEDGMENTS

The authors express their sincere gratitude to the Administrative Department of Science, Technology, and Innovation (COLCIENCIAS) and the Universidad del Norte for funding this study through the "Research Project 745/2016, Contract 037-2017, No. 1215-745-59105".

REFERENCES

AASHTO. (2018). T195: Standard Method of Test for Determining Degree of Particle Coating of Asphalt Mixtures. *American Association of State Highway and Transportation Officials Provisional Standards.* https://standards.globalspec.com/std/13053333/AASHTO T 195

Abudinen D., Fuentes L.G. and Carvajal J. (2017). Travel Quality Assessment of Urban Roads Based on International Roughness index: Case study in Colombia. *Transportation Research Record, 2612*, 1–10. https://doi.org/10.3141/2612-01

Alaloul W.S., Altaf M., Musarat M.A., Javed M.F. and Mosavi A. (2021). Systematic review of life cycle assessment and life cycle cost analysis for pavement and a case study. *Sustainability, 13*(4377), 1–38. https://doi.org/10.3390/su13084377

ASTM. (2015). D4402/D4402M-15: Standard Test Method for Viscosity Determination of Asphalt at Elevated Temperatures Using a Rotational Viscometer. *American Society for Testing and Materials (ASTM) International,* 1–4. https://doi.org/10.1520/D4402_D4402M-15

Bare J. (2011). TRACI 2.0: The Tool for the Reduction and Assessment of Chemical and Other Environmental Impacts 2.0. *Clean Technologies and Environmental Policy, 13*, 687–696. https://doi.org/10.1007/s10098-010-0338-9

Bare J., Young D. and Hopton M. (2012). EPA/600/R-12/554: Tool for the Reduction and Assessment of Chemical and Other Environmental Impacts (TRACI) - User's Manual. *U.S. Environmental Protection Agency.*

Behnood A. (2020). A Review of the Warm Mix Asphalt (WMA) Technologies: Effects on Thermo-mechanical and Rheological Properties. *Journal of Cleaner Production, 259*, 120817. https://doi.org/10.1016/j.jclepro.2020.120817

Cross S., Chesner W., Justus H. and Kearney E. (2011). Life-cycle Environmental Analysis for Evaluation of Pavement Rehabilitation Options. *Transportation Research Record, 2227*, 43–52. https://doi.org/10.3141/2227-05

Del Ponte K., Madras Natarajan B., Pakes Ahlman A., Baker A., Elliott E. and Edil T.B. (2017). Life-cycle Benefits of Recycled Material in Highway Construction. *Transportation Research Record, 2628*, 1–11. https://doi.org/10.3141/2628-01

Diab A., Sangiorgi C., Ghabchi R., Zaman M. and Wahaballa A. (2016). Warm Mix Asphalt (WMA) Technologies: Benefits and Drawbacks - a Literature Review. *4th Chinese European Workshop on Functional Pavement Design*, 1145–1154. https://doi.org/10.1201/9781315643274-127

Dong Y., Hossain M. U., Li H. and Liu P. (2021). Developing Conversion Factors of LCIA Methods for Comparison of LCA Results in the Construction Sector. *Sustainability, 13*(9016), 1–16. https://doi.org/10.3390/su13169016

Hamdar Y.S., Kassem H. A. and Chehab G.R. (2018). Using Different Performance Measures for the Sustainability Assessment of Asphalt Mixtures: Case of Warm Mix Asphalt in a Hot Climate. *Road Materials and Pavement Design*, 1–24. https://doi.org/10.1080/14680629.2018.1474795

Harvey J.T., Meijer J., Ozer H., Al-Qadi I., Saboori A. and Kendall A. (2016). FHWA-HIF-16-014: Pavement Life Cycle Assessment Framework. *Federal Highway Administration*.

Hasan M.R., You Z., and Yang X. (2017). A Comprehensive Review of Theory, Development, and Implementation of Warm Mix Asphalt Using Foaming Techniques. *Construction and Building Materials*, *152*, 115–133. https://doi.org/10.1016/j.conbuildmat.2017.06.135

Herrmann I.T., Moltesen A. (2015). Does it matter Which Life Cycle Assessment (LCA) Tool You Choose? - A Comparative Assessment of SimaPro and GaBi. *Journal of Cleaner Production, 86*, 163–169. https://doi.org/10.1016/j.jclepro.2014.08.004

INVIAS. (2013). INV-ART.450: Mezclas Asfalticas en Caliente de Gradacion Continua (Concreto Asfaltico). *Especificaciones Generales de Construcción de Carreteras*.

ISO. (2006a). ISO 14040: Environmental Management - Life Cycle Assessment - Principles and Framework. *International Organization for Standardization (ISO)*.

ISO. (2006b). ISO 14044: Environmental Management - Life Cycle Assessment - Requirements and Guidelines. International Organization for Standardization (ISO).

Lee S.I., Carrasco G., Mahmoud E. and Walubita L.F. (2020). Alternative Structure and Material Designs for Cost-Effective Perpetual Pavements in Texas. *Journal of Transportation Engineering, Part B: Pavements*, *146*(4), 04020071. https://doi.org/10.1061/jpeodx.0000226

Loureiro C., Moura C., Rodrigues M., Martinho F., Silva H., Oliveira J. (2022). Steel Slag and Recycled Concrete Aggregates: Replacing Quarries to Supply Sustainable Materials for the Asphalt Paving Industry. *Sustainability, 14*(5022), 1–31. https://doi.org/10.3390/su14095022

Martinez-Arguelles G., Acosta M., Dugarte M. and Fuentes L. (2019). Life Cycle Assessment of Natural and Recycled Concrete Aggregate Production for Road Pavements Applications in the Northern Region of Colombia: Case Study. *Transportation Research Record, 2673*(5), 397–406. https://doi.org/10.1177/0361198119839955

Martinez-Arguelles G., Dugarte M., Fuentes L., Sanchez E., Rondon H., Pacheco C., Yepes J. and Lagares R. (2019). Characterization of Recycled Concrete Aggregate as Potential Replacement of Natural Aggregate in Asphalt Pavement. *IOP Conference Series: Materials Science and Engineering, 471*(102045), 1–9. https://doi.org/10.1088/1757-899X/471/10/102045

NREL. (2009). U.S. Life Cycle Inventory Database Roadmap. *National Renewable Energy Laboratory*.

Plati C. (2019). Sustainability Factors in Pavement Materials, Design, and Preservation Strategies: A Literature Review. *Construction and Building Materials, 211*, 539–555. https://doi.org/10.1016/j.conbuildmat.2019.03.242

Polo-Mendoza R., Peñabaena-Niebles R., Giustozzi F. and Martinez-Arguelles G. (2022). Eco-Friendly Design of Warm Mix Asphalt (WMA) with Recycled Concrete Aggregate (RCA): A Case Study from a Developing country. *Construction and Building Materials, 326*, 126890. https://doi.org/10.1016/j.conbuildmat.2022.126890

PRé Sustainability. (2020). Simapro Database Manual: Methods Library. *SimaPro Website*.

Ryberg M., Vieira M.D.M., Zgola M., Bare J. and Rosenbaum R.K. (2013). Updated US and Canadian normalization factors for TRACI 2.1. *Clean Technologies and Environmental Policy, 16*(2), 329–339. https://doi.org/10.1007/s10098-013-0629-z

Salehi S., Arashpour M., Kodikara J., and Guppy, R. (2021). Sustainable Pavement Construction: A Systematic Literature Review of Environmental and Economic Analysis of Recycled Materials. *Journal of Cleaner Production, 313*, 127936. https://doi.org/10.1016/j.jclepro.2021.127936

Sanchez-Cotte E., Fuentes L., Martinez-Arguelles G., Rondon H., Walubita L. and Cantero J. (2020). Influence of Recycled Concrete Aggregates from Different Sources in Hot Mix Asphalt Design. *Construction and Building Materials, 259*, 120427. https://doi.org/10.1016/j.conbuildmat.2020.120427

Sanchez-Cotte E., Pacheco C., Ana F., Pineda Y., Mercado R., Yepes-Martinez J. and Lagares R. (2020). The Chemical-mineralogical Characterization of Recycled Concrete Aggregates from Different Sources and Their Potential Reactions in Asphalt Mixtures. *Materials, 13*(5592), 1–18. https://doi.org/10.3390/ma13245592

Vega D., Martinez-Arguelles G. and Santos J. (2019). Life Cycle Assessment of Warm Mix Asphalt with Recycled Concrete Aggregate. *IOP Conference Series: Materials Science and Engineering, 603*(052016), 1–9. https://doi.org/10.1088/1757-899X/603/5/052016

Vega D., Martinez-Arguelles G. and Santos J. (2020). Comparative Life Cycle Assessment of Warm Mix Asphalt with Recycled Concrete Aggregates: A Colombian Case Study. *Procedia CIRP*, *90*, 285–290. https://doi.org/10.1016/j.procir.2020.02.126

Vega D., Santos J. and Martinez-Arguelles G. (2020a). Carbon Footprint of Asphalt Road Pavements Using Warm Mix Asphalt with Recycled Concrete Aggregates: A Colombian Case study. *Pavement, Roadway, and Bridge Life Cycle Assessment* 2020, 333–342. isbn: 9781003092278

Vega D., Santos J. and Martinez-Arguelles G. (2020b). Life Cycle Assessment of Hot Mix Asphalt with Recycled Concrete Aggregates for Road Pavements Construction. *International Journal of Pavement Engineering*. https://doi.org/10.1080/10298436.2020.1778694

Vidergar P., Perc M. and Lukman R.K. (2021). A Survey of the Life Cycle Assessment of Food Supply Chains. *Journal of Cleaner Production*, *286*, 125506. https://doi.org/10.1016/j.jclepro.2020.125506

Weidema B.P., Bauer C., Hischier R., Mutel C., Nemecek T., Reinhard J., Vadenbo C.O. and Wenet G. (2013). Overview and Methodology: Data Quality Guideline for the Ecoinvent Database Version 3. *Ecoinvent Association*.

Recycled porous asphalt concretes: Pros and contras

Rosolino Vaiana*
Department of Civil Engineering, University of Calabria, Cosenza, Italy

Filippo G. Praticò
DIIES Department, University "Mediterranea" of Reggio Calabria, Italy

Teresa Iuele, Manuel De Rose & Giusi Perri
DINCI Department of Civil Engineering, University of Calabria, Cosenza, Italy

Rosario Fedele
DIIES Department, University "Mediterranea" of Reggio Calabria, Italy

Francesco De Masi
DINCI Department of Civil Engineering, University of Calabria, Cosenza, Italy

ABSTRACT: Porous European Mixes (PEMs) have been recognized as one of the most adequate solutions for pursuing asphalt pavement sustainability. In fact, apart from the well-known advantages in terms of skid resistance improvement in wet conditions, porous mixes are also useful for traffic noise reduction. Several studies show that the percentage of noise reduction for PEMs can reach a value of around 20% if compared with a traditional asphalt mix. Nevertheless, few research still deal with the identification of mix design solutions for noise reduction optimization. Another important issue for producing an eco-friendly asphalt concrete is related to the amount of RAP (Reclaimed Asphalt Pavement) used inside the mix.In the light of the above, the objectives of the study described in this paper are confined into the assessment of surface and acoustic performance of recycled porous asphalt concretes (PEMs). Acoustic properties were measured according to the ISO10534-2. Also surface texture was investigated according to the standards ISO 13473-1; ISO/CD TS 13473-4; ISO 13473-3. Moreover, the relationship between acoustic absorption coefficients and pavement surface performance was studied as a function of the degree of compaction of laboratory asphalt mixes (by gyratory compactor). Finally, recycled PEMs performance was compared with the one of a reference PEMS produced by virgin porous mix. Results show that both drainability and texture depth are well correlated with several parameters associated with the acoustic absorption spectra for both recycled and virgin mixes, regardless of the degree of compaction. Results can benefit both practitioners and researchers.

Keywords: Porous Asphalt, Sustainability, Recycling

1   INTRODUCTION AND OBJECTIVES

1.1   *Background*

Noise pollution is one of the main challenges to be faced in the current century. Matter of fact, noise exposure can lead to several kinds of health problems (Praticò 2014). Among many others, traffic noise is surely the most dominant source, especially in urban areas.

---

*Corresponding Author: rosolino.vaiana@unical.it

DOI: 10.1201/9781003429258-77

Therefore, over the last years, from a sustainable point of view, researchers seek to improve the acoustical properties of road pavements with the purpose of finding more cost-effective ways for the reduction of noise pollution. The generation and propagation of traffic noise is strongly related to the surface texture and the acoustic absorption properties of the road pavement. In particular, the major contribution to the traffic noise is due to the interaction between vehicle tires and road surface, also known as rolling noise (Praticò et al. 2014a). Thus, noise reduction must be pursued by acting primarily on the type of the surface (porous/non-porous) and the system of the interconnected voids of the course layer (mix porosity), on which the mechanism of acoustic impedance depends. Porous European Mix (PEM from here on) represents the most efficient and widespread pavement technology, over conventional dense graded asphalt mix, to pursue the mitigation of noise levels (Losa & Leandri 2011). Several terms are used to define them (e.g., permeable friction courses-PFC, open-graded friction courses-OGFC, porous asphalt-PA, etc.). PEM is a gap-graded asphalt mixture, obtained by reducing the fine aggregates in favor of coarse aggregates. Therefore, the resultant asphalt surface is characterized by a high percentage of air voids, generally of 18% to 22% (Alvarez et al. 2018; Praticò et al. 2018), most of them interconnected, that allows to obtain a porous structure capable of efficiently absorbing the band of frequencies range emitted at the tire-pavement interface.

Moreover, one of the obvious abilities of PEMs is the drainability. In fact, high air void contents make mixtures to channel water through the pavement structure (Xie et al. 2019). This results in a lower volume of water flowing on the surface and determines a significant improvement in skid resistance performance, especially in wet conditions, with a reduction of hydroplaning and splash and spray effect and the enhancement of the visibility (Gu et al. 2018). As a consequence, PEMs treatments determine a reduction in wet weather vehicle crashes or accident rates in comparison to dense graded mixtures.

However, several issues influence the performance and the durability of PEMs. Cloggins is one of the major distresses, which consists in the occlusion of the pores of the structure due to dirt and debris particles (Wu et al. 2020). Thus, the permeability of PEMs may gradually decrease, influencing the noise, friction, and drainability performance over time (Praticò & Vaiana 2012). As a result, regular maintenance should be planned to ensure the good performance of the surface pavement for a longer period of time (Praticò et al. 2012a), otherwise, the performance decay would require the milling of the existing surface.

Besides, the recent strategies promoted by the European Commission aim to create a system by 2050 based on the principles of an efficient resources consumption. In particular, the focus is on the reuse of waste materials and on the reduction of the energy and greenhouse emissions related to the extraction and processing of primary materials (European Commission 2011). Increasing recycling rates has become a central issue in the field of road pavement design and construction: the use of alternative materials is considered a suitable solution contributing to the reduction of virgin aggregates.

To date, research has mostly focused on the recycling of reclaimed asphalt pavement (RAP) in hot mix asphalt (HMA) contributing to minimize waste production and with the final goal of optimizing its reuse without reducing the pavement quality and performance. Little research has been documented pertaining to the sustainable rehabilitation of PEM and more specifically, to the recycling of PEMs back to PEMs. RAP, in fact, is widely implemented in dense asphalt concrete. Less frequently these waste materials are used in the production of porous asphalt concretes. Recycled PEMs lead to several benefits concerning the economic, social, and environmental sphere both during the construction stage and the long-term period (Praticò et al. 2011). As well as in every process that includes the use of RAP materials in new constructions, some aspects need to be addressed also in recycling PEMs due to RAP variability and the specific impact of RAP on pavement performance (such as bearing, skid resistance, acoustic, and permeability performance)

(Boscaino 2001). Factors that affect RAP variability, for example, are related to the inhomogeneity of the mixture (gradation and binder content) because of the previous maintenance and preservation operations or to the milling and transportation processes. (Praticò et al. 2012a). Furthermore, RAP mixtures variability increases with the addition of RAP (Li et al. 2008).

In their carefully designed study, Praticò et al. (2012b) found promising functional and mechanical performance of high-RAP content mixes. Around 82% of RAP (from PEM) was added for obtaining a two-layer porous asphalt. The experimental plan draws attention also to bearing and friction which are the main investigated properties of the rehabilitated wearing course (Praticò et al. 2011).

1.2 Objectives and scope

The objectives of the study described in this paper are confined to the assessment of surface and acoustic performance of recycled porous asphalt concretes. This research in the wake of a wider project of national interest (PRIN 2008 – "Drenante da Drenante"). The above-mentioned project involved two different laboratories (Lab MAST – University of Calabria and RAR-University of Reggio Calabria) and was mainly aimed at maximizing the recycling of RAP from OGFCs. The designed experimental plan consisted of several tests which were carried out in order to shed some light on the performance and sustainability level achieved for mixtures with high percentages of RAP from OGFC. Results on volumetric, mechanical, and functional performance of porous European mixes with high-RAP content are reported in previous studies related to the first task of the project (Praticò et al. 2011; Praticò & Vaiana 2012; Praticò et al., 2012a; 2013; 2014b). The paper is organized as follows: Section 2 presents the methodology (experimental plan, synthesis of tests carried out) and materials (analysis of RAP gradation and composition); Section 3 shows test results and discussion. Finally, conclusions are drawn in Section 4.

2 MATERIALS AND METHODS

2.1 Experimental plan

An experimental plan was designed and carried out in order to investigate on the surface and acoustic performance of a recycled PEM; in particular, the experimental plan includes 4 different phases:

1. Specimens manufacturing: samples of 150 mm of diameter were produced by using the Superpave Gyratory Compactor (Caputo et al. 2020). Samples were derived from two different sets of PEMs concretes, one produced with recycled material (M4) and the other, as a reference mixture (OGFC), made with virgin material;
2. Assessment of volumetric properties. Mixes volumetric properties were evaluated in terms of permeability (k, ASTM PS 129), bulk specific gravity (Gmb, cor: Vacuum sealing method according to ASTM D6752/ AASHTO T 331; dim: dimensional) and effective porosity (n, CoreLok method ASTM D6073) (Praticò & Vaiana 2013; Praticò et al. 2014c);
3. Texture Evaluation. Macrotexture measurements were carried out on each gyratory specimen by means of a Laser Profilometer (ISO 13473-3);
4. Acoustic Absorption Measurement. Samples with 100 mm diameter were cored from the previous specimens in order to measure the acoustic absorption by means of the Kundt's Tube (ISO 10534-2);

A further analysis of the overall data and the search for correlations between the measured quantities complete the experimental plan.

Figure 1.   Highlights of the experimental plan; a) Core drilling process; b) Cored specimen; c) Impermeabilization of the sample for acoustic absorption measurement; d) Kundt's Tube.

2.2   *Recycled and reference PEMs*

In particular, 84.2% of the reclaimed material was used for the production of a recycled single layer porous asphalt, M4 (Praticò *et al.* 2014b). The employed RAP, in turn, was obtained from the milling of an old porous asphalt pavement. Its dosage was determined in a standardized optimization process of a previous research project (Praticò *et al.* 2012, 2013, 2014b, 2015).

The OGFC reference mixture was designed with virgin materials (aggregates and binder) according to Italian Standard Specification for porous asphalt mixture (Anas 2001).

Mixes were compacted at 170°C by the gyratory compaction method, according to the EN 12697-31 by means of the Superpave Gyratory Compactor (SGC). Three different compaction levels were achieved: 10, 23 e 50 gyrations. In particular, 23 gyrations represent the number of gyrations which allows to obtain an intermediate percentage of air voids compared to the air voids recorded for 10 and 50 gyrations. For each level of compaction 4 samples were produced. Overall, considering two different mixes (M4 & reference OGFC), 24 samples were manufactured.

Figure 2 shows the gradation of M4 and the reference OGFC mixtures, represented by means of a 0.45 power gradation chart. The chart also shows the maximum density gradation (dotted straight line from the maximum aggregate size through the origin), the control points (of a dense-graded gradations), and restricted zone (through which gradations should

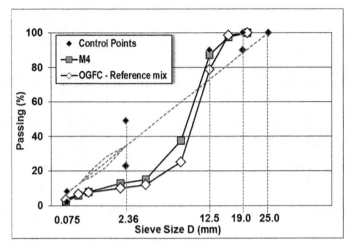

Figure 2.   Recycled porous asphalt (M4) and reference OGFC gradations.

not pass). Rejuvenating agents were added to the mix in order to improve the recovered binder performance.

3   RESULTS AND DISCUSSIONS

Table 1 provides an overview of the results from the tests carried out on both M4 and reference OGFC mix. Data are reported for each compaction level (@10, @23, and @50 (i..e., number of gyrations, 10, 23, and 50, respectively). Results demonstrate that recycled PEM undergoes a greater compaction than the reference mixture, which is reflected in a greater bulk specific gravity and a reduction of porosity. Regarding macrotexture, there is no big difference in terms of MPD results, which are even higher in mixture M4 for the compaction levels @23 and @50.

Table 1.   Summary of results of the tests carried out.

ID Sample@(N)	OGFC_@10	OGFC_@23	OGFC_@50	M4_@10	M4_@23	M4_@50
MPDiso (mm)	3.50	2.32	2.18	3.02	2.62	2.37
k_{20} (cm/sec)	0.38	0.18	0.11	0.19	0.09	0.02
CoreLok - % Porosity	18.36	15.39	12.79	14.56	11.79	9.83
CoreLok - Bulk Specific Gravity (g/cmc)	2.05	2.12	2.20	2.12	2.22	2.27

The comparative analysis between the mixes for each parameter is displayed in the following figures. As can be seen in Figure 3, besides the influence of the compaction rate, the permeability of the recycled mix is lower than the one of the reference porous mixes.

The permeability coefficient observed for M4 for 10 gyrations is comparable with the value obtained for the reference mix for 23 gyrations, also the value recorded for M4 for @23 is approximately comparable with the k20 obtained for OGFC and @50. The decrease in porosity between M4 and OGFC is comparable for each level of compaction and ranges from 21% (@10 gyrations) to 23% (@23 and @50 gyrations).

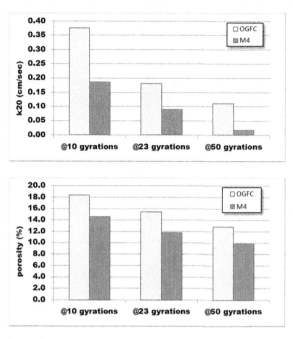

Figure 3. Permeability performance: variability of the permeability coefficient k_{20} and porosity for each level of compaction for the recycled and reference mixes.

Table 2. Acoustic absorption results.

Frequency (Hz)	a_0 | Acoustic Absorption Coefficient					
160	0.086	0.089	0.096	0.085	0.120	0.100
200	0.051	0.066	0.076	0.054	0.095	0.078
250	0.075	0.103	0.138	0.098	0.136	0.080
315	0.114	0.154	0.224	0.173	0.161	0.090
400	0.173	0.247	0.321	0.241	0.168	0.113
500	0.347	0.419	0.288	0.301	0.198	0.189
630	0.695	0.400	0.202	0.385	0.214	0.200
800	0.606	0.264	0.202	0.383	0.183	0.176
1000	0.254	0.272	0.265	0.304	0.253	0.167
1250	0.193	0.378	0.329	0.311	0.306	0.229
1600	0.353	0.339	0.271	0.315	0.280	0.336
$a_0$max	0.695	0.419	0.329	0.385	0.306	0.336

The measures of acoustic absorption were carried out on each sample through the use of the Kundt's Tube, according to the procedure defined by ISO 10534-2. This latter allows evaluating in the laboratory the acoustic absorption spectrum, recording the a_0 coefficients in the frequency range between 160 and 1,600 Hz, as reported in Table 2. The maximum values of the acoustic absorption coefficient a_0 max are reported below.

Values are plotted in Figure 4, which shows the acoustic absorption spectrum for each level of compaction, comparing the reference OGFC with the recycled porous asphalt M4. Focusing on the range between 600 and 830 Hz, which corresponds to part of the frequencies that refer to tire-pavement interaction of moving vehicles (Mahmud et al. 2021), OGFC

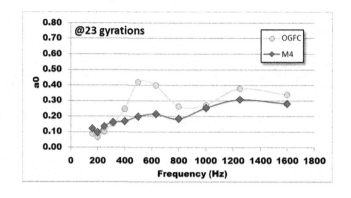

a)

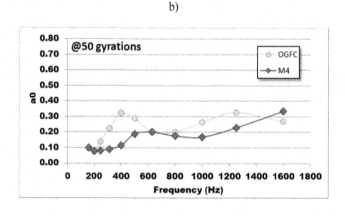

b)

c)

Figure 4. Acoustical performance: variability of the acoustic absorption coefficient a_0 for each level of compaction for the recycled and reference mixes.

shows a significant reduction in the absorption coefficient between the levels @10 and @23, from 0.695 to 0.400, and the same percentage decrease is noted in mixture M4 (42% and 40% of reduction, respectively). Therefore, although the reference mix exhibits better acoustic performance, M4 follows the same patterns, and this is true at @10 and @23 gyrations, whereas values tend to overlap at @50 gyrations.

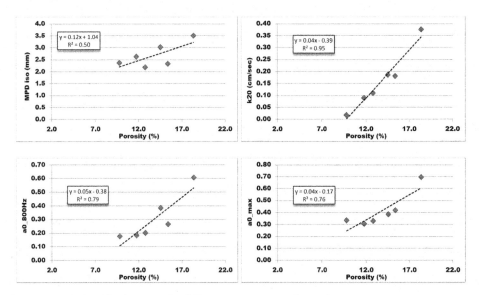

Figure 5.   Relationships between the different parameters and porosity.

Figure 5 reports the relationships between porosity and the main parameters analysed. As it is possible to see, the best linear regression of the experimental data was found for the relationship between k_{20} and porosity (R^2 = 0.95; the higher the porosity, the higher the permeability performance). The other investigated variables show a low dispersion of data and correlation coefficients always higher than 0.5. Significant trends are obtained for the acoustic absorption coefficient (R^2 = 0.79 and 0.76 for the maximum acoustic absorption coefficient a_0 and the 800 Hz frequency tested, respectively; higher porosity improves the performance of the pavement in the field of noise pollution reduction). For the relationship between porosity and MPD, it is noted that MPD is moderately affected by porosity. Taken together, these results suggest that the recycled mix shows performance which are coherent with the trends defined for the reference porous mix.

4   CONCLUSIONS

Several studies showed the importance of investigating the potential use of RAP within asphalt pavements, with the aim of pursuing the objectives of reducing the consumption of raw materials and reuse materials otherwise destined to landfill. Previous investigations confirm the adequacy of the recycled PEMs regarding the mechanical properties. The aim of the paper was the investigation on the surface performance of OGFC mixture obtained with high percentages of RAP coming from reclaimed porous asphalt. About 84% of RAP was used in a single layer porous asphalt and compared with an OGFC mixture made of virgin materials.

Although the milling process and aging impacts on the RAP characteristics and the recycled PEM is susceptible to a greater compaction, performance is promising and like the one of the reference mixes. Therefore, the possibility of using high percentages of RAP together with the good performance achieved could make this technique suitable for paving.

Further research is needed aimed at quantifying the corresponding economic and environmental sustainability using LCA and LCCA analysis.

REFERENCES

Alvarez A.E., Mora J.C. and Espinosa L.V., 2018. Quantification of Stone-on-stone Contact in Permeable Friction Course Mixtures Based on Image Analysis. *Construction and Building Materials*, 165, 462–471.

ANAS 2001. *Capitolato Speciale d'Appalto – Pavimentazioni.* ANAS, Ente Nazionale per le Strade, Italy.

Boscaino G. and Pratico F.G., 2001. A Classification of Surface Texture Indices of Pavement Surfaces [Classification et Inventaire des Indicateurs de la Texture Superficielle des Revêtements des Chaussées]. *Bulletin des Laboratoires des Ponts et Chaussees* (234), pp. 17–34+123+125+127.

Caputo P., Calandra P., Vaiana R., Gallelli V., De Filpo G. and Oliviero Rossi C. (2020). Preparation of Asphalt Concretes by Gyratory Compactor: A Case of Study with Rheological and Mechanical Aspects. *Applied Sciences*, vol. 10, ISSN: 2076–3417, doi: 10.3390/app10238567.

European Commission 2011. *A Resource-efficient Europe – Flagship Initiative Under the Europe 2020 Strategy, Brussels.* Available online: https://eur-lex.europa.eu/LexUriServ/LexUriServ.do?uri=COM:2011:0021: FIN:EN:PDF.

Gu F., Watson D., Moore J. and Tran N., 2018. Evaluation of the Benefits of Open Graded Friction Course: Case Study. *Construction and Building Materials*, 189, 131–143.

Li X., Marasteanu M.O., Christopher W. and Clyne T.R., 2008. Effect of RAP (Proportion and Type) and Binder Grade on the Properties of Asphalt Mixtures. *Transportation Research Board 86th Annual Meeting Compendium of Papers, National Research Council, Washington, DC.*

Losa M. and Leandri P., 2012. A Comprehensive Model to Predict Acoustic Absorption Factor Of Porous Mixes. *Materials and Structures*, 45(6), 923–940.

Mahmud M.Z., Hassan N.A., Hainin M.R., Ismail C.R., Jaya R.P, Warid M.N., Yaacob H. and Mashros N., 2021. Characterisation of Microstructural and Sound Absorption Properties of Porous Asphalt Subjected to Progressive Clogging. *Construction and Building Materials*. 283: 122654.

Praticò F.G., Vaiana R. and Giunta M., 2011. *Can You Really Recycle PEMs back to PEMs and be Confident Also on Surface Properties? AIPCR - XXIVth World Road Congress, Mexico City.*

Praticò F.G. and Vaiana R., 2012. *Improving Infrastructure Sustainability in Suburban and Urban Areas: Is Porous Asphalt the Right Answer? And how? WIT Transactions on the Built Environment.* 128: 673–684.

Praticò F.G., Vaiana R., Giunta M., Moro A. and Iuele T., 2012a. Permeable Wearing Courses by Recycling PEMs: Strategies and Technical Procedures. *Procedia: Social & Behavioral Sciences, vol. 53, p. 276–285, ISSN: 1877-0428, doi: 10.1016/j.sbspro.2012.09.880*;

Praticò F.G., Vaiana R. and Giunta M., 2012b. Sustainable Rehabilitation of Porous European Mixes. *ICSDC 2011: Integrating Sustainability Practices in the Construction Industry.* 535–541.

Praticò F.G. and Vaiana R., 2013. A Study on Volumetric Versus Surface Properties of Wearing Courses. *Construction and Building Materials*, vol. 38, p. 766–775, ISSN: 0950-0618, doi: 10.1016/j. conbuildmat.2012.09.021.

Praticò F.G., Vaiana R. and Giunta M. 2013. Pavement Sustainability: Permeable Wearing Courses by Recycling Porous European Mixes. *Journal of Architectural Engineering*, ISSN: 1076–0431, doi: 0.1061/ (ASCE)AE.1943-5568.0000127.

Praticò, F.G., 2014. On the Dependence of Acoustic Performance on Pavement Characteristics. *Transportation Research Part D: Transport and Environment*, 29, 79–87.

Praticò F.G., Vaiana R., Fedele R., 2014a. A study on the dependence of PEMs acoustic properties on incidence angle. *International Journal of Pavement Engineering*, vol. 16, p. 632–645, ISSN: 1029-8436, doi: 10.1080/10298436.2014.943215.

Praticò F.G., Vaiana R., Iuele T. and Puppala A.J. 2014b. HMA Sustainability: Producing a Recycled Permeable Mix that Performs as well as the Original Porous Mix. In: *Sustainability, Eco-efficiency, and Conservation in Transportation Infrastructure Asset Management.* p. 641–646, CRC Press, Taylor & Francis Group, Pisa, April 22-25 2014, doi: 10.1201/b16730-92.

Praticò F.G., Vaiana R. and Moro A., 2014c. Dependence of Volumetric Parameters of Hot Mix Asphalts on Testing Methods. *Journal of Materials in Civil Engineering*, vol. 26, p. 45–53, ISSN: 0899-1561, doi: 10.1061/(ASCE)MT.1943-5533.0000802.

Praticò F.G, Vaiana R and Iuele T. 2015. Permeable Wearing Courses from Recycling Reclaimed Asphalt Pavement for Low-Volume Roads. Optimization Procedures. *Transportation Research Record*, vol. 2474, p. 65–72, ISSN: 0361-1981, doi: http://dx.doi.org/10.3141/2474-08.

Praticò F.G., Vaiana R., Noto S., 2018. Photoluminescent Road Coatings for Open-graded and Dense-Graded Asphalts: Theoretical and Experimental Investigation. *Journal of Materials in Civil Engineering*, vol. 30, ISSN: 0899-1561, doi: 10.1061/(ASCE)MT.1943-5533.0002361.

Wu H., Yu J., Song W., Zou J., Song Q., and Zhou L., 2020. A Critical State-of-the-art Review of Durability and Functionality of Open-graded Friction Course Mixtures. *Construction and Building Materials*, 237, 117759.

Xie N., Akin M., and Shi X., 2019. Permeable Concrete Pavements: A Review of Environmental Benefits and Durability. *Journal of Cleaner Production*, 210, 1605–1621.

Roads and Airports Pavement Surface Characteristics – Crispino & Toraldo (Eds)
© 2023 The Author(s), ISBN 978-1-032-55149-4

Mix design of recycled asphalt concretes for sustainable rural mobility: Bicycle and/or pedestrian routes

Saverio Olita*, Donato Ciampa, Maurizio Diomedi & Francesco Paolo Rosario Marino
School of Engineering, University of Basilicata, Potenza, Italy

ABSTRACT: The ever-increasing demand for environmental sustainability in the production of bituminous asphalt concretes requires technical solutions to reduce pollutants emissions into the atmosphere and to reduce the consumption of natural raw materials, encouraging the recycling of secondary raw materials. This paper presents the results of a bituminous asphalt concrete mix design study to be used, in single layer and rural context, for construction of bicycle and/or pedestrian superstructures that are sustainable at the same time, in environmental and economic terms. In particular, Construction and Demolition Waste (CDW) recycled aggregates cold-bonded with bitumen emulsion were used. The first part of the study focused on the mix design of asphalt concretes made up exclusively of CDW aggregate and variable percentages of bitumen $(6 \div 10\%)$. The first results of mixes performance characterization (porosity, Marshall stability, indirect tensile stress, Cantabro index, etc.) showed quite modest results. To improve the mixtures performance, an addition of 2% by weight of artificial filler (pozzolanic cement) was provided. Laboratory performance characterization has made it possible to establish that these recycled asphalt concretes can be effectively used for the construction of bicycle and/or pedestrian superstructures, as well as for the environmentally sustainable construction of a secondary interconnecting road to serve the diffuse building of the territory in the suburban area. Since the obtained recycled asphalt concrete, by its nature, does not exhibit a good surface texture, in order to provide an aesthetically aspect and ensure the road surface regularity and durability, a surface finishing was experimented with a mixture of Portland cement and finely crushed glass that would provide the surface with a light color and a certain degree of refraction and reflectance. The research demonstrates that it is possible to achieve good results in terms of both reducing production costs and environmental impact, in the production of eco-sustainable asphalt concretes with "lower performance", in absolute terms, but which are fully compatible with those required for cycling and/or pedestrian superstructures.

Keywords: CDW (Construction and Demolition Waste), Sustainable Asphalt Concretes, Cycle/Pedestrian Routes, Sustainability, Circular Economy

1 INTRODUCTION

Environmental sustainability, cities liveability and the possibility of reducing activities with high environmental impact are now indispensable goals and foundational elements at the basis of development policies. In this context, sustainable territorial development and sustainable urban design are key themes of planning and an integral part of its lexicon (DETR 1998).

---

*Corresponding Author: saverio.olita@unibas.it

DOI: 10.1201/9781003429258-78

Uncontrolled, and often unbalanced, growth in production and consumption causes problems such as air pollution (Agostinacchio et al. 2014) and exhaustion of non-renewable natural resources, generating environmental incompatibilities and making new models of sustainable development increasingly necessary (Marino et al. 2019). The development of a recovery and regeneration system through design, innovation, waste and pollution management enables the construction of circularity models, with a holistic multidisciplinary approach to environmentally sustainable design. In these models interact operations that involves practical actions such as reducing land consumption and more efficient use of resources (*urban innovation*), design of nZEB (*nearly Zero Energy Building*) and positive energy balance buildings (*building innovation*), bicycle paths and sharing systems (*mobility innovation*), telecommunications and monitoring systems (*digital innovation*).

Even in road engineering, there is enormous potential, in terms of innovation, that by making use of the circular economy is available for sustainable development. Such as, for example, the strategy of reducing the use of raw materials, recovery-reuse-recycling with the minimization and valorisation of waste, useful life extension of materials to be reused also thanks to *eco-design* processes that improve or finalize their recyclability.

The contribution of this study is precisely in the recycling of end-of-life materials, otherwise destined for landfill, for the design of suitably packaged recycled asphalt mixes, whose laboratory performance characterization is aimed at making mixes suitable for use in the construction of bicycle and/or pedestrian superstructures, as well as, for the environmentally sustainable construction of a secondary interconnecting road system serving the widespread construction of the territory in the suburban area.

The research was carried out in the Road Construction and Construction Technology (La. Te.C.) laboratories of the School of Engineering of the University of Basilicata and was supported under the MIUR PON R&I 2014-2020 Program - project MITIGO, ARS01_00964 (MitiGO 2020).

2 STATE OF ART AND RESEARCH OBJECTIVES

Construction and demolition waste (CDW) constitutes a very wide and varied range of materials. A substantial fraction of them, particularly those containing concrete and bricks, are suitable to replace natural aggregates in a variety of applications (Agostinacchio et al. 2009; Bennert et al. 2000; Ciampa & Olita 2014; Ciampa et al. 2020; Olita & Ciampa 2021; Ossa et al. 2016), including the mix design of asphalt concretes for road pavement base layers or asphalt concretes intended for uses that do not require the achievement of particular performance levels.

According to the Statistical Office of the European Union - EUROSTAT, CDW is, in absolute terms, the most significant stream of special waste generated in Europe. Specifically for 2018, in the EU (European Union - 28 countries) CDW generation was 371.9 million tons (Mt), up from 2016 - 344.7 Mt and 2014 - 314.9 Mt (EUROSTAT 2022). In Italy, waste from construction and demolition activities produced in 2019 was 68.3 Mt, corresponding to 44.4 percent of total special waste (ISPRA 2021). Data on CDW reuse also confirm a continuous increasing trend. In Italy, the CDW reuse rate, calculated on the basis of data on the generation and management of this type of waste, stood at 76.2 percent in 2016, above the 70 percent target set by European legislation (Directive 2008/98/EC) for 2020.

The experimentation presented in this paper is marked precisely by the use of recycled aggregates, obtained from civil works construction and demolition activities (CDW), suitable to replace 100% virgin quarry aggregates. In addition, a low-impact cold recovery technique using bitumen emulsion was used to make the recycled asphalt concrete mixes. The experimental investigation, therefore, focused on the design of a recycled asphalt concrete mix, to be used in a single layer and suburban area, which can be used in an economical and low-impact bicycle/pedestrian superstructure in terms of materials used and construction techniques.

3 MIX-DESIGN OF ECO-SUSTAINABLE ASPHALT CONCRETE MIX

3.1 *Materials employed*

The CDW aggregate used for the experiments was supplied by INECO s.r.l. (Barile-PZ-Italy). The CDW was obtained from the selection and crushing, in a fixed plant, of recycled materials and specifically from demolition waste and scrap from construction and road works. Compliance with the leaching test (UNI 10802 and EN 12457-2) is met as evidenced by the results shown in Table 1.

Table 1. Results of leaching tests on CDW.

Parameters	Units of measure	Test Results	Threshold limit value
Nitrates	mg/l	5	50
Fluorides	mg/l	0.4	1.5
Cyanides	µg/l	2	50
Barium	mg/l	0	1
Copper	mg/l	0.01	0.05
Zinc	mg/l	0	3
Beryllium	mg/l	1	10
Cobalt	mg/l	6	250
Nichel	mg/l	4	10
Vanadium	µg/l	15	250
Arsenic	µg/l	6	50
Cadmium	µg/l	1	5
Chromium	µg/l	15	50
Lead	µg/l	17	50
Selenium	µg/l	1	10
Mercury	µg/l	0	1
COD	mg/l	17	30
Solfates	mg/l	35	750
Chlorides	mg/l	9	750
PH	pH Unit	8.3	5.5 <> 12.0

The used CDW, from the particle size point of view, consists of a d/D fraction equal to 0/31.5, characterized by aggregates with maximum diameter (D) equal to 31.5 mm and minimum (d) equal to 0. The particle size distribution, shown in Figure 1 and Table 2, was carried out by employing the sieves set related to the "Base Group" of EN 13242 suitably supplemented. Figure 1 and Table 2, shown the ANAS grading envelope (ANAS Spa 2016) for one base layer HMA (Hot Mix Asphalt) against which the possibility of using CDW was evaluated. According to EN 13242, the main constituents of CDW aggregate have been identified.

In particular it results in: Rc=7.36% (concrete, concrete products, concrete masonry elements), Ru=74.35% (unbound aggregate, hydraulically bound aggregate), Rb=17.76% (clay masonry elements - bricks, tiles), Rg=0.06% (glass) and X = 0.47% (other). Therefore, the content categories constituting CDW according to EN 13242 are: Rc7, Rcug70 and Rb30.

The analysis carried out showed that, the particle size distribution of the CDW aggregate falls largely within the reference grading envelope (Figure 1) despite the fact that a filler deficit is evident at the bottom and a slight surplus of coarse aggregate at the top. However, considering that CDW is largely made up of "hydraulically bound aggregate" i.e., plaster mortar (Ru=74.35%) it is reasonable to expect further disintegration of this constituent during the mixing phase of the recycled asphalt concrete mix resulting in the production of fine and simultaneous reduction of coarse material.

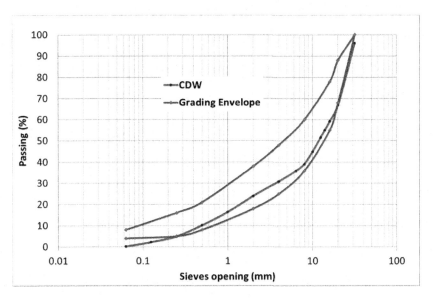

Figure 1.   CDW particle size distribution and ANAS reference grading envelope.

Table 2.   CDW particle size distribution and ANAS reference grading envelope.

Sieves opening (mm)	Passing (%)	Envelope lower limits (%)	Envelope upper limits (%)
40	100.0	100	100
31.5	95.9	100	100
20	66.7	68	88
16	59.1	55	78
14	54.9	-	-
12.5	51.5	-	-
10	44.8	-	-
8	39.0	36	60
6.3	35.7	-	-
4	30.7	25	48
2	23.9	18	38
1	16.3	-	-
0.5	10.1	8	21
0.25	4.8	5	16
0.125	2.1	-	-
0.063	0.1	4	8

Therefore, at an initial stage, it was considered reasonable not to carry out any aggregate particle size integration, except to conduct post-mixing verifications of its particle size distribution. These verifications conducted on some of the studied admixtures confirmed the particle size distribution tendency to fall almost entirely within the reference grading envelope, especially in the upper part, while a slight filler deficit persisted in the lower part of the curve. The choice made is related to the goal of as-is use of CDW in favour of greater economic sustainability of the recycled asphalt concrete mix.

The real volumic mass of CDW was found to be 2.56 g/cm^3, while the Los Angeles index (EN 1097-2) was found to be 30. According to EN13808, the used bitumen emulsion is characterized by code C55B4. It is a 55 percent unmodified bitumen cationic emulsion with a class 4 (fast) rupture behaviour.

3.2 Mix-design of cold recycled asphalt concrete: Phase 1

The first phase of mix-design study involved identifying the optimum moisture content of recycled aggregate. This determination was made by evaluating the workability of the aggregate and emulsion mixture for the purpose of mix homogeneity.

The dry CDW aggregate was wetted with an initial 4 percent water content, and the workability/homogeneity of mixing with 7 percent bitumen was evaluated. Then water content of aggregate was increased by 1 percent from time to time, keeping the bitumen percentage fixed. In this way, the aggregate wetting optimum value to meet the mixture workability/homogeneity requirement was identified. This value was found to be 10%.

3.3 Mix-design of cold recycled asphalt concrete: Phase 2

Having defined the recycled aggregate optimum moisture content, the optimum binder content was identified by evaluating the performance response of different mixtures in terms of porosity, Marshall stability (EN 12697-17), indirect tensile strength (EN 12697-23) and Cantabro index (EN 12697-17). Therefore, different mixtures were made with the following bitumen percentages: 6, 7, 8, 9 and 10%. It should be clarified that the strict implementation of Marshall mix-design methodology (EN 12697-34), let alone the volumetric one (EN 12697-31), was not deemed necessary, considering to this type of recycled mix, no special mechanical performance is required but rather maximum environmental/economic sustainability.

The authors assumed as minimum performance acceptability threshold, performance characteristic values equal to one-third of those normally required of a conventional base layer asphalt concrete mix. Therefore, Marshall stabilities >2.3 kN and indirect tensile strengths >0.25 N/mm^2 were considered acceptable. The EN 12697-17 (Cantabro test) evaluates the Marshall specimen mass loss after 300 rotations in Los Angeles device (EN 1097-2) without abrasive load. In this work, the mass loss at 50 and 100 rotations was evaluated.

3.4 Mix-design of cold recycled asphalt concrete: Phase 3

Table 3, which summarizes all mixtures composition and their performance characteristics, shows that most economically sustainable cold-recycled asphalt concrete mixes (mixes 1 and

Table 3. Mechanical characterization of recycled asphalt concrete mixtures.

Mix	Bitumen (%)	Hydraulic binder (%)	Porosity (%)	Marshall Stability (kN)	R_t(N/ mm^2)	Cantabro index@50 rotations (%)	Cantabro index@100 rotations (%)
1-CDW100%	6	-	19.8	2.54	0.31	25.42	42.39
2-CDW100%	7	-	19.9	2.60	0.33	24.45	40.15
3-CDW100%	8	-	19.1	2.63	0.38	23.05	38.23
4-CDW100%	9	-	18.6	2.72	0.38	22.66	32.76
5-CDW100%	10	-	18.9	2.79	0.39	15.08	28.39
6-CDW100%	6	2	17.5	3.85	0.49	10.42	17.83
7-CDW100%	8	2	16.4	3.96	0.58	8.76	14.54

2) exhibit modest performance overall. In order to keep low bituminous binder percentage and still improve the overall performance of the recycled asphalt concrete mix, 2 additional mixes were prepared with 6% and 8% bituminous binder by adding 2% hydraulic binder i.e., pozzolanic cement (Table 3, mixes 6 and 7).

The 2% by weight of pozzolanic cement (filler) not only improves adhesive properties with aggregates, but also, in synergy with bitumen emulsion, makes the recycled asphalt concrete mix more rigid. In addition, it resolves the filler deficit highlighted by the lower part of the mix's particle size distribution. The mixes made up, with the addition of hydraulic binder, consist of 100% CDW aggregate moistened to 10%, 6% (mix 6) and 8% (mix 7) bitumen and 2% cement.

Figure 2 shows, as an example, the configuration of a mix 6 Marshall specimen, before the Cantabro test and subsequently at 50 and 100 rotations. The images show the disintegration degree and relative mass loss experienced by the specimen, progressively, at 50 and 100 rotations.

3.5 *Considerations for recycled asphalt mixes*

The performance characterization of cold recycled concrete mix shown in Table 3 indicate an overall low performance mix when compared with a traditional base course HMA. However, as already clarified, the particular use of recycled asphalt concrete mix for construction of cycle/pedestrian routes makes it possible to set a performance target that the authors deemed permissible within one third of that normally required by Specifications for a base course HMA.

The comparative analysis of results shows a marked performance improvement of mixes 6 and 7, in which 2% pozzolanic cement was added, compared to their predecessors.

In particular, mix 6 compared with mix 1 shows indirect tensile strengths about 60% higher and Marshall stability 50% higher. The Cantabro index @100 rotations also shows a significantly lower mass loss of mix 6 compared to mix 1, due to the greater consistency conferred by the pozzolanic cement filler. Although mix 7 exhibits the highest performance parameters, it is less economically sustainable as it requires 2% more bituminous binder than mix 6. Therefore, mix 6 is definitely the more sustainable mix that balances the mechanical strength aspect and the economic aspect.

4 ECO-SUSTAINABLE ASPHALT CONCRETE SURFACE FINISHING

As already stated, the recycled asphalt mix design was conducted with reference to a base course layer mix of aggregates. By its nature, this asphalt concrete mix is quite open and when applied it produces a rough surface with a very pronounced macro-roughness.

In addition, the nature of the aggregate (100% CDW) in which the grains are very varied in shape and material type makes the compacted surface of the recycled asphalt concrete mix even more "uneven". Hence the need for a surface finishing treatment to make the rolling surface "smooth" and levelled. This treatment is required to meet several requirements:

1. to reduce the surface macro-roughness to low values and determine a "smooth" and regular surface;
2. to waterproof the underlying single layer of recycled asphalt concrete;
3. to provide the surface with a light coloring in order to make it less impactful from an environmental point of view;
4. give the surface a reflective characteristic to make path visible to bicycle headlights with environmental darkness;
5. be sustainable in terms of economics, technologies and materials to be used.

Leaving aside all commercial solutions, which have the fundamental flaw of having a generally high cost and the need, in some cases, for relatively complex construction technologies, a simple and sustainable solution was sought. The surface finishing tested in this study involves the use of Portland cement mixed with recycled colored glass granulate.

The finishing layer construction technique involves the preparation of a dry mix of Portland cement and crushed glass all passed through a 0.5 mm sieve in percentages varying

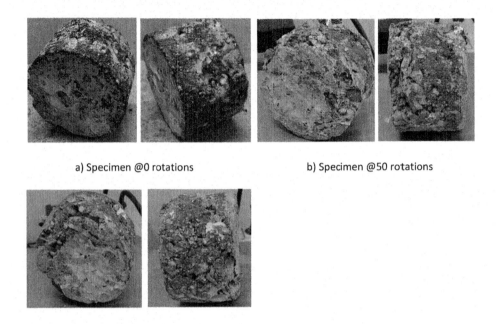

a) Specimen @0 rotations   b) Specimen @50 rotations

c) Specimen @100 rotations

Figure 2.   a) Mashall specimen before Cantabro test - b) @ 50 rotations - c) @ 100 rotations.

from 15 to 25% by weight depending on the desired level of reflectance. The mixture is distributed dry, with a rubber-coated rake, over the surface of the recycled asphalt layer, evenly filling all the intergranular voids associated with macrotexture. Subsequently, the surface is sprayed with water spray until the glass-cement mixture has completely hydrated.

The surface finishing study was conducted in laboratory by implementing on the surface of special Marshall specimens finishing layers with % glass varying from 0 to 30% (Figure 3a). This study made it possible to identify the average consumption per square meter of cement as well as the by weight percentages of glass most effective in making the surface reflective, which were found to vary from 15 to 25%. Figures 3b and 3c, as examples, show a surface finishing layer with 0 and 30% glass, respectively.

It should be noted that verifications were also conducted in accordance with the Decree of 11 January 2017 (G.U. no. 23 of 28/1/2017), by which the Italian Ministry of the Environment and Protection of Land and Sea established the adoption of minimum environmental criteria for construction (CAM), thus aligning with the environmental protection strategies already prevailing internationally. About the Solar Reflectance Index (SRI), section 2.2.6 of the cited decree, concerning the "*Reduction of the impact on the microclimate and atmospheric pollution*" establishes the following: for impermeable surfaces, the use of materials with a high solar reflectance index is foreseen, which, for external surfaces (e.g., pedestrian paths, pavements, squares, cycle paths, etc.) must guarantee an SRI index $\geq$ 29. The SRI index, which generally has a value between 0 and 100, the higher it is the more the surface exposed to solar radiation will remain "cool" (i.e., have a low temperature rise).

The solar reflectance and emissivity of a material, measured by means of a UV-VIS-NIR spectrophotometer and an emissometer, in accordance with ASTM E 903 and ASTM G17 standards, allowed the determination of the stationary surface temperature (TS) reached by the material under the environmental conditions of solar radiation and ventilation defined in ASTM E 1980-01. ASTM E 1980-01 standard requires the SRI index to be calculated under the following environmental conditions:

1. solar radiation 1000 W/m^2;
2. ambient temperature 37°C (310 K);
3. "sky" temperature 27°C (300 K);
4. ventilation: weak (0 ÷ 2 m/s); medium (2 ÷ 6 m/s); strong (6 ÷ 10 m/s).

The SRI index is calculated as:

$$SRI = \frac{(Tb - TS)}{(Tb - Tw)} \cdot 100 \tag{1}$$

where Tb and Tw are the temperatures reached respectively by a black surface (solar reflectance factor 0.05 - emissivity 0.9 - SRI = 0 under standard environmental conditions) and a white surface (solar reflectance factor 0.8 - emissivity 0.9 - SRI = 100 under standard environmental conditions) under the same environmental conditions for which TS was calculated.

The investigation revealed that the Portland cement surface finishing, tested according to ASTM C 1549 has a solar reflectance of 0.3 and an SRI of 34. With Portland cement surface finishing with 20 per cent crushed glass, a total SRI value of 42 was obtained, considering the color pigmentation shown in Figure 3c.

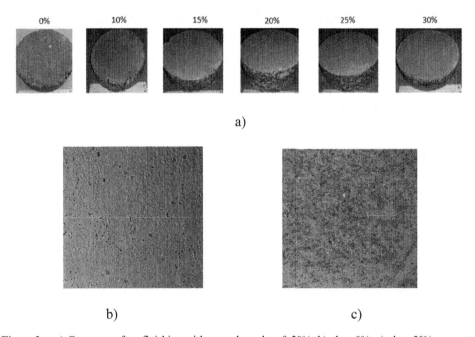

Figure 3. a) Cement surface finishing with granulate glass 0-30%; b) glass 0%; c) glass 30%.

5 MATERIALS APPLICATION AREAS AND ECONOMIC SUSTAINABILITY

As stated, the proposed experimental investigation focused on the mix design of a recycled asphalt concrete mix that could be used in an economical and environmentally sustainable cycle/pedestrian superstructure, in terms of materials and construction techniques.

In this light, designed recycled asphalt concrete will be employed for the construction of segments to complete the "*Lucanian Dolomites cycle route*" (Ciclovia delle Dolomiti Lucane) and, more generally, for the construction of eco-sustainable cycle/pedestrian routes within the so-called "*Quadrilateral of the Lucanian Dolomites*" defined by the municipal territories of Albano di Lucania Campomaggiore, Castelmezzano and Pietrapertosa (Potenza-Basilicata-Italy), all of which are characterized by remarkable landscape, environmental, historical and tourist significance. This area is part of the Lucanian Dolomites (Basilicatanet 2022) and can be accessible via E847-Basentana highway. The 28 km-long "*Lucanian Dolomites cycle route*" is part of this area (Figure 4). It is largely made up of bicycle paths shared with vehicular traffic, which is very limited on this route, rather than developing on its very own. The route has as its reference point the Gallipoli Cognato Regional Park, a natural reserve rich in forests, and is defined by a ring circuit around the two towns of Castelmezzano and Pietrapertosa (Figure 4).

About economic sustainability of recycled asphalt concrete and its associated surface finishing, a detailed analysis of production and installation costs was conducted. Assuming an application thickness of 10 cm, the production and installation costs, per square meter, of the single layer and the surface finishing are 19% lower than the alone cost, of the same thickness, of traditional base layer HMA.

Considering only the out-of-pocket costs, as first approximation, a saving is identified, which does not take into account the additional economic advantages related to the using 100% CDW environmental benefits. However, it is worth pointing out the lower performance of the material that will have to be evaluated in order to identify the overall economic sustainability.

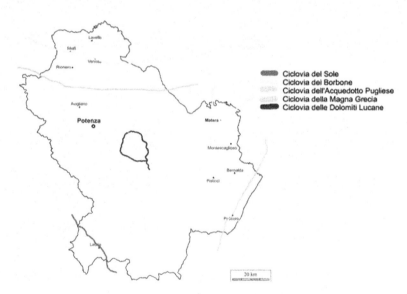

Figure 4. Cycle routes of Basilicata - Lucanian Dolomites cycle route.

6 CONCLUSIONS

This paper provides the results of a recycled cold asphalt concrete mix made of 100% CDW aggregates be used in a single-layer, extra-urban context for the construction of bicycle and/or pedestrian superstructures that is both environmentally and economically sustainable.

Given the recycled asphalt concrete lower performance and the performance threshold fixed at 1/3 of that of a traditional base course layer HMA, the asphalt concrete mixtures performance characterization (porosity, Marshall stability, indirect tensile strength, Cantabro index, etc.) showed results compatible with the thresholds set. In particular, a mix (mix 6) composed of: 100% CDW aggregate, 2% pozzolanic cement (filler) and 6% bituminous binder (obtained with a cationic bituminous emulsion with 55% unmodified bitumen), was identified which is more sustainable, among the studied recycled asphalt concrete mixes, and which balances the aspect of mechanical resistance and that of economic efficiency.

Since the obtained asphalt concrete, by its nature, does not exhibit a good surface finishing, in order to give an aesthetically pleasing appearance and guarantee the regularity and durability of the road surface, a surface finishing using Portland cement and finely crushed recycled glass (20%) was experimented with, providing the surface with a light color and a certain degree of refraction and reflectance (SRI = 42). The conducted experimentation shows how it is possible to achieve good results in terms of both lower production costs and environmental impact, in the production of environmentally sustainable asphalt concretes that exhibit, in absolute terms, "lower performance" but are fully compatible with those required for cycling and/or pedestrian superstructures.

AUTHOR CONTRIBUTIONS

All authors contributed equally to the research and the writing of this manuscript. All authors have read and agreed to the published version of the manuscript.

RESEARCH SUPPORT

This research has been supported by MIUR PON R&I 2014-2020 Program (project MITIGO, ARS01_00964).

ACKNOWLEDGMENTS

The authors are grateful to Arch. M.A. Schirò Administrator of INECO s.r.l. (Barile, Potenza, Italy) for supplying the CDW and technical support in the preparation of the article.

REFERENCES

Agostinacchio M., Ciampa D., Diomedi M. and Olita S., (2014). The Management of Air Pollution from Vehicular Traffic by Implementing Forecasting Models. In: *Massimo Losa and Tom Papagiannakis, Sustainability, Eco-efficiency, and Conservation in Transportation Infrastructure Asset Management.* p. 549–560, London: CRC Press 2014 - Taylor & Francis Group, ISBN: 9781138001473, doi: 10.1201/b16730-81.
Agostinacchio M., Diomedi M. and Olita S. (2009). "The Use of Marginal Materials in Road Constructions: Proposal of an Eco-compatible Section". In: *Andreas LOIZOS, Manfred PARTL, Tom SCARPAS, Imad*

AL-QADI. *Advanced Testing and Characterisation of Bituminous Materials*. Vol. 2, p. 1131–1142, London: Taylor & Francis Group, ISBN/ISSN: 978-0-415-55854-9.

ANAS Spa, 2016, Capitolato Speciale di Appalto—Pavimentazioni. Norme Tecniche per L'esecuzione del Contratto, Parte 2 (IT.PRL.05.21-Rev. 1.0). 2016. Available online: https://www.stradeanas.it/sites/default/files/CSA%20-%20NORME%20TECNICHE.pdf (accessed on 11 May 2022).

Basilicatanet, http://www.basilicatanet.com/ita/web/item.asp?nav=dolomitilucane (accessed on 11 May 2022).

Bennert T., Papp W.J., Jr., Maher A. and Gucunski N. Utilization of Construction and Demolition Debris Under Traffic-Type Loading in Base and Subbase Applications. *Transp. Res. Rec.* 2000, 1714, 33–39.

Ciampa D. and Olita S. (2014). Proposal of Eco-compatible Mixtures (C&D-EAF slag) for Road Constructions. In: *Massimo Losa and Tom Papagiannakis, Sustainability, Eco-efficiency, and Conservation in Transportation Infrastructure Asset Management*. p. 279–286, London: CRC Press 2014 - Taylor & Francis Group, ISBN: 9781138001473, doi: 10.1201/b16730-42.

Ciampa D.; Cioffi R.; Colangelo F.; Diomedi M.; Farina I. and OlitaS., 2020, Use of Unbound Materials for Sustainable Road Infrastructures. *Appl. Sci.* 2020, 10, 3465, https://www.mdpi.com/2076-3417/10/10/3465.

DETR – Department of the Environment Transport and the Regions 1998, *Planning for Sustainable Development. Towards Better Practice*. London, p. 12.

Directive 2008/98/EC on Waste and Repealing Certain Directives. Available online: https://eur-lex.europa.eu/legal-content/IT/LSU/?uri=celex:32008L0098 (accessed on 11 May 2022).

Eurostat, Generation of Waste-by-Waste Category, Hazardousness and NACE Rev 2 Activity. Available online: https://ec.europa.eu/eurostat/data/database (accessed on 11 May 2022).

ISPRA - Istituto Superiore per la Protezione e la Ricerca Ambientale, www.isprambiente.gov.it, Rapporti n. 344/2021 *"Rapporto Rifiuti Speciali – Edizione 2021"*, 2021, ISBN 978-88-448-1052-8.

Marino F.P.R., Lembo F., and Fanuele V. 2019. "Towards More Sustainable Patterns of Urban Development." *In SBE19 - Emerging Concepts for Sustainable Built Environment*, IOP Publishing - IOP Conference Series: Earth and Environmental Science 297 (2019) 012028, doi:10.1088/1755-1315/297/1/012028

MitiGO, 2020, MIUR PON R&I 2014-2020 Program (project MITIGO, ARS01_00964) - Azione II - Cluster tecnologici – dal titolo: *"MitiGO - Mitigazione dei Rischi Naturali per la Sicurezza e la Mobilità Nelle Aree Montane del Mezzogiorno"* https://www.mitigoinbasilicata.it/ (accessed on 11 May 2022).

Olita S. and Ciampa D. SuPerPave® Mix Design Method of Recycled Asphalt Concrete Applied in the European Standards Context. Sustainability 2021, 13, 9079, https:// doi.org/10.3390/su13169079

Ossa A., García J.L. and Botero E. Use of Recycled Construction and Demolition Waste (CDW) Aggregates: A Sustainable Alternative for the Pavement Construction Industry. *J. Clean. Prod.* 2016, 135, 379–386.

Pavements surfaces and urban heat islands (cool pavements, etc.)

Roads and Airports Pavement Surface Characteristics – Crispino & Toraldo (Eds)
© 2023 The Author(s), ISBN 978-1-032-55149-4

Colorimetry of pavements for urban planning

Laure Lebouc*
Routes de France, Paris, France
Light and Lighting Team, Cerema, Angers, France
Research Team, Spie batignolles malet, Portet-sur-Garonne, France

Sébastien Liandrat
STI Research Team, Cerema, Clermont-Ferrand, France

Romain Lafon
Routes de France, Paris, France
Research Team, Eurovia, Mérignac, France

Aurélia Nicolaï
Routes de France, Paris, France
Research Team, Spie batignolles malet, Portet-sur-Garonne, France

Fabrice Fournela & Florian Greffier
Light and Lighting Team, Cerema, Angers, France

ABSTRACT: In recent years, the use of light-coloured pavements has been encouraged internationally for urban design. In particular, colour is a lever that allows the designer to meet an objective of urban roads: the sharing of space, for uses and users. It is also a key factor for pavements in environmental and monumental contexts, where the aesthetics and appearance of the finished layer are essential characteristics.

The *Pavements and Lighting* working group has established a large library of current and innovative urban pavements available on the French market. The colour of these surfaces is notably part of the assessed properties. The objective of this study is to have a tool and an analysis method to understand the development of a project and its evolution according to the effects of colour change.

A set of 30 pairs of samples were characterised in their initial state and after 30 months of natural ageing in outdoor conditions, i.e. exposed to rain and sun. It included pavements of several categories: asphalt concrete (bituminous or synthetic binder), mastic asphalt, poured and precast cement concrete. The colour of each sample was measured using three different processes (devices and methods).

The results provide a first overview of the colorimetric coordinates per pavement family and show a wide range of achievable colorimetric characteristics. The impact of natural ageing on the colour of the road surface is studied by the comparison of the original with the 30 months measurements. The obtained results highlighted that not all pavements evolve in the same way, depending on their mixture and their finishing stage. Usually, the asphaltic pavements (concrete and mastic) tend to evolve more than the cement concrete samples (poured and precast). The colour of a pavement without surface treatment changes more than a treated one, as well as when dark aggregates are used. The same is true when the binder is synthetic, but it seems that the use of TiO_2 in synthetic asphalt pavements reduces the colour evolution.

Keywords: Road Surface, Urban Planning, Colorimetry

---

*Corresponding Author; laure.lebouc@spiebatignolles.fr

DOI: 10.1201/9781003429258-79

1 INTRODUCTION

The colour of road surfaces contributes to the perception of public spaces. By day, new types of materials with different colours can improve the legibility of the urban space by materializing on the ground the spaces dedicated to different uses. By night, a light-coloured pavement can reduce energy consumption for lighting. However, colour is very rarely considered in the design of urban development, there is no standard methodology for the assessment of the colour of pavement and its evolution over time is never evaluated.

The *Pavement and Lighting* working group (P&L group or in French "Revêtements et Lumière") is composed of project managers and public authorities[1], professional associations and unions of lighting designers[2] and road builders[3], public and private research organisations[4], and expert consultants. The group first organised and monitored demonstrators and operations on real sites to show the relevance of the challenges and concepts of optimal lighting (Abdo *et al.* 2010; Christory *et al.* 2014). It then elaborated a library of actual and innovative urban pavements available on the market to:

- facilitate the choice of decision makers,
- develop tools and methods for managers, lighting designers and road builders to optimize lighting both in interurban and urban areas.

This panel of urbans pavements contains asphaltic pavements, cement concrete pavements and natural stones. A complete characterization of this large pavement sample panel was conducted at the initial state (T0) and after 30 months of natural ageing (T30) (Liandrat *et al.* 2022; Muzet *et al.* 2021). The colour is notably part of the assessed properties.

This paper first presents the processes employed to measure the colour of each pavement. The results provide a first overview of the colorimetric coordinates per pavement family. Then, the results of the different processes are used to define a colour difference threshold. Finally, the impact of natural ageing on the colour of the road surface is studied by the comparison of the original with the 30 months measurements.

2 METHODOLOGY

2.1 *CIELAB colour space basics*

The 1976 CIE L*a*b* colour space, also referred to as CIELAB, is a colour space defined by the International Commission on Illumination (abbreviated CIE) in 1976 and particularly used for the characterisation of surface colours (CIE 2018). It characterises a colour with an intensity parameter L* corresponding to the luminance and two chrominance parameters, a* and b*, that describe the colour. CIELAB was intended as a perceptually uniform space, where a given numerical change corresponds to a similar perceived change in colour.

It is based on the opponent colour model of human vision, where red and green form an opponent pair, and blue and yellow form another opponent pair. The lightness value L*

---

[1] *AITF (Association des Ingénieurs Territoriaux de France)*
[2] *AFE (Association Française de l'Eclairage)*
[3] Office des Asphaltes, Routes De France, Specbea (Spécialistes de la Chaussée en Béton Et des Aménagements), CIMbéton (Centre D'information Sur le Ciment et ses Applications), Eurobitume
[4] Cerema (Centre d'Etudes et d'expertise sur les Risques, l'Environnement, la Mobilité et l'Aménagement), CERIB (Centre d'Etudes et de Recherches de l'Industrie du Béton), CTMNC (Centre Technique des Matériaux Naturels de Construction).

defines black at 0 and white at 100. The a* axis is relative to the green–red opponent colours, with negative values toward green and positive values toward red. The b* axis represents the blue–yellow opponents, with negative numbers toward blue and positive toward yellow (Figure 1).

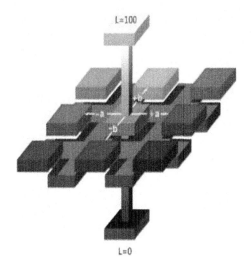

Figure 1.   Representation of CIELAB colour space.

The CIELAB coordinates (L*, a*, b*) in this colour space can be calculated from the tristimulus values XYZ (from CIEXYZ) with the following formulas (Eqs. 1–3). The subscript n denotes the values for the reference white point.

$$L^* = 116 \times f\left(\frac{Y}{Y_n}\right) - 16 \tag{1}$$

$$a^* = 500\left(f\left(\frac{X}{X_n}\right) - f\left(\frac{Y}{Y_n}\right)\right) \tag{2}$$

$$b^* = 200\left(f\left(\frac{Y}{Y_n}\right) - f\left(\frac{Z}{Z_n}\right)\right) \tag{3}$$

where $f(t) = \begin{cases} \sqrt[3]{t} & t > \delta^3 \\ \dfrac{t}{3\delta^2} + \dfrac{4}{29} & \text{otherwise} \end{cases}$ with $\delta = \dfrac{6}{29}$.

Knowing the coordinates of two surfaces, it is possible to evaluate the colour difference between them with a Euclidean distance as follows:

$$\Delta E_{ab}^* = \sqrt{\left(L_2^* - L_1^*\right)^2 + \left(a_2^* - a_1^*\right)^2 + \left(b_2^* - b_1^*\right)^2} \tag{4}$$

2.2   Methods and materials

To assess the colorimetry of the samples, several experiments were conducted using three Konica-Minolta colour measurement devices: two chromameters of the same model

(CR-410, Figure 2) and a spectrophotometer (CM-2300d, Figure 3). In this way, it will be possible to verify that the colour measurements do not differ from one instrument to the other and we will be able to define a valid colour difference threshold for these three processes.

Figure 3.   Konica-Minolta Spectrophotometer CM-2300d.

Figure 2.   Konica-Minolta Chromameter CR-410.

All the three instruments illuminate the sample with an illuminant D65 and have an integrating sphere that allows for homogeneous diffusion. They differ in the size of their measurement area: chromameters have a 50 mm aperture while the spectrophotometer has an 8 mm aperture. As a pavement is composed of aggregates whose size is usually between 2 and 14 mm, the measurement area of the chromameter is not sufficient to characterize a pavement in its entirety. To be more representative, for each of the protocols used, the experiment consisted of measuring the trichromatic characteristics L*, a* and b* on a road surface at several close areas.

For one sample, 20 measurements are made at different points with the chromameters as shown in Figure 4, while 56 measurements are needed with the spectrophotometer (because of its small aperture) (see Figure 5). The localisation of the measurement points could differ, depending on the shape of the sample (disk or rectangular). These measurement points are then averaged for each of the L*, a* and b* coordinates and assigned to the respective pavement.

2.3   Pavement samples

The first step was to establish a representative panel of urban and interurban surfacing materials including innovative French technologies. This panel includes 30 different pavements:

- 7 asphalt concrete (with bituminous or synthetic binder),
- 6 mastic asphalt (with bituminous or synthetic binder),
- 10 poured cement concrete,
- 7 precast cement concrete paving blocks.

For each type of pavement, several parameters have been differentiated to represent both conventional and innovative pavements used in cities. In particular, the following para-meters were variable, depending on materials:

- Formulation: nature and percentage of binder, aggregates granulometry, porosity;

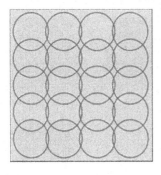

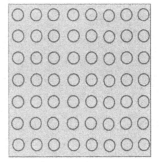

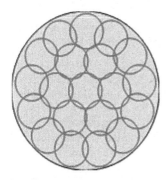

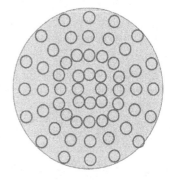

Figure 4.   Distribution of measurement areas
with chromameters.

Figure 5.   Distribution of measurement areas
with spectrophotometer.

- Poured or precast material;
- Aggregates colour: dark, clear depending on the colour of the used stones;
- Surface treatment: raw pavements, sandblasted, broomed, etc.

To work with values for T0 and T30 measured under the same conditions, two samples of each pavement were actually made and one of them was placed in the refrigerator for the 30 months in order to stop its ageing while the other one was placed outside.

3   RESULTS AND DISCUSSION

All the pavements were measured at their initial state (called T0) and then installed outside in order to undergo a natural ageing for 30 months (called T30). This ageing implies sun, rain, and any other weather condition exposition but not any mechanical influence such as traffic, or wear and tear or soiling. The ageing applied corresponds to what could be observed on both the central lanes of roads and on urban pavements not used by cars, such as cycle paths, sideways or squares.

The measurements show that a wide range of colours can be obtained within a single pavement family. This is achieved through the use of different coloured aggregates, different types of binders, pigments and surface treatments. In Figure 6, four different coloured asphalt concretes and four different coloured precast cement concrete paving blocks are shown.

3.1   *Definition of the colour difference threshold*

To investigate the effect of natural ageing on the colour and to be confident in our results, it is necessary to establish a threshold above which the colour difference is considered significant.

Figure 6. Illustration of the possible colour diversity of the pavements. On the first row, four examples of asphalt concretes. Below, four examples of precast cement concrete paving blocks.

As there is no reference threshold for pavements, we decided to construct it based on the results obtained by the three measurement processes. Thus, the overall variability is established from the individual variabilities obtained by the three processes. We remind here that each result of a process for a sample is itself derived from several measurements on the sample. This makes it possible to take into account that we are working with different instruments, of different natures and that the measurements are made by different people. The consensus standard is in fact defined on the principle of intercomparison tests between laboratories.

First, we calculate the average coordinates of each pavement from measurements obtained with these three devices (example for L* in Eq.5). Then we determine the colour difference between the coordinates of each process and the average coordinates (example for CM-2300d process in Eq.6). Finally, all calculated colour differences are grouped and sorted in ascending order and the threshold is set so that 95% of the ΔE_{ab}^* are lower. The resulting colour difference threshold is 4.4 (Figure 7). Below this threshold, we are in the uncertainty

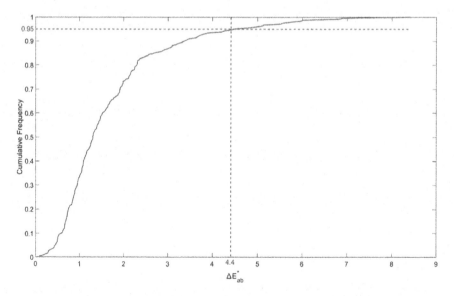

Figure 7. Cumulative frequency of the calculated colour differences.

range of the intercomparison method and therefore cannot conclude on the significance of the colour difference.

$$L_{mean} = \frac{(L_{mean_{CR1}} + L_{mean_{CR2}} + L_{mean_{CM}})}{3} \tag{5}$$

$$\Delta E^*_{ab_{CM}} = \sqrt{(L_{mean_{CM}} - L_{mean})^2 + (a_{mean_{CM}} - a_{mean})^2 + (b_{mean_{CM}} - b_{mean})^2} \tag{6}$$

3.2 Effect of natural ageing

To work with values for T0 and T30 measured under the same conditions, two samples of each pavement were actually made and one of them was placed in the refrigerator for the 30 months in order to stop its ageing while the other one was placed outside. Measurements were made at T0 and T30 on the samples left in the refrigerator. It was verified that the colour difference is less than 4.4, which is why the refrigerated samples are considered as T0 samples in the following.

The graphs of the Figure 8 present the evolution of the colorimetric coordinates during aging for each sample, the origin of the arrow is the T0 measurement, and the arrowhead is the T30 measurement.

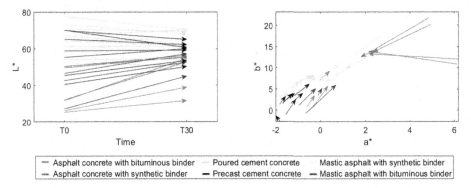

Figure 8. Representation of the colorimetric coordinates for all the pavements. The figure on the left shows the evolution of the lightness L* and the figure on the right shows the change in chrominance in the (a*, b*) plane. The initial state T0 measurement is the value at the origin of the arrow and the T30 measurement corresponds to the arrowhead. The asphalt concretes are in magenta and red, the mastic asphalts in cyan and blue, the poured cement concrete in green and the precast cement concrete in black.

The results of the colorimetric measurements of the pavements in their new condition at T0 show that a huge variety of colours can be obtained (Figure 8). After 30 months, the variety fades a little but remains important.

In the Figure 8 on the right, it can be seen that all the arrows seem to point to the same area, corresponding to a grey pavement. The evolution in the (a*, b*) plan is really important for the asphaltic pavements with synthetic binder (concrete and mastic) which were initially more coloured, whereas it remains more moderate for those with bituminous binder, just like for cement concrete samples (both poured and precast) (Figure 8, right). It can also be seen in the Figure 8 on the left that the asphaltic pavements (concrete and mastic) generally tend to lighten while the cement concrete samples (both poured and precast) darken.

To see if the colour changes are significant, we look for each pavement if the colour difference between T0 and T30 is above 4.4, the previously defined threshold. The Figure 9 presents each pavement type, and the minimum and maximum differences between T30 and T0. The colour of the asphalt concrete and mastic asphalt samples changes whatever the formulation. On the contrary, the difference in colour of the cement concrete samples between T0 and T30 is often not perceptible, especially for the poured ones. Also, we observe that the raw asphaltic samples evolve more than the surface treated ones. For example, asphalt concrete samples No. 2 and 3 are the same but No. 3 is sandblasted. The colour difference varies from one to three times. Moreover, Figure 9 shows that the colour of pavements with dark aggregates tend to change less than those with light aggregates, regardless of the pavement family. Finally, we find a previous result: the asphaltic pavements with synthetic binder appear to evolve more than the bituminous ones. For example, asphalt concrete samples No. 2 and 4 are the same but No. 4 is made with a synthetic binder.

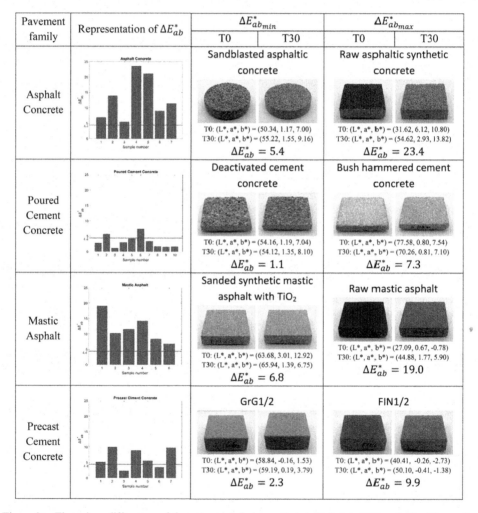

Figure 9.   The colour differences of the pavements between their initial state (T0) and after 30 months of natural outdoor ageing (T30). – Pavement family, colour difference (graph) with threshold in red, surface aspect (pictures), colorimetric characterization (L*, a* and b*) for T0 and T30 and colour difference.

However, we can notice that the use of TiO_2 for the asphaltic synthetic pavements, used to bleach and stabilise colour (Pane & Lecomte 2008; Ployeart & Van Audenhove 2010), seems to reduce the evolution of the colour.

4 CONCLUSION AND PERSPECTIVES

In this paper, the colorimetry of a panel of classic and innovative pavements was measured with three different processes. For reasons of material organisation, some results are not yet available, but in the short term all the selected coverings will be integrated, in particular those of the natural stone family.

The spectrophotometer CM-2300d allows the heterogeneity of the sample to be highlighted due to its smaller aperture but requires more time. The choice of an instrument is a compromise between precision and methodology. However, we have seen that all methods allow for acceptable colour definition.

Moreover, our study has confirmed that current urban pavements can cover a wide range of colorimetric characteristics. This is achieved using different coloured aggregates, different types of binder, pigments, and surface treatments. In this way, this diversity can improve the legibility of the urban space by marking on the ground the spaces dedicated to the different uses.

Also, the evolution of the colour over time differs greatly depending on the type of pavement. Usually, the asphaltic pavements (concrete and mastic) tend to evolve more than the cement concrete samples (poured and precast). The colour of a pavement without surface treatment changes more than a treated one, as well as when dark aggregates are used. The same is true when the binder is synthetic, but it seems that the use of TiO_2 in synthetic asphalt pavements reduces the colour evolution.

Finally, we also made measurements of a pseudo albedo on some samples using a spectroradiometer and a continuous spectrum halogen source. This preliminary study seems to show a link between L* and this pseudo albedo. Therefore, the next step will be to adapt the measurement method to obtain a real albedo. In this way, we will see if this link between L* and the albedo of the pavement really exists. In which case, recommendations concerning the effects of urban heat island can be made based on colorimetric coordinates only. Also, by knowing the effect of natural ageing, it will be possible to evaluate the effect of natural ageing on the albedo of a pavement.

ACKNOWLEDGEMENTS

We would like to thanks Joseph Abdo (CIMbéton), Sophie Banette (AITF/Montpellier 3 M), Frédérico Batista (CD 78), Vincent Bour (AFE/Comatelec), Jean-Pierre Christory, Jérôme Dehon (AFE/Comatelec), Brice Delaporte (Office des asphaltes), Jérôme Dherbecourt (Routes de France/EIFFAGE Routes), Thibaut Le Doeuff (CERIB), Christine Leroy (Routes de France), Valérie Muzet (Cerema), Didier Pallix (CTMNC), Florence Pero (SPECBEA) and Enoch Saint Jacques (Université Gustave Eiffel).

This project received financial and/or intellectual support of the *Pavements and Lighting* group members.

REFERENCES

Abdo J., Batista F., Carré D., Christory J.-P., Depetrinji A., Gandon-Léger P. and Peret M., 2010. Démarche Innovante '*Revêtements et Lumière' de l'idée à la Pratique*". RGRA. 885, pp. 49–53. (in french).
Christory J.-P., Batista F., Gandon-Léger P. and Talbourdet P., 2014. Pavements and Light for the Right Lighting: Contribution of Concrete Pavements. In: *Eupave Symposium on Concrete Roads*, Praha, CZ.

CIE 2018. CIE 015:2018 Colorimetry, 4th Edition. *International Commission on Illumination (CIE)*. https://doi.org/10.25039/TR.015.2018

Liandrat S., Muzet V., Bour V., Dehon J., Christory J.P., Delaporte B., Abdo J., Pero F. and Ledoeuff T. Sep 2022. Pavements, energy efficient actors in public lighting. *The 9th Symposium on Pavement Surface Characteristics*, Milan, Italy.

Muzet V., Liandrat S., Bour V., Dehon J., Christory J.P. Sep 2021. Is it Possible to Achieve Quality Lighting Without Considering the Photometry of the Pavements? *Conference CIE 2021, NC Malaysia*, CIE Kuala Lumpur, Malaysia. pp.11–25.

Pane C. and Lecomte M., Sep 2008. *Les Liants Clairs et Leur Rôle Dans la Construction Routière (In french)*. Publication Shell bitumen.

Ployeart C. and Van Audenhove P., 2010. *Vers une Composition Optimale des Bétons Routiers (In french)*. FEBELCEM.

Roads and Airports Pavement Surface Characteristics – Crispino & Toraldo (Eds)
© 2023 The Author(s), ISBN 978-1-032-55149-4

Systemic analysis of the thermal performance of road surfaces

Maxime Frere*
LIED Université de Paris Cité, Paris, France

Martin Hendel
Département SEN, Université Gustave Eiffel, ESIEE Paris, France

Julien Van Rompu & Simon Pouget
Eiffage Route, CORBAS, France

Laurent Royon
LIED Université de Paris Cité, Paris, France

ABSTRACT: Faced with heat waves that will be more and more frequent in the coming years, cities with a high mineral content will have to limit urban heat islands on the one hand and create cooler areas on the other. The regulation of heat transfers implies reflections on the nature of the infrastructures, in particular the roads. It is in this context that our work is integrated, where the objective is to be able to choose, as of the design of an urban space, the coatings presenting the thermal specificities adapted to these new problems. To this end, it is necessary to better understand the existing relationships between the properties of the constituents and those of the obtained asphalt. We present here an analysis of the thermal properties carried out on a wide range of pavement samples and on the different elements composing them in order to determine the different thermal interdependencies.

Keywords: ICU, Thermal Coatings, Albedo, Thermal Conductivity

1 INTRODUCTION

Bituminous roads, from their creation with Mac Adam to the present day, have been developed mainly in response to the need for mechanical resistance and durability. As traffic density has increased, new pavements have been developed that offer the durability and mechanical efficiency that we know today.

Today, in the face of climate change leading to more intense heat waves, it is becoming important to take into consideration the thermal properties of pavements. Current road materials are mostly heat accumulators and contribute strongly to the urban heat island (UHI) phenomenon (OKE 1982), leading to critical health situations due to thermal stress (Heaviside et al. 2016; Shahmohamadi et al. 2011; Tan et al. 2010), as was the case during the scorching summer of 2003. The need to create cool spaces has therefore arisen. This is based on various solutions, which can be coupled, such as the introduction of vegetation (Bowler et al. 2010), mitigation by evaporation (Hendel et al. 2016) and the design of coatings focused on the dual role of mechanical and thermal performance (Santamouris 2013). This is notably the case of doped coatings (Mizwar et al. 2019; Pan et al. 2014), with air spaces (Hassna et al. 2016; Kevern et al. 2012), with controlled granulometry (Jiaqi et al. 2015), or with reflective paints (Mulian et al. 2015).

---

*Corresponding Author: maxime.frere@u-paris.fr

DOI: 10.1201/9781003429258-80

The study presented here focuses on new pavement formulations, considering the percentage of binder, the nature of the rocks and the granulometry. The objective is to be able, through the measurement of the thermal properties (albedo, thermal conductivity, thermal capacity) of its various components, to compose/propose the most adapted coating to face the stakes of heat waves.

The data obtained through this panel of materials will ultimately allow, through the use of a multi-criteria analysis, to choose, depending on the location, the road infrastructures that best improve the thermal comfort of residents.

The various urban development operators will therefore be able to use this database to optimize the deployment of relevant solutions according to needs. For example, the uses of a square, depending on whether it is located in a residential or business district, will be developed in a very different way.

For example, a square surrounded by office buildings will be coolest during the day, with night-time discomfort being of secondary importance. On the other hand, a square surrounded by residential buildings must provide thermal comfort at night. This must be balanced by the fact that a portion of the population at risk, notably the retired, remains in these dwellings during the day.

Thus, we will first focus on the impact of the different components and formulations on the thermal performance of the final pavement.

2 METHODOLOGY

2.1 *Thermals properties*

The main thermal properties of materials are: albedo, thermal conductivity, heat capacity at constant volume and emissivity. The albedo determines the proportion of solar radiation that is reflected by a material. The thermal conductivity reflects its propensity to transmit heat within it. The heat capacity indicates the amount of thermal energy that the material is able to accumulate for a given temperature rise. The emissivity will indicate the power dissipated by radiation as a function of temperature, in correlation with the law of black bodies.

Our choice of physical parameter to measure was made on the basis of the volume balance presented in and Figure 1. We consider here that the coating is deployed as a thin layer and that its conduction flow is unidirectional.

$$L_+ + S_+ - L_- - S_- - H - C = \rho C_p v \frac{\delta T}{\delta t}\bigg|_V \tag{1}$$

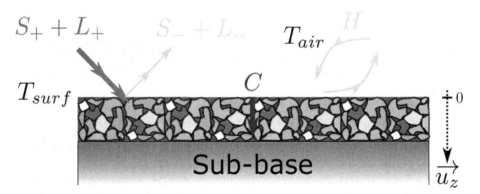

Figure 1.   Scheme of thermal equation.

with:

- L_+, (W/m^2): the infrared radiative flux through the surface, composed of solar and diffuse radiation.
- S_+, (W/m^2): the radiative flux visible by the surface, composed of solar and diffuse radiation.
- L_-, (W/m^2): the reflected and emitted flux on the long wavelengths.
- S_-, (W/m^2): the reflected flux on short wavelengths.
- $C = -\lambda. \overrightarrow{grand(T_z)}$, (W/m^2): the conduction flow to the deep soil.
- $H = h(T_{surf} - T_{air})$, (W/m^2): the convective flow with the air, where h the convection coefficient varies with the wind.
- C_p: Heat capacity at constant volume ($MJ.m^{-3}.k^{-1}$)

If the resultant of the balance for is positive, the surface layer will rise in temperature, which is generally the case during the day on an unshaded place. If the resultant is zero, it is the stationary case and the surface temperature will remain constant. Finally, the case of the night will correspond to a negative resultant, where the pavement will see its temperature decrease.

Thus with the thermal conductivity and albedo we are able to evaluate the different fluxes, coupled with the measurements of the thermal capacity at constant volume we are able to evaluate the impact of these flux variations on the temperature of the material.

2.2 Albedo measurement

The measurement of albedo is done with an Agilent Cary 5000 UV-Vis-NIR spectro-photometer with an integrating sphere allowing us to calculate the albedo according to the ASTM E903-12 standard. This dimensionless quantity ranges from 0 to 1, usually varying from 0.05 (conventional black asphalt) to 0.45 (stabilized sand). Although a high albedo seems to be the simplest solution to combat pavement heating, it is achieved in exchange for an increase in the radiosity of the material during the day, which is unfavorable to the thermal comfort of the pedestrian.

2.3 Conduction measurement and thermal capacity

Measurements for thermal conduction are made using the Hot-Disk 1500 and associated methodology (Gustafsson 1991). However, some samples will be carried out using a radia-tive method (Parison et al. 2021), giving a higher accuracy on materials with high roughness.

2.3 Materials and components

Having in our case the opportunity to study the exact elements that compose the final pavement, it will be possible to finely evaluate the covariant parameters.

Bituminous pavements, also called asphalt mixes, are usually made of crushed rocks, called aggregates. Their size is controlled by the use of sieves, giving a granulometric curve, in order to play on the physical properties of the whole.

The cohesion of the aggregates of an asphalt mix is generally ensured by a petroleum binder called bitumen, thus giving the whole an increased elasticity and durability.

Since pure binders are too liquid to be handled at room temperature, we study sealants, mixtures of bitumen with a certain percentage of fines, with a diameter lower than the micrometer.

For our study we have 8 bituminous concretes (BB) (Figure 3) which were elaborated from the crushed rocks represented in Figure 3 and the binders represented in Figure 2. We will use the denominations specified on these figures to designate the various samples.

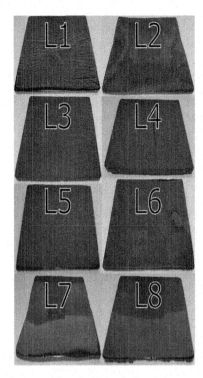

Figure 2. Illustration of the binders and sealants studied.

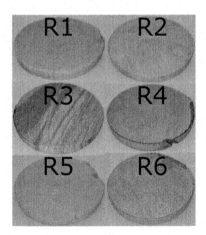

Figure 3. Illustration of rock samples.

2.4 *Binders and rocks*

As binders play a major role in the elasticity and durability of roads, we cannot afford to sacrifice these mechanical properties for thermal parameters.

We have at our disposal pure bitumens (L8, L3), sealants (L1, L2) and plant-based binders (L4, L5, L7) (Figure 2) which represent the thermal limits accessible while preserving the mechanical properties.

Figure 4. Samples of bituminous coatings obtained by assembling the binders and rocks presented above.

The objective is to obtain the greatest variation of results on the bituminous pavements of Figure 4, in terms of thermal behavior.

In order to characterize the thermal properties of aggregates, planar samples of different rock types are studied. The choice here was directed by the need to have the largest possible difference between albedos and conductivities.

2.5 *The assembled coatings*

The thermal properties of the asphalt will depend strongly on its components, but especially on the grading curves which will determine the presence of voids and the compactness of the whole [12].

Thus, with different combinations of rocks (Figure 3) and binders (Figure 2), different asphalt concretes are obtained as shown in Figure 4.

Some asphalt concrete samples, for example BB2, undergo stripping. This allows to simulate their ageing by removing part of the binder initially present on the surface of the asphalt aggregates.

3 PRELIMINARY RESULTS

At this stage, only the albedo of binders or sealants, flat rocks and coatings is studied. The albedo of binders is presented in Table 1, that of rocks in Table 2 and finally that of bituminous pavements in Table 3.

Table 1.   Binders and sealants albedo.

Binder	Albedo	Binder	Albedo
L1	0,05	L2	0,05
L3	0,06	L4	0,06
L5	0,17	L6	0,04
L7	0,08	L8	0,05

Table 2.   Albedo of rocks sample.

Rock	Albedo	Rock	Albedo
R1	0,43	R2	0,41
R3	0,29	R4	0,34
R5	0,63	R6	0,51

Table 3.   Albedo of asphalt concrete samples.

Asphalt Concrete	Albedo	Asphalt Concrete	Albedo
BB1	0,36	BB2	0,22
BB3	0,33	BB4	0,06
BB5	0,19	BB6	0,06
BB7	0,17	BB8	0,06

It can be seen that the binders studied offer a limited range of albedo between 0.05 and 0.17. On the other hand, there is a larger range for the rocks studied with values between 0.29 and 0.63.

Conventional black binders appear to have a greater impact on the final coatings. Indicating that the binders do not have zero transmission.

Figure 4 shows the spectral reflectivity measured for the rock R4 (Figure 3), the binder L7 (Figure 2) and the asphalt BB3 (Figure 4), obtained from these two components. The influence of the components on the final asphalt mix is clearly observed. Indeed, the albedo of the asphalt mix is a compound between the albedo of the binder and the albedo of the rocks, all of which seems to be weighted by the proportions of each component.

The comparison of the raw BB4 and stripped BB2 pavement (Table 3) illustrates the tendency of the pavements to approach the albedo of their aggregates as they age. Indeed, wear will mainly strip the binder present on the surface aggregates, thus reducing the proportion of binder exposed to solar radiation on the surface of the asphalt. In addition to this aging, the surrounding fines may be deposited if no rolling or maintenance is performed on the pavement.

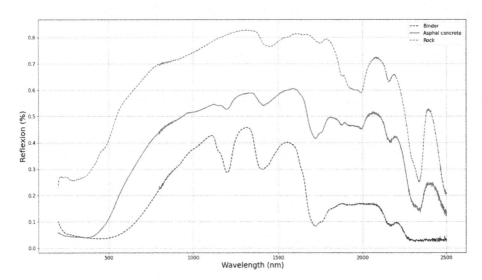

Figure 5.   Spectral reflectivity of the R4 rock, the L7 binder and the resulting draining asphalt BB3.

Future work will similarly focus on the analysis of thermal conductivity, as well as the heat capacity of aggregates, binders and asphalt.

4   CONCLUSION

Currently we have been able to determine the albedos for each element, the process of measuring conductivity and heat capacity is underway.

The hope of this work is to provide a better predictability of the thermal properties of coatings, including the influence of the properties of their constituents. This first study is a first step in the vast work of developing coatings with controlled thermal properties. The hope is to allow an advance in the deployment of thermal optimums, according to the needs and uses of the various urban places.

In order to reach this goal, this work is integrated to a research on a demonstrator at the scale of a square and a study of the different urban topologies on the effects of convections in hot weather.

ACKNOWLEDGEMENTS

This research was funded by the Eiffage Group and the Gustave Eiffel University in the framework of the E3S program. This program was supported by the "Investissement d'Avenir" program launched by the French government and implemented by the ANR, with the reference ANR-16-IDEX-0003.

The authors would like to thank Florian Moreaux and Luc Bonnel for their contribution to the preparation of the samples.

REFERENCES

Bowler D.E., Buyung-Ali L., Knight T.M. and Andrew S.P., Urban Greening to Cool Towns and Cities: A Systematic Review of the Empirical Evidence, *Landscape and Urban Planning*, vol. 97, n° 13, pp. 145–155, 2010.

Gustafsson S.E., Transient Plane Source Techniques for Thermal Conductivity and Thermal Diffusivity Measurements of Solid Materials, *Review of Scientific Instruments*, vol. 62, n° 13, p. 797, 1991.

Hassna A., Aboufoula M., Wub Y., Dawsona A. and Garcia A., Effect of Air Voids Content on Thermal Properties of Asphalt Mixtures, *Construction and Building Materials*, vol. 115, pp. 327–335, 2016.

Heaviside C., Sotiris V. and Xiao-Ming C, Attribution of Mortality to the Urban Heat Island During Heatwaves in the West Midlands, UK., *Environmental Health*, vol. 15, n° %11, pp. 49–59, 2016.

Hendel M., Pierre G., Morgane C., Youssef D. and Laurent R., Measuring the Effects of Urban Heat Island Mitigation Techniques in the Field: Application to the Case of Pavement-Watering in Paris, *Urban Climate*, vol. 16, pp. 43–58, 2016.

Jiaqi C., Miao Z., Hao W. and Liang L., Evaluation of Thermal Conductivity of Asphalt Concrete with Heterogeneous Microstructure, *Applied Thermal Engineering*, vol. 84, pp. 368–374, 2015.

Kevern J.T., Haselbach L. and Schaefer V.R., Hot Weather Comparative Heat Balances in Pervious Concrete and Impervious Concrete Pavement Systems, *Journal of Heat Island InstituteInternationa*, vol. 7, n° 12, pp. 231–237, 2012.

Mizwar I., Napiah M. and Sutanto M., Thermal Properties of Cool Asphalt Concrete Containing Phase Change Material, Chez *IOP Conference Series: Materials Science and Engineering*, 2019.

Mulian Z., Lili H., Fei W., Haichen M., Yifeng L. and Litao H., Comparison and Analysis on Heat Reflective Coating for Asphalt Pavement Based on Cooling Effect and Anti-skid Performance, *Construction and Building Materials*, vol. 93, pp. 1197–1205, 2015.

Oke T. R, The Energetic Basis of the Urban Heat Island, *Quarterly Journal of the Royal Meteorological Society*, vol. 108, n° %1455, pp. 1–24, 1982.

Pan P., Wu S., Xiao Y., Wang P. and Liu X., Influence of Graphite on the Thermal Characteristics and Anti-ageing Properties of Asphalt Binder, *Construction and Building Materials*, vol. 68, pp. 220–226, 2014.

Parison S., Hendel M., Grados A., Jurski K. and Royon L., A Radiative Technique for Measuring the Thermal Properties of Road and Urban Materials, *Road Materials and Pavement Design*, vol. 22, n° 15, pp. 1078–1092, 2021.

Santamouris M., Using Cool Pavements as a Mitigation Strategy to Fight Urban Heat Island—A Review of the Actual Developments, *Renewable and Sustainable Energy Reviews*, vol. 26, pp. 224–240, 2013.

Shahmohamadi P., Che-Ani A., Etessam I., Maulud K. and Tawil N., Healthy Environment: the Need to Mitigate Urban Heat Island Effects on Human Health, *Procedia Engineering*, vol. 20, pp. 61–70, 2011.

Tan J., Zheng Y., Xu T.C.G., Liping L., Guixiang S., Xinrong Z. *et al.* The Urban Heat Island and its Impact on Heat Waves and Human Health in Shanghai, *International Journal of Biometeorology*, vol. 54, n° 11, pp. 75–84, 2010.

Roads and Airports Pavement Surface Characteristics – Crispino & Toraldo (Eds)
© 2023 The Author(s), ISBN 978-1-032-55149-4

An analytical model for forecasting pavement temperature

Usama B. Ayasrah* & Laith Tashman
Department of Civil and Environmental Engineering, University of Jordan, Amman, Jordan

ABSTRACT: In this study, a temperature forecasting model is developed based on local conditions using an analytical method in Jordan. A pavement specimen was taken to compute the layers properties of flexible pavement. Temperature data were taken from surface, 6cm, 11cm, and 40cm depths to validate the developed model. The original assumption of the proposed model that is treating the pavement structure as a full-depth structure in terms of thermal properties is changed in order to accurately establish the pavement temperature profiles. Furthermore, the temperature profiles are computed after analyzing input parameters, and then, 24 hours temperature data is used to validate the calculated temperature profiles using MATLAB software for each stated depth. Three types of heat transfer modes were included. Rather than the heat conduction transfer mode, which flows the energy within solid objects, the radiation and heat convection modes were included by simulating the temperature values of the pavement surface. This assumption can facilitate the applicability of the temperature prediction model easily. The study results concluded that the prediction model predicted the field temperature profiles of the pavement structure reasonably. The maximum relative error of the analytical proposed model was less than 10 (%) at 6 cm, 11 cm, and 40 cm depths, respectively.

Keywords: Flexible Pavement, Heat Flux, Temperature Model

1 INTRODUCTION

The surrounding environment, vehicular loads and improper mix design are the main causes of asphalt pavement deformation and failures. There are various environmental conditions such as air temperature and seepage due to precipitations. In semi-infinite solid pavement systems, while the temperature of the pavement surface is controlled by the radiation and convection heat transfer modes, the heat conduction flux mode conveys the energy to the underlying pavement layers.

The succession of the temperature values during the day induces stresses in the pavement structures. Chen *et al.* (2016) and Wang and Al-Qadi (2013) concluded that the HMA layer responses including, stresses and strains can be accurately obtained only by considering the temperature profiles of the pavement structures because pavement responses highly depend on the temperature profiles within the structures. Also, the viscoelastic behavior of the asphalt surface layer contributes to the heterogeneity. For instance, the properties of the pavement materials are influenced by the temperature such as creep, stiffness and Young's modulus. The variation of temperature amplitude during the day will dramatically shift these properties. So that in the design phase, the pavement layers should be characterized adequately to resist the effect of temperature cycles and to account for the thermal effect, in which the importance of initiating anticipating model for the pavement structures arises.

---

*Corresponding Author: usama.qasim@eng.hu.edu.jo

DOI: 10.1201/9781003429258-81

Solaimanian and Kennedy (1993) proposed an analytical approach to forecast pavement temperature. Based on the net rate of the in and out energy of a given area, q_{net}, the absorbed energy by that area can be computed as shown in Eq. (1) through defined boundaries of the pavement surface. Additionally, Zhang *et al.* (2019) indicated a schematic diagram that included the heat exchange process in the pavement structure as shown in Figure 1.

$$q_{net} = q_s \pm q_{ln} \pm q_c \pm q_k \qquad (1)$$

Where:

q_s: energy absorbed by pavement, W
q_{ln}: the net long-wave heat transfer rate
q_c: convection heat flux, $J/(m^2.hr)$
q_k: the conduction heat flux, W/m^2

Many studies have shown the importance of the temperature effect on pavement performance. For instance, Xue *et al.* (2013) simulated the behavior of the asphalt pavement under temperature-stress accompanied by traffic loads. Including the viscoelastic behavior of the asphalt layer for the analytical temperature prediction model in China and taking the numerical method for pavement stresses and strains, the authors stated that the asphalt pavement behavior considerably depends on the pavement temperature. The horizontal stresses highly grew up when the temperature increased. More shear stresses manifested when the temperature climbed up. As a result, with high temperature induced on the pavement surface, shear failure may occur. By contrast, changing temperature values have little effect on vertical stresses.

This study focused on the accuracy improvement of prediction temperature profiles for the pavement structures, in which the first step in analyzing the pavement behavior could be established.

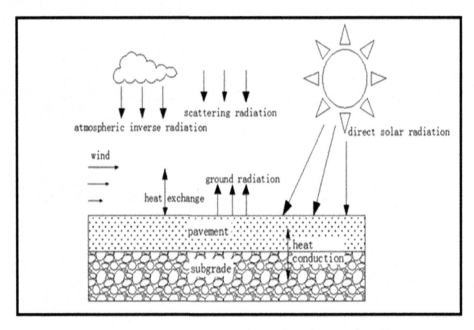

Figure 1. Field Pavement Temperature Heat Transfer Modes. (Zhang *et al.* 2019).

2 MODEL DEVELOPMENT

2.1 *Parameter input*

Actual temperature field data were derived from the pavement surface, 6, 11, and 40 (cm) depths. For the recording process, a data-logger temperature device was used. A pavement sample was tested to acquire the asphalt, base course and subgrade densities (ASTM D2726 2019; ASTM D1557-12e1 2012). Also, thermal properties of pavement layers were computed according to (ASTM D5334-08 2008; Nave 2010).

2.2 *Field location*

The field test of this study is located in Jordan. The material and thermal properties of pavement layers are listed in Table 1.

Table 1. Pavement Structure Scheme Included Material and Thermal Properties.

Layer	Information
Asphalt Layer	Thickness = 7 cm; Density = 2224 kg/m^3; Diffusivity = 0.002211 m^2/h
Base Course Layer	Thickness = 9 cm; Density = 2180 kg/m^3; Diffusivity = 0.0804 m^2/h
Natural Subgrade Soil	Thickness = ∞; Density = 2096 kg/m^3; Diffusivity = 0.0507 m^2/h

2.3 *The analytical model*

In this study, the analytical solution for a semi-infinite object is applied to the pavement structures. MATLAB software is used to solve the partial differential equation (PDE) of the heat conduction equation within the pavement layers, which can be shown in Eq. (2):

$$\frac{\partial^2 T}{\partial x^2} = \frac{1}{\alpha}\frac{\partial T}{\partial t} \tag{2}$$

Where:

T: is the temperature, °C
x: is the space dimension, m
t: is the time, h
α: is the thermal diffusivity. m^2/h, and can be calculated from Eq. (3)

$$\alpha = \frac{k}{\rho c} \tag{3}$$

Where:

k: thermal conductivity, J/(m.hr.°C);
ρ : material density, kg/m^3;
c: specific heat capacity, J/(kg.°C)

The separation of variables technique of (PDE) is used to establish the closed-form solution of the heat conduction equation throughout the pavement structures (Wang 2015):

$$T_{(x,t)} = \frac{1}{L}\left[c.x + (L - x).s_{(t)}\right] + W_{1(x,t)} + W_{2(x,t)} \tag{4}$$

Where: $T_{(x,t)}$: asphalt pavement temperature (°C); $s_{(t)}$: pavement surface temperature function (°C); α: thermal diffusivity $(m^2/h), n = 0, \quad 1, \quad 2, \ldots, W_{1(x,t)}$ and $W_{2(x,t)}$: infinite series equations; c = constant temperature (°C); L = finite depth where the Earth temperature is assumed to be = 20°C (m), measured from the top of the pavement surface.

Additionally, Previous studies have considered the radiation parameters to set up the top boundary conditions of the pavement structure, which is the pavement surface. For instance, (Dempsey & Thompson 1970) stated that the total net energy amount of radiation includes the net short-wave radiation, the long-wave radiation released by the atmosphere, and the reflected long-wave radiation by the pavement surface.

3   ANALYSIS AND DISCUSSION

3.1   *The analytical temperature model inputs*

Based on the analytical approach, the pavement temperature prediction model was done with MATLAB software aid. The pavement temperature values were measured at the mentioned depths for two consecutive days in March and May. The data sets were categorized such that the temperature data for the first 24 hours were used to analyze the model behavior, and the rest of the temperature data was used for the validation process of the anticipating model.

Figure 2 shows temperature profiles for each tested depth as well as the pavement surface. The temperature profiles peak after midday when the radiation reaches the maximum effect. The 6 cm profile, which lies in the asphalt layer, reaches its maximum value nearly at 15:00. If the depth increases, the peak of temperature seems likely to shift further in the day. The 11 cm curve approximately has higher values than the 6 cm profile at the beginning and the ending of the analysis period (i.e., at night). This is justified by the time dependency of the heat conduction equation.

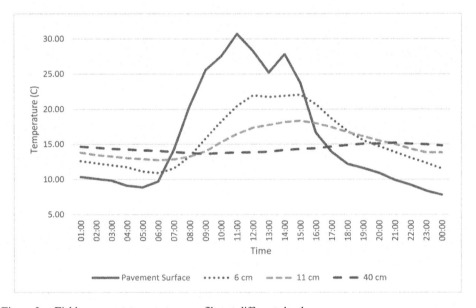

Figure 2.   Field pavement temperature profiles at different depths.

It is noticeable from Figure 2 that the temperature profile of the surface had a sudden drop in the temperature values which was starting at 11:00 and lasted for 2 hours. This abrupt reduction stems from the cloud cover effect, which relives the radiation intensity.

This study deemed the temperature profile at 6 cm in the asphalt layer to calibrate the proposed anticipated model.

3.2 *The analytical solution for the proposed model*

After optimizing the accuracy of the anticipating model inputs, the findings are used to solve the 1-D heat conduction formula. The developed model shows reasonable anticipating results. The maximum relative error (%) for the proposed anticipating model was less than 10 (%) for 6cm, 11cm, and 40cm depths, respectively, as shown in Figure 3. Also, the average relative error (%) for the proposed prediction model was less than 3 (%) at 6cm, 11cm, and 40cm depths, respectively. The temperature profiles at relatively deep depths, such as 40cm depth were approximately constant due to the assumption of a steady temperature $c = 20°C$ at a finite depth (L), which equals 3m.

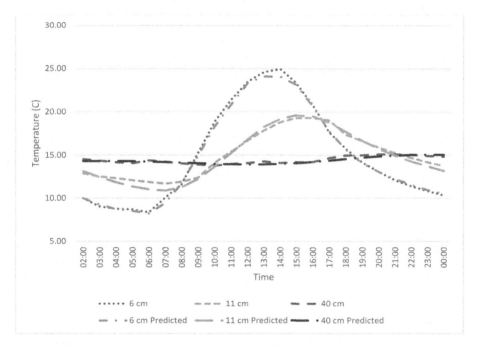

Figure 3. Pavement temperature profiles measured and predicted.

In this study, three input parameters were analyzed to lower the accumulated prediction error. Firstly, the number of n-terms in the infinite series equations, $W_{1(x,t)}$ and $W_{2(x,t)}$, was selected. Secondly, the proper assumption for the finite depth (L) was assumed to be 3 m. Finally, instead of using the asphalt layer thermal properties to predict the base course layer temperature profiles, the average thermal properties of the asphalt and the base course layer were used.

4 CONCLUSION

Plenty of previous studies have included the temperature as a major factor that distorts the behavior of the pavement, which leads to distresses. For that reason, predicting pavement temperature profiles accurately is a crucial step in the pavement design phase. In this paper, an analytical temperature prediction model was used to generate temperature profiles throughout the pavement layers for the pavement structure. Also, instead of focusing on deriving a new theoretical anticipating model, this paper aimed at enhancing the assumptions and the selected parameters of a previous prediction model to minimize the accumulated error.

The Superpave grading system is well-equipped with a grading system that is derived from extreme pavement temperature conditions. The Superpave grading system is generally obtained from empirical temperature profile models. Hence, it is important to have good anticipations of the pavement temperature profiles within the pavement structures in order to use. The inputs of the proposed model include actual temperature field values and thermal properties of the tested pavement material. In addition, the proposed model generally has the benefit of applicability in any region. In this method, the pavement surface temperature was simulated rather than including complex radiation parameters. So, this study can form the basis for further studies in Jordan and can be used to facilitate the application of Superpave technology.

This study divided up the analysis section into two steps. Firstly, the critical input parameters of the selected model were examined using actual field data sets, and then, the prediction process was run via the enhanced parameters and assumptions onto another data set. In this study, the earth's temperature was assumed to have a constant temperature ($^{\circ}$C) at a pre-defined depth (L), and the thermal properties of the base course layer were taken into consideration, unlike the previous study.

REFERENCES

ASTM International (2008). *Standard Test Method for Determination of Thermal Conductivity of Soil and Soft Rock by Thermal Needle Probe Procedure*, ASTM International: West Conshohocken, Pa, USA, ASTM D5334-08.

ASTM International (2012). *Standard Test Methods for Laboratory Compaction Characteristics of Soil Using Modified Effort (56,000 ft-lbf/ft3 (2,700 kN-m/m3))*, ASTM International: West Conshohocken, PA, D1557-12e1.

ASTM International (2019). *Standard Test Method for Bulk Specific Gravity and Density of Non-Absorptive Compacted Asphalt Mixtures ASTM International*, ASTM International: West Conshohocken, PA, ASTM D2726.

Chaudhry M. A. & Zubair S. M. (1992). Heat Conduction in a Semi-infinite Solid Subject to Time-dependent Boundary Conditions. *Presented at 1992 ASME/AIChE National Heat Transfer Conference*, San Diego: ASME.

Chen J.Q., Wang H., Li M. and Li L. (2016). Evaluation of Pavement Responses and Performance with Thermal Modified Asphalt Mixture, *Materials and Design* (111), 88–97.

Dempsey B., and M. Thompson. A Heat-Transfer Model for Evaluating Frost Action and Temperature Related Effects in Multilayered Pavement Systems. *Transportation Research Record: Journal of the Transportation Research Board*, 1970. 342: 39–56.

Shiguang X. and Yuansheng G. (2009). *Geothermal Foundation*. Beijing: Science Press.

Solaimanian M. & Thomas Kennedy W. (1993). Predicting Maximum Pavement Surface Temperature Using Maximum Air Temperature and Hourly Solar Radiation, *Transportation Research Records* (1417), 1–11.

Sun L. (2016). *Structural Behavior of Asphalt Pavements (1st ed.)*. The Boulevard, Langford Lane, Kidlington, Oxford OX5 1GB, United Kingdom. ISBN: 978-0-12-849908-5.

Nave R. (2010). *Specific Heat*. HyperPhysics. Georgia State University.

Wang D. (2015). *Simplified Analytical Approach to Predicting Asphalt Pavement Temperature*, ASCE, DOI: 10.1061/MT.1943-5533.0000826.

Wang H. and Al-Qadi I. L. (2013). Importance of Nonlinear Anisotropic Modeling of Granular Base for Predicting Maximum Viscoelastic Pavement Responses Under Moving Vehicular Loading, *Journal of Engineering Mechanics* 139 (1) 29–38.

Xue Q., Lei L., Ying Z., Yi-Jun C., & Jiang-Shan L. (2013). Dynamic Behavior of Asphalt Pavement Structure Under Temperature-stress Coupled Loading, *Applied Thermal Engineering Journal* (53), 1–7.

Zhang N., Wu G., Chen B., Cao C. Numerical Model for Calculating the Unstable State Temperature in Asphalt Pavement Structure. *Coatings*. 2019; 9(4):271. https://doi.org/10.3390/coatings9040271.

Weather conditions impact (snow, ice, etc.)

Roads and Airports Pavement Surface Characteristics – Crispino & Toraldo (Eds)
© 2023 The Author(s), ISBN 978-1-032-55149-4

Winter damage to wearing course

Heidi Kauffmann*
Pole of Winter Maintenance, Cerema, Nancy, France

Pascal Rossigny
Department of Transport, Infrastructure and Materials, Cerema, Sourdun, France

ABSTRACT: During the winter of 2009–2010, numerous deteriorations appeared on the road networks of north-eastern France. Following this episode, various expert reports were written, attempting to explain the phenomenon. Special monitoring of what has been called "winter damage" (disbonding by patches, potholes, cracking, tearing, etc.) has also been set up on the national road network. It is in this context that an **IDRRIM** working group (an institute working on road issues) was set up to try to better understand the technical origins of these particular degradations. This work identified certain recurring phenomena that affect the service life of wearing courses, particularly those subject to severe weather conditions. A technical note was produced in order to propose recommendations to road managers working in areas subject to severe to very severe winter conditions in order to limit these risks. The note is divided into four main areas:

1. Choice of materials: the different types of wearing course usually used in France are reviewed in order to identify the most suitable ones and define conditions of use that limit the risk.
2. Formulation: bitumen characteristics and mix composition in terms of richness modulus and water resistance.
3. Implementation conditions: quality of the substrate (treatment of cracks, elimination of lamination, sweeping, etc.), tack coat (quality, dosage and breakage), compliance of the manufacture with the formulation study (binder content, grading, etc.), implementation temperature, weather conditions, compactness. But also quality of longitudinal joints, treatment of transverse cracks, management of areas difficult to compact, quality of the start of work, macrotexture and management of sinuous areas at altitude.
4. Perspectives with information on techniques being tested or used by other countries such as surface dressing, Stone Mastic Asphalt or the integration of lime in asphalt.

A call for feedback has been launched in order to evaluate these techniques but also to discover new ones.

Keywords: Wearing Course, Winter Damage, Formulation

1 FRAMEWORK OF THE STUDY

1.1 *Context*

During the winter of 2009–2010, numerous unusual degradations appeared on the road networks of eastern France in the form of potholes, cracks, tears and especially disbonding

---

*Corresponding Author: heidi.kauffmann@cerema.fr

DOI: 10.1201/9781003429258-82

in patches with a very rapid rate of evolution that necessitated the closure of roads to traffic (Figure 1). This phenomenon was repeated in the following years to varying degrees depending on the severity of the winter and the geographical areas.

These degradations have been generically named "winter damage".

While the link with meteorological conditions such as alternating periods of freezing and thawing and heavy precipitation seems to be established, the precise reasons for their occurrence and the mechanisms involved in these phenomena can be varied.

Figure 1.   Winter damage on the RN4 in Lorraine – February 2010.

At the request of the Ministry of Ecology, Sustainable Development, Transport and Housing, two reports were drawn up in 2010 by the Conseil général de l'Environnement et du Développement durable [1] and by the Mission d'Appui au Réseau Routier National [2].

1.2   *Conclusion of the first report*

Reports in 2010 suggested the following causes:

- A greater than expected increase in heavy goods traffic not taken into account when designing the road structures;
- Ageing of the wearing courses;
- The choice of materials unsuited to the climatic conditions, such as ultra-thin asphalt concrete (**BBUM**), high modulus asphalt concrete (**BBME**) or draining asphalt concrete (**BBTM**);
- Insufficient control of the use of very thin wearing courses;
- Defects in implementation;
- Unsuitable maintenance methods resulting in a stacking of successive layers of asphalt (the "millefeuille" phenomenon).

Following this observation, the rapporteurs recommended a reflection on the policy of maintenance, management and monitoring of road infrastructures and advised continuing the analysis of the causes and updating the technical doctrine according to the results that would be obtained.

1.3   *Setting up of a working group*

A working group has been set up under the aegis of **IDRRIM** (Institut des Routes, des Rues et des Infrastructures pour la Mobilité).

It brought together members of the Ministry, managers of national and departmental roads, and public and private technical experts.

Its objective was to better understand the phenomena that cause winter damage and to propose technical solutions to reduce the risk.

2 WORKING APPROACH TO UNDERSTANDING THE PHENOMENA

2.1 *Areas of work*

The first area of work, which is more experimental, was dealt with in the framework of theme 1 of the DVDC research project on pavement deterioration mechanisms [3]. It will not be detailed in this article.

The second axis was based on feedback with, on the one hand, the carrying out of tests and auscultations on zones presenting winter damage and, on the other hand, a statistical analysis of the deteriorations observed on the road and motorway network of the DIR Est (Direction Interdépartementale des Routes Est).

2.2 *Testing areas with degradation described as winter damage*

Since 2010, the road ministry has set up a monitoring system for sections with winter damage on the national state managed road network.

The sections identified by the DIR Est following the winters of 2016–2017 and 2017–2018 were analysed on the basis of photographic surveys and monitoring of the evolution of the deterioration using historical data (results of periodic quality inspections (IQRN), history of works, etc.). For sections where the deterioration was particularly unusual and rapid (e.g. wearing course less than 7 years old), additional investigations (in situ visual inspections and core sampling) were carried out.

As the winters after 2015 were mild, the volume of damage was low and it is difficult to conclude how representative it is.

However, the following assessment was made:

- In a large number of cases, the degraded wearing courses were old (> 12 years);
- Some surface deterioration was in fact due to structural problems (cracking of semi-rigid pavements, poor bonding of base layers, etc.), with winter conditions only accelerating the phenomenon;
- In some cases, the technical choices made during maintenance were unsuitable (e.g. BBTM laid directly on a rigid material, stacking of thin layers without milling the degraded underlying layer).

2.3 *Statistical analysis of the condition of wearing courses in DIR Est*

The study was based on an analysis of photographic surveys of the entire DIR Est network, i.e. approximately 2,500 km of pavements, to assess the condition of the wearing courses and their evolution (comparison with previous photographs).

These elements were cross-referenced with the history of the works, the nature of the wearing courses, their age, the volume of heavy goods traffic, the climatic classification and the nature of the last works carried out between new (rehabilitation of the whole structure) and maintenance (repair of the wearing course only).

Figure 2 shows an example of the analysis graphs produced. It shows the condition of wearing courses over 10 years old according to the nature of the wearing course, for the highest traffic class (Ts: heavy goods vehicle traffic over 2 000 trucks per day).

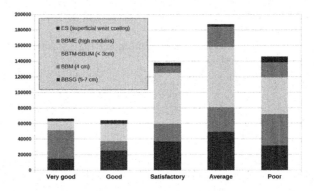

Figure 2. Visual condition of wearing courses over 10 years old according to the type of asphalt used – DIR Est network – traffic Ts – 2017.

One of the objectives of this analysis was to determine whether certain types of wearing course were less suitable for sections in severe climatic zones and subject to heavy truck traffic.

The question was raised in particular for very thin asphalt concrete (BBTM), for which fears were expressed about winter behaviours:

• This relatively porous material could allow water to penetrate the pavement;
• Its thinness could make the wearing course/binder course interface more susceptible to thermal shock and delamination.

The study showed that, statistically, in new structures, thin asphalt concrete (BBM) and semi-rough asphalt concrete (BBSG) were more resistant than BBTM. On the other hand, in maintenance, the evolution was equivalent between the different materials.

In a second step, a more targeted analysis of BBTM was carried out.

Figure 3 shows the visual condition of the DIR Est's BBTM according to their age.

In general, the condition is good or even very good for wearing courses less than 10 years old (which is considered to be the usual life span of a wearing course) and it remains acceptable in half of the cases for wearing courses 11 to 14 years old.

The fact that more than a third of the DIR Est's BBTM were over 15 years old in 2017 confirms the ageing nature of the network, with resources that do not allow structures to be renewed at the necessary frequency. It should be noted, however, that a reform of the maintenance policy and additional resources have since made it possible to accelerate the

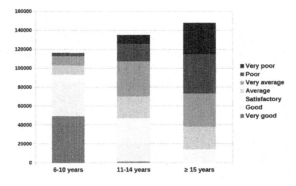

Figure 3. Visual condition of the DIR Est's BBTM wearing courses according to their age.

pace of repairs. The proportions presented here should therefore not be considered as representative of the current situation.

A zoom was carried out on the BBTM sections where the deterioration was abnormal. Recurrent elements linked to the implementation were identified:

• Stacking of thin layers during successive maintenance phases without systematic milling of damaged layers;
• Presence of longitudinal joints;
• Presence of transverse cracks linked to already degraded semi-rigid sub-base layers;
• Probable compaction deficiencies (areas difficult to compact, bridges, start of construction, etc.).

2.4 *Conclusion of the study*

The conclusions of this study are in line with those of the 2010 reports.

In the majority of cases, the winter damage is an acceleration of the damage that would have occurred in subsequent years due to winter conditions.

The high age of the wearing courses may explain the extent of the phenomenon on certain networks. The implementation of a new maintenance policy and the allocation of additional resources in recent years should make it possible to reduce this factor.

With the exception of some very specific materials (high modulus, draining or ultra-thin asphalt concrete), there is no reason to exclude certain types of wearing course, if normal lifetimes are respected.

Inappropriate maintenance techniques, such as thin layer piling, are a major factor in the occurrence of winter damage.

The installation conditions (compaction, management of longitudinal joints, treatment of cracks, etc.) have a strong influence on the speed of degradation.

3 ADVICE TO PROJECT OWNERS

3.1 *Information note for project owners*

On the basis of these results, an information note for project owners was produced [4].

It deals with the selection and application of surface layers in climatic zones with severe winter conditions.

This document, which is intended to be pragmatic, is based mainly on observations and feedback rather than on experiments or research. Indeed, even if many studies have been carried out, it is not possible to identify a single origin for what is known as "winter damage".

However, some recurring phenomena affecting the service life of wearing courses, particularly those subject to severe weather conditions, have been identified.

The aim of this note is to propose recommendations to owners and contractors working in areas subject to severe to very severe winter conditions in order to limit these risks.

It is divided into four main areas:

• The choice of materials;
• The formulation;
• The conditions of implementation;
• The perspective.

3.2 *Choice of materials*

The note indicates, for the coatings most commonly used in France, whether and under what conditions they ca be used in areas subject to severe winters.

3.2.1 *Materials usually used in areas with severe winter conditions*

Experience has shown that the materials listed below are well suited for areas with harsh winters:

BBSG (NF EN 13108-1, 5–7 cm):

These continuous granular matrix mixes are dense and have a more closed macrotexture. As a result, they are less susceptible to surface chipping and water infiltration.

Due to their formulation and application thickness, they also tolerate less stringent application conditions than thin techniques.

Under heavy traffic and aggressive winter conditions, they have slower degradation kinetics than discontinuous formulas, thus allowing more flexibility in the timing of layer renewals.

They can therefore be used without difficulty, provided that they are carefully applied and that they are not reloaded onto a BBM or a BBTM.

Asphalt concrete for flexible pavement (BBCS):

Due to its formulation (generally continuous matrix, high modulus of richness, "soft" grade 70/100 or even 160/220 bitumen), this technique is well suited to the maintenance of lightly trafficked and poorly structured pavements, including in areas with an aggressive winter.

3.2.2 *Materials suitable for harsh conditions, provided that the recommendations for use are followed*

Thin asphalt concrete (BBM) and very thin asphalt concrete (BBTM)

These are discontinuous granular matrix coatings implemented in thin or very thin layers ($\leq$ 4 cm).

Under heavy truck traffic, their degradation kinetics can accelerate significantly after 8 to 10 years of use. This generally requires renewal after 10 to 14 years of use.

They have particular sensitivities due to:

- their thinness which :
  - induces brittleness at the interface with the lower layer;
  - does not delay the rise of cracks due to the contraction/dilation effects of the rigid base layers;
  - make them sensitive to the conditions of implementation (temperature and hygrometry).
  - their porosity which favours the circulation of water. A cohesive, dense and waterproof support is therefore necessary.

In order to prevent abnormal deterioration, it is advisable to:

- not stack thin layers;
- ensure the quality of a support that must be:
  - not too rigid: do not lay directly on concrete, a material treated with hydraulic binder (MTLH) or a high modulus asphalt mix (EME). For BBTM, laying directly on a bituminous gravel is to be avoided;
  - sound: cracked or degraded layers must be milled off beforehand and the base must be waterproofed. The BBTM require a bonding layer at least 6 cm thick with a closed texture to waterproof the structure (e.g. BBSG);
  - homogeneous in terms of altimetry. A reprofiling may be necessary.

- treat the tack coat with a fast-breaking emulsion incorporating a high consistency residual binder. A dosage of 450 g/m^2 is recommended.

Subject to these reservations, BBM and BBTM can be used without difficulty as a wearing course in areas with severe winter conditions.

Surface coatings

The service life of wearing courses and cold mix asphalt materials is highly dependent on the mix design, the condition of the substrate, the quality of the constituents and their application.

As an early maintenance solution, on an existing clean wearing course with little cracking, this type of treatment can extend the service life of the wearing course by several years (typically 10 to 20 years for ESU and 7 to 10 years for MBCF).

On surfaces with active cracking (excluding thermal cracking), surface coatings shall be avoided because they do not stop the rise of cracks and they can become detached due to water infiltration.

The formulation of ESU on degraded substrates is all the more complicated as the traffic is high.

On the most heavily trafficked routes, ESU must therefore be laid on homogeneous substrates with little degradation. For MBCF, particular attention should be paid to the deformability of the substrate (deflection $< 50/100$th mm for traffic $\geq$ T1).

3.2.3 *Coatings to avoid*

Porous asphalt concrete (BBDr) should be avoided for winter maintenance reasons. Indeed, their high porosity can lead to an acceleration of the drop in surface temperature with an increased risk of icing and consequently more frequent winter maintenance operations. They are also more sensitive to damage caused by the passage of snow ploughs.

High modulus asphalt concrete (BBME) made with a grade of bitumen less than or equal to 20/30 generates so-called "top cracking" due to the cold brittleness of these bitumens. This anarchic cracking causes considerable difficulties in maintenance.

The use of this type of BBME as a surface layer should therefore be avoided in areas with a harsh winter. For the treatment of singular points with high stresses with respect to rutting (roundabouts, bus lanes, etc.), the use of BBSG with polymer-modified binder is a good alternative.

3.2.4 *Conclusion*

Table 1 summarizes the possibilities of using the various pavements in severe winter conditions.

Table 1. Potential use of different types of wearing course materials in severe winter conditions.

Matérial	Low traffic	Heavy traffic	Nature of the bottom layer
BBSG (NF EN 13108-1 / 5–7 cm)	Yes Pure bitumen > 50/70 or polymer modified	Yes Polymer modified bitument	All except thin layer (< 4 cm)
BBCS	Yes Pure bitumen > 70/100	No	All except thin layer (< 4 cm)
BBM (NF EN 13108-1 / 3–4 cm)	Yes, subject to the quality of the substrate and the bonding layer		GB or BBSG (binder course)
BBTM (NF EN 13108-2 / 2–3 cm)	Polymer modified bitument		BBSG binder course
ESU (NF EN 12271 – surface coating) MBCF (cold mastic asphalt)	Yes, subject to the quality of the substrate		Any clean, low-cracking surface course
BBDr (NF EN 13108-7 – draining)	No, for winter maintenance reasons		
BBME (NF EN 13108/1 – high modulus)	No, as hard bitumen generates cracking		

3.3 *Formulation*

To improve the resistance of road materials to severe winter conditions, the IDRRIM note sets out formulation principles.

3.3.1 *Characteristics of bituminous binders*

The objective is to have binders that are resistant to both low (as measured by the FRAAS test) and high (as measured by the ring and ball method) service temperatures while ensuring that this performance lasts over time.

It is therefore recommended, particularly for roads with more than 50 HGVs per day, to use binders modified by adding SBS polymers.

The performance targets for modified binders (according to NF EN 14023) are the following:

- Ring and ball temperature (NF EN 1427) $\geq$ 60°C
- FRAASS (NF EN 12593) $\leq$ – 12°C
- Elastic return at 10°C (NF EN 13398) $\geq$ 75 %
- Tensile test at 5°C (NF EN 13587 – NF EN13703) > 3 J/cm^2

After RTFOT curing at 163°C (NF EN 12607-1), the specifications are as follows (NF EN 12591):

- Mass variation (NF EN 12607-1) $\leq$ 0.5 %
- Increase of the softening point (NF EN 1427) $\leq$ 8°C
- Remaining penetrability (NF EN 1426) $\geq$ 60 %

After curing and ageing PAV at 100°C / 20 h (according to NF EN 12607-1 + NF EN 14769):

- Increase of the softening point (NF EN 1427) $\leq$ 16°C
- Compliance with the values declared by the producer for the tests
  - o DSR: Determination of complex shear modulus and phase angle – dynamic shear rheometer (according to NF EN 14770)
  - o BBR: Determination of the modulus of rigidity in bending – Bar bending rheometer (according to NF EN 14771)

For light truck traffic (< 50 HGV/day), the use of pure road bitumen conforming to standard NF EN 12591 is accepted. However, road bitumen grades lower than 35/50 should not be used for wearing courses.

Regardless of the traffic class, pure bitumen should not be used for BBTM and BBM.

For formulations incorporating asphalt aggregates, the above recommendations apply to the final binder (mixture of the new binder and the aggregate binder). The incorporation of aggregate changes the characteristics of the final asphalt binder. This consideration should be taken into account in the design study by using a suitable binder and/or reducing the aggregate content.

3.3.2 *Characteristics of binders for plasters and surface coatings*

The only performance to be imposed, within the framework of a performance-based approach, corresponds to the maximum of the pendulum cohesion test (according to NF EN 13588).

It should be greater than or equal to 1.2 J/cm^2 (on the stabilised binder), for heavy traffic and for the most severe areas from a climatic point of view.

The products will also be selected taking into account the usual values of the other characteristics of the stabilised binder (according to NF EN 13074-2):

- Softening point – ball and ring method (NF EN 1427) $\geq$ 48°C;
- FRAASS (NF EN 12593) $\leq$ – 16°C for ESU stabilised binder;
- FRAASS (NF EN 12593) $\leq$ – 12°C for the stabilised binder of MBCF.

For further details, please refer to the Cerema-IDRRIM guides on "Wearable Surface Coatings" [5] and "Cold-Poured Bituminous Materials" [6].

3.3.3   *Composition of asphalt mixes*

Premature aggregate stripping is one type of winter damage encountered. However, there is no relevant test to predict this phenomenon. Otherwise, particular attention must be paid to the binder/aggregate affinity.

In regions where water resistance problems are common, it is recommended to increase the water resistance requirements by: 0.10 (NF EN 12697-12 method A) and 10 (NF EN 12697-12 method B) compared to the minimum value of the standard.

For example, the use of additives (addition of tackifiers or hydrated lime) can be a solution.

Furthermore, the richness modulus K (as defined in standard NF P98-149) reflects the conventional thickness of the binder film coating the aggregate. It therefore has a direct influence on the water resistance and durability of asphalt performance.

For certain conditions of use or particular operations, it can be interesting to add specification of a value for the richness modulus.

3.4   *Implementation precautions*

As with all worksites, the construction of wearing courses in harsh winter zones requires a good command of the conditions of implementation like quality of the substrate (treatment of cracks, removal of lamination, sweeping, etc.), bond coat (quality, dosage and breakage), conformity of the formulation (binder content, granulometry, ...), application temperature, weather conditions and compactness.

But also quality of longitudinal joints, treatment of transverse cracks, management of hard-to-compact areas, quality of the start of the work, macrotexture and management of winding areas at altitude.

The IDRRIM note insists on these last six points because it has been noted that many deteriorations described as "winter damage" resulted from the lack of control of one of these elements. Table 2 recalls the main points of vigilance and the nature of the responses provided.

Table 2.   Reminder of the issues in terms of formulation and implementation set out in the IDRRIM note.

	Objective	Possible answers
Formulation	Choice of binder: high and low temperature resistance	Modified binders and specific characteristics
	Quality of asphalt: Resistance to tearing	Water resistance and richness modulus
Implementation	Longitudinal joints: avoid or reduce their impact	Instructions for heat treatment of joints
	Transverse cracks: limit the rise of cracks	Method of treating the substrate. Bridging, purging and/or geogrid.
	Sub-compactness: avoid material loss and potholes	Quality of compaction including difficult areas
	Material departure: avoid too open macrotextures	Maximum LMP thresholds
	Beginning of construction: an area of particular vigilance	Implementation of controls from the beginning of the work
	Winding areas: special stresses	Careful selection of materials and macrotexture

4 PERSPECTIVE

The following paragraphs deal with techniques or formulation recommendations for which encouraging experiments have been carried out but for which the results are currently insufficient to reach a definitive conclusion.

4.1 *Combined surface coating*

This is a complex combining a specific Superficial Coating (ES) (open mesh) and a Cold Mastic Asphalt Mixture (MBCF), called Combined Superficial Coating (RSC).

This new technique would make it possible to combine the advantages of wearing courses (ESU) and cold-cast bituminous materials (MBCF), particularly in terms of waterproofing, while minimising their drawbacks and increasing their durability.

IDRRIM's information note No. 35 of January 2018 [7] provides project owners with decision-making aids for gradually integrating this technique into their pavement maintenance policy.

4.2 *Stone mastic asphalt*

Stone Mastic Asphalt (NF EN 13108-5), traditionally used in Germany and Switzerland, are not widely used in France.

These are discontinuous graded asphalt mixes with a high binder content (of the order of 6.5%), a fairly high proportion of fines and with cellulose fibres embedded. Depending on the type of traffic, the binder used may be pure bitumen or polymer modified bitumen.

The thickness of application is about 4 cm (similar to a BBM).

The main advantages of SMAs are high compactness, high macrotexture and good water resistance. This makes them particularly suitable for use in harsh winter areas.

Since 2010, the department of Bas-Rhin (Alsace) has been testing this technique on roads of varying traffic and altitude. The results are very encouraging, including for roads with heavy traffic.

4.3 *Lime integration*

The addition of hydrated lime in the form of active filler has been tested to improve binder-aggregate affinity and reduce the risk of stripping. The results seem encouraging in areas with harsh winters. These experiments deserve to be continued to consolidate the process.

Experiments carried out in particular in the Vosges region [8] on the main road network or on the motorway network have shown that a solution consisting of adding 1.5 to 2% (but not more than 2%) of hydrated lime to the BBSG formulas was promising.

5 CONCLUSION

The technical advice proposed in the document should make it possible to greatly reduce the risk of winter damage occurring, even if the understanding of the phenomenon remains incomplete. As the winters of 2014 to 2021 were very mild, little damage of this type was observed. However, it is therefore advisable to continue investigations on this subject because, in the event of an exceptional winter, significant damage to the network cannot be ruled out. At the same time, an analysis of the impact of high temperatures on pavement structures should be carried out, as recent heat waves have revealed new types of deterioration.

REFERENCES

[1] Corté J-F. and Garnier P., 2010. *Rapport de la Mission Sur Les Dégâts Causés au RRN Durant l'hiver 2009–2010. Rapport n° 007210-01. Conseil Général de l'Environnement et du Développement Durable.*

[2] Pendarias D., 2010. *Dégâts Occasionnés au Réseau Routier National Durant l'hiver* 2009–2010. Rapport MARRN.

[3] DVDC – PN ANR MOVE – IREX (*Institut Pour la Recherche Appliquée et l'Expérimentation en Génie civil*) – https://dvdc.fr.

[4] Note d'information n°43 – *Choix et Mise en œuvre Des Couches de Surface Dans Les Zones Soumises à Des conditions Climatiques Hivernales Rigoureuses – décembre* 2020. IDRRIM.

[5] Enduits superficiels d'usure – *Guide Technique* – septembre 2017 – Cerema IDRRIM.

[6] Matériaux bitumineux coulés à froid – *Guide Technique* – septembre 2017 – Cerema IDRRIM.

[7] Revêtements superficiels combinés- Note d'Information n°35 – janvier 2018 – IDRRIM.

[8] Pibis P., Siegel J-B., El-Bedoui S. and Lesueur D. *Improving The Durability of Asphalt Mixes – Use of Hydrated Lime in the Vosges* – RGRA March 2019

Roads and Airports Pavement Surface Characteristics – Crispino & Toraldo (Eds)
© *2023 The Author(s), ISBN 978-1-032-55149-4*

Pavement heating system using electroconductive paint for preventing ice formation on the surface layer

Thomas Attia*, Flavien Geisler & Simon Pouget
Research and Innovation Department, Eiffage Infrastructures, Lyon, France

Dominique Le Blanc & Camille Thomasse
Eiffage Energie Systèmes-Clemessy, Mulhouse, France

ABSTRACT: Winter road maintenance is an important issue in countries located in mountainous areas or subjected to cold weather. Snowstorms and ice can cause major disturbances of the traffic, paralyzing vital activities. An innovative pavement heating system has been developed in order to keep strategic infrastructures like bridges or airports always running. It consists in a layer of electroconductive paint laid on the base course beneath the surface layer. Electricity is injected in the paint using metal electrodes evenly spaced in the length of the road and as the current spreads in the paint, it induces a homogeneous Joule heating of the pavement. The voltage used ensures electrical safety for users in accordance with European standards for electrical protection. The system is completed with an automatic system that predicts the surface temperature of the pavement and anticipates the possible formation of ice. An efficient prediction system is key to maintain the energy consumption as low as possible. Various studies were conducted to evaluate the efficiency of this system and its durability under mechanical loading or temperature and humidity variations. A full-scale demonstrator has been constructed on the access ramp of a car park near Lyon, France. This paper presents some of the results of the conception studies and the first ones obtained with the demonstrator.

Keywords: Pavement Heating, Deicing, Anti-Icing, Electroconductive Paint, Bituminous Mixtures, Automatic Control

1   INTRODUCTION

Heavy snowfalls and ice storms cause significant disturbances in the transport network of several regions of the world, some of them being important economic centres as well (Northern China, USA and Canada, Northern and Central Europe …). Road infrastructures of these areas are important links and their malfunctioning is detrimental to the welfare of numerous inhabitants. The main strategies to maintain roads serviceable are the mechanical removal of snow and the use of ice melting agents (salt brine and other chemicals). However, plowing the snow can damage the surface layer of pavements and melting agents are harmful for the environment (Hintz & Relyea 2019); they can also damage the infrastructures by corroding metallic elements (Vassie 1984) and deteriorating cement concrete (Sajid *et al.* 2022).

Engineers and researchers have designed alternative solutions to achieve winter maintenance. Some of these solutions prevent the accumulation of snow and the formation of ice and thus keep the roads serviceable without any interruption.

---

*Corresponding Author: thomas.attia@eiffage.com

DOI: 10.1201/9781003429258-83

Pipes embedded in the pavement can be used to circulate fluids (water, propylene glycol, ammonia ...) that heat the surface layer (Chapman 1952). Heat can come from various sources: geothermal energy (Eugster 2007; Lee *et al.* 1984), fossil fuel energy (Cress 1995) or even solar energy. There are indeed reversible systems that circulate fluid during summer, gathering the heat of the pavement warmed by the sun, and store the energy for winter use (Eugster & Schatzmann 2002; Morita & Tago 2000). These solutions are efficient but costly compared to the classical methods. Electrically conductive elements can also be placed in the pavement and heat the surface when a current is supplied, thanks to Joule heating. Among these elements are metal cables (Henderson 1963; Lai *et al.* 2015), carbon fiber tape (Yang *et al.* 2012) and conductive rubber composites (Wei *et al.* 2020). Conductive cement concretes (Wu *et al.* 2015; Xie & Beaudoin 1995; Yehia & Tuan 2000) and conductive bituminous mixtures (Wang *et al.* 2016; Wu *et al.* 2005) have been developed by adding conductive materials (steel shaving, steel fibres, carbon fibres or graphite) in the mixes. Other solutions like infrared heat lamps and microwave deicing (Ding *et al.* 2018) have also been studied.

Due to their cost or their high energy consumption the solutions presented above are intended for strategic areas that need constant practicability (steep slopes, car parks, pedestrian crosswalks, ...) or that are sensitive to melting agents (metallic structures, concrete bridges, ...).

This paper describes an innovative system of pavement heating system (Geisler & Olard 2020) that could be cheaper than the existing solutions. It consists in applying electroconductive paint beneath the pavement surface layer and supplying a current inducing homogeneous heating of the surface layer. Results from thermal and mechanical tests performed in laboratory are presented in this paper. The first feedbacks from a full-scale demonstrator will also be discussed.

2 MATERIALS AND PROCEDURE FOR LABORATORY TESTS

2.1 *Materials and sample fabrication process*

Electrically conductive paint is obtained by adding conductive particles to a non-conductive resin binder. It has already been used for heating purposes: applied on walls for domestic heating or applied on wind turbine blades for deicing. Two types of electroconductive paint were tested in this project, they have the same conductivity and are both water-based paint, but their resin binders are different. The company Rescoll developed them. They are called paint A and paint B in the following.

Laboratory samples (50x18x10 cm^3) consists of two bituminous mixtures layers with a layer of paint and electrodes at the interface. First, the lower bituminous mixture layer (50x18x5 cm^3) is compacted using a French wheel compactor (following the standard NF EN 12697-33 + A1:2007). The lower layer is made of a "GB3", a bituminous mixture commonly used as a base layer, with a Nominal Maximum Aggregate Size (NMAS) of 14 mm. Bitumen emulsion is applied on the lower layer with a residual dosage of 250 g/m^2 to act as a clean support for the paint. Tinned copper electrodes are then placed on the bitumen after the emulsion broke, the bitumen acting as a glue maintaining the electrodes. The paint is applied on the bitumen and on the electrodes with a bar coater to control the paint thickness. It is left to dry in ambient conditions for 24 h, drying being necessary to ensure optimal electrical properties. Bitumen emulsion is applied on the dry paint with a residual dosage of 250 g/m^2 to act as a tack coat for the upper layer. Finally, the upper layer of bituminous mixture (50x18x5 cm^3) is compacted. It is made of a "BB5", a bituminous mixture used in surface layers, with a NMAS of 14 mm. Pictures of a laboratory sample during its fabrication process are presented in Figure 1. Two samples were made following this fabrication process: sample A with paint A and sample B with paint B.

Figure 1.   Laboratory sample fabrication: a) bituminous mixture lower layer with bitumen emulsion and electrodes, b) electroconductive paint, c) complete sample with the bituminous mixture upper layer.

2.2   *Experimental procedures*

2.2.1   *Heating tests*

The objective of the heating tests was to evaluate the performance of the electroconductive paints. Samples A and B were placed in a climate chamber where the temperature was maintained at 10°C. As shown in Figure 2, on each sample, one thermocouple was placed on the top of the surface layer and one on the bottom of the lower layer to follow the evolution of temperature. An electrical power corresponding to 400 W/m^2 was then supplied in each sample for about 6 hours.

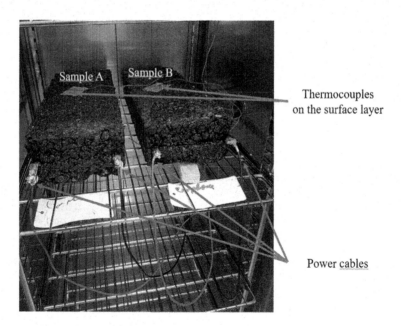

Figure 2.   Samples inside the climate chamber for heating tests.

2.2.2 Rutting tests

The aim of the rutting tests was to provide a first insight of the mechanical resistance of the electroconductive paint in the pavement under loads representative of the traffic. Samples A and B have the dimensions of standard samples used for rutting tests in France (NF EN 12697-22). The rutting machine, shown in Figure 3, applies 5 kN on a pneumatic wheel that performs a back-and-forth movement on the sample. The sample temperature is maintained at 60°C, and 30,000 cycles were applied at a frequency of 1 Hz. Measurements of the rut were performed during the test as well as measurements of the electrical resistance of the paint between the electrodes. The wheel rolled between the electrodes and not above them.

Figure 3.   Rutting machine.

3   LABORATORY TESTS RESULTS

3.1   Heating tests

The evolution of the temperature of the samples during the heating tests is presented on Figure 4. The tests were stopped when the temperature reached a constant value. The temperature on the upper layer of sample A rose by 3°C and the temperature below the lower layer by 5°C. For sample B, the temperature rose by 4°C on the surface layer and by 9°C below the base layer. The evolution of the temperature in the sample is highly dependent on boundary conditions. Notably, the airflow in the climate chamber keeping the temperature at a constant value induced high convection exchange that decreased the surface temperature measured during the experiment. An inhomogeneous ventilation in the chamber could explain the temperature difference between the top and the bottom of the samples. The humidity level was not controlled either. However, this test did show a significant rise of the temperature surface thanks to the electroconductive paint.

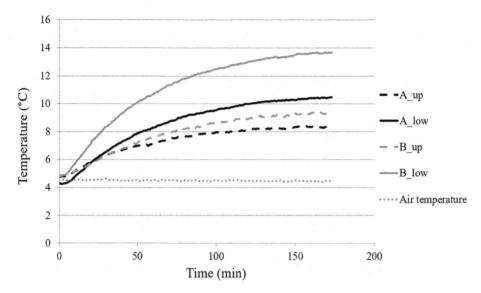

Figure 4. Evolution of the temperature of the samples during the heating tests.

3.2 *Rutting tests*

The electrical resistance between electrodes is an important parameter for industrial applications. Whatever the resistance, one can apply any power to the paint but the higher the
resistance, the higher the voltage needed to reach the same power. With the idea of working
at a low voltage to ensure user's safety, it is important to make sure that the resistance value
does not increase too much in order to be able to reach sufficient power values.

The evolution of the sample electrical resistance during the rutting tests is presented on
Figure 5. The rutting test was performed at a temperature of 60°C but measurements were
performed before the test began at the ambient temperature of about 20°C. It had been

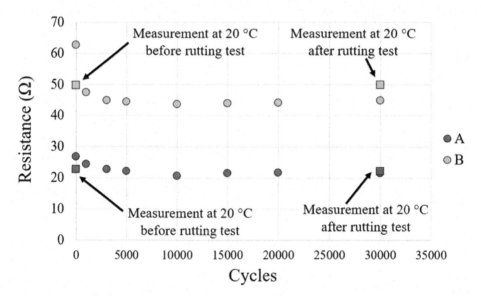

Figure 5. Evolution of the electrical resistance of the samples during the rutting tests.

previously observed that the electrical resistance depended on the temperature. The measurements at 60°C before the first rutting cycles confirmed that the resistance increased with the temperature. During the cycles, the resistance diminished possibly because of the compaction at the interface between the layers. After the cycles were applied, the resistances were also measured once the samples had cooled down. The resistance at ambient temperature before and after the 30,000 rutting cycles remained the same: 22 Ω in the end against 22.7 Ω in the beginning for sample A, 49.8 Ω against 49.7 Ω for sample B. It is a first indication that the electroconductive paint use in a pavement is resilient to traffic.

The rut size observed in the wheel path was also monitored. The mean rut size after 30,000 cycles was 3.5% of the sample height for sample A and 4.4% for sample B. Both samples met the highest requirements (inferior to 5%) as described in the French standard (NF EN 13108) for bituminous mixtures.

4 FULL-SCALE DEMONSTRATOR: CAR PARK ACCESS RAMP

4.1 *Construction works*

A full-scale demonstrator was built during autumn 2021 in the Eiffage Infrastructures central laboratory in Corbas (France), a city in the Lyon metropolitan area. The access ramp of a car park was equipped with the electroconductive paint solution on a 65 m^2 surface. As for the laboratory samples, bitumen emulsion was applied on the base course before the paint. As shown on Figure 6, the electrodes were evenly placed on the bitumen, perpendicularly to the vehicle movement direction. The paint was then applied manually on the electrodes and the bitumen (Figure 6). Electrodes were fixed with concrete nails to prevent stripping by the construction machines when applying the surface course. A tack coat made of bituminous emulsion was applied after the paint had dried for 48 h. This time can be reduced to 24 h if the weather is favourable, for instance during a warm and dry summer day. Finally, the upper layer of bituminous mixture was laid on the surface.

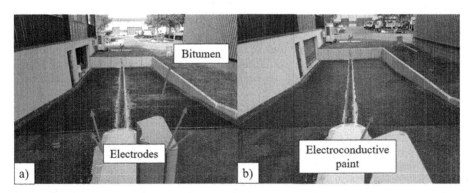

Figure 6. Construction steps of the demonstrator: a) electrodes and bitumen emulsion on the base course, b) electroconductive paint layer.

An automatic control system was also installed on the finished demonstrator to optimise the energy consumption as it can be seen on Figure 7. A weather station (measuring temperature, air humidity, wind speed, solar radiation, visibility, etc.) and sensors (measuring temperature at different depths in the pavement and the humidity surface state) were implemented to develop a model predicting the surface temperature of the pavement. The whole system is remote-controlled.

Figure 7. Demonstrator of pavement heating system using electroconductive paint: a) pavement and b) weather station and electrical cabinet hosting the automatic control system.

The maximum voltage between the electrodes is 120 V of direct current. The system satisfies the French electrical safety standards (NFC 15-100) without the need of additional protection. Moreover, the bituminous mixture of the surface layer is a very good electrical insulator. The electrical conductivity of the paint was selected to obtain a maximum heating power of 350 W/m^2, which is enough to maintain the pavement snow-free during winter in this area of France according to a climatic study and in agreement with the literature.

4.2 Heating tests

Heating tests were conducted on the demonstrator during the 2021–22 winter. These tests aimed to assess the efficiency of the system out of the laboratory and to gather data about its functioning. The transient behaviour of the heating was also evaluated. The tests were performed under various weather conditions (rainy, sunny, cloudy, windy, at night or during daytime) to account for all the parameters that influence the pavement surface temperature.

An example of test result is presented on Figure 8 where the thermographic image of half of the heated ramp can be seen. This test was performed from 18:00 on the 8th of December 2021 to 9:00 on the 9th of December 2021. The weather was cold (between 0 and 3°C), rainy (5.2 mm of precipitation during the night) and a little windy (average speed between 0 and 4 m/s). A constant power of 200 W/m^2 was supplied throughout the test. When the picture was taken at the end of the test, the temperature of the unheated pavement surface was about 3°C when the temperature of the heated surface had an average value between 9 and 10°C with a good homogeneity.

On Figure 8 can also be seen the effect of repairing the solution after it has been damaged (for instance to access pipes or networks under a pavement). Repair consists in recreating a base course on which the process described in section 4.1 is repeated. The initial electrodes must be kept and the new paint layer must be connected to the remaining paint layer. As it can be seen on Figure 8, the repaired zone is operational but with lower temperature uniformity.

Unfortunately, neither snowy nor icy weather conditions happened at the location of the demonstrator during the first winter's operation.

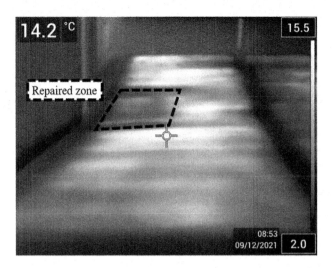

Figure 8.   Thermographic image of the ramp after heating.

5   CONCLUSIONS

An innovative heating system for pavements using electroconductive paint beneath the surface course has been developed. Laboratory samples representative of the system were made of two layers of bituminous mixture and a conductive interface with paint and electrodes. Heating tests performed in thermal chamber showed that the electroconductive paint significantly heated the samples when power was supplied. Rutting tests proved that the electrical properties of the paint remained unchanged when loads representative of traffic were applied to the samples.

The first full-scale demonstrator of the system was built in a car park access ramp. Multiple sensors were installed to monitor the climatic conditions and the temperature in the pavement. Heating tests were performed on the demonstrator and thermographic images of the ramp showed that the surface layer is heated with a good homogeneity.

This demonstrator is a powerful tool to develop the thermal models of the pavement that are used to anticipate the apparition of ice or snow on the surface. The durability of the solution when exposed to outdoor conditions will also be monitored in the following winters.

ACKNOWLEDGEMENTS

This work is part of the ICCAR project supported by the French government through ADEME and its Investments for the Future programme (in french, "*Investissements d'Avenir*"). The ICCAR consortium is composed of three partners: Rescoll, Eiffage Infrastructures and LIED UMR 8236 CNRS Université Paris Cité. The authors are grateful to Mathieu Oyharçabal (Rescoll), Prof. Laurent Royon (LIED), and Prof. Xiaofeng Guo (LIED) for their technical and writing suggestions to improve this manuscript.

REFERENCES

Chapman W.P., 1952. Design of Snow Melting Systems. *Heating and Ventilating* 49 (4), 97–102.
Cress M.D., 1995. *Heated Bridge Deck Construction and Operation in Lincoln*, Nebraska. IABSE Symposium, San Francisco, 1995, pp. 449–454.

Ding L., Wang X., Zhang W., Wang S., Zhao J., and Li Y., 2018. Microwave Deicing Efficiency: Study on the Difference between Microwave Frequencies and Road Structure Materials. *Applied Sciences* 8 (12). https://doi.org/10.3390/app8122360

Eugster W., and Schatzmann J., 2002. Harnessing Solar Energy for Winter Road Clearing on Heavily Loaded Expressways. *Proceedings XIth PIARC International Winter Road Congress*, Sapporo, Japan, January 2002.

Eugster W.J., 2007. Road and Bridge Heating Using Geothermal Energy. Overview and Examples. *Proceedings European Geothermal Congress*, Unterhaching, Germany.

Geisler F., and Olard F., 2020. *Chaussée à interface conductrice, son procédé de construction et un procédé de chauffage de cette chaussée* [in French]. EP 3,739,120,A1.

Henderson D.J., 1963. Experimental Roadway Heating Project on a Bridge Approach and Discussion. *Highway Research Record* 14, 14–23.

Hintz W.D., and Relyea R.A., 2019. A Review of the Species, Community, and Ecosystem Impacts of Road Salt Salinisation in Fresh Waters. *Freshwater Biology* 64 (6), 1081–1097. https://doi.org/10.1111/fwb.13286

Lai J., Qiu J., Chen J., Fan H., and Wang K., 2015. New Technology and Experimental Study on Snow-Melting Heated Pavement System in Tunnel Portal. *Advances in Materials Science and Engineering* 2015. https://doi.org/10.1155/2015/706536

Lee R.C., Sackos J.T., Nydahl J.E., and Pell K.M., 1984. Bridge Heating Using Ground-Source Heat Pipes. *Transportation Research Record* 962, 51–57.

Morita K., and Tago M., 2000. *Operational Characteristics of the GAIA Snow-melting System in Ninohe, Iwate, Japan*. Geo-Heat Center Bulletin 21 (4), 5–11.

Sajid H.U., Jalal A., Kiran R., and Al-Rahim A., 2022. A Survey on the Effects of Deicing Materials on Properties of Cement-based Materials. *Construction and Building Materials* 319. https://doi.org/10.1016/j.conbuildmat.2021.126062

Vassie P., 1984. Reinforcement Corrosion and the Durability of Concrete Bridges. *Proceedings of the Institution of Civil Engineers* 76 (3), 713–723. https://doi.org/10.1680/iicep.1984.1207

Wang H., Yang J., Liao H., and Chen X., 2016. Electrical and Mechanical Properties of Asphalt Concrete Containing Conductive Fibers and Fillers. *Construction and Building Materials* 122, 184–190. https://doi.org/10.1016/j.conbuildmat.2016.06.063

Wei H., Han S., Han L., and Li Q., 2020. Mechanical and Electrothermal Properties of Conductive Ethylene-Propylene-Diene Monomer Rubber Composite for Active Deicing and Snow Melting. *Journal of Materials in Civil Engineering* 32 (8). https://doi.org/10.1061/(ASCE)MT.1943-5533.0003266

Wu J., Liu J., and Yang F., 2015. Three-phase Composite Conductive Concrete for Pavement Deicing. *Construction and Building Materials* 75, 129–135. https://doi.org/10.1016/j.conbuildmat.2014.11.004

Wu S., Mo L., Shui Z., and Chen Z., 2005. Investigation of the Conductivity of Asphalt Concrete Containing Conductive Fillers. *Carbon* 43 (7), 1358–1363. https://doi.org/10.1016/j.carbon.2004.12.033

Xie P., and Beaudoin J.J., 1995. Electrically Conductive Concrete and Its Application in Deicing. *Advances in Concrete Technology Proceedings of the 2nd CANMET/ACI International Symposium*, Las Vegas, Nevada, 1995, 399–417.

Yang T., Yang Z. J., Singla M., Song G., and Li Q., 2012. Experimental Study on Carbon Fiber Tape-based Deicing Technology. *Journal of Cold Regions Engineering* 26 (2), 55–70. https://doi.org/10.1061/(ASCE)CR.1943-5495.0000038

Yehia S., and Tuan C. Y., 2000. Thin Conductive Concrete Overlay for Bridge Deck Deicing and Anti-Icing. *Transportation Research Record* 1698 (1), 45–53.

Surface heat budget of sixteen pavement samples on an experimental test site in the Parisian region

Sophie Parison*
Université Paris Cité, LIED, Paris, France

Maïlys Chanial
Université Paris Cité, LIED, Paris, France
Marie de Paris, Direction de la Voirie et des Déplacements, Paris, France

Frédéric Filaine
Université Paris Cité, LIED, Paris, France

Martin Hendel
Université Paris Cité, LIED, Paris, France
Département SEN, Université Gustave Eiffel, ESIEE Paris, Noisy-le-Grand, France

ABSTRACT: In order to limit the effect of heat on the population during heat waves, the City of Paris has recently focused research efforts on finding alternatives to existing road materials. In this regard, this paper presents a field experiment that consists of 16 paving samples of 4.25×3.95 m^2 each and 30 cm deep, filled with different traditional and alternative road structures. Each structure is instrumented with 5 thermofluxmeters at different depths and on the surface, measuring temperature and heat flux every 5 minutes and continuously during the summer of 2021. A weather station located near the experimental area collects air temperature, relative humidity, solar and thermal infrared radiation, as well as wind speed. This paper compares the results of the heat balance analysis between the 16 structures. Their daytime and nighttime thermal behavior during the summer is studied. The perspectives of this work concern the estimation of the microclimatic impact of materials on a pedestrian.

Keywords: Urban Microclimate, Pavement Materials, Urban Cooling, Heat-Waves, Surface Heat Budget

1   INTRODUCTION

When combined with the urban heat island phenomenon, heat waves pose significant health issues for the population (Lemonsu *et al.* 2013). In this respect, the City of Paris has been interested in the implementation of different passive strategies for cooling public spaces (Chaumont *et al.* 2021; Hendel *et al.* 2016; Parison *et al.* 2020a). Among the range of existing passive solutions (vegetation, energy efficiency, reflective materials, evaporative cooling, etc.), the use of alternatives to existing urban materials is a known strategy to fight against heat islands (Bowler *et al.* 2010; Guo & Hendel 2018; Santamouris 2015). Indeed, urban materials absorb part of the solar radiation and radiate infrared, which contributes to increase the thermal inertia of the city in a significant way, under certain meteorological conditions (Santamouris *et al.* 2011). Studies of the thermal behavior of urban materials in

---

*Corresponding Author: sophie.parison@u-paris.fr

DOI: 10.1201/9781003429258-84

the field can be found in the scientific literature, focusing on roofs, facades, and road surfaces (Akbari & Levinson 2011; Doya *et al.* 2012; Wang *et al.* 2021). Nevertheless, few of them take interest in the comparison of a large number of realistic pavement structures simultaneously undergoing to the same weather conditions (Li *et al.* 2013; Takebayashi & Moriyama 2012), or to the thermal stress experienced by a pedestrian.

In this paper, the thermal behavior of sixteen road material structures under field conditions is therefore examined. The evolution of temperatures and heat fluxes is observed at the surface and in depth with sensors integrated into the samples. The purpose of this work is to compare the heat balance of different materials during the summer period (Qin & Hiller 2014), in order to evaluate the effect of the conversion of an asphalt site (the most commonly found material in Paris, France). The surface heat budget of each sample is calculated (net radiation, atmospheric heat flux and conductive flux). The mitigation of the conveyed heat for each material depending on the period (day/night) is discussed.

2 MATERIALS AND METHOD

2.1 *Experimental test site and instrumentation*

The measurement site, located in Bonneuil-sur-Marne in the suburban Parisian region (France), consists of 16 samples measuring 4.25 m by 3.95 m and 30 cm deep (see Figure 1). Each sample is filled with a road structure in accordance with the prescriptions of the City of Paris' Road Division, thus consisting of several layers of materials.

Figure 1.   Visible (left) and infrared (right) photographs of the experimental test site and.

All the pavement samples were instrumented with thermofluxmeters (type T, with dimension 100 × 100 cm for 1 mm thickness), measuring simultaneously temperature and heat flux. In depth, 3 to 4 (depending on the sample) thermofluxmeters are placed at the center of the samples placed −25, −15 and −8 cm deep as well as at the interface between the two upper layers of material, for which the depth is variable depending on the paving structure and therefore on the sample. A sketch of the test site is shown in Figure 2, with photographs of the samples' surface taken during the summer of 2021. The details of the composition of the constitutive layers of materials of the samples, their thickness as well as the depths at which the thermofluxmeters are places, are presented in Table 1.

Before installation, each thermofluxmeter was covered with a bituminous coating to protect it from humidity. On the surface, the center of the samples is instrumented with a sensor prior painted with a matte black paint of emissivity 0.95 and albedo 0.05. On the surface, the bituminous coating is used to ensure a good contact between the bottom of the sensor and the surface of the sample.

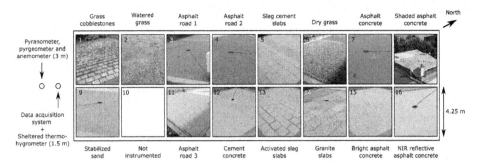

Figure 2. Design of the test site (to scale), with photographs of the surface of the samples.

Table 1. Composition and thickness (in cm) of the constitutive layers of materials form the samples, and depth (in cm) of the thermofluxmeters.

#	Thickness (cm) and layer materials	Instrument depth (cm)
1	Cobblestones (15), sand (5), Cement-non-treated base (10)	0 – 15 – 20 – 25
2	Vegetal soil (30)	0 – 4 – 8 – 15 – 25
3	Asphalt 1 (5), Cement-non-treated base (25)	0 – 5 – 8 – 15 – 25
4	Asphalt 2 (4), Asphalt (3), Cement-non-treated base (23)	0 – 4 – 7 – 15 – 25
5	Slag cement slab (10), sand (5), Cement-non-treated base (15)	0 – 15 – 20 – 25
6	Terre végétale (30)	0 – 4 – 8 – 15 – 25
7	Asphalt concrete (2), concrete (10), soil subgrade (18)	0 – 2 – 8 – 15 – 25
8	Shadehouse, asphalt concrete (2), concrete (10), soil subgrade (18)	0 – 2 – 8 – 15 – 25
9	Stabilized sand (8), soil subgrade(22)	0 – 4 – 8 – 15 – 25
10	Not instrumented	/
11	Asphalt 3 (4), Cement-non-treated base (26)	0 – 4 – 8 – 15 – 25
12	Concrete (10), soil subgrade (20)	0 – 4 – 8 – 15 – 25
13	Activated slag slab (10), sand (5), Cement-non-treated base (15)	0 – 10 – 15 – 25
14	Granite slabs (8), sand (7), Cement-non-treated base (15)	0 – 8 – 15 – 25
15	Bright asphalt concrete (2), concrete (10), soil subgrade (18)	0 – 2 – 8 – 15 – 25
16	NIR reflective asphalt concrete (2), concrete (10), soil subgrade (18)	0 – 2 – 8 – 15 – 25

In addition, a weather station as well as the data acquisition center are installed south of the site, about 1.5 m aside from the edge of the samples. The weather station includes a sheltered thermo-hygrometer 1.5 m high (air temperature and relative humidity), as well as a pyranometer (solar radiation 0.3 – 3 μm), a pyrgeometer (infrared radiation 3 – 100 μm) and an ultrasonic anemometer (wind speed) at 3 m height. The data acquisition system is powered by a battery coupled with a photovoltaic panel and acquires data continuously every 5 minutes (averaging data per minute).

The samples include both "traditional" materials (asphalt concrete sidewalk, concrete, granite slabs, etc.) and alternative surfaces (reflective, permeable surfaces, etc.). Three samples are vegetal: two grasses (one watered and one dry) and watered grass cobblestones. Four samples contain an asphalt concrete sidewalk structure, with various top layers: traditional, with a 1-m-high shadehouse, with bright-coloured asphalt, or with near-infrared radiation (NIR) reflective paint (respectively on samples 7, 8, 15, and 16).

Samples 5 and 13 both have slabs incorporating slag cement, which is a by-product of the metallurgical industry identified in literature as interesting compound for cool pavements (Anupam *et al.* 2021). Sample 9 is made of stabilized sand (a traditional pavement used in parks and gardens in Paris), and finally, samples 3, 4 and 11 are made of pervious asphalt.

2.2 Surface heat budget

The energy balance of a dry surface is described entirely by equation (1), with R_n the net radiation at the surface, H the atmospheric convective flux, and V the heat flux conducted to the surface (Parison et al. 2020b).

$$R_n = H + V \tag{1}$$

Equations (2) and (3) develop the expression for net radiation as a function of solar (S) and infrared (L) radiation (resp. S_{up} and L_{up} for the reflected and re-emitted terms by the surface), albedo α, emissivity ε, Stefan Boltzmann's constant σ and surface temperature T_s.

$$R_n = S + L - \left(S_{up} + L_{up}\right) \tag{2}$$

$$R_n = (1 - \alpha)S + L\left(\varepsilon - \sigma T_s^4\right) \tag{3}$$

The albedo is measured in the field with an albedometer following the ASTM E1918-16 standard (ASTM E1918-16). The emissivity values are either taken from the literature (Kotthaus et al. 2014), from previous measurements (Parison et al. 2020b) or from the manufacturer's data sheet for the NIR reflective paint. The albedo and emissivity values are listed in Table 2 below.

Table 2. Albedo and emissivity of the samples.

#	Albédo	Emissivité	#	Albédo	Emissivité
1	0,22	0,92	9	0,40	0,90
2	0,22	0,97	10	/	/
3	0,20	0,95	11	0,37	0,95
4	0,24	0,95	12	0,49	0,95
5	0,53	0,95	13	0,57	0,95
6	0,20	0,97	14	0,34	0,93
7	0,13	0,95	15	0,46	0,95
8	0,13	0,95	16	0,59	0,88

The atmospheric heat flux H (communicated to the ambient air by convection) is expressed with equation (4), with h the convective exchange coefficient and T_a the air temperature. The latter can be calculated using equation (5), with the surface sensor's measured heat flux φ_g, with controlled emissivity and albedo.

$$H = h(T_s - T_a) \tag{4}$$

$$H = R_n^{\varepsilon=0,95;a=0,05} - \varphi_g \tag{5}$$

From the pedestrian heat stress perspective, radiosity ($S_{up} + L_{up}$) as well as atmospheric heat flux (H) are two detrimental terms that should be reduced when implementing urban cooling strategies. For this paper, these two terms only will be considered.

2.3 Analysis days

The analysis focuses on days with meteorological conditions representative of a summer day or a fortiori heat-wave days. Such days are selected between early June and mid-September and present good insolation conditions (cloud cover <3 oktas), low wind speeds (<3 m/s), and minimum and maximum air temperatures respectively above 16°C and 25°C (Parison *et al.* 2020). Based on these criteria, 16 days were selected for the analysis of the summer 2021 data. These data are shown in Figure 3 for July the 20th of 2021 which met these criteria.

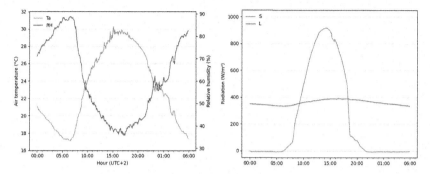

Figure 3.   Air temperature T_a, relative humidity RH, incident solar S and infrared L radiations for July, the 20th 2021.

3   RESULTS

3.1   *Convective exchanges coefficient*

To determine the convective exchange coefficient h, equation (5) is used to calculate the atmospheric heat flux H from the flux measurement of the sample surface sensor. The value of the exchange coefficient h is mainly influenced by the convection regime (free or forced). For this reason, and in order to increase the size of the sample used, the calculated flux values for H were merged between samples and between radiative days (i.e. with similar meteorological conditions, especially regarding wind speed) (i.e. 16 signals times 16 days).

Figure 4 represents H (calculated from φ_g) as a function of the temperature difference between the surface and the ambient air, as per equation (4). The surface temperature corresponds to the thermo-fluxmeter measurement of the considered sample, while the temperature is the same for all samples, the latter being measured only at the weather station, 1.5 m above ground level (Figures 1 and 2).

From these figures, two convection regimes are identified: one during the insolation period, approximately between 9am and 6 pm (UTC + 2) (Figure 4 left), and the second regime at night, from 0am to 6am and again from 9 pm to midnight the next day (Figure 4 right). A

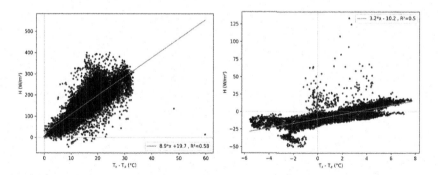

Figure 4.   Atmospheric heat flux H versus temperature difference between the surface and the air for the 16 samples and 16 analysis days all merged. Left: from 9am to 6 pm (UTC + 2). Right: from 0am to 6am and from 9 pm to 0 pm (UTC + 2).

linear regression of the data allows us to determine the coefficient h for the two regimes of "day" and "night", which is respectively equal to 8.9 W/m²/K and 3.2 W/m²/K. These "step" values are used afterwards regardless of the radiative day considered. For the periods excluded from the linear regression (i.e. the transition between the two regimes from 6am to 9am and from 6 pm to 9 pm), an intermediate value of h is used, calculated by smoothing out its value with that time window using a hyperbolic function. The convection coefficient is considered here as constant during the day or night. In reality, it should be noted that its value is subject to variations according to the time of day and the day itself, not taken into account here.

3.2 Surface heat budget

The atmospheric convection term H obtained from the heat balance for July the 20th 2021 is presented in Figure 3. In order to smooth the measurement fluctuations, the term H is recalculated with equation (5), knowing the exchange coefficient h (calculated in section 3.1). The atmospheric heat flux H is plotted against time in Figure 5 for each sample.

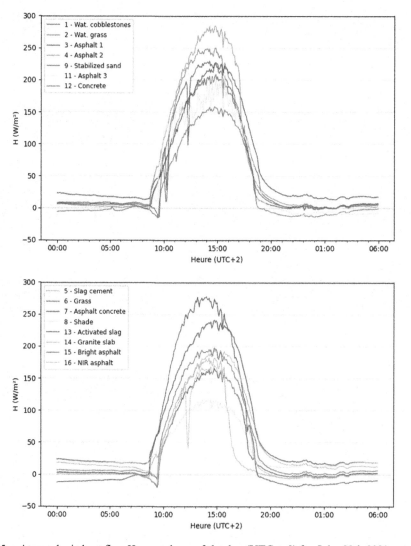

Figure 5. Atmospheric heat flux H versus hour of the day (UTC + 2) for July, 20th 2021.

During the insolation period, the ranking of structures follows that of the surface temperature, the air temperature and the coefficient h being unchanged between samples (as per equation 4). The structures that transfer the most heat to the ambient air are mainly those with a low albedo: asphalt concrete sidewalk, asphalt roads and vegetated samples. For the latters, this can be explained by the low thermal inertia of the vegetal soil compared to mineral structures. On the other hand, it may indicate a poor contact between the sensor and the surface, due to its vegetation. In-depth sensors will help clarify this point.

The samples that heat the air the least are the shaded sample (solar radiation is reduced by half) and then the concrete sample. Compared to traditional asphalt concrete sidewalk, the use of bright-, NIR reflective- and shaded- asphalt concretes can reduce H by a maximum of 50, 80 and 125 W/m², respectively. Slag cement slabs, having a lower thermal inertia than the majority of structures, heat up more quickly and reach their maximum at about solar noon, while the others reach it about 1.5 hours later. At night, these trends are generally

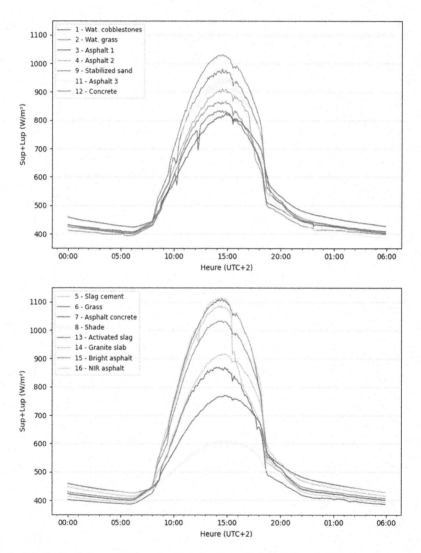

Figure 6.   Radiosity (solar and infrared reflected and emitted radiations) of the samples' surfaces versus hour of the day (UTC + 2) for July, 20th 2021.

maintained with some exceptions: grass samples become the least heating structures, their low inertia helping them to cool down very quickly. On the other hand, the shade becomes one of the least favourable structures at night, as it slows down the cooling of the surface due to the reduction of the sky view factor it causes.

The radiosity of the surfaces (re-emitted and reflected radiation, $S_{up} + L_{up}$) is now considered. Knowing the albedo and the emissivity of the surfaces (see Table 2), the net radiation is calculated according to equation (3), and radiosity is deducted from equation (2). The radiosity is plotted for each sample in Figure 6 for July 20th, 2021.

For radiosity, two competing effects are at work: on the one hand, visible radiosity will be greater the higher the albedo (during the day), and on the other hand, infrared radiosity is high if the surface temperature is also high. The tendencies observed in Figure 5 are thus globally reversed here during the day, the materials with high albedo being less favourable from this standpoint. On the other hand, the shade remains the most interesting structure from the pedestrian's standpoint (the incident solar radiation being reduced by half). The radiation from the shade itself is not taken into account here. At night, these observations are reversed: in the absence of solar radiation, the structures with the highest radiosity are those emitting the most infrared, i.e. the warmest structures, i.e. those with a low albedo as observed previously, all other things being equal.

4   CONCLUSION

An experimental site was instrumented to study the thermal behaviour of sixteen road structure samples during summer periods. Measurements during the summer of 2021 allowed the calculation of the thermal balance of the surface of the samples, in particular the atmospheric heat flux as well as the radiosity (radiation reflected and re-emitted by the surface towards the sky). The results allow us to discriminate between the structures and show competing effects from the pedestrian's heat stress standpoint. On the one hand, samples with a low albedo heat up more than the others and release more heat by convection with the ambient air, during the day but also at night, by thermal inertia. These structures will therefore have a more negative effect on the pedestrian.

On the other hand, regarding radiosity, the observed tendency is reversed during the day, with highly reflective structures (high albedo) radiating more on a pedestrian than less reflective structures. However, at night, the opposite occurs, as warmer structures (low albedo) radiate more infrared at night than structures with a lower temperature (high albedo). In any case, for the daytime, the shaded sample remains by far the best option to limit the thermal stress of the pedestrian, either from the standpoint of atmospheric convective heat or radiated heat. Future perspectives of this work concern the estimation of the impact of the presented materials on the pedestrian heat stress.

NOMENCLATURE

Symbols:

α	albedo, [-]
ε	emissivity, [-]
φ_g	global heat flux density, W/m^2
H	atmospheric convection, W/m^2
h	convective exchanges coefficient, W/m^2/K
L	infrared radiation (3 – 100 μm), W/m^2

R_n net radiation, W/m^2
S solar radiation $(0,3 - 3 \ \mu m)$, W/m^2
σ Stefan-Boltzmann's constant $(5.67 \ 10^{-8} \ W/m^2/K)$
T_a air temperature, °C
T_s surface temperature, °C
V conductive heat flux, W/m^2

Abbreviations:

Wat. watered
RH relative humidity
NIR near infrared $(3 - 100 \ \mu m)$
up upwards

ACKNOWLEDGMENTS

The authors acknowledge the technical support provided by the Public Space Laboratory of the City of Paris' Road Division.

REFERENCES

Akbari H., & Levinson R. (2008). Evolution of Cool-roof Standards in the US. *Advances in Building Energy Research*, 2(1), 1–32.

Anupam B.R., Sahoo U.C., Chandrappa A.K., & Rath P. (2021). Emerging Technologies in Cool Pavements: A Review. *Construction and Building Materials*, 299, 123892.

ASTM E1918-16: Standard Test Method for Measuring Solar Reflectance of Horizontal and Low-Sloped Surfaces in the Field. *American Society for Testing and Materials*. West Conshohocken, PA.

Bowler D.E., Buyung-Ali L., Knight T.M., & Pullin A.S. (2010). Urban Greening to Cool Towns and Cities: A Systematic Review of the Empirical Evidence. *Landscape and Urban Planning*, 97(3), 147–155.

Chaumont M., Parison S., Kounkou-Arnaud R., Long F., Bernik A., Da Silva M., & Hendel M. (2021). "Tierce Forêt": Greening a Parking Lot. *34rd Conference of the International Association of Climatology*, Casablanca, Morocco.

Doya M., Bozonnet E., & Allard F. (2012). Experimental Measurement of Cool Facades' Performance in a Dense Urban Environment. *Energy and Buildings*, 55, 42–50.

Guo X., & Hendel M. (2018). Urban Water Networks as an Alternative Source for District Heating and Emergency Heat-wave Cooling. *Energy*, 145, 79–87.

Hendel M., Gutierrez P., Colombert M., Diab Y., & Royon L. (2016). *Measuring the Effects of Urban Heat Island Mitigation Techniques in the Field: Application to the Case of Pavement-watering in Paris. Urban Climate*, 16, 43–58.

Kotthaus S., Smith T.E., Wooster M.J., & Grimmond C.S.B. (2014). Derivation of an Urban Materials Spectral Library Through Emittance and Reflectance Spectroscopy. *ISPRS Journal of Photogrammetry and Remote Sensing*, 94, 194–212.

Lemonsu A., Kounkou-Arnaud R., Desplat J., Salagnac J.L., & Masson V. (2013). Evolution of the Parisian Urban Climate Under a Global Changing Climate. *Climatic Change*, 116(3), 679–692.

Li H., Harvey J., & Kendall A. (2013). Field Measurement of Albedo for Different Land Cover Materials and Effects on Thermal Performance. *Building and Environment*, 59, 536–546.

Parison S., Hendel M., & Royon L. (2020a). A Statistical Method for Quantifying the Field Effects of Urban Heat Island Mitigation Techniques. *Urban Climate*, 33, 100651.

Parison S., Hendel M., Grados A., & Royon L. (2020b). Analysis of the Heat Budget of Standard, Cool and Watered Pavements Under Lab Heat-wave Conditions. *Energy and Buildings*, 228, 110455.

Qin Y., & Hiller J. E. (2014). Understanding Pavement-surface Energy Balance and its Implications on Cool Pavement Development. *Energy and Buildings*, 85, 389–399.

Santamouris M., Synnefa A., & Karlessi T. (2011). Using Advanced Cool Materials in the Urban Built Environment to Mitigate Heat Islands and Improve Thermal Comfort Conditions. *Solar Energy*, 85(12), 3085–3102.

Santamouris M. (2015). Regulating the Damaged Thermostat of the Cities—Status, Impacts and Mitigation Challenges. *Energy and Buildings*, 91, 43–56.

Takebayashi H., & Moriyama M. (2012). Study on Surface Heat Budget of Various Pavements for Urban Heat Island Mitigation. *Advances in Materials Science and Engineering*, 2012.

Wang C., Wang Z. H., Kaloush K. E., & Shacat J. (2021). Cool Pavements for Urban Heat Island Mitigation: A Synthetic Review. *Renewable and Sustainable Energy Reviews*, 146, 111171

Airport pavements

Roads and Airports Pavement Surface Characteristics – Crispino & Toraldo (Eds)
© 2023 The Author(s), ISBN 978-1-032-55149-4

Towards efficient airfield pavement surface condition monitoring using deep learning models

Ronald Roberts* & Fabien Menant
MAST-LAMES, Université Gustave Eiffel, IFSTTAR, Campus de Nantes, Bouguenais, France

Michael Broutin
Service Technique de l'Aviation Civile (STAC), Bonneuil-sur-Marne, France

ABSTRACT: Deep Learning techniques have shown value as possible solutions for automating road data collection and analysis, with many agencies now using them. They can help road managers in many situations where there is unavailable or subjective manual road data. The issues of insufficient data and instrumentations also plague airport managers who have airfield pavements to maintain, which need to be kept in good condition. This study proposes applying Deep Learning image detection models originally built using secondary road imagery in France to now be used for automating airfield pavement monitoring using a simple single camera setup onboard a vehicle. Whilst the structures are similar, the distresses and general surrounding context are different. The study, therefore, tests the efficiency of the models on a runway in France to understand how prebuilt models can be adapted. It focuses on cracking detection and the impacts of the changing pavement surroundings. These are evaluated using practical field tests and experiments. Results from the tests show the models can provide a good overview of runway conditions and offer managers vital input to help plan maintenance actions.

Keywords: Airfield pavements, Deep learning, Pavement distresses, Pavement management

1 INTRODUCTION: CONTEXT AND OVERVIEW

The new and evolving challenge facing airfield pavement managers focuses on asset management. Specifically, in the context of combining aging pavements, increasing air traffic (which involves ever more aggressive aircrafts in terms of single wheel loads and contact pressures), and progressively more sustainable development concerns to optimize maintenance and rehabilitation works.

At present, there are more than 160 airport platforms in France, in metropolitan France or overseas, which represent more than 100 million airfield pavement square meters to maintain, divided into flexible and composite pavements (75%) and rigid ones (25%).

In-service regular assessment of pavement conditions is the key to good asset management, since it allows understanding pavement behavior, and anticipating and optimizing maintenance. In France, the majority of airfield pavement managers rely on a visual distress survey to accomplish this. All relevant distresses, listed in the French Aviation authority's distress catalogue (Service Technique de l'Aviation Civile 2007), which contains 20 distresses for flexible pavements and 11 for rigid ones, are recorded in terms of the number of occurrences, sizes or densities, and severities. Indicators called Indice de Service (IS) are computed

---

*Corresponding author: ronald.roberts@univ-eiffel.fr

DOI: 10.1201/9781003429258-85

from these data – [Méthode IS]: a global indicator (IS) computed from all distresses, a structural index computed from distresses representative of structural problems (IS$_{Struct}$) and a surface index computed from distresses indicating surface problems only (IS$_{Surf}$) (Service Technique de l'Aviation Civile 2013). The IS index is used in the French airfield pavement management system with a range of 0 to 100, where 100 represents no degradation and 0 represents the need to close the runway.

The usual implementation of the IS survey is currently manual. The studied area is divided into 500m^2 unit sections for flexible pavement, and the distresses are visually characterized and reported by a pedestrian team for each unit section (flexible pavements) or each concrete slab (rigid pavements). This way of proceeding is time-consuming and implies heavy operational constraints for the airport, and it may raise subjectiveness problems (and thus repeatability concerns). This also leads to infrequent conditional surveys. This is the reason why it is compelling to develop automated distress survey systems and associated analysis tools.

1.1 Airfield pavement maintenance needs and goals of the study

It is widely understood that airport Managers need to ensure their pavements are in good conditions. Whilst it can be noted that many airports have high-end systems that can produce detailed and accurate conditional surveys using expensive systems such as the Laser Crack Measurement System (LCMS) (Ragnoli et al. 2018), several airports across France do not have the resources or access to these systems. As a result, they require cheaper solutions to obtain pavement condition data as at present many rely on manual methods to carry out pavement condition assessments similar to road pavements where only 40% of French departments utilize automated methods (L'Observatoire National de la Route 2019). This is especially the case for French overseas territories, where it is not feasible to transport survey vehicles back and forth from mainland France. The possibility of utilizing Deep learning (DL) for this presents an opportunity for these airfield pavement managers as DL systems have seen significant increases in their use for detecting, diagnosing and predicting defects (Liu et al. 2017) even if the quality of the input images is not high. Particularly in the field of pavement distress detection, DL models have shown merit in distress detection using images (Gopalakrishnan 2018). However, as most of these models are primarily focused on normal road surveys, it is imperative to understand how these models can be adapted for airfield pavements. To this end, this study aims to utilize existing DL models originally trained for secondary roads and test their transferability to the airfield pavement context. The study uses these models, adapting them to the particular context and compares them to models trained directly on airfield pavement images. This provides an understanding of how models can perform and may be adapted in case of need. Furthermore, using the DL models, it is here proposed a simple instrumentation including a single camera, which could be easily implemented on vehicles such as the inspection vehicles, which are already used daily for FOD (Foreign Object Debris) detection. This would enable regular surveys to be carried out by airfield managers without extra costs to the airport nor further complicating the current operational constraints.

2 INSTRUMENTATION AND SURVEY METHODOLOGY

2.1 Workflow to test deep learning models

As stated, the purpose of the study is to test the transferability of existing DL models for the detection of distresses initially developed by Université Gustave Eiffel for secondary roads to use on runways. In the first instance, the previous models were run and detections were produced followed by an assessment of the results with respect to airfield conditional survey requirements. The next step consisted of retraining models with new data in order to improve the results and take special features of airfield pavements into account. The schema shown in Figure 1 displays the workflow of the study.

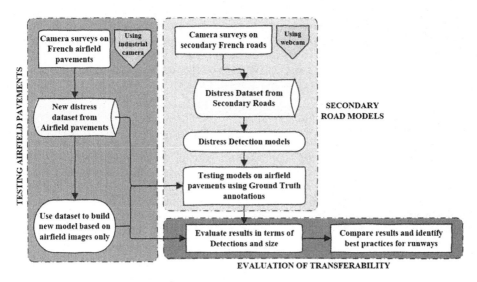

Figure 1.   Workflow to utilize previous DL models for runway analysis.

2.2   *Camera features and setup*

For the surveys on the airfield pavements, images were captured using an industrial camera affixed to a vehicle equipped at a vertical height of 2.5 m above the surface to cover a large surface with an acceptable ground resolution (~1.5 mm/pixel in this study). Surveys were carried out with GPS-located images captured from a top-down perspective as shown in Figure 2 with drivers using runway marks to guide the path of the survey vehicle whilst driving at slow speeds to control the survey. The color camera equipped with a 12 mm lens provides successive images (4096 × 2160 px), with a relatively regular distance sampling; each image represents an actual on-ground distance of 3 m in the transverse direction and 1.5 m in the longitudinal direction. The relationship between the image size and the corresponding ground dimension is important as it allows for an estimation of the real size of the

Figure 2.   Instrumentation used to capture images during the survey.

distress. It must be noted here that the images used for the original model were captured with a webcam affixed to the survey vehicle with a resolution of 640 × 480 px, which is much less than the resolution of images captured on the airfield pavement. This is one important difference and is a scenario that is likely to occur with many typical road distress models as the level of details required of those managers will be substantially less than those required of airfield pavements. This relates to the traffic expected on airfield pavements and the need to have runways maintained at a much higher level of standard than typical secondary roads. The camera setup here could be easily replicated in the FOD surveys previously mentioned, thus providing an opportunity for frequent surveys on the runway if a detection model could prove adequate using the images collected in these surveys.

2.3 *Pre-processing of images from surveys*

In order to produce a complete image covering the runway, the acquisition software was set in such a way as successive images share the same part of the pavement surface (30 cm at least in the longitudinal direction). This image overlap ensures no distresses are neglected but may cause an overestimation of the total amount of distresses on the other hand. Consequently, a cropping operation to each image was applied to remove the common parts between images. Each couple of successive images was examined and cropped according to the corresponding GPS coordinates. As the latter are not exact, the result is not perfect but acceptable to go ahead in the process and avoid an inflation in the total amount of distresses.

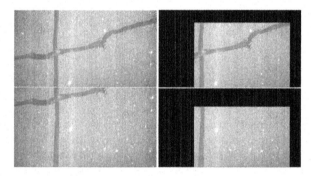

Figure 3.   Example of cropped images from surveys (right – images before and left – images after crop).

The images captured were labelled to identify the distresses for both, a ground truth for comparisons to the detections made by the DL models, and for training new models tailored to airfield pavements. To do this, the open-source software, LabelImg (Tzutalin 2015) was utilized, which allows annotating images in the required DL training format (Everingham *et al.* 2014).

3   DEEP LEARNING MODELS EMPLOYED

For the analysis, the study initially considered DL models that were developed based on imagery from secondary roads in France. These models were developed using TensorFlow (Abadi *et al.* 2016) using open-source models from their object detection model zoo (TensorFlow 2020). The model zoo provides base models that were previously trained on millions of images and can be repurposed for new tasks using transfer learning. Transfer learning involves reusing existing base models with new images to retrain the model's final layers to detect different objects than the original model was trained for (Canziani *et al.* 2017; Pan & Yang 2010) and this allows achieving high degrees of performance without the need to train an entirely new model.

3.1 *Typical deep learning setup*

For the model development, a typical DL workflow was applied as shown in Figure 4. This involved preparing the images for training and including a comprehensive annotation procedure to label images with distresses. The model development included optimization and sensitivity tests considering different hyperparameters and augmentation strategies to yield the best model results. Figure 4 highlights the key stages of the DL process considering the TensorFlow environment. Within it, it is important to state that the process splits the dataset into three subsets for training, valuation and testing (in a ratio of 70:20:10). The largest subset is for training the model, and then the valuation is used for evaluating the model during training and the final subset, the test set is used for simulating a real-world test by applying the model on these images (never before seen by the model) after training has been completed. Models were developed using a Windows 10 PC with an NVIDIA Quadro P4000 GPU (8 GB ram) and total CPU memory of 32 GB @ 3.7GHz and a 64bit processor. The general speed of the models on the workstation is approximately 0.6 seconds per training step with an average of 100,000 steps used in the training process.

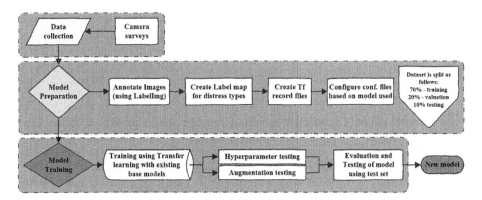

Figure 4.   Workflow to utilize Tensorflow models.

3.2 *Deep learning models used*

The focus of the study was utilizing models that were previously trained for flexible pavements roads by the study team (Roberts *et al.* 2020) and adapted for secondary roads in France (Roberts 2021) using a database of over 6,000 images. As previously mentioned, these images were of a lower quality than the ones captured over the airfield pavements but for the purposes of secondary road managers, however, they are sufficient. Within this work, many models were analysed but the one with the highest performance was the faster_rcnn_inception_v2_coco model which uses the Faster Regions with convolution Neural Network (Faster R-CNN) (Ren *et al.* 2017). The model uses the inception architecture as a backbone (Szegedy *et al.* 2016) and is based on the COCO (common objects in context) dataset (Lin *et al.* 2014). This model will be the focus of the study. Respectively, other models considered used bases with variations of the Single Shot Detection (SSD) network (Liu *et al.* 2016) and the yolo network (Redmon *et al.* 2016). However, these models had poorer performance in detecting pavement distresses compared to the Faster R-CNN based model. The Faster R-CNN based model is slower in real-time detection than the others but has an advantage in attaining higher accuracies when detecting smaller objects (Li *et al.* 2021). This is important as pavement distresses are quite small and thus it is understandable that this model's performance is higher. A reason for the additional accuracy is the fact that the faster-RCNN model takes a two-step approach to the detection task wherein the model essentially proposes regions in the first step then extract features from these regions and classifies the regions based on the features. The other two model bases mentioned are quicker to make detections because they essentially look at the model only once. The full architectures of these models are well known and thus will not be elaborated on here as the aim of the paper is using the models for a new purpose specifically for airfield pavements.

The models developed for secondary roads were trained to detect five distress categories in line with requirements from the French Manual for distress collection and identification (Laboratoire Central des Ponts et Chaussées (LCPC) 1998). These were Ravelling (RV), Transverse cracking (TC), Longitudinal cracking (LC) and Alligator/Block cracking (BC). These categories represent the most relevant to French secondary road managers. For assessing the performance of DL models, the typical and most commonly used metrics of precision, recall and f1 score were utilized (Han *et al.* 2014). Precision in this case refers to the ratio of accurate distresses detected to the total number of predicted distresses with higher precision values indicating limited false positives. Meanwhile, Recall refers to the ratio of accurate distresses detected to the total number of actual distresses with higher recall values indicating limited false negatives. The f1 score is a combination of the two where recall and precision are reflected equally. The use of a particular metric largely corresponds to the use of the model in real life as different stakeholders may try to avoid certain occurrences such as false detections, which would see them preferring a higher precision.

4 APPLICATION OF MODELS

4.1 *Description of the test site*

For the analysis of the DL models, condition surveys were carried out at the Lognes Aerodrome in Lognes, France. For the survey, four corridors were utilized along the runway as shown in Figure 5. Each corridor has a width of 2m with an approximate length of 550m and the overlap correction was done to ensure that the images captured in each corridor reflect only the width of that corridor. The Vehicle was aligned so that the camera would capture the entire corridor's width and with the four corridors the entire width of the runway. This allowed for a total of 1819 images to be captured when driving at 30km/hr (the speed represents the best compromise between operational constraints the ability to follow a predefined trajectory and picture quality).

Figure 5. Left – Aerial airport view and right – Overview of how images were captured during the survey.

Once the images were captured, the DL model was initially run over all the survey images. For the model application, python scripts were developed wherein the model could be run on any image dataset to produce output images of the detections and also a text file containing the images with detected distresses, distress type and the size and location of the distress within the image (this is represented by a rectangular-shaped box). This result allows post-treatment wherein the distress size can be calculated by using the dimensions of the box containing the detection. This is done considering the actual dimensions represented by the pixels in an image.

4.2 *Results*

Before assessing the performance of the models on airfield pavements, the initial performance metrics of the Faster-RCNN model as tested on secondary roads were considered and this is shown in Figure 6, showing only the results from the cracking distress categories.

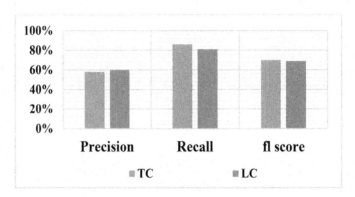

Figure 6.   Overview of the performance of the Faster R-CNN model for distresses on secondary roads.

Within the figure, it can be seen that the recall values are a bit higher than the precision but overall, the f1 scores are adequate in detecting the distresses. For a comparison to distresses required in an airfield conditional survey, there is no split of cracking with respect to its presence in the wheel path as opposed to the secondary road requirements which require distinguishing between cracking within and out of the wheel path. The other distresses from the previous model considering Ravelling and Block cracking do exist for flexible airfield pavements but in most maintained airfield pavements, their occurrence is less frequent with cracking being the most commonly found distress. The pavement condition is determined by the estimation of the service index (IS) as computed by the encountered distress types, their dimensions and severities. In this study, the focus was on measurements of the cracking distress as the survey revealed that this was the main distress pattern on the surveyed runway. This study aimed to utilize the detections to provide an understanding, particularly of the total length of cracks that appear in the survey. Nevertheless, for future considerations, it is important to consider other distress patterns so as to satisfy the full practical standard requirements. An example of the detections produced by the model on pavement images is shown in Figure 7.

In applying the model to the airfield images, visually it was noted that the model was able to detect the longitudinal and transverse cracking. This can be seen in Figure 8. Notably, the model did not have any major issues distinguishing cracking from the repair section (as shown in the figures) which can appear as distresses and which traditional methods have difficulties differentiating. An explanation for this is that the original model was trained on images with a variety of instances of flexible pavements that had elements that could be misconstrued as distresses such as shadows, pavement marks and repairs. This emphasizes

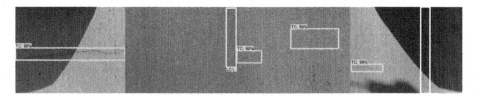

Figure 7.   Example of distress detection on secondary road images.

Figure 8.   Resulting detections on the airfield pavement images.

the need for DL models to be trained with a dataset containing many background elements that could appear in real life.

Given the visual performance, the next step was to train an entirely new model considering only the airfield images and compare the performance metrics. This was done using the Lognes survey images, annotating them to represent two distress categories – longitudinal and transverse cracking. This produced a dataset comprising 2660 instances of longitudinal cracking and 1454 instances of transverse cracking. A comparison of the metrics achieved when testing the airfield distress dataset for both models (the one trained with secondary road images and the one trained with airfield images) is shown in Figure 9. The performance of the models is in fact higher than that achieved for the secondary road images but this is expected given the higher resolution of the images, which allows more image details to be seen by the models.

From the figure, it can be seen that when trained with images directly from the same dataset, the results improve particularly for the recall. Improved recall translates to lower chances of missed detections. This is in line with other research work where the precision for airfield pavement distress detection is in the range of 85% and recall – 65% (Li *et al.* 2022). However, the precision remains consistent in terms of values for both models indicating they both will produce low false distress detections, which is key. This suggests that whilst it is preferable to have a dataset trained on exact images, once models are based on images with similar characteristics, they can be effective. As the surveyed secondary roads in France are also asphaltic in nature (as the runway in Lognes), this can explain the level of performance. However, the models tailored to the airfield pavements present a greater opportunity to yield more accurate distress dimensions as opposed to the models based on the secondary road images.

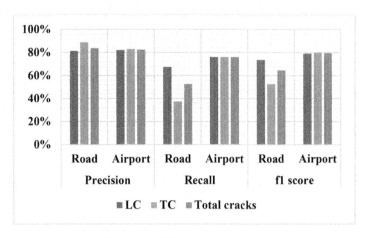

Figure 9.   Comparison of the performance of DL models.

Using the detections and the ground truth boxes, the lengths of the cracks detected were determined by calculating the length and widths of the detected boxes depending on whether the crack was longitudinal or transversal. This is shown in Table 1 with comparisons to the ground truth. From this, it was shown that the model was able to achieve an adequate performance level in identifying runway cracks, which is critical towards computing the required service index. Additionally, the crack lengths were generated automatically, saving time and resources.

Table 1.   Detected crack lengths and comparison to ground truth labels.

	Detected crack lengths (m)	Ground Truth crack lengths (m)	Difference to detections	% cracks detected
Longitudinal cracking (LC)	744.75	1080.33	−31%	69%
Transverse cracking (TC)	632.76	638.47	−1%	99%
Total cracks	1377.51	1718.8	−20%	80%

Further tests were made to consider different image characteristics. This included changing the image brightness and using greyscale images. The greyscale tests in particular produced noteworthy results in that when using these images, the model was able to produce 13% more detections. This allowed for the total percentage of cracks detected when compared to the ground truth, to be 90%. This is substantial and therefore it could be inferred that the greyscale change helps the model by filtering out colour allowing it to focus on the cracks, which appear easier to see in greyscale. Further tests will be carried out to establish coherence to this hypothesis. Nevertheless, what is important to the process is understanding the needs of the condition survey and requirements and this study was able to do that to yield valuable data for the airfield manager.

5   OUTLOOK AND FUTURE WORK

The work presented in this study represents an initial overview of a larger project considering automated detection systems for airfield pavement distresses. It establishes that DL detection models can be readily used once trained in similar circumstances even though

877

direct training on exact images yields better results. It must be stressed however that training and optimization of the models are required to achieve the best results and models should not simply be transferred from scenario to scenario without testing. The next steps of the wider project involve considering applications on rigid pavements (as they are also widely used in the area of airfield pavements) and to determine how models need to be adapted for this change. Additionally, whilst the current study was able to establish crack lengths, the widths are also needed to establish severities and this will be considered using the Mask R-CNN DL segmentation model given the performance of the detection model based on the Faster R-CNN network. The possibility of considering different camera perspectives will also be considered including the use of drones for image analysis. With the combination, the full IS index could be established which will provide substantial relief to airfield pavement managers. The study, therefore, presents an important base for future work.

ACKNOWLEDGEMENTS

The authors would like to acknowledge and thank Aéroport de Paris for offering experiments on the runway of the Lognes Aiport. Additionally, the authors would like to thank the French FEREC foundation, which funded the project.

REFERENCES

Abadi M. et al. 2016. TensorFlow: A System for Large-Scale Machine Learning, in: *Proceedings of 12th USENIX Symposium on Operating Systems Design and Implementation*. USENIX Association, Savannah, Georgia, USA, pp. 265–283.

Canziani A., Culurciello E. and Paszke A., 2017. *An Analysis of Deep Neural Network Models for Practical Applications*. ArXiv.

Everingham M., Eslami S.M.A., Van Gool L., Williams C.K.I., Winn J. and Zisserman A., 2014. The Pascal Visual Object Classes Challenge: A Retrospective. *Int. J. Comput. Vis.* 111, 98–136.

Gopalakrishnan K., 2018. Deep Learning in Data-driven Pavement Image Analysis and Automated Distress Detection: A review. *Data* 3, 28.

Han J., Kamber M. and Pei J., 2014. Data Mining: Data Mining Concepts and Techniques, Third Edit. ed, *The Morgan Kaufmann Series in Data Management Systems (Selected Titles)*. Elsevier, Waltam, MA.

L'Observatoire National de la Route, 2019. *Rapport Observatoire National de la Route (ONR) 2019* [WWW Document]. Rapport. URL https://www.idrrim.com/ressources/documents/11/7102-IDRRIM_Rapport_ONR-2019.pdf (accessed 1.12.22).

Laboratoire Central des Ponts et Chaussées (LCPC), 1998. *Catalogue des dégradations de surface des chaussées: version 1998* [Online]. Catalogue. URL http://www.ifsttar.fr/fileadmin/user_upload/editions/lcpc/MethodeDEssai/MethodeDEssai-LCPC-ME52.pdf (accessed 1.12.22).

Li, H., Jing, P., Huang, R., Gui, Z., 2022. *Algorithm for Crack Segmentation of Airport Runway Pavement under Complex Background based on Encoder and Decoder* 1706–1711.

Li W., Feng X.S., Zha K., Li S. and Zhu H.S., 2021. Summary of Target Detection Algorithms, in: *Proceedings of International Conference on Computer Big Data and Artificial Intelligence (ICCBDAI 2020)*. Changsha, China.

Lin T.Y., Maire M., Belongie S., Hays J., Perona P., Ramanan D., Dollár P. and Zitnick C.L., 2014. Microsoft COCO: Common objects in context. *Lect. Notes Comput. Sci.* (including Subser. Lect. Notes Artif. Intell. Lect. Notes Bioinformatics) 8693 LNCS, 740–755.

Liu W., Anguelov D., Erhan D., Szegedy C., Reed S., Fu C.Y. and Berg A.C., 2016. SSD: Single Shot Multibox Detector, in: *Proceedings of European Conference on Computer Vision ECCV 2016*. Amsterdam, The Netherlands, pp. 21–37.

Liu W., Wang Z., Liu X., Zeng N., Liu Y. and Alsaadi F.E., 2017. A Survey of Deep Neural Network Architectures and Their Applications. *Neurocomputing* 234, 11–26.

Pan S.J. and Yang Q., 2010. A Survey on Transfer Learning. *IEEE Trans. Knowl. Data Eng.* 22, 1345–1359.

Ragnoli A., De Blasiis M. and Di Benedetto A., 2018. Pavement Distress Detection Methods: A Review. *Infrastructures* 3, 58.

Redmon J., Divvala S., Girshick R. and Farhadi A., 2016. You Only Look Once: Unified, Real-Time Object Detection, in: *Proceedings of 2016 IEEE Conference on Computer Vision and Pattern Recognition (CVPR)*. IEEE, Las Vegas, Nevada, USA, pp. 779–788.

Ren S., He K., Girshick R. and Sun J., 2017. Faster R-CNN: Towards Real-Time Object Detection with Region Proposal Networks. *IEEE Trans. Pattern Anal. Mach. Intell.* 39, 1137–1149.

Roberts R., 2021. A New Approach to Road Pavement Management Systems by Exploiting Data Analytics, *Image Analysis and Deep Learning*. University of Palermo.

Roberts R., Giancontieri G., Inzerillo L., Di Mino G., 2020. Towards Low-Cost Pavement Condition Health Monitoring and Analysis Using Deep Learning. *Appl. Sci.* 10, 319.

Service Technique de l'Aviation Civile, 2007. *État de la Surface des Chaussées Aéronautiques Catalogue Pour la Réalisation des Relevés Visuels de Dégradations sur Chaussées Aéronautiques*. Paris, France.

Service Technique de l'Aviation Civile, 2013. *Méthode indice de service – Guide Méthodologique*. Paris, France.

Szegedy C., Vanhoucke V., Ioffe S., Shlens J. and Wojna Z., 2016. Rethinking the Inception Architecture for Computer Vision. *Proc. IEEE Comput. Soc. Conf. Comput. Vis.* Pattern Recognit. 2016-Decem, 2818–2826.

TensorFlow, 2020. *TensorFlow 1 Detection Model Zoo* [Online]. URL https://github.com/tensorflow/models/blob/master/research/object_detection/g3doc/tf1_detection_zoo.md (accessed 2.21.22).

Tzutalin, 2015. LabelImg.

Roads and Airports Pavement Surface Characteristics – Crispino & Toraldo (Eds)
© *2023 The Author(s), ISBN 978-1-032-55149-4*

Aviation risk assessment: Analyzing the critical boundary conditions that can affect the risk of runway-related accidents

Misagh Ketabdari*, Claudia Nodari, Maurizio Crispino & Emanuele Toraldo
Department of Civil and Environmental Engineering, Politecnico di Milano, Milan, Italy

ABSTRACT: The safety of runway-related operations has been always the main scope in airport risk assessment. The importance of the runway safety is getting even more significant with the increase in the air transport demands over the last decade. In this regard, this study focuses on identifying most influencing boundary conditions on the safety of runway-related operations by adopting a practical aviation risk assessment model developed by Airport Cooperative Research Program (ACRP). One International Airport, as the case study, is selected and simulated through adopted model to assess the safety of the aircraft operations in excursion events including Landing Overruns (LDOR), Landing Undershoots (LDUS), Landing Veer-offs (LDVO), Takeoff Overruns (TOOR), and Takeoff Veer-offs (TOVO). For this matter, the operating aircraft characteristics, Runway Safety Area's (RSA's) geometry, one year of daily operation data, and the real-time weather data for the same year of the selected airport are collected. As an outcome of this simulation, four different sensitivity analyses based on changes in runway geometry, weather conditions, runway end safety area infrastructure (e.g., presence of arresting system), and airport annual movements are conducted to evaluate the impacts of these parameters on the probability of occurrence of runway-related accidents. These sensitivity analyses clarify the dependency of aircraft ground operations' safety on airport's boundary conditions, which help airport operators to increase their runways safety by strengthening their weaknesses.

Keywords: Aviation Risk Assessment, Runway-Related Accidents, Runway Boundary Conditions.

1   INTRODUCTION

Civil Aviation represents all the non-military air operations, including private and commercial flights. Basically, civil aviation can be classified into scheduled air transport and general aviation. The scheduled air transport, which forms a major part of civil aviation, includes the passenger flights and cargo flights. This study focuses on the passenger flights, which are mainly commercial, while the general aviation can be private or commercial (ICAO 2001).

The number of the passengers transported by scheduled flights has increased globally in the last 10 years. According to International Civil Aviation Organization (ICAO)'s 2018 annual global statistics report, the total number of the passengers carried on scheduled service rose to 4.3 billion in 2018, which is 6.4 per much higher than 2017. Regarding to the departure number, it increased 33.5 per to 37.8 million in 2018. For the past 2019, the number of passengers is recorded as 4.5 billion (ICAO 2018). Figure 1 shows the detailed statistics of passengers from 2009 to 2018, at the global scale.

---

*Corresponding Author: misagh.ketabdari@polimi.it

                                              DOI: 10.1201/9781003429258-86

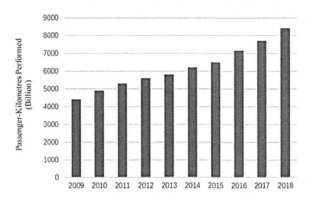

Figure 1. Increasing trend of passenger volumes at globe scale from 2009 to 2018 (ICAO, 2018).

Air transport makes a great contribution to the global economy. It improves the human mobility all around the world. It's essential for the global tourism and business and it accelerates the global trade. It helps countries to participate into the global distribution of each industry. According to International Air Transport Association (IATA), 52,2 million metric tons of goods valued at USD 5.6 trillion were transported in 2015 by airlines, which is estimated as 35% of all global trades in the same year (Shepherd *et al.* 2016). Moreover, air transport industry also supports a total of 62.7 million jobs globally. It provided 9.9 million jobs directly (ICAO 2018). Therefore, it can be declared that the aviation is the most global industry which connect different people, culture, and business.

With the increase in air traffic volume over the years, guarantying the safety of operations should be the highest priority for the aviation industry. Airlines from all over the world, carried 3.3 billion passengers in 2014 with 12 fatal accidents and 641 fatalities (Tyler 2014). According to IATA Annual Review, the aviation safety is decreasing year by year since the numbers of flights are increasing annually, therefore, continuous updated safety measures are required to be applied to the aviation industry (Tyler 2014).

Airport is playing a key role in aviation safety, in particular runway safety is one of the most significant parameters to be guaranteed. Runway-related accidents can be defined as Landing Overrun (LDOR), Take-off Overrun (TOOR), Landing Under-shoot (LDUS), Landing Veer-off (LDVO), and Take-off Veer-off (TOVO) (Ketabdari *et al.* 2018). The summary of runway-related accidents and incidents from 1980 to 2009 are presented in Figure 2.

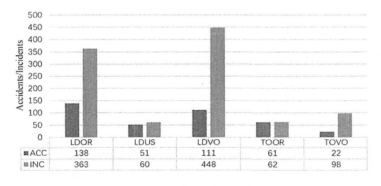

	LDOR	LDUS	LDVO	TOOR	TOVO
ACC	138	51	111	61	22
INC	363	60	448	62	98

Figure 2. History of the runway-related events from 1980 to 2009 (Ayres *et al.* 2013).

It can be seen from the figure that landing incidents/accidents accounted for approximately 83% of the events, and takeoffs for the remaining 17%. Overruns, including landing and takeoffs, accounted for 44% of the events, veer-offs 48%, and undershoots only 8% of the total number of events. Incidents are the events without any causalities, on the contrary, accidents include events with injuries, fatalities, and any type of causalities.

To mitigate the probability of occurrence and/or the severity of the consequences of each runway-related accidents, the level of impact of different operation boundary conditions should be initially assessed. In this regard, this study focuses on identifying most influencing boundary conditions on the safety of runway-related operations by adopting a risk model developed by Airport Cooperative Research Program (ACRP).

2   LITERATURE REVIEW

Many studies have been conducted over the past decades to evaluate the risk of runway-related accidents and in general, the safety of runway operations. In 2003, the required length of Runway End Safety Area (RESA) rose from 90 m to 240 m according to ICAO, which was based on the crude analysis of historical overrun events without considering the environmental situation and runway suitability (Kirkland *et al.* 2003).

In 2008, ACRP released Report 03 regarding the analysis of aircraft overruns and undershoots for Runway Safety Areas (RSAs). This report introduced a probabilistic approach to evaluate the degree of protection offered by a specific RSA and provided a risk-based assessment procedure in terms of different risk variables associated with aircraft overruns and undershoots. Furthermore, three sub-models of probability of occurrence, location, and consequences for LDOR, LDUS, and TOOR were developed. This study is based on Historical Operation Data (HOD), which requires a large database of accidents records (Hall 2008).

In 2009, Wong *et al.*, developed a more risk-sensitive and flexible airport safety area strategy. This research overcomes the limitations regarding the necessity of large database of accidents by using the Normal Operation Data (NOD), including non-accident flights. In addition, a general approach, which could analyze different accident events including take-off and landing overruns, undershoots, veer-offs, was developed (Wong *et al.* 2009).

In 2011, ACRP released Report 50 regarding improved models for risk assessment of runway safety areas. This report expands the research presented in ACRP report 3. Although it was still based on sub-models of accident frequency, location probability model and consequence, all runway-related events of LDOR, LDUS, LDVO, TOOR, and TOVO were considered in the methodology. Both HOD and NOD are involved in this approach, which make the results more relevant and reliable. Moreover, an analytical software of Runway Safety Area Risk Analysis (RSARA) was developed and verified to quantify the risk of mentioned runway-related events and support runway safety planning. Thanks to this software, it is possible to study and prioritize the alternatives for RSA improvements (e.g., application of Engineered Material Arresting System - EMAS, declared distances, etc.) (Ayres 2011).

In 2014, ACRP released Report 107 regarding the development of a runway veer-off location distribution risk assessment model and reporting template. In this report, the location models for both lateral and longitudinal accident distances in the RSA were improved with higher precision. In addition, a new consequence model was developed by integrating the new location models and the existing frequency models from the previous ACRP reports. Last but not the least, an analytical tool of Lateral Runway Safety Area Risk Analysis (LRSARA) was developed and verified to estimate the veer-offs probability of occurrence, the full-stop aircraft location after veer-off, and the severity of the consequences in presence of obstacles in the vicinity of runway (Ayres *et al.* 2014).

In 2015, Trucco *et al.* developed a two-step procedure which converts probability and severity results obtained from ACRP methodology into topological grid forms. Thanks to this procedure, it is possible to characterize the terrain surrounding the runway and therefore giving the possibility to compare them with the underlying infrastructure. This approach is a qualitative visual tool that can be adopted for airport runway planning and runway safety management (Trucco et al. 2014).

3 GOALS AND METHODOLOGY

The main goal of the research discussed in this paper is to assess the critical boundary conditions that can affect the risk of runway-related accidents. The research method adopted in the study is based on the model proposed by ACRP in report 50 (Ayres 2011). As discussed in detail in the following sub-sections, the model is divided in three different sub-models of:

- Accident probability: a sub-model to calculate the probability of occurrence of runway-related accidents (e.g., overruns, undershoots, veer-offs);
- Wreckage location: a sub-model to evaluate the probability of occurrence of runway-related accidents at locations in the proximity of the runway (e.g., RESA, lateral RSA);
- Accident consequences: a sub-model to assess the severity of consequences in case of accidents (e.g., negligible, minor, major, hazardous, and catastrophic).

3.1 Accident probability sub-model

This sub-model assesses the probability of occurrence of different runway-related accidents by considering several independent variables (e.g., runway geometry, aircraft characteristics, weather conditions, etc.). This sub-model is applicable to LDOR, LDUS, LDVO, TOVO, and TOOR. The logistic regression approach was adopted to carry out the multivariate analysis since its more relevant to the dichotomous characteristic of the possible outcomes (i.e., accident or normal operation), as demonstrated in the Equation 1.

$$P_{accident\ occurrence} = \frac{1}{1 + e^{-(b_0 + b_1 x_1 + b_2 x_2 + b_3 x_3 + ...)}} \tag{1}$$

Where, $P_{accident\ occurrence}$ is the probability of a specific type of runway-related accident occurring (e.g., LDOR) in certain operational conditions; X_i are independent variables (e.g., ceiling, visibility, crosswind, precipitation, aircraft type, criticality factor), and b_i are the regression coefficients (Ayres 2011). All the independent variables are transformed into binary format which means it can only be 0 (absence of boundary condition) or 1 (presence of boundary condition) so that it can avoid the non-linear relationship between the variables.

3.2 Wreckage location sub-model

This sub-model assesses the likelihood of an aircraft leaving the runway area and stops in RSA. Like the accident probability sub-model, the wreckage location is also specific for the event type. Therefore, five models can be developed. These wreckage location models are based on historical accident data for all five types of associate events. In formulating these models, many factors would affect separately for each type of accident, as demonstrated in Equation 2. For instance, in the case of overruns, the wreckage location would be altered by the type of terrain.

$$P_{location > x} = e^{-ax^n} \tag{2}$$

Where, $P_{location > x}$ is the probability the overrun/undershoot distance along the runway centerline beyond the runway end is greater than x; x is a certain location or distance beyond the runway end; and a and n are regression coefficients (Ayres 2011).

3.3 Accident consequence sub-model

This sub-model depends on parameters such as the kinetic energy of crash, quantity of fuel and the quantity of fuel and the local type of terrain. In most cases, the consequences are related to the speed that the aircraft hit an obstacle and the type of the obstacle. The higher the speed is, the more serious consequence happens. Based on the location model, the terrain type, and the deceleration model developed by Kirkland *et al.* in 2003, the probability the aircraft moving above a certain speed when hitting the obstacle can be estimated. This sub-model develops a qualitative method to evaluate the severity of consequence according to different types of obstacles, aircraft speed and size, and location.

To conclude, the combination of these three sub-models to evaluate the risk of runway-related accidents is demonstrated in Figure 3.

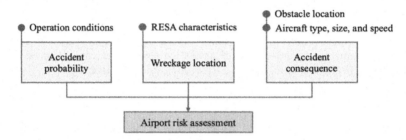

Figure 3.   Airport risk assessment sub-models (Ayres 2011).

4   CASE STUDY

One international airport from Italy is selected as case study. The airport has two runways of 17/35 (601 m length) and 18/36 (2442 m length). The runway 18/36 is generally used for commercial traffic with a preference of utilizing Runway 36 since the tailwind is not too strong. The runway 17/35 runway is mainly used for general aviation. Therefore, regarding the goals of this study, runway 18/36 is selected in the simulations.

The selected airport is simulated through the RSARA software developed by ACRP to evaluate the risk of its runway operations. According to the ACRP Report 50, the airport historical and weather data needed as inputs of the simulation process should be collected for at least one year. Therefore, one year of HOD and the related Historical Weather Data (HWD) for the selected case study are collected from 23rd of February 2020 to 23rd of February 2021. The parameters that should be covered in HOD are flight number, FAA code, category and type, and runway designation and bound.

The period for the weather data must match the period of operational data to determine the actual weather conditions for each operation. In order to accurately match HWD with the flight operations, HWD are collected for every 30 minutes. The weather parameters that should be covered in HWD are visibility, wind direction, wind speed, air temperature, ceiling, and presence of different weather conditions (e.g., rain, snow, snow showers, rain showers, thunderstorms, ice crystals, ice pellets, snow pellets, pellet, showers, freezing rain, freezing drizzle, wind gusts and fog, etc.).

In order to define the RSA geometric layout of each runway ends, first the geometry of undershoot RSA should be drawn in a spreadsheet and then the overrun RSA layout of runway should be designed. It should be noted that in the designing process on the spreadsheet, the runway is not on the actual scale. It is the representation to facilitate the position of the runway. Subsequently, these designs should be imported into the RSARA.

The next step is to define the side Obstacle Free Area (OFA) distance. The lateral OFA distance is the clearance from the edge of runway to the nearest obstacle, fixed or movable.

In most cases, it will be an aircraft located in a parallel taxiway. However, in some cases, a hanger or another fixed object may be the obstacle.

By simulating the collected data through RSARA, the average probability for each event can be achieved for the selected case study. The average number of years is calculated between two different accidents and incidents. For instance, the LDOR incident happens one time over 100 years. Although this means the probability of occurrence of LDOR is improbable, considering the widespread cancellation of flights caused by the COVID-19 in 2020, the number of aircraft operations at the airport is much smaller than usual. Therefore, the probability of an aircraft accident can be also lower than other operational years. The percentage above the Target Level of Safety (TLS) is 3% means only 3% movement in Runway 36 has the higher risk than 1.0E-06.

5 SENSITIVITY ANALYSIS OF OPERATION'S BOUNDARY CONDITIONS

Assessing the safety of runway-related operations in different boundary conditions can help to analyze and quantify the influence of each parameter on the final risk probability. Moreover, it can determine the importance of each boundary condition on the probability of occurrence of each event and the severity of the related consequences. In this regard, four different boundary conditions of runway length, weather condition, RESA infrastructure (e.g., presence of EMAS), and airport annual movement are selected to be varied in order to perform sensitivity analyses of the risk of airport accidents.

5.1 Runway length

To evaluate the influence of runway length on the safety of runway-related operations, four length scenarios with increase of 1000 ft (305 m) and 2000 ft (610 m) and decrease of 1000 ft (305 m) and 2000 ft (610 m) are defined and the rest of the boundary conditions remain unchanged. The risks of these newly defined scenarios are assessed and compared to the current length of the runway, which is 8012 ft (2442 m), as presented in Figure 4.

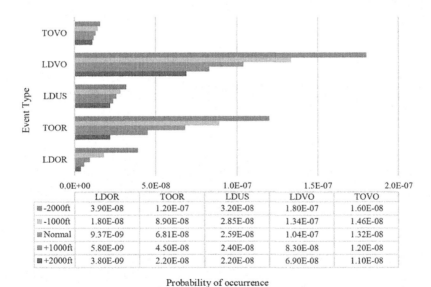

	LDOR	TOOR	LDUS	LDVO	TOVO
■ -2000ft	3.90E-08	1.20E-07	3.20E-08	1.80E-07	1.60E-08
▨ -1000ft	1.80E-08	8.90E-08	2.85E-08	1.34E-07	1.46E-08
▨ Normal	9.37E-09	6.81E-08	2.59E-08	1.04E-07	1.32E-08
▨ +1000ft	5.80E-09	4.50E-08	2.40E-08	8.30E-08	1.20E-08
■ +2000ft	3.80E-09	2.20E-08	2.20E-08	6.90E-08	1.10E-08

Probability of occurrence

Figure 4. Probability of occurrence of runway-related events in different runway lengths.

It can be noticed that the probability of **TOOR**, respect to normal/current runway length, is decreased by 34% and 67% in case of increases in runway length of 1000 ft (305 m) and 2000 ft (610 m), respectively, and it is increased by 31% and 76% in case of decreases in runway length of 1000 ft (305 m) and 2000 ft (610 m), respectively.

The probability of LDOR is decreased by 38% (for +1000 ft), 60% (for +2000 ft), and it is raised by 92% (for -1000 ft) and 316% (for -2000 ft). According to the results, variation in runway length has noticeable influence on TOOR and LDOR events with very small impacts on LDVO and TOVO events, since veer-off events usually happen along the runway area, and they are mainly affected by the runway surface situation and crosswind. Moreover, LDUS, regarding its nature, does not experience any significant changes by modifying runway length.

5.2 Weather condition

To evaluate the impact of weather parameters on the safety of runway-related operations, two weather scenarios of best and worst operational weather conditions are defined. In this regard, operational weather parameters (e.g., visibility, wind direction, wind speed, air temperature, ceiling, and presence of rain, snow, snow showers, rain showers, thunderstorms, ice crystals, ice pellets, snow pellets, pellet, showers, freezing rain, freezing drizzle, wind gusts and fog, etc.) should be defined carefully (Ketabdari *et al.* 2020a).

ICAO developed an application to define bad-weather and good-weather conditions by investigating how the weather phenomena can affect the aircraft performance. This application provides information about weather conditions at international airdrome worldwide using METeorological Aerodrome Reports (METARS). The algorithm is largely based on a technical note from Eurocontrol on Algorithm to describe weather conditions at European airports (Eurocontrol 2011). According to this application, bad-weather scenario should include critical weather parameters (e.g., low ceiling, poor visibility, high freezing rate, heavy winds, extreme precipitation, etc.). On the contrary, good-weather condition should be free of any critical weather parameters that can compromise the safety of maneuvers.

The rest of the boundary conditions remain unchanged. The risks of these newly defined scenarios are assessed and compared to the existing weather condition of the case study, as presented in Figure 5.

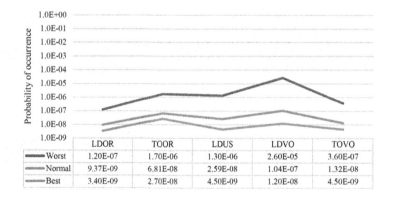

	LDOR	TOOR	LDUS	LDVO	TOVO
Worst	1.20E-07	1.70E-06	1.30E-06	2.60E-05	3.60E-07
Normal	9.37E-09	6.81E-08	2.59E-08	1.04E-07	1.32E-08
Best	3.40E-09	2.70E-08	4.50E-09	1.20E-08	4.50E-09

Event Type

Figure 5.   Probability of occurrence of runway-related events in different weather conditions.

It can be noticed that variation in weather parameters can affect noticeably the probability of occurrence of all runway-related events, with great order of magnitude.

5.3 RESA infrastructure

To evaluate the impact of RESA materials on the safety of runway-related operations, a new scenario of installment of Engineered Materials Arresting System (EMAS), instead of asphalt pavement at the runway end safety area, is defined. EMAS is passive safety intervention in case of overrun event, which is normally adopted for the airport with limited area (e.g., landlocked airports located in residential areas or surrounded by mountains/seas) (Ketabdari *et al.* 2020b).

EMAS, as demonstrated in Figure 6, is a bed of customized cellular cement material, designed to crush under the weight of an aircraft. It is an officially recognized and promoted method by Federal Aviation Administration (FAA).

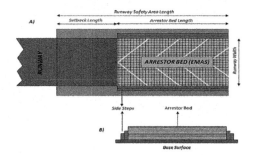

Figure 6. EMAS interface with the aircraft and typical installation of EMAS on RESA.

According to ICAO, it is recommended to extend the length of standard RESA from 90 m to 240 m (ICAO annex 14 2013), but the available fields on some airports are limited due to their location and surrounding areas. Therefore, it is impossible to increase the length of the RESA on these airports to respect the ICAO's recommendation. Moreover, from economic point of view, this activity can be highly expensive and unaffordable for the airports' concessionaires. In these situations, EMAS can be considered as an alternative solution (Ketabdari *et al.* 2021).

According to FAA AC 150/5220-22B, EMAS must be capable of safely stopping an aircraft that leaves the runway, traveling at 70 knots (FAA 2005). EMAS beds can be vary in length from 60 m ($\cong$ 200 ft) to 120 m ($\cong$ 400 ft), which for this study, the 90 m ($\cong$ 300 ft) is set as the EMAS bed length. The rest of the boundary conditions remain unchanged. The risk of this newly defined scenario is assessed and compared to the existing paved RESA of the case study, as presented in Figure 7.

It can be noticed that there are no changes in the probabilities of occurrence of LDUS, LDVO, and TOVO, since EMAS affects only overrun events. Moreover, installation EMAS at RESA will decrease the severity of consequences of LDOR and TOOR, therefore, it is very reasonable that the probabilities of occurrence of LDOR and TOOR do not change noticeably respect to the paved RESA scenario.

5.4 Airport annual movement

To evaluate the impact of annual movement on the rate of runway-related accidents, a new scenario of lower annual movement, as a result of Covid-19 outbreak, is defined for the selected case study. As reported by ASSAEROPORTI - *Associazione Italiana Gestori Aeroporti*, the selected airport operated with an average of 115,000 movements before Covid-19, compared to average of 34,000 movements after the virus outbreak, which means it dropped down nearly 70% of the normal volume. The rest of the boundary conditions

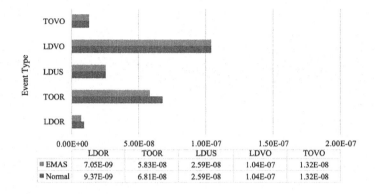

	LDOR	TOOR	LDUS	LDVO	TOVO
EMAS	7.05E-09	5.83E-08	2.59E-08	1.04E-07	1.32E-08
Normal	9.37E-09	6.81E-08	2.59E-08	1.04E-07	1.32E-08

Probability of occurrence

Figure 7.   Probability of occurrence of runway-related events for different RESA materials.

remain unchanged. The risk of this newly defined scenario is assessed and compared to the normal annual movement of the case study, as presented in Table 1.

Table 1.   Comparison of the number of years between events, before and after Covid-19.

Type of accident	Average probability (before Covid-19)	Average number of years to critical accident (before Covid-19)	Average probability (after Covid-19)	Average number of years to critical accident (after Covid-19)
LDOR	9.37E-09	>100	9.37E-09	>100
TOOR	6.81E-08	57	6.81E-08	83
LDUS	2.59E-08	78	2.59E-08	>100
LDVO	1.04E-07	49	1.00E-07	74
TOVO	1.32E-08	93	1.32E-08	>100

It can be noticed that there are no changes in the probabilities of occurrence of runway-related accidents, but the frequency of accidents drops down noticeably, since in the Covid-19 outbreak period, the selected case study has experienced dramatical reduction in daily and annual aircraft movements.

6   CONCLUSION

Risk assessment methods in aviation greatly rely on the knowledge of the factors influencing risk and safety during aircraft operations. One of the weak points of the common airport risk assessment approaches is their focus on qualitative assessment to support decisions rather than quantitative risk evaluation. This study focuses on identifying most influencing boundary conditions on the safety of runway-related operations by adopting a practical aviation risk assessment (both quantitative and qualitative) model, developed by ACRP. In this regard, one Italian airport is selected as the case study to assess the risks of its runway-related accidents (i.e., LDOR, TOVO, LDUS, LDVO and TOVO) through RSARA tool, which is based on the ACRP model.

To analyze and quantify the influence of each boundary condition on the safety of airport operations, new scenarios respect to current situation at the selected airport are defined by varying the runway length, weather condition, RESA infrastructure (e.g., presence of

EMAS), and airport annual movement. The risks of these newly defined scenarios are assessed and compared. It can be noticed from the results that runway length extension can significantly lower the probabilities of occurrence of overrun accidents. Moreover, weather conditions can alter remarkably the likelihood of all five runway-related accidents. Furthermore, the presence of EMAS at RESA can decrease the severity of the consequences of overrun accidents. Finally, airport annual movement can affect the frequency of accidents over the years and not their probabilities of occurrence over the operations.

These sensitivity analyses clarify the dependency of the safety of aircraft ground operations on each assessed boundary condition, which help airport operators and managers to increase their runways safety in a more controlled environment.

REFERENCES

Ayres M., 2011. *Improved Models for Risk Assessment of Runway Safety Areas* (Vol. 50). Transportation Research Board.

Ayres Jr M., Shirazi H., Carvalho R., Hall J., Speir R., Arambula E., and Pitfield D., 2013. Modelling the Location and Consequences of Aircraft Accidents. *Safety Science*, 51(1), 178–186.

Ayres Jr M., Carvalho R., Shirazi H. and David R. E., 2014. *Development of a Runway Veer-off Location Distribution Risk Assessment and Reporting Template* (No. Project 04-14).

Eurocontrol, 2011. *Algorithm to Describe Weather Conditions at European Airports, ATMAP Weather Algorithm* (Version 2.3), technical note.

Federal Aviation Administration (FAA), 2005. Engineered Materials Arresting System (EMAS) for Aircraft Overrun. *Tech. Rep. Advisory Circular* 150/5200-22A.

Hall J.W., 2008. Analysis of Aircraft Overruns and Undershoots for Runway Safety Areas (Vol. 3). *Transportation Research Board*.

International Civil Aviation Organization (ICAO), 2001. *Annex 6. Operation Of Aircraft-Part I-International Commercial Air Transport-Aeroplanes.*

International Civil Aviation Organization (ICAO), 2013. *Convention on International Civil Aviation – Annex 14: Aerodromes – Aerodrome Design and Operations.* Vol. 1. 6th ed.

International Civil Aviation Organization (ICAO), 2018. The World of Air Transport in 2018. Annual Report 2018. [cited 2022 May 11]. Available from: https://www.icao.int/annual-report-2018/Pages/the-world-of-air-transport-in-2018.aspx.

Ketabdari M., Giustozzi F. and Crispino M., 2018. Sensitivity Analysis of Influencing Factors in Probabilistic Risk Assessment for Airports. *Safety Science*, 107, 173–187.

Ketabdari M., Toraldo E. and Crispino M., 2020a. Numerical Risk Analyses of the Impact of Meteorological Conditions on Probability of Airport Runway Excursion Accidents. *In International Conference on Computational Science and Its Applications* (pp. 177–190). Springer, Cham.

Ketabdari M., Toraldo E., Crispino M. and Lunkar V., 2020b. Evaluating the Interaction Between Engineered Materials and Aircraft tyres as Arresting Systems in Landing Overrun Events. *Case Studies in Construction Materials*, 13, e00446.

Ketabdari M., Toraldo E. and Crispino M., 2021. Analyzing the Bearing Capacity of Materials used in Arresting Systems as a Suitable Risk Mitigation Strategy for Runway Excursions in Landlocked Aerodromes. *In Eleventh International Conference on the Bearing Capacity of Roads*, Railways and Airfields, Volume 1 (pp. 226–235). CRC Press.

Kirkland I., Caves R.E., Hirst M., and Pitfield D.E., 2003. The Normalisation of Aircraft Overrun Accident Data. *Journal of Air Transport Management*, 9(6), 333–341.

Shepherd B., Shingal A. and Raj A., 2016. Value of Air Cargo: Air Transport and Global Value Chains. Montreal: *The International Air Transport Association (IATA)*.

Trucco P., De Ambroggi M. and Leva M.C., 2015. Topological Risk Mapping of Runway Overruns: A Probabilistic Approach. *Reliability Engineering & System Safety*, 142, 433–443.

Tyler T., 2016. IATA Annual Review 2016. Recuperado de https://www. iata. org/en/publications/annual-review.

Wong D.K., Pitfield D.E., Caves R.E. and Appleyard A.J., 2009. The Development of a More Risk-sensitive and Flexible Airport Safety Area Strategy: Part I. The Development of an Improved Accident Frequency Model. *Safety Science*, 47(7), 903–912.

Roads and Airports Pavement Surface Characteristics – Crispino & Toraldo (Eds)
© 2023 The Author(s), ISBN 978-1-032-55149-4

Airport noise emission: Models' comparison and application on a real case study

Misagh Ketabdari*, Fazel Nasserzadeh, Maurizio Crispino & Emanuele Toraldo
Department of Civil and Environmental Engineering, Politecnico di Milano, Milan, Italy

ABSTRACT: In recent years, due to increase in travel demands of principal transportation infrastructures such as road, railway and airport, the corresponding air and noise pollutions are escalated. Regarding the latter, scientific literatures provide several models capable of computing the parameters related to noise emissions. Some of them are strictly related to a certain infrastructure (e.g., roads, or airports), other are more general. Independently of this, the noise assessment results are strictly linked to the definition of the infrastructure location.

In this regard, the research described in this paper is devoted to compare the leading international noise models, analyzing the impact of key parameters included in each model on the noise emission response. Later, these models, which can be stochastic or deterministic, are compared to provide a comprehensive and applicable noise model to the airport infrastructures. This first stage of the research allowed to define the model that best fitted to the airport infrastructure needs in terms of noise.

As a second stage, the noise emission of an International Airport, which is selected as the case study, is assessed by means the model defined during the first stage of the research. In this assessment, the airport noise emissions for two different years are modeled to interpret how and why the noise patterns change over the infrastructure lifetime. In the paper, the outputs are presented as Noise Contours overlaid with land use map of the selected case study.

Keywords: Aviation Noise Emission, Noise Modeling, Noise Contour Maps

1 INTRODUCTION

The assessment of the effect of traffic noise on the surrounding areas is critical in the context of urban settlements. Human life's quality is negatively affected by constant exposure to acoustic transport noise that exceeding the tolerable threshold (Ouis 2001).

Most researchers perform an environmental assessment of the noise of transportation through a calculation campaign or a software simulation. The latter needs a very reliable mathematical structure modelling of the transport infrastructure, the sources of noise emission, and the sound propagation in the environment (Quartieri et al. 2009). The advantages of using a Traffic Noise Model (TNM) can be divided in two parts:

- adopting in the construction of new/modern transportation networks for acoustical impact evaluation and post-construction, while avoiding noise mitigation techniques that often present a higher cost,

---

*Corresponding Author: misagh.ketabdari@polimi.it

                                        DOI: 10.1201/9781003429258-87

- adopting in the existing transportation network to evaluate the related traffic noise emission and consequently to define mitigation strategies that best fitted the actual boundary conditions.

Increase in the number of passengers of road, rail, and air networks over the recent years, as presented in Figure 1, led to higher investigation demands of noise effects on the human health by researchers.

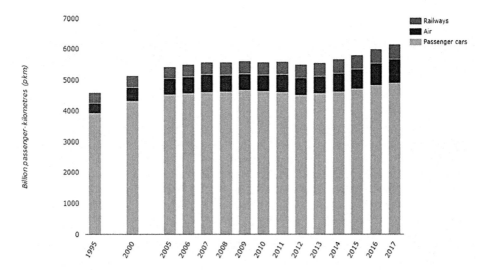

Figure 1.   Increasing trend of passenger volumes in EU (G. s. institute 2013).

World Health Organization (WHO), in the latest publication of environmental noise guidelines for the European region (WHO 2018), strongly recommends that the average noise level produced by each means of transport, should stay under a certain value (Table 1), as traffic noise above this level is associated with adverse health effects. Where, L_{den} is equivalent noise exposure level during the day, evening, nighttime intervals; and L_{night} is Equivalent noise exposure level during only the night interval.

Table 1.   Average noise exposure, source by source, recommended by WHO (2018).

Noise Source	L_{den} (dB)	L_{night} (dB)
Road Traffic	53	45
Railway Traffic	54	44
Aircraft	45	40

According to Table 1, the recommended noise level produced by aircraft (for both day and night) is the lowest among the other. This means airport operations can have more critical impacts on human health respect to other means of transportation. Therefore, the scope of this study is to evaluate the noise exposure levels at one selected airport (case study) and its surrounding areas, through a well-known international TNM.

2 LITERATURE REVIEW

In the last decade, several aircraft noise modelling at the vicinity of the airfield are proposed by different territories such as AzB by Germany (G. s. institute 2013), CNOSSOS-EU by Europe (Kephalopoulos *et al.* 2012), and Aviation Environmental Design Tool (AEDT) by US Federal Aviation Administration (FAA) (Lee *et al.* 2020). Most of these aviation noise models comply with regulations determined by International Civil Aviation Organization (ICAO), European Civil Aviation Conference (ECAC), and European Union Aviation Safety Agency (EASA).

In 1992, ANCON (Aircraft Noise Contour Model) has been released by the UK's Civil Aviation Authority (CAA) (Ollerhead 1992). In 1999, the second edition of ANCON was released, which is completely different from the first edition in terms of noise source modeling parameters (Ollerhead *et al.* 1999). Moreover, for the second edition of ANCON, a complete noise database of all aircraft operated in Heathrow, Gatwick, and Stansted airports was collected from the continuous measurements and radar tracks data around these airports.

In 2016, ECAC has published the new edition of doc29, which is discussed in the next chapter, with a slight difference respect to the previous editions regarding the aircraft noise performance data, but most of the principals are remain constant. This model uses the international aircraft noise database, which were provided previously by FAA, EUROCONTROL, and EASA, and extracts a wide spectrum of data such as aircraft weight during take-off and landing, aircraft power engine in each performance, aircraft aerodynamic parameters, etc., to model the noise source (ECAC 2016a). Different modifications are applied to the base noise source level for non-reference conditions like different distances from the observer to the source, different flight procedures, aircraft speeds, lateral attenuation effects, and noise directivity effects. Following the European Noise Directive (END 2002), the metrics used in this model are L_{den} and L_{night} as well (ECAC 2016b).

In 2018, ICAO has published the second edition of doc9911, which is the repetition of ECAC doc29 with no fundamental changes (ICAO 2018). Therefore, TNM proposed by ECAC doc 29 can still be the main reference point for modelling the airfield noise emissions.

3 COMPARISON OF PRINCIPAL AVIATION NOISE MODELS

In general, noise models are divided into two categories: stochastic models and deterministic models. Stochastic models appeal to simplicity and acceptability, and the deterministic models represent better the physical nature of the problem (Kumičák 2004). The focus of this study is on assessing two reference deterministic models, as described in the following, because of their higher level of precision in modeling the noise emissions.

3.1 *The standard method of computing noise contours around civil airports by ECAC*

The approach of this method is to calculate the noise around the receiver points at the airport's vicinity, for landing and departure procedures, or any operations with a different flight path. Basically, recent editions of this method follow the international Aircraft Noise and Performance database (ANP) (ECAC 2016a). This database is the function of Noise Power Distance (NPD) relationship for various aircraft in different conditions and the noise propagation distances (ECAC 2016b). Thanks to this model it is possible to calculate the noise contours, which are based on the measurement of long-term noise levels, in five steps, as demonstrated in Figure 2.

The airport data, mentioned in Figure 2, includes the runway geometry, airport meteorology data, airport surrounding topography, land cover, radar data, Air Traffic Control

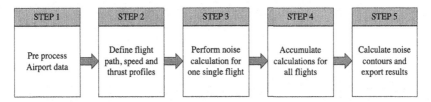

STEP 1	STEP 2	STEP 3	STEP 4	STEP 5
Pre process Airport data	Define flight path, speed and thrust profiles	Perform noise calculation for one single flight	Accumulate calculations for all flights	Calculate noise contours and export results

Figure 2. Calculation steps of airport noise contour based on ECAC doc29 (ECAC 2016a).

(ATC) data. Each airport has a reference point coordinate; each runway has its reference coordinate, which is usually considered in the middle of the runway. Each aircraft's movement is described by its 3D flight path and the varying engine power and speed along the path. The aircraft flight path should be georeferenced compared to a reference point. Moreover, the local aircraft coordinate should be defined.

By splitting the infinite flight path length into several finite-length segments, each segment's noise level is calculated, then their contributions are considered to calculate the overall maximum or exposure noise level. Equation 1 represents the formula to obtain the maximum noise level for each segment.

$$L_{max,seg} = L_{max}(P,d) + \Delta_I(\varnothing) - \Lambda(\beta,l)\ [dB(A)] \tag{1}$$

Where, $L_{max,seg}$ is the maximum noise level for each segment, in dB(A); L_{max} (P,d) is the single event maximum noise level obtained from the NPD table, in dB(A); $\Delta_I(\varnothing)$ is the engine installation correction, which describes a variation in lateral directivity, in dB(A); and $\Lambda(\beta,l)$ is air to ground lateral attenuation, in dB(A).

The exposure noise level for each segment is calculated by Equation 2.

$$L_{AE,seg} = L_{AE,\infty}(P,d) + \Delta_I(\varnothing) - \Lambda(\beta,l) + \Delta_V + \Delta_F\ [dB(A)] \tag{2}$$

Where, $L_{AE,seg}$ is exposure noise level for finite length segment, in dB(A); $L_{AE,\infty}$ (P,d) is the single event noise exposure level obtained from the NPD table, in dB(A); Δ_V is duration (speed) correction, which applicable for non-reference speed, in dB(A); and Δ_F is finite segment correction, which describes noise level fraction for each finite segment, in dB(A).

If the considered segment belongs to the takeoff or landing ground segments, some corrections are added to the above equations (e.g., correction due to the Start Of Roll – SOR, correction due to the finite arrival or departure segment corresponds to the flight procedure, etc.).

3.2 Aircraft noise contour model (ANCON II) by CAA

ANCON II is the empirical and deterministic model used to produce the annual aircraft noise exposure contours, for the airports of Heathrow, Gatwick, and Stansted (Ollerhead et al. 1999). It is summing the Sound Exposure Level (SEL) induced by all passing aircraft to measure long-term equivalent noise level (L_{eq}) at a point on the ground (Ollerhead 1992). One aircraft's SEL is determined by its flight path (in three dimensions), the amount of noise it produces along that path, and how sound propagates from the aircraft to the ground.

The fact that the flight path and noise emission are related for each aircraft is a critical factor governing SEL (both are dependent on how the aircraft is flown, i.e., the operating procedure, particularly changes in engine power). Therefore, noise levels produced by different movements of the same or similar aircraft can vary dramatically.

This model uses NPD data consistent with the FAA NPD table, but most of the UK's aircraft are not in the FAA NPD table. This is the main reason for using the national NPD database, and this database composes of many measurements performed around the London airports. The procedure to calculate the noise level by ANCON II is presented in Figure 3.

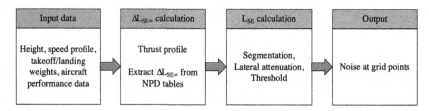

Input data	$\Delta L_{SE\infty}$ calculation	L_{SE} calculation	Output
Height, speed profile, takeoff/landing weights, aircraft performance data	Thrust profile Extract $\Delta L_{SE\infty}$ from NPD tables	Segmentation, Lateral attenuation, Threshold	Noise at grid points

Figure 3. Calculation procedure of ANCON II (Ollerhead *et al.* 1999).

For any single aircraft operation, the flight path is divided into finite-length segments, and for each segment, the exposure noise level is calculated. Combining the contribution of each noise-significant segment is used to calculate an operation noise exposure.

The straight segment noise exposure level is the core of the calculation to obtain an equivalent noise level, as described by Equation 3.

$$\Delta L_{SE} = \Delta L_{SE,\infty} - 10 \times \log F - \Lambda \, [dB(A)] \tag{3}$$

Where, ΔL_{SE} is a noise-significant segment exposure noise level for a single operation, in dB (A); $\Delta L_{SE,\infty}$ is the basic exposure noise level for an infinite segment that obtained from NPD, in dB(A); F is the correction due to the energy fraction (finite-length segment correction) in dB(A); and Λ is the correction due to the lateral attenuation, in dB(A).

The main differences between the described TNMs are provided in Table 2.

Table 2. Principal aircraft noise models.

Noise Model	Country	Category	Year	Vehicle types	Noise metric	Input parameters	Reference Source-receiver distance/receiver height
ECAC doc29-ICAO doc 9911	Worldwide	Aircraft	2016–2018	Civil and commercial aircraft	L_{Aeq}, L_{max}, L_{DEN}, L_{night}	Airport data, ATC data (aircraft, routes, altitude, speed), aircraft weight and performance, meteorological data	Slant distances at 200, 400, 630, 1000, 2000, 4000, 6300, 10000, 16000, 25000 (foot)
ANCON II	England	Aircraft	1999	Airbus, Boeing, MD, Fokker, Concord, DC	L_{aeq} (16hour), L_{max}	Aircraft altitude, speed profile, take-off, and landing weights, aircraft performance data	Slant distance at 61, 122, 192, 305, 610, 1220, 1922, 3050, 4880, 7625 (meter)

4 METHODOLOGY: ADOPTED NOISE MODEL

An Italian airport is selected as the case study to assess the noise patterns in the vicinity of the airport. The noise analysis is performed according to the model proposed by ECAC in doc29, 4th edition, through IMMI2020 software. Noise indicators are evaluated according to national regulation to best fit the boundary conditions.

The Italian Ministry of Environment determined the national noise indicator around the airports (L_{VA}). In this regard, the methodology proposed in Italian Ministerial Decree D. M.31/10/1997 (ENAC 1997) is adopted in this study to calculate the noise indicator and consequent isolation curves.

$$L_{VA} = 10 \times \log \left[\frac{1}{N} \sum_{j=1}^{N} 10^{\frac{L_{VAj}}{10}} \right] [dB(A)] \tag{4}$$

Where, L_{VA} is an index of Italian airport noise level in dB(A); N is the number of days in the observation period; and L_{VAj} is the daily value of the airport noise level in dB(A).

The observation period (N) is equal to 21 days of the three weeks of the greatest air traffic at the single airport, i.e., the period with the most significant number of movements (take-offs and landings), which are chosen from the periods of 1 February – 31 May, 1 June – 30 September, 1 October – 31 January. The value of L_{VAj} is determined by applying Equation 5.

$$L_{VAj} = 10 \times \log \left(\frac{17}{24} \times 10^{\frac{L_{VAd}}{10}} + \frac{7}{24} \times \log 10^{\frac{L_{VAn}}{10}} \right) [dB(A)] \tag{5}$$

Where, L_{VAj} is the daily noise level in dB(A); L_{VAd} is the average noise level during the day period (6:00-23:00) in dB(A); and L_{VAn} is the average noise level during the night period (23:00-6:00) in dB(A).

L_{VAd} and L_{VAn}, which refer respectively to the day (6:00-23:00) and night (23:00-6:00), can be calculated as follows,

$$L_{VAd} = 10 \times \log \left(\frac{1}{T_d} \sum_{i=1}^{N_d} 10^{\frac{SEL_i}{10}} \right) [dB(A)] \tag{6}$$

$$L_{VAn} = 10 \times \log \left(\frac{1}{T_n} \sum_{i=1}^{N_n} 10^{\frac{SEL_i}{10}} \right) [dB(A)] \tag{7}$$

Where, T_d and T_n are the durations of the day and night period in seconds, respectively (i.e., 61200 seconds and 25200 seconds); N_d and N_n are the numbers of day and night flights respectively; and SEL_i (Sound Exposure Level) is the sound level of the ith event associated with the single movement, normalized to one second in dB(A).

According to D.M.31/10/1997, the airport commissions, set up for each airport, define the airport's respective areas' boundaries (ENAC 1997). Within the areas thus defined, the following L_{VA} limits and intended use requirements:

• Zone A: $L_{VA} < 65$ dB(A), there are no restrictions on use,
• Zone B: $65 < L_{VA} < 75$ dB(A), agricultural activities and livestock breeding, industrial, commercial, office, tertiary activities are allowed, subject to the adoption of adequate soundproofing measures,
• Zone C: $L_{VA} > 75$ dB(A), only activities related to the use and services of airport infrastructures are allowed.

Outside the airport surroundings, the L_{VA} index must be less than 60 dB(A), and the absolute emission limit should be respected, which is the maximum noise value measured near the receptors from one or more sound sources, including the airport infrastructure (ENAC 1997).

With Legislative Decree 194/2005, the new acoustic descriptors (L_{den}, L_{day}, $L_{evening}$, L_{night}) have been incorporated into Environment Noise Directive 2002/49/EC (ERA 2020), which constitute the indicators for the acoustic mapping of the main traffic infrastructures (roads, railways, and airports), borne by the managing bodies and agglomerations with more than 100,000 inhabitants.

In this regard, on each day of the year, three periods are defined as: daytime period (from 06:00 AM to 08:00 PM – 14 hours), evening period (from 08:00 PM to 10:00 PM – 2 hours), and night period (from 10:00 PM to 06:00 AM – 8 hours). Moreover, 5 and 10 additional dB (A) penalty values are considered for the evening and night period flight noise, respectively. Finally, the day-evening-night noise value for a long period (e.g., one year) can be computed by Equation 8.

$$L_{den} = 10 \times \log\left(\frac{14}{24} \times 10^{\frac{L_{day}}{10}} + \frac{2}{24} \times 10^{\frac{L_{evening}+5}{10}} + \frac{8}{24} \times 10^{\frac{L_{night}+10}{10}}\right) [dB(A)] \qquad (8)$$

Where, L_{den} is day-evening-night A-weighted noise level in dB(A); and L_{day}, $L_{evening}$, L_{night} are the long-term equivalent continuous A-weighted levels, determined on the set of respective periods of competence over the entire calendar year.

4.1 Scenario selection and data collection

The noise levels of the selected case study are computed for different annual movements of 2019 (scenario A), and 2023 (scenario B). The considered annual movement for scenario B is 15% higher than scenario A, with the same aircraft types, due to the airport's extension. Noise data is extracted from different Noise Monitoring Terminals (NMT) in 2019 and weekly, monthly, and annual equivalent noise levels are calculated for each terminal. In IMMI2020, the exact positions of NMTs are modelled to compare the modelled receivers' values and NMT's measured data. The exact positions of NMTs (receivers) are located on the population density map, as presented in Figure 4.

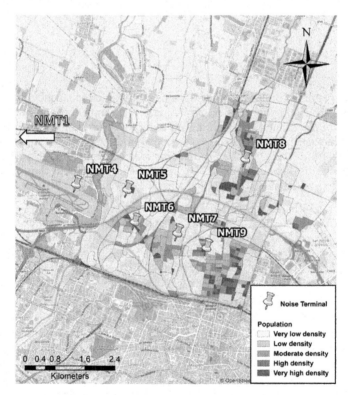

Figure 4.   Population density map of the districts surrounding the case study airport with NMTs' locations.

5 RESULTS AND DISCUSSION: COMPARISON BETWEEN MODELED RESULT AND MEASUREMENT

As explained before, the receiver points are defined in the IMMI2020 with the exact coordinates of the NMTs. The extracted measured data for 2019 are compared with the modelled results to validate the modelling procedures. As an example of this comparison, the measured data and modelled results for scenario A are summarized in Table 3.

Table 3. Model validation with measurement data.

Receiver Name	X Coordinate (UTM)	Y Coordinate (UTM)	Model (IMMI 2020) dB(A)	Measurement dB(A)	Relative Difference (%)
NMT 1	679455.00	4935116.00	68.76	65.65	4.70%
NMT 4	682674.00	4933787.00	68.71	61.95	10.90%
NMT 5	684025.00	4933691.00	55.26	55.34	0.10%
NMT 6	683973.00	4932962.00	64.45	62.86	2.50%
NMT 7	685420.46	4932558.49	58.72	55.05	6.70%
NMT 8	686882.81	4934360.80	48.73	53.03	8.10%
NMT 9	686125.00	4932170.00	56.52	55.18	2.40%

The average relative difference between the model and measured values is 5%, which has a good agreement. The differences can be due to the assumption in modelling because of the lack of input data.

The modelled noise contour maps for both scenarios are presented in Figures 5 and 6. In the same manner the authors developed overlapping runway-related risk contour maps applicable to the selected case study in the previous studies (Ketabdari *et al.* 2019, 2021a, 2021b).

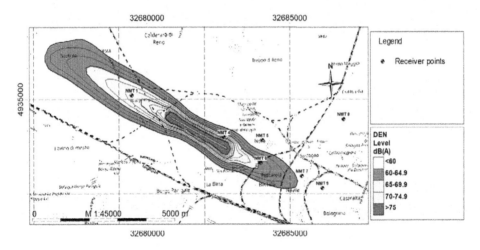

Figure 5. Airport modelled noise map for scenario 2019.

As seen, the noise contour map for scenario 2023 is more expanded in residential areas compared to scenario 2019, due to higher annual movements. Therefore, mitigation measures are required to be applied inside these residential areas, to reduce the people's exposure

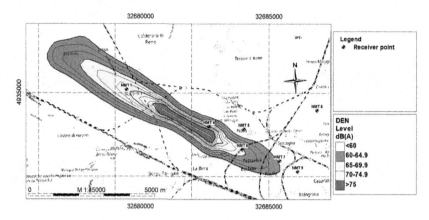

Figure 6.   Airport modelled noise map for scenario 2023.

to excessive noise emissions. Although the noise direction is extended towards the NMT1 for scenario 2023, since that area is industrial, there is no need to do any special mitigation measurement.

For a better overview concerning the noise spreading compared to the land use around the airport, the land use map is overlaid with the noise contour maps for both scenarios, as shown in Figure 7.

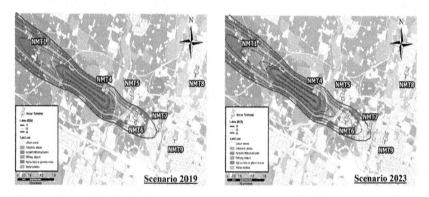

Figure 7.   Overlaying the land use map on the modelled noise contour maps for scenarios A and B.

As it can interpreted from the figure, the land use planning around the airport respect to the levels of emitted noise are consistent with the limits and requirements mentioned in D. M.31/10/1997 (ENAC 1997), therefore, no land use planning revision is required. In detail, in 2023, the noise level at the NMT1, which is an industrial zone, will be 71 dB(A), which refers to zone B of land cover according to D.M.31/10/1997 (ENAC 1997). In zone B, agriculture and industrial activities are allowed, therefore, the land cover rule around the airport for NMT1 is respected. The NMT4 is in a low-populated residential area, that it is possible to reduce the noise level with some mitigation barriers. The NMTs 5 and 6 are in the vicinity of the dense-populated area. The noise level at NMT5 will be 57 dB(A), which refers to zone A, therefore there is no restriction in this zone for any activity. The noise level in 2023 at NMT6 will be 66 dB(A), which refers to the border of zone A and zone B of land cover limitation, therefore some further mitigation measures can be needed. Finally, the noise levels at NMTs 7,8, and 9 are in the acceptable decibel ranges for both scenarios.

7 CONCLUSION

The research activities performed in this study consist of two stages. The first stage, which is dedicated to evaluating the existing NMTs from the literature, allowed to select a comprehensive model that best fitted to the airport infrastructure needs in terms of noise. According to the output from this stage, the standard method of computing noise contours around civil airports proposed by ECAC in doc29, 4th edition (2016), which can cover longer slant distances of noise receivers respect to other models, is selected.

At the second stage, the noise emissions of annual operations of an international airport (the case study) are simulated through the selected model. In this regard, two different operating years are modeled to interpret how and why the noise patterns change over the infrastructure lifetime, and the outputs are presented as Noise Contours maps. According to the results, there is a good agreement of the modelled results and the measured ones, therefore, it is validated that the selected methodology has sufficient precision.

REFERENCES

Academy of European Law (ERA), 2020. *Environment Noise Directive 2002/49/EC. European Parliament, Council of the European Union.* [Online]. Available: https://eur-lex.europa.eu/eli/dir/2002/49/oj/eng.

European Civil Aviation Conference (ECAC), 2016a. Doc.29, 4th Edition. *Report on Standard Method of Computing Noise Contours Around Civil Airports.* Volume1: Applications guide.

European Civil Aviation Conference (ECAC), 2016b. Doc29, 4th Edition. *Report on Standard Method of Computing Noise Contours Around Civil Airports.* Volume 2: Technical Guide.

G. s. institute, 2013. DIN 45684-1: *Acoustics – Determination of Aircraft Noise Exposure at Airfields – Part 1: Calculation Method.*

International Civil Aviation Organization (ICAO), 2018. *Doc 9911, 2nd Edition. Recommended Method for Computing Noise Contours Around Airports.*

Italian Civil Aviation Authority (ENAC), 1997. *Airport Noise Measurement Methodology. Decreto Ministeriale* (D.M.3/10/1997).

Kephalopoulos S., Paviotti M. and Anfosso-Lédée F., 2012. *Common Noise Assessment Methods in Europe (CNOSSOS-EU). Common Noise Assessment Methods in Europe* (CNOSSOS-EU), 180-p.

Ketabdari M., Millán I.P., Toraldo E., Crispino M. and Pernetti M., 2021a. Analytical Optimization Model to Locate and Design Runway-Taxiway Junctions. *The Open Civil Engineering Journal*, 15(1).

Ketabdari M., Millán I.P., Crispino M., Toraldo E., and Pernetti M., 2021b. Numerical Prediction Model of Runway-Taxiway Junctions for Optimizing the Runway Evacuation Time. In *International Conference on Computational Science and Its Applications* (pp. 298–308). Springer, Cham.

Ketabdari M., Crispino M., and Giustozzi F., 2019. Probability Contour Map of Landing Overrun Based on Aircraft Braking Distance Computation. In *Pavement and Asset Management* (pp. 731–740). CRC Press.

Kumičák J., 2004. Stochastic and Deterministic Models of Noise. In *Advanced Experimental Methods for Noise Research in Nanoscale Electronic Devices* (pp. 61–68). Springer, Dordrecht.

Lee C., Thrasher T., Hwang S., Shumway M., Zubrow A., Hansen A. and Solman G., 2020. *Aviation Environmental Design Tool (AEDT) User Manual Version 3c* (No. DOT-VNTSC-FAA-20-04).

Ollerhead J.B., 1992. *The CAA aircraft noise contour model: ANCON version 1.* Cheltenham, UK: Civil Aviation Authority.

Ollerhead J. B., Rhodes D. P., Viinikainen M. S. and Mo D. J., 1999. *R&D REPORT 9842: The UK Civil Aircraft Noise Contour Model ANCON: Improvements in Version 2.* Civil Aviation Authority (CAA), London.

Ouis D., 2001. Annoyance from Road Traffic Noise: A Review. *Journal of Environmental Psychology*, 21(1), 101–120.

Quartieri J., Mastorakis N.E., Iannone G., Guarnaccia C., D'ambrosio S., Troisi A. and Lenza T.L.L., 2009. A Review of Traffic Noise Predictive Models. *In Recent Advances in Applied and Theoretical Mechanics, 5th WSEAS International Conference on Applied and Theoretical Mechanics (MECHANICS'09) Puerto De La Cruz, Tenerife, Canary Islands, Spain December* (pp. 14–16).

World Health Organization (WHO), 2018. *Environmental Noise Guidelines for the European region.*

Roads and Airports Pavement Surface Characteristics – Crispino & Toraldo (Eds)
© 2023 The Author(s), ISBN 978-1-032-55149-4

A prediction model of the coefficient of friction for runway using artificial neural networks

Danilo Rinaldi Bisconsini* & Mariana Myszak
Department of Civil Engineering, Federal University of Technology – Paraná, Pato Branco, Brazil

José Breno Ferreira Quariguasi, Lucas Moreira Magalhães &
Francisco Heber Lacerda de Oliveira
Department of Transportation Engineering, Federal University of Ceará, Fortaleza, Brazil

Jorge Braulio Cossio Durán & José Leomar Fernandes Júnior
Department of Transportation Engineering, São Carlos School of Engineering at the University of São Paulo, São Carlos, Brazil

ABSTRACT: Monitoring the pavement surface is essential to guarantee operational safety during landing and takeoff operations of aircrafts. Data collection is essential for the generation of performance prediction models applied in the Airport Pavement Management System (APMS), which help administrators to make early decisions about pavement Maintenance and Rehabilitation (M&R) strategies. Despite the importance of the friction coefficient for operational safety of landing and takeoff operations of aircrafts, there are few models for predicting friction performance in Brazil, mainly applied to airport pavements. This paper will be designed for the friction performance prediction model at Afonso Pena International Airport (SBCT), located in Curitiba, South of Brazil. The models will be based on information collected during the period from January 2015 to December 2019, from friction coefficient data provided by the National Civil Aviation Agency (ANAC), number of flight operations obtained by the Department of Airspace Control (DECEA), relative humidity obtained by the Airspace Control Institute (ICEA) and data from maintenance and rehabilitation services (M&R) applied over the period of analysis provided by the administration of SBCT. Using these data, a friction performance prediction model based on Artificial Neural Networks (ANN) was developed for the SBCT. During the training stage, different models were investigated, obtaining an R^2 of 81.34% for the best model during the training phase, and an R^2 of 68.99% for the test phase. It is expected that the study will contribute to the operational safety of landing and takeoff operations of aircrafts and to the advancement of scientific research that seeks to develop and improve pavement performance prediction models applied to APMS.

Keywords: Runway, Friction Coefficient, Airport Pavement Management System

1 INTRODUCTION

The activities of monitoring, maintenance and conservation of airport pavements, which are part of the routine of Airport Pavement Management Systems (APMS), are essential to ensure the operational safety of aircraft during operations on taxiways and runways. Among the most important pavement performance parameters for the operational safety of aircraft

---

*Corresponding Author: bisconsini@utfpr.edu.br

DOI: 10.1201/9781003429258-88

is the coefficient of friction, measured on the pavement surface. Reducing the coefficient of friction of the runway can lead to serious accidents and incidents, being critical mainly for the runways, where the interaction between the aircraft's landing gear tire and the pavement surface is more prominent. Despite its importance, decision-making related to airport pavement maintenance and rehabilitation measures is often subjective, based on experience and not on data science.

According to the International Civil Aviation Organization (ICAO 2002), the coefficient of friction is defined as the ratio between the tangential force required to maintain uniform relative motion between the tire-pavement contact surfaces and the perpendicular force acting on the contact surface, that is, the weight distributed over the area of the aircraft tires.

Friction between the runway and the aircraft tires is what allows for safe braking and acceleration in take-off and landing operations (Wells & Young 2004). Runways with low friction values lead to the need for a longer braking distance, as well as having greater chances of skidding and aircraft sliding (Bezerra Filho & Oliveira 2013). In addition, the tire-pavement adhesion is of great importance for the operational safety of aircraft, especially on short runways, where the available length is close to the operating length required by certain aircraft operating at the airport (Silva 2008).

It is the responsibility of the aerodrome operator to judge the need to measure the coefficient of friction after the execution of new pavements or after maintenance services, taking into account the type, location and extent of the intervention (ANAC 2019). The coefficient of attrition values obtained must be above the minimum levels established by regulatory authorities in the airline industry. In Brazil, the limits for this and other performance parameters are established by ANAC (National Civil Aviation Agency). The threshold values for the coefficient of friction are defined in item 153.205(g) of the Brazilian Civil Aviation Regulation (RBAC) No. 153 (ANAC 2018).

The development of friction performance prediction models contributes to early decision making on pavement maintenance and rehabilitation (M&R) strategies, contributing to increased operational safety in aircraft ground operations. However, the development of pavement performance prediction models is not always a simple task, since the variation of the infrastructure condition is a complex phenomenon, depending on several variables that affect the pavement, such as climatic factors, number of landings and takeoffs. and pavement structure. Therefore, it is essential to continuously monitor different information and apply robust statistical techniques to obtain useful models for pavement management applications.In Brazil, there are few friction performance prediction models, mainly applied to airport pavements.

In this work, a friction performance prediction model was developed for the Afonso Pena International Airport (SBCT), located in Curitiba, capital of the state of Paraná, in the southern region of Brazil. The model was based on information collected during the period from September 2015 to September 2019, from friction coefficient data provided by the National Civil Aviation Agency (ANAC), number of takeoff and landing operations obtained by the Department of Airspace Control (DECEA), relative humidity obtained by the Airspace Control Institute (ICEA) and data from maintenance and rehabilitation services (M&R) applied throughout the analysis period provided by the SBCT administration. The model was developed from the application of Artificial Neural Networks (ANN).

1.1 *Factors involved in the variation of the coefficient of friction*

The value of the tire-pavement friction coefficient varies according to loading characteristics, tire types and pavement surface conditions. One of the most critical factors for reducing the coefficient of friction is the presence of thick water depths on the surface of the pavement, which can occur in rainy periods, especially during heavy rainfall. The volume of water that accumulates under the pavement can cause the phenomenon of hydroplaning, which causes loss of traction, impairing the directional control of the aircraft and, consequently, increasing the probability of accidents (Silva 2008).

The number of aircraft take-off and landing operations also has a great influence on the variation of the coefficient of friction. Busier airports, where there is a greater flow of flights, require more of the support capacity of the pavement, as well as the ability of the pavement surface to maintain minimum values of coefficient of friction, especially at the runway thresholds that receive the first touch of the aircraft, where there is greater wear on the pavement surface.

Another factor that contributes to the reduction of the friction coefficient of the runways is the accumulation of rubber in aircraft tires caused by the partial melting of the rubber, which occurs due to the increase in temperature generated by the friction between the tires and the pavement surface. For this reason, aviation regulatory agencies require the execution of rubber removal services with pre-established frequencies, according to the flow of aircraft that the airport receives.

The macro-texture and micro-texture of the pavement surface are factors of great influence on the values of the friction coefficient of the track. The macro-texture depends on the type of surface layer defined for the pavement, which varies according to the type and content of binder, as well as the origin of the aggregates, their diameters and granulometric range. The macrotexture is the main responsible for the surface drainage of rainwater. The more open the macrotexture, the greater its ability to prevent the formation of water sheets on the pavement. Microtexture, on the other hand, is related to the degree of individual texture of the surface of the aggregate particles that make up the asphalt or Portland Cement Concrete mixture of the surface layer (Rodrigues Filho 2006).

The coefficient of friction values also varies according to the method used for its measurement. The measurement procedures vary according to certain characteristics of the equipment used for the evaluation, being common the use of dynamic measurement systems, generally towable, that allow the measurement of the friction of one or more tires that are braked and dragged, simulating the braking of a vehicle.

In some equipment, the friction between the tire and the pavement is caused not by blocking the tire, but by using a certain angle in relation to the direction of movement of the tire. Measurements must be carried out on a wet runway, as this is the most critical condition for hydroplaning to occur. For this reason, the surface of the pavement is artificially moistened by the equipment. The amount of water released by each equipment, for the formation of the water layer, also varies, as well as the speed of operation (Bernucci et al. 2006).

Due to these variables, the value of the coefficient of friction measured by different equipment may vary, although there are models for obtaining standard friction values suggested by researchers and regulatory agencies. In Brazil, friction measurements are reported in 100-meter segments, with limit values defined according to the type of equipment and measurement speed. For the friction performance model developed in this work, friction data measured using the Skiddometer equipment, with a test speed of 65 km/h and a water depth of 1 mm, were used. In this case, the maximum value of friction coefficient established for carrying out maintenance services is 0.60 and the minimum is 0.50 (ANAC 2019).

1.2 *Application of artificial neural networks in performance prediction models*

Fwa et al. (1997) developed a model based on Artificial Neural Networks (ANN) with the objective of predicting the need for rubber removal services on airport runways. The model achieved a success rate of 90.0% during the testing phase. Yao et al. (2019) developed models to predict pavement deterioration based on different performance parameters, among which the coefficient of friction stands out. The model achieved a coefficient of determination of 86.1% during the testing phase.

Moura (2021) developed a friction coefficient prediction model using ANN with the objective of creating a systematic maintenance schedule for Natal Airport, in the State of Rio Grande do Norte, Brazil. The author tested different models, with the final architecture configured from four hidden layers composed of a layer of 128 neurons, two of 64, and one containing a neuron. The weights were optimized using the RMSprop(0.001) function, limiting the processing to 1000 iterations. The author also applied an overfitting fix that was applied to the function via the EarlyStopping function.

Midtfjord, Bin and Huseby (2021) used the XGBoost method which provides approximations with SHAP (SHapley Additive Explanations) to generate a model for predicting slippery conditions for airport runways. The models were trained with meteorological and friction coefficient data. The models identify slippery road conditions with a metric for measuring model performance using ROC (Receiver Operating Characteristics) curves to calculate the area under the curve (AUC), obtaining a value of 0.95, predicting the coefficient of friction with a MAE (Mean Absolute Error) of 0.0254.

Quariguasi *et al.* (2021) developed a friction coefficient prediction model using Artificial Neural Networks (ANN) for Fortaleza International Airport (SBFZ), in the state of Ceará, Brazil. The model was generated based on the following input variables: (a) measurement distance; (b) measurement side; (c) rubber removal; (d) relative humidity; (e) age of the pavement; and, (f) number of operations between rubber removals, during the period from 2015 to 2019. Among the parameters tested for the development of the model, the authors highlighted the use of up to two hidden layers, with a range from 1 to 100 neurons; the use of sigmoid, hyperbolic tangent and rectified linear curves for the activation function; variation of alpha values from 0.001 to 1, this being a regularization term to avoid excessive adjustments; limitation of iterations in the range of 200 and 1000; two weight optimization methods, the Stochastic Gradient Descent and the L-BFGS, a quasi-Newton family optimizer. The authors presented a viable model with a coefficient of determination (R^2) of 0.775 to implement and monitor operational safety conditions.

Quariguasi *et al.* (2021) point out that the use of the model developed for other airports could generate errors, as it was created from specific boundary conditions, such as the equipment and procedures used in the measurements (Grip tester at 65 km/h) and measurements carried out at 3 m of the track axis. In addition, it cites that the use of data with values much lower or higher than those used during the training phase of the model also lead to errors.

Moura (2021) emphasizes the importance of the existence of a reliable database for the development of models, pointing out some limitations related to the data used for the development of the model, such as the lack of detailed information about flight operations and aircraft characteristics, and the non-collection of climatic data obtained by the airport base. The author also highlights that variations in weather conditions, traffic and the type of mix of the surface layer of the pavement can make it difficult to build a universal model, proposing a framework for the development of ANN-based friction coefficient prediction models.

2 METHOD

This section presents the procedures used for the development of the performance prediction model of friction applied to the SBCT. The methodology applied can be divided into four steps. The first step refers to the literature review about features that may affect the friction coefficient and its prediction models. Then, in the second step, the friction coefficient data were collected from 35 reports from the National Civil Aviation Agency (ANAC), the number of flight operations obtained by the Department of Airspace Control (DECEA), relative humidity obtained by the Airspace Control Institute (ICEA) and data from maintenance and rehabilitation services (M&R) applied over the period of analysis provided by the administration of SBCT. All the friction coefficient measurements reports were performed by a Skiddometer at 65 km/h at 3 metres from the centreline.

After, during the third step, the dataset was organized and randomly divided into two different sets, 80% for training and 20% for testing. Then, the model was trained and tested using Python programming language and Scikit-Learn library. It was tested several hyperparameters, such as one and two hidden layers, number of neurons ranging from 8 to 100, and learning rate ranging from 0.001 to 0.1. Due to the random initialization of the weights, each model was run three times and its results were saved to compute their mean and standard deviation to find the best architecture. In the fourth step, the results were analyzed, including the errors. The results were analyzed by Coefficient of Determination (R^2), Mean Absolute Error (MAE), and Mean Squared Error (MSE).

The following features were chosen to predict the friction coefficient: 1) distance from the main threshold, 2) relative humidity from a climatological station located at the airport, 3) the pavement surface age, 4) a categorical feature to represent if a rubber removal was carried out between the friction measurements (it receives 0 if it was not performed and 1, otherwise), and 5) the number of flights operations that occurred between the rubber removal procedures.

3 RESULTS AND DISCUSSIONS

The best ANN architecture found presented a Coefficient of Determination (R^2) mean of 67.67% and a standard deviation of 1.17. Its hyperparameters are two hidden layers composed of 88 and 72, respectively, linear rectified as activation function, and a learning rate of 0.01. Then, the results shown in Table 1 refer to the best result among the three analyzed.

Table 1.   Results of the best architecture.

	Training	Test
R^2	81.34%	68.99%
MAE	0.026	0.033
MSE	0.001	0.002

Figure 1 shows the scatter plot from the training phase and Figure 2 shows the scatter plot from the test phase. The scatter plots show a trend between the observed and predicted

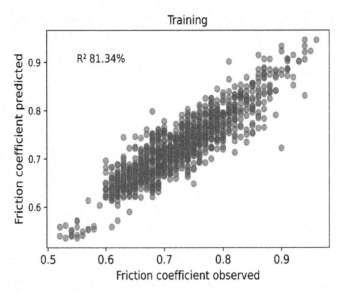

Figure 1.   Results of the training phase.

values. However, there is still a significant dispersion. Figure 3 shows the error histogram between the observed and predicted values from the training phase and Figure 4 shows the error histogram between the observed and predicted values from test phase. According to Figure 3, most of the error in the training phase ranges from −0.05 and 0.05. In Figure 4 it is observed that most of the test phase error ranges from −0.05 and 0.05, mainly between −0.025 and 0.025.

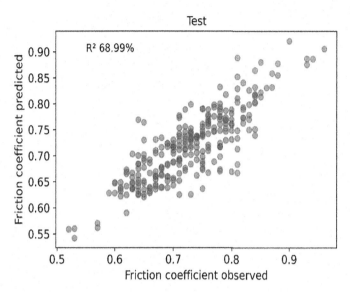

Figure 2. Results of the test phase.

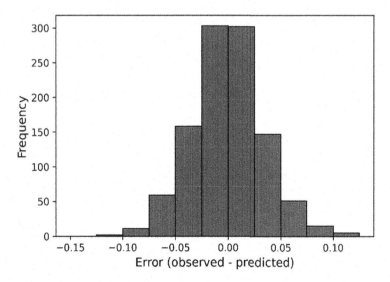

Figure 3. Error histogram from training phase.

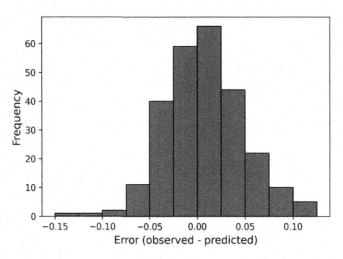

Figure 4. Error histogram from test phase.

4 CONCLUSIONS

In this work, a friction performance prediction model was developed for the Afonso Pena International Airport, in the city of Curitiba, Brazil. Data of friction coefficient, number of flight operations, relative humidity and data from maintenance and rehabilitation (M&R) services were used. The model was developed through the use of Artificial Neural Networks (ANN). During the training stage, different models were investigated, obtaining an R^2 of 81.34% for the best model during the training phase, and an R^2 of 68.99% for the test phase. It is expected that the study will contribute to the operational safety of aircraft take-off and landing operations and to the advancement of scientific research that seeks to develop and improve pavement performance prediction models applied to APMS.

The model can help the airport operator to make decisions regarding the need to measure the coefficient of friction or to perform maintenance and rehabilitation services, especially preventive measures, reducing the subjectivity of the process and favoring the operational safety of landing and take-off operations.

The developed model will undergo new tests in order to improve its predictive power. New tests include the use of data from other airports to obtain a sample with a greater number of data and greater dispersion. Other variable variables should be applied, such as the type of equipment used to measure the coefficient of friction, temperature and the presence of grooving in the pavement.

ACKNOWLEDGEMENTS

To the administrators of Afonso Pena International Airport (SBCT) for providing data and to the National Civil Aviation Agency (ANAC) for the financial resources for the development of the research.

REFERENCES

ANAC 2018. Superintendência de Infraestrutura Aeroportuária – SIA. Regulamento Brasileiro da Aviação Civil, RBAC n. 153. Aeródromos - Operação, Manutenção e Resposta à Emergência, EMENDA n. 02. *Agência Nacional de Aviação Civil.* Available from: https://www.anac.gov.br/assuntos/legislacao/legislacao-1/boletim-de-pessoal/2018/8s1/anexo-i-rbac-no-153-emenda-no-02.

ANAC 2019. Superintendência de Infraestrutura Aeroportuária – SIA. Regulamento Brasileiro da Aviação Civil, RBAC n. 153. Aeródromos - Operação, Manutenção e Resposta à Emergência, EMENDA n. 03. *Agência Nacional de Aviação Civil.* Available from: https://www.anac.gov.br/assuntos/legislacao/legislacao-1/boletim-de-pessoal/2019/6s1/anexo-v-rbac-no-153-emenda-03.

Bezerra Filho C.I.F. and e Oliveira F.H.L., 2013. Análise da Correlação Entre a Macrotextura e o Coeficiente de Atrito em Pavimentos Aeroportuários. *Congresso de Pesquisa e Ensino em Transportes — ANPET 2013 (27), Belém-PA, Brasil.*

Fwa T.F., Chan W.T. and Lim C.T., 1997. Decision Framework for Pavement Friction Management of Airport Runways. *Journal of Transportation Engineering* 123 (6), 429–435. DOI: 10.1061/(ASCE)0733-947X(1997)123:6(429).

ICAO 2002. Manual de Servicios de Aeropuertos. Parte 2. Estado de la Superficie de los Pavimentos. Cuarta Edición. *Organización de Aviación Civil Internacional, Lima, Peru.*

Midtfjord A.D., Bin R.de and Huseby A.B.A, 2021. Machine Learning Approach to Safer Airplane Landings: Predicting Runway Conditions Using Weather and Flight Data. *ArXiv - CS - Computers and Society (IF), DOI: arxiv-2107.04010.*

Moura I.R.D., 2021. Proposta de um Modelo de Avaliação de Pavimentos Aeroportuários para Determinação de Estratégias de Manutenção. *Dissertação (mestrado) - Universidade Federal do Rio Grande do Norte, Centro de Tecnologia, Programa de Pós-Graduação em Engenharia de Produção, Natal-RN, Brasil, 185 p.*

Oliveira F.H.L., 2008. Considerações Sobre a Prática dos Serviços de Remoção de Borracha em Pavimentos Aeroportuários. Reunião Anual de Pavimentação - RAPv (39), *Encontro Nacional de Conservação Rodoviária – ENACOR (13), Recife-PE, Brasil.*

Quariguasi J.B.F., Oliveira F.H.L., de Reis S.D.S.e, 2021. *A Prediction Model of the Coefficient of Friction for Runway Using Artificial Neural Network, Transportes, 29 (2), 9–11.* DOI:10.14295/transportes.

Ribeiro A.J.A, da Silva C.A.U. and Barroso S.H.D.A., 2018. *Metodologia de Baixo Custo Para Mapeamento Geotécnico Aplicado à Pavimentação. Transportes, 26 (2), 84–100.* DOI:10.14295/transportes.v26i2.1491.

Rodrigues Filho O.S., 2006. Características de Aderência de Revestimentos Asfálticos Aeroportuários – Estudo de caso do Aeroporto Internacional de São Paulo/Congonhas. *Dissertação de Mestrado. Escola Politécnica da Universidade de São Paulo.* São Paulo-SP, Brasil.

Silva J.P.S., 2008. Aderência Pneu-Pavimento em Revestimentos Asfálticos Aeroportuários. *Dissertação de Mestrado. Departamento de Engenharia Civil e Ambiental.* Universidade de Brasília. Brasília-DF, Brasil.

Wells A.T. and Young S.B., 2004. *Airport Planning & Management. 6th Edition. McGraw-Hill. New York, USA.*

Yao L., Dong Q., Jiang J. and Ni F., 2019. Establishment of Prediction Models of Asphalt Pavement Performance based on a Novel Data Calibration Method and Neural Network. *Transportation Research Record.,* 2673 (1), 66–82. DOI:10.1177/0361198118822501.

Author index

Salt, G. 243
Sandberg, U. 313, 651
Sanders, P.D. 223, 590
Santos, J. 600
Sartori, C. 768
Sasaki, K. 355
Sato, M. 265
Schmidt, B. 201
Schroedter, T. 14
Schwanen, W. 313
Segundo, I.R. 346
Silva, A. 759
Sirvio, K. 189
Sivapatham, P. 14, 213
Soares, S. 759
Sollazzo, G. 64, 567
Spielhofer, R. 294, 662
Spinelli, P. 74
Stein, L. 107
Stöckner, M. 520, 557
Stöckner, U. 520, 557

Sund, E.K. 678
Suomela, S. 189
Sýkora, M. 54

Tashman, L. 829
Tetley, S. 201
Than, N.T. 168
Thomas, C. 335, 641
Thomasse, C. 848
Tomiyama, K. 265, 355
Toraldo, E. 23, 97, 146, 880, 890
Tozzo, C. 74
Tucker, M. 304, 304
Tullio, C. 34, 748
Turbay, E. 778

Uchôa 759
Unitt, R.P. 365

Vaiana, R. 788
van Aalst, W. 382

van Antwerpen, G. 382
van Antwerpen, M. 382
Vanhooreweder, B. 696
Viner, H.E. 590
Vreugdenhil, B. 382

Waligora, J. 323
Wayman, M. 725
Wehr, R. 662
White, G. 3, 737
Williams, R. 759
Wix, R. 469
Wright, A. 335
Wright, M. 725

Yamaguchi, K. 82
Yamaguchi, Y. 355

Zachrisson, B. 159
Zhuoyuan, C. 168
Zuo, W. 725